Dispersal Ecology and Evolution

EDITED BY

Jean Clobert

Station d'Ecologie Expérimentale du CNRS à Moulis, France

Michel Baguette

Muséum National d'Histoire Naturelle, France

Tim G. Benton

Institute of Integrative and Comparative Biology, University of Leeds, UK

and

James M. Bullock

Centre for Ecology & Hydrology, UK

OXFORD
UNIVERSITY PRESS

OXFORD
UNIVERSITY PRESS

Great Clarendon Street, Oxford, OX2 6DP,
United Kingdom

Oxford University Press is a department of the University of Oxford.
It furthers the University's objective of excellence in research, scholarship,
and education by publishing worldwide. Oxford is a registered trade mark of
Oxford University Press in the UK and in certain other countries

First Edition published in 2012

Published in the United States of America by Oxford University Press
198 Madison Avenue, New York, NY 10016, United States of America

British Library Cataloguing in Publication Data
Data available

Library of Congress Control Number: 2012940355

ISBN 978–0–19–960889–8

Contents

Preface

Jean Clobert, Michel Baguette, Tim G. Benton, and James M. Bullock

At an unprecedented speed, the biosphere is undergoing alteration at all levels of its organization. Pollutants, habitat fragmentation, global warming, species' invasion, and land management changes are creating considerable changes in the functioning of meta-ecosystems. This is occurring through changing aspects of the physical/chemical environment (e.g. temperature, humidity, erosion, etc.), changing habitat availability (e.g. habitat loss, fragmentation, and degradation), and by changing biodiversity dynamics (invasive species, changing species composition, etc.). For millennia, organisms have evolved ways of tackling spatial and temporal fluctuations in their abiotic and biotic environments. Local adaptation, selection, and phenotypic plasticity are all well-known processes to adapt to environmental change or fluctuation. However, such processes have limits especially when the environmental changes are too rapid or too big as is currently the case. In many cases, the only mechanism left to cope with large scale modifications of the local environment will be to leave it and settle in a better one; i.e. to disperse.

Dispersal has long been recognized to be an important life-history trait. Dispersal evolves in response to any kind of alteration in the biotic and abiotic environment, and dispersal profoundly changes the potential for local adaptation and population genetic differentiation. It plays a prominent role in meta-population (and in meta-community and meta-ecosystem) dynamics, species invasion, range shifting, and hence in conservation biology and population management. With the rapid changes that most ecosystems are undergoing, there has been a renewed interest about the causes, consequences, and mechanisms of dispersal. Over the last twenty years there has been a steady increase of studies dealing with dispersal and/or migration. In particular, two important books on dispersal, edited by two of the current editors, were published about ten years ago (Clobert *et al.* 2001, Bullock *et al.* 2002). The great theoretical and empirical advances over even that short period are testament to the research interest in this subject. Despite this interest and although the importance of this particular life-history trait is widely recognized, it is perhaps understudied with respect to its overall importance in spatial ecology. There are several reasons for this. Dispersal may occur at spatial scales that are often difficult to cover. Dispersal is also a complex behaviour (see Part I) which is the result of an integration of genetic, morphological, physiological, and, sometimes, behavioural factors with an individual's specific environment (see Parts II and III, Benton and Bowler 2005, Clobert *et al.* 2009). Dispersal is often confused with migration (which are periodic movements), and this confusion is furthered as some geneticists also call dispersal migration). Dispersal is basically a three-step process which includes departure from the local environment, the movement across a more or less hostile matrix, and settlement in a new habitat. Processes at work within each of these stages might be different (Part IV) and each stage entails different costs (Bonte *et al.* 2011). Furthermore, the proximal causes of departure, the path across the matrix, and the criteria for settling in a new habitat might strongly interact, creating as many potential strategies as the dimension of these interactions. It follows that, depending on the type of environmental change considered (Parts V and

VI), the consequences of dispersal for dynamics (population and evolutionary) can be complex and variable, and therefore be contrary to the predictions which are derived from simplest models.

The book has been organized in seven parts in which we have tried to cover most of the hottest topics in the study of dispersal. The first part deals with the many interacting causes of the dispersal process. The second deals with the genetic structure of the trait, and the third with its association with other life-history traits. Part III considers the covariation between dispersal and other life-history traits, and stresses reality of dispersal syndromes. Part IV recognizes that recent years have seen the development of conceptual frameworks to integrate movement into dispersal. The importance of dispersal for spatial dynamics is the theme developed in the fifth part. Because the biosphere is experiencing unprecedentedly rapid alterations, the two last parts focus on two important aspects of these alterations: habitat fragmentation and global (in particular climatic) changes.

Each section is composed of five chapters: the first reviews the recent research, the second either reviews relevant theory or presents theoretical work pertaining to the section, and finally chapters 3, 4, and 5 describe case studies on vertebrates, invertebrates, and plants respectively. In some cases, the example chapters are mini-reviews of systematic, and particularly well-studied taxonomic groups. The only exception to this structure is the 'genetic' section for which no real theory is yet available on the way the genetic structure underpinning the dispersal process has evolved.

The book concludes with a chapter which is an attempt to bridge human and non-human dispersal. We deliberately give some details about human expansion and the likely reasons for it such that everyone might build their own opinion about the possible parallels.

To help readers, we provide a Glossary with definitions of the most important terms used throughout the book. In most cases, a given term has the same meaning through all chapters. In order to avoid repetition, we also provide four initial introductory boxes for the most well-studied, or iconic, model systems.

We are grateful to leading authors in the field of dispersal for their time and dedication to the production of this book. We dedicate our efforts with this volume to our respective partners and children.

Acknowledgements

Jean Clobert is funded by the CNRS. The book has also benefitted from a 'Plant-Etat-Région' supported by the Ariège Department, the Communauté des Communes of Saint-Girons, the Midi-Pyrénées region, and the FEDER European program. It has also been supported by a European program BIODIVERSA, a FP7 European program SCALES, and two French ANR grants; DIAME and MOBIGEN. I also wish to express gratitude to all members of the informal group 'Dispersal', since many of the key inspirations to construct this book were coming from the discussions we had in our successive meetings. I also want to thank all the authors for the willingness to participate and their hard work to produce up to date work, and A.P. Moller, F. Blanquart, and R. Huttel for chapter reviews.

Michel Baguette is at the Muséum National d'Histoire Naturelle, Paris, seconded to the CNRS USR 2936 Experimental Ecology Station of the CNRS in Moulis. My work is supported by two European programs from the FP6 BIODIVERSA ERA-net (TenLamas and CLIMIT), by the FP7 European program SCALES, and by the French Agence Nationale de la Recherche (ANR open call DIAME and 6th extinction MOBIGEN). I wish to thank Jean for inviting me to edit this book, and Tim and James for our fruitful collaboration. I wish also to thank the authors of the sections 'The Genetics of Dispersal' and 'Dispersal and Habitat Fragmentation' for their enthusiasm and their decisive contributions. Blandine Doligez, Gabriel Nève, and several chapter authors provided expert and constructive reviews that were really helpful in the editing process. Helen Eaton and Ian Sherman were patient and positive editors at OUP. Finally, I express my gratitude to Virginie and the members of the CaLauMaTo team for their presence and daily support.

Tim Benton's interest in dispersal was stimulated by a variety of collaborators over the years. Amongst the first was Justin Travis, and other early inspirations came from discussions with many people including Jean Clobert, Calvin Dytham, and Robert Payne. Despite being spectacularly unsuccessful at raising funding for research in dispersal, Diana Bowler's PhD project on dispersal in the mite system provided an excellent testing ground for many ideas and development of a world view. Karen Mustin, with Justin and Calvin, provided another excellent project for developing model-based ideas. To all these people, plus old and new collaborators, thanks are due. To Ian Sherman and, especially, Helen Eaton at OUP, I record my personal thanks for maintaining faith and impetus when things stalled. In another life, having been a book editor, I appreciate their efforts mightily. Finally, to the family, thanks for helping with the whole process—evenings, weekends and helping to host editorial meetings at home. I also want to thank Atle Mysterud for chapter review.

James Bullock works at the Centre for Ecology and Hydrology, UK. My work on the spatial ecology of plants and the restoration and conservation of species' populations and ecosystems is supported by a wide range of funders, including the Natural Environment Research Council, the Economic and Social Research Council, the Department for Environment and Rural Affairs, Natural England, the European Commission, and the European FP7 project SCALES. I thank Jean, Michel, and Tim for inviting me to be involved in this exciting project, and acknowledging that plants are important too! I thank the authors of the Dispersal Kernels section for their excellent and timely contributions. Several chapter authors along with Merel Soons, Nils Bunnefeld, Olav Skarpaas, and Roger Cousens helped me immensely with constructive reviews. Finally, my gratitude goes to Joanna who has encouraged and inspired me with all my endeavours.

Glossary

Condition-dependent refers to *a factor that depends on an individual's condition* (often used in terms of individual characteristics like size or nutritional status). Condition-dependent implies a trait that is not necessarily fixed but could have responded to the environment or maternal state (unlike sex or age). Sex-dependent, age-dependent, or stage-dependent (e.g. larva versus adult) is not condition-dependent.

Context-dependent refers to a *behaviour or trait that depends on the environment external to the individual* (whether abiotic environment, like temperature or habitat, or the biotic environment, including population density or social environment). Context-dependence is therefore different from condition dependence, although the environmental context may influence an individual's condition.

Dispersal is *movement leading to spatial gene flow* (whatever the behaviour or movement mode giving rise to it). Dispersal does not require movement followed by reproduction (as gene flow occurs by an individual moving its own genes across space) but clearly movement followed by reproduction is typical.

(Dispersal) **decision** is where an organism changes movement status (e.g. from residency to transience, or transience to residency). This does not necessarily imply a conscious decision as it could be entirely probabilistic or it could be a neurological or physiological integration of many factors internal and external. There is a considerable body of evolutionary theory, with empirical support, that decision rules (e.g. 'if you are big then disperse' or 'disperse with probability *p*') can be adaptive, and therefore have arisen through natural selection.

Dispersal syndrome describes co-variation of multiple traits (whether behavioural and/or life-history traits) associated with dispersal. A dispersal syndrome is therefore equivalent to a multivariate **dispersal phenotype**.

Informed dispersal occurs where dispersal decisions are affected by information or cues (at any stage of the dispersal process: emigration, movement, immigration into a new patch). Informed dispersal does not imply conscious decisions; for example, high density can prompt emigration from a patch via a range of proximal mechanisms (visual or chemical cues, increased interference competition, reduction in per-capita resources, or reduction in condition).

A **meta-population** is a specific form of structured population driven by colonisation/extinction dynamics. Not all structured populations are meta-populations.

An animal's **personality** is consistent co-variation in a range of behavioural traits across individuals.

Structured population is a population that has (spatial) structure, in that it is non-uniform in density or genetics. Spatial structure may typically arise from habitat that is spatially autocorrelated, though it can arise through social interactions.

List of contributors

Michel Baguette, Muséum National d'Histoire Naturelle, Dept Ecologie & Gestion de la Biodiversité, Station d'Ecologie Expérimentale du CNRS à Moulis, CNRS USR 2936, Moulis, 09200, Saint-Girons, France
baguette@mnhn.fr

William E. Banks, Univ. Bordeaux, CNRS UMR 5199 PACEA, Equipe Préhistoire, Paléoenvironnement, Patrimoine, Bâtiment B18, Avenue des Facultés, 33405 Talence, France & Biodiversity Institute, University of Kansas, 1345 Jayhawk Boulevard, Lawrence, KS 66045-7562, USA
w.banks@pacea.u-bordeaux1.fr

Tim G. Benton, Institute of Integrative & Comparative Biology, Faculty of Biological Sciences, University of Leeds, Leeds LS2 9JT, UK
t.g.benton@leeds.ac.uk

Dries Bonte, Terrestrial Ecology Unit, Ghent University, Dept. Biology, KL Ledeganckstraat 35, BE-9000 Ghent, Belgium
dries.bonte@ugent.be

Luca Börger, Centre d'Etudes Biologiques de Chizé – CNRS, Villiers-en-Bois, 79360 Beauvoir-sur-Niort France
lborger@uoguilph.ca

Diana E. Bowler, Biodiversity and Climate Research Centre, Senckenberganlage 25, D-60325 Frankfurt am Main, Germany
diana.e.bowler@gmail.com

Jennifer A. Brisson, School of Biological Sciences, University of Nebraska, Lincoln, NE 68588, USA
jbrisson@unl.edu

James M. Bullock, Centre for Ecology & Hydrology, Benson Lane, Wallingford OX10 8BB, UK
jmbul@ceh.ac.uk

Pierre-Olivier Cheptou, UMR 5175 CEFE-CNRS, 1919 route de Mende, 34293 Montpellier Cedex 5, France
pierre-olivier.cheptou@cefe.cnrs.fr

Jean Clobert, Station d'Ecologie Expérimentale du CNRS à Moulis, CNRS USR 2936, Moulis, 09200, Saint-Girons, France
jean.clobert@ecoex-moulis.cnrs.fr

Julien Cote, Laboratory of Evolution and Biological Diversity, CNRS & Université Paul Sabatier, UMR 5174, 118 Route de Narbonne, 31062 Toulouse, Cedex 9, France
julien.cote@univ-tlse3.fr

Aurélie Coulon, Muséum National d'Histoire Naturelle, Dept Ecologie & Gestion de la Biodiversité, 1 Avenue du Petit Château - 91800 Brunoy, France
acoulon@mnhn.fr

Francesco d'Errico, Univ. Bordeaux, CNRS UMR 5199 PACEA, Equipe Préhistoire, Paléoenvironnement, Patrimoine, Bâtiment B18, Avenue des Facultés, 33405 Talence, France & Department of Archaeology, History, Cultural Studies and Religion, University of Bergen, Postboks 7805, NO-5020 Bergen, Norway
f.derrico@pacea.u-bordeaux1.fr

Kathleen Donohue, Department of Biology, Duke University, Box 90338 Durham, NC 27708, USA
k.donohue@duke.edu

Antoine Dornier, UMR 5175 CEFE-CNRS, 1919 route de Mende, 34293 Montpellier Cedex 5, France
antoinedornier@gmail.com

Alex Douglas, School of Biological Sciences, University of Aberdeen, Zoology Building,

Tillydrone Avenue, Aberdeen AB24 2TZ, Scotland, UK
a.douglas@abdn.ac.uk

Renée A. Duckworth, University of Arizona, Department of Ecology & Evolutionary Biology, Tucson, AZ, USA
rad3@email.arizona.edu

Simon Ducatez, Department of Biology, McGill University, 1205 avenue Docteur Penfield Montréal, Québec, Canada
simon.ducatez@mail.mcgill.ca

Calvin Dytham, Department of Biology (area 18), University of York, York, YO10 5DD, UK
cd9@york.ac.uk

Hélène Fréville, Muséum National d'Histoire Naturelle, Dept Ecologie & Gestion de la Biodiversité, CEFE CNRS UMR 5175, 1919 route de Mende, 34293 Montpellier Cedex 5, France
helene.freville@cefe.cnrs.fr

John Fryxell, Department of Integrative Biology, University of Guelph, Guelph, Ontario, Canada
jfryxell@uoguelph.ca

Mats Gyllenberg, Department of Mathematics and Statistics, PO Box 68, FIN-00014 University of Helsinki, Finland
mats.gyllenberg@helsinki.fi

Jocelyn C. Hall, Department of Biological Sciences, University of Alberta, Canada
jocelyn.hall@ualberta.ca

Olivier Honnay, Biology Department, Laboratory of Plant Ecology, University of Leuven, Kasteelpark Arenberg 31, 3001 Heverlee, Belgium
olivier.honnay@bio.kuleuven.be

Ilkka Hanski, Metapopulation Research Group, Department of Biosciences, PO Box 65 (Viikin kaari 1), FI-00014 University of Helsinki, Finland
ilkka.hanski@helsinki.fi

Thomas Hovestadt, Field Station Fabrikschleichach, Glashüttenstr. 5 Fabrikschleichach D- 96181 Rauhenebrach, Germany
hovestadt@biozentrum.uni-wuerzburg.de

Hans Jacquemyn, Biology Department, Laboratory of Plant Ecology, University of Leuven, Kasteelpark Arenberg 31, 3001 Heverlee, Belgium
hans.jacquemyn@bio.kuleuven.be

Eva Kisdi, Department of Mathematics and Statistics, PO Box 68, FIN-00014 University of Helsinki, Finland
kisdi@mappi.helsinki.fi

Etienne Klein, INRA PACA. Domaine Saint Paul, Site Agroparc 84914 Avignon Cedex 9, France
etienne.klein@avignon.inra.fr

Hanna Kokko, Centre of Excellence in Biological Interactions, Division of Evolution, Ecology and Genetics, Research School of Biology, Australian National University, Canberra, ACT 0200, Australia
hanna.kokko@anu.edu.au

Xavier Lambin, School of Biological Sciences, University of Aberdeen, Zoology building, Tillydrone Avenue, Aberdeen AB24 2TZ, Scotland, UK
x.lambin@abdn.ac.uk

Diane Le Bouille, School of Biological Sciences, University of Aberdeen, Zoology building, Tillydrone Avenue, Aberdeen AB24 2TZ, Scotland, UK
diane.lebouille@gmail.com

Jean-François Le Galliard, CNRS—UMR 7625, Ecologie et Evolution, Université Pierre et Marie Curie, Case 237, Bâtiment A, 7 Quai St Bernard 75005 Paris, France
galliard@biologie.ens.fr

Delphine Legrand, Muséum National d'Histoire Naturelle, Dept Ecologie & Gestion de la Biodiversité, Station d'Ecologie Expérimentale du CNRS à Moulis, CNRS USR 2936, Moulis, 09200, Saint-Girons, France
delphine.legrand@ecoex-moulis.cnrs.fr

Valérie Lehouck, Terrestrial Ecology Unit, Ghent University, Dept. Biology, KL Ledeganckstraat 35, BE-9000 Ghent, Belgium
valerie.lehouck@ugent.be

Luc Lens, Terrestrial Ecology Unit, Ghent University, Dept. Biology, KL Ledeganckstraat 35, BE-9000 Ghent, Belgium
luc.lens@ugent.be

Manuel Massot, CNRS—UMR 7625, Ecologie et Evolution, Université Pierre et Marie Curie, Case 237, Bâtiment A, 7 Quai St Bernard 75005 Paris, France
mmassot@snv.jussieu.fr

Erik Matthysen, Evolutionary Ecology Group, Department of Biology, University of Antwerp, B-2020 Antwerpen, Belgium
erik.matthysen@ua.ac.be

Ran Nathan, Movement Ecology Lab, Department of Ecology, Evolution and Behavior, The Hebrew University of Jerusalem, Givat Ram, Jerusalem 91904, Israel
rnathan@cc.huji.ac.il

Matthew K. Oliver, School of Biological Sciences, University of Aberdeen, Zoology building, Tillydrone Avenue, Aberdeen AB24 2TZ, Scotland, UK
m.k.oliver@abdn.ac.uk

Henrik Pärn, Centre for Conservation Biology, Department of Biology, Realfagbygget, NO-7491 Trondheim, Norway
henrik.parn@bio.ntnu.no

Hans Joachim Poethke, Field Station Fabrikschleichach, Glashüttenstr. 5 Fabrikschleichach D- 96181 Rauhenebrach, Germany <http://www.evolutionaryecology.biozentrum.uni-wuerzburg.de/service/how_to_reach_us/>
poethke@biozentrum.uni-wuerzburg.de

Eloy Revilla, Department of Conservation Biology, Estacion Biologica de Donaña CSIC, Calle Americo Vespucio s/n, E-41092 Sevilla, España
revilla@ebd.csic.es

Juan J. Robledo-Arnuncio, Departamento de Ecología y Genética Forestal, Centro de Investigación Forestal (CIFOR)—INIA, Ctra. de la Coruña Km 7.5, 28040 Madrid, Spain
jjrobledo@gmail.com

Ophélie Ronce, Université Montpellier 2, CNRS, Institut des Sciences de l'Evolution, CC65, Place Eugène Bataillon, 34095, Montpellier Cedex 5, France
ophelie.ronce@univ-montp2.fr

François Rousset, Université Montpellier 2, CNRS, Institut des Sciences de l'Evolution, CC65, Place Eugène Bataillon, 34095, Montpellier Cedex 5, France
francois.rousset@univ-montp2.fr

Rafael Rubio de Casas, National Evolutionary Synthesis Center, NESCent, 2024 W. Main Street, Suite A200 Durham, NC27705, USA
rafa@nescent.org

Marjo Saastamoinen, Metapopulation Research Group, Department of Biosciences PO Box 65 (Viikinkaari 1), FI-00014 University of Helsinki, Finland
marjo.saastamoinen@helsinki.fi

Bernt-Erik Sæther, Centre for Conservation Biology, Department of Biology, Realfagbygget, NO-7491 Trondheim, Norway
bernt-erik.sather@bio.ntnu.no

Nicolas Schtickzelle, Université catholique de Louvain, Biodiversity Research Centre—Earth and Life Institute, Place Croix du Sud, 4–1348 Louvain-la-Neuve, Belgium
nicolas.schtickzelle@uclouvain.be

Frank M. Schurr, Plant Ecology and Nature Conservation, University of Potsdam, Maulbeerallee 3,14469 Potsdam, Germany
frank.schurr@uni-potsdam.de

Toon Spanhove, Terrestrial Ecology Unit, Ghent University, Dept. Biology, KL Ledeganckstraat 35, BE-9000 Ghent, Belgium
toon.spanhove@ugent.be

Jostein Starrfelt, Dept. of Biosciences, University of Helsinki, Finland and Norwegian Institute for Water Research (NIVA), Gaustadalléen 21, NO-0349 Oslo, Norway
jostein.starrfelt@niva.no

Virginie M. Stevens, Station d'Ecologie Expérimentale du CNRS à Moulis, CNRS USR 2936, Moulis, 09200, Saint-Girons, France
virginie.stevens@dr14.cnrs.fr

Chris Sutherland, School of Biological Sciences, University of Aberdeen, Zoology building, Tillydrone Avenue, Aberdeen AB24 2TZ, Scotland, UK
r02css9@abdn.ac.uk

Edoardo Tedesco, School of Biological Sciences, University of Aberdeen, Zoology building, Tillydrone Avenue, Aberdeen AB24 2TZ, Scotland, UK
edoted@hotmail.com

Justin M. J. Travis, University of Aberdeen, Institute of Biological & Environmental Science, Zoology Building, Tillydrone Avenue, Aberdeen, AB24 2TZ, Scotland
justin.travis@abdn.ac.uk

Camille Turlure, Université catholique de Louvain, Biodiversity Research Centre—Earth and Life Institute, Place Croix du Sud, 4–1348 Louvain-la-Neuve, Belgium
camille.turlure@uclouvain.be

Margarete Utz, Department of Mathematics and Statistics, PO Box 68, FIN-00014 University of Helsinki, Finland and Ludwig-Maximilians-

University Munich, Department of Biology II, Großhaderner Str. 2, D-82152 Martinsried-Planegg, Germany
margarete.utz@helsinki.fi

Hans Van Dyck, Université catholique de Louvain, Biodiversity Research Centre—Earth and Life Institute, Place Croix du Sud, 4–1348 Louvain-la-Neuve, Belgium
hans.vandyck@uclouvain.be

Christopher W. Wheat, Department of Biological and Environmental Sciences, PL 65, Viikinkaari 1, 00014 University of Helsinki, Finland; and Department of Zoology, SE-106 91 Stockholm, Sweden
christopher.wheat@helsinki.fi; chris.wheat@zoologi.su.se

Charles G. Willis, Department of Biology, Duke University, Box 93308 Durham, NC 27708 US
charlie.willis@duke.edu

Anthony J. Zera, School of Biological Sciences, University of Nebraska, Lincoln, NE 68588, USA
azera1@unlnotes.unl.edu

CASE STUDY I

The common lizard (*Zootoca vivipara*, anciently *Lacerta vivipara*)—a model system for the study of the causes, mechanisms, and consequences of dispersal

Jean Clobert

Lacertidae is an old world lizard family belonging to the infra-order Autharcoglossa (order Squamata) (Figure I.I). The common lizard belongs to the single-species genus *Zootoca* (from the Greek 'to give birth'). *Zootoca vivipara* (formally *Lacerta vivipara*) is a small reptile (on average 60 mm body length, 20 for juvenile) inhabiting peat bogs, humid grasslands, or heath land, and is widely distributed across Eurasia (from Great Britain to the Sakhalin Island, and from the polar circle to north of Spain, north of Italy, and the Balkans). The species gives birth to an average clutch of five (range 1–13) soft-shelled eggs from which independent juvenile emerge within one hour. The species feeds on small insects, and adults and juveniles utilize prey size ranges that overlap, leading to some competition (Massot *et al.* 1992; Lecomte *et al.* 1994). Females and males have colourful ventral faces; males are orange/red to white, while females are orange/dark yellow/pale yellow. Distinct reproductive or competitive strategies are found in association with these ventral colour morphs in females (Vercken *et al.* 2007; Vercken *et al.* 2010). Back patterns are also variable in this species with reticulated individuals being more adapted to warm climatic conditions (Lepetz *et al.* 2009), having higher clutch size but smaller juvenile survival than linear individuals (with more melanin).

In the population studied (Mont Lozere, south-east of the Massif Central, France), the species spends six months in hibernation (from October until March). Males emerge first followed by juveniles and adult females one month later. Mating takes place immediately after female emergence, while yolking takes place in early May. After two months of gestation, females lay eggs in the mid-to-end of July, the date of parturition, due to global warming, is now about a month earlier than it used to be (Chamaillé-Jammes *et al.* 2006). Juveniles are dark at birth and turn dark to medium maroon one month later. Juvenile annual survival is 0.25 on average, adult females 0.55, and adult males 0.3. After four years of age, adult survival drops to about half that of younger adults (Ronce *et al.* 1998; Massot *et al.* 2011).

Dispersal has been studied in a variety of ways in this species. Natural dispersal patterns (as well as a host of other life-history traits) have been recorded over 25 years in two populations (Massot and Clobert 2000; Massot *et al.* 2008). In particular, we found that reticulated individuals disperse less than linear individuals (Lepetz *et al.* 2009), and that juveniles issued from poorly fed mothers during gestation (June–July) disperse either more or less than juveniles issued from well-fed mothers (Massot and Clobert 1995; Massot *et al.* 2008) depending on weather conditions after birth (temperature in

Dispersal Ecology and Evolution. First Edition. Edited by Jean Clobert *et al.*

Figure I.1 Left panel: picture of two newly born individuals with a bit of the tail removed for paternity assessment (photo used with permission of Sandrine Meylan); Right panel: picture of an adult pregnant female with a linear back pattern (photos used with permission of Olivier Calvez). S. Meylan and O. Calvez are both members of the lizard research team.

August; Massot *et al.* 2008). Dispersal propensity and most probably distance was found to correlate negatively with the temperature in June (Massot *et al.* 2008). Transplant experiments in other natural populations have been conducted three times to study the interplay between dispersal, density, habitat characteristics (humidity), and environmental influences (humidity and temperature) during egg development (Massot *et al.* 2002; Meylan *et al.* 2007). These experiments suggest a multifactorial influence of biotic and abiotic factors on dispersal behaviour (Clobert *et al.* 2009). To unravel this complexity, semi-natural enclosed meta-populations (two patch systems, and then more than two patches) have been used to study the interplay between the causes of dispersal, population processes, and individual characteristics (Le Galliard *et al.* 2005). Population density and kinship both increases dispersal (Léna *et al.* 1998; Le Gaillard *et al.* 2003; Cote *et al.* 2007). However, most of these responses were dependent on individual characteristics such as individual size (Cote and Clobert 2007a; Cote and Clobert 2010), individual physiology (de Fraipont *et al.* 2000; Meylan *et al.* 2002), or individual personality (Cote and Clobert 2007b). They were also dependent on the information available to the resident individuals provided by immigrants to these populations (Lecomte *et al.* 2004; Cote and Clobert 2007a). In particular, density was found to have a positive effect on dispersal only when the population was in contact with other populations through immigrants (Clobert *et al.* 2009). This model system has provided evidence for a complex, multifactorial, determination of the dispersal process, and the role of past and present information in shaping individual dispersal decisions.

References

Chamaillé-Jammes, S., Massot, M., Aragon, P. *et al.* (2006). 'Global warming induces only positive individual responses in mountain populations of common lizards *Lacerta vivipara*', *Global Change Biology* **12:** 392–402.

Clobert, J., Le Galliard, J.-F., Cote, J. *et al.* (2009). 'Informed dispersal, heterogeneity in animal dispersal syndromes and the dynamics of spatially structured populations', *Ecology Letters* **12:** 197–209.

Cote, J. and Clobert, J. (2007). 'Social information and emigration: lessons from immigrants', *Ecology Letters* **10:** 411–17.

Cote, J. and Clobert, J. (2007). 'Social personalities influence natal dispersal in a lizard', *Proceedings of the Royal Society B* **274:** 383–90.

Cote, J., Clobert, J., Fitze, P.S. (2007). 'Kin competition promotes colonization success', *Proceedings of the National Academy of Sciences of the USA* **104:** 9703–8.

de Fraipont, M., Clobert, J., John-Alder, H. *et al.* (2000). 'Pre-natal stress increases offspring philopatry', *Journal of Animal Ecology* **69:** 404–13.

Lecomte, J., Boudjemadi, K., Sarrazin, F. *et al.* (2004). 'Connectivity and homogenisation of population sizes: an experimental approach in *Lacerta vivipara*', *Journal of Animal Ecology* **73:** 179–89.

Lecomte, J., Clobert, J., Massot, M. *et al.* (1994). 'Spatial and behavioral consequences of a density manipulation in the common lizard', *Ecoscience* **1:** 300–10.

Le Gaillard, J.-F., Ferriére, R., Clobert, J. (2005). 'Immigration within a metapopulation: Demographic and behavioral consequences in the common lizard', *Journal of Animal Ecology* **74:** 241–9.

LeGaillard, J.-F., Ferrière, R., Clobert, J. (2003). 'Mother-offspring interactions affect natal dispersal in a lizard', *Proceedings of the Royal Society B* **270:** 1163–9.

Léna, J.-P., Clobert, J., de Fraipont, M. *et al.* (1998). 'The relative influence of density and kin competition on dispersal in the common lizard', *Behavioral Ecology* **9:** 500–7.

Lepetz, V., Massot, M., Chaine, A. *et al.* (2009). 'Climate warming and evolution of morphotypes in a reptile', *Global Change Biology* **15:** 454–66.

Massot, M., Clobert, J. (1995). 'Influence of maternal food availability on offspring dispersal', *Behavioral Ecology and Sociobiology* **37:** 413–18.

Massot, M. and Clobert, J. (2000). 'Process at the origin of ressemblance of natal dispersal among offspring of a same litter', *Journal of Evolutionary Biology* **13:** 707–19.

Massot, M., Clobert, J., Ferrière, R. (2008). 'Climate warming, dispersal inhibition and extinction risk', *Global Change Biology* **14:** 461–9.

Massot, M., Clobert, J., Lorenzon, P. (2002). 'Condition-dependent dispersal, and the ontogeny of dispersal behaviour: an experimental approach', *Journal of Animal Ecology* **71:** 253–61.

Massot, M., Clobert, J., Montes-Poloni, L. *et al.* (2011). 'An integrative study of ageing in a wild population of common lizards (*Lacerta vivipara*)', *Functional Ecology* **25:** 848–58.

Massot M., Clobert, J., Pilorge, T. *et al.* (1992). 'Density dependence in the common lizard (*Lacerta vivipara*): demographical consequences of a density manipulation', *Ecology* **73:** 1742–56.

Meylan, S., Belliure, J., Clobert, J. *et al.* (2002). 'Determinants of dispersal in juveniles of the common lizard (*Lacerta vivipara*): stress removes the influence of maternal condition', *Hormones and Behaviour* **42:** 319–26.

Meylan, S., Clobert, J., Sinervo, B. (2007). 'Adaptive significance of maternal induction of density-dependent phenotypes', *Oikos* **116:** 650–61.

Ronce, O., Clobert, J., Massot, M. (1998). 'The evolution of offspring dispersal with mother age, some theory and one example', *Proceedings of the National Academy of Sciences of the USA* **95:** 600–5.

Vercken, E., Clobert, J., Sinervo, B. (2010). 'Frequency dependent reproductive success in female common lizards: a real-life hawk-dove-bully game?', *Oecologia* **162:** 49–58.

Vercken, E., Massot, M., Sinervo, B. (2007). 'Colour polymorphism and alternative reproductive strategies in females of the common lizard *Lacerta vivipara*', *Journal of Evolutionary Biology* **20:** 221–32.

CASE STUDY II

Spiders as a model in dispersal ecology and evolution

Dries Bonte

Spiders are important predators of insect pests that often 'parachute' into new areas on single strands of silk. This behaviour was first described by Martin Lister in 1684 in a remarkable publication in the Proceedings of the Royal Society and was also noted by Charles Darwin during his expeditions with *The Beagle* (Bell *et al.* 2005). Silk-assisted aerial dispersal is known from almost all spider families and phylogenetic analysis suggests that it most likely evolved in parallel with the raise of grasses in the Cretaceous era (135–65 mya), and the subsequent temporal and spatial heterogeneities in habitat structure due to, for example, herbivory and trampling (Bell *et al.* 2005).

Aerial dispersal in spiders is mediated by distinct behaviours preceding transfer, so-called tiptoe behaviour (i.e. stretching of legs, raising abdomen, and producing silk threads; see Figure II.Ia). By means of this behaviour, spiders initiate either long-distance dispersal when silk threads are transported in the air (*ballooning*), or short-distance dispersal when silk threads are used as bridges (*rappelling*). Due to aerodynamic constraints, ballooning is restricted to smaller individuals. In temperate regions, these comprise predominantly the adults of smaller money spiders (Linyphiidae) or the juveniles of larger wolf- and crabspiders (Lycosidae, Thomisidae). In the latter, routine cursorial movements (Bonte *et al.* 2004b) or maternally assisted spreading of offspring (Bonte *et al.* 2007b) exist as alternative adult and natal dispersal mechanisms, respectively. Ballooning and cursorial movements have a differential impact on metapopulation structure and species distribution (Bonte *et al.* 2003b, 2004a; Lambeets *et al.* 2010; Pétillon *et al.* in press). The costs of ballooning are directly related to the availability of suitable habitat within the dispersal perimeter, while short-distance dispersal success is more likely to be correlated with the spatial configuration of resources and habitat. Ballooning, but not rappelling is partly under genetic control with heritability estimates ranging from 0.1–0.6 in both a wolf spider and a money spider, depending on the prevailing environmental conditions (Bonte *et al.* 2003a, 2009, Bonte and Lens 2007)

Because of stereotypic behaviours preceding transfer, aerial dispersal behaviour in spiders provides a unique opportunity to study dispersal initiation before the onset of the actual transfer. As such, behavioural reaction norm approaches can be applied under standardized laboratory conditions. Research on spider dispersal was initiated in the 1970s by Richter (1970, 1971) and Vughts and van Wingerden (1976), who studied tiptoe behaviour in relation to meteorological conditions and habitat characteristics. A similar approach was followed by Weyman and Duffey (Weyman 1995) to unravel the importance of sex and age as condition-dependent factors. From 2003 onwards, Bonte and collaborators studied species belonging to the genus *Erigone* which are predominantly associated with disturbed habitats and wolf spiders from spatially structured habitats like dune grasslands and salt marshes. They demonstrated evolutionary patterns related to species traits (Bonte *et al.* 2003c), landscape structure (Bonte *et al.* 2006, 2007a), and plasticity in relation to prey availability (2008a, De Meester and Bonte 2010; Pétillon *et al.* in press), wind velocity

Dispersal Ecology and Evolution. First Edition. Edited by Jean Clobert *et al.*

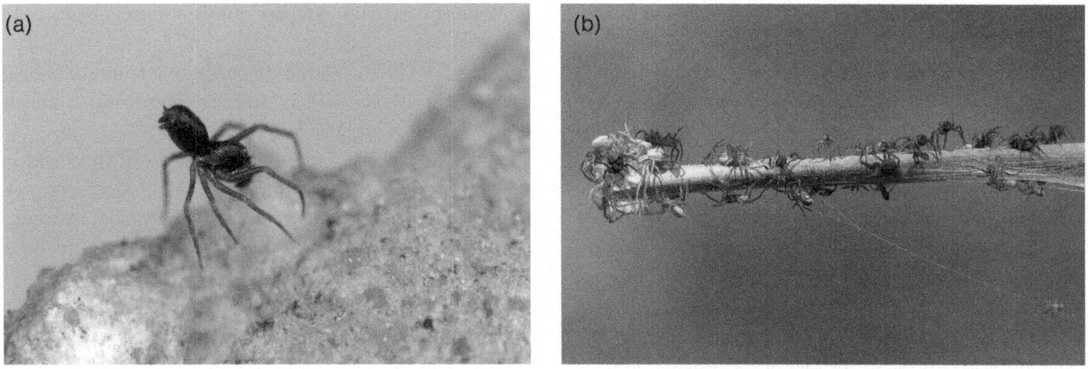

Figure II.1 (a) Onset of ballooning by the performance of stereotypic tiptoe-behaviour in *Dictyna uncinata* (P.Oger/ARABEL image bank). (b) Mass ballooning during spring by juvenile wolf- and crabspiders on an elevated structure in a meadow (D. Appels/ARABEL image bank).

(Bonte *et al.* 2007a), and thermal conditions during development (Bonte *et al.* 2008b). Inbreeding (Bonte 2009) and the presence of specific endosymbiontic bacteria (Goodacre *et al.* 2009) constrain dispersal in spiders. Recently, Entling *et al.* (2011) used this approach to document the importance of disturbance as an important driver of aerial dispersal, while Larrivée and Buddle (2011) demonstrated showed high dispersal ballooning rates in North American forest spiders.

References

Bell, J.R., Bohan, D.R., Shaw, E.M. *et al.* (2005). 'Ballooning dispersal using silk: world fauna, phylogenies, genetics and models', *Bulletin of Entomological Research* **95**: 69–114.

Bonte, D. (2009). 'Inbreeding depresses short and long distance dispersal modes in three congeneric spiders', *Journal of Evolutionary Biology* **22**: 1429–4.

Bonte, D., Baert, L., Lens, L. *et al.* (2004a). 'Effects of aerial dispersal, habitat specialisation, and landscape structure on spider distribution across fragmented grey dunes', *Ecography* **27**: 343–9.

Bonte, D., Bossuyt, B., Lens, L. (2007a). 'Aerial dispersal plasticity under different wind velocities in a salt marsh wolf spider', *Behavioral Ecology* **18**: 438–43.

Bonte, D., De Blauwe, I., Maelfait, J.-P. (2003a). 'Environmental and genetic background of tiptoe-initiating behaviour in the dwarfspider *Erigone atra*', *Animal Behaviour* **66**: 169–74.

Bonte, D., Lens, L. (2007). 'Heritability of spider ballooning motivation under different wind velocities', *Evolutionary Ecology Research* **9**: 817–27.

Bonte, D., Lens, L., Maelfait, J.P. (2004b). 'Lack of homeward orientation and increased mobility result in high emigration rates from low-quality fragments in a dune wolf spider', *Journal of Animal Ecology* **73**: 643–50.

Bonte, D., Lens, L., Maelfait, J.-P. *et al.* (2003b). 'Patch quality and connectivity influence spatial dynamics in a dune wolfspider', *Oecologia* **135**: 227–33.

Bonte, D., Lukac, M., Lens, L. (2008a). 'Starvation affects two aerial dispersal decisions in *Erigone*-spiders in a different way', *Basic and Applied Ecology* **9**: 308–15.

Bonte, D., Swertvaeger, I., De Clercq, N. (2009). 'Repeatability of dispersal behavior in a common dwarf spider: evidence for different mechanisms behind short and long distance dispersal', *Ecological Entomology* **34**: 271–6.

Bonte, D., Travis, J., Swertvaeger, I. *et al.* (2008b). 'Thermal conditions during juvenile development affect adult dispersal in a spider', *Proceedings of the National Academy of Sciences of the United States of America* **105**: 17000–5.

Bonte, D., Van Belle, S., Maelfait, J.-P. (2007b). 'Maternal-care and reproductive-state dependent mobility determine initial natal dispersal in a wolf spider', *Animal Behaviour* **74**: 63–9.

Bonte, D., Van den Borre, J., Lens, L. (2006). 'Geographic variation in wolfspider dispersal behaviour is related to landscape structure', *Animal Behaviour* **72**: 655–62.

Bonte, D., Vandenbroecke, N., Lens, L. (2003c). 'Low propensity for aerial dispersal in specialist spiders from fragmented landscapes', *Proceedings of the Royal Society London, Series B* **270**: 1601–7.

De Meester, N., Bonte, D. (2010). 'Information use and density-dependent dispersal in an agrobiont spider', *Behavioral Ecology* **21**(5): 992–8.

Entling, M.H., Stämpfli, K., Ovaskainen, O. (2011). 'Increased propensity for aerial dispersal in disturbed habitats due to intraspecific variation and species turnover', *Oikos* **120**: 1099–09.

Goodacre, S., Martin, O., Bonte, D. *et al.* (2009). 'Microbial modification of host long-distance dispersal capacity', *BMC Biology* **7**: 32.

Lambeets, K., Breyne, P., Bonte, D. (2010). 'Spatial genetic variation of a riparian wolf spider *Pardosa agricola* (Thorell, 1856) on lowland river banks: the importance of functional connectivity in linear spatial systems', *Biological Conservation* **143**: 660–8.

Larrivée, M., Buddle, C.M. (2011). 'Ballooning propensity of canopy and understorey spiders in a mature temperate hardwood forest', *Ecological Entomology* **36**: 144–51.

Pétillon, J., Decae, A., Deruyter, D. *et al.* (in press). 'Habitat use, but not dispersal limitation, as the mechanism behind the aggregated population structure of the mygalomorph species Atypus affinis', *Animal Biology* (in press).

Richter, C.J. (1970). 'Aerial dispersal in relation to habitat in eight wolf spider species (Pardosa, Araneae, Lycosidae)', *Oecologia* **5**: 200–14.

Richter, C.J. (1971). 'Some aspects of aerial dispersal in relation to habitat in different populations of wolf spiders with particular reference to *Pardosa amentata* (Araneae, Lycosidae)', *Miscellaneous Papers Landbouw Hogeschool, Wageningen* **8**: 77–88.

Vughts, H.F., Van Wingerden, W.K.R.E. (1976). 'Meteorological aspects of aeronautic behaviour of spiders', *Oikos* **27**: 433–44.

Weyman, G.S. (1995). 'Laboratory studies of the factors stimulating ballooning behavior by Linyphiid spiders', *Journal of Arachnology* **23**: 75–84.

CASE STUDY III

Spatial structure and dynamics in the Glanville fritillary (*Melitaea cinxia*) metapopulation

Ilkka Hanski

The Glanville fritillary butterfly (*Melitaea cinxia*) has a large metapopulation in a network of 4000 dry meadows in the Åland Islands in Finland (Figure III.1A and C). The meadows are small, with an average area of 0.17 ha and none greater than 10 ha. The corresponding local populations are also small, mostly < 10 full-sib larval groups in the autumn and < 100 adult butterflies in the summer (females lay their eggs in clutches of 150–200 eggs and the larvae remain gregarious until the final instar in the following spring). The small populations are prone to go locally extinct for many reasons (Hanski 1998), but at the landscape level local extinctions are compensated for by the establishment of new populations by dispersing females. The numbers of extinctions and recolonizations vary from one year to another (Figure III.1E) depending on the environmental conditions, but the size of the metapopulation as a whole has remained relatively stable over the past 20 years, consisting of a few thousand larval groups (Figure III.1D) in 500–800 local populations per year.

Figure III.1B shows genetically homogeneous clusters of local populations based on allele frequencies in neutral markers. Note that these clusters typically consist of tens of discrete local populations within small regions. Genetic similarity of nearby local populations is due to substantial gene flow: empirical and modeling studies indicate that roughly half of the butterflies in a population inhabiting an average-sized meadow disperse to at least one other meadow during their life-time, up to a distance of 2–3 km from the natal population (Chapter 23). However, though there is no or is only limited genetic differentiation in neutral markers among nearby local populations, the small-scale spatial structure has important consequences for the dynamics. For instance, inbreeding in small local populations increases their risk of extinction (Saccheri *et al.* 1998), and there may be significant non-neutral genetic variation among nearby populations. The gene phosphoglucose isomerase provides a prime example of the latter.

Phosphoglucose isomerase (*Pgi*) encodes for a glycolytic enzyme and is highly polymorphic in the Glanville fritillary. Orsini *et al.* (2009) identified a SNP in the coding region of *Pgi* as the key genetic variant at the DNA level. This SNP, *Pgi_111*, which corresponds to the allozyme allele *f* (Hanski and Saccheri 2006), shows strong associations with a range of life history traits, including clutch size and longevity. The AC heterozygotes in *Pgi_111* typically exhibit superior performance to the AA homozygotes (Table 23.1 in Chapter 23), while the CC homozygotes are very uncommon (for the latter see Orsini *et al.* 2009). For metapopulation dynamics, it is especially noteworthy that AC heterozygotes have roughly twice the flight metabolic rate (Niitepõld 2010) and fly roughly twice the distance in the field under commonly occurring low ambient temperatures than the AA homozygotes (Niitepõld *et al.* 2009). Based on their superior dispersal capacity, one could expect that new populations are often established by the AC heterozygotes,

Dispersal Ecology and Evolution. First Edition. Edited by Jean Clobert *et al.*

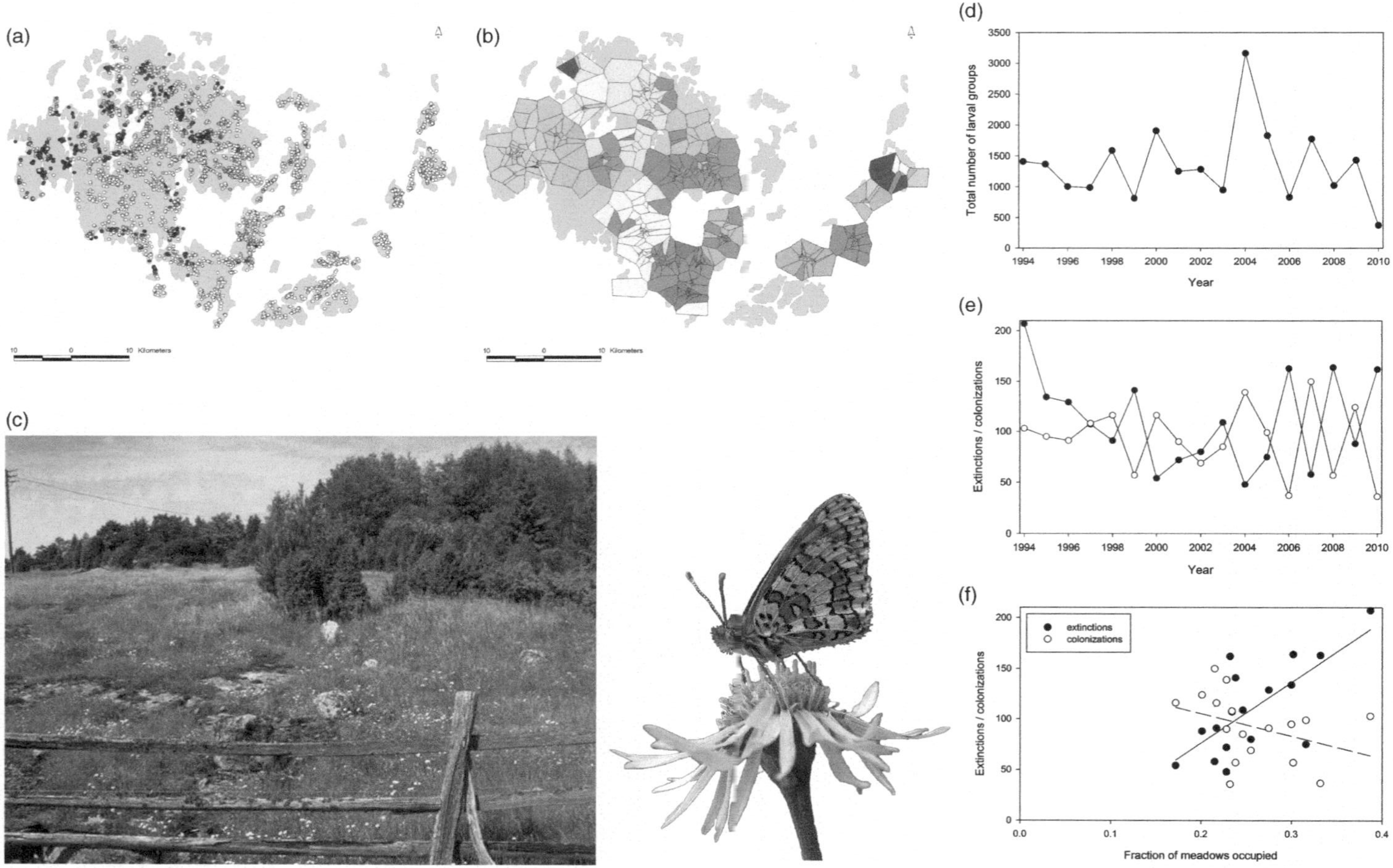

Figure III.1 (a) Map of the Åland Islands in Finland showing the network of dry meadows. Open (unoccupied) and closed dots (occupied) give a snapshot of patch occupancy in one year. (b) Genetic clustering of local populations based on 4 microsatellites and 10 SNPs. Clusters of local populations with homogeneous allele frequencies are indicated by different grey intensities. (c) Photograph of a meadow (photo by Marko Nieminen). (d) Total number of larval groups in the metapopulation in the years 1993–2010, and (e) the corresponding numbers of local extinctions and recolonizations per year. (f) The annual numbers of extinctions and recolonizations plotted against the fraction of meadows occupied. To have comparable data for the different years, results in panels (d) to (f) were calculated only for those meadows that have been surveyed in every year in 1993–2010; the real size of the metapopulation is ca 50% greater (from Hanski, 2011; butterfly image by Niclas Fritzén, modified by Sami Ojanen).

and hence it is not surprising that the frequency of the AC heterozygotes is higher in newly established than old populations. For further discussion see Chapter 23.

References

Hanski, I. (1998). 'Metapopulation dynamics', *Nature* **396**: 41–9.

Hanski, I. (2011). 'Eco-evolutionary spatial dynamics in the Glanville fritillary butterfly', *Proceedings of the National Academy of Sciences of the USA* **108**: 14397–14404.

Hanski, I. and Saccheri, I. (2006). 'Molecular-level variation affects population growth in a butterfly metapopulation', *PLoS Biology* **4**: 719–26.

Niitepõld, K. (2010). 'Genotype by temperature interactions in the metabolic rate of the Glanville fritillary butterfly', *Journal of Experimental Biology* **213**: 1042–8.

Niitepõld, K., Smith, A.D., Osborne, J.L. *et al.* (2009). 'Flight metabolic rate and *Pgi* genotype influence butterfly dispersal rate in the field', *Ecology* **90**: 2223–32.

Orsini, L., Wheat, C.W., Haag, C.R. *et al.* (2009). 'Fitness differences associated with *Pgi* SNP genotypes in the Glanville fritillary butterfly (*Melitaea cinxia*)', *Journal of Evolutionary Biology* **22**: 367–75.

Saccheri, I.J., Kuussaari, M., Kankare, M. *et al.* (1998). 'Inbreeding and extinction in a butterfly metapopulation', *Nature* **392**: 491–4.

CASE STUDY IV

Heterocarpy in *Crepis sancta* (Asteraceae) as a model system to study dispersal

Pierre-Olivier Cheptou

Everyone knows the difference in the ray and central florets of, for instance, the daisy, [...] in some Compositous plants, the seeds also differ in shape and sculpture;[..] But in regard to the differences [two types of] seeds, which are not always correlated with any differences in the flowers, it seems impossible that they can be in any way advantageous to the plant.

—On the origin of species (Darwin 1859)

Crepis sancta is an annual allogamous Mediterranean weed occupying disturbed habitats such as wastelands, vineyards, and urban areas. As in many composite plants (see Imbert 2002), the species *C. sancta* produces two types of achenes (Figures IV.Ia and IV.Ib): achenes without pappus at the periphery of the capitulum (10–15 per head), and achenes with a pappus at the center of the capitulum (80–100 per head), typically dispersed by wind. Achenes with pappus are light (0.1 +/– 0.01 mg), and achenes without pappus are heavy (0.27 +/– 0.02 mg), which results in a very contrasted rate of falling, 0.23 +/– 0.01 $m.sec^{-1}$ and 1.48 +/– 0.08 $m.sec^{-1}$, respectively (Imbert *et al.* 1999). Neither of these types exhibit dormancy. Because of the two identifiable forms of diaspores, the individual phenotype of dispersal can be easily measured as the proportion of each type, which is known to be heritable in the case of *Crepis sancta*. This makes the plant a manageable model to study adaptive processes of dispersal.

Urban fragmentation and the short-term evolution of dispersal

While *C. sancta* forms large and continuous populations in the countryside, urban populations are highly fragmented. Specifically, in cities it grows in small patches around trees on the pavements in the concrete matrix, suitable habitat representing less than 1% of the total surface in Montpellier, France (Figures IV.Ic and IV.Id). Taking advantage of this simplistic habitat, urban habitat have revealed colonization/extinction dynamics within the matrix (Dornier *et al.* 2011), and a high mortality of dispersing achenes due to the extreme fragmentation of urban habitat. Thanks to substantial inheritance of dispersal traits, the proportion of non-dispersing achenes has evolved from 10% in the countryside to about 15% in urban patches in a dozen of generations (Cheptou *et al.* 2008).

References

Cheptou, P.O., Carrue, O. *et al.* (2008). 'Rapid evolution of seed dispersal in an urban environment in the weed Crepis sancta', *Proceedings of the Nationl Academy of Sciences of the United States of America* **105**: 3796–9.

Darwin, C. (1859). On the origin of species by means of natural selection or the preservation of favoured races in the struggle for life. John Murray, London.

Dornier, A., Pons, V., Cheptou, P.-O. (2011). 'Colonization and extinction dynamics of an annual plant metapopulation in an urban environment', *Oikos* **120**: 1240–6.

Dispersal Ecology and Evolution. First Edition. Edited by Jean Clobert *et al.*

Figure IV.1 (a) The species *Crepis sancta*. (b) Non-dispersing achene (left), and dispersing achene (right). (c) Urban patchy environment in Montpellier, south of France (here, Antigone district; photo courtesy G. Przetak). (d) Mean (and SE) for the proportion of non-dispersing seeds estimated in greenhouse (R_d) in unfragmented and fragmented 'patchy' populations.

Imbert, E. (2002). 'Ecological consequences and ontogeny of seed heteromorphism', *Perspectives in Plant Ecology, Evolution and Systematics* **5**: 13–36.

Imbert, E. *et al.* (1999). 'The effects of achene dimorphism and among-population on the dispersal in time and in space in *Crepis sancta* (Asteraceae)', *Canadian Journal of Biology* 77: 508–13.

PART I

The Multiple Causes of the Dispersal Process

CHAPTER 1

Multicausality of dispersal: a review

Erik Matthysen

1.1 Introduction

This chapter aims to review the main causes leading to dispersal, with an emphasis on proximate factors. I will consider causality in dispersal as any factor that explains variation among individuals (within or across species) in any component of dispersal. For a discussion on ultimate causes and on different types of causation and its terminology, I refer to Chapter 2. I will focus in particular on the multicausal nature of dispersal and the interactions between causal factors, a topic that has so far received rather limited attention (Bowler and Benton 2005). First I will introduce the concept of dispersal, briefly discuss its ultimate explanations, and give an overview of different possible mechanisms. Then after a general introduction on multicausality, I will review the different types of causal factors that may explain dispersal patterns. I will distinguish between causation at the individual level and at the level of the environment, but also consider interactions between these levels; for example, when individuals differ in their sensitivity to environmental cues. A recurring theme in the review will be the different ways in which causal factors affect the three stages of dispersal (emigration, transfer, and immigration).

1.1.1 Definitions of dispersal

Dispersal is the movement of an individual from site of birth to site of reproduction (natal or pre-breeding dispersal) or its movement between successive sites of reproduction (post-breeding or simply breeding dispersal). It is the main mechanism leading to gene flow within and between populations. In this review I focus on natal dispersal which is typically much larger in extent than breeding dispersal, and therefore has the largest impact on both individual fitness and population structure. Note that especially at population level, dispersal is often referred to as 'migration', but I will avoid the latter term because in ecological research it also refers to regular movements by animals between different locations (e.g. seasonal or diurnal migration). Dispersal is typically defined in terms of sites or locations. In organisms that are sessile or bound to a nest in the reproductive stage, these locations can be easily defined. However, in viviparous organisms or those that scatter their eggs over multiple sites, the exact locations of birth and reproduction may be more difficult to define, and therefore often proxies for the site of reproduction are used such as a territory, home-range, habitat patch, or membership of a social group. This also applies to male organisms where the site of reproduction can be alternatively defined as the nest site where parental care is provided, or the area where matings take place (e.g. in a lek system, Dunn and Braun 1985).

Dispersal is often seen as a three-stage movement with a distinction between emigration, transfer, and immigration stages (also referred to as 'departure', 'transience' or 'vagrant', and 'settllement' stages, Baguette and Van Dyck 2007; Ronce 2007; Clobert et al. 2009). Note that some authors consider exploration or prospecting behaviour prior to dispersal as a component of the emigration stage (e.g. Bowler and Benton 2005). While this distinction is helpful, it should be remembered that these definitions can vary with the type of organism and mode of dispersal, as well as the scale considered. For example,

Dispersal Ecology and Evolution. First Edition. Edited by Jean Clobert *et al.*

'emigration' may refer to the act of leaving the parental organism (seeds being released or consumed by a seed-dispersing frugivore, Chapter 5), leaving a familiar environment such as the parental territory, or leaving a patch of habitat. Similarly the 'immigration' stage may refer to entering a habitat patch, joining a social group within a patch, or the actual establishment of a nesting site. The three-stage concept is also more difficult to apply to iteroparous organisms that keep on moving throughout their life, such as nomadic birds or primates that switch between social groups after reaching maturity (Section 1.3).

1.1.2 Multiple functions of dispersal

Dispersal by definition takes organisms away from their site of birth. This has a number of possible consequences that in turn determine the selective pressures on dispersal as a life-history trait, and are generally phrased in terms of fitness costs and benefits. The evolutionary forces shaping differences in dispersal patterns are dealt with more extensively in Chapter 2 but are briefly summarized here:

1. Dispersal generally reduces the likelihood of interacting with kin; therefore a major benefit is the avoidance of inbreeding and of competition with kin (kin selection theory), but a possible cost is the loss of cooperation with kin, as well as the potential costs of outbreeding.
2. Dispersal increases the variance in expected fitness by distributing offspring from the same parents over different conditions ('bet hedging'); depending on the nature of spatio-temporal variability in fitness prospects this can be a strong selective force on dispersal.
3. As a plastic or conditional trait, dispersal allows the organism to escape local unfavourable conditions including (over)crowding, and/or to prospect actively for more favourable conditions (including availability and quality of mates). A special case which is also linked with the previous point is dispersal as a strategy to exploit ephemeral or successional habitats which predictably deteriorate in quality over time.
4. Dispersal also has costs which can operate in all stages of dispersal. Costs may be associated with the energetic and physiological cost of movement itself, and/or the development of locomotory structures allowing movement. There are also risks associated with movement itself, often reinforced by moving through unfamiliar or less suitable habitat, such as exposure to predators or risk of starvation.

1.1.3 Mechanisms of dispersal

Before reviewing the multiple causes leading to dispersal (or lack of dispersal, i.e. philopatry), we have to consider the many different mechanisms involved. These are highly varied for the simple reason that any mechanism affecting an organism's physical location, both active and passive, can eventually contribute to its dispersal. On the other hand, the simple representation of dispersal as a change of location may be misleading, as this change can be the result of processes that vary enormously in complexity (Figure 1.1). This can range from a single act of movement such as a seed falling from a tree or a bird undertaking a single flight from one forest to another, to much more complex processes involving a sequence of behaviours and decisions, even including pre-emigration behaviours such as exploration and social interactions with group members.

The most general and intuitively obvious distinction between mechanisms is between active and passive dispersal. Active dispersal means that the organism controls its own locomotion leading to, or contributing to, the dispersal process. This may involve complex decision making in relation to a number of cues from the physical and social environment, and in all three stages of dispersal. Variation in active dispersal can be attributed to differences in capacity to move (morphology, size, endurance), and in dispersal propensity or motivation (Benard and McCauley 2008). A particular point of discussion is to what extent dispersal is the result of 'special behaviour' of which the main function is to disperse (Van Dyck and Baguette 2005); or whether it is a by-product of other movements. In some cases, organisms enter a recognizable locomotory or behavioural mode which allows the observer to predict the onset of dispersal (e.g. ballooning in

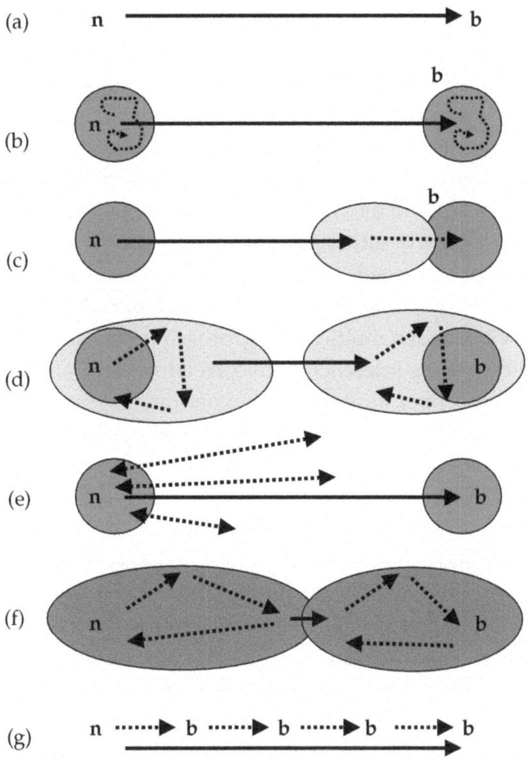

Figure 1.1 Schematic representation of different processes which may lead to dispersal from the natal site (N) to one or multiple sites of breeding (B). In each case the full arrow represents what may be considered the actual dispersal event, while dotted arrows are other types of movements—e.g. within home-ranges or territories—that are typically not classified as dispersal. (a) An individual undertaking a single movement from the location of birth to the site of reproduction, as in a seed being blown away from the mother plant and germinating at the site where it lands. (b) An individual that lives in a natal territory (dark grey area) before dispersing to another area and establishing a breeding territory. This applies to many species with permanent pair or family territories. (c) The dispersing individual first establishes a non-breeding home-range (light grey area) from where it moves into its final breeding home-range. Examples are amphibia or birds that spend the non-breeding season outside their reproductive habitat. (d) The dispersing individual first establishes a non-breeding home-range including the natal area, then leaves for another home-range, which later contracts to the breeding territory. Examples can be found in many bird species where territoriality is confined to the breeding season. (e) Dispersal is preceded by exploration movements (dotted arrows) out of the natal area, as observed in several small mammals (Selonen and Hanski 2006). (f) Dispersal happens when an individual leaves its natal social group when it encounters the boundary of the group territory in order to join a neighbouring group. An example is primates with large permanent group territories (Jack and Fedigan 2004). (g) An individual continues to move throughout its lifetime and reproduces intermittently without settling in a specific breeding area; an example is oviposition in multiple sites by butterflies (Doak 2000).

spiders, Bonte *et al.* 2007a; or dispersal 'restlessness' in birds, Ritchison *et al.* 1992). In other cases dispersal can be recognized by animals moving out of their usual habitat into less suitable areas. More complicated situations can exist when the dispersal process is not achieved in a single episode of movement. For example, individuals may move away from their natal area towards a non-breeding area, and at a later stage disperse to the proper breeding site (or return to the natal site; Figure 1.1). Another example is the movement of primates between neighbouring social groups, where the actual dispersal event may be an individual joining a neighbouring group when they meet at a territory boundary; an even more complex event is aggressive group take-over (Jack and Fedigan 2004). In these cases it may be difficult to discriminate between movement phases of which the primary function is dispersal, and other types of movement such as migration, foraging, or exploration (Van Dyck and Baguette 2005). Exploratory movements in which animals periodically return to their site of origin before actually undertaking dispersal (Figure 1.1) are variously considered as pre-dispersal behaviour (in which case it can be seen as a potential causal factor in dispersal), or as part of the dispersal process itself (Mabry and Stamps 2008). In one of the very few studies on the link between exploration and subsequent dispersal, Selonen & Hanski (2006) somewhat surprisingly found that the two were negatively correlated, suggesting that lack of exploration is part of a long-distance dispersal 'syndrome'.

Passive dispersal implies that movement is largely outside the organism's immediate control and depends on external forces such as gravity, wind, currents, and other organisms. Passive dispersal predominates in plants, microbial organisms, and some small invertebrates with low mobility. Agents of passive dispersal are highly varied, including gravity, wind (anemochory), water currents (hydrochory; rafting), and animals (zoochory). In particular cases, mobile organisms can actively transport their offspring (Bonte *et al.* 2007b) which can variously be regarded as active or passive dispersal. Note that even highly mobile 'active' dispersers can occasionally experience passive dispersal, such as birds displaced by storms or animals 'rafting' on ocean currents. On the other hand, passive dispersers can exert some control on movement; e.g. by choosing the time of release in relation to wind conditions (Bonte *et al.* 2007a). Aside from these cases, passive (and also active) dispersers can adjust their movement to external conditions on an evolutionary or a developmental timescale by either natural selection or phenotypic plasticity (Donohue 1999; Bonte *et al.* 2008b).

Importantly, different mechanisms of dispersal can coexist in a population. Examples are the combination of dispersal by wind and animal vectors in plants (Bullock *et al.* 2002); the combination of active movement and passive 'ballooning' in spiders (Bonte 2009); and the combination of short-distance 'territory budding' and long-distance 'jump dispersal' in cooperative woodpeckers (Kesler *et al.* 2010). While the contribution of these different dispersal modes to dispersal patterns has been widely recognized (Fraser *et al.* 2001; Higgins *et al.* 2003; Kesler *et al.* 2010), relatively few studies have examined the causation of different modes of dispersal within the same population (Bonte 2009; Castro *et al.* 2010).

Note that in this chapter, I will not consider variation in mortality as a causal factor in explaining dispersal patterns. There is general agreement that dispersal is costly, and that selective mortality associated with dispersal can contribute to patterns of realized dispersal (Bowler and Benton 2009). Unfortunately, empirical data on mortality during dispersal are still scarce, undoubtedly due to the large spatial scales involved which make it even harder to detect deaths during dispersal than it is to find surviving dispersers. Despite advances in remote detection technologies, the large sample sizes required for mortality analyses are still unavailable for the majority of taxa.

1.2 Multicausality in dispersal

Dispersal is a process with a complex causation, for a number of reasons. First of all dispersal is—like many life-history traits—only expressed through an often complex interaction of the organism with its environment, therefore dispersal is likely to be driven by a combination of individual characteristics and environmental effects (Lambin *et al.* 2001; Bowler and Benton 2005; Benard and McCauley 2008). Given the heterogeneity and complexity of dispersal mechanisms, and the variety of proximal factors involved, different aspects of the environment are likely to act together in altering the costs and benefits of movement. If we want to test hypotheses properly either on functional aspects (evolution of dispersal strategies) and/or mechanisms (plasticity, condition-dependent dispersal) it is obvious that a multicausal approach is necessary to allow strong inference from correlative data, and to set up proper multifactorial experiments. An excellent example of such multicausal experimental approach is given in Chapter 3. Two other examples are the studies on spider dispersal by Bonte and colleagues (Bonte *et al.* 2007a; Bonte *et al.* 2008b; Bonte 2009) and on multiple determinants of seed dispersal by Donohue (1999 and references therein).

Secondly, and again similarly to other life-history traits, costs and benefits differ among individuals, therefore not all individuals respond in the same way to cues about their environment (physical or social) and/or cues about their own state. In other words, dispersal is often phenotype- or condition-dependent (Bowler and Benton 2005; Clobert *et al.* 2009). For example, sex differences in the extent of dispersal are ubiquitous and particularly well studied in birds and mammals (Clarke and Saether 1997; Handley and Perrin 2007) but sexes can also differ in their response to external conditions (examples are given further on).

Thirdly, dispersal is inherently a spatial process, and therefore it is likely to be influenced by causal factors or cues acting at different spatial scales (Clobert *et al.* 2009) including both biotic and abiotic components. Many of the costs and benefits of dispersal are scale-dependent, but not always in the same way. For example, benefits obtained through avoidance of inbreeding and/or kin competition diminish very quickly at relatively short distances from the natal area. On the other hand, effects of landscape structure only come into play once a disperser moves outside the natal habitat patch, and continue to operate throughout dispersal. Another aspect of spatial scale is related to information. At the onset of dispersal individuals have better information on local cues than on distant cues (Bowler and Benton 2005), but information accumulates during dispersal, depending among other things on dispersal mode and cognitive abilities (Mabry and Stamps 2008). This inevitably creates heterogeneity in the use of information and experience in immigration, transfer, and emigration decisions.

Finally, dispersal includes a large variety of mechanisms as already mentioned earlier These mechanisms or modes of dispersal also contribute differentially to the three stages of dispersal, and therefore may affect short- and long-distance movements in a different way (Higgins *et al.* 2003; Ronce 2007; Long *et al.* 2008; Castro *et al.* 2010; Kesler *et al.* 2010). While from a population dynamical and genetic perspective it may be justifiable to represent the entire dispersal process with a single distribution or kernel, for a better understanding of the factors that drive variation in dispersal, we need to be able to unravel the causal influences on these different components of the dispersal process (Nathan *et al.* 2008).

1.3 Causation at the individual level: genetics and development

1.3.1 Individual variation

Variation in dispersal can to a large extent be explained by variation in morphological, physiological, or behavioural traits that affect individual movement and orientation capacity. These include body size, locomotory appendages, seed morphology, and sensory and cognitive capacities (Bowler and Benton 2005; Benard and McCauley 2008). Many if not most of these traits are genetically determined to a large degree (Clobert *et al.* 2001; Ronce 2007). A full review of variation in dispersal-related morphology is beyond the scope of this review. However, it should be noted that the requirements for successful dispersal can clearly differ between the main dispersal stages. A typical trade-off is caused by allocation of limited resources to either locomotory structures or reproductive investment; well-known examples are the dispersal-germination trade-off affecting seed size and morphology in plants (Eriksson and Jakobsson 1999), and the flight-fecundity trade-offs in insects (Roff and Fairbairn 2001). Thus, enhanced success in the transfer stage may actually compromise settling success. Still, correlations need not always be negative, as in some cases characteristics for successful transfer and immigration may also co-evolve leading to 'superdisperser' phenotypes (Bonte and de la Pena 2009).

In active dispersers, differences in dispersal can be caused not only by variation in movement capacity or competitive ability, but also by dispersal behaviour. These factors are not necessarily independent because propensity and dispersal ability may be correlated (Anholt 1990; Verhulst *et al.* 1997) though not necessarily due to a genetic basis (Coll and Yuval 2004). A few studies have now shown that dispersal propensity can indeed be inherited (Sinervo *et al.* 2006; Doligez *et al.* 2009) while others have found no such evidence (Pasinelli and Walters 2002; Laloi *et al.* 2009). Dispersal behaviour can be triggered by environmental factors (see following) and/or by ontogenetic switches in physiology (Ritchison *et al.* 1992; Nunes *et al.* 1999; Einum *et al.* 2006). Recently a lot of attention has been paid to the role of heritable 'personalities'. Many studies have now shown consistent associations between dispersal and other behavioural characteristics such as aggressiveness or exploration behaviour, suggesting the existence of innate behavioural syndromes (Clobert *et al.* 2009; Cote *et al.* 2010). While the mechanisms behind these associations are not yet known in detail, personalities are generally conceived as different strategies of coping with stress, which includes differences in the use of information. Therefore we should not expect uniform and linear relationships between personality and dispersal, but rather differential responses to environmental cues leading to more complex relationships (see further in Chapter 10).

1.3.2 Condition dependence

Aside from genetic variation, internal state variables can strongly affect dispersal decisions (condition-dependent dispersal) (Dufty and Belthoff 2001; Clobert *et al.* 2009). Condition and genetic variation may also interact as shown by Sinervo *et al.* (2006), who found interactions between maternal effects such as egg size and paternal genotype on lizard dispersal. Internal state can provide an individual with information on the fitness prospects of its environment and thereby affects its decision to leave, which we will discuss further as part of influences of the natal habitat (Section 4.2). This information can be acquired by the organism by monitoring its own success in resource acquisition or the predation risk in the environment. However, information can also be transferred in the pre-natal stage through differential allocation of resources and/or hormonal influences by the parents, which in turn can be related to parental condition, size, or age (Dufty and Belthoff 2001). Such effects have been shown, for example, to influence wingedness in insects and plant seeds (Sloggett and Weisser 2002; Wender *et al.* 2005), size-mediated variation in locomotory capacity in insects and amphibia (Anholt 1990; Beck and Congdon 2000) and dispersal propensity of lizards and mammals (Ronce *et al.* 1998; Massot *et al.* 2002; Zedrosser *et al.* 2007). Tschirren *et al.* (2007) showed experimentally that female birds modulate offspring dispersal through deposition of yolk androgens in a sex-specific manner, which appears to result in an adaptive sex difference in dispersal. It should be noted, however, that parental manipulation of dispersal may not necessarily lead to an optimal decision for the offspring, since optimal dispersal rates may differ for parents and offspring (Starrfelt and Kokko 2010). Thus, hormone- or allocation-mediated maternal effects may also be a mechanism for parents to optimally disperse their offspring and/or to reduce direct parent-offspring competition (Le Galliard *et al.* 2003; Section 4.1). An additional aspect that has hardly received attention is the effect of parentally transmitted endosymbionts on offspring dispersal, as recently demonstrated in *Rickettsia*-infected spiders (Goodacre *et al.* 2009).

Aside from providing information on its fitness prospects, condition can also affect an individual's ability to compete for access to high-quality habitats, known as the 'silver spoon' effect (Forero *et al.* 2002; van Oort and Otter 2005; Stamps 2006). Thus, individuals raised in good conditions may be more likely to stay in their natal environment through a combination of low dispersal propensity and high competitive ability. On the other hand, high-quality individuals may also be better competitors and/or better able to bear the costs of dispersal, in which case the quality of the rearing environment might have a positive, rather than negative effect on dispersal. Many studies have indeed shown that individuals in good condition disperse at higher rates (disregarding the effect of habitat quality), depart earlier, or disperse over longer distances (Wahlstrom and Liberg 1995; Nunes *et al.* 1999; Massot *et al.* 2002; Bonte 2009; Delgado *et al.* 2010). High-quality individuals can also afford to be more selective and therefore may spend more time searching for high-quality vacancies, which may provide an alternative mechanism for a silver spoon effect (Botello and Krug 2006; Delgado *et al.* 2010). Thus, effects of phenotypic quality on dispersal may involve different mechanisms, and the exact shape of the relationship may depend on the relative importance of competition, costs of dispersal, and variation in habitat quality.

1.3.3 Age and experience

In many taxa, dispersal is related to specific ages (or developmental stages), typically occurring shortly after behavioural independence or release from the parent. Nevertheless, delayed dispersal also occurs (as is typical in cooperative breeders, e.g. Ekman and Griesser 2002), and different timing strategies can coexist in a population (Walls *et al.* 1999; Ekman and Griesser 2002; Kesler *et al.* 2010). Differences in age at onset of dispersal may reflect behavioural strategies connected to dispersal ability (e.g. size) and fitness prospects, whereby early departure is often seen as an indicator of high dispersal ability and/or motivation (Walls *et al.* 1999; Byers 2000; Einum *et al.* 2006; Delgado *et al.* 2010). Indeed, when dispersal is peaked in time and settlement involves strong competition, the timing of dispersal may reflect a trade-off between the increasing ability to disperse, and decreasing prospects for successful settling. The latter can also be induced by seasonal time constraints such as the need for pre-hibernation fattening (Nunes *et al.* 1998). Dispersal propensity and timing may therefore be influenced by different factors, as in ground squirrels where they are mediated by hormones and condition, respectively (Nunes *et al.* 1998; Nunes *et al.* 1999).

As they age, individuals accumulate information which may alter their dispersal decisions, and this can affect different stages of dispersal. For example, several studies have shown that breeding site selection of experienced breeders differs from that of inexperienced immatures (Kenward *et al.* 2001; Parejo *et al.* 2007). Experience may be especially important for dispersal during the vulnerable transfer stage (Frair *et al.* 2007), but few studies have examined how dispersers accrue information during transfer and how this affects their choice of travel paths and destinations (but see Doerr and Doerr 2005; Mabry and Stamps 2008). For example, Kenward *et al.* (2001) found a correlation between natal (pre-breeding) and post-breeding dispersal distances in buzzards, but they were not able to decide whether this points towards a carry-over effect of previous experience, or can be attributed to consistent individual variation in personality or habitat quality. I will come back to this issue when discussing habitat selection in Section 1.4.2.

1.4 Causation at the level of environment

1.4.1 Parents and kin

An important component of the natal environment affecting dispersal patterns is the presence of parents and kin. Parents can directly influence dispersal of offspring through aggression (Lambin *et al.* 2001), by moving offspring around during parental care (Bonte *et al.* 2007b; Matthysen *et al.* 2010), or by territory bequethal (Berteaux and Boutin 2000). These different behaviours may all help to resolve parent-offspring competition and risk of inbreeding. In fact, the occurrence of intersexual parent-offspring aggression followed by dispersal is often seen as evidence for active inbreeding avoidance under parental control (Handley and Perrin 2007). Offspring also increase their dispersal rate as a direct response to the presence of kin (Lambin *et al.* 2001; Pasinelli and Walters 2002; Le Galliard *et al.* 2003; Moore *et al.* 2006) and/or preferentially join groups containing non-kin (Handley and Perrin 2007). Because relatedness diminishes quickly with distance, kinship can be expected to affect the emigration process much more than the transfer or immigration stages (Kenward *et al.* 2001). In feral horses, for example, sibling competition affected dispersal from the natal group while dispersal from the natal area was related to non-kin group size (Kaseda *et al.* 1997). Somewhat similarly, changes in population composition that decreased inbreeding risk but increased mate competition in deer led to a reduction in initial short-distance dispersal in the spring, but an increase in subsequent dispersal in the autumn (Long *et al.* 2008). Note, however, that the effect of dispersal distance on post-dispersal relatedness may be very different if kin do not disperse independently (Sinervo and Clobert 2003; Matthysen *et al.* 2005). Dispersers may also respond to cues about the parent's likelihood of surviving, leading to a reduction in dispersal with increasing parental age or decreasing parental condition (Ronce *et al.* 1998; Zedrosser *et al.* 2007). Few studies have examined variation in individual responses to presence of kin. A notable exception is an experimental study on lizards by Cote and Clobert (2010) who showed that when kin competition was important, larger individuals dispersed more often regardless of patch connectivity, while in an environment with non-kin competition, dispersal was related to both individual size and patch connectivity. The suggested explanation is that kin competition promotes dispersal regardless of competitive ability, and larger individuals are better able to bear the cost.

1.4.2 Habitat quality and population density

Dispersal is closely associated with habitat selection (Stamps 2001; Doligez *et al.* 2002; Stamps *et al.* 2005), and therefore we expect dispersers to respond to the quality of habitats and availability of resources in the natal as well as the external environment. Many studies have shown that dispersers are more likely to leave habitats of lower quality, caused by shortage of resources (Imbert and Ronce 2001; Bonte *et al.* 2008a; Mathieu *et al.* 2010), elevated levels of predation or parasitism (Sorci *et al.* 1994; McIntosh *et al.* 2002), or physical disturbance (Bates *et al.* 2006). These effects can lead to higher dispersal propensity, often mediated through conditional effects which trigger dispersal in the individual or its offspring (see Section 3), and/or by an increase in strategic investment in dispersal-related traits such as wings (Benard and McCauley 2008).

Effects of habitat quality can be expected to interact closely with population density because density may modulate the relationship between habitat quality and expected fitness. Indeed, many studies have shown that population density can lead to increased dispersal through competition (Lambin *et al.* 2001; Matthysen 2005). Plant studies have also shown how maternal architecture, which in turn reflects environmental conditions during growth, interacts with plant density to affect seed dispersal patterns (Donohue 1999; Wender *et al.* 2005). Nevertheless, dispersal can also respond negatively to population density both in the emigration and immigration stages due to Allee effects and/or conspecific attraction (Roland *et al.* 2000; Kim *et al.* 2009). High densities may also prevent dispersal because of the high cost of social interactions (the social fence effect; Lambin *et al.* 2001), or by reducing body condition and thereby dispersal (Wahlstrom and Liberg 1995). Density-dependent dispersal responses may in turn be modulated by individual characteristics. For example, males and females may respond differently to cues of local density (Albrectsen and Nachman 2001; Eikenaar *et al.* 2008; Loe *et al.* 2009; De Meester and Bonte 2010), group size (Kaseda *et al.* 1997), or habitat quality (Sorci *et al.* 1994; Matter and Roland 2002). In several taxa, individuals with a competitive advantage (e.g. in better condition, older, or hatched earlier) respond more strongly to density variation in the natal habitat, possibly because they are better able to withstand moderate competition levels (Hanski *et al.* 1991; Byers 2000; Forero *et al.* 2002; Einum *et al.* 2006). As a consequence, dispersal may be biased towards low-quality individuals at low but not high densities (Hanski *et al.* 1991). Two studies have shown interactions between competitive ability and habitat quality or density in the receiving habitat rather than the habitat of departure. In lizards, populated patches received heavy yearling dispersers while empty patches were colonized by relatively lean animals (Le Galliard *et al.* 2005). In great tits, larger juveniles were more successful in entering high-quality habitats (Verhulst *et al.* 1997; Garant *et al.* 2005).

Variation in both habitat quality and population density can be expected to affect emigration and immigration processes in similar ways. Dispersers are expected to preferentially settle in high-quality habitat with low competition (Wright and Boxshall 1999), but in the case of conspecific attraction, they may actually avoid low-density habitats (Valimaki and Itamies 2003). However, competition effects can also differ between the three stages of dispersal and/or vary with spatial scale (Matthysen 2005). In red deer, high densities led to reduced emigration on the one hand, but longer distances among emigrants on the other hand (Loe *et al.* 2009). In Aleppo pines, high plant density caused a change in the dispersal distribution with lower median distances but more frequent long-distance dispersal (Schurr *et al.* 2008). Habitat conditions in the transfer stage are likely to have very different effects compared to conditions in the natal or post-dispersal habitat, because successful transfer depends strongly on short-term survival factors such as food and shelter, but does not require habitat suitable for mating, nest-building, or raising offspring (Palomares *et al.* 2000). While in transfer, high densities or group sizes may actually promote safety during movement and thereby enhance dispersal (Fletcher 2006; Schmidt *et al.* 2010).

An important scale-related issue related to habitat selection is the availability of information. Even though individuals may collect information on their environment prior to or during dispersal (Selonen and Hanski 2006; Mabry and Stamps 2008), and may even use information on remote areas from immigrants (Cote and Clobert 2010), we can expect dispersers in general to have better information on their natal surroundings than on distant areas (Bowler and Benton 2005). Thus, dispersers may well use different cues for emigration, transfer, and immigration decisions. A well-known study on songbirds showed that immigrants use more superficial information on local habitat quality than emigrants (Doligez *et al.* 2002). The opposite was found in butterflies where male immigration, but not emigration, was related to habitat quality (Matter and Roland 2002). However, very few studies have directly addressed the availability of information at different scales. Mayland *et al.* (2000) showed that aphids responded differently to local and distant cues on prey availability. Kenward *et al.* (2001) found that local cues promoted long-distance dispersal in buzzards, while cues at larger distances from the nest were more likely to prevent dispersal. Cues of local habitat quality or density may have different effects in relation to development or age. This was shown in an experimental study on lizards where dispersal was differentially affected by temperature and humidity in the prenatal versus the postnatal environment (Massot *et al.* 2002; Chapter 3). Somewhat similarly, De Meester and Bonte (2010) showed that density during early development only affected short-distance dispersal in spiders, while long-distance movements were also affected by density at the onset of dispersal. The use of information may also be limited by condition or time constraints. Owls in poor condition followed straighter paths, indicating that they invested less in collecting information and hence probably made less efficient settlement choices (Delgado *et al.* 2010). In time-limited dispersers, habitat selectivity of settlers may decrease throughout the transfer stage, known as the 'desperate larva hypothesis' (Botello and Krug 2006; Stamps 2006).

A final point is that habitat quality can affect dispersal in more complex ways if phenotypes differ in their suitability to live in specific habitats. Such differences may arise genetically through local adaptation (which in turn is counteracted by dispersal; see Hanski *et al.* 2002 for an example), but also through phenotypic plasticity. The latter can include the acquisition of habitat-specific foraging skills or resistance to parasites, known as 'habitat training' (Davis and Stamps 2004; Stamps *et al.* 2009). This can strongly affect decision rules of active dispersers by inducing them to cue for natal habitat rather than evaluating overall habitat quality. In organisms with less effective habitat selection, selection in favour of phenotype-environment matching can strongly affect post-dispersal settlement patterns (Marshall *et al.* 2010).

1.4.3 Landscape structure

Many organisms experience their landscape as a mosaic of patches of varying quality. The size and quality of patches, the distances between them, and the biotic and abiotic conditions in the intervening matrix habitat are likely to affect emigration and immigration, and especially the transfer stage. Many studies have documented the effect of landscape isolation on emigration rates and dispersal distance distributions (Coulon *et al.* 2004; Baguette and Van Dyck 2007; Englund and Hamback 2007; Schtickzelle *et al.* 2007; Ockinger and Smith 2008). In general, dispersal between habitat patches—as opposed to dispersal within the breeding habitat—is perceived as costly because of the cost of locomotion as well as risks associated with a 'hostile' matrix that may lack critical resources and shelter. Therefore shorter distances between patches and a more permeable matrix (including the presence of corridors or stepping stones) are expected to favour higher dispersal rates and longer dispersal distances. However, landscape structure should not necessarily affect different components of dispersal in the same way. For example, larger distances between habitat patches (or colonies) have repeatedly been shown to have a neutral or negative effect on emigration propensity, but a positive effect on dispersal distances (Matthysen *et al.* 1995; Serrano *et al.* 2001; Mennechez *et al.* 2003; Long *et al.* 2005). This effect is understandable from a functional perspective because a higher patchiness increases the potential costs of leaving suitable habitat, but at the same time increases the distance to be travelled towards other suitable patches. Mechanistically this can be explained by a difference in movement behaviour inside habitat and in the matrix (Schtickzelle *et al.* 2007). Effects of landscape characteristics on emigration and immigration may also differ. For example, 'hard' boundaries with strong contrasts between habitat and matrix are likely to reduce emigration, while they may actually increase immigration rates due to improved detectability for dispersers (Englund and Hamback 2007). Landscape effects on dispersal will be further discussed in Chapter 30.

In passive dispersers as well, landscape structure can have an important effect by altering the abiotic and biotic conditions that affect movement (the 'extended landscape', see Chapter 5). Examples are the lower effect of density on aquatic invertebrate dispersal under high current flow rates (Fonseca and Hart 1996), or the complex interactions between weather variables and patch quality on aerial dispersal decisions in spiders (Bonte *et al.* 2003; Bonte *et al.* 2007a). The influence of weather conditions on active dispersers has received much less attention (but see Walls *et al.* 2005; Rizkalla and Swihart 2007; Delgado *et al.* 2010). Landscape structure can also affect zoöchorous dispersal by affecting the presence and/or movement patterns of biotic agents responsible for dispersal (see further in Chapter 5).

In the short term, landscape structure may directly affect dispersal propensity and movement during transfer. In the longer term, it is likely to affect adaptive variation in dispersal-related traits which again can be differentially expressed according to the stage of dispersal. Several studies have shown that dispersal capacity and/or propensity tend to decrease over time in isolated populations (Cody and Overton 1996; Ahlroth *et al.* 2010; Desrochers 2010). Hanski *et al.* (2004) further showed that dispersal is positively related to connectivity in old, but not in new populations. An experimental study on woodland butterflies reared in different conditions showed that individuals from more fragmented landscapes were less likely to leave shaded habitat but had higher wing loadings (indicating higher flight capacity) and were better at orienting towards wooded habitat (Merckx *et al.* 2003; Merckx and Van Dyck 2007).

Note that the structure of the landscape may not only affect dispersal rates or distances as such, but can also have more qualitative effects on dispersal. Variation in landscape permeability may channel dispersers towards destinations with higher connectivity, for example by using corridors or stepping stones (e.g. Haddad *et al.* 2003; Gilbert-Norton *et al.* 2010). This in turn may lead to non-random dispersal patterns and distribution of kin even at a fine-scale level (Matthysen 2002). Landscape structure can also hamper the efficiency of collecting information prior to or during dispersal (Fletcher 2006), which in turn can result in suboptimal settlement patterns with respect to resource or mate availability (Matthysen and Currie 1996; Matter *et al.* 2009).

1.5 Conclusions and perspectives

Several concluding points can be highlighted from this review. A first point is that dispersal is highly diverse in its mechanisms, and therefore inevitably invokes a large number of potential causes. In terms of mechanistic processes, the downstream drifting of an aquatic invertebrate larva can hardly be compared with the transfer of a young primate to a neighbouring group, yet both may contribute to separating the individual from its kin and/or the colonization of undersaturated habitats; thus functionally they may be more similar than at first recognized. Nevertheless, since any process changing an organism's location may eventually contribute to its dispersal, the factors that explain variation in dispersal are inevitably extremely varied and heterogeneous. Thus, similar dispersal patterns may derive from very different mechanisms and causes, while seemingly different outcomes may be due to the same causal factor (Massot *et al.* 2002). Theories explaining the evolution and consequences of dispersal will have to take these widely divergent causal factors into account, yet the majority of existing models still treat dispersal as a simple process characterized by a single rate or kernel. The increasing attention to modelling movement trajectories in a more realistic fashion (Nathan *et al.* 2008; Barton *et al.* 2009) offers perspectives towards a more mechanism-driven modelling approach.

A second point is that multicausality of dispersal can to a large extent be linked to the availability and use of information. The flow of information from the environment to the dispersing (or philopatric) individual, often channelled through the parental phenotype, is strongly linked with spatial scale and highly uneven with respect to the origin and destination of dispersal. As individuals depart from their site of origin they accumulate information, but this information changes in content and in quality, depending on their movement strategy and cognitive abilities (Clobert *et al.* 2009). The question of how dispersal decisions are influenced by the accumulation of information, especially in the transfer and settling stage, has so far received little attention. Few studies have analysed characteristics of movement paths in detail in the light of information, despite the fact that new developments in remote observation technology, data logging, and analytical techniques are making this increasingly more feasible (Doerr and Doerr 2005; Nathan *et al.* 2008; Urbano *et al.* 2010).

A third recurring point in this review is that causal factors may differ greatly in their effects between the three main stages of dispersal. Figure 1.2 is an attempt to generalize these findings in terms of overall importance of effects, obviously grossly ignoring variation across taxa and contexts, and not based on any quantitative analysis. Nevertheless, some general points seem to emerge from this review. Innate variation in dispersal ability (wingedness, size, endurance, navigation ability), whether genetically or developmentally determined, plays a major role during transfer but less so in emigration and immigration decisions. Simply put, the act of leaving or entering a habitat is generally less demanding in terms of movement capacity, than transferring between habitats. On the other hand, innate behaviours (dispersal 'propensity', perhaps linked to 'personality') are clearly important in emigration decisions and probably also affect behaviour during transfer, but it is less clear to what extent different strategies exist with respect to settling decisions. The impact of parents and kin is obviously most important in the emigration stage, but with a few exceptions is unlikely to have a direct impact (other than through persisting maternal effects) beyond the initial transfer stage. Habitat quality and density have strong effects—though not always in the same way—in the emigration and immigration stage, but less so during transfer. Conversely, the physical environment has its strongest potential effect in the transfer stage, although in passive dispersers it is likely to affect emigration as well. Finally, landscape structure is undoubtedly highly important in the transfer stage, followed by a large but somewhat lesser impact in the emigration stage (mainly through responses to boundaries), and a much smaller effect in the immigration stage.

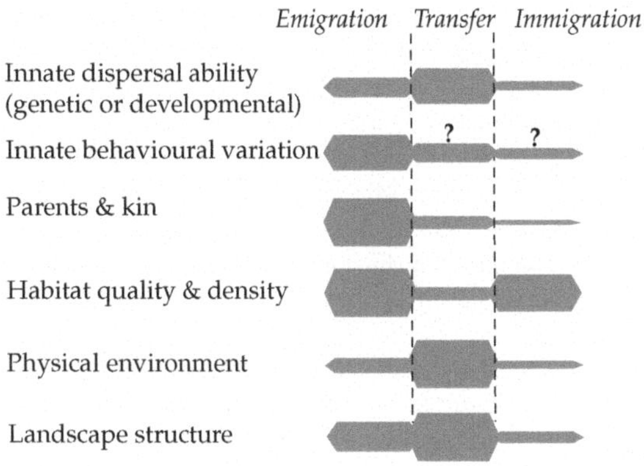

Figure 1.2 General overview of the importance of different causal factors in the dispersal process. Bar width indicates the relative importance of each factor in each of the three stages of the dispersal process. See text for further explanation.

Clearly the simplified scheme in Figure 1.2 may look quite different for taxa using very different mechanisms of dispersal. The main point here is to draw attention to the strongly divergent profiles of major causal agents across the three dispersal stages, implying that results from studies focusing on a particular stage (and/or on a subset of the dispersal distance distribution) cannot be easily generalized to other stages.

A fourth and final point is that interactions between causal factors still deserve more attention. Where possible, I have highlighted known interaction effects between causal variables, but while the number of cases documenting such interactions is steadily increasing, they are often limited to a specific taxon and/or study area, and rarely have such interactions been studied in a systematic fashion across all dispersal stages within the same model system. Interactions between causal factors are probably a major reason why different studies on the proximate causes of dispersal have repeatedly come up with diverging or even contradictory results.

References

Ahlroth, P., Alatalo, R.V. and Suhonen, J. (2010). 'Reduced dispersal propensity in the wingless waterstrider Aquarius najas in a highly fragmented landscape', *Oecologia* 162: 323–30.

Albrectsen, B. and Nachman, G. (2001). 'Female-biased density-dependent dispersal of a tephritid fly in a fragmented habitat and its implications for population regulation', *Oikos* 94: 263–72.

Anholt, B.R. (1990). 'Size-biased dispersal prior to breeding in a damselfly', *Oecologia* 83: 385–87.

Baguette, M. and Van Dyck, H. (2007). 'Landscape connectivity and animal behavior: functional grain as a key determinant for dispersal', *Landscape Ecology* 22: 1117–29.

Barton, K.A., Phillips, B.L., Morales, J.M. *et al.* (2009). 'The evolution of an 'intelligent' dispersal strategy: biased, correlated random walks in patchy landscapes', *Oikos* 118: 309–19.

Bates, A.J., Sadler, J.P. and Fowles, A.P. (2006). 'Condition-dependent dispersal of a patchily distributed riparian ground beetle in response to disturbance', *Oecologia* 150: 50–60.

Beck, C.W. and Congdon, J.D. (2000). 'Effects of age and size at metamorphosis on performance and metabolic rates of Southern Toad, Bufo terrestris, metamorphs', *Functional Ecology* 14: 32–8.

Benard, M.F. and McCauley, S.J. (2008). 'Integrating across life-history stages: Consequences of natal habitat effects on dispersal', *American Naturalist* 171: 553–67.

Berteaux, D. and Boutin, S. (2000). 'Breeding dispersal in female North American red squirrels', *Ecology* 81: 1311–26.

Bonte, D. (2009). 'Inbreeding depresses short and long distance dispersal in three congeneric spiders', *Journal of Evolutionary Biology* 22: 1429–34.

Bonte, D., Bossuyt, B. and Lens, L. (2007a). 'Aerial dispersal plasticity under different wind velocities in a salt marsh wolf spider', *Behavioral Ecology* 18: 438–43.

Bonte, D. and de la Pena, E. (2009). 'Evolution of body condition-dependent dispersal in metapopulations', *Journal of Evolutionary Biology* 22: 1242–51.

Bonte, D., Deblauwe, I. and Maelfait, J.P. (2003). 'Environmental and genetic background of tiptoe-initiating behaviour in the dwarfspider Erigone atra', *Animal Behaviour* 66: 169–74.

Bonte, D., Lukac, M. and Lens, L. (2008a). 'Starvation affects pre-dispersal behaviour of Erigone spiders', *Basic and Applied Ecology* 9: 308–15.

Bonte, D., Travis, J.M.J., De Clercq, *et al.* (2008b). 'Thermal conditions during juvenile development affect adult dispersal in a spider', *Proceedings of the National Academy of Sciences of the USA* 105: 17000–5.

Bonte, D., Van Belle, S. and Maelfait, J.P. (2007b). 'Maternal care and reproductive state-dependent mobility determine natal dispersal in a wolf spider', *Animal Behaviour* 74: 63–9.

Botello, G. and Krug, P.J. (2006). '"Desperate larvae" revisited: age, energy and experience affect sensitivity to settlement cues in larvae of the gastropod Alderia sp.', *Marine Ecology Progress Series* 312: 149–59.

Bowler, D.E. and Benton, T.G. (2005). 'Causes and consequences of animal dispersal strategies: relating individual behaviour to spatial dynamics', *Biological Reviews* 80: 205–25.

Bowler, D.E. and Benton, T.G. (2009). 'Variation in dispersal mortality and dispersal propensity among individuals: the effects of age, sex and resource availability', *Journal of Animal Ecology* 78: 1234–41.

Bullock, J.M., Kenward, R.E., Hails, R.S. *et al.* (2002). *Dispersal ecology*. Cambridge University Press, Cambridge.

Byers, J.E. (2000). 'Effects of body size and resource availability on dispersal in a native and a non-native estuarine snail', *Journal of Experimental Marine Biology and Ecology* 248: 133–50.

Castro, S., Ferrero, V., Loureiro, J. *et al.* (2010). 'Dispersal mechanisms of the narrow endemic Polygala vayredae: dispersal syndromes and spatio-temporal variations in ant dispersal assemblages', *Plant Ecology* 207: 359–72.

Clarke, A.L. and Saether, B.-E. (1997). 'Sex biases in avian dispersal: a reappraisal', *Oikos* 79: 429–38.

Clobert, J., Danchin, E., Dhondt, A.A. *et al.* (2001). *Dispersal*. Oxford University Press, New York.

Clobert, J., Le Galliard, J.F., Cote, J. *et al.* (2009). 'Informed dispersal, heterogeneity in animal dispersal syndromes and the dynamics of spatially structured populations', *Ecology Letters* 12: 197–209.

Cody, M.L. and Overton, J.M. (1996). 'Short-term evolution of reduced dispersal in island plant populations', *Journal of Ecology* 84: 53–61.

Coll, M. and Yuval, B. (2004). 'Larval food plants affect flight and reproduction in an oligophagous insect herbivore', *Environmental Entomology* 33: 1471–76.

Cote, J. and Clobert, J. (2010). 'Risky dispersal: avoiding kin competition despite uncertainty', *Ecology* 91: 1485–93.

Cote, J., Clobert, J., Brodin, T. *et al.* (2010). 'Personality-dependent dispersal: characterization, ontogeny and consequences for spatially structured populations', *Philosophical Transactions of the Royal Society B* 365: 4065–76.

Coulon, A., Cosson, J.F., Angibault, J.M. *et al.* (2004). 'Landscape connectivity influences gene flow in a roe deer population inhabiting a fragmented landscape: an individual-based approach', *Molecular Ecology* 13: 2841–50.

Davis, J.M. and Stamps, J.A. (2004). 'The effect of natal experience on habitat preferences', *Trends in Ecology and Evolution* 19: 411–16.

De Meester, N. and Bonte, D. (2010). 'Information use and density-dependent emigration in an agrobiont spider', *Behavioral Ecology* 21: 992–8.

Delgado, M.D., Penteriani, V., Revilla, E. *et al.* (2010). 'The effect of phenotypic traits and external cues on natal dispersal movements', *Journal of Animal Ecology* 79: 620–32.

Desrochers, A. (2010). 'Morphological response of songbirds to 100 years of landscape change in North America', *Ecology* 91: 1577–82.

Doak, P. (2000). 'Population consequences of restricted dispersal for an insect herbivore in a subdivided habitat', *Ecology* 81: 1828–41.

Doerr, E.D. and Doerr, V.A.J. (2005). 'Dispersal range analysis: quantifying individual variation in dispersal behaviour', *Oecologia* 142, 1–10.

Doligez, B., Danchin, E. and Clobert, J. (2002). 'Public information and breeding habitat selection in a wild bird population', *Science* 297: 1168–70.

Doligez, B., Gustafsson, L. and Part, T. (2009). '"Heritability" of dispersal propensity in a patchy population', *Proceedings of the Royal Society B* 276: 2829–36.

Donohue, K. (1999). 'Seed dispersal as a maternally influenced character: Mechanistic basis of maternal effects and selection on maternal characters in an annual plant', *American Naturalist* 154: 674–89.

Dufty, A.M., Jr. and Belthoff, J.R. (2001). 'Proximate mechanisms of natal dispersal: the role of body condition and hormones', in J. Clobert, E. Danchin, A.A. Dhondt *et al.* eds, *Dispersal*. Oxford University Press, New York, 203–16.

Dunn, P.O. and Braun, C.E. (1985). 'Natal Dispersal and Lek Fidelity of Sage Grouse', *Auk* 102: 621–7.

Eikenaar, C., Richardson, D.S., Brouwer, L. *et al.* (2008). 'Sex biased natal dispersal in a closed, saturated population of Seychelles warblers Acrocephalus sechellensis', *Journal of Avian Biology* 39: 73–80.

Einum, S., Sundt-Hansen, L. and Nislow, K.H. (2006). 'The partitioning of density-dependent dispersal, growth and survival throughout ontogeny in a highly fecund organism', *Oikos* 113: 489–96.

Ekman, J. and Griesser, M. (2002). 'Why offspring delay dispersal: experimental evidence for a role of parental tolerance', *Proceedings of the Royal Society B* 269: 1709–13.

Englund, G. and Hamback, P.A. (2007). 'Scale dependence of immigration rates: models, metrics and data', *Journal of Animal Ecology* 76: 30–5.

Eriksson, O. and Jakobsson, A. (1999). 'Recruitment tradeoffs and the evolution of dispersal mechanisms in plants', *Evolutionary Ecology* 13: 411–23.

Fletcher, R.J. (2006). 'Emergent properties of conspecific attraction in fragmented landscapes', *American Naturalist* 168: 207–19.

Fonseca, D.M. and Hart, D.D. (1996). 'Density-dependent dispersal of black fly neonates is mediated by flow', *Oikos* 75: 49–58.

Forero, M.G., Donazar, J.A. and Hiraldo, F. (2002). 'Causes and fitness consequences of natal dispersal in a population of black kites', *Ecology* 83: 858–72.

Frair, J.L., Merrill, E.H., Allen, J.R. *et al.* (2007). 'Know thy enemy: Experience affects elk translocation success in risky landscapes', *Journal of Wildlife Management* 71: 541–54.

Fraser, D.F., Gilliam, J.F., Daley, M.J. *et al.* (2001). 'Explaining leptokurtic movement distributions: Intrapopulation variation in boldness and exploration', *American Naturalist* 158: 124–35.

Garant, D., Kruuk, L.E.B., Wilkin, T.A. *et al.* (2005). 'Evolution driven by differential dispersal within a wild bird population', *Nature* 433: 60–5.

Gilbert-Norton, L., Wilson, R., Stevens, J.R. *et al.* (2010). 'A Meta-Analytic Review of Corridor Effectiveness', *Conservation Biology* 24: 660–8.

Goodacre, S.L., Martin, O.Y., Bonte, D. *et al.* (2009). 'Microbial modification of host long-distance dispersal capacity', *BMC Biology* 7, 1–8.

Haddad, N.M., Bowne, D.R., Cunningham, A. *et al.* (2003). 'Corridor use by diverse taxa', *Ecology* 84: 609–15.

Handley, L.J.L. and Perrin, N. (2007). 'Advances in our understanding of mammalian sex-biased dispersal', *Molecular Ecology* 16: 1559–78.

Hanski, I., Breuker, C.J., Schops, K. *et al.* (2002). 'Population history and life history influence the migration rate of female Glanville fritillary butterflies', *Oikos* 98: 87–97.

Hanski, I., Eralahti, C., Kankare, M. *et al.* (2004). 'Variation in migration propensity among individuals maintained by landscape structure', *Ecology Letters* 7: 958–66.

Hanski, I., Peltonen, A. and Kaski, L. (1991). 'Natal dispersal and social dominance in the common shrew *Sorex araneus*', *Oikos* 62: 48–58.

Higgins, S.I., Nathan, R. and Cain, M.L. (2003). 'Are long-distance dispersal events in plants usually caused by nonstandard means of dispersal?', *Ecology* 84: 1945–56.

Imbert, E. and Ronce, O. (2001). 'Phenotypic plasticity for dispersal ability in the seed heteromorphic Crepis sancta (Asteraceae)', *Oikos* 93: 126–34.

Jack, K.M. and Fedigan, L. (2004). 'Male dispersal patterns in white-faced capuchins, Cebus capucinus Part 1: patterns and causes of natal emigration', *Animal Behaviour* 67: 761–9.

Kaseda, Y., Ogawa, H. and Khalil, A.M. (1997). 'Causes of natal dispersal and emigration and their effects on harem formation in Misaki feral horses', *Equine Veterinary Journal* 29: 262–6.

Kenward, R.E., Walls, S.S. and Hodder, K.H. (2001). 'Life path analysis: scaling indicates priming effects of social and habitat factors on dispersal distances', *Journal of Animal Ecology* 70: 1–13.

Kesler, D.C., Walters, J.R. and Kappes, J.J. (2010). 'Social influences on dispersal and the fat-tailed dispersal distribution in red-cockaded woodpeckers', *Behavioral Ecology* 21: 1337–43.

Kim, S.Y., Torres, R. and Drummond, H. (2009). 'Simultaneous positive and negative density-dependent dispersal in a colonial bird species', *Ecology* 90: 230–9.

Laloi, D., Richard, M., Federici, P. *et al.* (2009). 'Relationship between female mating strategy, litter success and offspring dispersal', *Ecol. Lett.* 12: 823–9.

Lambin, X., Aars, J. and Piertnet, S.B. (2001). 'Dispersal, intraspecific competition, kin competition and kin facilitation: a review of the empirical evidence', in J. Clobert, E. Danchin, A.A. Dhondt *et al.*, eds, *Dispersal*. Oxford University Press, New York, 261–72.

Le Galliard, J.F., Ferriere, R. and Clobert, J. (2003). 'Mother-offspring interactions affect natal dispersal in a lizard', *Proceedings of the Royal Society B* 270: 1163–9.

Le Galliard, J.F., Ferriere, R. and Clobert, J. (2005). 'Effect of patch occupancy on immigration in the common lizard', *Journal of Animal Ecology* 74: 241–9.

Loe, L.E., Mysterud, A., Veiberg, V. *et al.* (2009). 'Negative density-dependent emigration of males in an increasing red deer population', *Proceedings of the Royal Society B* 276: 2581–7.

Long, E.S., Diefenbach, D.R., Rosenberry, C.S. *et al.* (2008). 'Multiple proximate and ultimate causes of natal dispersal in white-tailed deer', *Behavioral Ecology* 19: 1235–42.

Long, E.S., Diefenbach, D.R., Rosenberry, C.S. *et al.* (2005). 'Forest cover influences dispersal distance of white-tailed deer', *Journal of Mammalogy* 86: 623–9.

Mabry, K.E. and Stamps, J.A. (2008). 'Searching for a New Home: Decision Making by Dispersing Brush Mice', *American Naturalist* 172: 625–34.

Marshall, D.J., Monro, K., Bode, M. *et al.* (2010). 'Phenotype-environment mismatches reduce connectivity in the sea', *Ecology Letters* 13: 128–40.

Massot, M., Clobert, J., Lorenzon, P. *et al.* (2002). 'Condition-dependent dispersal and ontogeny of the dispersal behaviour: an experimental approach', *Journal of Animal Ecology* 71: 253–61.

Mathieu, J., Barot, S., Blouin, M. *et al.* (2010). 'Habitat quality, conspecific density, and habitat pre-use affect the dispersal behaviour of two earthworm species, Aporrectodea icterica and Dendrobaena veneta, in a mesocosm experiment', *Soil Biology and Biochemistry* 42: 203–9.

Matter, S.F., Ezzeddine, M., Duermit, E. *et al.* (2009). 'Interactions between habitat quality and connectivity affect immigration but not abundance or population growth of the butterfly, Parnassius smintheus', *Oikos* 118: 1461–70.

Matter, S.F. and Roland, J. (2002). 'An experimental examination of the effects of habitat quality on the dispersal and local abundance of the butterfly Parnassius smintheus', *Ecological Entomology* 27: 308–16.

Matthysen, E. (2002). 'Boundary effects on dispersal between habitat patches by forest birds (Parus major, P-caeruleus)', *Landscape Ecology* 17: 509–15.

Matthysen, E. (2005). 'Density-dependent dispersal in birds and mammals', *Ecography* 28: 403–16.

Matthysen, E., Adriaensen, F. and Dhondt, A.A. (1995). 'Dispersal distances of nuthatches, *Sitta europaea*, in a highly fragmented landscape', *Oikos* 72: 375–81.

Matthysen, E. and Currie, D. (1996). 'Habitat fragmentation reduces disperser success in juvenile nuthatches *Sitta Europaea*: evidence from patterns of territory establishment', *Ecography* 19: 67–72.

Matthysen, E., Van de Casteele, T. and Adriaensen, F. (2005). 'Do sibling tits (Parus major, P-caeruleus) disperse over similar distances and in similar directions?', *Oecologia* 143: 301–7.

Matthysen, E., Van Overveld, T., Van de Casteele, T. *et al.* (2010). 'Family movements before independence influence natal dispersal in a territorial songbird', *Oecologia* 162: 591–7.

Mayland, H., Margolies, D.C., Charlton, R.E. (2000). 'Local and distant prey-related cues influence when an acarine predator leaves a grey patch. *Entomologia Experimentalis et Applicata* 96: 245–52.

McIntosh, A.R., Peckarsky, B.L. and Taylor, B.W. (2002). 'The influence of predatory fish on mayfly drift: extrapolating from experiments to nature', *Freshwater Biology* 47: 1497–513.

Mennechez, G., Schtickzelle, N. and Baguette, M. (2003). 'Metapopulation dynamics of the bog fritillary butterfly: comparison of demographic parameters and dispersal between a continuous and a highly fragmented landscape', *Landscape Ecology* 18: 279–91.

Merckx, T. and Van Dyck, H. (2007). 'Habitat fragmentation affects habitat-finding ability of the speckled wood butterfly, Pararge aegeria L.', *Animal Behaviour* 74: 1029–37.

Merckx, T., Van Dyck, H., Karlsson, B. *et al.* (2003). 'The evolution of movements and behaviour at boundaries in different landscapes: a common arena experiment with butterflies', *Proceedings of the Royal Society Series B* 270: 1815–21.

Moore, J.C., Loggenberg, A. and Greeff, J.M. (2006). 'Kin competition promotes dispersal in a male pollinating fig wasp', *Biology Letters* 2: 17–19.

Nathan, R., Getz, W.M., Revilla, E. *et al.* (2008). 'A movement ecology paradigm for unifying organismal movement research', *Proceedings of the National Academy of Sciences of the USA* 105: 19052–9.

Nunes, S., Duniec, T.R., Schweppe, S.A. *et al.* (1999). 'Energetic and endocrine mediation of natal dispersal behavior in Belding's ground squirrels', *Hormones and Behavior* 35: 113–24.

Nunes, S., Ha, C.D.T., Garrett, P.J. *et al.* (1998). 'Body fat and time of year interact to mediate dispersal behaviour in ground squirrels', *Animal Behaviour* 55: 605–14.

Ockinger, E. and Smith, H.G. (2008). 'Do corridors promote dispersal in grassland butterflies and other insects?', *Landscape Ecology* 23: 27–40.

Palomares, F., Delibes, M., Ferreras, P. *et al.* (2000). 'Iberian lynx in a fragmented landscape: Predispersal, dispersal, and postdispersal habitats', *Conservation Biology* 14: 809–18.

Parejo, D., White, J. and Danchin, E. (2007). 'Settlement decisions in blue tits: difference in the use of social information according to age and individual success', *Naturwissenschaften* 94: 749–57.

Pasinelli, G. and Walters, J.R. (2002). 'Social and environmental factors affect natal dispersal and philopatry of male red-cockaded woodpeckers', *Ecology* 83: 2229–39.

Ritchison, G., Belthoff, J.R. and Sparks, E.J. (1992). 'Dispersal Restlessness—Evidence for Innate Dispersal by Juvenile eastern Screech-Owls?', *Animal Behaviour* 43: 57–65.

Rizkalla, C.E. and Swihart, R.K. (2007). 'Explaining movement decisions of forest rodents in fragmented landscapes', *Biological Conservation* 140: 339–48.

Roff, D.A. and Fairbairn, D.J. (2001). 'The genetic basis of dispersal and migration, and its consequences for the evolution of correlated traits', in J. Clobert, E. Danchin, A.A. Dhondt *et al.*, eds, *Dispersal.*Oxford University Press, New York, 191–202.

Roland, J., Keyghobadi, N. and Fownes, S. (2000). 'Alpine Parnassius butterfly dispersal: Effects of landscape and population size', *Ecology* 81: 1642–53.

Ronce, O. (2007). 'How does it feel to be like a rolling stone? Ten questions about dispersal evolution', *Annual Review of Ecology Evolution and Systematics* 38: 231–53.

Ronce, O., Clobert, J. and Massot, M. (1998). 'Natal dispersal and senescence', *Proceedings of the National Academy of Sciences of the USA* 95: 600–5.

Schmidt, K.A., Dall, S.R.X. and van Gils, J.A. (2010). 'The ecology of information: an overview on the ecological significance of making informed decisions', *Oikos* 119: 304–16.

Schtickzelle, N., Joiris, A., Van Dyck, H. *et al.* (2007). 'Quantitative analysis of changes in movement behaviour within and outside habitat in a specialist butterfly', *BMC Evolutionary Biology* 7, 1–15.

Schurr, F.M., Steinitz, O. and Nathan, R. (2008). 'Plant fecundity and seed dispersal in spatially heterogeneous environments: models, mechanisms and estimation', *Journal of Ecology* 96: 628–41.

Selonen, V. and Hanski, I.K. (2006). 'Habitat exploration and use in dispersing juvenile flying squirrels', *Journal of Animal Ecology* 75: 1440–9.

Serrano, D., Tella, J.L., Forero, M.G. *et al.* (2001). 'Factors affecting breeding dispersal in the facultatively colonial lesser kestrel: individual experience vs. conspecific cues', *Journal of Animal Ecology* 70: 568–78.

Sinervo, B., Calsbeek, R., Comendant, T. *et al.* (2006). 'Genetic and maternal determinants of effective dispersal: The effect of sire genotype and size at birth in side-blotched lizards', *American Naturalist* 168: 88–99.

Sinervo, B. and Clobert, J. (2003). 'Morphs, dispersal behavior, genetic similarity, and the evolution of cooperation', *Science* 300: 1949–51.

Sloggett, J.J. and Weisser, W.W. (2002). 'Parasitoids induce production of the dispersal morph of the pea aphid, Acyrthosiphon pisum', *Oikos* 98: 323–33.

Sorci, G., Massot, M. and Clobert, J. (1994). 'Maternal parasite load increases sprint speed and philopatry in female offspring of the common lizard', *American Naturalist* 144: 153–64.

Stamps, J.A. (2001). 'Habitat selection by dispersers: integrating proximate and ultimate approaches', in J. Clobert, E. Danchin, A.A. Dhondt *et al.*, eds, *Dispersal.* Oxford University Press, New York, 110–22.

Stamps, J.A. (2006). 'The silver spoon effect and habitat selection by natal dispersers', *Ecology Letters* 9: 1179–85.

Stamps, J.A., Krishnan, V.V. and Reid, M.L. (2005). 'Search costs and habitat selection by dispersers', *Ecology* 86: 510–18.

Stamps, J.A., Krishnan, V.V. and Willits, N.H. (2009). 'How Different Types of Natal Experience Affect Habitat Preference', *American Naturalist* 174: 623–30.

Starrfelt, J. and Kokko, H. (2010). 'Parent-Offspring Conflict and the Evolution of Dispersal Distance', *American Naturalist* 175: 38–49.

Tschirren, B., Fitze, P.S. and Richner, H. (2007). 'Maternal modulation of natal dispersal in a passerine bird: An adaptive strategy to cope with parasitism?', *American Naturalist* 169: 87–93.

Urbano, F., Cagnacci, F., Calenge, C. *et al.* (2010). 'Wildlife tracking data management: a new vision', *Philosophical Transactions of the Royal Society B* 365: 2177–85.

Valimaki, P. and Itamies, J. (2003). 'Migration of the clouded Apollo butterfly Parnassius mnemosyne in a network of suitable habitats—effects of patch characteristics', *Ecography* 26: 679–91.

Van Dyck, H. and Baguette, M. (2005). 'Dispersal behaviour in fragmented landscapes: Routine or special movements?', *Basic and Applied Ecology* 6: 535–45.

van Oort, H. and Otter, K.A. (2005). 'Natal nutrition and the habitat distributions of male and female black-capped chickadees', *Canadian Journal of Zoology-Revue Canadienne De Zoologie* 83: 1495–501.

Verhulst, S., Perrins, C.M. and Riddington, R. (1997). 'Natal dispersal of Great Tits in a patchy environment', *Ecology* 78: 864–72.

Wahlstrom, L.K. and Liberg, O. (1995). 'Patterns of Dispersal and Seasonal Migration in Roe Deer (Capreolus-Capreolus)', *Journal of Zoology* 235: 455–67.

Walls, S.S., Kenward, R.E. and Holloway, G.J. (2005). 'Weather to disperse? Evidence that climatic conditions influence vertebrate dispersal',. *Journal of Animal Ecology* 74: 190–7.

Walls, S.S., Manosa, S., Fuller, R.M. *et al.* (1999). 'Is early dispersal enterprise or exile? Evidence from radio-tagged buzzards', *Journal of Avian Biology* 30: 407–15.

Wender, N.J., Polisetty, C.R. and Donohue, K. (2005). 'Density-dependent processes influencing the evolutionary dynamics of dispersal: A functional analysis of seed dispersal in Arabidopsis thaliana (Brassicaceae)', *American Journal of Botany* 92: 960–71.

Wright, J.R. and Boxshall, A.J. (1999). 'The influence of small-scale flow and chemical cues on the settlement of two congeneric barnacle species', *Marine Ecology Progress Series.* 183: 179–87.

Zedrosser, A., Stoen, O.G., Saebo, S. *et al.* (2007). 'Should I stay or should I go? Natal dispersal in the brown bear', *Animal Behaviour* 74: 369–76.

CHAPTER 2

The theory of dispersal under multiple influences

Jostein Starrfelt and Hanna Kokko

2.1 Introduction

In this book, dispersal is defined as movement that leads to gene flow. This is a fairly wide definition of dispersal, as it is not limited to what type of entities that move (adults, juveniles, seeds, or pollen), mechanism of movement (passive dispersal, active locomotion, etc.), and whether or not the movement is seen as an evolutionary adaptation. To survey models that deal, in some sense, with movement that leads to gene flow is therefore difficult. Here we will briefly chart the established ultimate factors affecting dispersal evolution, as well as the proximate factors that can be of importance in individuals' decisions to disperse or to stay philopatric.

A prerequisite in understanding the theory of dispersal is that not every model is built with the same scope and purpose. A classic in the modelling literature is the trade-off between generality, realism, and precision, presented by Levin (1966). Though this view has been criticized for being somewhat unclear (see, for instance, Odenbaugh 2006), it does point to the fact that not all models yield the same *kind* of insight or knowledge. The scope of some models is to fit a specific biological system (e.g. Zheng *et al.* 2009), while others can be seen as proof-of-principle models, where a specific mechanism, pattern, or causality can be shown to be, in principle, possible; for example, that optimal distances or rates of dispersal can differ between mothers and offspring (Starrfelt and Kokko 2010; Motro 1983). In this chapter we focus mainly on mathematical models where we are interested in the consequences of a set of fairly reasonable assumptions more than in making accurate predictions; i.e. models that favour generality and realism over precision. Also, we focus on reasons behind dispersal and their interactions (including feedback effects), while spending less time exploring the consequences of dispersal in non-evolutionary models. Lastly we highlight the importance of incorporating models where not only the mean effects are taken into account.

2.2 Dispersal and its consequences: a feedback loop

The collection of models of dispersal can reasonably be divided into those models where dispersal is primarily seen as an effect (i.e. something to be explained) and those where dispersal is seen as a cause of something else. In the first case, we are attempting to explain why the patterns of dispersal are the way they are, instead of some different pattern, by manipulating some variable other than dispersal. Models attempting to explain dispersal *per se* range from models that are tailored towards analysing movement and dispersal (e.g. Turchin 1998) to evolutionary models attempting to find general mechanisms selecting for dispersal-enhancing traits.

In the second type of model, the goal is not to explain dispersal but to examine the consequences of it; i.e. by manipulating dispersal propensities and studying their effect on something else. In such models, the ecological or evolutionary pattern to be explained can be, for example, a large-scale ecological pattern such as the species distributions described by island biogeography (MacArthur and Wilson 1967). Other examples include the study of species ranges (e.g. Holt and Keitt 2005; Holt 2003),

Dispersal Ecology and Evolution. First Edition. Edited by Jean Clobert *et al.*

or population persistence (e.g. Hanski 2001); both can be studied either in a temporally invariant world or in a changing one—the latter often referring to climate change. Finally, spatial distributions of populations (e.g. Mueller and Fagan 2008) can also be the pattern to be explained, and often such work considers the evolutionary dynamics of traits other than dispersal (e.g. Rice and Papadopoulos 2009). Dispersal in such contexts is often assumed to be a population-level trait that does not evolve: for example, in their study of pollen and seed flow and their effect on species range expansion, Hu and He (2006) considered selection on viability loci but dispersal was a non-evolving trait—for other similar studies of local adaptation and plasticity, see for example, Sultan and Spencer (2002), and for effects on other traits such as helping and harming conspecific individuals, see for example, Johnstone and Cant (2008).

This leads us to our third and, in our opinion, extremely important viewpoint. Dispersal can produce ecological patterns, but these patterns can again influence the selective pressures on dispersive traits. Only by closing this loop, that is, by realizing that dispersal can at the same time act as both a cause and an effect, will we get the full picture of the evolutionary ecology of dispersal. This principle applies whether or not dispersal is considered the only evolving trait or whether it co-evolves with other traits such as cooperation (Le Galliard *et al.* 2005) or some trait influencing fitness in a spatially and temporally heterogeneous landscape (Blanquart and Gandon 2011). This, of course, makes the study of interacting causes of dispersal challenging, and most models attempt to improve our understanding of the feedback by considering one or a few factors at a time.

A good example is the study of Blanquart and Gandon (2011), who considered two traits (two loci), the first of which influences a fitness-related trait such as the immune response to a parasite, and the second controls dispersal. Spatial variation in the environment then creates local adaptation (since locally unfit individuals produce fewer offspring). The effect of local adaptation that is easiest to comprehend is that it selects against dispersal: the very existence of an individual in its natal patch implies that its parents must have been fit enough to produce young. Given that traits are heritable and environmental challenges are autocorrelated spatially as well as temporally, dispersing elsewhere is likely to lead to lower fitness than staying in the natal patch (Hastings 1983; McNamara and Dall 2011).

However, when there is temporal variation, the story is considerably more complex, as some dispersal can be selected for despite local adaptation (Blanquart and Gandon 2011). Simultaneously, dispersal itself can increase the population-wide level of local adaptation, measured as the difference between the mean fitness of subpopulations in their native environment *versus* elsewhere (i.e. expected fitness if these subpopulations were transplanted to another deme at random). Dispersal has this effect because it provides a flux of individuals with different genotypes that local selection can act on; without genetic variation there can be no response to selection, and thus no adaptation. The net result is complicated, but in a nutshell, selection for more dispersal can increase local adaptation until the linkage disequilibrium between the two alleles (philopatric individuals are more adapted to local conditions) is strong enough to select against further dispersal, leading to a predicted rate of dispersal that depends on the timescale of environmental fluctuations (Blanquart and Gandon 2011).

This is a beautiful example of eco-evolutionary feedback, but as noted by the authors themselves, their analysis ignores many forces known to promote dispersal: for example, another type of determinant of local fitness is the degree of inbreeding of the offspring, which obviously can depend on the distance dispersed (assuming that mating occurs after dispersal). Since the relatedness structure of a population depends on dispersal, selection for dispersal will co-evolve with this structure (e.g. Gandon and Rousset 1999; Lehmann and Perrin 2003). What happens when 'everything occurs simultaneously and interacts with everything'? This is a difficult question to answer, because our understanding of complex systems is not much advanced by throwing every conceivable causality into a model and staring at the outcome. There is probably no alternative to first understanding each mechanism on its own, before joining two (and then maybe more) into a joint model to see how predictions change (Gandon and Michalakis 2001). To some extent, this is how the field indeed has progressed.

2.3 Ultimate and proximate factors in explaining dispersal

There are two very distinct answers to the question 'what causes dispersal'. These are most easily viewed by recapping Niko Tinbergen's (1963) four questions which he instructed scientists to ask about any animal behaviour. The questions, loosely based on Aristotle's four types of causes, can be exemplified as:

1. Proximate:
   a. Mechanism. What stimuli elicit the behaviour?
   b. Development or ontogeny. How does the behaviour change with age or develop in a given individual?
2. Ultimate:
   a. Phylogeny. How does the particular behaviour compare with similar behaviours in similar species?
   b. Adaptation or function. How does the behaviour affect the individuals' chances of survival and reproduction?

The major part of the theory of explaining dispersal has dealt with the last ultimate question. Dobzhansky's statement that 'nothing in evolution makes sense, except in the light of evolution' (Dobzhansky 1973) is an example of how ultimate thinking should always guide our quest for answering all of these questions. We would like to emphasize, however, that in evolutionary models the converse is also important; understanding how ultimate causes influence dispersal evolution may require some knowledge about how the dispersal strategy develops and is triggered. For instance, whether the development of a dispersive morph is under parental or offspring control (i.e. how dispersive behaviour develops), changes the predictions of ultimate models of dispersal rates (Motro 1983) and distances (Starrfelt and Kokko 2010). Likewise, since dispersal strategies can be dependent on the state of individuals such as age (Bowler and Benton 2009) or condition (see Chapter 11), evolutionary trajectories may depend on how the dispersive traits develop in the individual organism. Dispersal strategies that are conditional on some attribute of the subpopulation in which individuals find themselves can also show markedly different patterns of movement (particularly invasions) than unconditional strategies (e.g. Travis *et al.* 2009). Thus, a too explicit focus on ultimate questions only, without information of proximate processes, could fail, and we therefore briefly review proximate factors now before continuing to ultimate factors.

2.4 Proximate factors

In the modelling literature, proximate factors are included in both studies where dispersal is seen as an explanation and in the literature on explaining dispersal (see, for example, Tables 1 and 2 in Bowler and Benton 2005). Proximate factors must induce movement of an individual, and often the organism makes use of these factors in a way that makes adaptive sense: for example, female parasitoid wasps *Cotesia glomerata* are observed to leave their natal patch disproportionately often when the local mating opportunities are confined to inbreeding (Ruf *et al.* 2011). Similarly, ontogenetic changes in dispersiveness are often argued to be adaptive; e.g. the greater mobility of young versus old individuals can be seen to be an example of the 'asset-protection principle' (Clark 1994)—the young should be less risk averse than old individuals, since they do not yet have an established home range or territory, they have less to lose by moving. But dispersal can also be triggered at any age, e.g. by breeding failure of self and/or neighbours (Rioux *et al.* 2011). This is called the 'win-stay, lose-switch' strategy (Switzer 1993). There appears to be little work that attempts to connect such adaptive rules of breeding dispersal to a more general understanding of why, ontogenetically, individual dispersal propensities decline with age: early work by Johst and Brandl (1999) addresses this issue, suggesting that the age structure of a population is important, but this modelling work has been followed up only scantily.

Thus we do not know much about what factors determine the relative frequency of natal *versus* breeding dispersal.

Against the general impression that proximate cues of dispersal often serve to trigger movements that prove adaptive, it is also healthy to remember that dispersal of organisms often is not 'intentional' in this sense. A flock of birds displaced by a storm during migration can potentially establish a new breeding area; fieldfares breeding in Iceland are a potential example. If all non-intentional movement (currents, winds, seed carrying by animals, etc.) occurred completely irrespectively of traits of individuals, the evolutionary impact of such effects might remain limited, although even then the spatio-temporal structure of the populations would be influenced by displacements, and this will have feedback effects back to evolutionary traits. Importantly, and adding a direct component to such indirect effects, physical forces often displace individuals differentially depending on the morphology as well as behaviour of individuals. Migrant bird species, for example, show longer dispersal distances than year-round residents living in the same area (Paradis *et al.* 1998). Likewise, within a species, more mobile individuals might forage and search for mates more efficiently, but such individuals can also end up exiting the local habitat patch more often, forcing them become dispersers (especially in small organisms such as butterflies that are easily displaced by wind; Haag *et al.* 2005; Niitepõld *et al.* 2009).

It may be as difficult to disentangle 'voluntary' and 'involuntary' dispersal decisions in such cases as it is to ask whether the shape of the dispersal kernel is the precise 'intended' (i.e. adaptive) result when temporally varying wind conditions interact with the wing shape of seeds. For the purpose of deriving evolutionary and ecological predictions, luckily, this does not matter. The spread of organisms over space results from the trait distribution interacting with the conditions found in the environment, and the evolution of traits proceeds according to the expected dispersal kernel predicted by a trait (see Chapter 16). As the above examples show, it is important to be aware of the proximate mechanisms by which a certain trait produces a particular type of dispersal kernel. Ultimate explanations (selection) will be based on this kernel (the outcome of dispersal; see Chapter 16), while the evolving trait may be a morphological or behavioural characteristic. Most models of dispersal evolution lump these two together. The newly emerging field of 'movement ecology' shows some promise in making the link much more explicit (Jordano 2011; Nathan *et al.* 2008; Mueller and Fagan 2008; Kesler *et al.* 2010), but progress in this area is yet to be linked to evolutionary theory on dispersal.

2.5 Ultimate factors

As stated earlier, evolutionary models of dispersal typically do not let the morphological or behavioural characteristics and the consequent movement rules evolve. Instead, they simplify the world by considering that the evolving trait is either a rate of dispersal (usually defined as the proportion of young that leave their natal patch) or a parameter describing the dispersal kernel (e.g. the mean distance in an exponential distribution). The focus away from proximate explanations is that this allows the models to concentrate on the basic and general conundrum: why do individuals show adaptations that enhance movement even though movement is risky? Risks may involve a combination of energetic costs and risks of moving through unfamiliar matrix habitat, as well as not easily finding a new suitable place to settle and breed. In addition, under spatial variation, natal patches yielding many potential dispersers will be better breeding sites than average, which selects against dispersal (McNamara and Dall 2011).

Since Hamilton and May (1977), it has been known that kin selection offers a powerful explanation: by alleviating competition for resources and thus enhancing the reproductive success of kin that do not move, it can select for dispersal even if the environment is temporally and spatially invariant. Of course, spatio-temporal variation often exists in the environment and it too can select for dispersal (Crespi and Taylor 1990; McPeek and Holt 1992; Doebeli and Ruxton 1997; Heino and Hanski 2001); this category includes the study and effects of range margins (Dytham 2009). Density dependence similarly creates spatio-temporal variation among habitat patches that can favour dispersal (Olivieri *et al.* 1995). Compared to these factors, inbreeding avoidance is a relatively distinct mechanism that can select for dispersal. It is often argued to lead to a sex bias in dispersal, because for inbreeding to be avoided, it is sufficient for one sex to disperse (Motro 1991; Perrin and Mazalov 1999; Lehmann and Perrin 2003).

Naturally, all these factors can interact, and much theoretical progress has focused on predicting evolutionary trajectories when several causal routes to dispersal are considered simultaneously. To mention examples, Ozaki (1995) and Gandon and Michalakis (1999) consider kin competition in scenarios that include spatio-temporal variation, and Parvinen *et al.* (2003) include density dependence, demographic stochasticity, and temporal variation in habitat suitability in their model. An interesting feature of such work in general is that verbal explanations of models may focus on only a subset of factors that are at work in the model. A good example is provided by kin competition. This potential route to positive dispersal rates is present whenever related individuals have the chance to interact locally: for example, they may potentially breed in the same patch but there is within-patch density-dependence, which predicts either that not all attempts to gain a breeding site within a patch will be successful, or the breeding success of all local breeders declines with local density. Clearly, there then exists a component of indirect fitness that increases by dispersing. However, a model that is not explicitly framed in terms of inclusive fitness—it might be formulated as an individual-based simulation of births and deaths—may have its author emphasizing the effects of density-dependence or temporal or spatial variation, although kin effects have implicitly (and correctly) been incorporated. This makes it somewhat challenging to read the literature on the various causes of dispersal, although there are attempts to tease apart the relative influence of different causes (Poethke *et al.* 2007).

As stated earlier in this chapter, there may be no shortcut to the hard work of first considering each effect in isolation and building a more complex system of interacting forces thereafter (Gandon and Michalakis 2001). In this endeavour it may prove helpful to strive towards a conceptual framework where all the factors that favour the evolution of dispersive traits are seen as manifestations of a single cause of dispersal as an adaptation. This single cause is variation in expected genotypic fitness over space and/or time. Such variability, in turn, can come about through four different mechanisms (or factors): variation in inclusive fitness over space (avoidance of kin competition), externally varying environmental conditions (and hence fitness) over space and time, variability in conditions over space and time driven by the population itself (e.g. demographic stochasticity), or variation in fitness due to relatedness structure of the population. The last fact may include kin interactions in a non-mating context (cooperation among kin) but also reduced survival or fecundity of inbred offspring, leading to avoidance (or tolerance, Kokko and Ots 2006) of inbreeding.

The conditions for these general mechanisms to select for dispersive traits differ slightly. For dispersal to be favoured over philopatry through the first three mechanisms, we minimally require an assumption of local competition for resources. If competition is global (that is, if competition between individuals is independent of spatial position), there is no general incentive to move at all, since it will by definition not affect an individual's chance of reproductive success. For dispersal to evolve to avoid negative fitness consequences of inbreeding, competition need not be local, but mating must be, so that moving decreases the chance of mating with kin, and hence the probability of reduced fitness of inbreed offspring.

Any adaptation will be affected by both costs and benefits, and there are a number of factors that will reduce the fitness of a dispersing individual/genotype compared to a philopatric one. Earlier, we have commented on the fact that evolutionary dispersal models tend to lump the entire movement process into one variable such as 'dispersal rate'. While this has the advantage of keeping the model simple, a better understanding of selection on actual dispersal-enhancing traits will require considering the movement process in more detail. It can be instructive to divide dispersal into the three different phases of emigration, inter-patch movement, and immigration together with the corresponding potential fitness-reducing factors. For the emigration phase, there might be costs associated with developing the phenotypic mechanism of dispersal; often models posit a trade-off between dispersal abilities and survival or fecundity, though this might not be empirically grounded in all systems (Hanski *et al.* 2006). There are also several factors that can increase the expected fitness in the natal patch compared to other patches, therefore favouring philopatry: loss of kin cooperation and familiarity with the natal territory are potential examples that can be viewed as costs paid by dispersers that relate to the decision to emigrate. The transfer phase (i.e. movement of individuals through unsuitable matrix habitat) has its own costs, as individuals may experience a direct mortality risk while dispersing, in addition to a probability of not finding a suitable habitat at all. This last factor relates to success as an immigrant, although the difficulties of finding suitable vacant breeding habitat obviously can interact with mortality risks as such difficulties can prolong the phase of moving through matrix habitat (e.g. McCarthy 1999).

In addition to understanding the effects of these ultimate factors alone and in concert, there is yet another perspective on the multicausal nature of dispersal, and in fact evolutionary change in general (Rice 2008; Rice and Papadopoulos 2009); these factors can affect all moments (i.e. means, variances, skewness, etc.) of fitness distributions. For instance, the degree of kin competition among individuals in a patch will not be exactly equal for all individuals, but can be seen as a distribution of competitive impacts of kin, with a mean, variance, etc. In most analytical models, assumptions are made so as to capture the effects of the mean of such distributions, mostly by assuming infinite population sizes or number of patches making the higher moments negligible. Another example is that of mortality risk during the movement phase of dispersal. Assuming that the number of dispersers is very large, one disregards the probabalistic nature of this risk: a given fraction of the dispersers are assumed to die instead of focusing on how this risk leads to a distribution of individuals either surviving or dying, with an accompanying degree of variation. In general it is very difficult to derive how these fitness distributions at the level of the individuals have an effect on the fitness at the level of genotype(s), and one of the major challenges of modelling the evolution of dispersal lies here; many approaches (e.g. the direct-fitness method) do not take this variability into account and only deal with the mean fitness of individuals (and genotypes) (Ronce *et al.* 2001; Ronce 2007). This also contributes to the difficulty of linking individual-based models, where the effects of the full distribution of individual fitnesses will be taken into account, to more analytical methods focusing on mean effects only.

2.6 Dispersal homeostasis

The distribution of individual fitness, particularly the higher moments, depends on what proximate mechanisms of dispersal decisions are incorporated in a model. As an example of how these higher moments can affect the evolution of dispersal, we will now construct a simple individual-based model mimicking the classic direct-fitness treatments of Frank (1986, 1998). For a related model, see Lehmann and Balloux 2007.

The model considers dispersal homeostasis, by which we mean that for a given dispersal trait d, a focal mother will divide her clutch into a dispersive fraction (d), and a philopatric fraction ($1-d$). This differs from letting each offspring develop into a dispersive morph with probability d not only in that it assumes a different developmental process, but also in that this latter option introduces variance among broods in the number of dispersers. We refer to these two options as homeostatic and randomized dispersal, respectively. Our aim now is to mimic the simple direct-fitness model of dispersal to evaluate if there is potential for dispersal homeostasis to perform better than randomized dispersal in our simple system, despite the mean dispersal rate being the same for both.

We posit P patches, wherein one asexual individual can give rise to offspring whose dispersal phenotype she determines. Under the homeostatic interpretation of d, all individuals of genotype d will always produce $n \times d$ dispersing offspring (as

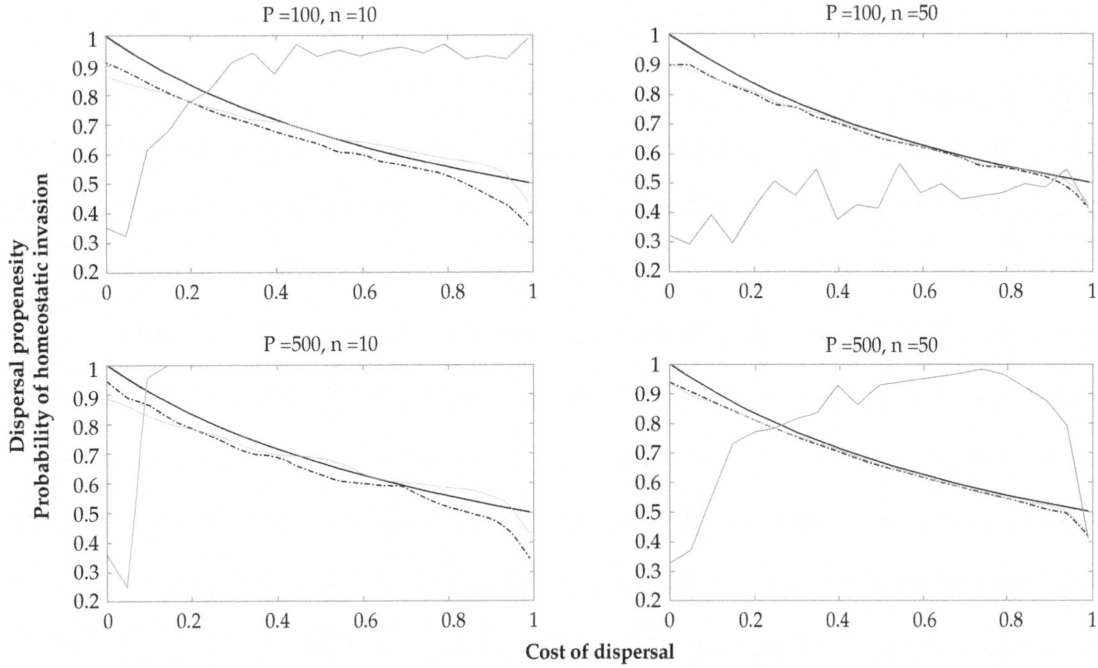

Figure 2.1 Consider *P* patches where one sole individual can breed. This individual is every generation chosen randomly among all the philopatric individuals from last generation and immigrants from other patches. Each reproducing individual gives rise to *n* offspring, each of which either develops into a dispersive phenotype and disperses, or stays philopatric. The mother dies after reproduction. The maternal allele at a haploid locus determines either the probability that each of her offspring will disperse (randomized model) or the fraction of her offspring that will disperse (homeostatic model) to a randomly chosen different patch. We therefore only consider allelic values $d \in \left\{\frac{0}{n}, \frac{1}{n}, \frac{2}{n}, \ldots \frac{n}{n}\right\}$. For each birth there is a small probability (1/500) that the allele will mutate to a randomly chosen different value. For the dispersing phenotype there is a probability of mortality (c) during the movement phase which varies between simulations from 0 to 0.99. The mean allelic values for 100 simulations for each cost setting are shown for the randomized model (dotted line) and the homeostatic (dash-dotted line). The predictions from the analytical treatment of Frank (1986, 1998) are given by the full black line.

For comparing the two different types of phenotype determination or development, the population is initialized with randomized dispersal monomorphic for the optimal dispersal propensity predicted by the direct-fitness model and there are no mutations in the dispersal propensity. After 500 generations, individuals have a slight probability (1/500) of changing to a homeostatic determination for 50 generations. We then let the simulations run until one of the types have outcompeted the other. The grey line in the figure shows the fraction of 100 replicates in which the homeostatic dispersal strategy outcompeted the randomized one, despite the fact that the mean fitness of these two strategies are identical.

well as n ×(1–*d*) non-dispersing offspring). Under a randomized view, each individual of genotype *d* will produce a distribution of dispersing offspring that has the mean $n \times d$ and variance $n \times d \times (1-d)$. Since the mean number of offspring, and therefore the mean fitness of mothers, is the same for both of these setups, we can detect the effect of higher moments of the distribution by comparing evolution under randomized and homeostatic determination of dispersal.

Our individual-based model has the same setup as the direct-fitness model of Frank (1986, 1998). In our first runs of the simulations, we vary the cost of dispersal from 0 to 0.99 under both homeostatic and randomized dispersal (see legend for Figure 2.1 for details). After 2000 generations, the mean dispersal propensity under both assumptions conform fairly nicely to the direct-fitness predictions (Frank, 1986, 1998), at least for moderate costs (Figure 2.1). The simulation results are also closer to these predictions when we increase both the number of patches and the number of offspring, which is an expected pattern since both are assumed to be large in the direct-fitness approach. Thus, these simulations provide reassurance that our stochastic individual-based implementation can reproduce the basic results from the direct-fitness model.

To investigate the potential for selection for homeostatic dispersal, we can now initialize simulations with randomized strategies where *d* equals the predicted optimal dispersal propensity from the direct-fitness model, and see if a homeostatic strategy with the same dispersal propensity *d* will invade and take over the population.

When comparing the two different proximate mechanisms (homeostatic and randomized dispersal), a useful baseline is provided by setting the mortality cost of dispersal to zero and the dispersal propensity to 1. With this setting, there is no difference between the two different proximate mechanisms of dispersal, since 'dispersing with probability 1' and 'dispersing the whole clutch' both imply that all offspring develop as the dispersive morph. In this particular setup, a neutral allele (use of homeostasis instead of randomized dispersal) grows to fixation with probability between 0.3 and 0.4. For all the other dispersal propensities, the invasion probability of a homeostatic determination of dispersal morphs is higher (Figure 2.1). This is a clear indication that, even though both proximate interpretations of the dispersal propensity in the direct-fitness model are valid, they can be teased apart in our model, and that homeostatic (i.e. with completely deterministic proximate causality) dispersal strategies are favoured over randomized ones (i.e. with probabilistic proximate causality). Since the mean (or expected) fitness of the parents in these models are identical, homeostatic dispersal strategies can evolve through their effect of the higher moments of the fitness distributions. In general, all moments of a fitness distribution contribute to evolutionary change (Rice, 2008), and thus the fact that different proximate mechanisms of dispersal can imply clear differences in these distributions appears as understudied as the linking of movement rules to evolutionary change in general.

2.7 Summary

Our chapter highlights that dispersal can be seen as multicausal in many different perspectives. It is multicausal in the sense that researchers may be interested in what causes dispersal, or dispersal as a cause of other ecological or evolutionary patterns and processes—or, ideally, the feedback between these two. It is also multicausal in the sense that we often both choose to, and have to, deal with different notions of causality. The decision for a particular individual to emigrate, alter tactics in the movement phase, or when and where to settle, can all occur both in a probabilistic and a deterministic way. Dispersal is also multicausal in the sense that the ultimate 'factors' identified as selecting for dispersive traits (spatio-temporal variability, avoidance of kin competition, and inbreeding) affects not only the mean fitness of individuals and genotypes, but also the higher moments. Finally, the multicausal nature of dispersal relates to how we explain dispersal patterns; we can do so by referring to either proximate and ultimate factors, both of which can be either deterministic or probabilistic. We suggest that a tighter link between the proximate and ultimate two would bring the field forward. We thus urge more explicit treatment of proximate mechanisms (developmental and mechanistic) in ultimate models, both to study the effects of including different proximate causes of dispersal, but also to clarify the full impact on the fitness distributions and evolutionary consequences of these.

References

Blanquart, F. and Gandon, S. (2011). 'Evolution of Migration in a Periodically Changing Environment', *American Naturalist* 177: 188–201.

Bowler, D.E. and Benton, T.G. (2005). 'Causes and consequences of animal dispersal strategies: relating individual behaviour to spatial dynamics', *Biological Reviews* 80: 205–25.

Bowler, D.E. and Benton, T.G. (2009). 'Variation in dispersal mortality and dispersal propensity among individuals: the effects of age, sex and resource availability', *Journal of Animal Ecology* 78: 1234–41.

Clark, C.W. (1994). 'Antipredator behavior and the asset-protection principle', *Behavioral Ecology* 5, 159–70.

Crespi, B.J. and Taylor, P.D. (1990). 'Dispersal Rates Under Variable Patch Density', *American Naturalist* 135: 48–62.

Dobzhansky, T. (1973). 'Nothing in Biology Makes Sense Except in the Light of Evolution', *The American Biology Teacher* 35: 125–9.

Doebeli, M. and Ruxton, G.D. (1997). 'Evolution of dispersal rates in metapopulation models: Branching and cyclic dynamics in phenotype space', *Evolution* 51: 1730–41.

Dytham, C. (2009). 'Evolved dispersal strategies at range margins', *Proceedings of the Royal Society B* 276: 1407–13.

Frank, S.A. (1986). 'Dispersal Polymorphisms in Subdivided Populations', *Journal of Theoretical Biology*, 122: 303–9.

Frank, S.A. (1998). *Foundations of social evolution*. Princeton University Press, Princeton NJ.

Gandon, S. and Michalakis, Y. (1999). 'Evolutionary Stable Dispersal Rate in a Metapopulation with Extinctions and Kin Competition. *Journal of Theoretical Biology* 199: 275–90.

Gandon, S. and Michalakis, Y. (2001). 'Multiple causes of the evolution of dispersal', in J. Clobert, J.D. Nichols, E. Danchin *et al.*, eds, *Dispersal*. Oxford University Press, Oxford.

Gandon, S. and Rousset, F. (1999). 'Evolution of stepping-stone dispersal rates', *Proceedings of the Royal Society of London Series B* 266: 2507–13.

Haag, C.R., Saastamoinen, M., Marden, J.H. *et al.* (2005). 'A candidate locus for variation in dispersal rate in a butterfly metapopulation', *Proceedings of the Royal Society B* 272: 2449–56.

Hamilton, W.D. and May, R.M. (1977). 'Dispersal in Stable Habitats', *Nature* 269: 578–81.

Hanski, I. (2001). 'Population dynamic consequences of dispersal in local populations and metapopulations', in J. Clobert, J.D. Nichols, E. Danchin *et al.* eds, *Dispersal*. Oxford University Press, Oxford: 283–98.

Hanski, I., Saastamoinen, M. and Ovaskainen, O. (2006). 'Dispersal-related life-history trade-offs in a butterfly metapopulation', *Journal of Animal Ecology* 75: 91–100.

Hastings, A. (1983). 'Can spatial variation alone lead to selection for dispersal', *Theoretical Population Biology*, 24: 244–51.

Heino, M. and Hanski, I. (2001). 'Evolution of migration rate in a spatially realistic metapopulation model', *American Naturalist* 157, 495–511.

Holt, R.D. (2003). 'On the evolutionary ecology of species' ranges', *Evolutionary Ecology Research* 5: 159–78.

Holt, R.D. and Keitt, T.H. (2005). 'Species' borders: a unifying theme in ecology', *Oikos* 108: 3–6.

Hu, X.S. and He, F.L. (2006). 'Seed and pollen flow in expanding a species' range', *Journal of Theoretical Biology*, 240: 662–72.

Johnstone, R.A. and Cant, M.A. (2008). 'Sex Differences in Dispersal and the Evolution of Helping and Harming', *American Naturalist* 172: 318–30.

Johst, K. and Brandl, R. (1999). 'Natal versus breeding dispersal: Evolution in a model system', *Evolutionary Ecology Research* 1: 911–21.

Jordano, P. (2011). 'Pollen, seeds and genes: the movement ecology of plants', *Heredity* 105: 329–30.

Kesler, D.C., Walters, J.R. and Kappes, J.J. (2010). 'Social influences on dispersal and the fat-tailed dispersal distribution in red-cockaded woodpeckers', *Behavioral Ecology* 21: 1337–43.

Kokko, H. and Ots, I. (2006). 'When not to avoid inbreeding', *Evolution*, 60: 467–75.

Le Galliard, J.F., Ferriere, R. and Dieckmann, U. (2005). 'Adaptive evolution of social traits: Origin, trajectories, and correlations of altruism and mobility', *American Naturalist* 165: 206–24.

Lehmann, L. and Balloux, F. (2007). 'Natural selection on fecundity variance in subdivided populations: Kin selection meets bet hedging', *Genetics* 176: 361–77.

Lehmann, L. and Perrin, N. (2003). 'Inbreeding avoidance through kin recognition: Choosy females boost male dispersal', *American Naturalist* 162: 638–52.

Levins, R. (1966). 'Strategy of Model Building in Population Biology', *American Scientist* 54: 421.

MacArthur, R.H. and Wilson, E.O. (1967). *The theory of island biogeography*. Princeton University Press, Princeton.

McCarthy, M.A. (1999). 'Effects of competition on natal dispersal distance', *Ecological Modelling* 114: 305–10.

McNamara, J.M. and Dall, S.R.X. (2011). 'The evolution of unconditional strategies via the "multiplier effect"', *Ecology Letters*, 14: 237–43.

McPeek, M.A. and Holt, R.D. (1992). 'The Evolution of Dispersal in Spatially and Temporally Varying Environments', *American Naturalist* 140: 1010–27.

Motro, U. (1983). 'Optimal Rates of Dispersal 3. Parent Offspring Conflict', *Theoretical Population Biology* 23: 159–68.

Motro, U. (1991). 'Avoiding inbreeding and sibling competition—The evolution of sexual dimorphism for dispersal', *American Naturalist* 137: 108–15.

Mueller, T. and Fagan, W.F. (2008). 'Search and navigation in dynamic environments—from individual behaviors to population distributions', *Oikos* 117: 654–64.

Nathan, R., Getz, W.M., Revilla, E. *et al.* (2008). 'A movement ecology paradigm for unifying organismal movement research', *Proceedings of the National Academy of Sciences of the United States of America* 105: 19052–9.

Niitepold, K., Smith, A.D., Osborne, J.L. *et al.* (2009). 'Flight metabolic rate and Pgi genotype influence butterfly dispersal rate in the field', *Ecology*, 90: 2223–32.

Odenbaugh, J. (2006). 'The strategy of "The strategy of model building in population biology"', *Biology and Philosophy* 21: 607–21.

Olivieri, I., Michalakis, Y. and Gouyon, P.H. (1995). 'Metapopulation Genetics and the Evolution of Dispersal', *American Naturalist* 146: 202–28.

Ozaki, K. (1995). 'Intergall migration in Aphids—A model and a test of ESS dispersal rate', *Evolutionary Ecology* 9: 542–9.

Paradis, E., Baillie, S.R., Sutherland, W.J. and Gregory, R. D. (1998). 'Patterns of natal and breeding dispersal in birds', *Journal of Animal Ecology* 67: 518–36.

Parvinen, K., Dieckmann, U., Gyllenberg, M. *et al.* (2003). 'Evolution of dispersal in metapopulations with local density dependence and demographic stochasticity', *Journal of Evolutionary Biology* 16: 143–53.

Perrin, N. and Mazalov, V. (1999). 'Dispersal and inbreeding avoidance', *American Naturalist* 154: 282–92.

Poethke, H.J., Pfenning, B. and Hovestadt, T. (2007). 'The relative contribution of individual and kin selection to the evolution of density-dependent dispersal rates', *Evolutionary Ecology Research* 9: 41–50.

Rice, S.H. (2008). 'A stochastic version of the Price equation reveals the interplay of deterministic and stochastic processes in evolution', *BMC Evolutionary Biology* 8: 262 DOI:10.1186/1471-2148-8-262.

Rice, S.H. and Papadopoulos, A. (2009). 'Evolution with Stochastic Fitness and Stochastic Migration', *Plos One:* 4 (10): e7130. DOI:10.1371/journal.pone.0007130.

Rioux, S., Amirault-Langlais, D.L. and Shaffer, F. (2011). 'Piping Plovers make decisions regarding dispersal based on personal and public information in a variable coastal ecosystem', *Journal of Field Ornithology* 82: 32–43.

Ronce, O. (2007). 'How does it feel to be like a rolling stone? Ten questions about dispersal evolution', *Annual Review of Ecology Evolution and Systematics* 38: 231–53.

Ronce, O., Olivieri, I., Clobert, J. *et al.* (2001). 'Perspectives on the study of dispersal evolution', in J. Clobert, J.D. Nichols, E. Danchin *et al.*, eds, *Dispersal.* Oxford University Press, Oxford 341–57.

Ruf, D., Dorn, S. and Mazzi, D. (2011). 'Females leave home for sex: natal dispersal in a parasitoid with complementary sex determination', *Animal Behaviour* 81: 1083–9.

Starrfelt, J. and Kokko, H. (2010). 'Parent-Offspring Conflict and the Evolution of Dispersal Distance', *American Naturalist* 175: 38–49.

Sultan, S.E. and Spencer, H.G. (2002). 'Metapopulation Structure Favors Plasticity over Local Adaptation', *American Naturalist* 160: 271–83.

Switzer, P.V. (1993). 'Site fidelity in predictable and unpredictable habitats', *Evolutionary Ecology* 7: 533–55.

Tinbergen, N. (1963). 'On Aims and Method in Ethology', *Zeitschrift für Tierpsychologie* 20: 410–33.

Travis, J.M.J., Mustin, K., Benton, T.G. and Dytham, C. (2009). 'Accelerating invasion rates result from the evolution of density-dependent dispersal', *Journal of Theoretical Biology* 259: 151–8.

Turchin, P. (1998). *Quantitative analysis of movement: measuring and modeling population redistribution in animals and plants.* Sunderland, Mass, Sinauer.

Zheng, C.Z., Ovaskainen, O. and Hanski, I. (2009). 'Modelling single nucleotide effects in phosphoglucose isomerase on dispersal in the Glanville fritillary butterfly: coupling of ecological and evolutionary dynamics', *Philosophical Transactions of the Royal Society B* 364: 1519–32.

CHAPTER 3

Multi-determinism in natal dispersal: the common lizard as a model system

Jean Clobert, Manuel Massot, and Jean-François Le Galliard

3.1 Introduction

Theoretical models suggest that dispersal is an omnibus response to many biotic and abiotic factors (Hamilton and May 1977; McPeek and Holt 1992; Travis and Dytham 1998; reviewed in Clobert *et al.* 2004), and multi-determinism in dispersal behaviour has been indeed reported (Chapter 1). There are, however, open questions about the extant of multi-determinism in dispersal and, most importantly, about how information on environmental factors is processed and integrated to produce optimal dispersal behaviours (Massot *et al.* 2002). For example, many questions can be raised. Is the presence of predators more important than the presence of conspecifics? Is food abundance more relevant than the presence of genetically related partners? Only a few models and empirical studies have looked at the relative importance of different factors acting on dispersal. For example, Gandon and Michalakis (1999) found that environmental stochasticity was driving the evolution of dispersal more strongly than competition or inbreeding, while Perrins and Malazov (2000) found that only a mixture of causes (inbreeding and kin competition) could explain the evolution of sex-biased dispersal. However, we do not know if these factors act independently, synergistically or, on the contrary, suppress each other's effects (Massot *et al.* 2002). In fact, few observational and even less experimental studies have been designed to address these critical issues.

There are several reasons for the lack of thorough studies of multi-determinism in dispersal behaviour (Clobert *et al.* 2001, 2004). Long term studies of vertebrates (mostly mammals and birds) have been designed mainly to answer demographic questions about the role of density-dependence and climate variations, while similar studies of insects most often did not record individual identity. The spatial scale of many studies is often shorter than the typical scale of dispersal of the study species, with the consequence that few dispersal attempts are recorded and dispersal is considered as a variable of secondary interest. An additional reason is that empirical studies of dispersal often failed to ascribe unambiguously dispersal variation to a particular cause, especially for species with complex social systems. Experiments as well as detailed behavioural studies with appropriate model systems then constitute essential tools to disentangle the processes and patterns of dispersal. Like many other small vertebrates, the common lizard (*Zootoca vivipara*) is a suitable model to investigate the multi-determinism of natal dispersal (Box 1). Long-term studies can be initiated in the field where individuals are marked at birth and virtually all dispersal attempts of surviving individuals can be recorded (Clobert *et al.* 1994). In addition, experiments could be conducted with small groups of individuals in artificial settings, but also with entire populations in natural or semi-natural (fenced populations) conditions. Studies on individually marked animals were essential to gather data on genetic, physiological, and behavioural correlates of dispersal. Long-term studies enabled us to understand processes of natal dispersal as well as correlation between natal dispersal and some environmental factors, while short-term field studies and experiments were used to decipher the main determinants of natal dispersal.

Dispersal Ecology and Evolution. First Edition. Edited by Jean Clobert *et al.*

We crossed experimentally some of these factors to question their interactive effects and tested them during several life stages to identify when they influence the decision to disperse.

This chapter presents an overview of the results of this integrated approach of natal dispersal. Our work will focus on two major issues. First, we summarize knowledge on all proximate determinants of dispersal in the common lizard, focusing on main effects of several abiotic and biotic factors that can promote the evolution of dispersal. Second, we present evidence that the main effects of some factors interacted with others or with maternal and individual factors. More specific data on individual correlates of dispersal, especially dispersal personalities, are reported in Chapter 12.

Figure 3.1 The common lizard (*Zootoca vivipara*). An adult pregnant female with a linear back pattern. (Photos used with permission of Olivier Calvez.)

3.2 The common lizard as a model system

The common lizard (Figure 3.1) is a small live-bearing reptile (60 mm of snout–vent length) widely distributed across Eurasia. It feeds on small invertebrates such as homopters and spiders, and inhabits bogs, heathlands, or wet grasslands, where arthropod abundance and habitat humidity are important factors for the abundance of this species (Lorenzon *et al.* 2001). We studied several French populations where sexual maturity is reached at one or two years of age, and mating takes place as soon as the females emerge from hibernation, which occurs in March or April depending on local climate conditions. After approximately two months of gestation, females lay in June–July on average five soft-shelled eggs from which juveniles usually hatch within less than two hours. Since our study populations are ovoviviparous, hormonal exchanges between the mother and her embryos are possible, and factors that influence the mother until the late gestation period can impact on offspring development and natal dispersal behaviour (maternal effects, Massot and Clobert 2000; Massot *et al.* 2002). In October–November, all individuals enter the hibernation phase. Like other ectothermic vertebrates, temperature is an important determinant of physiological performances and life-history decision. An observed rise in the mean maximal daily temperature in May, June, and August over the last 30 years has been especially influential on demography and dispersal in our species (Chamaillé-Jammes *et al.* 2006; Massot *et al.* 2008). Our studies focused on natal emigration, a fundamental step of natal dispersal defined here as the movement out of the natal or maternal range, but we have also studied natal immigration and movements. Dispersal involves short distance movements in the common lizard (between 30–100 meters), but natal dispersal is much more prevalent than breeding dispersal (breeding dispersal rate average 1/10 of natal dispersal rate; Massot 1992). Based on this natural history, we hypothesized at the beginning of the study that temperature and humidity might act in a different way at the different stages (before, during, and after gestation) of the life cycle, or from one year to another. In the same way, because the species had a limited dispersal distance of a few hundred meters (Clobert *et al.* 2004), local competition was also expected to be a critical factor of dispersal decisions (Massot *et al.* 1992).

3.3 General observational and experimental procedures

During more than 20 years, we gathered information on natal dispersal by either following individuals in natural populations or by using experiments where we modified environmental factors in natural or semi-natural populations. Our general monitoring procedures consisted in capturing young and adults in populations around May–June, and juveniles around August–September. Upon capture, individuals were geo-referenced, marked, and described (body length and weight, back and ventral colour pattern, parasites) following standard protocols. They were then released at their capture point. One month before parturition, we removed pregnant females from the studied populations and maintained them in laboratory until parturition. At birth, juveniles were individually marked, and their body length and weight were measured. In some cases, behavioural measures such as 'boldness', 'social inclination', and thermoregulation were also collected on adult females and their offspring (de Fraipont *et al.* 2000; Cote and Clobert 2007a, 2007b). In addition, field experiments were conducted to manipulate environmental conditions experienced by adult females and their offspring. This was achieved either by manipulating conditions from the field, by reciprocal transplants of females across populations with different density, social, or humidity conditions (Lorenzon *et al.* 2001; Vercken *et al.* 2009), or by changing the laboratory conditions during gestation (de Fraipont *et al.* 2000; Massot *et al.* 2002).

To achieve more involved and replicated experiments, we set up in 1993 several two-patch systems within a natural habitat (Lecomte and Clobert 1996). The two-patch systems consisted of two 10 m × 10 m enclosures connected by two-way corridors ending with pitfall traps to capture and identify dispersers. By manipulating population composition (sex, genetic structure) in these enclosures, we were able to investigate in great details and in a realistic situation the subtle effects of several social factors on natal dispersal (Le Galliard *et al.* 2003; Cote and Clobert 2007a, 2007b). Measurements and monitoring procedures were done during these experiments as described above for natural populations.

3.4 Proximate factors of natal dispersal: main effects

Dispersal evolves under the action of several factors, ranging from spatio-temporal stochasticity in the environment, to competition, to inbreeding avoidance, and to interspecific interactions (Chapter 1).We illustrate here that our study species is sensitive to several biotic and abiotic factors (see Table 3.1 for an overview).

Table 3.1 Review of studies testing determinants of natal dispersal in the common lizard *Zootoca vivipara*. The study designs (Sd.) were observational (O) and experimental (E). 'Pattern observed' reported non significant tests (NS), the direction of significant relationships with the dispersal rate, if significant results involved main or interactive effects, and if effects where conditional to offspring characteristics and years. 'Ontogeny' reports if effects were postnatal, prenatal (when known, indicated as short-term from mid to end of gestation or long-term from previous years to mid-gestation) or genetic. We reported some 'Undefined factors' that may be due to different types of factors.

Abiotic factors	Sd.	Pattern observed	Ontogeny	Ref.
temperature	O,E	Dispersal inhibition by the climate warming, interaction with the maternal feeding	Short-term prenatal effect & interaction with temperature of the previous year	1
temperature during gestation	E	Interaction with postnatal humidity	Short-term prenatal effect	2,3
humidity at the mother habitat	O,E	Interaction with prenatal humidity	Long-term prenatal or genetic effect	3
humidity at the natal site	E	Main effect of higher dispersal rates in drier sites	Postnatal effect	3
humidity during gestation	E	Interaction with humidity at the mother habitat modulated by an interaction with prenatal temperature	Short-term prenatal effect	3

Table 3.1 *Continued*

Heterospecific factors	Sd.	Pattern observed	Ontogeny	Ref.
Food availability during gestation	E	Effect that depends on year, interaction with temperature	Short-term prenatal effect & interaction with temperature of the previous year	1,2,4,5
predator attack at mid-gestation	E	NS	–	2
maternal parasite load	E	Negative effect in offspring females, positive effect in offspring males	Short-term prenatal effect	6
Conspecific factors	**Sd.**	**Pattern observed**	**Ontogeny**	**Ref.**
total density	E	Positive or negative density dependence, no interaction with the presence of the mother, interaction with social tolerance at birth, body size at birth and attributes of immigrants	Postnatal effect	7,8,9,10,11
density of adult females	E	Positive density dependence	Postnatal effect	12
density of adult males	E	Negative density dependence	Postnatal effect	12
adult sex-ratio	E	NS	–	13
connectivity with another population	E	Change in attributes of candidates for dispersal, interaction with the presence of the mother and body mass	Postnatal effect	14,15
corpulence of adult females	E	NS	–	12
corpulence of familiar adult females	E	Positive effect	Postnatal effect	12
proximity of an adult female during gestation	E	NS	–	2
attributes of immigrants	E	Interaction with total density and body size at birth	Postnatal effect	8
Maternal factors	**Sd.**	**Pattern observed**	**Ontogeny**	**Ref.**
maternal stress	E	Main negative effect, interactions with mother body size, mother corpulence, litter sex-ratio and stress duration	Prenatal effect	16,17,18,19
mother body size	O,E	NS or positive relationship, interaction with maternal stress	–	2,17,18
mother corpulence	O,E	Positive effect, interaction with maternal stress	Prenatal or postnatal effect	12,18
mother age	O	Negative relationship in offspring females, NS in offspring males	Prenatal effect	20
maternal reproductive strategy	O	Interaction with litter success, no interaction with mother body size, mother corpulence and litter size	Prenatal effect	21
litter size	O	NS	–	2,22
litter success	O	NS as a main effect, interaction with reproductive strategy	–	2,21

litter sex-ratio	O	Relationships that depend on sex and year, interactions with maternal stress	Prenatal effect	2,16,22,23
laying date	O	NS	–	2,23
presence of the mother after birth	E	Positive effect for offspring females in a large experimental system, NS in males, NS in a small system, no interaction with total density interaction with habitat connectivity and body mass,	Postnatal effect	9,12,14
mother-offspring competition at the population scale	E	Interaction with body size at birth	Postnatal effect	22
Undefined factors	**Sd.**	**Pattern observed**	**Ontogeny**	**Ref.**
juvenile stress	E	NS	–	18
natal site	E	Study area effects	Prenatal & postnatal effects	2,3,4,10,11,15
Family effect	O	Sibling resemblance in dispersal behaviour, interaction with the maternal reproductive strategy	Prenatal & postnatal effects, no or weak genetic effect	2,3,4,12,14,21

References

1. Massot, M., Clobert, J., Ferrière, R. (2008). *Global Change Biology* 14:461–469;
2. Massot, M. Clobert, J. (2000). *Journal of Evolutionary Biology* 13:707–719;
3. Massot, M., Clobert, J., Lorenzon, P., Rossi, J.M. (2002). *Journal of Animal Ecology* 71:253–261;
4. Clobert, J., Massot, M., Lecomte, J., Sorci, G., de Fraipont, M., Barbault, R. (1994); In *Lizard ecology: historical and experimental perspectives* (L.J. Vitt, R. Pianka, R., eds). Princeton University Press, 183–206;
5. Massot, M., Clobert, J. (1995). *Behavioral Ecology and Sociobiology* 37:413–418;
6. Sorci, G., Massot, M., Clobert, J. (1994). *American Naturalist* 144:153–164;
7. Cote, J., Clobert, J. (2007). *Proceedings of the Royal Society B* 274:383–390.
8. Cote, J., Clobert, J. (2007). *Ecology Letters* 10:411–417;
9. Le Galliard, J.F., Ferrière, R., Clobert, J. (2003). *Proceedings of the Royal Society B* 270:1163–1169;
10. Meylan, S., Clobert, J, Sinervo, B. (2007). *Oikos* 116:650–661;
11. Lecomte, J., Boudjémadi, K., Sarrazin, F., Cally, K., Clobert, J. (2004). *Journal of Animal Ecology* 73:179–189;
12. Léna, J.P., Clobert, J., de Fraipont, M., Lecomte, J., Guyot, G. (1998). *Behavioural Ecology* 9:500–507;
13. Le Galliard, J.F., Fitze, P.S., Ferrière, R., Clobert, J. (2005). *Proceedings of the National Academy of Sciences USA* 102:18231–18236;
14. Cote, J., Clobert, J. (2010). *Ecology* 91:1485–1493;
15. Boudjémadi, K., Lecomte, J., Clobert, J. (1999). *Journal of Animal Ecology* 68:1207–1224;
16. Vercken, E., de Fraipont, M., Dufty, A.M., Clobert, J. (2007). *Hormones and Behavior* 51:379–386;
17. Meylan, S., de Fraipont, M., Clobert, J. (2004). *Ecoscience* 11:123–129;
18. Meylan, S., Belliure, J., Clobert, J., de Fraipont, M. (2002). *Hormones and Behavior* 42:319–326;
19. de Fraipont, M., Clobert, J., John-Alder, H. S. (2000). *Journal of Animal Ecology* 69:404–413;
20. Ronce, O., Clobert, .J, Massot, M. (1998). *Proceedings of the National Academy of Sciences USA* 95:600–605;
21. Laloi, D., Richard, M., Federici, P., Clobert, J, Teillac-Deschamps, P., Massot, M. (2009). *Ecology Letters* 12:823–829;
22. Cote, J., Clobert, J., Fitze, P.S. (2007). *Proceedings of the National Academy of Sciences USA* 104:9703–9708;
23. Massot, M, Clobert, J., Chambon, A., Michalakis, Y. (1994). *Oikos* 70:172–176.

3.4.1 Abiotic factors

Two of the main determinants of population density in the common lizard are temperature and humidity (Clobert *et al.* 1994). These factors also generally determine habitat quality and limit the distribution of many other reptiles (Le Galliard *et al.* 2011). In line with this, we found that higher temperature during gestation generally reduced natal dispersal (Massot *et al.* 2002, 2008), while humidity influenced natal dispersal at all stages, namely before, during, and after *in utero* development (Table 3.1, Massot *et al.* 2002). A higher humidity could increase dispersal at one life stage but decrease it at another stage (Massot *et al.* 2002). Microhabitat structure had also some impact on transience and settlement, since dense forest or more rocky areas repulsed dispersers or induced longer distance dispersal, while open areas tended to attract lizards (Boudjemadi *et al.* 1999; Zajitschek *et al.*, unpub. data).

3.4.2 Heterospecific factors

The availability of food had a great influence on natal dispersal. In long-term food manipulation during gestation, Massot and Clobert (1995, 2000; Massot *et al.* 2008) demonstrated that supplementary maternal feeding can increase or decrease offspring dispersal. In addition, juveniles that received supplementary food a few days after birth dispersed less frequently when back in their habitat of origin (Massot, unpublished). Although tail breaks (a surrogate measure of predation pressure in this species) did not correlate with natal dispersal (Massot and Clobert 2000), high maternal parasite loads during gestation correlated with more offspring philopatry (Sorci *et al.* 1994).

3.4.3 Non-genetic social factors

Intraspecific competition shapes importantly population dynamics in the common lizard (Massot *et al.* 1992; Lecomte *et al.* 1994). Indeed natal dispersal was enhanced by population density in most cases (Lecomte *et al.*; 2004; Le Galliard, Ferrière and Clobert 2003; Le Galliard Ferrière and Clobert 2005; Massot and Clobert 2000; Cote and Clobert 2007b). However, other factors interacted with population density and modified the sign of the correlation between dispersal and density (see next section), even reversing the sign of density-dependence in some cases (Le Galliard *et al.* 2003, Meylan *et al.*; 2007). Natal dispersal was also sensitive to population composition and structure. The adult sex ratio (Léna *et al.* 1998, but see Le Galliard *et al.* 2005) and the proportion of yellow-coloured females in the population (ventral colouration indicates social competition in females, Vercken *et al.* 2010, 2012) increased natal dispersal (Table 3.1). These non-genetic social effects are therefore complex, probably reflecting the fact that congeners may provide different cues about the costs and benefits of natal dispersal (Clobert *et al.* 2009). In particular, density may either act as a cue for competition or as a cue for habitat quality depending on the context (Clobert *et al.* 2009; Cote *et al.* 2010). Two studies examined if density-dependence involved maternal effects and found no effect of population density or social interactions during gestation, while effects of non-genetic social factors at the postnatal stage were obvious (Table 3.1). Density during gestation might not be a good descriptor of juvenile future competitive environment because either juveniles are entering mostly in competition among themselves, or social conditions at gestation time are poor predictors of social conditions later on (rather random fluctuating densities overall, Massot *et al.* 1992).

3.4.4 Genetic social factors

Hamilton and May (1977) and Comins *et al.* (1980) demonstrated that genetic relatedness with congeners are important drivers of the evolution of dispersal. Indeed, we found that kin competition, and specifically mother-offspring competition (Ronce *et al.* 1998; Le Galliard *et al.* 2003), is involved in natal dispersal in the common lizard. The likelihood of kin competition can be assessed at the prenatal stage (through changes in maternal age and/or condition, Ronce *et al.* 1998; de Fraipont *et al.* 2000; Meylan *et al.* 2002, 2004), but the presence of the mother and its health status are also paramount to trigger natal dispersal at the postnatal stage (Léna *et al.* 1998; Le Galliard *et al.* 2003; Meylan *et al.* 2007; Cote *et al.* 2007). Thus, the mechanisms involving the genetic and non-genetic social factors are probably different. In addition, offspring males and females of the same litter reacted in an opposite way to the absence of their mother in an experiment on kin competition (Le Galliard *et al.* 2003), possibly indicating that brothers avoided inbreeding risks with their sisters. However, other facts indicate a moderate relationship between inbreeding avoidance and natal dispersal. First, sex-biased dispersal is not pronounced (Massot et Clobert 2000), and offspring males and females tend to disperse to similar rather than opposite locations in nature (Zajitschek *et al.*; unpub. data). For example, out of 45 pairs of sister and brother, more than 75% settled at a distance less than 20 meters, and therefore shared substantial part of their home range. Second, the limited avoidance of inbreeding by natal dispersal is also confirmed by the presence of a significant genetic structure at a fine scale (Stevens *et al.*; unpublished data). Third, close inbreeding can be avoided by kin recognition and mate choice as juveniles discriminate their mother, brothers, and sisters at birth (Léna and de Fraipont 1998), and females select sexual partners based on genetic proximity (Richard *et al.* 2009). Thus, inbreeding avoidance is not a major determinant of natal dispersal. Rather, sexual differences in dispersal behaviour are probably due to sexual differences in competition and life-history strategies (e.g. mate versus resource competition, Le Galliard *et al.* 2003).

3.4.5 Maternal factors

We also found that maternal factors (see Table 3.1) and phenotypic traits of the offspring correlate with dispersal behaviour. This includes maternal body size and body condition (de Fraipont *et al.* 2000; Léna *et al.* 1998), maternal levels of corticosterone (de Fraipont *et al.* 2000; Meylan *et al.* 2004), maternal dorsal colour patterns (Lepetz *et al.* 2009), and offspring or mother behaviour (de Fraipont *et al.* 2000; Cote and Clobert 2007a, 2007b). These results are reported in Chapters 10 and 12, and will not be described further here.

3.4.6 Summary and main conclusions

Natal dispersal is an omnibus response to alterations in the environment of the common lizard since many social and non-social factors were influential (Table 3.1). Natal dispersal was influenced by the environment experienced by mothers before mating and during gestation and also by the environment experienced by juveniles after birth (Massot and Clobert 2000; Dufty *et al.* 2002). Young common lizards appear to collect numerous sources of information to take a dispersal decision. It remains however unclear if these results imply a combined action of many ultimate factors. Indeed, in some cases at least, some proximate factors (for example, habitat vegetation structure and density) could be proximate cues for several ultimate factors (habitat quality or competition). Also, it is difficult at this stage to draw a clear hierarchy among proximate determinants of dispersal since most factors were not tested in the same study or in the same context (natural habitats or experimental enclosures). Our feeling however is that the effects of two environmental factors (temperature and habitat humidity) and two social factors (density and mother-offspring interactions) are dominant over the effects of heterospecific factors, inbreeding avoidance, and mate choice. For example, increased temperatures during the summer caused a reduction of 50% of natal dispersal in the long term (Chapter 25). Local density and the likelihood of mother-offspring interactions had strong (over 30%) quantitative effect on dispersal in artificial settings (Léna *et al.* 1998; Le Galliard *et al.* 2003) but also in wild populations (Ronce *et al.* 1998; Meylan *et al.* 2007).

3.5 Combined effects of proximate factors

We were able to consider the consistency of some proximate factors through time and life stages, as well as the additive and interactive effects of several proximate factors on natal dispersal. The results of our studies highlight the complexity of natal dispersal behaviour in the common lizard.

3.5.1 The same factor exhibits different responses

Starting in 1990, for 18 years we manipulated food levels delivered to pregnant females during late gestation in the laboratory. Supplementary maternal feeding had little effect on juvenile morphology or condition (Le Galliard *et al.* 2006), but had a strong effect on natal dispersal (Massot and Clobert 1995, 2000; Massot *et al.* 2008). In several years, supplementary feeding enhanced dispersal, which is in line with other results showing that natal dispersal increases when maternal condition improves (Table 3.1). However, a more recent analysis demonstrated that this effect is observed only when the temperature in August is high; otherwise, supplementary feeding decreased dispersal (Massot *et al.* 2008). The magnitude of positive and negative effects of maternal feeding on natal dispersal was similar. There is here a significant interaction between the prenatal effects of food, possibly implying health of the mother and kin competition, and the postnatal effects of temperature during early life, possibly implying habitat quality. Another example of such interaction is the effect of humidity and temperature (Massot *et al.* 2002). In a reciprocal transplant experiment (Lorenzon *et al.* 2000), we tested the effects of humidity at different developmental stages including before, during, and after *in utero* development. Humidity had a negative effect on natal dispersal at the postnatal stage, while humidity during gestation had a positive effect in litters from dry habitat only. These results were also year-dependent. This inter-annual variation demonstrates our lack of understanding of humidity effects and could be due to the action of a non-recorded factor. It is nevertheless possible that the same cue is integrated over different developmental stages to trade off the different messages coming from the environment, a bit like corticosterone over different stages of the embryos' development (see following, and Vercken *et al.* 2007).

Indeed, a third example of interactive effects involves the effects of corticosterone, a hormone involved in stress responses and energetic expenditure. In at least three different experiments, chronically corticosterone-treated females produced more philopatric offspring (Table 3.1), which is in line with the idea that low maternal condition promotes natal philopatry in this species. Yet we also found that corticosterone-treated offspring did not change their natal dispersal behaviour when tested in the absence of their mother (Meylan *et al.* 2002), and that the duration of the maternal corticosterone treatment was important; i.e. that the stress should be chronic (Vercken *et al.* 2007). Philopatric juveniles were produced when maternal exposure to corticosterone lasted at least a month (chronic stress), otherwise dispersing offspring were produced. A likely explanation is that chronic stress indicates a female's health degradation (a cue for mother-offspring competition), while acute stress is more related to variation in the mother's environment (food abundance, predation, etc., Vercken *et al.* 2007). Thus, a common result of our experiments is the presence of interactions among manipulated factors that plague some biological interpretations, in particular interactions involving social and abiotic factors. Some interactions were found only once or involved manipulated and non-manipulated factors, and might be spurious or the result of some unexplored factors.

3.5.2 Contrasted responses across families

A recurrent finding in our studies is that juvenile dispersal is strongly family-dependent (Massot and Clobert 2000). Family effects might be generated by many environmental factors, such as living in a common environment, and are not necessarily related to maternal traits. However, some proximate factors had contrasting effects on natal dispersal depending on maternal traits. For example, corticosterone-treated females produced philopatric offspring only when females were large (Meylan *et al.* 2004). Not only was natal dispersal dependent on female's traits, but also the dispersal juvenile phenotype (Meylan *et al.* 2002; Chapter 10). In the same way, the female ventral colour (a signal of their competitive strategy) influenced differently offspring dispersal depending on the proportion of female colour scores into the population (increased with the proportion of yellow females; Vercken *et al.* 2012). Both environmental and genetic effects can explain why these maternal traits influence natal dispersal.

3.5.3 Interactions among factors

Interactions among factors have been found regularly (Table 3.1). For example, we have seen that temperature and humidity influence jointly natal dispersal (Massot *et al.* 2002). Dispersal was more frequent in dry habitats, but relatively less when the temperature during gestation was high (Massot *et al.* 2002). A synergy between temperature and humidity is expected since high temperatures combined with low humidity generate strong evaporative water loss and act as an important physiological stress (Chapter 25). However, the observed interaction between temperature and humidity was more difficult to predict. Surprisingly, interactions among non-genetic social factors are less numerous (Table 3.1). In a reciprocal transplant experiment to study the role of density at different developmental stages (Meylan and Clobert 2005; Meylan *et al.* 2007), no interaction was found and only the density at the postnatal stage influenced natal dispersal (Meylan *et al.* 2007). In addition, interactions among biotic factors and attributes of dispersers were numerous and involved body size, body corpulence, and behavioural traits in juveniles (see Table 3.1, and Chapters 10 and 12). An interesting interaction was found between social factors and habitat connectivity (Lecomte *et al.* 2004; Cote and Clobert 2007a): density tended to increase dispersal attempts when populations were connected, while it had no effect on dispersal attempts in the absence of a connection. The effect of the presence of a corridor, as described earlier, might reflect a social effect: the return of unsuccessful dispersers to their population might indicate a lack of connectivity, which might imply more generally that individuals exchanging among populations can influence natal dispersal if residents are sensitive to social interactions with immigrants (Cote and Clobert 2007a, 2007b). Thus young individual common lizards adopt different dispersal strategies with respect to biotic factors, which sometimes might blur the main effect of the same biotic factors at the population level.

3.5.4 Interactions between non-genetic and genetic social factors

The evidence for an interaction between non-genetic and genetic social factors is ambiguous. In the common lizard, kin competition occurs primarily between the mother and her daughters (Ronce *et al.* 1998; Le Galliard *et al.* 2003). We therefore manipulated kin competition by releasing offspring in natal habitats with their mother or with an unrelated post-parturient female. The total number of individual dispersing was similar in the presence or absence of the mother, but a major difference was found in the sex ratio and body condition of dispersing individuals (Léna *et al.* 1998; Le Galliard *et al.* 2003). Many of the results reported earlier have been found in interactions with the juvenile phenotype, either with respect to non-genetic and genetic social factors (Chapter 12). Clearly, individuals perceive the quality of the social habitat in different ways and take the dispersal 'decision' according to their phenotype.

3.5.5 Summary and main conclusion

Studies to unravel interactions among stages and factors require fully factorial experimental designs and are therefore often difficult to achieve. Yet using such experiments, we found some significant interactions among some factors. In addition, we found much inter-familial and inter-individual variation in dispersal behaviour as indicated by the interactions between individual or maternal factors and some proximate social factors. These results should call for caution when interpreting univariate or correlative approaches.

3.6 Synthesis: an informed base theory of factors integration

Clearly, natal dispersal behaviour, defined here as the decision to leave a natal range, is influenced by several factors in the common lizard, including both abiotic and biotic conditions. Maternal characteristics and maternal environments experienced at different developmental stages influence dispersal decisions. Finally, the offspring phenotype, sometimes produced in response to maternal effects, is also different between dispersing and non-dispersing lizards (see Chapter 12). Natal dispersal events occur over small distances and happen very early in life (Clobert *et al.* 1994). In such a situation, strategic information gathering prior to dispersal should be used to reduce dispersal costs and optimize natal dispersal decisions, but this is likely to depend on multiple factors like those described here and in several other species (see Chapters 1 and 4). Which cue is relevant, and when to be sensitive to a given cue, should depend on the content and reliability of the information provided by that cue. For example, the non-genetic social environment can be reliably sampled after birth (Massot *et al.* 1992), while competition between mother and her offspring can be assessed at a prenatal stage using cues providing information on maternal health. We provided several examples of maternal effects by which offspring integrate information from their environments and assess their fitness prospects. In addition, the prenatal environment often interacts with the postnatal environment as if maternal effects shaped the offspring dispersal reaction norm. This interaction could be achieved though the organizational effects of hormones (Dufty *et al.* 2002) or by genome-wide epistasis (Sinervo *et al.* 2008). The study of the genotype-by-environment interactions is therefore the most promising avenue for unravelling the complexity of the dispersal-decision process and its potential consequences on population dynamic and evolution. The existence of dispersal-dedicated phenotype (see Chapter 12) and the numerous interactions between, on the one hand, abiotic and biotic factors and, on the other, the individual phenotype, strongly militate for a more detailed investigation of multiple causal approaches to the study of dispersal.

What do these results tell us about the evolution of dispersal? Firstly, they demonstrate that dispersal decisions are condition-dependent, and that condition-dependence can inform us about the ultimate cause of dispersal. This feature has some strong consequences on dispersal evolution and metapopulation dynamics. Second, dispersal seems to be an individual-based trait; i.e. there is some strong variation among individuals about the motivation and cue sensitivity in dispersal decision. These individual 'strategies' are not the simple reflection of the 'best of a bad job' strategy, but more probably correspond to evolutionary trade-offs as the variety of dispersal syndromes found within population of this species seems to indicate (Chapter 10 and 12). Third, the evolution of dispersal is probably compounded because the variety of factors influencing it makes it unlikely that a simple mechanism might operate for all (for example, density and kinship, social, and non-social factors). We believe that dispersal is the aggregated result of different components which might or might not co-evolve depending of the species; i.e. there are several 'dispersals' within a species. To what extent our results extend to other species remains an open question, although results reported in Chapters 1, 4, 12, 13, and 14 strongly militate for their generality.

References

Belliure, J., Meylan, S. and Clobert J. (2004). 'Pre-natal and post-natal effects of corticosterone on behavior in juveniles of the common lizard, Lacerta vivipara', *Journal of Experimental Zoology: Part A Comparative and Experimental Biology* **301**: 401–10.

Chamaillé-Jammes S., Massot M., Aragon P. *et al.* (2006). 'Global warming and positive fitness response in mountain populations of common lizards *Lacerta vivipara*', *Global Change Biology* **12**: 392–402.

Clobert J., Danchin E., Nichols, J.D., *et al.* eds, (2001). *Dispersal*. Oxford University Press, Oxford.

Clobert J., Ims R.A. and Rousset F. (2004). 'Causes, mechanisms and consequences of dispersal', in I. Hanski and O.E. Gagiotti, eds, *Ecology, Genetics and Evolution of Metapopulations* . Elsevier, Academic Press, London, 307–35.

Clobert, J., Massot, M., Lecomte, J. *et al.* (1994). 'Determinants of dispersal behavior: the common lizard as a case study', in L.J. Vitt and E.R. Pianka, eds, *Lizard ecology. Historical and experimental perspectives*. Princeton University Press, Princeton, 183–206.

Comins, H.N., Hamilton, W.D., and May, R.M. (1980).'Evolutionarily stable dispersal strategies', *Journal of Theoretical Biology* **82:** 205–30.

Cote, J., Boudsocq, S. and Clobert, J. (2008). 'Density, social information and space use in the common lizard (*Lacerta vivipara*)', *Behavioral Ecology* **19:** 163–8.

Cote, J. and Clobert, J. (2007a). 'Social information and emigration: lessons from immigrants', *Ecology Letters* **10:** 411–17.

Cote, J. and Clobert, J. (2007b). 'Social personalities influence natal dispersal in a lizard', *Proceedings of the Royal Society B* **274:** 383–90.

Cote, J. and Clobert, J. (2010). 'Risky dispersal: avoiding kin competition despite uncertainty', *Ecology* **91**: 1485–93.

Cote, J., Clobert, J. and Fitze, P. S. (2007). Mother-offspring competition promotes colonization success. *Proceedings of the National Academy of Sciences USA* **104:** 703–8.

de Fraipont, M., Clobert, J., John-Alder, H. *et al.* (2000). 'Increased pre-natal maternal corticosterone promotes philopatry of offspring in common lizards *Lacerta vivipara*', *Journal of Animal Ecology* **69:** 404–13.

Dufty, A.M., Clobert, J. and A.P. Møller (2002). 'Hormones, develomental plasticity and adaptation', *Trends in Ecology and Evolution* **17:** 190–6.

Gandon, S. and Michalakis, Y. (1999). 'The evolution of dispersal in a metapopulation with extinction and kin competition', *Journal of Theoretical Biology* **199:** 275–90.

Hamilton, W.D. and May, R.M. (1977). 'Dispersal in stable habitats', *Nature* **269:** 578–81.

Lecomte, J., Boudjemadi, K., Sarrazin, F. *et al.* (2004).'Connectivity and homogenisation of population sizes: an experimental approach in the common lizard', *Journal of Animal Ecology* **73:** 179–89.

Lecomte, J. and Clobert, J. (1996). 'Dispersal and connectivity in populations of the common lizard (*Lacerta vivipara*): an experimental approach', *Acta Oecologica* **6:** 585–98.

Lecomte, J., Clobert, J. and Massot, M. (1993). 'Shift in behaviour related to pregnancy in *Lacerta vivipara*', *Revue d'Ecologie (Terre et vie)* **48:** 99–107.

Lecomte, J., Clobert, J., Massot, M. *et al.* (1994). 'Spatial and behavioral consequences of a density manipulation in the common lizard', *Ecoscience* **1:** 300–10.

Le Galliard, J.-F., Ferrière, R. and Clobert, J. (2003). 'Mother-offspring interactions affect natal dispersal in a lizard', *Proceedings of the Royal Society of London B* **270:** 1163–9.

Le Galliard, J.-F., Ferrière, R. and Clobert, J. (2005). 'Effect of patch occupancy on immigration in the common lizard', *Journal of Animal Ecology* **74:** 241–9.

Le Galliard J.-F., Massot M., Baron J.-P. & Clobert J. (2011, in press). Ecological effects of climate change on European reptiles', in J. Brodie, E. Post and D. Doak, eds, *Conserving wildlife populations in a changing climate.* University of Chicago Press, Chicago, USA.

Le Galliard, J.-F., Massot, M., Landys, M. *et al.* (2006). 'Ontogenic sources of variation in the sexual size dimorphism of a viviparous lizard', *Journal of Evolutionary Biology* **19:** 690–704.

Léna, J.P., Clobert, J., de Fraipont, M. *et al.* (1998). 'The relative influence of density and kinship on dispersal in the common lizard', *Behavioral Ecology* **9:** 500–7.

Léna, J.P. and de Fraipont, M. (1998). 'Kin recognition in the common lizard', *Behavioural Ecology and Sociobiology* **42:** 341–7.

Lepetz V., Massot M., Chaine A. *et al.* (2009). 'Climate warming and evolution of morphotypes in a reptile', *Global Change Biology* **15:** 454–66.

Lorenzon, P., Clobert, J., and Massot, M. (2001). 'The contribution of phenotypic plasticity to adaptation in *Lacerta vivipara*', *Evolution* **55:** 392–404.

Massot, M. (1992). 'Movement patterns of the common lizard (*Lacerta vivipara*) in relation to sex and age', In Z. Korsos and I. Kiss, eds, Proceedings of the Sixth Ordinary General Meeting of the Societas Europea Herpetologica. Budapest, Hungary Natural History Museum.

Massot, M. and Clobert, J. (1995). 'Influence of maternal food availability on offspring dispersal', *Behavioral Ecology and Sociobiology* **37:** 413–18.

Massot, M. and Clobert, J. (2000). 'Processes at the origin of similarities in dispersal behaviour among siblings', *Journal of Evolutionary Biology* **13:** 707–19.

Massot, M., Clobert, J. and Ferrière, R. (2008). 'Climate warming, dispersal inhibition and extinction risk', *Global Change Biology* **14:** 1–9.

Massot, M., Clobert, J., Lorenzon, P. *et al.* (2002). 'Condition-dependent dispersal and ontogeny of the dispersal behaviour: an experimental approach', *Journal of Animal Ecology* **71:** 253–61.

McPeek, M.A. and Holt, R.D. (1992). 'The evolution of dispersal in spatially and temporally varying environments', *American Naturalist* **140:** 1010–27.

Meylan, S., Belliure, J., Clobert, J. *et al.* (2002). 'Determinants of dispersal in juveniles of the common lizard (*Lacerta vivipara*): stress removes the influence of maternal condition', *Hormones and Behavior* **42:** 319–26.

Meylan, S., Belliure, J., Clobert, J. *et al.* (2002). 'Stress and body condition as prenatal and postnatal determinants of dispersal in the common lizard (*Lacerta vivipara*)',. *Hormones and Behavior* **42:** 319–26.

Meylan, S. and J. Clobert (2005). 'Is Corticosterone mediated phenotype development adaptive?', *Hormones and Behavior* **48:** 44–52.

Meylan, S., Clobert, J. and Sinervo, B. (2007). 'Adaptive significance of maternal induction of density-dependent phenotypes', *Oikos* **116:** 650–61.

Meylan, S., de Fraipont, M. and Clobert, J. (2004). 'Maternal size and stress and offspring philopatry: an experimental study in the common lizard (*Lacerta vivipara*)' *Ecoscience* **11:** 123–9.

Perrin, N. and Mazalov, V. (2000). 'Local competition, inbreeding, and the evolution of sex-biased dispersal', *American Naturalist* **155:** 116–27.

Richard, M., Losdat, S., Lecomte, J., *et al.* (2009). 'Optimal level of inbreeding in the common lizard', *Proceeding of the Royal Society B* **276:** 2779–86.

Ronce, O., Clobert, J. and Massot, M. (1998). 'Natal dispersal and senescence', *Proceedings of the National Academy of Sciences USA* **95:** 600–5.

Sinervo, B., Clobert J., Miles D.B. *et al.* (2008). 'The role of sender-receiver gene epistasis in life history trade-offs: dissecting the genomic architecture of organismal design in social systems', *Heredity* **101**: 197–211.

Sorci, G., Massot, M. and Clobert, J. (1994). 'Maternal parasite load increases sprint speed and philopatry in female offspring of the common lizard', *American Naturalist* **144:** 153–64.

Sorci, G., Swallow, J.G., Garland, T. Jr *et al.* (1995). 'Quantitative genetics of locomotor speed and endurance in the lizard *Lacerta vivipara*', *Physiological Zoology* **68**: 698–720.

Travis, J.M.J. and Dytham, C. (1998). 'The evolution of dipsersal in a metapopulation: a spatially explicit, individual-based model', *Proceedings of the Royal Society B* **265:** 17–23.

Vercken, E, Clobert, J. and Sinervo, B. (2010). 'Frequency dependent reproductive success in female common lizards: a real-life hawk-dove-bully game?', *Oecologia* **162**: 49–58.

Vercken, E., Sinervo, B., Clobert, J. (2012). 'The importance of a good neighbourhood: dispersal decisions in juvenile common lizards are based on social environment', *Behavioural Ecology* (in press).

Vercken, E., de Fraipont M., Dufty, A. *et al.* (2007). 'Maternal corticosterone exposure and the determinism of natal dispersal in the common lizard (*Lacerta vivipara*)', *Hormone and Behavior* **51**: 379–86.

CHAPTER 4

Dispersal in invertebrates: influences on individual decisions

Tim G. Benton and Diana E. Bowler

4.1 Introduction

Dispersal is *'any movement of individuals or propagules with potential consequences for gene flow across space'* (Ronce 2007). By this definition, dispersal is defined by its outcome (movement in space), not by its mechanism (behavioural mode). Using this definition, it becomes clear that any definition of dispersal as a particular behaviour will become system-dependent. For some systems, dispersal will arise from a particular behaviour, happening only once per generation of movement from a natal area to another area (e.g. sessile organisms or ones with permanent home ranges or territories). For such systems it is easy to define dispersal as a rare act of movement. For other systems, defining a unique dispersal behaviour is rarely possible. Many insects may have broadly similar patterns of behaviour: laying eggs in a patch, refuelling and moving between patches and reproducing again (e.g. blowflies lay eggs and forage on carrion, and may reproduce multiple times as adults; mosquitoes lay multiple broods and feed between; *Drosophila* move between fruit; butterflies may move between patches of host-plants; aphidophagous syrphids move across the agricultural landscape seeking aphid colonies to oviposit near, and feeding from nectar and pollen (Wratten *et al.* 2003)). In these systems, there is a net displacement over time, but the dispersal across the landscape is a result of what would traditionally be seen as foraging or oviposition-site-selection behaviour rather than 'dispersal'. In other organisms, resources may be ephemeral. For example, soil mites, *Sancassania berlesei*, are general detritivores and are thought to live on local patches of organic matter—such as decomposing plant material or soil macrofauna—which may persist for a few generations until depleted entirely. When the resources in the patch become exhausted, the animals leave and seek alternative resources and will move until food is encountered. This leads to inter-patch movement that is behaviourally related to foraging and yet also functionally dispersal. In the literature, inter-patch movements have alternatively been seen as dispersal, foraging or oviposition-site selection and sometimes just 'movement'. Occasionally, the same behaviour or motivation is called different things depending on the distance moved: short distance inter-patch movements are 'foraging behaviour', but rarer long distance ones are 'dispersal'. Clearly, this lack of clarity can lead to confusion in some situations; defining dispersal in a simple way as movement leading to gene flow removes this confusion, and has potential to unite the three literatures of 'movement', 'behaviour', and 'dispersal' into a functional whole.

The decision to disperse is not a simple one. Firstly, it involves a range of developmental processes (e.g. the decision whether to develop wings, the allocation of resources to lipid stores versus growth, etc). It involves behavioural decisions (to leave the natal patch, to continue or turn back, to settle). None of these decisions are 'conscious' in the sense we may decide, but they are decisions in the sense that evolution by natural selection will shape resource allocation rules in certain ways, and behavioural rules to respond to certain stimuli. There are many proximate factors involved in dispersal decisions and there is clearly insufficient

Dispersal Ecology and Evolution. First Edition. Edited by Jean Clobert *et al.*

space in this short chapter to review the topic in the millions of species of invertebrates occupying every conceivable habitat on the globe. However, we briefly list some of the generic factors that differentiate dispersal in invertebrates from vertebrates. Invertebrates can have considerable flexibility in their life histories (which can sometimes be very complex) and often can develop specific machinery (e.g. wings) or specific life-history stages for movement (e.g. dispersal morphs, hypopi, in mites; dispersive planktonic larvae) or a combination of life-history stage and behaviour (aerial dispersal by juvenile spiders on silk threads, or a parasite's manipulation of host behaviour to encourage transmission to the next host). For some species it is therefore possible to differentiate long-distance dispersal from short-distance routine movements. Sexual size dimorphism can be significant in invertebrates, often leading to sex-biased dispersal strategies. Invertebrates sit on the 'fast-slow life-history continuum', from very short lived to very long lived and may have long juvenile stages (e.g. 17 year cicadas, which are juveniles for nearly 17 years and adults for a couple of weeks) or short juvenile and long adult stages (e.g. giant clams, which are free-living for days and sessile for up to centuries). The overall context of the life history is therefore crucial in determining the scope for dispersal, which is often constrained developmentally into a short period. Invertebrates often have low levels of investment per offspring, leading to high fecundity. Clutches of eggs may commonly be laid, leading to the potential for strong kin competition early in life, and this is often a driver of juvenile dispersal away from the natal patch in addition to later, longer-distance dispersal by the flying adults.

In the last decade there has been an increasing recognition that variability between individuals is both ubiquitous and important in ecology (Benton *et al.* 2006). Given that dispersal is a key process underlying spatial dynamics (Chapters 20–24), individual variation in dispersal propensity and ability will be important for how dispersal affects dynamics (Bowler and Benton 2005). Some of this individual variation is driven by 'fixed' factors like genes, age/stage, or sex, while some is driven by plastic responses to resources or other environmental factors, such as predators or social factors like the presence of conspecifics or kin (Chapter 12). We term the sensitivity of dispersal to external factors context-dependence, and restrict the term condition-dependence to the influence of an animal's size, physiological, or nutritional condition *sensu stricto.*

Our purpose here is to focus on the proximal factors that influence an individual's decisions, using examples from invertebrates. We first address the environmental factors, such as the quality of the local patch and the information that an individual may use to inform decisions. We then address the role of individuals' age, stage, sex, and condition on dispersal decisions, before closing with some comments on dispersal adaptations.

4.2 Environmental- or context-dependent factors influencing dispersal

Population size, or density, along with resource availability, plays a key role in determining the overall intensity of competition within a population. There are many empirical studies that show a positive relationship between dispersal and population density in a range of invertebrate taxa such as butterflies (Baguette *et al.* 1996), spiders (De Meester and Bonte 2010), fly larvae dispersing in water (Fonseca and Hart 1996), and thrips (Rhainds *et al.* 2005). Density may deplete resources and cause greater competition, which pushes the smaller or weaker competitors to leave (butterflies, Brunzel 2002; odonates, McCauley 2010) or, in the case of locusts, physical jostling is the proximal cue causing the change from a solitary form to a swarming form (Simpson *et al.* 1999). Density-dependent dispersal may also be related to inbreeding probability as has been empirically demonstrated in bees (Kukuk *et al.* 2005). Not all organisms increase dispersal with density; there are examples, for instance, in moths (Menendez *et al.* 2002) and damselfies (Chaput-Bardy, Gregoire *et al.*, 2010), when individuals are more likely to disperse from a patch with low numbers of individuals. This may arise from Allee effects, which reduce mating opportunities, or to the benefits of living with conspecifics, or because low-density patches are low quality.

Environmental impacts on individuals' dispersal decisions are not necessarily restricted to the external environment experienced by individuals during their lifetime. For some organisms, there may be parental effects on dispersal. In many taxa, parents can influence their offspring development by changing the resources allocated to their offspring, or the way the resources can be used, such as via epigenetic modification of the genome, or changing a hormonal titre in an egg, or the competitive environment an offspring will experience (Benton *et al.* 2008). In such a way, parents can influence the life-history strategy of their offspring, including their potential to disperse. For example, in gypsy moths, the offspring of mothers who had been nutritionally stressed, caused by living in trees that had been stripped of leaves, are more likely to disperse (Diss *et al.* 1996). Furthermore, by altering the number of eggs laid in a clutch and the spatial distribution of clutches, mothers directly impact the strength of kin competition and risk of inbreeding in their offspring, which influences the offspring's need to disperse.

Other ways in which the environment can either be expected to affect dispersal, or affect phenotype that in turn affects dispersal, include the presence of predators or diseases. The presence of predators increases the proportion of winged offspring in pea aphids (Kunert *et al.* 2005), and fish predators increase dispersal in water beetles (McCauley and Rowe 2010). In the case of parasites and pathogens, activation of the immune system may also signal that one is in a risky habitat, and make disperse more likely, as recently shown in a dragonfly (Suhonen *et al.* 2010). Alternatively, infected animals may be less likely to move if infection impairs movement. For example, genetic analysis of populations of a pest vector, the leafhopper *Scaphoideus titanus*, indicate that populations carrying the phytoplasma, are less dispersive than ones uninfected (Papura *et al.* 2009). Infected animals may also be more likely to move if this is a parasitic adaptation to increase the parasites' transmission. For example, malarial-infected mosquitoes feed more often and therefore disperse the parasite (Koella *et al.* 1998).

Environmental factors also play an important role in how far individuals can move. Successful dispersal generally declines with increasing distance, reflecting the distance-related cost of dispersal. In the desert amphipod, *Hemilepistus reamuri*, the costs of dispersal are related to dehydration, which in turn is related to both the distance and speed moved (Baker and Rao 2004), and thus costs depend on both sex and body condition (e.g. size) (Baker 2004).

Typically, in the past, dispersal models have assumed a fixed rate of dispersal and that individuals have either random movement or perfect knowledge of the environment and optimal dispersal (Bowler and Benton 2005). Neither of these assumptions is realistic and for most organisms, dispersal is potentially a risky process where the risk can be offset if individuals can collect information on the quality of the environment to allow adaptive decision-making. As some external factors directly affect an individual's phenotype, changes in traits such as lipid reserves can be seen as providing a source of information that is based on direct experience. However, other external factors may not directly affect phenotype but nonetheless be perceived and carry information that can influence dispersal decisions. Some of these cues (such as density) may be sampled by individuals before dispersal, reflecting the quality of the current patch, and so shape their initial motivation, or timing of, dispersal. While other cues may be sampled once individuals start dispersal, and may be used in navigation or settlement, in this case the use of such cues may depend on dispersal mode, with individuals using wind-borne methods less able to control their movement once started. Increasingly, we recognize that dispersal decisions are influenced by an individual's external environment, and their internal state (Clobert *et al.* 2009; Schjorring 2002; Stamps and Krishnan 2005); in other words, the decisions around dispersal integrate information from a variety of sources. There are many unresolved questions on how an individual integrates the different sources of information available, and how this interacts with individual traits and their initial motivation to disperse.

Some sources of external information were discussed earlier (crowding and harassment, the presence of parasites and predators). Other sources

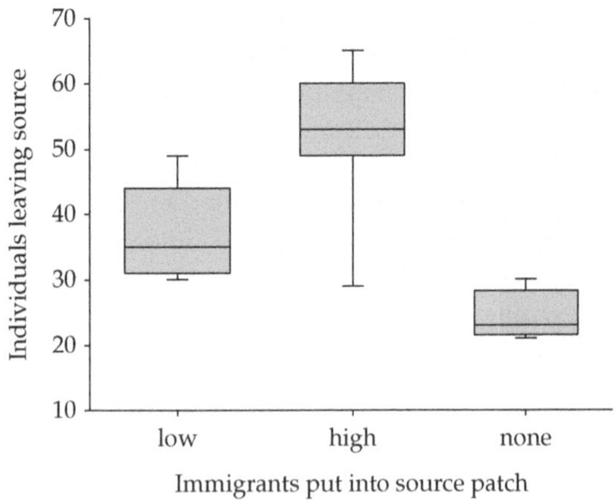

Figure 4.1 Informed dispersal in soil mites. Equally sized populations were seeded with a low or high density of immigrants (and equal number of residents removed, to maintain density). The number of successful dispersers to a connected 'empty' patch was counted over the following week. Where more immigrants were added to the source population, more individuals subsequently dispersed: Kruskal-Wallis one way anova, $H = 28.26$ $DF = 2$ $P < 0.0005$ (Moore, Cameron, Shroeder and Benton, unpublished data).

include volatile cues that can be detected from a distance and indicate the existence of nearby patches, reducing the risk of dispersal. For example, the predatory mite *Phytoselius* is more likely to leave a patch if there are chemical cues of nearby patches (Nachappa *et al.* 2006). Soil mites are more likely to disperse from a patch if there is a turnover of individuals in the patch (irrespective of any change in density), perhaps signalling the presence of other patches nearby via chemical means (Figure 4.1).

Similarly, cues can be gained about the locality outside the patch of residency through temporary prospecting trips, as shown in butterflies (Conradt *et al.* 2001). During dispersal itself, cues may be gained about a patch's quality prior to settlement (Stamps and Krishnan 2005). For example, in *Hemilepistus*, experience of the home patch is used as a relative measure to assess potential new patches (Baker 2005), though exactly how is unknown. Chemical cues can sometimes indirectly signal patch quality, for example fruitflies use aggregation pheromones to estimate the density of individuals in the patch and use this as a proxy for patch condition (de Gee *et al.* 2008; Lof *et al.* 2009).

4.3 Stage- and sex-dependent influences on dispersal

An animal's ability and propensity to move is often related to their life-history stage and sometimes also their sex. Many invertebrates are effectively sessile for parts of their life cycle, necessitating a motile phase associated with dispersal. In many marine invertebrates, the motile stage is planktonic. In insects (especially the endopterygotes), the motile stage is typically the winged adults, while larvae are restricted to the natal site. Although this is a rule of thumb, even in insects, many larvae move and, to an extent, disperse to avoid kin competition. Larval dispersal is sometimes far from insignificant; in some sawflies, many lepidopterans like armyworms and processionary caterpillars, the larvae can disperse considerable distances via walking or ballooning on silk threads (Floater 1996; Johns *et al.* 2009; McDonald 1991; Steinbauer 2009). In the case of the winter moth, larval dispersal is associated with reduced adult movement. Nonetheless, when all stages or instars are motile, there is typically variation in their movement ability. For example, in mites, movement between patches carries both metabolic and mortality costs (related to dehydration) and adults, which can move faster, are more likely to disperse than juveniles (Bowler and Benton 2009). In a similar vein, fully reproducing adult females, which are essentially swollen 'egg machines', have reduced movement ability and also reduced dispersal propensity. This means that, overall, pre-reproductive adult females and males disperse most easily, compared to both juveniles and mature females (Bowler and Benton 2009). That female dispersal may become limited due to their investment of resources in eggs may be common in many insects (the 'Oogenesis-flight syndrome', see Lorenz 2007). Differential allocation of resources between the sexes will contribute to sex-biased dispersal, which may generally arise through two routes. Firstly, if there is sexual size dimorphism then the different sexes may travel different distances if movement is affected by size. For example, in the scolytid beetle *Ips typographus*, long-distance dispersal is commonly observed and in these cases it is the males that go further than females because they are larger (Salle *et al.* 2007). Second, males and females may differ in their propensity to move for reasons related to reproduction. In some cases, females may be more likely to disperse as they are seeking oviposition sites, while males may disperse less if they are territorial; this happens, for example, in the tephritid fly *Paroxyna plantagenis* (Albrectsen and Nachman 2001), and in the lekking damselfly *Calyopterx spendens* (Chaput-Bardy *et al.* 2010). In other cases, for non-territorial males, males may be more likely to disperse to increase their access to 'new' females that they have not mated with yet (Lawrence 1988). In a range of 'tethered flight' experiments, the effort put into flight has been shown to depend on sex and mating status (Bruzzone *et al.* 2009; Elliott and Evenden 2009).

4.4 An individual's size or physiological 'condition' can influence dispersal

An animal's size or physiological condition is typically an integration of the developmental processes over its lifetime (and sometimes, via parental effects, previous generations). Many studies have shown that the availability of food resources, or the effects of food availability on body size or body condition, influences dispersal. In mites, there is considerable plasticity in development depending on food, leading to differences in size and age at maturity (Plaistow *et al.* 2004), which impact on subsequent life-history traits such as fecundity and survival (Beckerman *et al.* 2003; Plaistow *et al.* 2006). This also affects dispersal propensity and the survival of individuals that attempt dispersal (Bowler and Benton 2009): well-fed individuals are larger, which causes males to disperse more but inhibits movement in females, probably because their size slows their movement. In the moth *Choristoneura conflictana*, lipid stores influence individual's flight times, as determined by laboratory tethered experiments, in a way that depends on sex, mating status, and age (Elliott and Evenden 2009).

There are many other examples where body size or body condition has been shown to affect dispersal propensity, or ability to undertake successful dispersal, as reflected by distance travelled, or where individuals eventually settle (Baker 2004; Baker and Rao 2004; Bruzzone *et al.* 2009; Burgess *et al.* 2009). Often, there are positive correlations (larger animals disperse more), which may reflect an increased capacity to disperse, but sometimes the reverse is true. For example, in the dragonfly *Pachydiplax longipennis*, small males are more likely to disperse as they are weak competitors (McCauley 2010); while in horned beetles, small males are also more likely to move but this is because they can disperse more easily: large horns are a handicap to movement (Okada *et al.* 2007; Yamane *et al.* 2010).

4.5 Dispersal decisions are influenced by dispersal phenotypes

As discussed elsewhere in this work, the concept of a dispersal phenotype is much debated. In one sense, any generic locomotory adaptation—whether morphological, physiological, or behavioural—can be broadly considered a 'dispersal phenotype'. However, there may be inter-individual variation in dispersal caused by context- or condition- dependence such that the dispersal phenotype is preferentially expressed in some individuals rather than others. Often dispersal behaviour is associated with covariation in a number of phenotypic traits (e.g. size, sex, condition), creating a 'dispersal syndrome' (Chapters 10–14).

The most apparent mechanism for dispersal in insects is via flight (though not all insects disperse by flight!), so wings are perhaps the most obvious dispersal phenotype. However, in some species there is a clear context-dependent expression of this phenotype in that individuals develop into winged forms in set ecological circumstances (e.g. time of year). Non-flying terrestrial invertebrates also have a range of dispersal adaptations that allow aerial dispersal. For example, several taxa (spiders, some lepidopterans) disperse via 'ballooning', where an individual stands at an exposed site, such as the tip of a plant, lets out a line of silk to catch the wind, and then travels as 'aerial plankton'. In spiders, the propensity to balloon in a given context has been shown to have a heritable component (as has the propensity to move shorter-distances by rappelling between neighbouring plants) (Bonte and Lens 2007, Bonte *et al.* 2008). Another adaptive route to dispersal is via phoresy, which is when an organism 'hitch-hikes' a ride on another organism that has a greater movement capacity. Many mites have a facultative phoretic morph, called a hypopus, that hitch-hikes on beetles (Houck and O'Connor 1991). Hypopi are non-feeding and capable of prolonged existence, with a specialized 'sucker' to allow it to adhere to the beetle carapace. As with spiders' ballooning, development into a long-distance dispersing phoretic mite is context-dependent (they will typically chose to develop this way when patches decline in quality), and a common observation is that the propensity to develop into hypopi under similar conditions typically declines in laboratory populations with time, suggesting that there is also a heritable component in this dispersal adaptation (Benton, unpublished data).

In some systems, a dispersal phenotype is not obviously related to dispersal *per se*. As discussed elsewhere (Case Study III, Chapters 8 and 23), in Glanville fritillaries, molecular genetic analysis indicated that this dispersal phenotype was associated with variation in a key glycolytic enzyme glucose-6-phosphate isomerize (also called phosphoglucose isomerize, or *pgi*) During flight, the supply of ATP to the flight muscles can become a rate-limiting step, especially under certain temperature conditions (Hanski *et al.* 2006), so variation in the *pgi* gene therefore influences flight performance; however, because it is a general metabolic enzyme, it has also additional affects on physiological performance throughout life (creating a dispersal syndrome, Chapter 13). The *pgi* gene is known to be important in a range of taxa and on different traits including *Drosophila* flight performance (Eanes *et al.* 2006), crickets (where genetic variation is maintained by balancing selection (Broughton and Harrison 2003)), beetles and their cold tolerance and movement rate (Dahlhoff *et al.* 2008, Rank *et al.* 2007), crustaceans and their heat tolerance and survival (Patarnello and Battaglia 1992), and even different ecotypes in *Arabidopsis*, perhaps also associated with temperature tolerance (Kawabe *et al.* 2000). These studies illustrate that there can be relatively simple genetic factors underlying not only movement behaviour but the way this relates to an individual's life-history in general and creating co-variation in many traits. In this case, having a higher metabolic rate changes the propensity and ability to move between patches along with a suite of other traits.

4.6 Conclusions

The empirical literature on invertebrates strongly indicates a number of consistent messages. Firstly, the decision to leave a patch, move between patches, or settle in a new patch differs between individuals and is not a fixed probability. Second, as dispersal is movement creating gene flow, there are perhaps as many dispersal behaviours as species and to define it in operational terms as a single behavioural mode is impossible. Dispersal is sometimes associated with leaving a natal patch, sometimes associated with seeking a place to reproduce, or a mate, sometimes associated with finding food or shelter, and sometimes associated with just moving (as may occur by drifting in wind or water currents); indeed, some individuals may disperse throughout life, using multiple modes of movement. Third, movement ability and the underlying proximal drivers of dispersal are very dependent on individual characteristics: genes, size/age/stage, sex, and condition (including disease state, perception of risk, and even maternal condition). Fourth, although convenient to model dispersal as a random, diffusion, term, organisms typically have evolved a range of ways of using information to make informed decisions (Clobert *et al.* 2009). This suggests that individual movement paths are far from random walks across the landscape (Smouse *et al.* 2010).

The evolutionary function of dispersal is to move genes around the landscape. Movement, however, arises from a number of motivations and a single movement can fulfil more than one function (e.g. moving to find food is also functionally dispersal). So, although some behaviours are specifically 'for' dispersal, others may be for dispersal and finding food, shelter, mates, etc. The definition of dispersal as movement only by a specific mode has created much confusion, as it ignores the 'non-dispersal' movements that, over a lifetime, may create as much or more spatial gene flow. By being able to track individuals more than we have in the past, we are both observing individual movement more finely than we have previously, and understanding the condition- and context-dependence of those movements. We are increasingly in a position to unite 'movement ecology', the study of movements across a landscape, (Morales *et al.* 2010; Nathan *et al.* 2008; Smouse *et al.* 2010) with 'dispersal ecology', which has traditionally focused on a single class of movement, 'dispersal'.

References

Albrectsen, B., Nachman, G. (2001). 'Female-biased density-dependent dispersal of a tephritid fly in a fragmented habitat and its implications for population regulation', *Oikos* **94**: 263–72.

Baguette, M., Convie, I., Neve, G. (1996). 'Male density affects female spatial behaviour in the butterfly Proclossiana eunomia', *Acta Oecologica-International Journal of Ecology* **17**: 225.

Baker, M.B. (2004). 'Sex biased state dependence in natal dispersal in desert isoponds, Hemilepistus reaumuri', *Journal of Insect Behavior* **17**: 579–98.

——(2005). 'Experience influences settling behaviour in desert isopods, Hemilepistus reaumuri', *Animal Behaviour* **69**: 1131–8.

Baker, M.B., Rao S. (2004). 'Incremental costs and benefits shape natal dispersal: Theory and example with Hemilepistus reaumuri', *Ecology* **85**: 1039–51.

Beckerman, A.P., Benton, T.G., Lapsley, C.T. *et al.* (2003). 'Talkin' 'bout my generation: Environmental variability and cohort effects', *American Naturalist* **162**: 754–67.

Benton, T.G., Plaistow, S.J., Coulson, T.N. (2006). 'Complex population dynamics and complex causation: devils, details and demography', *Proceedings of the Royal Society B* **273**: 1173–81.

Benton, T.G., St Clair, J.J.H., Plaistow, S.J. (2008). 'Maternal effects mediated by maternal age: from life histories to population dynamics', *Journal of Animal Ecology* **77**: 1038–46.

Bonte, D., Lens, L. 2007. 'Heritability of spider ballooning motivation under different wind velocities', *Evolutionary Ecology Research* **9**: 817.

Bonte, D., Travis, J.M.J., De Clercq, N. *et al.* (2008). 'Thermal conditions during juvenile development affect adult dispersal in a spider', *Proceedings of the National Academy of Sciences of the USA* **105**: 17000.

Bowler, D.E., Benton, T.G. (2005). 'Causes and consequences of animal dispersal strategies: relating individual behaviour to spatial dynamics. *Biological Reviews* **80**: 205–25.

——(2009). 'Variation in dispersal mortality and dispersal propensity among individuals: the effects of age, sex and resource availability', *Journal of Animal Ecology* **78**: 1234–41.

Broughton, R.E., Harrison, R.G. (2003). 'Nuclear gene genealogies reveal historical, demographic and selective factors associated with speciation in field crickets', *Genetics* **163**: 1389–401.

Brunzel, S. (2002). 'Experimental density-related emigration in the cranberry fritillary Boloria aquilonaris', *Journal of Insect Behavior* **15**: 739–50.

Bruzzone, O.A., Villacide, J.M., Bernstein, C. *et al.* (2009). 'Flight variability in the woodwasp Sirex noctilio (Hymenoptera: Siricidae): an analysis of flight data using wavelets', *Journal of Experimental Biology* **212**: 731–7.

Burgess, S.C., Hart, S.P., Marshall, D.J. (2009). 'Pre-Settlement Behavior in Larval Bryozoans: The Roles of Larval Age and Size', *Biological Bulletin* **216**: 344–54.

Chaput-Bardy, A., Gregoire, A., Baguette, M. *et al.* (2010). 'Condition and Phenotype-Dependent Dispersal in a Damselfly, Calopteryx splendens', *PLoS One* 5 (5): e10694. DOI 10.1371/journal.phone.0010694.

Clobert, J., Le Galliard, J.F., Cote, J. *et al.* (2009). 'Informed dispersal, heterogeneity in animal dispersal syndromes and the dynamics of spatially structured populations', *Ecology Letters* **12**: 197–209.

Conradt, L., Roper, T.J., Thomas, C.D. (2001). 'Dispersal behaviour of individuals in metapopulations of two British butterflies', *Oikos* **95**: 416–24.

Dahlhoff, E.P., Fearnley, S.L., Bruce, D.A. *et al.* (2008). 'Effects of Temperature on Physiology and Reproductive Success of a Montane Leaf Beetle: Implications for Persistence of Native Populations Enduring Climate Change', *Physiological and Biochemical Zoology* **81**: 718–32.

de Gee, M., Lof, M.E., Hemerik, L. (2008). 'The effect of chemical information on the spatial distribution of fruit flies: II parameterization, calibration, and sensitivity', *Bulletin of Mathematical Biology* **70**: 1850–68.

De Meester, N., Bonte, D. (2010). 'Information use and density-dependent emigration in an agrobiont spider', *Behavioral Ecology* **21**: 992–8.

Diss, A.L., Kunkel, J.G., Montgomery, M.E. *et al.* (1996). 'Effects of maternal nutrition and egg provisioning on parameters of larval hatch, survival and dispersal in the gypsy moth, Lymantria dispar L.', *Oecologia* **106**: 470–7.

Eanes, W.F., Merritt, T.J.S., Flowers, J.M. *et al.* (2006). 'Flux control and excess capacity in the enzymes of glycolysis and their relationship to flight metabolism in Drosophila melanogaster', *Proceedings of the National Academy of Sciences of the USA* **103**: 19413–18.

Elliott, C.G., Evenden, M.L. (2009). 'Factors influencing flight potential of Choristoneura conflictana', *Physiological Entomology* **34**: 71–8.

Floater, G.J. (1996). 'Estimating movement of the processionary caterpillar Ochrogaster lunifer Herrich-Schaffer (Lepidoptera: Thaumetopoeidae) between discrete resource patches', *Australian Journal of Entomology* **35**: 279–83.

Fonseca, D.M., Hart, D.D. (1996). 'Density-dependent dispersal of black fly neonates is mediated by flow', *Oikos* **75**: 49.

Hanski, I., Saastamoinen, M., Ovaskainen, O. (2006). 'Dispersal-related life-history trade-offs in a butterfly metapopulation', *Journal of Animal Ecology* **75**: 91–100.

Houck, M.A., O'Connor, B.M. (1991). 'Ecological and Evolutionary Significance of Phoresy in the Astigmata', *Annual Review of Entomology* **36**: 611–36.

Johns, R.C., Quiring, D.T., Ostaff, D.P. (2009). 'Sex biased intra-tree oviposition site selection and larval foraging behavior of a specialist herbivore. *Entomologia Experimentalis et Applicata* **132**: 264–74.

Kawabe, A., Yamane, K., Miyashita, N.T. (2000). 'DNA polymorphism at the cytosolic phosphoglucose isomerase (PgiC) locus of the wild plant Arabidopsis thaliana', *Genetics* **156**: 1339–47.

Koella, J.C., Sorensen, F.L., Anderson, R.A. (1998). The malaria parasite, Plasmodium falciparum, increases the frequency of multiple feeding of its mosquito vector, Anopheles gambiae', *Proceedings of the Royal Society Series B* **265**: 763–8.

Kukuk, P.F., Bitney, C., Forbes, S.H. (2005). 'Maintaining low intragroup relatedness: evolutionary stability of nonkin social groups', *Animal Behaviour* **70**: 1305–11.

Kunert, G., Otto, S., Rose, U.S.R. *et al.* (2005). 'Alarm pheromone mediates production of winged dispersal morphs in aphids', *Ecology Letters* **8**: 596–603.

Lawrence, W.S. (1988). 'Movement ecology of thre red milkweed beetle in relation to population-size and structure', *Journal of Animal Ecology* **57**: 21–35.

Lof, M.E., de Gee, M., Hemerik, L. (2009). 'Odor-mediated aggregation enhances the colonization ability of Drosophila melanogaster', *Journal of Theoretical Biology* **258**: 363–70.

Lorenz, M.W. (2007). 'Oogenesis-flight syndrome in crickets: Age-dependent egg production, flight performance, and biochemical composition of the flight muscles in adult female Gryllus bimaculatus', *Journal of Insect Physiology* **53**: 819–32.

McCauley, S.J. (2010). 'Body size and social dominance influence breeding dispersal in male Pachydiplax longipennis (Odonata)', *Ecological Entomology* **35**: 377–85.

McCauley, S.J, Rowe, L. (2010). Notonecta exhibit threat-sensitive, predator-induced dispersal', *Biology Letters* **6**: 449–52.

McDonald, G. (1991). 'Oviposition and larval dispersal of the common armyworm, *Mythimna-convecta* (Walker) (Lepidoptera: Noctuidae)', *Australian Journal of Ecology* **16**: 385–93.

Menendez, R., Gutierrez, D., Thomas, C.D. (2002). 'Migration and Allee effects in the six-spot burnet moth Zygaena filipendulae', *Ecological Entomology* **27**: 317–25.

Morales, J.M., Moorcroft, P.R., Matthiopoulos, J. *et al.* (2010). 'Building the bridge between animal movement and population dynamics', *Philosophical Transactions of the Royal Society B* **365**: 2289–301.

Nachappa, P., Margolies, D.C., Nechols, J.R. (2006). 'Resource-dependent giving-up time of the predatory mite, Phytoseiulus persimilis', *Journal of Insect Behavior* **19**: 741–52.

Nathan, R., Getz, W.M., Revilla, E. *et al.* (2008). 'A movement ecology paradigm for unifying organismal movement research', *Proceedings of the National Academy of Sciences of the USA* **105**: 19052.

Okada, K., Nomura, Y., Miyatake, T. (2007). 'Relations between allometry, male-male interactions and dispersal in a sap beetle, Librodor japonicus', *Animal Behaviour* **74**: 749–55.

Papura, D., Delmotte, F., Giresse, X. *et al.* (2009). 'Comparing the spatial genetic structures of the Flavescence doree phytoplasma and its leafhopper vector Scaphoideus titanus', *Infection Genetics and Evolution* **9**: 867–76.

Patarnello, T., Battaglia, B. (1992). 'Glucosephosphate Isomerase and Fitness—Effects of Temperature on Genotype Dependent Mortality and Enzyme-Activity in Two Species of the Genus Gammarus (Crustacea: Amphipoda)', *Evolution* **46**: 1568–73.

Plaistow, S.J., Lapsley, C.T., Beckerman, A.P. *et al.* (2004). 'Age and size at maturity: sex, environmental variability and developmental thresholds', *Proceedings of the Royal Society B* **271**: 919–24.

Plaistow, S.J., Lapsley, C.T., Benton, T.G. (2006). 'Context-dependent intergenerational effects: The interaction between past and present environments and its effect on population dynamics', *American Naturalist* **167**: 206–15.

Rank, N.E., Bruce, D.A., McMillan, D.M. *et al.* (2007). 'Phosphoglucose isomerase genotype affects running speed and heat shock protein expression after exposure to extreme temperatures in a montane willow beetle', *Journal of Experimental Biology* **210**: 750–64.

Rhainds, M., Shipp, L., Woodrow, L. *et al.* (2005). 'Density, dispersal, and feeding impact of western flower thrips (Thysanoptera: Thripidae) on flowering chrysanthemum at different spatial scales', *Ecological Entomology* **30**: 96–104.

Ronce, O. (2007). 'How does it feel to be like a rolling stone? Ten questions about dispersal evolution', *Annual Review of Ecology Evolution and Systematics* **38**: 231.

Salle, A., Arthofer, W., Lieutier, F. *et al.* (2007). 'Phylogeography of a host-specific insect: genetic structure of Ips typographus in Europe does not reflect past fragmentation of its host', *Biological Journal of the Linnean Society* **90**: 239–46.

Schjorring, S. (2002). 'The evolution of informed natal dispersal: Inherent versus acquired information', *Evolutionary Ecology Research* **4**: 227–38.

Simpson, S.J., McCaffery, A.R., Hagele, B.F. (1999). 'A behavioural analysis of phase change in the desert locust', *Biological Reviews of the Cambridge Philosophical Society* **74**: 461–80.

Smouse, P.E., Focardi, S., Moorcroft, P.R. *et al.* (2010). 'Stochastic modelling of animal movement', *Philosophical Transactions of the Royal Society B* **365**: 2201–11.

Stamps, J., Krishnan, V.V. (2005). 'Nonintuitive cue use in habitat selection', *Ecology* **86**: 2860–7.

Steinbauer M.J. (2009). 'Thigmotaxis maintains processions of late-instar caterpillars of Ochrogaster lunifer', *Physiological Entomology* **34**: 345–9.

Suhonen, J., Honkavaara, J., Rantala, M.J. (2010). 'Activation of the immune system promotes insect dispersal in the wild', *Oecologia* **162**: 541–7.

Wratten, S.D., Bowie, M.H., Hickman, J.M. *et al.* (2003). 'Field boundaries as barriers to movement of hover flies (Diptera: Syrphidae) in cultivated land', *Oecologia* **134**: 605–11.

Yamane, T., Okada, K., Nakayama, S. *et al.* (2010). 'Dispersal and ejaculatory strategies associated with exaggeration of weapon in an armed beetle', *Proceedings of the Royal Society B* **277**: 1705–10.

CHAPTER 5

Integrating context- and stage-dependent effects in studies of frugivorous seed dispersal: an example from south-east Kenya

Valérie Lehouck, Dries Bonte, Toon Spanhove, and Luc Lens

5.1 Introduction

Endozoochory refers to plant dispersal by which viable seeds are defaecated or regurgitated by animals (Van der Pijl 1969). Grazing herbivores can disperse seeds through the ingestion of—apparently unassisted—seeds along with plant foliage (Janzen 1984). This contrasts with distinctively coloured, fleshy, or arillated fruits that are attractive to frugivorous dispersers—mainly bats, monkeys, and birds (Howe and Smallwood 1982; Terborgh 1986; Fleming *et al.* 1987). Because of the biotic interaction involved, endozoochorous seed dispersal strongly depends on various behavioural components of the dispersing vectors, such as foraging and seed handling (including ingestion and gut treatment), movement away from the parent tree, and defaecation or regurgitation (Schupp 1993; Wang and Smith 2002). As a consequence, and in contrast to most abiotically dispersed seeds, endozoochorous seed dispersal is likely to be clumped (Clark *et al.* 1998; Russo *et al.* 2006) and non-random (Wenny 2001; Russo *et al.* 2006; Carlo and Morales 2008). Because endozoochorous seed dispersal involves a network of potential dispersing agents (Bascompte and Jordano 2007), a wide array of frugivores that may or may not act in sequence and vary in their effectiveness as a seed disperser, jointly shape the resulting dispersal kernel (Jordano *et al.* 2007; Spiegel and Nathan 2007; Martínez *et al.* 2008 ; Lehouck *et al.* 2009b). As such, changes in landscape composition may affect endozoochorous seed dispersal through changes in plant habitat configuration and vector community composition and behaviour (Holl 1999; Cordeiro and Howe 2001; Levey *et al.* 2005). In dynamic systems where dispersal effectiveness is highly context-dependent, a disperser guild consisting of multiple species increases ecosystem resilience (Schupp *et al.* 2010).

Main selective forces driving the evolution of dispersal-related traits include (i) avoidance of kin competition by escaping high seed and seedling mortality near the parent tree, (ii) directed dispersal towards (un)suitable habitat, and (iii) colonization of new habitat as a risk spreading strategy (Chapter 1; Howe and Smallwood 1982; Rousset and Gandon 2002). While these ultimate factors can be expected to shape the overall (seed) dispersal strategy, a variety of proximate factors related to the abiotic and biotic environment determine variation in dispersal kernels. Abiotic conditions are mainly related to structural components of the landscape, biotic ones to those of the dispersing community and the prevalence of diseases or predators. Because the latter can be expected to be affected by the landscape context too, both sets of conditions strongly interact. Whereas studies of landscape effects usually refer to the direct effects of (a)biotic factors in a given location on single processes, they often ignore such indirect or cascading effects. We therefore introduce the term 'extended landscape', in parallel with the principle of the 'extended phenotype'

Dispersal Ecology and Evolution. First Edition. Edited by Jean Clobert *et al.*

introduced by Dawkins (1982), to apply a holistic view on landscape effects that comprise all biotic and abiotic components, either direct or indirect.

5.2 Seed dispersal of *Xymalos monospora* in the Taita Hills: a case study

Xymalos monospora is a small, dioecious tree indigenous to African submontane moist forest (Beentje 1994; Aerts *et al.* 2011) and common in the highly fragmented cloud forests of the Taita Hills, part of the Eastern Afromontane biodiversity hotspot (Mittermeier *et al.* 2005). Indigenous forest remnants of the Taita Hills are 1–120 ha in size, often surrounded by exotic plantations consisting of *Eucalyptus*, *Pinus*, and *Cupressus* species and/or embedded in a matrix of smallholder agriculture (mainly banana, maize, and cabbage). Spatial recruitment of *X. monospora* is mainly affected by avian seed dispersal due to the virtual absence of a soil seed bank and very limited seed predation and secondary dispersal (Lehouck *et al.* 2009a; Lehouck *et al.* 2009c). Three frugivorous bird species, the Taita thrush (*Turdus helleri*), stripe-cheeked greenbul (*Andropadus milanjensis*), and Hartlaub's turaco (*Tauraco hartlaubi*), jointly remove 80% of the fleshy, yellow fruits (Lehouck *et al.* 2009a). These bird species represent contrasting functional groups based on their main ecological characteristics relevant for dispersal (see Dennis and Westcott 2006 for an overview of these characteristics). *T. helleri* is a medium-sized, highly sedentary forest specialist, while *T. hartlaubi* is a large-sized, mobile species that regularly ventures outside the boundaries of indigenous forest. *A. milanjensis* occupies an intermediate position in size, mobility, and level of forest dependency.

5.2.1 Stage and context dependency of frugivorous seed dispersal

Frugivorous seed dispersal is typically affected by multiple biotic and abiotic factors that may either act separately or in concert. Moreover, seed dispersal comprises a multistage process in which not only the movement phase (transience) *per se*, but also the pre- (departure) and post-dispersal phases (settlement) may be affected by the interaction between plants, their dispersers, and the (a)biotic environment (Figure 5.1). While the importance of taking into account pre- and post-dispersal phases in seed dispersal studies is widely recognized (see, for example, Nathan and Muller-Landau 2000; Wang and Smith 2002), studies considering the multiphase process of seed dispersal in a spatio-temporally variable context are extremely rare (Schupp *et al.* 2010). We aim to focus here on the holistic effect of the (a)biotic environment ('extended landscape') on each of the seed dispersal phases.

During the *departure phase*, fruit intake by the dispersal agents (the animal vectors) not only depends on the spatial and temporal fruit distribution, but also on the attractiveness of fruits to dispersers which is mediated by crop size, fruit size, colour, or odor (Jordano 1987; Schmidt *et al.* 2004). Figure 5.2A shows that in the Taita Hills, *X. monospora* bearing large crops of ripe fruits have more fruits removed, which in this case can be attributed to more visits by frugivorous birds (Lehouck *et al.* 2009a). The presence of other fruiting *X. monospora* trees within the neighbourhood increases the competition for fruit removal by frugivores, resulting in reduced per capita removal rates. The visitation length and number of fruits eaten per visit also differ among bird species (Lehouck *et al.* 2009a), and as such, seed dispersal vectors vary in their quantitative and qualitative contribution towards fruit removal (Lehouck *et al.* 2009a, 2009d).

During the *transience phase*, animal vectors determine the distance and direction of seed dispersal through gut transit times and post-foraging movements, both of which may vary among dispersers. For instance, gut passage or regurgitation rates generally depend on the behaviour, morphology, and physiology of the frugivores as well as on properties of seeds, such as size, pulp-to-seed ratio, and the presence of secondary compounds (Traveset 1998). Mobility and habitat use of frugivores, in turn, may not only depend on their behaviour and morphology (Jordano *et al.* 2007), but may also vary with season (Bowen *et al.* 2007), time of the day (Westcott *et al.* 2005), landscape structure (Levey *et al.* 2008), and/or the spatio-temporal distribution of critical resources such as food and water

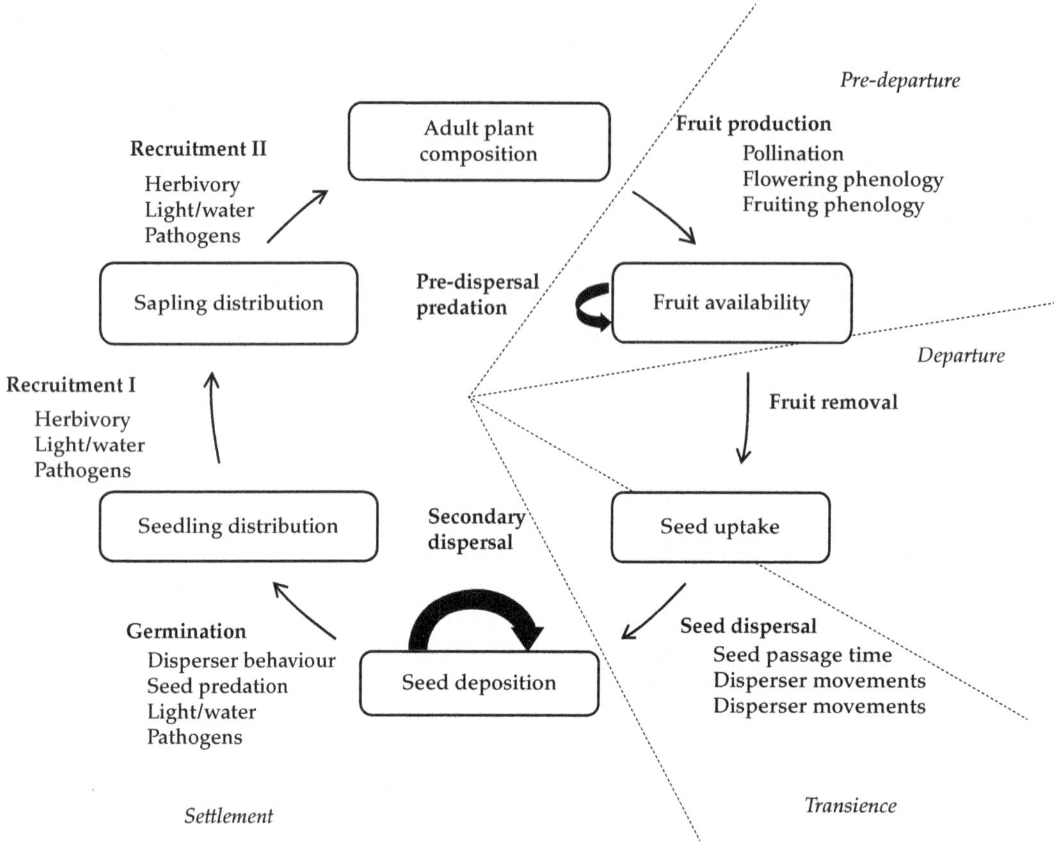

Figure 5.1 Schematic representation of the life cycle of fleshy fruited plant species, with indication of four phases in the dispersal process. Processes (bold text accompanying arrows) lead to patterns (rectangles). Each of the sub-processes is affected by the interaction among fruiting plants, their disperser guild, and the environment.

(Thies *et al.* 2006; Lehouck *et al.* 2009d). As a result, species-specific gut passage or regurgitation times of swallowed seeds in combination with the movement behaviour of frugivores carrying seeds, will proximately determine the dispersal distances of animal-dispersed seeds in a given spatio-temporal context (e.g. Murray 1988; Schupp 1993). This will ultimately affect the fate of individual seeds, plant populations, communities, and ecosystems. Although transit times of *X. monospora* seeds do not differ among the three main dispersers in the Taita Hills (median transit times are 34, 33, and 40 minutes for *T. helleri, A. milanjensis* and *T. hartlaubi* respectively; Lehouck *et al.* 2009b; Lehouck *et al.* 2011), both the distances covered during transit and the microhabitats visited when depositing the seeds vary considerably. The endemic forest specialist *T. helleri* mainly disperses seeds at a local scale; i.e. within the boundaries of single indigenous forest patches (mean displacement ± SE during median transit time of seeds through the gut: 45 ± 4 m, Lehouck *et al.* 2009b). As a result, density-dependent mortality rates related to decreased host-specific pathogen loading, seed predation, herbivory, or sibling competition away from the parent tree are reduced (Lehouck *et al.* 2009a). The larger and more mobile *T. hartlaubi* disperses seeds at intermediate spatial scales (115 ± 51 m), whereby seeds may be carried into nearby new habitats such as exotic plantations bordering indigenous forest patches (Lehouck *et al.* 2009b). The behaviour of the medium-sized *A. milanjensis* is intermediate: while this species mainly disperses seeds over short distances (61 ± 5 m), it has been observed to deposit seeds in exotic plantations and the agricultural matrix immediately bordering indigenous forest fragments. Since foraging *A. milanjensis* cross indigenous forest boundaries more often in small than in large fragments (Lehouck *et al.* 2009b), areas surrounding the former are exposed to higher levels of seed rain. Yet none of the three frugivores have been observed to move between indigenous forest fragments during the restricted time window of seed gut passage. Because forest specialist birds often fail to persist in small degraded fragments due to their limited mobility (Lens *et al.* 2002; Devictor *et al.* 2008), interior trees depending on this avian guild for dispersal may experience dispersal failure if forest generalists cannot function as equivalent substitutes. Although some disturbed fragments in the Taita Hills appear to benefit from visits of non-forest dependent seed dispersers, such 'resilience' is neither predictable nor reliable in time or space, and does not counterbalance the seed dispersal loss by the more efficient forest-dependent *T. helleri, A. milanjensis* and *T. hartlaubi* (Figure 5.2b).

During the *settlement phase,* frugivores may directly affect the success and speed of germination of gut-passed seeds by removing fruit pulp, by increasing the permeability of the seed coat for water and nutrients, or by embedding the seed in nutrient-rich faecal material (Traveset 1998). In addition, seed survival, seed germination, seedling establishment, and seedling performance may all depend on how and where seeds are deposited. For example, seed fate may differ when deposited clumped versus scattered (Russo and Augspurger 2004), in open versus covered microhabitat (Wenny and Levey 1998), or under versus away from the parent tree (Harms *et al.* 2000). In the Taita Hills, the germination rate *X. monospora* seeds is six to ten times higher, and seeds germinate up to twice as fast, after bird ingestion compared to intact fruits (Figure 5.2c). Furthermore, germination and seedling establishment benefit by being removed from the parent tree where density-dependent mortality is high (Figure 5.2d), although seeds may end up under conspecific trees in high numbers due to directed dispersal (see Carlo and Morales 2008; Spiegel and Nathan 2010).

Density-dependent seed predation can generate different recruitment patterns depending on the host specificity, mobility, and social behaviour of the predator/pathogen guild (Nathan and Casagrandi 2004). Provided that the average distance of predator activity is smaller than the average seed-dispersal distance, the expected fitness of seeds in homogeneous landscapes is presumed to be optimal at intermediate distances. At this range of distances, trade-offs between increasing direct costs, related to investment in dispersal-related morphological structures, and decreasing indirect costs, related to density-dependent mortality, are balanced (Janzen-Connell hypothesis; Rousset and Gandon 2002; Nathan and Casagrandi 2004). However, in the case of increased deposition of seeds in specific sites ('directed dispersal'), density-dependent mortality of dispersed seeds may still be high, and these costs may not outweigh the advantages of being dispersed to otherwise 'safe sites' (Spiegel and Nathan 2010). Moreover, changes in the environmental context, such as elevated predation rates or habitat heterogeneity, may alter cost-benefit balances in either direction (Schupp *et al.* 2010).

5.2.2 Effects of the extended landscape on seed dispersal and seedling establishment

When fragmentation causes a decrease in abundance of fruit-eating animals, this will almost inevitably lead to reduced seed dispersal rates and lower net average seed dispersal distance, as a higher proportion of seeds is predicted to land under the parent tree (Cordeiro and Howe 2001; Cordeiro and Howe 2003). Concomitantly, poorer recruitment occurs if density-dependent mortality exists in these plant species (Cordeiro *et al.* 2009; Lehouck *et al.* 2009c). However, when ignoring the large amount of seeds that fall below the parent crown as intact fruits, the dispersal distance and recruitment success of seeds that are effectively removed by a frugivore are not necessarily reduced. Actually, these 'effectively dispersed seeds' (hereafter 'dispersed seeds') are often the only relevant ones contributing to gene flow, especially in case of endozoochorous-dispersed fleshy fruits that often fail to germinate when they remain covered by fruit pulp (Lehouck *et al.* 2011; Figure 5.2b). The fate of dispersed seeds strongly depends on the characteristics of the disperser guild in interaction with the environment. For example, if

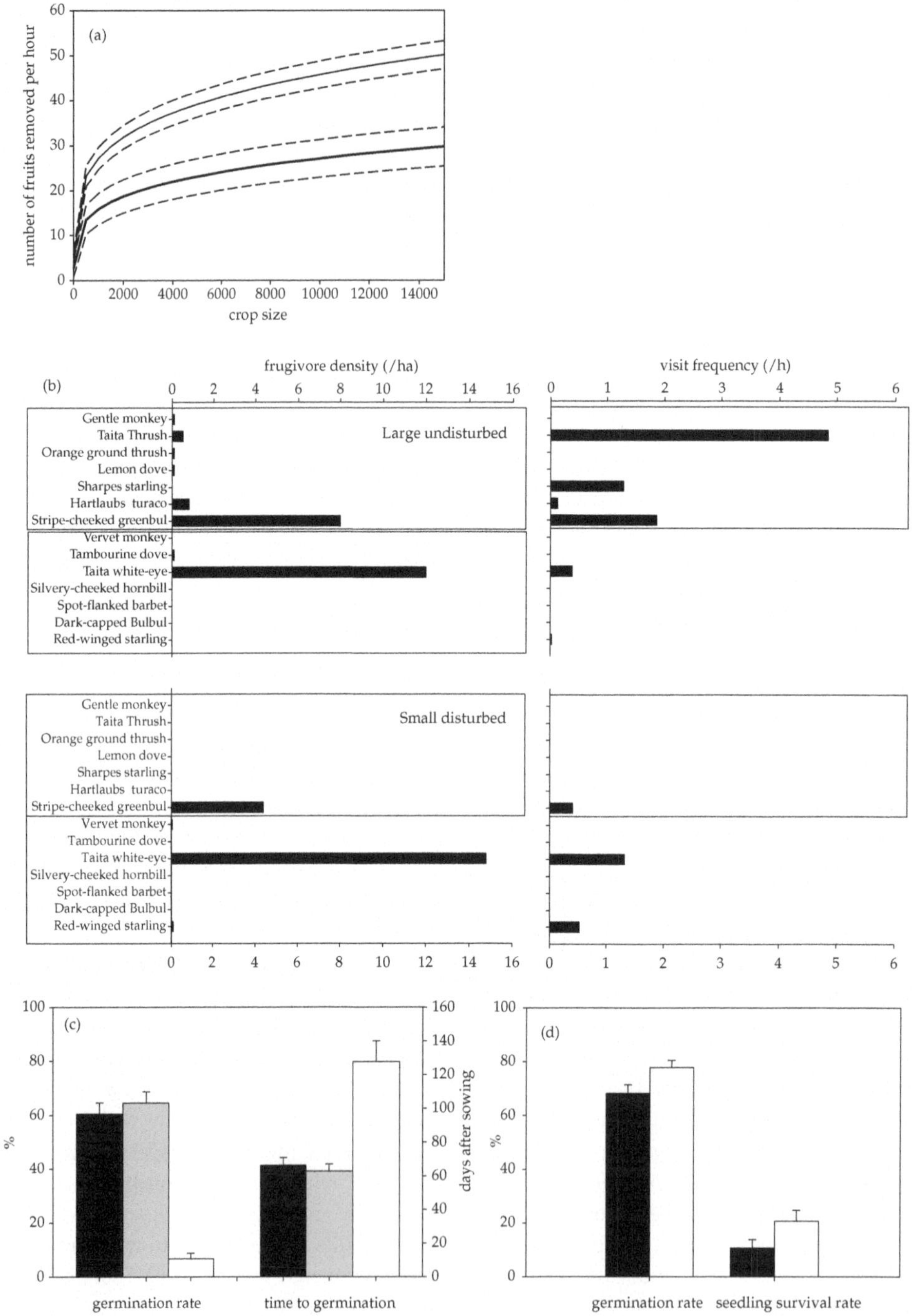

(a)
number of fruits removed per hour
crop size
0
10
20
30
40
50
60
2000
4000
6000
8000
10000
12000
14000
(b)
frugivore density (/ha)
visit frequency (/h)
Large undisturbed
Small disturbed
Gentle monkey
Taita Thrush
Orange ground thrush
Lemon dove
Sharpes starling
Hartlaubs turaco
Stripe-cheeked greenbul
Vervet monkey
Tambourine dove
Taita white-eye
Silvery-cheeked hornbill
Spot-flanked barbet
Dark-capped Bulbul
Red-winged starling
0 2 4 6 8 10 12 14 16
0 1 2 3 4 5 6
(c)
%
days after sowing
germination rate
time to germination
0 20 40 60 80 100
0 20 40 60 80 100 120 140 160
(d)
%
germination rate
seedling survival rate

the remaining frugivorous species in fragmented habitats altogether drop more seeds under the parent tree and/or are less mobile, then a reduced seed-dispersal distance and recruitment rate of dispersed seeds can be predicted. Yet empirically linking spatial patterns of a plant's offspring to the behaviour and performance of their main seed dispersers in a heterogeneous landscape remains difficult (Wang and Smith 2002). Mechanistic simulation models provide a powerful alternative to such complex biological systems (De Angelis and Mooij 2005; Levey *et al.* 2005; Russo *et al.* 2006; Carlo and Morales 2008).

We combined high-resolution data on seed removal, gut passage, and dispersal by the three key avian frugivores from the Taita Hills with experimental data on post-dispersal seed germination and seedling establishment to predict how realized distribution patterns of established seedlings of a forest interior tree may be affected by the 'extended landscape'. This encompasses (i) shifts in the disperser species composition, (ii) shifts on environmental conditions and their effects on the disperser behaviour and on recruitment success, and (iii) the interaction between (i) and (ii). Figure 5.3 shows a flow chart with the sequential modelling steps. Local extinction of the highly mobile *T. hartlaubi* is expected to lead to a reduction of average and maximum distances over which removed seeds are dispersed, because seeds are no longer dispersed over larger distances (Figure 5.4a). The less mobile forest specialist *T. helleri*, however, deposits seeds relatively close to the source tree and consistently so within indigenous forest boundaries. Hence, local extinction of this species is predicted to cause an increase, rather than decrease, of mean seed dispersal distances (Figure 5.4a and b). Probability functions describing the dispersal distances of seeds in tiny fragments are more prone to changes in frugivore community than those in large ones (Figure 5.4b). This is because movement distances of *T. helleri* are limited by indigenous forest boundaries in small forests. Compared to probability functions of seeds, patterns of realized distances of two year-old seedlings are less sensitive to shifts in frugivore community (results not shown), but small forest patches are less buffered against changes in the matrix surrounding the forest, and show more variation in realized distance probability than seedling distribution patterns in larger forests (variable curve shape and less smooth curves; Figure 5.4c). This results from contrasting seed germination and seedling survival rates in farmland versus plantation rather than from contrasting bird movements that would limit seed dispersal in one of these matrix habitats. The strong context dependence of seed dispersal effectiveness in terms of both quantity and quality, can nicely be summarized using a seed dispersal effectiveness landscape (Schupp *et al.* 2010, Figure 5.4D). This can be a powerful tool to predict the changes in recruitment under different land-use changes, such as the transformation of small forest fragments into larger ones, or the elimination of exotic plantations bordering small forest fragments in favour of agriculture (Figure 5.4d).

5.3 Conclusions and future directions

Our study on seed dispersal and recruitment of the tropical forest tree *X. monospora* in the fragmented setting of the Taita Hills' cloud forests shows that seed dispersal by frugivorous animals is stage- and context-dependent. More importantly, this highly simplified ecological model shows that landscape

Figure 5.2 Stage- and context-dependent effects on various dispersal-related aspects of *X. monospora* seeds in the Taita Hills. (a) Increase of *X. monospora* fruit removal (number of fruits removed/hour) as a function of tree crop size (absolute number of ripe fruits per tree) and conspecific fruit availability in the large neighbourhood (within 3 ha). *Per capita* fruit removal rates per hour are reduced and large crops less attractive with high (50 000 fruits/ha; thick line) compared to low (500 fruits/ha; thin line) fruit density surrounding the focal tree, an effect due to competition of fruiting trees for frugivores. Dashed lines show values ± SE (Lehouck *et al.* unpubl.). (b) Frugivore density and visit frequency in a large undisturbed (upper) and a small disturbed (lower) forest fragment. Avian forest specialists (black rectangular) render disproportionately more visits and remove more fruits than forest generalists (grey rectangular). However, this functional group is largely absent in small, highly disturbed forest fragments, resulting in reduced seed dispersal rates in the latter (after Lehouck *et al.* 2009a.). (c) Average germination rate (in % of germinated seeds; left) and time to germination (in days after sowing; right) of *X. monospora* seeds that are defaecated by their main frugivores (black) compared to manually depulped fruits (grey) and intact fruits (white). Error bars depict standard errors (after Lehouck *et al.* 2010). (d) Average percentage of germinated seeds and survived seedlings of *X. monospora* under (black) and away from (white) conspecific trees. Error bars depict standard errors (after Lehouck *et al.* 2009c).

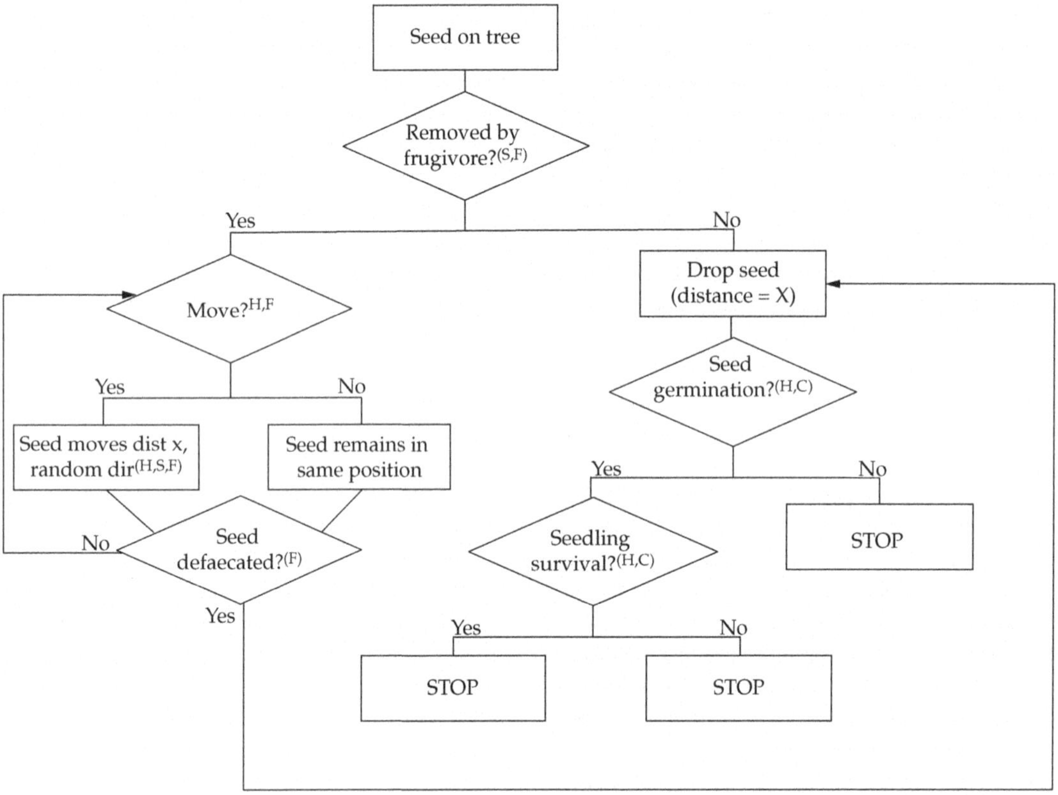

Figure 5.3 Flow diagram showing the algorithm for the spatial explicit simulation of seed dispersal and subsequent seedling establishment. Squares indicate individual events (decisions), while diamonds indicate yes/no questions (data sampled from empirical data, see text for details). Letters between brackets refer to the dependence of the data on habitat (matrix type and/or microhabitat) (H), forest size (S), frugivore species (F), and climate (rainfall; C). The algorithm is repeated until all seeds on the trees are fallen or removed (64 000 seeds in total, randomly distributed over the trees). Forest fragments are simulated as squares with side length 2^x m, with x = 2 – 10.

effects extend beyond the effects of fragment size and matrix type *per se*, and that indirect effects of landscape changes via altered frugivore community as well as combined effects of these three parameters, may have important consequences for seed dispersal and plant recruitment. Such a highly dynamic nature of seed-dispersal effectiveness may have important implications for the balance among selective pressures driving evolution of seed dispersal features and may change our intuitive ideas on the resilience of an ecosystem (see Schupp *et al.* 2010). Evidently, a more holistic approach encompassing all biotic and abiotic components involved in an ecological process, such as this seed dispersal system, must be pursued to grasp fully the idea of the 'extended landscape' effect on ecosystem functioning. High-resolution empirical studies such as the one of the Taita Hills, may be an important step forward to bridge our current knowledge in this field.

Given the dual nature of the plant-disperser mutualism and the species-specific responses of both frugivores and fleshy-fruit producing plants towards landscape changes, it can be expected that context-dependency of zoochorous seed dispersal is stronger compared to other dispersal mechanisms. In addition, compared to temperate zones, the pressure for tropical tree seeds to be instantly dispersed to suitable sites for establishment, is expected to be higher. This is because dispersal limitation, i.e. the failure of seeds to arrive at favourable sites, is pervasive in the tropics (Clark *et al.* 1999), where density-dependent seed and seedling mortality seem to be more prevalent as well (Hyatt *et al.* 2003), and where seed dormancy, a complementary risk-reduction mechanism to dispersal (Snyder 2006) seems to be rather scarce (Vázquez-Yanes and Orozco-Segovia 1993; Baskin and Baskin 2001). Tropical animal-dispersed plant species may therefore be more dependent on their dispersal agent than their temperate counterparts (often with soil seed bank) or than non-animal dispersed species (being less dispersal limited), although this remains to be tested. If the high sensitivity of biotic functional networks (resulting from tight or diffuse mutualisms) in species-rich, stable tropical systems to changes in the extended landscape can be generalized, then the functioning of tropical ecosystems may be even more affected by global changes that previously anticipated.

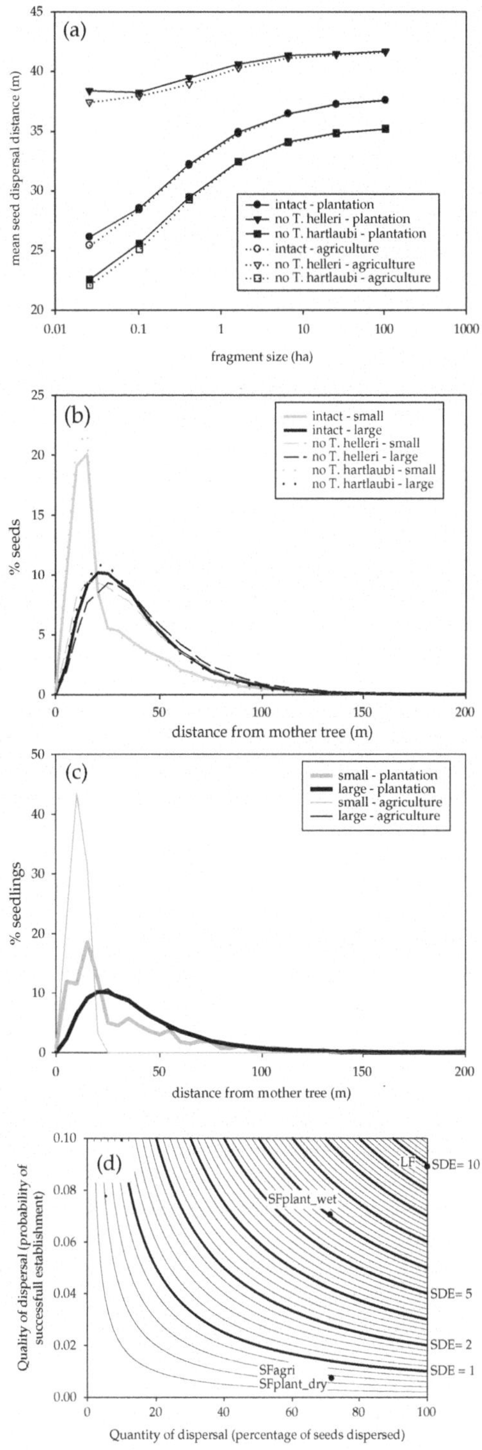

Figure 5.4 Effects of the extended landscape on seed dispersal characteristics of *X. monospora* in the Taita Hills, as simulated by IBM (Lehouck *et al.* unpublished). (a) Mean seed dispersal distance as a function of fragment size in forest fragments surrounded by exotic plantations (black symbol, solid lines), or agriculture (white symbol, dotted lines) under different frugivore communities. (b) Distance probability function of avian dispersed seeds in small (0.25 ha—grey) and large (100 ha—black) forest patches (in this case surrounded by smallholder agriculture). Different line types indicate the change in frugivore composition. Probabilities of seed arrival are expressed per unit distance. (c) Distance probability function of seeds reaching the seedling stage (two years old) in small (0.25 ha—grey) and large (100 ha—black) forest patches surrounded by exotic plantations (thick lines) or farmland (thin lines). The frugivore community is kept intact in all cases. The curves in the large patch show > 90% overlap, such that they appear as one single line. Probabilities of seedling distribution are expressed per unit distance. (d) Seed dispersal effectiveness (SDE) landscape illustrating context-dependent shift in SDE for *X. monospora* in the Taita Hills. SFagri: small fragment surrounded by farmland (no difference according to rainfall); SFplant_dry/wet: small fragments surrounded by exotic plantation in dry versus wet years; LF: large forest fragment (no difference according to matrix type and/or rainfall). Seed quality is expressed as the probability of a seed to establish successfully as a two year old seedling, whereas quantity refers to the percentage of seeds dispersed relative to LF. Lines indicate isoclines (SDE constant). We here suppose an intact frugivore community in LF and a single-species community in SF (only *A. milanjensis* remaining). We note that quality is a stronger factor distinguishing the SDE of *X. monospora* in different fragmentation contexts than quantity.

References

Aerts, R., Thijs, K.W., Lehouck, V. *et al.* (2011). 'Woody plant communities of isolated Afromontane cloud forests in Taita Hills, Kenya', *Plant Ecology* **212**: 639–49.

Bascompte, J. and Jordano, P. (2007). 'Plant-animal mutualistic networks: the architecture of biodiversity', *Annual Review of Ecology, Evolution and Systematics* **38**: 567–93.

Baskin, C.C. and Baskin, J.M. (2001). *Seeds. Ecology, biogeography, and evolution of dormancy and germination*. Academic Press, San Diego. Beentje, H. (1994). *Kenya Trees, Shrubs and Lianas*. National Museum of Nairobi, Nairobi.

Bowen, L.T., Moorman, C.E. and Kilgo, J.C. (2007). 'Seasonal bird use of canopy gaps in a bottomland forest', *Wilson Journal of Ornithology* **119**: 77–88.

Carlo, T.A. and Morales, J.M. (2008). 'Inequalities in fruit-removal and seed dispersal: consequences of bird behaviour, neighbourhood density and landscape aggregation', *Journal of Ecology* **96:** 609–18.

Clark, J.S., Macklin, E. and Wood, L. (1998). 'Stages and spatial scales of recruitment limitation in Southern Appalachian forests', *Ecological Monographs* **68**: 213–35.

Clark, J.S., Silman, M., Kern, R. *et al.* (1999). 'Seed dispersal near and far: patterns across temperate and tropical forests', *Ecology* **80**: 1475–94.

Cordeiro, N.J. and Howe, H.F. (2001). 'Low recruitment of trees dispersed by animals in African forest fragments', *Conservation Biology* **15**: 1733–41.

Cordeiro, N.J. and Howe, H.F. (2003). 'Forest fragmentation severs mutualism between seed dispersers and an endemic African tree', *Proceedings of the National Academy of Sciences of the USA* **100**: 14052–6.

Cordeiro, N.J., Ndangalasi, H.J., McCentee, J.P. *et al.* (2009). 'Dispersal limitation and recruitment of an endemic African tree in a fragmented landscape', *Ecology* **90**: 1030–41.

Dawkins, R. 1982. *The Extended Phenotype: The Gene as the Unit of Selection*. W.H. Freeman and Company, Oxford.

Deangelis, D.L. and Mooij, W.M. (2005). 'Individual-based modeling of ecological and evolutionary processes', *Annual Review of Ecology Evolution and Systematics* **36**: 147–68.

Dennis, A.J. and Westcott, D.A. (2006). 'Reducing complexity when studying seed dispersal at community scales: a functional classification of vertebrate seed dispersers in tropical forests', *Oecologia* **149**: 620–34.

Devictor, V., Julliard, R. and Jiguet, F. (2008). 'Distribution of specialist and generalist species along spatial gradients of habitat disturbance and fragmentation', *Oikos* **117**: 507–14.

Fleming, T.H., Breitwisch, R. and Whitesides, G.H. (1987). 'Patterns of tropical vertebrate frugivore diversity', *Annual Review of Ecology and Systematics* **18**: 91–109.

Harms, K.E., Wright, S.J., Calderon, O. *et al.* (2000). 'Pervasive density-dependent recruitment enhances seedling diversity in a tropical forest', *Nature* **404**: 493–5.

Holl, K.D. (1999). 'Factors limiting tropical rain forest regeneration in abandoned pasture: seed rain, seed germination, microclimate, and soil', *Biotropica* **31**: 229–42.

Howe, H.F. and Smallwood, J. (1982). 'Ecology of seed dispersal', *Annual Review of Ecology and Systematics* **13**: 201–18.

Hyatt, L.A., Rosenberg, M.S., Howard, T.G. *et al.* (2003). 'The distance dependence prediction of the Janzen-Connell hypothesis: a meta-analysis', *Oikos* **103**: 590–602.

Janzen, D.H. (1984). Dispersal of small seeds by big herbivores—Foliage is the fruit', *American Naturalist* **123**: 338–53.

Jordano, P. (1987). 'Avian fruit removal—Effects of fruit variation, crop size, and insect damage', *Ecology* **68**: 1711–23.

Jordano, P., Garcia, C., Godoy, J.A. *et al.* (2007). 'Differential contribution of frugivores to complex seed dispersal patterns', *Proceedings of the National Academy of Sciences of the USA* **104:** 3278–82.

Lehouck, V., Spanhove, T., Colson, L. *et al.* (2009a). 'Habitat disturbance reduces seed dispersal of a forest interior tree in a fragmented African cloud forest', *Oikos* **118**: 1023–34.

Lehouck, V., Spanhove, T., Demeter, S. *et al.* (2009b). 'Complementary seed dispersal by three avian frugivores in a fragmented Afromontane forest', *Journal of Vegetation Science* **20**: 1110–20.

Lehouck, V., Spanhove, T., Gonsamo, A. *et al.* (2009c). 'Spatial and temporal effects on recruitment of an Afromontane forest tree in a threatened fragmented ecosystem', *Biological Conservation* **142**: 518–28.

Lehouck, V., Spanhove, T. and Lens, L. (2011). 'Avian fruit ingestion differentially facilitates seed germination of four fleshy-fruited plant species of an Afrotropical forest', *Plant Ecology and Evolution* **144**: 96–100.

Lehouck, V., Spanhove, T., Vangestel, C. *et al.* (2009d). 'Does landscape structure affect resource tracking by avian frugivores in a fragmented Afrotropical forest?' *Ecography* **32**: 789–99.

Lens, L., Van Dongen, S., Norris, K. *et al.* (2002). 'Avian persistence in fragmented rainforest', *Science* **298**, 1236–8.

Levey, D.J., Tewksbury, J.J. and Bolker, B.M. (2008). 'Modelling long-distance seed dispersal in heterogeneous landscapes', *Journal of Ecology* **96:** 599–608.

Mittermeier, R., Gil P. R., Hoffman, M. *et al.* (2005). *Hotspots revisited. Earth's biologically richest and most endangered terrestrial ecoregions*. Conservation International, Washington DC.

Murray, K.G. (1988). 'Avian seed dispersal of 3 Neotropical gap-dependent plants', *Ecological Monographs* **58**: 271–98.

Nathan, R. and Casagrandi, R. (2004). 'A simple mechanistic model of seed dispersal, predation and plant establishment: Janzen-Connell and beyond', *Journal of Ecology* **92**: 733–46.

Nathan, R. and H.C. Muller-Landau (2000). 'Spatial patterns of seed dispersal, their determinants and consequences for recruitment', *Trends in Ecology and Evolution* **15**: 278–85.

Rousset, F. and Gandon, S. (2002). 'Evolution of the distribution of dispersal distance under distance-dependent cost of dispersal', *Journal of Evolutionary Biology* **15**: 515–23.

Russo, S.E. and Augspurger, C.K. (2004). 'Aggregated seed dispersal by Spider monkeys limits recruitment to clumped patterns in *Virola calophylla*', *Ecology Letters* **7**: 1058–67.

Russo, S.E., Portnoy, S. and Augspurger, C.K. (2006). 'Incorporating animal behavior into seed dispersal models: implications for seed shadows', *Ecology* **87**: 3160–74.

Schmidt, V., Schaefer, H.M. and Winkler, H. (2004). 'Conspicuousness, not colour as foraging cue in plant-animal signalling', *Oikos* **106**: 551–7.

Schupp, E.W. (1993). 'Quantity, quality and the effectiveness of seed dispersal by animals', *Vegetatio* **108**: 15–29.

Schupp, E.W., Jordano, P. and Gomez, J.M. (2010). 'Seed dispersal effectiveness revisited: a conceptual review', *New Phytologist* **188**: 333–53.

Snyder, R.E. (2006). 'Multiple risk reduction mechanisms: can dormancy substitute for dispersal?' *Ecology Letters* **9**: 1106–14.

Spiegel, O. and Nathan, R. (2007). 'Incorporating dispersal distance into the disperser effectiveness framework: frugivorous birds provide complementary dispersal to plants in a patchy environment', *Ecology Letters* **10**: 718–28.

Spiegel, O. and Nathan, R. (2010). 'Incorporating density dependence into the directed-dispersal hypothesis', *Ecology* **91**: 1538–48.

Terborgh, J. (1986). 'Community aspects of frugivory in tropical forests', in A. Estrada, and T.H. Fleming, eds, *Frugivores and seed dispersal*. Junk Publishers Group, Dordrecht/Boston/Lancaster, 371–84.

Thies, W., Kalko, E.K.V. and Schnitzler, H.U. (2006). 'Influence of environment and resource availability on activity patterns of *Carollia castanea* (Phyllostomidae) in Panama', *Journal of Mammalogy* **87**: 331–8.

Traveset, A. (1998). 'Effect of seed passage trough vertebrate frugivores' guts on germination: a review', *Perspectives in Plant Ecology, Evolution and Systematics* **1**: 151–90.

Van der Pijl, L. (1969). *Principles of seed dispersal in higher plants*. Springer Verlag, Berlin.

Vázquez-Yanes, C. and Orozco-Segovia, A. (1993). 'Patterns of seed longevity and germination in the tropical rainforest', *Annual Review of Ecology and Systematics* **24**: 69–87.

Wang, B.C. and Smith, T.B. (2002). 'Closing the seed dispersal loop', *Trends in Ecology and Evolution* **17**: 379–85.

Wenny, D.G. (2001). 'Advantages of seed dispersal: a re-evaluation of directed dispersal', *Evolutionary Ecology Research* **3**: 51–74.

Wenny, D.G. and Levey, D.J. (1998). 'Directed seed dispersal by Bellbirds in a tropical cloud forest', *Proceedings of the National Academy of Sciences of the USA* **95**: 6204–7.

Westcott, D.A., Bentrupperbaumer, J., Bradford, M.G. *et al.* (2005). 'Incorporating patterns of disperser behaviour into models of seed dispersal and its effects on estimated dispersal curves', *Oecologia* **146**: 57–67.

PART II

The Genetics of Dispersal

CHAPTER 6

Quantitative, physiological, and molecular genetics of dispersal/migration

Anthony J. Zera and Jennifer A. Brisson

6.1 Introduction and overview

Dispersal, which can be defined broadly as 'any movement of individuals or propagules with potential consequences for gene flow across space' (Ronce 2007; see Section 6.2, and glossary for definitions) can profoundly influence many aspects of population biology, such as the evolution of locality-specific adaptation, the magnitude of genetic differentiation among populations and rates of speciation, rates of population extinction/colonization, and range expansion (Dingle 1996; Clobert *et al.* 2001; Ronce 2007). Adaptive differences in dispersal are common among genotypes, populations, or species (reviewed in Dingle 1996; Pulido 2007), and thus dispersal itself can evolve. The conditions under which, and mechanisms by which, this occurs have been of long-standing interest to evolutionary biologists and ecologists, which has naturally led to interest in the genetic basis of dispersal variation.

Dispersal is a complex syndrome of traits, consisting of the integrated expression of many morphological, physiological, and behavioural aspects of dispersal as well as other life-history traits (e.g. timing of reproduction) (Dingle 1996; Zera and Denno 1997; Roff and Fairbairn 2001; Duckworth 2008; Chapters 10–14). Thus, important issues in dispersal genetics have been (1) the genetic basis of variation in individual traits that contribute to the capacity to disperse (e.g. production of wings and flight fuel in insects), and the overall degree of dispersal movement *per se*, and (2) the genetic mechanisms that coordinate the adaptive expression of multiple dispersal traits with other organismal features (e.g. life-history traits) in the laboratory or field.

The 1980s and 1990s saw the first extensive genetic analyses of dispersal, focusing on whole-organism morphological or behavioural components, and correlated life-history traits (for reviews, see Harrison 1980; Berthold 1991; Gatehouse 1989; Dingle 1985, 1996; Roff and Fairbairn 1991, 2001, 2007; and Pulido 2007). These studies clearly established that phenotypic variation for whole-organism dispersal traits often has a strong, polygenic component, and that dispersal components often exhibit strong genetic correlations among themselves and with various life-history traits.

During the past decade, the field of dispersal genetics has matured and advanced on a number of fronts which will be discussed in detail in this chapter. Previous genetic studies were largely restricted to a few species of insects or birds. Quantitative-genetic analyses have since been applied to a wider range of organisms and dispersal traits, have become more sophisticated and better controlled, and have focused on more complex, biologically important aspects of dispersal (e.g. genetically variable reaction norms). Probably the most important recent advance has been the in-depth focus on physiological-genetic aspects of dispersal. Physiology is used here in a broad sense, to denote function at any level, molecular, biochemical, systemic, or whole-organismal. For example, some studies have identified genetically variable components of intermediary metabolism that are important in the energetics of dispersal, or aspects of endocrine regulation that control the expression of dispersal or life-history

Dispersal Ecology and Evolution. First Edition. Edited by Jean Clobert *et al.*

traits. Other studies have identified molecular- or biochemical-genetic polymorphisms that are associated with dispersal behaviour or performance in the field. Finally, various 'omics' approaches (e.g. transcriptome profiling) have begun to identify associations between global patterns of gene expression, or specific candidate genes, and phenotypes differing in aspects of dispersal. These physiological studies are revolutionizing the field of dispersal genetics by providing the first detailed information on the mechanisms by which genetic variation is translated into variation in dispersal phenotypes, the nature of trait correlations, and the nature of dispersal adaptation.

In this review we focus on what we consider to be the most important empirical advances in the genetics of dispersal in animals and plants during the past decade. Because the literature on this topic is extensive, we will focus on selected examples rather than being exhaustive.

6.2 Background: traits studied and methods of analysis

Definition of dispersal. There is considerable variation and debate regarding usage of the terms 'dispersal' and 'migration' (for example, see Dingle, 1996; Dingle and Drake 2007; Ronce 2007; Nathan *et al.* 2008; Hawkes 2009). These terms are sometimes used interchangeably, or are used to describe different types of movement; definitions often vary within or across sub-disciplines. Here we use the term dispersal broadly, as defined by Ronce (2007), to simply mean movement between habitat patches or breeding sites, irrespective of spatial scale, that potentially gives rise to gene flow. This term includes 'migration', which is defined behaviourally by some insect workers to denote active, persistent, straightened-out (and often long-distance) movement that is not undertaken to locate resources (Dingle 1996; Dingle and Drake 2007). We also briefly discuss 'seasonal migration' in birds. Where appropriate, we identify the specific type of movement when discussing its genetic basis.

Types of traits studied. Table 6.1 illustrates the diversity of dispersal components that have been investigated in genetic studies. Some traits comprise whole-organism aspects of morphology, life history, or behaviour, and involve aspects of dispersal capability as well as actual locomotion. These traits have been the foci of traditional studies of dispersal genetics, and continue to be investigated in a wider range of organisms. More recently, functional-genetic correlates of dispersal, such as aspects of endocrinology, intermediary metabolism, DNA sequence, as well as global transcriptome profiling, have been increasingly investigated.

Genetic analyses: quantitative-genetics. Traditional 'whole-organism' morphological, physiological, behavioural, or life-history aspects of dispersal, that have been subjected to genetic analysis, are classic 'quantitative traits'. These traits typically exhibit continuous phenotypic variation, and have a polygenic mode of inheritance, and thus have been analysed using standard quantitative-genetic approaches described below. Several prominent dispersal polymorphisms in insects (e.g. wing polymorphism) and plants (seed heteromorphism), although exhibiting discontinuous variation, also are under polygenic control (with a few notable exceptions; see Roff and Fairbairn 2007; Braendle *et al.* 2005a). These polymorphisms also have been analysed using quantitative-genetic protocols modified for polymorphic but polygenic traits (i.e. threshold traits; Roff 1996; Roff and Fairbairn 2001). Key genetic parameters estimated in these studies have been heritability (i.e. the proportion of phenotypic variation that is due to genetic variation), absolute or standardized additive genetic variance, genetic correlation, and various aspects of plasticity (e.g. norms of reaction, genotype X environment interaction).

Detailed discussion of standard quantitative-genetic designs can be found in Falconer and McKay (1996), Roff (1997), Lynch and Walsh (1998), and in reviews focusing on dispersal genetics: Merilä and Sheldon 2001; Roff and Fairbairn 2001; van Noordwijk *et al.* 2006; and Pulido 2007. Only a few salient points are made here regarding these protocols. Basically, traditional quantitative genetic analyses can be divided into two general classes: individual pair crosses, typically conducted over one generation (e.g. parent-offspring regression or full-sib/half-sib analyses), and multigenerational artificial selection conducted on populations. The former approach involves quantifying the co-variance between phenotype and relatedness in a series

Table 6.1 Examples of genetically variable components or correlates of dispersal

Type	Example	Technique	Lab/Field	Taxon	References
Morphology	Size of wings, flight muscles, whole body	h^2, r, DR, CR, Sib	L	Insect	Roff and Fairbairn 2001, 2007
	Seed and fruit size	h^2, r, Sib	L	Plant	Venable and Burquez 1989
	Maternal plant architecture	h^2, r, Sib	L	Plant	Wender *et al.* 2005
Life history	Ovarian mass, onset of egg production, duration of juvenile development	h^2, r, DR, CR, Sib	L	Insect	Roff and Fairbairn 2001, 2007; Zera and Harshman 2009
Behaviour	'Migratory Restlessness'	Sib	L	Bird	Berthold and Terrill 1991; Pulido 2007
	Exploratory behaviour	h^2, Sib	L, F	Bird/Mammal	Dingemanse *et al.* 2002; Krackow 2003
	Foraging behaviour	Mendelian analysis	L	Insect/Worm	de Bono & Bargmann 1998; Osborne *et al.* 1997
Physiology	Flight performance, duration, and distance	h^2, Sib	L, F	Insect	Gatehouse 1989,1997; Saastamoinen 2008
	Flight fuel concentration	CR	L	Insect	Zera and Larsen 2001
	Hormonal regulation of wing and muscle devel.	CR	L	Insect	Zera and Harshman 2009
	Levels of reproductive and flight hormones	CR	L, F	Insect	Zhao and Zera 2004; Zera *et al.* 2007; Zera, 2009
Biochemistry	Flight-fuel (lipid) biosynthesis	CR	L	Insect	Zhao and Zera 2002
	Activities of lipogenic enzymes	CR	L	Insect	Zera and Zhao 2003
	Enzyme electromorph and flight performance	Association mapping	F	Insect	Saastamoinen and Hanski 2008; Wheat *et al.* 2011
Molecular Biology	Expression of lipogenic genes	CR	L	Insect	Schilder *et al.* 2011
	Metabolic gene sequence and flight performance	Association mapping	F	Insect	Wheat *et al.* 2011
	Neurotransmitter gene sequence and movement	Association mapping	L, F	Bird/Mammal	Krackow and König 2008; Fidler *et al.* 2007; Trefilov *et al.* 2000
	Seed-shattering (i.e. dispersal) gene sequences	QTL Mapping	L	Plants	Li *et al.* 2006, Konishi *et al.* 2006
Functional Genomics	Transcriptome and dispersal morph	Genomics	L	Insect	Brisson *et al.* 2007
	Transcriptome and dispersal propensity	Genomics	F	Insect	Wheat *et al.* 2011

h^2 = heritability; r = genetic correlation; DR = direct response to artificial selection; CR = correlated response to artificial selection; Sib = sib analysis (Full-sib or half-sib breeding designs or parent offspring regression); L = measured in the laboratory; F = measured in the field; see text for description of Association and QTL mapping.

of crosses to identify the genetic component(s) of trait variation. For example, in parent-offspring regression, the average phenotype of offspring is regressed on the average phenotype of parents obtained in a series of pair crosses. The slope of the regression line is equal to the heritability. When multiple traits are studied, genetic correlations can be estimated from trait co-variances. When sibships are split between different environments, genotype X environment interactions and reaction norms can be measured. One persistent problem with commonly used full-sib and parent-offspring regression is the potential conflation of environmental with genetic effects on phenotypic similarity. For example, common environmental effects due to maternal effects, or to parents and offspring, or sibs, being raised in the same environment can contribute to the increased phenotypic resemblance among relatives, thus potentially substantially inflating the heritability estimate. Such conflation has been a major issue in many genetic analyses of dispersal traits, notably dispersal in birds in the field (van Noordwijk *et al.* 2006; Merilä and Sheldon 2001). Maternal characteristics arguably play an even larger role in plant relative to animal dispersal, since maternal plant traits such as height, branching patterns, and fruit strongly influence seed dispersal (Donohue 1999; Donohue *et al.* 2005; Wender *et al.* 2005; Chapter 14).

Because a half-sib design largely eliminates this problem, it is preferred over full-sib or parent-offspring regression. However, a half-sib design is more difficult to perform, especially in the field, and has rarely been used (see Sinervo *et al.* 2006, for a notable exception). Methods such as cross-fostering progeny or conducting parent-offspring regression on each parent separately can be used to reduce/identify effects of common environment or maternal effects (Weigensberg and Roff 1996; Merilä and Sheldon 2001). If quantitative genetic studies are conducted in the laboratory, as has been the case for nearly all studies of dispersal polymorphism in insects, greater control over environmental variation can be obtained, but extrapolation of results to the field becomes problematic. Heritabilities for most traits are lower in the field compared to the laboratory (Riska *et al.* 1989; Simmons and Roff 1994) and genotype X environmental interactions, which are common ((Weisser and Braendle 2001; Hazell *et al.* 2005), can be problematic. However, laboratory estimates of heritabilities appear to be at least roughly comparable to those in the field (Simons and Roff 1994; Weigenberg and Roff 1996; however, see Hoffmann 2000).

The second major quantitative-genetic approach is artificial selection, in which individuals from a population are selectively bred (e.g. parents are chosen who have a higher or lower than average phenotype) to produce the next generation, with this process being repeated over several generations. The magnitude of change (e.g. increase or decrease) in the mean phenotype of the selected population over successive generations provides an estimate of the heritability. The degree to which the average phenotype of other, non-selected traits changes over time provides an estimate of the genetic correlation. Artificial selection can be done either in the laboratory or field.

The production of populations genetically divergent for traits of interest by artificial selection is particularly useful for functional-genetic studies (Zera and Harshman 2009). Genetic correlations between the selected trait (e.g. morphology) and 'lower level' physiological traits (e.g. blood hormone level) identify candidate traits and ultimately genes whose variation potentially gives rise to the selected phenotype. Correlations between traits at the same hierarchical level (e.g. wing and flight muscle morphology; or morphology and egg production) identify variable traits that are inherited together because they are potentially influenced by the same variable genes with pleiotropic effects. Lines raised in different environments provide information on genotype X environment interaction and reaction norms.

'Animal Models' provide a powerful new statistical approach to investigate the genetics of complex traits (Merilä and Sheldon 2001; Kruuk 2004), such as dispersal (Duckworth and Badyaev 2007). The power of this approach is that it can use all of the information in a complex pedigree to estimate various quantitative-genetic parameters, rather than using only a subset of the data that conforms to a traditional quantitative-genetic breeding design.

This approach also can handle incomplete or unbalanced data sets, and can control for factors that influence traits under study. 'Animal Models' will most likely become increasingly popular in genetic studies of dispersal, especially in vertebrates where estimation of pedigrees in the field using molecular markers is becoming increasingly common. Finally, common-garden experiments and transplants, especially when used in conjunction with crossing studies, are important tools in field studies of dispersal genetics (van Noordwijk *et al.* 2006).

Molecular genetics and genomics. A major limitation of traditional quantitative-genetic analyses is that they do not allow an accurate assessment of the number or identity of individual variable genes that contribute to trait variation. Molecular-genetic and genomic approaches can be used in conjunction with, or independent of, standard quantitative analyses to address this issue. For example, if a high density linkage map is available then crosses between lines that vary in the phenotype of interest can be used to identify the contribution of specific gene regions (QTLs, quantitative trait loci) to variation in the phenotype by measuring the extent to which a phenotype co-segregates with the markers. This kind of genetic mapping approach has revealed the genetic architecture underlying dispersal phenotypes (e.g. Caillaud *et al.* 2002; Konishi *et al.* 2006). However, few studies have successfully taken the next step of identifying the gene or genes underlying QTLs. Even rarer are studies that have identified the nucleotide variants that contribute to phenotypic variation (e.g. Arnaud *et al.* 2011). Alternatively, if variable molecular (e.g. DNA sequence polymorphism) or biochemical (allozyme, electromorph) markers are available, then it is possible to determine the extent to which molecular markers are associated with, and potentially contribute to, variation in a phenotype of interest (association mapping; e.g. Trefilov *et al.*, 2000; Fidler *et al.*, 2007).

Finally, genomic approaches, such as transcriptome profiling via the use of microarrays have resulted in the first functional genomic investigations of dispersal. These approaches have only recently been applied to dispersal genetics (see studies of the pea aphid *Acyrthosiphon pisum*, and the Glanville butterfly, following). As discussed in detail in this volume by Wheat (Chapter 11), these studies involve the simultaneous measurement of differences between dispersal phenotypes in expression (i.e. transcript abundance) of thousands of identified genes. However, data from these studies must be interpreted with caution, since in the absence of additional functional information, gene expression differences are simply correlative. Further, transcriptional differences do not necessarily give rise to corresponding differences at the protein level (Zera 2011; Storz and Zera 2011) and thus may not be functionally significant. Given the explosive increase in research in this area, this aspect of research on dispersal genetics will certainly increase exponentially in the future.

6.3 Genetic analysis of dispersal traits

6.3.1 Dispersal capability

Genetic studies of traits involved in the capability to disperse, as opposed to dispersal movement *per se*, have been predominantly studied in wing-polymorphic insects, and to a lesser degree in seed heteromorphic plants.

Wing polymorphism in insects. Wing polymorphism consists of discontinuous variation in a suite of traits involved in the capability for flight and various correlated life-history traits. The flight-capable morph (long-winged, LW), typically has large functional wings and large functional flight muscles at the adult molt, a high concentration of flight fuel, and delayed and reduced egg production compared to the flightless (short-winged, SW, or wingless) morph. Dispersal polymorphism is phylogenetically widespread, occurring commonly in many insect groups, especially crickets, grasshoppers, beetles, true bugs, aphids, and planthoppers (Harrison 1980, Zera and Denno 1997; Roff and Fairbairn 2007; Zera 2009; Guerra 2011). Because the flightless morph can be easily identified in the field, this polymorphism has been well-studied with respect to the ecological correlates of dispersal. Wing morph can be produced by variation in genotype, environment, or most commonly, variation in both (Harrison 1980; Zera and Denno 1997; Roff and Fairbairn 2001, 2007; Guerra 2011).

Studies of dispersal polymorphism prior to the 1990s mainly investigated the mode of inheritance of morphological (wings, flight muscles) and correlated life-history traits (reviewed in Rankin and Burchsted 1992; Dingle 1996; Zera and Denno 1997; Roff and Fairbairn 2001, 2007; and Feder *et al.* 2010). The most intensive studies have focused on the cricket, *Gryllus firmus* (reviewed in Roff and Fairbairn 2001 2007), with additional studies on a variety of waterstrider, planthopper, and other cricket species (e.g. Zera *et al.* 1983, Zera and Denno 1997; Masaki and Shimuzu 1995; Matsumura 1996). With a few exceptions (see Roff and Fairbairn 2001, 2007, and following), these studies collectively demonstrated a polygenic mode of inheritance and high heritability for various dispersal traits (e.g. wing length measured by a variety of different protocols such as half-sib analysis and artificial selection). Very strong positive genetic correlations between wing length and muscle mass/functionality (close to 1.0), strong negative genetic correlations between these traits and ovarian mass/egg number (close to 1.0), and a strong effect of various environmental variables on morph expression are also typical. A number of studies have also documented reaction norms and genotype X environment interactions for dispersal traits, making wing polymorphism the most extensively studied model in dispersal genetics with respect to the genetics of plasticity and genotype X environment interaction (Matsumura 1996; Masaki and Shimizu 1995; Zera and Larsen 2001; Zera and Zhao 2003; Roff and Gelinas 2003; also see aphid studies, following). Morph expression is best viewed as a polygenic threshold trait (Roff 1996; Roff and Fairbairn 2001), the threshold level of which is determined by both multiple loci and environmental inputs. These basic studies of wing polymorphism have contributed importantly to the following key ideas: (1) genetic variation for dispersal is common, (2) the negative coupling between dispersal-related traits and life history traits also is genetically based, and (3) construction and maintenance of the flight apparatus has significant energetic and fitness costs which strongly influence the evolution of flightlessness in insects (Harrison 1980; Dingle 1996; Zera and Denno 1997; Roff and Fairbairn 2001, 2007; Zera 2009).

One of the most significant recent advances in the genetics of dispersal polymorphism has been in the area of physiological genetics (reviewed in Zera 2006, 2009; Zera and Harshman 2009, 2011). Integrative, multidisciplinary studies have provided new information on two long-standing topics in dispersal genetics: (1) genetic variation for hormone levels that underlie expression of various dispersal traits, and which co-ordinate traits in the dispersal syndrome, and (2) the biochemical genetics of energy allocation, and the trade-off between flight capability and egg production. The most detailed studies have been conducted on replicate, artificially selected lines of *Gryllus firmus* and *G. rubens* that are essentially pure-breeding for the dispersing morph or the flightless morph in the laboratory (i.e. correlated or indirect responses to selection). In some cases, reaction norms have been measured for these physiological traits across environments (Zera and Larsen 2001; Zera and Zhao 2003; Roff and Gelinas 2003). Because these studies have been reviewed in detail recently (Zera 2004, 2006, 2009; Zera and Harshman 2001, 2009, 2011), only a brief sampling of major findings will be given here.

First, large magnitude differences in the blood levels of several endocrine regulators (ecdysteroids, the enzyme juvenile hormone esterase) have been rigorously demonstrated between LW- and SW-selected lines during the juvenile stage, implicating genetic variation of these regulators as important causes of genetic variation and co-variation in expression of key morphological traits (Zera 2006, Zera *et al.* 2007). Second, during adulthood, LW- and SW-selected lines differ dramatically with respect to the blood level of juvenile hormone, a regulator of many aspects of flight, reproduction, and behaviour in insects. LW-selected lines exhibit a dramatic circadian rhythm for this hormone, while SW-selected lines are largely arrhythmic (Zhao and Zera 2004; Zera 2009). This unexpected finding contrasts with earlier speculative models that proposed a simple threshold concentration difference in JH level between the morphs (Fairbairn, 1994). Genetic variation for circadian traits, which has been largely ignored in dispersal studies, may be an important functional component of dispersal polymorphism.

Third, large magnitude genetic differences in aspects of lipid metabolism, most notably rates of flux through pathways of lipid biosynthesis, activities of lipogenic enzymes, and expression of lipogenic genes, have been identified between LW- and SW-selected lines. These are the first data on genetic modifications of intermediary metabolism contributing to variation in dispersal, and the genetic trade-off (negative genetic correlation) between dispersal capability and egg production. Importantly, crosses and backcrosses between LW- and SW-selected lines resulted in very strong co-segregation among morphological, biochemical, and reproductive traits (Zera and Zhao 2003) demonstrating strong genetic correlations among these traits. Fourth, large magnitude differences in whole-organism respiration and nutrient allocation to organs of flight and reproduction have been experimentally verified by feeding studies, respirometry, and measurement of masses of energy reserves, and organs of flight and reproduction (Mole and Zera 1993; Zera and Larsen, 2001; Crnokrak and Roff 2002; Roff and Gelinas, 2003; reviewed in Zera and Denno 1997; Zera and Harshman 2001, 2009).

A limitation of these physiological-genetic studies is the paucity of information on molecular variation that underlies systemic genetic variation in endocrinology and metabolism (however, see Zera and Harshman 2011; Schilder *et al.* 2011). In addition, nearly all studies have been undertaken in the laboratory. However, very similar phenotypic patterns exist in the field for morph-specific differences in JH and ecdysteroid titres, muscle phenotype, and ovarian mass. Laboratory stocks raised in the field also produce very similar morph frequencies and reproductive differences between morphs as are produced in the laboratory (Zera *et al.* 2007; and unpublished data). These data suggest that results of genetic studies undertaken in the laboratory can likely be extrapolated to the field.

Another significant recent advance in genetic studies of dispersal capability is the first functional-genomic study of genetic wing polymorphism in aphids [(Brisson *et al.* 2007); although the genomic basis of a dispersal polyphenism was explored in another aphid species, *Myzus persicae* (Ghanim *et al.* 2006)]. The pea aphid (*Acyrthosiphon pisum*) displays a genetic wing polymorphism in males in addition to an environmental wing polyphenism in females. In males, wing morph is controlled by allelic variation at a single locus, called *aphicarus* (*api*), that is located on the X chromosome (Smith and MacKay 1989; Caillaud *et al.* 2002; Braendle *et al.*, 2005a). Interestingly, among F_2 females segregating for the *api* polymorphism, there is an association between *api* allele and the polyphenic response in females (Braendle *et al.* 2005b); that is, variation among clones in the production of the dispersing female morph in response to an environmental variable. This suggests that the male genetic polymorphism and female environmental polyphenism may share mechanistic components. Zera and Tiebel (1989) previously reported an analogous situation regarding common components of the endocrine regulation of wing polymorphism and wing polyphenism in *Gryllus*.

A recent cDNA microarray study found differential gene expression, in a parallel manner, between the winged and wingless morphs of both the male genetic polymorphism and the female morphs of the environmental polyphenism (Brisson *et al.* 2007). Specifically, genes associated with energy production and flight musculature were expressed at higher levels in the winged morph. Thus, despite the difference in control of the morph switch between the polyphenism and polymorphism (i.e. environmental versus genetic), the downstream gene expression events leading to alternate morph production/function are fundamentally similar. This study sets the stage for detailed molecular-genetic analyses of candidate genes and global genomic analyses of differential morph development, trade-offs, etc.

Wing polymorphism field studies. Two notable field studies have identified adaptive microgeographical differences in genotype frequencies of dispersing and sedentary morphs (Denno *et al.* 1996; Simmons and Thomas 2004). Most recently, Simmons and Thomas (2004) showed that the frequency of the dispersing long-winged morph of two-wing polymorphic grasshopper species was higher in recently colonized habitats at an invasion front compared with the central, longer-established part of the range. A common-garden study in the laboratory showed that the LW phenotype had a genetic basis (but maternal effects could not be eliminated). Interestingly, the threshold response to density for production of the LW morph also was lower in the invasion front compared with central populations, suggesting an evolutionary change in the polygenic threshold for morph indication. In addition, the LW morph from the marginal population also flew a much longer distance in the laboratory compared with the LW morph from central populations. This is one of the few cases in which flight duration *per se* has been studied in a wing-polymorphic species. This study also shows that the syndrome of traits that define a morph can vary among populations, a point emphasized by Roff (1994) in earlier genetic studies. As is typical for wing-polymorphic species, dispersal ability traded off with egg production in these two grasshopper species, which likely explains why the frequency of the LW morph is reduced in central, more sedentary populations. In an analogous study, Denno (1996) showed that natural populations of a planthopper species from a more ephemeral habitat produced a higher frequency of the dispersing morph in response to hopper density, than did populations from a more stable habitat. A common-garden experiment provided evidence that these differences had a genetic basis (again, maternal effects could not be eliminated) which involved reduction in the threshold response to density for production of the winged morph in the ephemeral population. A recurring theme in recent genetic studies of dispersal is the functional importance of genotype X environment interaction, which allows expression of phenotypes in appropriate environments.

Aphid studies also have provided direct evidence on genetic variation, genotype X environment interaction, and differences in population in the frequency of dispersal genotypes in the field. The existence of clonal reproduction in this group has been especially useful for investigating reaction norms and genotype X environment interaction for dispersal traits. There is considerable genetic (i.e. clonal) variation in the proportion of winged versus wingless progeny produced by individual clones in response to environmental cues (Markkula 1963; Lamb and MacKay 1983; Groeters 1989; Weisser and Braendle 2001; Hazell *et al.* 2005; see following). However, it is unclear why clones differ in their tendency to produce dispersal morphs. Like the hopper studies, higher dispersal propensity has been observed from clones collected from shorter relative to longer-lived host plants (Groeters 1989), but generally even clones collected from the same type of host plant exhibit considerable genetic variation for the dispersal polyphenism (Lamb and MacKay 1983; Braendle *et al.* 2005b).

In addition to these ecological-genetic studies of the dispersal morph *per se*, a number of studies of waterstriders and planthoppers have shown that the frequency of the dispersal morph can have a strong influence on the magnitude of gene flow among populations. In general, genetic differentiation is most pronounced in species that contain a very low frequency of the dispersing morph (e.g. Zera 1981; Peterson and Denno 1998), although some dramatic exceptions are known (e.g. Peterson *et al.* 2001; and references therein).

Seed heteromorphism. Because plants are sessile, dispersal is dependent upon the production of propagules such as seeds. Many plants exhibit discontinuous variation for seed traits that affect dispersal capability. This polymorphism, termed 'seed heteromorphism', has been one of the best-studied aspects of natural genetic variation in seed morphology. A seed heteromorphic plant produces two or more different kinds of seeds, often with different dispersal attributes, on the same individual plant (Venable and Levin 1985). Seed heteromorphism is a broad term, sometimes more specifically referenced as heterocarpy, for differentiation of the fruits, and heterospermy for differentiation of the seeds. Seed heteromorphism is taxonomically widespread, occurring in 18 angiosperm families, with particularly high abundance in the Asteraceae and the Chenopodiaceae (Imbert 2002). It allows dispersal bet-hedging in response to changing environments (Venable 1985; Venable and Levin 1985; Imbert 2002).

Like insect wing dimorphism, seed heteromorphism exhibits heritable variation (Venable and Burquez 1989; Imbert 2001). As an example, *Heterosperma pinnatum* (Asteraceae) produces three types of achenes (single-seeded fruits) on their fruiting head. The more dispersable central achenes are long and barbed while the more dormant peripheral achenes are unbarbed, wide, and short; the other achenes are intermediate in location and phenotype (Venable 1987). Variation in the proportion of central versus peripheral achenes has high broad-sense heritability and among-population genetic variation, most likely to be due to spatially and temporally varying selection (Venable and Burquez 1989). Further, significant genetic correlations exist among morphological and life-history traits comprising seed heteromorphism (Venable and Burquez 1990), such as achene traits and seed head traits, suggesting a negative relationship between dispersal and dormancy.

Several studies have examined the evolution of dispersal in the field in seed-heteromorphic plants. The hawk's beard weed *Crepis sancta* (Asteraceae) is seed heteromorphic, with heritability of the ratio of dispersing to non-dispersing seeds being approximately 0.25 (Imbert 2001). In highly fragmented urban environments in southern France, the success of dispersing seeds is 55% lower than non-dispersing seeds because of the patchiness of suitable habitats (Cheptou *et al.* 2008; Case Study IV; Chapter 14). Surviving offspring exhibited a higher proportion of non-dispersing seeds relative to their parents after a generation of dispersal in this environment (Cheptou *et al.* 2008). Further, plants from fragmented urban populations of *C. sancta* have significantly higher proportions of non-dispersing seeds relative to plants from unfragmented populations when raised in a common-garden experiment (Cheptou *et al.* 2008). Thus, the relative proportion of dispersing seeds can rapidly respond to natural selection.

Seed shattering in domesticated rice. Extensive genetic characterization of domesticated plants presents an opportunity to examine the loci underlying variation in morphological traits contributing to dispersal capability, such as seed-shattering in rice. In wild rice species such as *Oryza rufipogon*, seeds are released from the plant by seed shattering at maturation. During the process of domestication, seed shattering and its associated loss of harvestable seeds was selected against with the end result being cultivated rice lines (*Oryza sativa*) retaining seeds on their stems. The seed-shattering phenotype has been intensively studied from a genetic perspective, with differences between wild and domesticated lines mainly due to two dominant genes on chromosomes 1 and 4 (Cai and Morishima 2000; Kennard *et al.* 2002; Thomson *et al.* 2003). Using a QTL analysis, Konishi *et al.* (2006) showed that a large portion of variation in the trait mapped to a region called *QTL of seed shattering in chromosome 1* (*qSH1*) locus, and subsequently to a single nucleotide substitution positioned 12 kb upstream of the open reading frame of a BEL1-type homeobox gene. (This same regulatory mutation underlies phenotypic variation in a seed dispersal structure in the Brassicaceae family; see Arnaud *et al.* 2011). Also using a QTL mapping approach, Li *et al.* (2006) demonstrated that the dominant shattering gene on chromosome 4 is a QTL previously called *Shattering 4 (sh4)*. They localized the mutation to a single nucleotide substitution in the DNA binding domain region of the trihelix family of plant-specific transcription factors. Both *qSH1* and *sh4* are expressed in the abscission layer that causes the separation of the rice grain from the pedicel, suggesting that the shattering phenotype emerges from changes in this abscission layer.

6.3.2 Genetics of dispersal movement

Genetic analysis of movement *per se* during dispersal has been investigated in many organisms, most notably insects, various vertebrates, and, to a lesser degree, plants, and a variety of other organisms. Different aspects of movement have been investigated in different groups. For example, in insects, dispersal/migratory flight has mainly been studied in the laboratory while studies of vertebrate dispersal have exclusively been undertaken in the field. Insect studies also have focused on correlated traits, similar to studies of wing polymorphic insects.

Insect dispersal flight in the laboratory and under semi-natural conditions. Genetic variation for flight itself, related to dispersal/migration, has been investigated in a number of wing-monomorphic insects (i.e. all individuals have wings) that fly short or long distances (100 kilometers per day). These studies have been reviewed extensively (Gatehouse 1987, 1997; Rankin and Burchsted 1992;

Dingle, 1996; Roff and Fairbairn 2001; Kent and Rankin, 2001), and will only be discussed briefly here. Significant heritabilities for flight duration, distance, or incidence have commonly been found in these species. In addition, genetic correlations between flight duration and a variety of reproductive (age at first reproduction, clutch size), morphological (wing length, body size), or physiological traits (lipid reserves) have been reported. Notable examples are the following, all of which were measured in long-distance colonists (referred to as long-distance migrants in the insect literature): *Oncopelus fasciatus* (Dingle 1996), *Melanoplus sanguinipes* (Kent Jr, and Rankin 2001), and *Spodoptera exempta* (Gatehouse 1987, 1997). Functionally important genotype X environment interactions have been identified in several cases (Gatehouse 1997; Dingle 1996). Studies of the migratory African armyworm, *Spodoptera exempta*, are good examples of this type of investigation. Individuals of this species can fly more than several hundred kilometers over one or more nights. Selection experiments on moths reared in the laboratory under high density demonstrated a strong genetic component with respect to duration of tethered laboratory flight (realized h^2 = ca. 0.5). However, when larvae of selected lines were raised under low density, flight duration of adult moths was very low, and did not differ between selected lines. This genotype X environment interaction appears to be adaptive in that it causes the expression of long-duration flight under the appropriate environmental conditions (Gatehouse 1987, 1997; Dingle 1996). High larval density occurs after onset of the rainy season in East Africa, and flight during this period allows moths to escape competition and colonize new habitats made available by the rains where they subsequently reproduce.

Similar results (e.g. significant heritability, genetic correlations with life-history traits, genotype X environment for flight duration) are found in the temperate migrant *Oncopeltus fasciatus* (reviewed in Dingle 1996). Interestingly, flight duration and fecundity are positively correlated phenotypically or genetically in some of these colonizing species (e.g. *Oncopeltus*, Dingle 1996; *Melanoplus sanguinipes*, Ranken and Burchsted 1992; also see the Glanville butterfly discussed following) but negatively correlated in wing-polymorphic species. This positive correlation is thought to be important for colonizing species, which must both fly to a new habitat and reproduce there (Rankin and Burchsted 1992). Thus the negative genetic association between flight capability and reproductive output that is almost universally found in wing-polymorphic species does not represent an inherent physiological-genetic constraint; different combinations of dispersal and other life-history traits can apparently be moulded by natural selection depending upon the biology of the species.

In contrast to laboratory studies of flight in the wing-monomorphic species discussed earlier, genetic analysis of dispersal movement in the Glanville fritillary butterfly (*Melitaea cinxia*) has been undertaken under semi-natural conditions in the field. This example will only be discussed briefly because it is the focus of Chapters 14 and 23. The Glanville fritillary exists as a meta-population in fragmented habitat patches interconnected by dispersal. Female progeny from newly colonized patches differ genetically from those of longer-duration patches in a variety of dispersal and life-history traits. Significant heritabilities for mobility, as well as several reproductive traits (age at first reproduction and body-mass-corrected egg mass) have been quantified in a large cage (32 × 26 × 3 metres) in the field by parent-offspring (mother-on-daughter) regression (Saastamoinen 2008) or full-sib analyses (Klemme and Hanski 2009). However, as is often the case with field heritability estimates, the extent to which maternal effects contributed to these heritabilities is unknown (see Saastamoinen 2008). Subsequent studies have identified a number of interesting correlations between allozymes (electophoretic variants) or DNA sequence variants of metabolic enzymes and dispersal rate, flight metabolic rate, and a variety of life-history traits (e.g. Saastamoinen and Hanski 2008; Klemme and Hanski 2009; Zheng *et al.* 2009). Most recently, in a landmark functional-genomic study, Wheat *et al.* (2011) found dramatic differences in expression in many important classes of genes between females from newly colonized versus older populations. To our knowledge, this is the first study comparing global gene expression in phenotypes that differ in dispersal in the field, analogous to the functional-genomic study of flightless and flight-capable morphs of the pea aphid (Brisson *et al.* 2007).

Genetic studies of dispersal movement in vertebrates. Genetic studies of dispersal/migration in vertebrates have largely focused on natal dispersal involving distances in the tens of kilometers or less in birds and reptiles. However, some classic studies have investigated long-distance seasonal migration in birds, which involves return trips between overwintering and breeding areas. Surprisingly, only a few limited studies of dispersal genetics have been conducted on mammals (Waser and Johnes 1989; Boonstra *et al.* 1997).

Long-distance seasonal migration in birds has been studied extensively from an evolutionary-genetic perspective in several passerines, most notably the blackcap, *Sylvia atricapilla* (for reviews, see Berthold and Terrill 1991; Berthold, 2003; Pulido 2007). Early classic work on this species, using artificial selection and various types of sib analysis, demonstrated a genetic basis to 'migratory restlessness' or *Zugunruhe*, a laboratory behaviour indicative of endogenous propensity to migrate. Common-garden experiments, and heritability estimates derived from both parent-offspring regression and artificial selection, have subsequently demonstrated a genetic basis for differences in the timing of migration within and between several species including *S. atricapilla* (Berthold and Terrill 1991; Pulido *et al.*, 2001; Pulido 2007 and references therein). Similar to the situation for wing-polymorphic insects discussed above, genetic control of migratory behaviour in these birds is best explained by the polygenic threshold model: genetically encoded factors, above or below some threshold, expressed under a specific photoperiod, collectively specify migratory versus non-migratory behaviour (Pulido *et al.* 1996; Pulido 2007). Marked geographic variation in migration in *S. atricapilla* is thought to result from geographic variation in the threshold that elicits migration, similar to the situation for differences among populations with respect to induction of the dispersing morph in planthoppers and grasshoppers (Denno *et al.* 1996; Simmons and Thomas 2004; see earlier). Limited information suggests that behavioural and physiological traits of the migratory syndrome, but not life-history traits, are genetically correlated (reviewed in Pulido 2007). Thus, genetic effects of migratory variation are well-established, but only in a few species of passerines to date.

The extent to which genetic variation contributes to variation in natal dispersal in vertebrates has been increasingly studied during the past decade. Earlier studies claiming to demonstrate heritability for natal dispersal in the great tit (Greenwood *et al.* 1979) were criticized on methodological grounds (van Noordwijk 1984), and some workers argued that variation in environment or condition-dependent behaviour are more important causes of variability in natal dispersal than genotype (e.g. Massot and Clobert 2000). The relative importance of genetic versus environmental influences on dispersal variability remains an open question, in large part due to the difficulty in undertaking sufficiently controlled quantitative-genetic studies in the field. Nevertheless, several well-controlled, intensive studies in birds provide strong evidence for a significant genetic component to dispersal variation in the field. For example, analysis of capture-recapture data of male great-reed warblers (*Acrocephalus arundinaceus*) obtained from the species whole Swedish breeding range over several years demonstrated a high heritability (ca. 0.5) for dispersal behaviour (i.e. philopatry versus interpopulation-dispersal; Hansson *et al.* 2003). Similarly, a significant heritability for natal dispersal distance was reported for the red-cockaded woodpecker (*Picoides borealis*), a study involving data collected over more than 15 years (Pasinelli *et al.*, 2004). Finally, as discussed below, Dingemanse *et al.* (2003) reported that genetic differences in personality influenced dispersal distance in the great tit, *Parus major.* Although it is not possible to eliminate inflation of these heritability estimates due to parental effects or common environment, authors in each study provide additional information indicating that, at the very least, genetic effects on dispersal were significant. Also, some authors report no evidence supporting the commonly held assumption that parental effects inflate avian heritability estimates in birds (Weigensberg and Roff 1996; Merilä and Sheldon 2001).

Several recent studies of lizards (e.g. Massot *et al.* 2003) also have provided evidence supporting a genetic component to dispersal in the field. In a particularly noteworthy study, Sinervo *et al.* (2006) demonstrated morph (genotype)-dependent dispersal in the side-blotched lizard, *Uta stansburiana*, a species which exhibits three genetically determined male colour morphs that differ in territorial behaviour. They followed dispersal of mature progeny, of known paternity, of the three male throat-coloured morphs produced in the field or laboratory (using a half-sib breeding design). They also experimentally assessed maternal effects on morph-specific dispersal acting through egg size on progeny size by experimentally manipulating egg size (e.g. by yolk removal). Progeny dispersal of both field- and laboratory-produced individuals was significantly affected by sire genotype, and a strong interaction between maternal effect and sire genotype on progeny dispersal distance was also noted. This study represents the most controlled investigation of dispersal genetics undertaken in the field to date.

Genetics of dispersal in plants. Few studies have examined the genetics of seed dispersal patterns in plants. Donohue *et al.* (2005) found significant heritability for seed dispersal distance and genetic correlations among maternal and seed dispersal traits between *Arabidopsis thaliana* ecotypes. This study was unique for plants in that it documented that these heritabilities and genetic correlations are density-dependent. Seed dispersal can, therefore, respond to selection and evolve, either via seed or maternal characters. Further, phenotypic plasticity can either directly or indirectly affect seed dispersability (Acosta *et al.* 1997; Donohue 1999; Imbert & Ronce 2001; Galloway 2002; Wender *et al.* 2005). As mentioned previously, a persistent theme in recent studies of dispersal genetics is the adaptive importance of environmentally contingent gene expression.

6.3.3 Behavioural and molecular genetics of movement

An important new area of study in dispersal genetics is the influence of genetic variation for behavioural traits (e.g. 'personality') on movement and dispersal. This work combines traditional quantitative genetics with molecular-genetic analyses of candidate genes.

Behavioural genetics of dispersal in vertebrates. In the great tit, *Parus major*, Dingemanse *et al.* (2002) and Drent *et al.* (2003) reported a significant heritability (0.3–0.6) for exploratory behaviour, a trait that is correlated with natal dispersal measured in the field (Dingemanse *et al.* 2003). These data suggest potential genetic co-variation between exploratory behaviour and dispersal. Several studies have gone on to investigate correlations between DNA sequence variation of a dopamine-receptor gene (*Drd4*) and exploratory behaviour. This polymorphic gene is involved in dopaminergic-mediated motivational behaviours, and DNA haplotype is associated with novelty-seeking behaviour in humans as well as a variety of non-human animals (Fidler *et al.* 2007). A single nucleotide polymorphism (SNP) in the coding region of this locus was genetically correlated with exploratory behaviour in artificially selected lines of *P. major*, and with exploratory behaviour in free-living birds (Fidler *et al.* 2007). However, a subsequent study (Korsten *et al.* 2010) did not find a similar association in other populations of *P. major*.

Analogous studies have been undertaken in mammals focusing on variation in DNA sequence of genes involved in aggressive behaviour. Male dispersal is often related to the onset of aggressive behaviour in mammalian natal groups. In free-ranging rhesus macaques, age of male natal dispersal was associated with DNA sequence in the promoter region of the serotonin transporter gene, with genotypes differing in age of natal dispersal by one year (Trefilov *et al.* 2000), as well as differing in timing of reproduction (Krawczak *et al.* 2007). Similarly, alleles of microsatellites located near an orthologue of this gene in house mice (*Mus musculus domesticus*) were associated with antagonistic behaviour between young male littermates (Krackow and König 2008). The onset of antagonistic behaviour is significantly related to dispersal propensity in this species, which, in turn, exhibits a heritable genetic component (Krackow 2003).

In a remarkable series of studies (Chapter 7), genetic co-variation was demonstrated between aggressive behaviour and dispersal ability in the western bluebird (Duckworth and Badyaev 2007; Duckworth 2008; Duckworth and Kruuk 2009). This coupling results in a higher frequency of aggressive individuals at an invasion front, which, in turn, contributes to the ability of the western bluebird to displace the less aggressive mountain bluebird. Duckworth (2008) argues that this dispersal syndrome was maintained before range expansion due to metapopulation dynamics of local extinction and recolonization, thus providing the genetic variation for rapid changes in dispersal phenotype during range expansion.

Behavioural genetics of locomotion in *Drosophila melanogaster* and *Caenorhabdites elegans*. Several behavioural polymorphisms involving locomotion have been studied in genetic model organisms allowing detailed analyses of the underlying molecular basis of the polymorphism. One of these is the rover-sitter polymorphism in *Drosophila melanogaster*. Natural populations exhibit a bimodal frequency distribution for the propensity of *D. melanogaster* to move in the presence of food (reviewed in Sokolowski 1998, 2001). The opposite ends of the spectrum of behaviour are called rovers and sitters and much of this variation is due to a single gene called *foraging*. The rover allele is dominant to the sitter allele (Sokolowski, 1980), which are found approximately in a 70:30 ratio within a single natural population (Sokolowski, 1980). Fitness effects of the alleles are density-dependent (Sokolowski *et al.* 1997), and the polymorphism is maintained by frequency-dependent selection (Fitzpatrick *et al.* 2007). The *for* gene has been mapped and identified: it encodes a cGMP-dependent protein kinase involved in signal transduction, with higher expression in rovers (Osborne *et al.* 1997).

C. elegans exhibits a behavioural polymorphism analogous to the *Drosophila* rover/sitters discussed above, although the behavioural variants are referred to as solitary versus gregarious foragers (Hodgkin and Doniach 1997; de Bono and Bargmann 1998) where the latter are more likely to disperse (Gloria-Soria and Azevedo 2008). Variation in this behaviour is due to a single amino acid substitution at the G-protein coupled receptor *npr-1* (de Bono and Bargmann 1998). The alternative alleles may be maintained as a trade-off between dispersal and competitive ability in heterogeneous environments—the solitary worms outcompete the gregarious worms within a food patch, but the gregarious worms can disperse away from a patch (Gloria-Soria and Azevedo 2008). Indeed, Friedenberg (2003) used a different mutant genotype (*rol-1*) in *C. elegans* to demonstrate that environmental stochasticity in space and time, including increased rates of food patch extinction, results in the evolution of increased dispersal propensity. More recently, Bendesky *et al.* (2011) reported natural variation in another G-protein coupled receptor (*tyra-3*) in *C. elegans* that influences the decision to leave a food patch.

6.4 Summary, synthesis, and future directions

Dispersal genetics has advanced significantly during the past decade in five ways:

(1) Fundamental studies of dispersal genetics have expanded considerably, both phylogenetically and methodologically. In some cases, such as vertebrates, better controlled studies have provided the first strong evidence for the existence of genetic variation for dispersal in the field (Hansson *et al.* 2003; Sinervo *et al.* 2006). Previously, the very existence of genetic variation for dispersal in this group had been questioned (Massot and Colber 2000). Similarly, detailed studies of dispersal in the field have been undertaken in plants, another understudied group (Imbert 2001; Cheptou *et al.* 2008; Donohue *et al.* 2005), and the first studies of genetic variation in dispersal movement in insects in the field have been reported (Saastamoinen 2008; Klemme and Hanski 2009), a group that had previously only been studied in the laboratory. Finally, investigations of dispersal genetics have expanded to previously unstudied laboratory organisms that allow new experimental approaches (e.g. *Tetrahymena*, Fjerdingstad *et al.* 2007; *C. elegans*, Friedenberg 2003; Kiontke 2008).

(2) The existence of genetic variation for plasticity (genetically variable reaction norms; genotype X environment interaction) has now been found in a number of cases where looked for, including density-dependent dispersal-flight or production of dispersal morphs in insects in the laboratory (Gatehouse 1987; Masaki and Shimizu 1995; Matsumura 1996; Braendle *et al.* 2005b), or field (Denno *et al.* 1996; Simmons and Thomas 2004), and seed dispersal in plants (Donohue *et al.* 2005). Adaptive genetic variation for dispersal plasticity allows the expression of genetically specified dispersal traits in the appropriate environment and will likely prove to be common and exceedingly important.

(3) The first in-depth physiological-genetic studies of dispersal capability have been conducted, mainly in wing-polymorphic insects. These studies have led to the first detailed insights concerning the identity of genetically variable endocrine regulators that control the expression of dispersal traits, genetic correlations between dispersal traits, and genetic correlations between dispersal and life-history traits (Zera and Harshman 2009). Another notable finding has been the biochemical-molecular genetics of intermediary metabolism and resource allocation that underlies genetic trade-offs between dispersal and other life-history traits (Zhao and Zera 2002; reviewed in Zera and Harshman 2009, 2011).

(4) The first studies of the molecular and genomic correlates of genetically variable dispersal traits have been reported. Notable examples include molecular and genomic aspects of dispersal capability (Brisson 2007; Konishs *et al.* 2006), dispersal behaviour in model organisms in the laboratory (Osborne *et al.* 1997; Freidenberg 2003) and field (Trefilov *et al.* 2000; Krackow and König 2008), and dispersal biochemical-genetics and genomics in the field (Zheng *et al.* 2009; Wheat *et al.* 2011).

(5) A number of notable studies (Simmons and Thomas 2004; Duckworth 2008) have also demonstrated the importance of dispersal genotypes in the field, such as during range expansion.

The field of dispersal genetics is now entering a new functional-genetic phase, moving beyond traditional quantitative genetic investigations that have largely focused on statistical components of dispersal variance. Although traditional quantitative genetics continues to be important in baseline studies that establish the existence of genetic variation and co-variation, it does not provide a useful perspective for understanding evolutionary mechanisms based on gene action (Hansen 2006; Dykuizen and Dean 2009). Recent functional-genetic studies are beginning to provide detailed information on the causal relationships among variation in gene sequences, physiological traits and whole-organism dispersal traits necessary for a deeper understanding of the mechanisms of dispersal evolution. For example, endocrine studies of genetic correlations among traits comprising the dispersal syndrome will allow a better understanding of the functional basis of these correlations, and hence the extent to which these correlations represent fundamental physiological constraints versus evolved couplings. Functional information will also allow investigations of such fundamental issues as the nature of the polygenic threshold, and the basis of plasticity, which in turn will allow a more mechanistic understanding of evolutionary changes in these traits (e.g. during range expansion).

Future studies of dispersal genetics should focus on the following:

(1) Detailed, controlled quantitative-genetic investigations of dispersal traits are still available for only a relatively few species and should continue to be a focus of future research. Fundamental issues such as the relative contribution of variation in environment (e.g. condition, maternal effects) versus genotype on dispersal variation in the field are still largely unresolved. Careful studies in lizards (Sinervo *et al.* 2006) indicate that both maternal effects and genotype are important contributors to dispersal variation. As mentioned above, the growing list of environmentally contingent, genetically variable dispersal traits indicates that a greater emphasis should be placed on investigating these traits using a reaction-norm

perspective, as advocated by van Noordwijk *et al.* (2006).

(2) The influence of molecular genetics and genomics is just beginning to be felt in dispersal genetics. Genomics will allow investigations of whole pathways and networks (cell signaling, systemic-endocrine, metabolic) underlying dispersal phenotypes. Importantly, although a growing number of interesting molecular correlates of dispersal have been identified, the extent to which these represent significant, functional relationships remains almost completely unknown. In-depth functional studies of these correlates, which is a difficult task in and of itself (Storz and Zera 2011), should be a top priority of future work on dispersal genetics.

(3) Physiological-genetic (e.g. endocrine) investigations have been undertaken in any detail in only a handful of traits in a very few species, and should also be a priority in future studies. As discussed above, physiology provides key information on the mechanisms by which variable genes are translated into variable whole-organism phenotypes. In addition, biochemical and physiological information will becomes increasingly important to evaluate the functional significance of rapidly accumulating genomic data, such as on transcript abundance (Feder and Walser 2005; Zera 2011).

(4) Finally, there is a need to undertake more multilevel, integrative investigations of dispersal genetics similar to those in wing-polymorphic *Gryllus* (e.g. Zera and Denno 1997; Roff and Fairbairn 2007; Zera 2009, 2011; Zera and Harshman 2009), the Glanville fritillary (Saastamoinen 2008; Wheat 2011; Hanski, 2011; Case Study III; Chapters 8, 23), *P. major* (Dingemanse *et al.* 2002; Fidler *et al.* 2007) and the mountain bluebird (Duckworth and Badyaev 2007; Duckworth 2008; Chapter 7). Integrative studies are especially powerful with respect to identifying the chain of causality from DNA sequence to whole-organism phenotype; they simultaneously provide information on the functional significance and mechanistic underpinnings of various traits at several levels of biological organization.

Acknowledgements

AJZ gratefully acknowledges support from the National Science Foundation (most recently IOS-05116973, IBN-0212486, IBN-0130665, and IBN-9808249). JAB acknowledges the generous support of the National Institute of Environmental Health Sciences (R00ES017367) and the School of Biological Sciences, University of Nebraska-Lincoln.

References

Acosta, F.J., Delgado, J.A., Lopez, E. (1997). 'Functional features and ontogenetic changes in reproductive allocation and partitioning strategies of plant modules', *Plant Ecology* **132**: 71–6.

Arnaud, N., Lawrenson, T., Ostergaard, L. (2011). 'The same regulatory point mutation changed seed-dispersal structures in evolution and domestication', *Current Biology* **21**: 1215–19.

Bendesky, A., Tsunozaki, M., Rockman, M.V. *et al.* (2011). 'Catecholamine receptor polymorphisms affect decision-making in *C. elegans*', *Nature* **472**: 313–19.

Berthold, P. (1991). 'Genetic control of migratory behavior', *Trends in Ecology and Evolution* **6**: 254–7.

Berthold, P. (2003). 'Genetic basis and evolutionary aspects of bird migration', *Advances in the Study of Behavior* **33**: 175–229.

Berthold, P., and Terrill, S.B. (1991). 'Recent advances in studies of bird migration', *Annual Review of Ecology and Systematics* **22**: 357–78.

Boonstra, R., and Hochachka, W.M. (1997). 'Maternal effects and additive genetic inheritance in the collard lemming *Dicrostonyx groenlandicus*', *Evolutionary Ecology* **11**: 169–82.

Braendle, C., Caillaud, M.C. and Stern, D.L. (2005a). 'Genetic mapping of *aphicarus*—a sex-linked locus controlling a wing polymorphism in the pea aphid (*Acyrthosiphon pisum*)', *Heredity* **94**: 435–42.

Braendle, C., Davis, G.K., Brisson, J.A. (2006). 'Wing dimorphism in aphids', *Heredity* **97**: 192–9.

Braendle, C., Friebe, I., Caillaud, M.C. (2005b). 'Genetic variation for an aphid wing polyphenism is genetically linked to naturally occurring wing polymorphism', *Proceedings of the Royal Society B* **272**: 657–65.

Brisson, J.A., Davis, G.K. and Stern, D.L. (2007). 'Common genome-wide transcription patterns underlying the wing polyphenism and polymorphism in the pea aphid (*Acyrthosiphon pisum*)', *Evolution & Development* **9**: 338–46.

Cai, H.W. and Morishima, H. (2000). 'Genomic regions affecting seed shattering and seed dormancy in rice', *Theoretical and Appliced Genetics* **100**: 840–6.

Caillaud, M.C., Boutin, M., Braendle, C. *et al.* (2002). 'A sex-linked locus controls wing polymorphism in males of the pea aphid, *Acyrthosiphon pisum* (Harris)', *Heredity* **89**: 346–52.

Cheptou, P.O., Carrue, O., Rouifed, S. (2008). 'Rapid evolution of seed dispersal in an urban environment in the weed *Crepis sancta*', *Proceedings of the National Academy of Sciences USA* **105**: 3796–9.

Clobert, J., Danchin, E., Dhondt, A.A. (2001). *Dispersal.* Oxford University Press, Oxford.

Crnokrak, P., and Roff, D. (2002). 'The trade-offs to flight capability in *Gryllus firmus*: the influence of whole-organism respiration rate on fitness', *Journal of Evolutionary Biology* **15**: 388–98.

de Bono, M., and Bargmann, C. (1998). 'Natural variation in neuropeptide Y receptor homolog modifies social behavior and food response in *C. elegans*', *Cell* **94**: 679–89.

Denno, R.F., Roderick, G.K., Peterson, M.A., *et al.* (1996). 'Habitat persistence underlies intraspecific variation in the dispersal strategies of planthoppers', *Ecological Monographs* **66**, 389–408.

Dingemanse, N.J., Both, C., Drent, P.J., van Oers, K., Noordwijk, A. J. (2002). 'Repeatability and heritability of exploratory behavior in great tits from the wild', *Animal Behaviour* **64**: 929–38.

Dingemanse, N.J., Both, C., van Noordwijk, A.J. *et al.* (2003). 'Natal dispersal and personalities in great tits (*Parus major*)', *Proceedings of the Royal Society B* **270**: 741–7.

Dingle, H. (1985). 'Migration', in G.A. Kerkut and L.I. Gilbert, eds, *Comprehensive Insect Physiology, Biochemistry and Pharmacology.* Pergamon, Oxford, 375–415.

Dingle, H. (1996). *Migration, the Biology of Life on the Move.* Oxford University Press, Oxford.

Dingle, H., and Drake, V.A. (2007). 'What is migration?', *BioScience* **57**: 113–21.

Donohue, K. (1999). 'Seed dispersal as a maternally influenced character: Mechanistic basis of maternal effects and selection on maternal characters in an annual plant', *American Naturalist* **154**: 674–89.

Donohue, K., Polisetty, C.R. and Wender, N.J. (2005). 'Genetic basis and consequences of niche construction: Plasticity-induced genetic constraints on the evolution of seed dispersal in *Arabidopsis thaliana*', *American Naturalist* **165**: 537–50.

Drent, P.J., van Oers, K. and van Noordwijk, A.J. (2003). 'Realized heritability of personalities in the great tit (*Parus major*)', *Proceedings of the Royal Society B* **270**: 45–51.

Duckworth, R. (2008). 'Adaptive dispersal strategies and the dynamics of range expansion', *American Naturalist* **172**: S4–17.

Duckworth, R.A., and Badyaev, A.V. (2007). 'Coupling of dispersal and aggression facilitates the rapid range expansion of a passerine bird', *Proceedings of the National Academy of Sciences USA* **104**: 15017–22.

Duckworth, R.A., and Kruuk, L.E.B. (2009). 'Evolution of genetic integration between dispersal and colonization ability in a bird', *Evolution* **63**: 968–77.

Dykhuizen, D.E., and Dean, A.M. (2009). 'Experimental evolution from the bottom up. In T. Garland, Jr and MR. Rose, eds, *Experimental Evolution. Concepts, Methods, and Applications of Selection Experiments.* University of California Press, Berkeley, 67–87.

Fairbairn, D.J. (1994). 'Wing dimorphism and the migratory syndrome: correlated traits for migratory tendency in wing dimorphic insects', *Researches on Population Ecology* **36**: 157–63.

Falconer, D.S., and Mackay, T.F.C. (1996). *Introduction to Quantitative Genetics.* Longman, Essex.

Feder, M.E., Garland Jr, T., Marden, J.H. *et al.* (2010). 'Locomotion in response to shifting climatic zones: Not so fast', *Annual Review of Physiology* **72**: 167–90.

Fidler, A.E., van Oers, K., Drent, P.J. *et al.* (2007). 'Drd4 gene polymorphisms are associated with personality variation in a passerine bird', *Proceedings of the Royal Society B* **274**: 1685–91.

Fitzpatrick, M.J., Feder, E., Rowe, L. & Sokolowski, M. B. (2007). 'Maintaining a behaviour polymorphism by frequency-dependent selection on a single gene', *Nature* **447**: 210–13.

Fjerdingstad, E.J., Schtickzelle, N., Manhes, P., (2007). 'Evolution of dispersal and life history strategies-*Tetrahymena* ciliates', *BMC Evolutionary Biology* **7**: 133.

Friedenberg, N.A. (2003). Experimental evolution of dispersal in spatiotemporally variable microcosoms. *Ecology Letters* **6**: 953–9.

Galloway, L.F. (2002). 'The effect of maternal phenology on offspring characters in the herbaceous plant *Campanula americana. Journal of Ecology* **90**: 851–8.

Gatehouse, A.G. (1989). 'Genes, environment and insect flight', in G.J. Goldworthy and C.H. Wheeler, eds, *Insect Flight.* CRC Press, Boca Raton, 115–38.

Gatehouse, A.G. (1997). 'Behavior and ecological genetics of wind-borne migration by insects', *Annual Review of Entomology* **42**: 475–502.

Gloria-Soria, A., and Azevedo, R.B.R. (2008). '*npr-1* Regulates foraging and dispersal strategies in *Caenorhabditis elegans*', *Current Biology* **18**: 1694–9.

Greenwood, P.J., Harvey, P.H. and Perrins, C.M. (1979). 'The role of dispersal in the great tit (*Parus major*): The causes, consequences and heritability of natal dispersal', *Journal of Animal Ecology* **33**: 175–229.

Groeters, F.R. (1989). 'Geographic and clonal variation in the milkweed-oleander aphid, *Aphis nerii* (Homoptera: Aphididae), for winged morph production, life history, and morphology in relation to host plant permanence', *Evolutionary Ecology* **3**: 327–41.

Guerra, P.A. (2011). 'Evaluating the life history trade-off between dispersal capability and reproduction: a meta-analysis', *Biological Reviews* **86**: 813–35.

Guerra, P.A. and Pollack, G.S. (2009). 'Flight behavior attenuates the trade-off between flight capability and reproduction in a wing polymorphic cricket', *Biology Letters* **5**: 229–31.

Hansen, T.F. (2006). 'The evolution of genetic architecture', *Annual Review of Ecology, Evolution and Systematics* **37**, 123–57.

Hansson, B., Bensch, S. and Hasselquist, D. (2003). 'Heritability of dispersal in the great reed warbler', *Ecology Letters* **6**: 290–4.

Harrison, R.G. (1980). 'Dispersal polymorphisms in insects', *Annual Review of Ecology and Systematics*, **11**: 95–118.

Hawkes, C. (2009). 'Linking movement behavior, dispersal and population processes: is individual variation a key?', *Journal of Animal Ecology* **78**: 894–906.

Hazell, S.P., Gwynn, D.M., Ceccarelli, S. (2005). 'Competition and dispersal in the pea aphid: clonal variation and correlations across traits', *Ecological Entomology* **30**: 293–8.

Hodgkin, J., and Doniach, T. (1997). 'Natural variation and copulatory plug formation in *Ceanorhabditis elegans*', *Genetics* **146**: 149–64.

Hoffmann, A.A. (2000). 'Laboratory and field heritabilities: Some lessons from *Drosophila*', in T.A. Mousseau, B. Sinervo, and J. Endler, eds, *Adaptive Genetic Variation in the Wild*. Oxford University Press, New York, 200–18.

Imbert, E. (2001). 'Capitulum characters in a seed heteromorphic plant, *Crepis sancta* (Asteraceae), variance partitioning and inference for the evolution of dispersal rate', *Heredity* **86**: 78–86.

Imbert, E. (2002). 'Ecological consequences and ontogeny of seed heteromorphism', *Perspectives in Plant Ecology, Evolution, and Systematics* **5**: 13–36.

Imbert, E. and Ronce, O. (2001). 'Phenotypic plasticity for dispersal ability in the seed heteromorphic *Crepis sancta* (Asteraceae)', *Oikos* **93**: 126–34.

Kennard, W.C., Phillips, R.L. and Porter, R.A. (2002). 'Genetic dissection of seed shattering, agronomic, and color traits in American wildrice (*Zizania palustris* var. interior L.) with a comparative map', *Theoretical and Applied Genetics* **105**: 1075–86.

Kent Jr, J.W., and Rankin, M.A. (2001). 'Heritability and physiological correlates of migratory tendency in the grasshopper *Melanoplus sanguinipes*', *Physiological Entomology* **26**: 371–80.

Kiontke, K. (2009). 'Evolutionary biology: patchy food may maintain a foraging polymorphism', *Current Biology* **18**: R1017–19.

Klemme, I., and Hanski, I. (2009). 'Heritability of and strong single gene (*Pgi*) effects on life-history traits in the Glanville fritillary butterfly', *Journal of Evolutionary Biology* **22**: 1944–53.

Konishi, S., Izawa, T., Lin, S.Y. (2006). 'A SNP caused loss of seed shattering during rice domestication', *Science* **312**: 1392–96.

Korsten, P., Mueller, J.C., Hermannstadter, C. *et al.* (2010). 'Association between DRD4 gene polymorphism and personality variation in great tits: a test across four wild populations', *Molecular Ecology* **19**: 832–43.

Krackow, S. (2003). 'Motivational and heritable determinants of dispersal latency in wild male house mice (*Mus musculus musculus*)', *Ethology* **109**: 671–89.

Krackow, S., and Konig, B. (2008). 'Microsatellite length polymorphisms associated with dispersal-related agonistic onset in wild house mice (*Mus musculus domesticus*)', *Behavioral Ecology and Sociobiology* **62**: 813–20.

Krawczak, M., Trefilov, A., Beard, J., *et al.* (2007). 'Male reproductive timing in rhesus macaques is influenced by the 5HTTLPR promoter polymorphism of the serotonin transporter gene', *Biology of Reproduction* **72**: 1109–13.

Kruuk, L.E.B. (2004). 'Estimating genetic parameters in natural populations using the 'animal model', *Philosophical Transactions of the Royal Society B* **359**: 873–90.

Lamb, R.J. & Mackay, P.A. (1983). 'Micro-evolution of the migratory tendency, photoperiodic response and developmental threshold of the pea aphid, *Acyrthosiphon pisum*', in V.K. Brown and I. Hodek, eds, *Diapause and Life Cycle Strategies in Insects*. Junk Publishers, The Hague, 209–17.

Li, C., Zhou, A. and Sang, T. (2006). 'Rice domestication by reducing shattering', *Science* **311**: 1936–9.

Lin, Z.W., Griffith, M.E., Li, X.R. *et al.* (2007). Origin of seed shattering in rice (*Oryza sativa* L.)', *Planta* **226**: 11–20.

Lynch, M., and Walsh, B. (1998). *Genetics and Analysis of Quantitative Traits*. Sinauer Associates, Inc., Sunderland.

Markkula, M. (1963). 'Studies of the pea aphid, *Acyrthosiphon pisum* Harris (Hom., Aphididae), with special reference to the differences in the biology of the green and red forms', *Annales Agriculturae Fenniae* **2**: 1–30.

Masaki, S. and Shimizu, T. (1995). 'Variability in wing form of crickets', *Researches on Population Biology* **37**: 119–28.

Massot, M., and Clobert, J. (2000). Processes at the origin of similarities in dispersal behavior among siblings', *Journal of Evolutionary Biology* **13**: 707–19.

Massot, M., Huey, R.B, Tsuji, J. *et al.* (2003). 'Genetic, prenatal, and postnatal correlates of dispersal in hatchling fence lizards (*Sceloporus occidentalis*)', *Behavioral Ecology* **14**: 650–5.

Matsumura, M. (1996). 'Genetic analysis of a threshold trait, density-dependent wing dimorphism in *Sogatella furcifera* (Horvath) (Hemiptera, Delphacidae), the white-backed planthopper', *Heredity* **76**, 229–37.

Merilä, J., and Sheldon, B.C. (2001). 'Avian quantitative genetics', *Current Ornithology*, **16**: 179–255.

Mole, S. and Zera, A.J. (1993). 'Differential allocation of resources underlies the dispersal-reproduction trade-off in the wing-dimorphic cricket, *Gryllus rubens*', *Oecologia* **93**: 121–7.

Nathan, R., Getz, E. Revilla, M. *et al.* (2008). 'A movement ecology paradigm for unifying organismal movement research', *Proceedings of the National Academy of Sciences USA* **105**: 19052–9.

Osborne, K.A., Robichon, A., Burgess, E., *et al.* (1997). 'Natural behavior polymorphism due to a cGMP-dependent protein kinase of *Drosophila*', *Science* **277**: 834–6.

Pasinelli, G., K. Schiegg, and Walters, J.R. (2004). 'Genetic and environmental influences on natal dispersal distance in a resident bird species', *American Naturalist* **164**: 660–9.

Peterson, M.A., and Denno, R.F. (1998). 'The influence of dispersal and diet breadth on patterns of genetic isolation by distance in phytophagous insects', *American Naturalist* **152**: 428–46.

Peterson, M.A., Denno, R.F., and Robinson, L. (2001). 'Apparent widespread gene flow in the predominantly flightless planthopper *Tumidagena minuta*', *Ecological Entomology* **26**: 629–37.

Pulido, F. (2007). 'The genetics and evolution of avian migration. *BioScience* **57**: 165–74.

Pulido, F., Berthold, P., Mohr, G. *et al.* (2001). 'Heritability of the timing of autumn migration in a natural bird population', *Proceedings of the Royal Society B* **268**: 953–9.

Pulido, F., Berthold, P., and van Noordwijk, A.J. (1996). 'Frequency of migrants and migratory activity are genetically correlated in a bird population: evolutionary implications', *Proceedings of the National Academy of Sciences USA* **93**: 14642–47.

Rankin, M.A., and Burchsted, J.C.A. (1992). 'The cost of migration in insects', *Annual Review of Ecology and Systematics* **37**: 533.

Riska, B., Prout, T. and Turelli, M. (1989). 'Laboratory estimates of heritabilities and genetic correlations in nature', *Genetics* **123**: 865–71.

Roff, D.A. (1994). 'Evidence that the magnitude of the trade-off in a dichotomous trait is frequency dependent', *Evolution* **48**: 1650–6.

Roff, D.A. (1996). The evolution of threshold traits in animals', *Quarterly Review of Biology* **31**: 243–51.

Roff, D.A. (1997). *Evolutionary Quantitative Genetics*. Chapman and Hall, New York.

Roff, D., and Fairbairn, D.J. (1991). 'Wing dimorphisms and the evolution of migratory polymorphisms among insects', *American Zoologist* **31**: 243–51.

Roff, D.A., and Fairbairn, D.J. (2001). 'The genetic basis of dispersal and migration, and its consequences for the evolution of correlated traits', in J. Clobert, E. Danchin, A.A. Dhondt, *et al.* Eds, *Dispersal*. Oxford University Press, Oxford, 191–202.

Roff, D.A., and Fairbairn, D.J. (2007). 'The evolution and genetics of migration in insects', *BioScience* **57**: 155–64.

Roff, D.A., and Gelinas M.B. (2003). 'Phenotypic plasticity and the evolution of trade-offs: the quantitative genetics of resource allocation in the wing-dimorphic cricket, *Gryllus firmus*', *Journal of Evolutionary Biology* **16**: 55–63.

Ronce, O. (2007). 'How does it feel to be like a rolling stone? Ten questions about dispersal evolution', *Annual Review of Ecology, Evolution and Systematics* **38**: 231–53.

Saastamoinen, M. (2008). 'Heritability of dispersal rate and other life history traits in the Glanville fritillary butterfly', *Heredity* **100**: 39–46.

Saastamoinen, M., and Hanski, I. (2008). 'Genotypic and environmental effects on flight activity and oviposition in the Glanville fritillary butterfly', *American Naturalist* **171**: E701–12.

Schilder, R., Zera, A.J., Black, C., Hoidal, M., and Wehrkamp, C. (2011). 'The biochemical basis of life history adaptation: Molecular and enzymological causes of NADP $^+$ -isocitrate dehydrogenase activity differences between morphs of *Gryllus firmus* that differ in lipid biosynthesis and life history', *Molecular Biology and Evolution* **28**: 3381–93. DOI: 10.1093/molbev/msr171.

Simmons, A.D., and Thomas, C.D. (2004). 'Changes in dispersal during species range expansion', *American Naturalist* **164**: 378–95.

Simons, A.M., and Roff, D.A. (1994). 'The effect of environmental variability on the heritabilities of traits of a field cricket', *Evolution* **48**: 1637–49.

Sinervo, B., Calsbeek, R., Commandant, A.T., Both, C., Adamopoulou, C., and Clobert, J. (2006). 'Genetic and maternal determinants of effective dispersal, The effect of sire genotype and size at birth in side-blotched lizards', *American Naturalist* **168**: 88–99.

Smith, M.A.H. and Mackay, P.A. (1989). 'Genetic variation in male alary dimorphism in populations of the pea aphid, *Acyrthosiphon pisum*', *Entomologia Experimentalis et Applicata* **51**: 125–32.

Sokolowski, M.B. (1980). 'Foraging strategies of *Drosophila melanogaster*, a chromosomal analysis', *Behavior Genetics* **10**: 291–302.

Sokolowski, M.B. (1998). 'Genes for normal behavioral variation: recent clues from flies and worms', *Neuron* **21**: 463–6.

Sokolowski, M.B. (2001). '*Drosophila*: Genetics meets behaviour. *Nature Reviews Genetics* **2**: 879–90.

Sokolowski, M.B., Pereira, H.S. and Hughes, K. (1997). 'Evolution of foraging behavior in *Drosophila* by density dependent selection', *Proceedings of the National Academy of Sciences USA* **94**: 7373–7.

Storz, J., and Zera, A.J. (2011). 'Experimental approaches for evaluating the contribution of amino acid mutations to adaptive phenotypic evolution', in V. Orgogozo and M.V. Rockman, eds, *Methods for Evolutionary Genetics*. Springer-Verlag, New York. 377–96.

Thomson, M.J., Tai, T.H., Mcclung, A.M., *et al.* (2003). 'Mapping quantitative trait loci for yield, yield components and morphological traits in an advanced backcross population between *Oryza rufipogon* and the *Oryza sativa* cultivar Jefferson', *Theoretical and Applied Genetics* **107**: 479–3.

Trefilov, A., Berard, J., Krawczak, M., and Schmidtke, J. (2000). 'Natal dispersal in rhesus macaques is related to serotonin transporter gene promoter variation', *Behavior Genetics*, **30**: 295–301.

van Noordwijk, A.J. (1984). 'Problems in the analysis of dispersal and a critique on its 'heritability' in the great tit', *Journal of Animal Ecology* **53**: 533–44.

van Noordwijk, A.J., Pulido, F., Helm, B., *et al.* (2006). 'A framework for the study of genetic variation in migratory behavior', *Journal of Ornithology* **147**: 221–33.

Venable, D.L. (1985). 'The evolutionary ecology of seed heteromorphism', *American Naturalist* **126**: 577–95.

Venable, D.L. (1987). 'The ecology of seed heteromorphism in *Heterosperma pinnatum* in central Mexico', *Ecology* **68**: 65–76.

Venable, D.L. and Burquez, A. (1989). 'Quantitative genetics of size, shape, life-history, and fruit characteristics of the seed heteromorphic composite *Heterosperma pinnatum*. 1. Variation within and among populations', *Evolution* **43**: 113–24.

Venable, D.L. and Burquez, A. (1990). 'Quantitative genetics of size, shape, life-history, and fruit characteristics of the seed heteromorphic composite *Heterosperma pinnatum*. 2. Correlation structure', *Evolution* **44**: 1748–63.

Venable, D. L. and Levin, D. A. (1985). Ecology of achene dimorphism in *Heterotheca latifolia*. 1. Achene structure, germination and dispersal', *Journal of Ecology* **73**: 133–45.

Waser, P., and Jones, T. (1989). 'Heritability of dispersal in banner-tailed kangaroo rates, *Dipodomys spectabilis*', *Animal Behaviour* **37**: 987–91.

Weigensberg, I., and Roff, D.A. (1996). 'Natural heritabilities, can they be reliably estimated in the laboratory?', *Evolution* **50**: 2149–57.

Weisser, W.W. and Braendle, C. (2001). 'Body colour and genetic variation in winged morph production in the pea aphid', *Entomologia Experimentalis et Applicata* **99**: 217–23.

Wender, N.J., Polisetty, C.R. and Donohue, K. (2005). 'Density-dependent processes influencing the evolutionary dynamics of dispersal: A functional analysis of seed dispersal in *Arabidopsis thaliana* (Brassicaceae)', *American Journal of Botany* **92**: 960–71.

Wheat, C.W., Fescemeyer, H.W., Kvist, J. *et al.* (2011). 'Functional genomics of life history variation in a butterfly metapopulation', *Molecular Ecology* **20**: 1813–28.

Zera, A.J. (1981). 'Genetic structure of two species of waterstriders (Gerridae, Hemiptera) with differing degrees of winglessness', *Evolution* **35**: 218–25.

Zera, A.J. (2004). 'The endocrine regulation of wing polymorphism: State of the art, recent surprises, and future directions', *Integrative and Comparative Biology* **43**: 607–16.

Zera, A.J. (2005). 'Intermediary metabolism and life history trade-offs: Lipid metabolism in lines of the wing-polymorphic cricket, *Gryllus firmus*, selected for flight capability vs. early age reproduction'. *Integrative and Comparative Biology* **45**: 511–24.

Zera, A.J. (2006). 'Evolutionary genetics of juvenile hormone and ecdysteroid regulation in *Gryllus*, A case study in the microevolution of endocrine regulation', *Comparative Biochemistry and Physiology, Part A* **144**: 365–79.

Zera, A.J. (2009). 'Wing polymorphism in *Gryllus* (Orthoptera, Gryllidae): Proximate endocrine, energetic and biochemical mechanisms underlying morph specialization for flight vs. reproduction', in D.W. Whitman and T.N. Ananthakrishnan, eds, *Phenotypic Plasticity of Insects. Mechanisms and Consequences*. Science Publishers, Enfield (NH), 609–53.

Zera, A.J. (2011). 'Microevolution of intermediary metabolism, Evolutionary genetics meets metabolic biochemistry', *Journal of Experimental Biology* **214**: 179–90.

Zera, A.J., and Denno, R.F. (1997). 'Physiology and ecology of dispersal polymorphism in insects', *Annual Review of Entomology* **42**: 207–31.

Zera, A.J., and Harshman, L.G. (2001). Physiology of life history trade-offs in animals', *Annual Review of Ecology and Systematics* **32**: 95–126.

Zera, A.J., and Harshman, L.G. (2009). 'Laboratory selection studies of life-history physiology in insects', in T. Garland Jr and M. R. Rose, eds, *Experimental evolution. Methods and Applications.* University of California Press, Berkeley, 217–62.

Zera, A.J., and Harshman, L.G. (2011). 'Intermediary metabolism and the biochemical-molecular basis of life-history variation and trade-offs in two insect models', in T. Flatt and A. Heyland, eds, *Mechanisms of Life History Evolution.* Oxford University Press, Oxford, 311–28.

Zera, A.J., Innes, D., and Saks, M.E. (1983). 'Genetic and environmental determinants of wing polymorphism in the waterstrider *Limnoporous canaliculatus*', *Evolution* **37**: 513–22.

Zera, A.J., and Larsen, A. (2001). 'The metabolic basis of life history variation: Genetic and phenotypic differences in lipid reserves among life history morphs of the wing-polymorphic cricket, *Gryllus firmus*', *Journal of Insect Physiology* **47**: 1147–60.

Zera, A.J. and K.C. Tiebel. (1989). 'Differences in juvenile hormone esterase activity between presumptive macropterous and brachypterous *Gryllus rubens*: Implications for the hormonal control of winglength polymorphism', *Journal of Insect Physiology* **35**: 7–17.

Zera, A., and Zhao, Z. (2003). 'Life-history evolution and the microevolution of intermediary metabolism: Activities of lipid-metabolizing enzymes in life-history morphs of a wing-dimorphic cricket', *Evolution* **57**: 568–96.

Zera, A.J., Zhao, Z. and Kaliseck, K. (2007). 'Hormones in the field: Evolutionary endocrinology of juvenile hormone and ecdysteroids in field populations of the wing-dimorphic cricket *Gryllus firmus*', *Physiological and Biochemical Zoology* **80**: 592–606.

Zhao, Z., and Zera, A.J. (2002). 'Differential lipid biosynthesis underlies a tradeoff between reproduction and flight capability in a wing-polymorphic cricket', *Proceedings of the National Academy of Sciences USA* **99**: 16829–34.

Zhao, Z., and Zera, A.J. (2004). 'The hemolymph JH titer exhibits a large-amplitude, morph-dependent, diurnal cycle in the wing-polymorphic cricket, *Gryllus firmus*', *Journal of Insect Physiology* **50**: 93–102.

Zheng, C., Ovaskainen, O., and Hanski, I. (2009). 'Modeling single nucleotide effects in *phosphoglucose isomerase* on dispersal in the Glanville fritillary butterfly: Coupling of ecological and evolutionary dynamics,' *Philosophical Transactions of the Royal Society B* **364**: 1519–32.

CHAPTER 7

Evolution of genetically integrated dispersal strategies

Renée A. Duckworth

7.1 Spatio-temporally varying environments and the evolution of dispersal

Dispersal is costly, so all else being equal, it is beneficial for an individual to remain in its natal population. Yet if habitat quality varies over time and space—as is often the case for species that depend on successional or ephemeral habitat—then the benefits of escaping habitat of declining quality to find higher quality habitat more than compensates for the cost of dispersal. Thus a key factor leading to the evolution of dispersal is a spatio-temporally varying environment (Roff 1975; Johnson and Gaines 1990; McPeek and Holt 1992; Paradis 1998).

Extensive spatio-temporal environmental variation means that parents and offspring are likely to experience different environmental conditions, and thus the costs and benefits of dispersal are also likely to be distinct across generations. This should favour environmental contingency of dispersal and result in a lack of correspondence between generations in individual dispersal decisions. While some environmental contingency of dispersal is often observed (Gaines and McClenagham 1980; Hanski 1999; Ronce 2007), many studies also find a strong resemblance between parents and offspring and corresponding significant heritability of dispersal behavior (Roff and Fairbairn 2001; Hansson *et al.* 2003; Pasinelli *et al.* 2004; Sinervo *et al.* 2006; see Chapter 6 for a review). How can such constancy in dispersal behavior across generations be reconciled with selection for context-dependency of dispersal that should result from spatio-temporal environmental variation?

In this chapter, I explore the conditions that might favour both constancy and context-dependency in dispersal phenotypes across generations for species that experience spatio-temporally varying habitat. I focus on spatio-temporally varying environments because theoretical models suggest that environmental variation is the most common evolutionary cause of dispersal, and because spatial and temporal environmental variation is ubiquitous in natural systems. I first suggest that across-generation constancy in dispersal arises from consistent natural selection for close integration of dispersal with a suite of behavioral, morphological, and life-history traits that enable successful dispersal. I then explore the conditions that produce correlational selection on dispersal and other traits, and suggest that such correlational selection may ultimately lead to their genetic integration. Finally, I use western bluebirds as a model system to show how integrating ecological, quantitative genetic, and behavioral information can provide novel insights into the evolution of dispersal phenotypes.

7.2 Setting the stage for environment-independent expression of dispersal

If the extrinsic environment is the main factor influencing the costs and benefits of dispersal and there is temporal variation in environmental quality, then individuals should make their dispersal decision based solely on their assessment of current environmental conditions. However, in many organisms, an individual's own phenotype can also influence the costs and benefits of dispersal because individuals vary in a suite of morphological, behavioral, and

Dispersal Ecology and Evolution. First Edition. Edited by Jean Clobert *et al.*

life-history traits that influence dispersal and/or colonizing ability (Gaines and McClenagham 1980; Dingemanse *et al.* 2003; Rehage and Sih 2004; Garant *et al.* 2005; Phillips *et al.* 2006, Duckworth and Badyaev 2007; Skjelseth *et al.* 2007; Chapter 12). If traits that alter the costs and benefits of dispersal are heritable, then these costs and benefits are likely to be similar across generations, and parents and offspring should make similar dispersal decisions. Assuming dispersal is also heritable, this sets the stage for evolution of genetic integration of these traits.

Interactions among traits that increase organismal fitness or performance can lead to selection for functional integration and trait co-expression (Lande 1979, 1980; Cheverud 1996). Consistent correlational selection on an evolutionary timescale can lead to the evolution of genetic correlations between traits at the level of the population (Houle 1991; Atchley *et al.* 1994; Cheverud 1996). Thus, one explanation for a significant genetic component to dispersal strategies is that consistent natural selection for close integration of dispersal with a suite of other behaviors and morphologies that enable successful dispersal has favoured across-generation constancy in the expression of dispersal.

7.3 Correlational selection for phenotype-dependent dispersal

Dispersal polymorphisms are frequently observed in species that depend on ephemeral or successional habitat, especially when habitat patches are moderately stable allowing the persistence of multiple generations, and thus a benefit to a philopatric strategy, before the patch disappears (Harrison 1980; Crespi and Taylor 1990; Johnson and Gaines 1990; Roff 1994). These species must continually recolonize new habitat in order to survive, and older and newer habitat patches are expected to differ in density of conspecifics because new populations are typically colonized by just a few individuals, but population density often increases rapidly once a population is established (Figure 7.1A). These predictable changes in density over time mean that the earliest colonizers experience conditions that are distinct from what later generations will experience. Because highly dispersive individuals are most likely to colonize new habitat (Hanski *et al.* 2004, Phillips *et al.* 2006) and philopatric individuals are more likely to breed in older, well-established populations, this sets the stage for correlational selection on dispersal and the traits that influence the ability to breed in different density conditions (Olivieri *et al.* 1995). Dispersers should display traits that adapt them to colonizing new areas and to breeding in low-density conditions, and philopatric individuals should display traits that adapt them to remaining near their natal site and to breeding in high-density conditions (Olivieri *et al.* 1995, Hanski 1999; Figure 7.1B). Under these conditions, correlational selection will only be detectable when comparing fitness of individuals of different dispersal strategies across populations that vary in their stage of colonization.

Some of the most dramatic examples of phenotype-dependent dispersal occur in insects where wing dimorphism is common (Harrison 1980; Roff 1986; Zera and Denno 1997). These dimorphic species depend on ephemeral habitat, and winged individuals are able to disperse over long distances to colonize new habitat patches; however, the possession of wings and the flight muscles that power them carries a significant fitness cost in terms of female fecundity (Roff 1984; Langellotto *et al.* 2000). Extensive work on wing dimorphic sand crickets by Derek Roff and Daphne Fairbairn has shown that wing morphology is genetically correlated to fecundity, muscle histolysis, and flight propensity (for an overview, see Roff and Fairbairn 2001; Roff and Fairbairn 2007).

In vertebrates, such distinct locomotor morphs within a species are rare; however, dispersal is often correlated with behavioral, physiological, or life-history traits, such as aggression, cooperative behavior or reproductive investment, rather than morphological traits (Gaines and McClenagham 1980; Dingemanse *et al.* 2003; Duckworth and Badyaev 2007; Skjelseth *et al.* 2007; Chapter 12). Thus in vertebrates, correlational selection, rather than targeting co-expression of dispersal behavior and locomotory performance, may more commonly target co-expression of dispersal and traits that influence either colonizing ability or the ability to

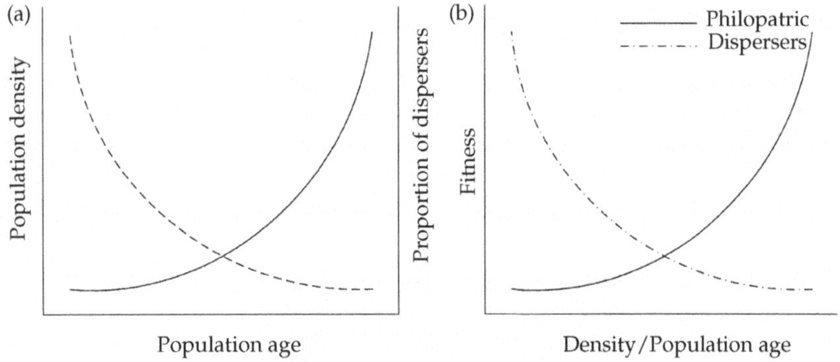

Figure 7.1 Conceptual framework for understanding the link between correlational selection on dispersal and other traits in the context of species that depend on ephemeral habitat. (a) During the process of colonizing new habitat patches, density of conspecifics should increase with population age (solid line) and proportion of dispersers should decrease with population age (dashed line) because new populations are colonized solely by dispersers and the proportion of dispersive individuals in the population decreases as offspring from the initial colonizing generations are recruited into the population as breeders. This sets up a situation where dispersive and philopatric individuals experience predictable differences in population density during the process of colonization. (b) Correlational selection for distinct dispersal strategies depends on population density and age. The link between population age and density means that dispersers should evolve a strategy (a suite of integrated behavioral, life-history, and morphological traits) that enables them to have high fitness in new populations where density of conspecifics is low, and philopatric individuals should evolve a strategy that enables them to have the high fitness in high density conditions that are commonly found in older populations. Trade-offs between dispersal-related traits and fecundity (see text) should prevent individuals from performing well in all density conditions.

perform in a high- or low-density population. Lack of distinct, easily observable, dispersal morphs in vertebrates may explain why phenotype-dependent dispersal is less well-documented in these groups compared to insect taxa. Moreover, the difficulty of recognizing and measuring the behavioral and life-history traits associated with dispersal in vertebrates has made studies of the genetic basis of phenotype-dependent dispersal difficult. The most detailed examination of the ecology and genetics of distinct dispersal strategies in a wild vertebrate has been carried out in western bluebirds (*Sialia mexicana*). We have used large-scale field experiments, measures of lifetime fitness, and pedigree-based quantitative genetic studies to show that western bluebirds have evolved distinct genetically integrated dispersal strategies that enables them to colonize new habitats continually while maintaining populations in older habitat patches.

7.4 Ecological context for evolution of distinct dispersal strategies in western bluebirds

Western bluebirds depend on tree cavities to breed—a limiting resource that historically was patchily distributed and ephemeral. Nest cavities occur at high densities following forest fire which creates suitable habitat for bluebirds by opening up understory vegetation and creating dead snags. Eventually, as the forest regrows, bluebirds are no longer able to breed in these habitat patches because snag density decreases and regrowth of the forest eliminates the open meadows bluebirds depend on to forage for insect prey (Power and Lombardo 1996; Guinan *et al.* 2000). Western bluebird's sister species, mountain bluebirds (*S. currucoides*), are frequently among the earliest colonizers following forest fires (Hutto 1995), whereas western bluebirds often show delayed patterns of colonization (Kotliar *et al.* 2007; Saab *et al.* 2007). Competition for nest cavities among these and other secondary cavity nesting species is intense and often involves aggressive displacement (Gowaty 1984, Newton 1994, Merilä and Wiggins 1995; Duckworth 2006b). Western bluebirds, while less dispersive and slower to find new habitat, are on average more aggressive than mountain bluebirds, and rapidly displace them when they colonize newly available habitat (Duckworth and Badyaev 2007).

The maintenance of these cycles of species replacement (Figure 7.2) depends at least partly on the evolution of two distinct dispersal strategies in western bluebirds where highly aggressive males tend to leave their natal populations and disperse to new areas to breed; whereas, non-aggressive males tend to remain in their natal population and eventually acquire a territory near relatives (Duckworth and Badyaev 2007; Duckworth 2008). Aggression and dispersal are functionally integrated because colonization of new populations by dispersing males requires the ability to outcompete earlier arriving heterospecific competitors for nesting sites and territories. Yet there is a trade-off between aggression and investment in parental care that results in a fecundity cost (Duckworth 2006b). Because of this trade-off, an individual's

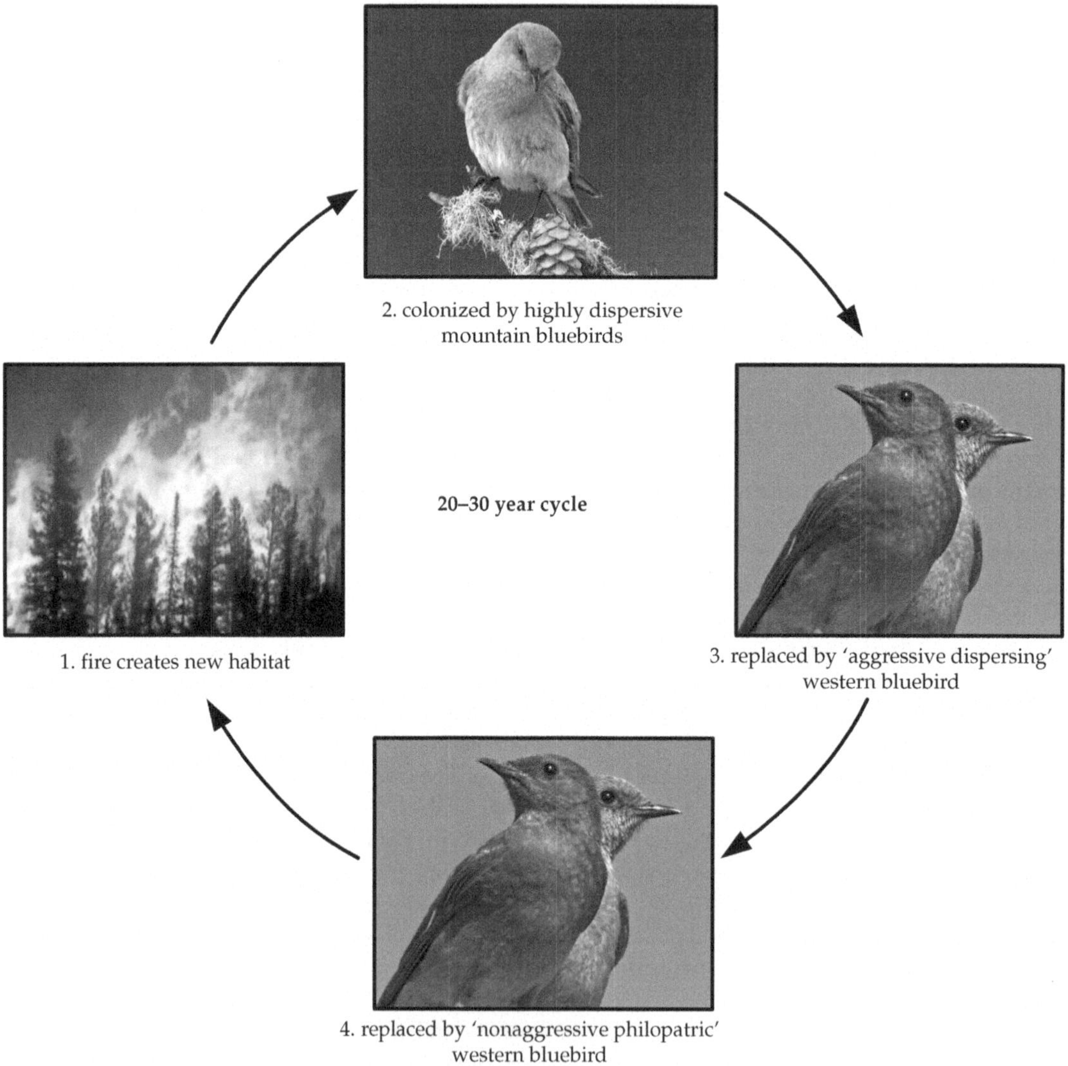

Figure 7.2 Ephemeral habitat and the evolution of both inter- and intra-specific differences in dispersal strategy produce cycles of species replacement. New habitat is created by fire (1), and is colonized first by highly dispersive mountain bluebirds (2). Mountain bluebirds are eventually replaced by aggressive dispersing type of western bluebird (3). In turn, aggressive dispersing western bluebirds are eventually replaced by non-aggressive philopatric type (4). Eventually, forest succession resets the cycle (steps 4 to 1). Burned forests can provide habitat for bluebirds for up to 30 years, approximately 20 bluebird generations.

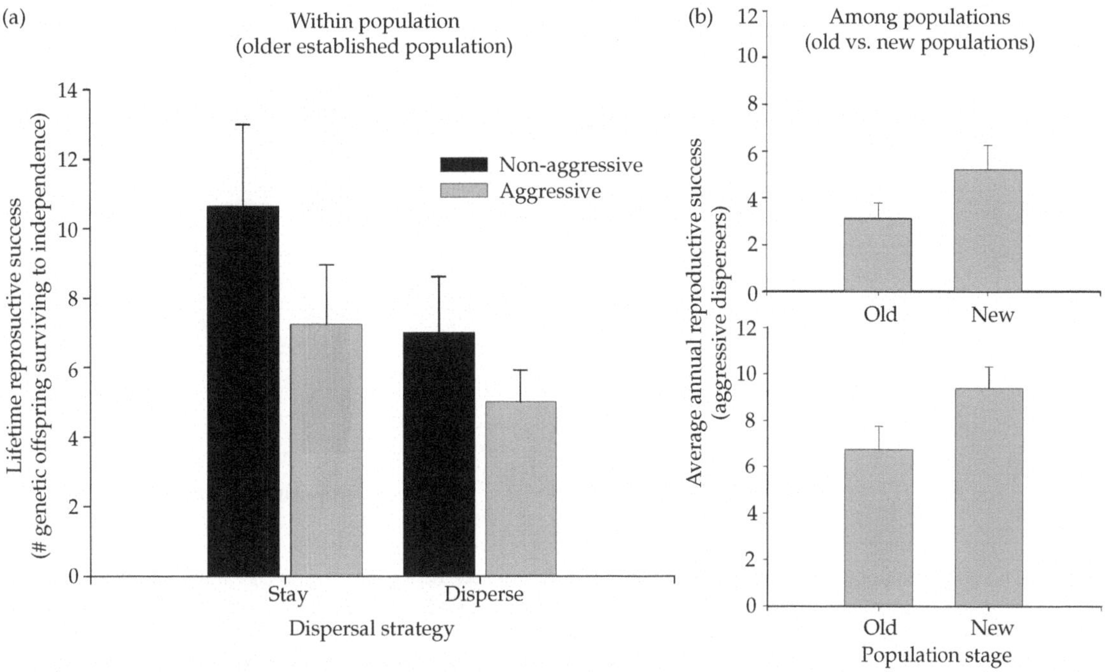

Figure 7.3 Correlational selection on aggression and dispersal is detectable when comparing populations of differing colonization stages. (a) Within an older, well-established population non-aggressive males that were born in the population and stay there to breed have the highest fitness (measured as lifetime genetically verified reproductive success); whereas aggressive dispersing males have the lowest fitness in this context. (b) When comparing fitness of aggressive dispersers across populations of differing colonization stage, they have higher fitness (measured as average annual reproductive success) when colonizing new areas. Data show replicated results where a new population was experimentally created within a few kilometers of an older well-established population in two geographically distinct regions. Only aggressive dispersers colonize new populations, so this is the only strategy compared in this population level comparison. Figure modified from Duckworth 2008.

performance in a particular population depends on both their aggressive phenotype and population stage (Duckworth 2006a, 2008). Non-aggressive males, despite being poor competitors, benefit from remaining in their natal population where they can gain a territory by cooperating with relatives. These males have higher fitness than aggressive males in older, well-established populations but rarely, if ever, are observed colonizing new populations (Figure 7.3A; Duckworth 2008); whereas aggressive males have the highest fitness when dispersing to new populations where density of conspecifics is low (Figure 7.3B; Duckworth 2008). The exact mechanism for higher fitness of aggressive males breeding in new populations is not known; however, they are likely to perform better in these populations because they acquire territories that are twice larger than in the older, highly dense, populations and thus have access to more resources (Duckworth 2008). In addition, the higher density of conspecifics in older populations means that aggressive males have more opportunities for agonistic encounters and thus may invest more in aggressive interactions at the expense of parental care in this context. The ecological importance of integration of these behaviors is emphasized by their role in facilitating the recent expansion of western bluebirds' range across the northwestern United States over the last 40 years (Duckworth and Badyaev 2007).

In concordance with these fitness costs and benefits, aggression and dispersal of western bluebird populations shift rapidly during the process of colonization (Duckworth 2008). Newly colonized populations comprise highly aggressive dispersers, and older well-established populations comprise less aggressive philopatric individuals (Figure 7.4).

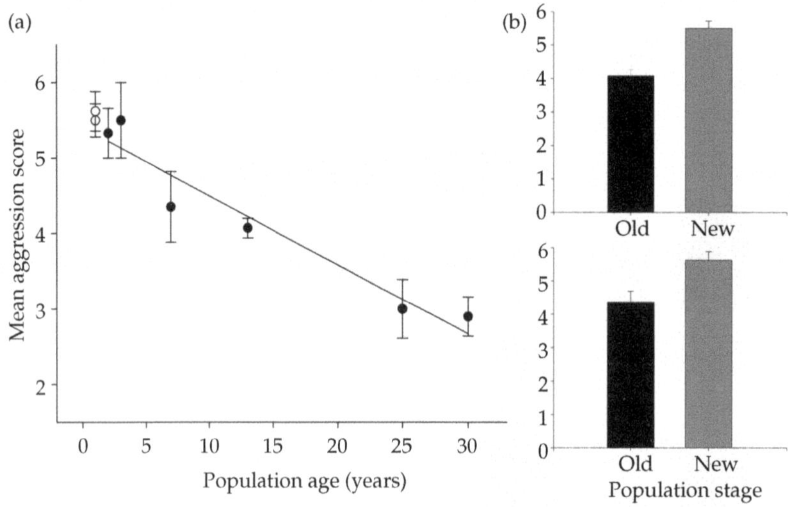

Figure 7.4 Newly colonized western bluebird populations were more aggressive than older, well-established populations. (a) In populations that were colonized naturally across a recent range expansion (black circles), younger populations were more aggressive than older populations. The new experimental populations in (a) (white circles) did not differ in aggression from newly colonized populations at the edge of the range. (b) Paired replicates of older, well-established, and newly created experimental populations show that new populations were more aggressive. Figure modified from Duckworth 2008.

These shifts are due to the non-random settlement of new populations by highly aggressive individuals followed by rapid decreases within these populations to lower levels of aggressive behavior (Duckworth and Badyaev 2007). Changes in aggression within populations are not due to flexibility in the expression of aggression as it does not systematically decrease with age and is not influenced by the current competitive environment (Duckworth 2006b; Duckworth and Badyaev 2007). Rapid phenotypic changes across populations are most likely due to a combination of nonrandom sorting of individuals during the initial stages of colonization in conjunction with natural selection favouring non-aggressive individuals once populations are established. Thus, these observations are concordant with the metapopulation model prediction that dispersal ability should shift rapidly as populations age (Olivieri *et al.* 1995; Hanski *et al.* 2004).

7.5 Proximate basis of distinct dispersal strategies: reconciling genetic variation and maternal effects

Consistent natural selection for co-expression of traits should lead to developmental and genetic integration of their components. Thus, if selection for co-expression of aggression and dispersal has been consistent for many generations, then quantitative genetic models predict that this should lead to the evolution of genetic integration of aggression and dispersal (Houle 1991; Atchley *et al.* 1994; Cheverud 1996).

To determine whether aggression and dispersal were genetically integrated in western bluebirds, we used use a complex multi-generational pedigree of individually marked western bluebirds to assess heritability of dispersal and aggression, and to determine whether the link between these traits is the result of their common dependence on genetic or environmental factors. The pedigree contained 873 individuals, spanned six generations, and contained both maternal and paternal half-sibs (due to both extra-pair paternity and divorce), as well as full and half-sibs reared in different nests (within and across breeding seasons), enabling estimates of heritability as well as genetic and environmental correlations. Analysis indicated that there was significant genetic variation for these behaviors and that they are positively genetically correlated (Duckworth and Kruuk 2009; Table 7.1), such that dispersive fathers were more likely to produce

Table 7.1 Components of variance (V_A, additive genetic variance and V_R, residual variance), heritability (h^2), and the phenotypic (r_P), genetic (r_G), and environmental correlations (r_E) for aggression and dispersal using a bivariate restricted maximum likelihood model. Standard errors are in parentheses. (*$P < 0.05$, **$P < 0.01$, P values estimated from likelihood ratio tests). Dispersal was considered a 'threshold trait' which assumes that it is underlain by a continuous distribution. Modified from Duckworth and Kruuk 2009.

Trait	Mean	V_A	V_R	h2	r_P	r_G	r_E
Aggression	4.02 (0.09)	0.89** (0.32)	0.82 (0.28)	0.52 (0.17)**	0.22 (0.09) *	0.55 (0.24)*	−0.21 (0.29)
Dispersal	0.57 (0.04)	0.13** (0.05)	0.08 (0.04)	0.60 (0.20)**			

aggressive offspring and philopatric fathers were more likely to produce non-aggressive offspring. This result is consistent with the idea that long-term functional integration of these traits has lead to their genetic integration.

Genetic integration of aggression and dispersal makes sense in the context of long-term selection due to functional integration of these traits; however, because the local availability of nest cavities is dynamic across years, environmental contingency in the expression of dispersal should also be favoured. For example, as density of bluebirds in newly colonized areas increases over time, resource availability decreases which means that the conditions that aggressive colonizers experience are different from the conditions experienced by later generations. Thus, even though there is selection for co-expression of aggression and dispersal which favours across-generation constancy in the link between these behaviors, there should be some environmental contingency in the specific strategy an individual pursues. Yet previous work has shown that expression of aggression in adulthood is not influenced by changes in resource availability (Duckworth 2006a). Moreover, the decision to disperse occurs during the juvenile period in the first few months of an individual's life. These observations point toward a possible role for maternal effects in the expression of dispersal strategies because if environmentally contingent expression occurs, it should have an effect early in ontogeny.

Maternal effects can enable breeding females to influence offspring phenotype in response to changes in current environmental conditions, and may be particularly important in species that experience temporal or spatial environmental variation (e.g. Sutherland 1969; Donohue 1999; Galloway 2005; Duckworth 2009). Pedigree-based quantitative genetic analysis did not detect a significant maternal effect on aggression or dispersal; however, in subsequent analyses, we showed that this was due to a significant effect of offspring laying order that increases variance in dispersal strategy among siblings of the same nest such that males from early-laid eggs were more aggressive and dispersive in adulthood compared to males from later-laid eggs (Duckworth 2009; Figure 7.5). In quantitative genetic models, maternal effects are typically estimated by looking for a greater similarity among offspring of the same mother over and above what one would expect from their genetic relatedness (Kruuk 2004). As a consequence, maternal effects that increase variance among offspring cannot be detected using an animal model approach.

The mechanism underlying the correlation between laying order and aggression is currently not known; however, other studies suggest that hormones or other substances deposited in the egg are likely to be important. In birds, breeding females lay one egg a day and changes in female behavior and physiology during oogenesis can lead to a gradient in hormones, carotenoids and morphogens among sequentially laid eggs in a clutch (Schwabl 1993; Lipar *et al.* 1999; Reed and Vleck 2001; Badyaev *et al.* 2006; 2008) which in turn has the potential to affect oocyte sex determination (Rutkowska and Badyaev 2008) and offspring phenotype (Schwabl 1996; Forstmeier *et al.* 2004; Groothius *et al.* 2004; Badyaev 2005; Amdan *et al.* 2006). Most importantly, it is thought that changing environmental conditions can influence fluctuations in hormone levels that occur when females are transitioning between oogenesis and egg laying, and can lead to differential allocation of maternal resources to oocytes of different ovulation order (Schwabl 1993; Groothius and Schwabl 2002; Müller *et al.* 2002; Badyaev *et al.* 2006; Rutkowska and Badyaev 2008).

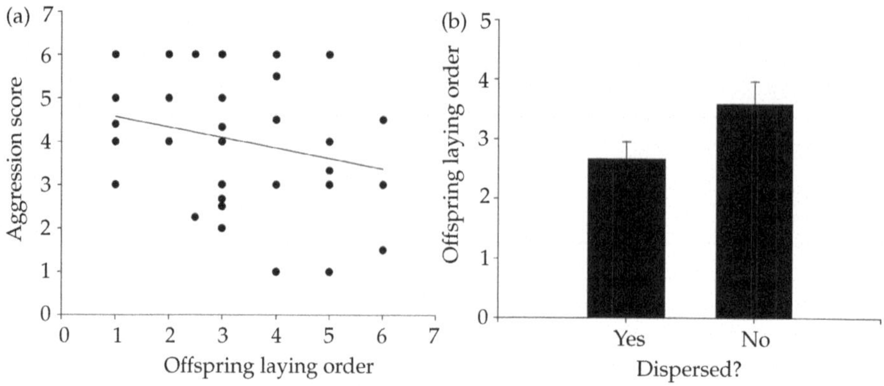

Figure 7.5 Maternal effect on expression of offspring dispersal ability. (a) Male aggression in adulthood is significantly correlated to his laying order, and (b) Males from earlier laid eggs were more likely to disperse away from their natal territory compared to males from later laid eggs. Figure modified from Duckworth 2009.

These observations raised the possibility that early ontogeny maternal effects on offspring dispersal strategy might be adaptive in the context of the dynamic process of population colonization. To test this idea, we carried out a series of field experiments to mimic different stages of colonization within a single population (Figure 7.6). In newly colonized populations, density of conspecifics is low and there are many open territories where non-aggressive individuals can acquire a territory near their parents. Under these conditions, producing male offspring that are non-aggressive and remain in their natal population is adaptive. In contrast, in older, well-established populations, density of bluebirds is saturated and there are few, if any, territories available for male offspring to acquire near their parents. Under these conditions, producing aggressive male offspring that are good at colonizing new areas is adaptive. From the mother's perspective, the difference between these two colonization stages is the local availability of nest cavities. Thus to determine whether the link between aggression and laying order was adaptive, we experimentally mimicked these different colonization stages by manipulating the local availability of nest boxes. For one group of birds, we increased the number of nest boxes on their territory, and for the other group, we left only a single nest box.

Results of this experiment showed that females adaptively adjusted their son's dispersal phenotype depending on the local availability of resources (Duckworth 2009). Females breeding under conditions of low nest cavity availability produced sons early in the laying order (i.e. they produced aggressive dispersers), whereas females breeding under conditions of high nest cavity availability produced sons later in the laying order (i.e. they produced non-aggressive stayers; Figure 7.7A). We also acquired data from a population in which nest cavity availability varied naturally and found the same pattern of female adjustment of offspring laying order in relation to the number of nest cavities on their territories (Figure 7.7B). The results of these studies suggested that maternal effects act as a bridge that enables the parental generation of colonizers to produce offspring that are less aggressive and more likely to remain philopatric and take advantage of the surplus of resources acquired by their parents. At the same time, by over-producing aggressive males when conditions are crowded or declining in quality, females ensure that their sons are prepared to leave their natal population, and search for and colonize a new habitat patch. Thus, maternal effects on male dispersal phenotype, by enabling close tracking of resource availability, are likely to have played a key role, not only in this species' recent range expansion, but also in this species' persistence.

Our long-term studies of western bluebirds show that the expression of dispersal strategies is

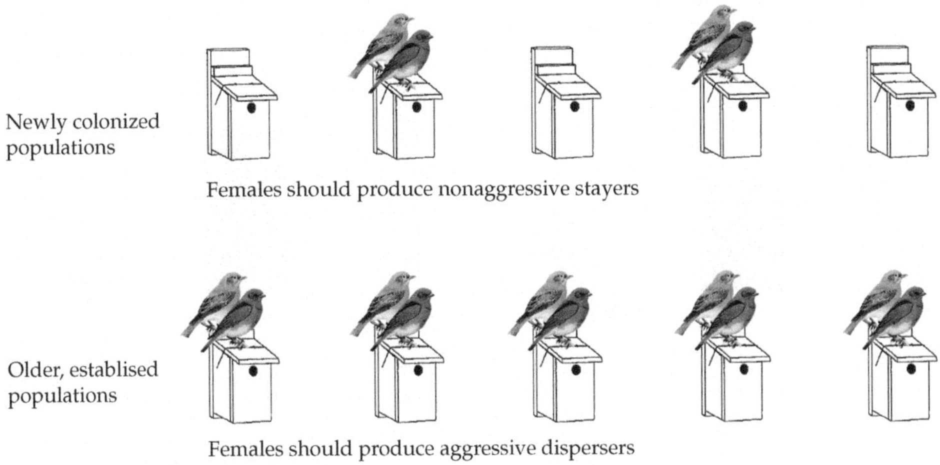

Figure 7.6 Experimental rationale for mimicking different colonization stages. Each box represents a territory and boxes with birds on them represent occupied territories. In newly colonized populations, density of bluebirds is low and there are many open territories—in this situation females would benefit by producing nonaggressive sons that can acquire a territory nearby. In older, established populations, density is high and territories are saturated—in this situation, females should produce aggressive sons that disperse.

influenced by genetic variation, but is also contingent on local environmental conditions. Such simultaneous across-generation stability and contingency of expression suggests that while the general propensity to disperse and be aggressive may be set by genetic variation, there is some flexibility of expression such that maternal effects can act to shift offspring toward one strategy or another depending on current environmental conditions. Thus genetic integration of the strategies ensures that the link between the appropriate combination of expression is maintained across generations while maternal effects fine-tune the expression of strategies to match environmental context.

To what extent can these findings inform our understanding of the evolution of dispersal strategies in other vertebrate taxa? The link between dispersal and aggression in this system is analogous in several ways to distinct dispersal polymorphisms described in many insect and plant species (Sorensen 1978; Harrison 1980; Venable and Levin 1985; Zera and Denno 1997). In these other taxa, dispersal is linked to traits that increase either mobility or colonization ability and is not just integrated with functional traits (e.g. winged versus non-winged insect morphs), but because of a trade-off between dispersal-related traits and fecundity (Roff 1984) is also linked to distinct life-history strategies. In western bluebirds, aggression is costly in terms of fecundity because highly aggressive males invest less in parental care (Duckworth 2006b) suggesting that the link between aggression and dispersal is but one axis of a suite of interrelated behaviors that have evolved as a consequence of the fundamental life-history trade-offs associated with dispersal. Finally, similar to our findings in western bluebirds, species that display distinct dispersal morphologies often depend on ephemeral habitat and variation in dispersal strategy is maintained in these systems through spatial and temporal variation in habitat availability (Roff 1986; Johnson and Gaines 1990; McPeek and Holt 1992). Such spatio-temporal variation in selection pressures can not only maintain variation in these dispersal-related traits, but on a regional scale can also produce correlational selection for their co-expression (Olivieri *et al.* 1995). The similarities between distinct behavioral strategies of western bluebirds and dispersal dimorphisms of insect taxa suggest that the evolution of phenotype-dependent dispersal is a common solution to the life-history trade-offs associated with dispersal in species that depend on ephemeral habitat.

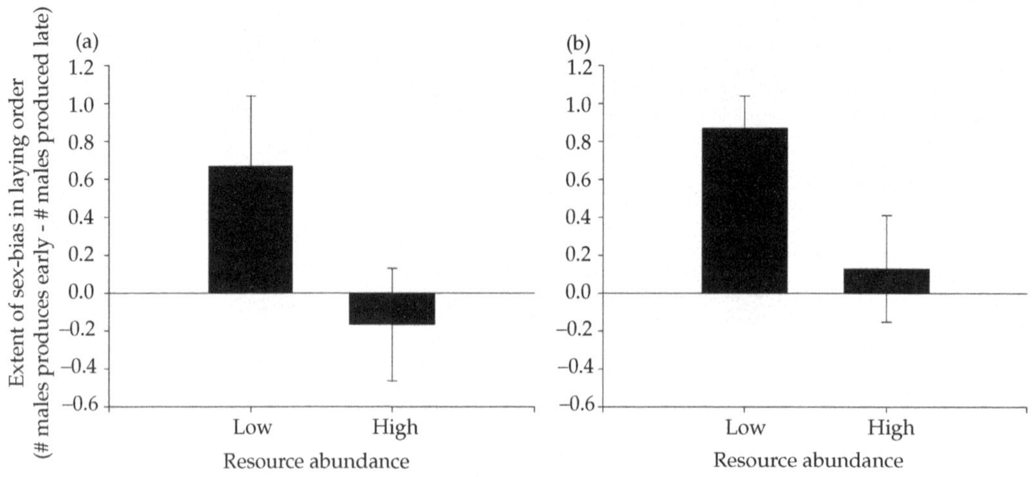

Figure 7.7 Evidence for adaptive adjustment of offspring dispersal ability. In both (a) experimental, and (b) naturally varying populations, the order in which females produced males depended on local nest cavity availability. When nest cavity availability was low, females produced males earlier, and when nest cavity availability was high, females produced males later. This sex-biased adjustment of offspring laying order was adaptive because earlier laid males are adapted to disperse; whereas, later-laid males are adapted to remain in their natal population. Figure modified from Duckworth 2009.

7.6 Concluding remarks

In this chapter, I suggest that the key to understanding both across generation constancy and contingency in dispersal is the recognition that individuals vary in an array of traits that influence dispersal ability and thus often express phenotype-dependent dispersal. Because an individual's own phenotype can influence the costs and benefits of dispersal, over the long term, this may favour genetic integration of dispersal and other traits.

The widespread occurrence of phenotype-dependent dispersal brings up the question of whether dispersal is a trait in its own right or whether it is an emergent property of individuals matching their phenotypes to the environmental contexts where they will perform best. In the latter scenario, the dispersal decision is reliably transmitted across generations because inheritance of particular traits makes offspring assessment of current environmental conditions similar to their parents' assessment. If this is true, then measures of genetic variation for dispersal behavior might reflect genetic variation for the suite of traits that influence dispersal rather than dispersal behavior *per se*. This is an important distinction because it determines how we investigate the proximate basis of variation in dispersal; i.e. whether we focus on the propensity of individuals to undertake directed movements, or focus on how behavioral, morphological and life-history traits interact to lead to a particular dispersal decision. Regardless, it is important to recognize that like most traits, dispersal tendency comprises multiple components, and thus understanding the developmental, functional, and genetic integration among these components is crucial to understanding its evolution.

Acknowledgements

I am grateful to A. Badyaev and an anonymous reviewer for comments which improved this chapter. This work was supported by funding from National Science Foundation grant 0918095.

References

Amdan, G.V., Csondes, A., Fondrk, M.K. (2006). 'Complex social behaviour derived from maternal reproductive traits', *Nature* **439**:76–8.

Atchley, W.R., Xu, S., and Vogl, C. (1994). 'Developmental quantitative genetic models of evolutionary change', *Developmental Genetics* **15**:92–103.

Badyaev, A.V. (2005). 'Maternal inheritance and rapid evolution of sexual size dimorphism: passive effects or active strategies?', *American Naturalist* **166**:S17–30.

Badyaev, A.V., Acevedo Seaman, D.A., Navara, K.J. (2006). 'Evolution of sex-biased maternal effect in birds: III. Adjustment of ovulation order can enable sex-specific allocation of hormones, carotenoids, and vitamins', *Journal of Evolutionary Biology* **19**:1044–57.

Cheverud, J.M. (1996). 'Developmental integration and the evolution of pleiotropy', *American Zoologist* **36**:44–50.

Crespi, B.J. and Taylor, P.D. (1990). 'Dispersal rates under variable patch density', *American Naturalist* **135**:48–62.

Dingemanse, N.J., Both, C., van Noordwijk, A.J. (2003). 'Natal dispersal and personalities in great tits (*Parus major*)', *Proceedings of the Royal Society B* **270**:741–7.

Donohue, K. (1999). 'Seed dispersal as a maternally influenced character: mechanistic basis of maternal effects and selection on maternal characters in an annual plant', *American Naturalist* **154**:674–89.

Duckworth, R.A. (2006a). 'Aggressive behavior affects selection on morphology by influencing settlement patterns in a passerine bird', *Proceedings of the Royal Society B* **273**:1789–95.

Duckworth, R.A. (2006b). 'Behavioral correlations across breeding contexts provide a mechanism for a cost of aggression', *Behavioral Ecology* **17**:1011–19.

Duckworth, R.A. (2008). 'Adaptive dispersal strategies and the dynamics of a range expansion', *American Naturalist* **172**:S4–17.

Duckworth, R.A. (2009). 'Maternal effects and range expansion: A key factor in a dynamic process?' *Philosophical Transactions of the Royal Society B* **364**:1075–86.

Duckworth, R.A. and Badyaev, A.V. (2007). 'Coupling of aggression and dispersal facilitates the rapid range expansion of a passerine bird', *Proceedings of the National Academy of Sciences USA* **104**:15017–22.

Duckworth, R.A. and Kruuk, L.E.B. (2009). 'Evolution of genetic integration between dispersal and colonization ability in a bird', *Evolution* **63**:968–77.

Forstmeier, W., Coltman, D.W., and Birkhead, T.R. (2004). 'Maternal effects influence the sexual behavior of sons and daughters in the zebra finch', *Evolution* **58**:2574–83.

Gaines, M.S. and McClenagham, L.R. (1980). 'Dispersal in small mammals', *Annual Review of Ecological Systems* **11**:163–96.

Galloway, L.F. (2005). 'Maternal effects provide phenotypic adaptation to local environmental conditions', *New Phytologist* **166**:93–100.

Garant, D., Kruuk, L.E.B., Wilkin, T.A. (2005). 'Evolution driven by differential dispersal within a wild bird population', *Nature* **433**:60–4.

Gowaty, P.A. (1984). 'House sparrows kill eastern bluebirds', *Journal of Field Ornithology* **55**:378–80.

Groothius, T.G.G., Müller, W., von Engelhardt, N. (2004). 'Maternal hormones as a tool to adjust offspring phenotype in avian species', *Neuroscience and Biobehavioural Reviews* **29**:329–52.

Groothius, T.G.G. and Schwabl, H. (2002). 'Determinants of within- and among-clutch variation in levels of maternal hormones in Black-Headed Gull eggs', *Functional Ecology* **16**:281–9.

Guinan, J.A., Gowaty, P.A., and Eltzroth, E.K., eds (2000). *Western Bluebird* (*Sialia mexicana*). Birds of North America, Inc., Philadelphia, PA.

Hanski, I. (1999). *Metapopulation Ecology*. Oxford University Press, Oxford.

Hanski, I., Erälahti, C., Kankare, M. *et al*. (2004). 'Variation in migration propensity among individuals maintained by landscape structure', *Ecology Letters* **7**:958–66.

Hansson, B., Bensch, S. and Hasselquist, D. (2003). 'Heritability of dispersal in the great reed warbler', *Ecology Letters* **6**:290–4.

Harrison, R.G. (1980). 'Dispersal polymorphisms in insects', *Annual Review of Ecological Systems* **11**:95–118.

Houle, D. (1991). 'Genetic covariance of fitness correlates: what genetic correlations are made of and why it matters', *Evolution* **45**:630–48.

Hutto, R.L. (1995). 'Composition of bird communities following stand-replacement fires in northern rocky mountain (USA) conifer forests', *Conservation Biology* **9**:1041–58.

Johnson, M.L. and Gaines, M.S. (1990). 'Evolution of dispersal: theoretical models and empirical tests using birds and mammals', *Annual Review of Ecological Systems* **21**:449–80.

Kotliar, N.B., Kennedy, P.L. and Ferree, K. (2007). 'Avifaunal responses to fire in southwestern montane forests along a burn severity gradient', *Ecological Applications* **17**:491–507.

Kruuk, L.B. (2004). 'Estimating genetic parameters in natural populations using the 'animal model', *Philosophical Transactions of the Royal Society B* **359**:873–90.

Lande, R. (1979). 'Quantitative genetic analysis of multivariate evolution, applied to brain: body size allometry', *Evolution* **33**:402–16.

Lande, R. (1980). 'The genetic covariance between characters maintained by pleiotropic mutations', *Genetics* **94**:203–14.

Langellotto, G.A., Denno, R.F. and Ott, J.R. (2000). 'A trade-off between flight capability and reproduction in males of a wing-dimorphic insect', *Ecology* **81**:865–75.

McPeek, M.A. and Holt, R.D. (1992). 'The evolution of dispersal in spatially and temporally varying environments', *American Naturalist* **140**:1010–27.

Merilä, J. and Wiggins, D.A. (1995). 'Interspecific competition for nest holes causes adult mortality in the collared flycatcher', *Condor* **97**:445–50.

Müller, W., Eising, C.M., Dijkstra, D. *et al.* (2002). 'Sex differences in yolk hormones depend on maternal social status in Leghorn chickens (*Gallus gallus domesticus*)', *Proceedings of the Royal Society B* **269**:2249–55.

Newton, I. (1994). 'The role of nest sites in limiting the numbers of hole-nesting birds—a review', *Biological Conservation* **70**:265–76.

Olivieri, I., Michalakis, Y. and Gouyon, P.-H. (1995). 'Metapopulation genetics and the evolution of dispersal', *American Naturalist* **146**:202–28.

Paradis, E. (1998). 'Interactions between spatial and temporal scales in the evolution of dispersal rate', *Evolutionary Ecology* **12**:235–44.

Pasinelli, G., Schiegg, K. and Walters, J.R. (2004). 'Genetic and environmental influences on natal dispersal distance in a resident bird species', *American Naturalist* **164**:660–9.

Phillips, B.L., Brown, G.P., Webb, J.K. *et al.* (2006). 'Invasion and the evolution of speed in toads', *Nature* **439**:803.

Power, H.W. and Lombardo, M.P. (1996). 'Mountain Bluebird'. in A. Poole and F. Gill, ed, *The Birds of North America.* Birds of North America, Inc, Philadelphia, PA, 1–21.

Rehage, J.S. and Sih, A. (2004). 'Dispersal behavior, boldness, and the link to invasiveness: a comparison of four *Gambusia* species', *Biological Invasions* **6**: 379–91.

Roff, D.A. (1975). 'Population stability and the evolution of dispersal in a heterogeneous environment', *Oecologia* **19**: 217–37.

Roff, D.A. (1984). 'The cost of being able to fly: A study of wing polymorphism in two species of crickets. *Oecologia* **63**:30–7.

Roff, D.A. (1986). 'The evolution of wing dimorphism in insects', *Evolution* **40**:1009–20.

Roff, D.A. (1994). 'Habitat persistence and the evolution of wing dimorphism in insects', *American Naturalist* **144**: 772–98.

Roff, D.A. and Fairbairn, D.J. (2001). 'The genetic basis of dispersal and migration and its consequences for the evolution of correlated traits', in J. Clobert, E. Danchin, A.A. Dhondt, eds, *Dispersal.* Oxford University Press, New York, 191–202.

Roff, D.A. and Fairbairn, D.J. (2007). 'The evolution and genetics of migration in insects', *Bioscience* **57**: 155–64.

Ronce, O. (2007). 'How does it feel to be like a rolling stone? Ten questions about dispersal evolution', *Annual Review of Ecological Systems* **38**:231–53.

Rutkowska, J. and Badyaev, A.V. (2008). 'Meiotic drive and sex determination: molecular and cytological mechanisms of sex ratio adjustment in birds', *Philosophical Transactions of the Royal Society B* **363**:1675–86.

Saab, V.A., Russell, R.E. and Dudley, J.G. (2007). 'Nest densities of cavity-nesting birds in relation to postfire salvage logging and time since wildfire', *Condor* **109**: 97–108.

Schwabl, H. (1993). 'Yolk is a source of maternal testosterone for developing birds', *Proceedings of the National Academy of Sciences USA* **90**:11446–50.

Schwabl, H. (1996). 'Maternal testosterone in the avian egg enhances postnatal growth', *Comparative Biochemistry and Physiology Part C* **114A**:271–6.

Sinervo, B., Calsbeek, R., Comendant T. (2006). 'Genetic and maternal determinants of effective dispersal: the effect of sire genotype and size at birth in side-blotched lizards', *American Naturalist* **168**:88–99.

Skjelseth, S., Ringsby, T.H., Tufto, J. (2007). 'Dispersal of introduced house sparrows Passer domesticus: an experiment', *Proceedings of the Royal Society B* **274**:1763–71.

Sorensen, A.E. (1978). 'Somatic polymorphism and seed dispersal', *Nature* **276**:174–6.

Sutherland, O.R.W. (1969). 'The role of crowding in the production of winged forms by two strains of the pea aphid, *Acyrthosiphon pisum*', *Journal of Insect Physiology* **15**:1385–410.

Venable, D.L. and Levin, D.A. (1985). 'Ecology of achene dimorphism in *Heterotheca latifola*: I. Achene structure, germination and dispersal', *Journal of Ecology* **73**: 133–45.

Zera, A.J. and Denno, R.F. (1997). 'Physiology and ecology of dispersal polymorphism in insects', *Annual Review of Entomology* **42**:207–30.

CHAPTER 8

Dispersal genetics: emerging insights from fruitflies, butterflies, and beyond

Christopher W. Wheat

8.1 Introduction

In our current post-genomics era, how can we gain general insights into the genomic underpinnings of dispersal variation and its ramifications upon ecological and evolutionary dynamics? One route is to identify and study the genetic variation affecting dispersal across a range of diverse species, thereby gathering the understanding needed for synthesis and generalization. Recent work demonstrates how genetic variation that correlates with individual metabolic rate and locomotion propensity also scales up to correlate with dispersal differences among demes, and even eco-evolutionary dynamics (Hanski 2011). Thus if genetic effects on locomotion are shared across species, these genes may also play an important role in dispersal variation among diverse taxa. In this light, those interested in the genetics of dispersal may have much to gain by delving into the literature focused on the genetics of locomotion in model genomic species. Here I review some of the functional genomic studies focused on locomotion differences in the model genomic system of *Drosophila melanogaster* and some other species. These findings are compared with single locus effects on locomotion and dispersal differences in butterflies. Important insights into the architecture of genetic variation and causality emerge, providing perspective for future studies of locomotion in relation to dispersal.

Knowledge of genetic polymorphism with measurable effects on dispersal variation provides a means of distinguishing among individuals for this complex phenotype that in many cases may be morphologically cryptic and difficult to quantify. Even a nascent genetic understanding provides a means to explore the range of possible trade-offs and interactions dispersal phenotypes exhibit (e.g. Chapters 10–14), as well as insights into the selection regime acting on the variation, and the age of the variation under selection (Storz and Wheat, 2010; Watt, 2000). In this chapter, I argue that a candidate gene approach, which can provide these insights, could greatly facilitate general advances in dispersal research (following much of the same line of argument for the study behavioural ecological traits; see Fitzpatrick *et al.* 2005).

Genetic polymorphisms with identified effects on specific phenotypes, whether through strong correlations or established causations, and which are likely to have similar effects in other species, are called candidate genes in recognition of their likely candidate effect across divergent, unstudied taxa. For example, bone morphology gene Bmp4, originally identified as affecting chicken beak morphology, has been shown to affect beak morphology variation in Darwin's finches (Abzhanov *et al.* 2004), as well as jawbone morphology in fish (Parsons and Albertson 2009). Similar generality of effect by a single gene (*Mcr1*) on fur and feather colour variation is seen across diverse taxa (Majerus and Mundy 2003). Candidate gene studies of morphologically conspicuous phenotypes are greatly facilitated by accurate genotype and phenotype relationships and the observation that developmental traits are likely to be very conserved among taxa (Shubin *et al.* 2009).

Morphologically cryptic phenotypes are inherently more difficult to study, especially when environmental effects are strong, but genetic understanding of

Dispersal Ecology and Evolution. First Edition. Edited by Jean Clobert *et al.*

such phenotypes can change this. For example, although variation in disease resistance remains cryptic until infection, candidate gene studies of MHC variation have revolutionized the study of vertebrate ecological and conservation genetics (Piertney and Oliver 2006). Dispersal phenotypes range from morphologically conspicuous to inconspicuous, leading to many proxies of dispersal quantification. Morphologically inconspicuous dispersal phenotypes complicate field and molecular studies because robust prediction and identification of those having higher or lower genetic predisposition to disperse, prior to, or after their dispersal, can be difficult. However, recent work in *M. cinxia* has gained important insights that facilitate the study of morphologically inconspicuous dispersal phenotypes (Hanski, 2011).

These developments in *M. cinxia* rest upon the pioneering work by W.B. Watt and colleagues in *Colias* butterflies over the last three decades, which documented strong correlations among genetic variation at the phosphoglucose isomerase (*Pgi*) gene, biochemical performance differences, flight performance variation, and adult fitness parameters, as well as developed functional genomic insights into the likely basis of these correlations (Watt 1977; Watt 1983; Watt *et al.* 1985; Watt *et al.* 2003; Wheat *et al.* 2006; Watt 2003). Using *Pgi* as a candidate gene, studies of *M. cinxia* have identified a strong correlation between allelic variation at *Pgi* and peak metabolic rate during flight while stimulating butterflies to fly in a small glass jar (Haag *et al.* 2005). Importantly, these findings also correlate with radio tracking measures of adult flight behaviour in the field (Niitepõld 2010; Niitepõld *et al.* 2009) and lead to sustained predictions of how *Pgi* alleles should be distributed among demes of a metapopulation based upon their expected effects on dispersal (Hanski and Saccheri 2006; Haag *et al.* 2005). Recently these findings have been even further extended to eco-evolutionary dynamics (Hanski, 2011). Importantly, the work on *Colias* butterflies strongly predicted PGI effects on peak metabolic rate metabolism during flight, which is exactly what has been observed in *M. cinxia*, even though the two species are approximately 80 million years divergent (Wheat *et al.* 2010). Thus, if genetic effects on locomotion are shared across diverse species in much the same way that morphological or behavioural traits are (Fitzpatrick *et al.* 2005), these genes may also play an important role in dispersal variation among taxa. In this light, those interested in the genetics of dispersal have much to gain by delving into the literature focused on the genetics of locomotion in model genomic species.

8.2 Functional genomic insights into locomotion

The study of dispersal has a long history in *Drosophila* flies, with estimates of dispersal rates dating back to the 1940s (Dobzhansky and Wright 1943; Powell *et al.* 1976). Over the past decade, studies have further focused on dispersal and related non-sexual behaviour across several *Drosophila* species (Sisodia and Singh 2005; Reaume and Sokolowski 2006; Sokolowski 2001), including documentation of temperature-dependent differences in emigration (Mikasa and Narise 1986). However, whether *Drosophila* locomotion differences are actually a good proxy for their dispersal is an open question as there is little, if any, field data on the subject. Genetic effects on realized dispersal differences may be obscured by many conspecific or environmental factors, e.g. see Baguette *et al.* 1996; Baguette *et al.* 1998, and certainly both active and passive (wind blown) movement can affect dispersal (Johnston 1982; Dobzhansky and Wright 1943). Rather than wade into these complexities arising from the small size of the fruit fly, my object here is to assess whether the empirical data on locomotion differences in *Drosophila* and other animals could be consistent with the gene to locomotion to dispersal relationship documented in butterflies. Beginning with a potentially ideal candidate gene for locomotion differences, attention then turns to QTL and microarray studies, followed by investigations of metabolic rate, flight performance, and central metabolism. Theses latter issues are of interest given the recent findings of allelic variation at *Pgi* being correlated with peak flight metabolic rate and dispersal differences in the butterfly *M. cinxia* (Case Study III, Chapter 23).

8.2.1 Candidate gene: the foraging gene example

D. melanogaster larvae when feeding on yeast paste exhibit natural and heritable polymorphism in how much they travel. This difference has been tracked down to allelic variation in a gene called *foraging* (*for*), which is a cyclic GMP-dependent protein kinase (PKG) (Osborne *et al.* 1997; Sokolowski 2001). Protein kinases modify other proteins by chemically adding phosphate groups to specific amino acid residues, which alters those proteins' function, and therefore PKG plays a role in signal transduction and regulation, although its exact role is little understood (Fitzpatrick and Sokolowski 2004). Larvae with the dominant rover allele (for^{R}) express more of the PKG enzyme, which causes them to travel further while feeding and be more likely to leave a given food patch in search of another, compared to larvae with the recessive sitter allele (for^{s}) (Figure 8.1) (Osborne *et al.* 1997; Sokolowski 2001). After feeding upon a drop of sugar, for^{R} adults walk around more compared to for^{s} adults (Pereira and Sokolowski 1993). These differences appear to be behavioural and foraging related, as neither larvae nor adults exhibit any difference in distance travelled in the absence of food, as they both travel far (Figure 8.1) (Pereira and Sokolowski 1993; Osborne *et al.* 1997). They also appear to be very pleiotropic, as recent work in *D. melanogaster* reveals that gene expression patterns in metabolic pathways differ between $for^{s\ R}/for^{s}$ individuals in response to food stress (Kent *et al.* 2009).

The molecular level understanding of *for* is impressive, but it shadows the important strides that were made before much was known about this gene. Earlier studies of apple orchard populations found a 70:30 rover:sitter phenotype frequency ratio (Sokolowski 1982), and uncovered behaviour patterns suggesting maintenance of the polymorphism by fluctuating selection (Sokolowski *et al.* 1986). Subjecting independent collections of equal frequency rover:sitter founders to high and low density conditions, Sokolowski and colleagues found that such conditions repeatedly selected for either the rover or the sitter phenotypes respectively (Sokolowski *et al.* 1997). Dissecting larval competition and selection dynamics even farther, recent work has studied the action of negative density-dependent selection, where a given allele is favoured only when rare. Larvae reared at equal densities but with different relative frequencies of rover versus sitter alleles revealed that each morph did best when rare. Importantly, these differences were environment dependent, as they were only observed under low nutrient conditions (Fitzpatrick *et al.* 2007).

Once the genetic basis of the rover/sitter phenotype was discovered, it became an ideal candidate gene for comparative study. PKG genes with *for*-like effects have been now been identified and studied in bees, ants, and nematodes with important results (Ingram *et al.* 2005; Kodaira *et al.* 2009; Lucas and Sokolowski 2009; Fujiwara *et al.* 2002; Ben-Shahar *et al.* 2002). Honeybee (*Apis mellifera*) nurses, who stay in the hive and tend larvae, have lower levels of PKG compared to foraging honeybees, which leave the hive in search of nectar and pollen. Honeybee nurses can be induced to forage by a pharmacological manipulation of the cGMP pathway which increases PKG (Ben-Shahar *et al.* 2002). Similar findings have been recently reported in the bumblebee (Kodaira *et al.* 2009). These results indicate that rather than allelic variation having an effect upon larvae, as is observed in *D. melanogaster*, changes in the expression of PKG gene can affect foraging related behaviour in adult stages of other insects. Related observations have been reported that identify yet another change in the mechanistic basis of PKG effects on foraging related phenotypes. Study of the worker-caste dimorphic ant *Pheidole pallidula* finds that the 'major' morphs, which are large and specialized in nest defence, have higher PKG levels than 'minor' morphs that are smaller and forage. However, environmental conditions can induce 'major' morphs to forage, resulting in a decrease in their PKG levels. Pharmacological manipulation that caused a lowering of PKG expression increased foraging in this case (Lucas and Sokolowski 2009). Similar results of lower PKG associated with increased foraging behaviour are

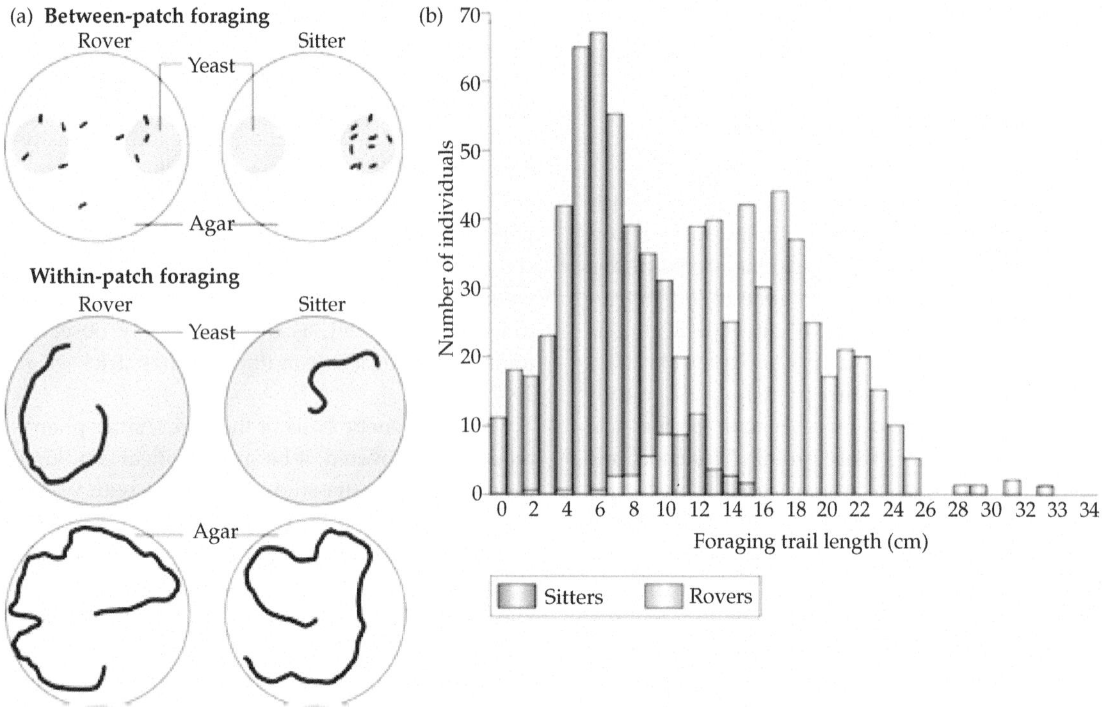

Figure 8.1 Foraging behaviour of rover and sitter phenotypes. (a) *Drosophila melanogaster* larvae feed upon patches of yeast paste spaced out on agar plate. Rover and sitter phenotypes differ in their between-patch foraging, as rovers have a high probability of leaving a food patch, while sitters will most likely remain within patches. Within-patch foraging is context dependent. Rovers travel farther than sitters when foraging within yeast patches, but when larvae are travelling on agar in search of food, both rover and sitter phenotypes move with equal speed over similar distance. (b) Distribution of foraging trail lengths for within yeast patch feeding shows a minimal overlap between rover and sitter phenotypes.

also found in the harvester ant (Ingram *et al.* 2005) and the nematode *Caenhorhabditis elegans* (Fujiwara *et al.* 2002). Thus, while higher expression levels of PKG are not always associated with increased foraging, variation upon some axis affecting PKG functional performance remains associated with foraging related behaviour. Many protein kinases are also post-translationally regulated, but currently such regulation is not known for these PGK genes.

These diverse observations suggest a conservation of function across diverse phyla (Lucas and Sokolowski 2009). Yet in none of these systems, though they differ in mechanism (allelic versus expression) and directionality (high or low levels can induce foraging) of effect, are the PKG effects absolute. In each example individuals can alter their foraging behaviour in response to environmental stimuli (Lucas and Sokolowski 2009). Thus, the pleiotropic nature of this gene and its environmental sensitivity are important considerations for experimental designs in novel species.

In summary, variation at PKG affects the locomotion behaviour of individuals within populations, but whether such PKG variation is actually involved in dispersal variation, where the movement of individuals results in gene flow, is an open question. Related open questions are how much dispersal variation could be driven by foraging stimulus, and whether foraging stimulated dispersal could be a general phenomenon. There is strong support for foraging stimulated dispersal in insects, with dispersal morphs appearing during conditions of high-density and low food (Zera and Denno 1997). If such density-dependent phase polyphenism is related to PKG gene expression, PKG variation could certainly be related to dispersal. Such an example is emerging from recent work on the desert locust (*Schistocerca gregaria*) that reports significantly higher PKG expression in the locust's gregarious, high-density form (Lucas *et al.* 2010). In its gregarious form, the desert locust swarms, and the resulting locus swarms are biblical in proportion and impact, having plagued human civilization since the time of the pharaohs. Locust plagues, originating in arid regions of Africa and the Middle East, can decimate regional crops, reach densities of up to 80 million insects per km^2, and travel between 5–200 km per day. There are even documented dispersal events from north-west Africa to the British Isles in 1954, and across the Atlantic to the Caribbean in 1988, covering a distance of 6000 km (Cressman 2009). Thus, variation at the PKG gene is associated not only with locust density-dependent phase polyphenism, but with what is arguably one of the largest, recurrent, and most devastating dispersal events in recorded history.

8.2.2 Top-down approaches: QTLs and microarrays

Dispersal can also be studied by trying to identify genetic variation and genomic regions associated with locomotion differences, thereby generating functional understanding, candidate genes, and insights into the genetic architecture of a complex trait. Working from the phenotype down to the genotype, top-down approaches use quantitative trait locus (QTL) mapping, gene expression patterns, and association studies to query genomic variation for correlations with the phenotype of interest (for an excellent review of top-down versus bottom-up approaches, see Vasemagi and Primmer 2005).

Quantitative genetic study of locomotion in Drosophila exemplifies top-down approaches to studying locomotion, with important implications for dispersal study. *D. melanogaster* lines selected over 25 generations for high or low locomotion, founded from crosses of a wild and lab strain, identified four QTL for locomotor behaviour (Jordan *et al.* 2006). Locomotor behaviour was scored by recording the duration of flight over 45 sec by an individual fly in a jar after two taps of the jar upon a table (Figure 8.2a). The total chromosomal region identified by these four QTL was 18,200 kbp long and contained 3207 genes (i.e. potential candidate genes). High-resolution mapping using deficiency complementation tests was used to investigate each of these four QTL regions (Jordan *et al.* 2006). Deficiency complementation tests assess locomotor behaviour in F1 flies derived from crosses of mutant flies, having either high or low locomotor behaviour (i.e. this study), with deficiency line flies, which have unique and large chromosomal deletions whose boundaries are known. For this study, deficiency lines were chosen that had deletions within the studied QTL regions. F1 offspring that failed to restore wild-type locomotor performance presumably resulted from the chromosomal deletion in the deficiency line, since when these genes are absent (i.e. deleted), they cannot complement, and thereby mask, the genes in the mutant fly that cause either the high or low locomotor behaviour. Using this approach with 61 different deficiency lines (each covering a large fraction of the QTL region), the four QTL were found to fractionate into 12 QTL of smaller size that each harbored on average 102 candidate genes.

These smaller QTL were then further studied by focusing upon 26 specific genes located within them as determined from genome sequence (Jordan *et al.* 2006). These 26 genes were selected based on their being previously associated with stress or behaviour in other studies. Of note is that these genes were found in only five of the 12 smaller QTL. Additional detailed complementation tests were then conducted using a new series of mutation fly lines having knockout mutations specific to each of the genes. Locomotor effects were demonstrated for 13 of these 26 genes, for which only three had previously been demonstrated to affect locomotor behaviour. A SNP association study upon a wild population sample was then conducted for one of these 13 genes, *Dopa decarboxylase* (*Ddc*), revealing a region of LD, extending through the gene and into the transcriptional control region, that was associated with locomotor differences. *Ddc* effects on locomotion are well known in the *D. melanogaster* literature, having either a positive or negative effect dependent upon genetic background (Jordan *et al.* 2006), and has been found to be associated with

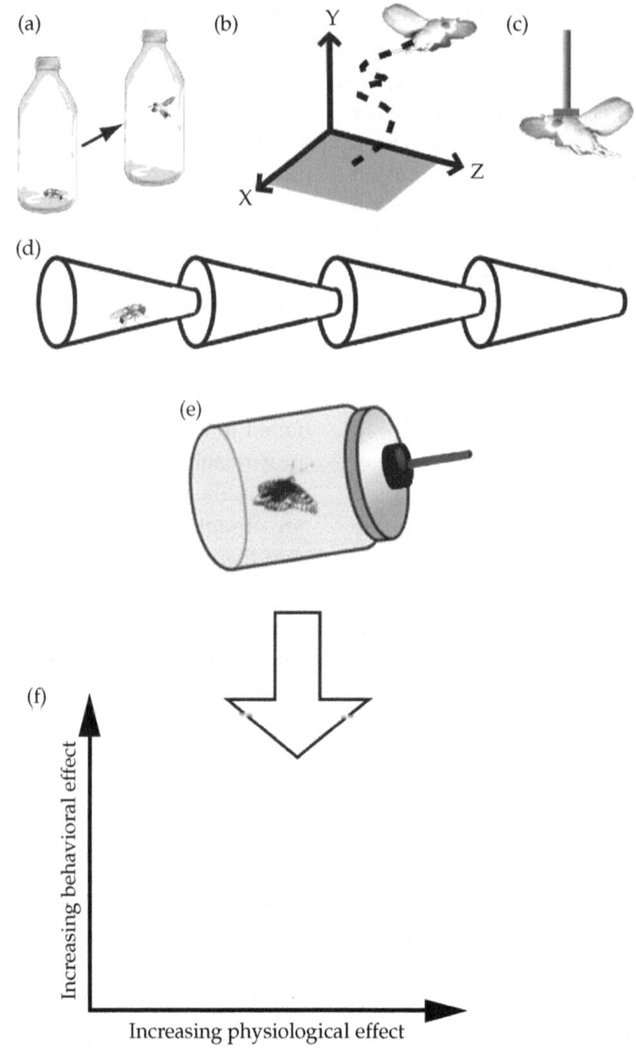

Figure 8.2 Measures of locomotion and a graph of locomotion or dispersal phenotype space. (a) An individual fly in a jar was disturbed by tapping the jar on the table, causing the fly to alight. Flight time was recorded over 45 seconds (Jordan *et al.* 2006, 2007). (b) Flight path in three dimensions was recorded after take-off using high-speed cameras, providing an estimate of mean path velocity and take-off angles (Montooth *et al.* 2003). (c) Individual flies were glued to a post and provided a perch to stand upon. When the perch was removed flight behaviour began and high-speed cameras recorded wing beat frequency (Eanes *et al.* 2006). (d) Flies introduced at one end of a linear series of funnel shaped tubes (n = 20) were allowed unidirectional travel through the tubes over 30 minutes, with their final tube location used as a locomotion score (van Dijken *et al.* 1979). (e) Individual butterflies were placed in a sealed jar and supplied with CO_2 free air. By jar tapping and shaking, butterflies were stimulated to fly over ten minutes, with the resulting peak in the rate of CO_2 production recorded (e.g. Niitepõld 2010). (f) Graph of a hypothetical locomotion or dispersal phenotype space, where a given phenotype can be placed based on its relative contribution from behavioural and physiological effects. This is intended as a starting point for discussion of the potential genetic basis of a given phenotype, illustrating that there are no simple answers regarding where measures (a) through (e), or even others such as mark-release-recapture data, should be located.

lifespan (DeLuca *et al.* 2003). A growing body of literature also documents an effect of *Ddc* on feeding and goal related behaviour in mice (Palmiter 2008).

Microarrays have also been employed to the study locomotion behaviour in locomotion selected lines (Jordan *et al.* 2007). Locomotor behaviour was scored identically to the previous QTL study (i.e. recording flight duration after taping jar on table; see Jordan *et al.* 2006) (Figure 8.2a). Unlike the previous QTL work that studied selected lines founded from a single wild and lab strain, this microarray study utilized population level variation as founding stock to establish their selection lines. Narrow-sense heritability for high versus low line locomotion behavioural differences was 0.16, with whole genome expression analysis of these lines identifying 1790 genes differentially expressed between them (at a false discovery rate of 0.001). Most of these expression differences arise due to some genes affecting the expression variation of many others (i.e. pleiotropy). In order to assess the quality of this information, ten of these genes were then screened using deficiency mapping, which showed that seven of them did indeed have impacts on locomotion (Jordan *et al.* 2007).

Using the same starting population, selection lines were also created and similarly studied for aggression (Edwards *et al.* 2006) and mating behaviour (Mackay *et al.* 2005). Interestingly, 311 differentially expressed genes were found to be in common across all three studied behaviours (locomotion, aggression, copulation latency). However, none of the correlations among them were significant or large (all $R^2 < 0.06$), suggesting that although different complex behaviours are affected by many of the same genes, the magnitude and direction of expression are different (Jordan *et al.* 2007). That is, behavioural differences appear to use the same building blocks, although they are arranged differently.

The final study investigated replicate selection lines of high and low locomotor activity at the whole chromosome level (van Dijken *et al.* 1979). Locomotor activity was scored, and subsequently selected upon, by counting the number of tubes (70 mm long) that individual flies travelled through within 30 minutes. Tubes were in a linear sequence and cone-shaped, which forced unidirectional travel of flies between tubes (Figure 8.2d). Crossing and mapping these lines identified the majority of differences due to the X chromosome. Also, low-activity alleles tended to be dominant over high locomotor-activity alleles.

These top-down approaches that have investigated the genomic architecture of locomotion variation in *D. melanogaster* provide insights into the genomic variation likely to give rise to dispersal variation. While initial QTL study identified only a small handful of chromosomal regions likely to affect locomotion, finer scale analysis uncovered that there are many independent loci underlying these large QTLs. Analysis of global gene expression similarly uncovered many genes having expression differences associated with locomotion variation. However, it is difficult to predict the number of genetic variants giving rise to those expression differences, as a handful of trans-acting regulatory genes may give rise to a large number of gene expression differences. Two studies identify the pervasiveness of pleiotropic effects, with many genes that appear to affect locomotor variation having also been found to be involved in stress, aggression, mating, and other behaviours. Finally, the observed dominance of low activity over high suggests that it is perhaps easier for low activity phenotypes to evolve, potentially due to deleterious mutations that decrease metabolic performance. This is a very important observation, suggesting that not all low-activity phenotypes are likely to have the same genetic basis, while high-activity phenotypes might. Thus it is likely that many genes with pleiotropic effects underlie variation in dispersal among individuals, and low activity many arise via deleterious mutations.

8.2.3 Metabolic rate, flight performance, and central metabolism

Two other studies have investigated the genetic basis of metabolic rate, flight performance, and central metabolism in *D. melanogaster*. They are highly relevant to the study of dispersal genetics given the recent series of studies that find a strong association between the *Pgi* gene and dispersal (Case Study III, Chapter 23). PGI catalyzes the second step of glycolysis, converting glucose-6-phosphate to fructose-6-phosphate. This is referred to as a branchpoint enzyme, as glucose-6-phosphate can also enter two other pathways (pentose phosphate shunt and glycogen biosynthesis). Montooth *et al.* (Montooth *et al.* 2003) investigated the genetic architecture of functionally related physiological traits by QTL mapping the enzymatic activity of metabolic enzymes (N = 11), metabolite storage (N = 2), basal metabolic rate, and free-flight performance. Free-flight performance was quantified for individual flies by tracking the first quarter-second of take-off flight using CCD cameras to capture 3D coordinate movement, providing an estimate of mean flight velocity (Figure 8.2b).

Variation in an enzyme's activity never mapped to its enzymatic locus, indicating extensive trans-acting regulation of metabolism. Even though epistatic effects were observed among QTLs affecting in flight velocity, metabolic rate, and enzyme activity, individual QTLs were found that could explain between 9% and 35% of the phenotypic variance in metabolic rate and flight performance. Of interest was a QTL associated with the expression of the branchpoint enzymes sharing glucose-6-phosphate (PGI, PGM, G6PD), metabolic rate, and flight performance, suggesting their coordinated regulation and potential pleiotropic effects on whole-animal performance. Montooth *et al.* (Montooth *et al.* 2003) also found a significant negative association between *Pgi* and metabolic rate explaining 9% of the variance in metabolic rate, and their association significantly mapped to a particular QTL. Of additional interest is the observation that although two QTLs were observed for mean flight velocity, which together accounted for 51% of its genetic variance, mean flight velocity was not correlated with any of the metabolic traits measured, including basal metabolic rate (Montooth *et al.* 2003).

These interesting findings are consistent with previous studies on *Pgi*, as they find an effect of *Pgi* and central metabolism on metabolic rate. However, this *D. melanogaster* work identifies a negative association between PGI activity and metabolic rate, while in *M. cinxia*, this relationship is assumed to be positive, as is found in *Colias* butterflies (Haag *et al.* 2005). There are likely to be many reasons for these differences, the foremost being that the *D. melanogaster* work quantified basal metabolic rate (Montooth *et al.*, 2003), while the butterfly work focuses on peak metabolic rate during stimulated flight (Figure 8.2e). Basal and peak metabolic rate demands are expected to have dramatically different effects on glycolytic flux and thereby differentially influenced by variation in central metabolism (Watt and Dean 2000). However, mean flight velocity should be a much closer approximation to peak metabolic rate, and thus the findings of no association between this measure and any of the metabolic traits is surprising (Montooth *et al.* 2003).

Next is a detailed investigation of the impact of enzymatic variation at specific steps in central metabolism on metabolic rate (Eanes *et al.* 2006). By generating lines of flies having partial knockouts of specific enzymatic steps in and around glycolysis, *in vivo* activity levels of these biochemical steps were quantitatively reduced and the resulting impact on flight performance assessed (see Figure 8.1; Eanes *et al.* 2006). Altering enzyme levels did have significant impacts at three enzymatic steps (phosphoglucomutase, glycogen phosphorylate, and hexokinase), two of which are considered classical regulatory steps. However, no effect on flight performance was observed for variation in other genes of glycolysis, even after reducing the PGI enzyme to 10% normal levels. These results were found to suggest that flight performance is robust to naturally occurring variation in enzyme activity across many of the enzymes of glycolysis, and that variation in PGI enzymatic step in particular should have little impact on flight performance. These latter findings are consistent with previous attempts to link natural variation in central metabolic enzyme levels with measures of flight performance in *D. melanogaster* (Laurie-Ahlberg *et al.* 1985).

How can we reconcile the findings indicating no effect of variation in glycolytic enzyme activity, especially in PGI enzyme activity, with previous work on *Colias* and *Melitaea* butterflies that have found large PGI effects? Drosophila melanogaster and these butterflies solely use glycolysis to fuel flight, but their flight dynamics are certainly different. Eanes *et al.* (2006) suggested that the synchronous flight muscles of butterflies might exert a lower flux demand upon glycolysis during free flight than in Diptera, thereby placing a lower demand upon high enzyme activity at intermediate steps. Maintaining high enzyme levels in the absence of metabolic benefits could be costly, resulting in trade-offs and potential fitness consequences (Eanes *et al.* 2006). Indeed flux demand is an important consideration, since only at high levels of demand may variation in enzyme activity impact flux itself (Watt and Dean 2000). With this in mind, it is important to note that flight performance in the Eanes *et al.* (2006) study was measured by quantifying voluntary wing beat frequency of *D. melanogaster* glued to a hook (Figure 8.2c). This measure is not equivalent to the metabolic demands and conditions of peak metabolic rate measured during free-flight, as in the *M. cinxia* studies (e.g. Niitepõld 2010) (Figure 8.2e). Rather, the metabolic demands of voluntary tethered flight lie somewhere above basal activity and substantially lower than that of free flight. For example, in the moth *Manduca sexta*, rates of oxygen consumption for tethered flight are below half that measured for free flight, although both have similar wing beat frequencies (Heinrich 1971). Therefore, the previously mentioned *D. melanogaster* studies, which used only tethered flight, placed only a weak demand upon ATP pools and their rephosphorylation mechanisms.

An additional point of consideration focuses on the biochemical 'defences' that work to maintain ATP levels upon their initial rapid depletion during flight (Nation, 2008). During the initiation of free flight, flight muscles consume the standing levels of ATP, with the subsequent seconds of ATP hydrolysis supplied by the rephosphorylation of ADP to ATP by arginine phosphokinase, the adenylate kinase (AK) reaction (2 ADP ↔ ATP + AMP), and the citric acid cycle. After these mechanisms have come into play, and complimentary to them, comes ATP generated by glycolysis. Certainly at the initiation of flight, glycolysis and the citric acid cycle are primed for increased flux due to the rapid generation of AMP and other allosteric interactions, but only after sufficient demand will ATP from glycolysis be used in direct support of flight (Sacktor and Hurlbut 1966; Sacktor and Wormser-Shavit 1966). Thus, given this 'defence' of ATP mechanisms and the lack of a dramatic demand upon central metabolism in the absence of lift during the flight assay, Eanes *et al.* (2006) may have drastically underestimated the actual impacts of reduced enzyme levels on glycolytic performance. In fact, when free flight assays are used to assess glycolytic enzyme effects on flight, much smaller changes in glycolytic enzyme levels have been found to have effects, even at PGI, and these were discussed earlier (Montooth *et al.* 2003). Consistent with the findings of *Pgi* effects on flight performance, the *Pgi* gene does show excess amino acid polymorphism in *D. melanogaster* (Moriyama and Powell 1996), as well as strong associations with flight performance and fitness across diverse arthropods (Wheat 2009), along with evidence of long-term balancing selection in butterflies (Wheat *et al.* 2010; Wheat *et al.* 2006; Watt 2003).

Establishing a causal understanding of enzymatic effects on flight performance is very difficult in the absence of transgenic manipulation. Corroborative evidence for causality can be obtained by investigating associations and molecular level predictions across independent samples and across higher levels of biological organization (Mackay *et al.* 2009). With this in mind, the classic enzyme kinetic studies of the genetic variants at PGI in *Colias* butterflies provide the best evidence to-date of *Pgi* effects, with rank order biochemical performance among genotypes correlating with those genotype's relative performance and fitness in the wild (Watt 2003). The recent findings of *Pgi* effects in the 80 million year divergent butterfly *M. cinxia*, which also exhibits associations between *Pgi* genotype and phenotype, provides important support for the effects of *Pgi* variation on flight performance. However, more work is needed in order to synthesize the diverse observations from *Drosophila* and butterflies, and other taxa (Wheat 2009).

8.3 Conclusions

In summary, dispersal variation is a complex phenotype, and no single candidate gene or pathway is likely to capture all of this complexity across diverse taxa. Simply, there is no candidate gene like *Bmp4* or *Mcr1* for dispersal. However, some genes affecting locomotion can also affect dispersal, via behavioural and/or physiological effects. Two candidate genes are emerging that appear to affect insect dispersal through such effects on locomotion, the *Pgi* and *for* genes. As such, these represent an untapped resource in the quest to develop a general understanding of dispersal. Candidate gene approaches focusing on these or other genes can benefit from the main issues illuminated in this brief comparative review, which are summarized below.

Firstly, predicting the exact mechanism and phenotypic effect of a candidate gene in a novel species is difficult since the mechanism, dominance, magnitude, direction, type of selection, and environmental context of such effects are likely to vary significantly across species. Yet for some candidate genes such as *foraging*, a similar role across divergent species can be found (Fitzpatrick *et al.* 2005). This is a very important perspective when searching for and studying a candidate gene that is relevant to the dispersal characteristics of a novel species (Figure 8.1; Fitzpatrick *et al.* 2005).

Second, insight into the likely genetic architecture underlying dispersal differences is critical when designing and interpreting studies without genomic resources. Findings suggest that trans-regulation, pleiotropy, and epistastic interactions among many genes of small effect may be a common component of dispersal genetics (Flint and Mackay 2009). If this is generally the case, finding and studying many genes having a measurable effect on dispersal variation in the wild will be very difficult.

Currently there are not enough studies of the genetic architecture of adaptive, complex phenotypes to predict how common genes of large effect are likely to be in novel taxa. Certainly a large number of studies in model genomic systems indicate that complex phenotypes are likely to adhere to an infinitesimal model (Mackay *et al.* 2009), having many genes of small effect. However, as studies reviewed here indicate, genes of large effect do occur in wild populations. Assessing the relative frequency of genes of large effect versus the infinitesimal model underlying dispersal variation will require studies specifically focused upon adaptive phenotypes in wild populations.

Third, the strategy employed by Jordan *et al.* (2007), who assayed natural variation using selection lines and microarray analysis, appears to be a good way to identify genetic variation affecting complex behaviours such as locomotion. Especially important is the relevance to population level phenomena this design affords, as the observed expression differences arise due to allelic variation from the sampled population. Note however, Jordan *et al.* (2007) did not identify the underlying causes of the observed expression variation, although they did verify that some of the observed expression differences could affect locomotion. Nevertheless, knowledge of the genes involved in the connection between genotype and phenotype provides a means of characterizing the functional basis of locomotion and its variation. Such understanding could be used to assess the role of locomotion variation in dispersal.

Finally, what should be the relative location of the different measures of locomotion performance covered in this chapter (Figure 8.2a–e), within the two dimension space of increasing behavioural or physiological contribution in Figure 8.2f? Where would the results of even a direct measure of dispersal be placed in Figure 8.2f, such as a mark-release-recapture experiment? There is no simple consensus among scientists, which reflects the complexity and reality of dispersal as a phenotype. As a result, many dispersal measures commonly confound behaviour and physiological contributions. While such confounding of contributions most likely has little consequences for field studies and modelling, when considering the genetic basis of dispersal, this confounding can severely complicate mechanistic insights. However, separating behavioural and physiological contributions may greatly facilitate the study of dispersal, exemplified by studies of *for* and *Pgi*. Therefore, trying to isolate behavioural and physiological effects may indeed be a promising approach to the study of dispersal genetics.

In closing, increased integration of mechanistic and ecological approaches appears necessary for significant progress to be realized in understanding the genetics of dispersal. Insights into the genetics of dispersal could greatly benefit from 1) mechanistic studies of locomotion phenotypes that have greater ecological relevance, and 2) ecological studies of dispersal that try to assess the role of candidate gene contributions to the physiological or behavioural aspects of dispersal in the field.

Acknowledgements

The author thanks W.B.Watt, J.H. Marden, R. Schilder and K. Niitepõld for their productive and insightful conversations over the years that have stimulated much of this article. Thanks also to M. Sokolowski and helpful comments on this manuscript from WBW, JHM, MS, and RS. Thanks also to Mathias Moser of the Fry group (Swiss Federal Institute of Technology (ETH)/Rhine-Waal University of Applied Sciences; http://fly.ini.uzh.ch) who provide the original *D. melanogaster* in flight photo, which I manipulated for Figure 8.2.

References

Abzhanov, A., Protas, M., Grant, B.R. *et al.* (2004). 'Bmp4 and morphological variation of beaks in Darwin's finches', *Science* 305: 1462–5.

Baguette, M., Convié, I. and Nève, G. (1996). 'Male density affects female spatial behaviour in the butterfly Proclossiana eunomia', *Acta Oecologica* 17: 225–32.

Baguette, M., Convié, I., Vansteenwegen, C. *et al.* (1998). 'Sex-biased density-dependent dispersal in a metapopulation of the butterfly Proclossiana eunomia', *Acta Oecologica* 19: 17–24.

Ben-shahar, Y., Robichon, A., Sokolowski, M.B. *et al.* (2002). 'Influence of gene action across different time scales on behavior', *Science* 296: 741–4.

Cressman, K., ed (1999). *Monitoring Desert Locusts in the Middle East: An Overview.* Yale University, 123–140.

Dobzhansky, T. and Wright, S. (1943). 'Genetics of natural populations. X. Dispersion rates in D. pseudoobscura', *Genetics* 28: 304–40.

Eanes, W.F., Merritt, T.J., Flowers, J.M. *et al.* (2006). 'Flux control and excess capacity in the enzymes of glycolysis and their relationship to flight metabolism in Drosophila melanogaster', *Proceedings of the National Academy of Sciences of the USA* 103: 19413–18.

Edwards, A.C., Rollmann, S.M., Morgan, T.J. *et al.* (2006). 'Quantitative genomics of aggressive behavior in Drosophila melanogaster', *PLoS Genetics* 2: e154.

Fitzpatrick, M.J., Ben-Shahar, Y., Smid, H.M. *et al.* (2005). 'Candidate genes for behavioural ecology', *Trends in Ecology and Evolution* 20: 96–104.

Fitzpatrick, M.J., Feder, E., Rowe, L. *et al.* (2007). 'Maintaining a behaviour polymorphism by frequency-dependent selection on a single gene', *Nature* 447: 210–12.

Fitzpatrick, M.J. and Sokolowski, M.B. (2004). 'In Search of Food: Exploring the Evolutionary Link Between cGMP-Dependent Protein Kinase (PKG) and Behaviour', *Integrative and Comparative Biology* 44: 28–36.

Flint, J. and Mackay, T.F.C. (2009). 'Genetic architecture of quantitative traits in mice, flies, and humans', *Genome Research* 19: 723–33.

Fujiwara, M., Sengupta, P. and McIntire, S.L. (2002). 'Regulation of body size and behavioral state of C. elegans by sensory perception and the EGL-4 cGMP-dependent protein kinase', *Neuron* 36: 1091–102.

Haag, C.R., Saastamoinen, M., Marden, J.H. *et al.* (2005). 'A candidate locus for variation in dispersal rate in a butterfly metapopulation', *Proceedings of the Royal Society B* 272: 2449–56.

Hanski, I. (2011). 'Dispersal and eco-evolutionary dynamics in the Glanville fritillary butterfly', in J. Clobert, M. Baguette, T. Benton *et al.* eds, *Informed dispersal and spatial evolutionary ecology*. Oxford University Press, Oxford.

Hanski, I. and Saccheri, I. (2006). 'Molecular-level variation affects population growth in a butterfly metapopulation', *Plos Biology* 4: 719–26.

Heinrich, B. (1971). 'Temperature Regulation of the Sphinx Moth, Manduca Sexta: I. Flight Energetics and Body Temperature During Free and Tethered Flight', *Journal of Experimental Biology* 54: 141.

Ingram, K.K., Oefner, P. and Gordon, D.M. (2005). 'Task-specific expression of the foraging gene in harvester ants', *Molecular Ecology* 14: 813–18.

Johnston, J.S. (1982). 'Genetic variation for anemotaxis (wind-directed movement) in laboratory and wild-caught populations of Drosophilia', *Behavior Genetics* 12: 281–93.

Jordan, K., Morgan, T. and Mackay, T. (2006). 'Quantitative Trait Loci for Locomotor Behavior in Drosophila melanogaster', *Genetics* 174: 271–84.

Jordan, K.W., Carbone, M.A., Yamamoto, A. *et al.* (2007). 'Quantitative genomics of locomotor behavior in Drosophila melanogaster', *Genome Biology* 8: R172.

Kent, C.F., Daskalchuk, T., Cook, L. *et al.* (2009). 'The Drosophila foraging gene mediates adult plasticity and gene-environment interactions in behaviour, metabolites, and gene expression in response to food deprivation', *PLoS Genetics* 5: e1000609.

Kodaira, Y., Ohtsuki, H., Yokoyama, J. *et al.* (2009). 'Size-dependent foraging gene expression and behavioral caste differentiation in Bombus ignitus', *BMC Research Notes* 2: 184.

Laurie-Ahlberg, C.C., Barnes, P.T., Curtsinger, J.W. *et al.* (1985). 'Genetic variability of flight metabolism in Drosophila melanogaster. II. Relationship between power output and enzyme activity levels', *Genetics* 111: 845–68.

Lucas, C., Kornfein, R., Chakaborty-Chatterjee, M. *et al.* (2010). 'The locust foraging gene', *Archives of Insect Biochemistry and Physiology* 74: 52–66.

Lucas, C. and Sokolowski, M.B. (2009). 'Molecular basis for changes in behavioral state in ant social behaviors', *Proceedings of the National Academy of Sciences of the USA* 106: 6351–6.

Mackay, T.F.C., Heinsohn, S.L., Lyman, R.F. *et al.* (2005). 'Genetics and genomics of Drosophila mating behavior', *Proceedings of the National Academy of Sciences of the United States of America* 102: Suppl. 1, 6622–9.

Mackay, T.F.C., Stone, E. and Ayroles, J.F. (2009). 'The genetics of quantitative traits: challenges and prospects', *Nature Reviews Genetics* 10: 565–77.

Majerus, M.E. and Mundy, N.I. (2003). 'Mammalian melanism: natural selection in black and white', *Trends in Genetics* 19: 585–8.

Mikasa, K. and Narise, T. (1986). 'Genetic variation of temperature-influenced emigration behavior of Drosophila melanogaster in a natural population', *Japanese Journal of Genetics* 61: 233–40.

Montooth, K.L., Marden, J.H. and Clark, A. (2003). 'Mapping determinants of variation in energy metabolism, respiration and flight in Drosophila', *Genetics* 165: 623–35.

Moriyama, E.N. and Powell, J.R. (1996). 'Intraspecific nuclear DNA variation in Drosophila', *Molecular Biology and Evolution* 13: 261–77.

Nation, J.L. (2008). *Insect Physiology and Biochemistry*. CRC Press, Boca Raton.

Niitepõld, K. (2010). 'Genotype by temperature interactions in the metabolic rate of the Glanville fritillary butterfly', *Journal of Experimental Biology* 213: 1042–8.

Niitepõld, K., Ovaskainen, O., Smith, A.D. *et al.* (2009). 'Flight metabolic rate and Pgi genotype influence butterfly dispersal rate in the field', *Ecology* 90: 2223–32.

Osborne, K.A., Robichon, A., Burgess, E. *et al.* (1997). 'Natural behavior polymorphism due to a cGMP-dependent protein kinase of Drosophila', *Science* 277: 834–6.

Palmiter, R. D. (2008). 'Dopamine signaling in the dorsal striatum is essential for motivated behaviors: lessons from dopamine-deficient mice. *Annals of the New York Academy of Sciences* 1129: 35–46.

Parsons, K.J. and Albertson, R.C. (2009). 'Roles for Bmp4 and CaM1 in shaping the jaw: evo-devo and beyond', *Annual Review of Genetics* 43: 369–88.

Pereira, H.S. and Sokolowski, M.B. (1993). 'Mutations in the larval foraging gene affect adult locomotory behavior after feeding in Drosophila melanogaster', *Proceedings of the National Academy of Sciences of the USA* 90: 5044–6.

Piertney, S. and Oliver, M. (2006). 'The evolutionary ecology of the major histocompatibility complex', *Heredity* 96: 7–21.

Powell, J.R., Dobzhansky, T., Hook, J.E. *et al.* (1976). Genetics of natural populations. XLIII. 'Further studies on rates of dispersal of Drosophila pseudoobscura and its relatives', *Genetics* 82: 293–506.

Reaume, C. and Sokolowski, M.B. (2006). 'The nature of Drosophila melanogaster', *Current Biology* 16: R623–8.

Sacktor, B. and Hurlbut, E.C. (1966). 'Regulation of metabolism in working muscle in vivo. II. Concentrations of adenine nucleotides, argining phosphate, and inorganic phosphate in insect flight muscle during flight', *Journal of Biological Chemistry* 241: 632–4.

Sacktor, B. and Wormser-Shavit, E. (1966). 'Regulation of metabolism in working muscle in vivo. I. Concentrations of some glycolytic, tricarboxylic acid cycle, and amino acid intermediates in insect flight muscle during flight', *Journal of Biological Chemistry* 241: 624–31.

Shubin, N., Tabin, C. and Carroll, S. (2009). 'Deep homology and the origins of evolutionary novelty', *Nature* 457: 818–23.

Sisodia, S. and Singh, B. (2005). Behaviour genetics of Drosophila: Non-sexual behaviour. *Journal of Genetics* 84: 195–216.

Sokolowski, M.B. (1982). 'Drosophila larval foraging behavior: Digging', *Animal Behavior* 30: 1252–3.

Sokolowski, M.B. (2001).' Drosophila: genetics meets behaviour', *Nature Reviews Genetics* 2: 879–90.

Sokolowski, M.B., Bauer, S.J., Wai-Ping, V. *et al.* (1986). 'Ecological genetics and behaviour of Drosophila melanogaster larvae in nature', *Animal Behavior* 34: 403–8.

Sokolowski, M.B., Pereira, H.S. and Hughes, K. (1997). 'Evolution of foraging behavior in Drosophila by density-dependent selection', *Proceedings of the National Academy of Sciences of the USA* 94: 7373–7.

Storz, J.F. and Wheat, C.W. (2010). 'Integrating evolutionary and functional approaches to infer adaptation at specific loci', *Evolution* 64: 2489–509.

Van Dijken, F.R., Van Sambeek, M.P. sand Scharloo, W. (1979). 'Divergent selection on locomotor activity in Drosophila melanogaster. III. Genetic analysis', *Behavior Genetics* 9: 563–70.

Vasemagi, A. and Primmer, C.R. (2005). 'Challenges for identifying functionally important genetic variation: the promise of combining complementary research strategies', *Molecular Ecology* 14: 3623–42.

Watt, W.B. (1977). 'Adaptation at specific loci. I. Natural selection on phosphoglucose isomerase of *Colias* butterflies: biochemical and population aspects', *Genetics* 87: 177–94.

Watt, W.B. (1983). 'Adaptation at specific loci. II. Demographic and biochemical elements in the maintenance of the *Colias* PGI polymorphism', *Genetics* 103: 691–724.

Watt, W.B. (2000). 'Avoiding paradigm-based limits to knowledge of evolution', *Evolutionary Biology* 32: 73–96.

Watt, W.B. (2003). 'Mechanistic studies of butterfly adaptations', in Boggs, C.L., Watt, W.B. and Ehrlich, P.R. eds, *Ecology and Evolution Taking Flight: Butterflies as Model Systems*. University of Chicago Press, Chicago.

Watt, W.B., Carter, P.A. and Blower, S.M. (1985). 'Adaptation at specific loci. IV. Differential mating success among glycolytic allozyme genotypes of *Colias* butterflies', *Genetics* 109: 157–75.

Watt, W.B. and Dean, A.M. (2000). 'Molecular functional studies of adaptive genetic variation in prokaryotes and eukaryotes', *Annual Review of Genetics* 34: 593–622.

Watt, W.B., Wheat, C.W., Meyer, E.H. *et al.* (2003). 'Adaptation at specific loci. VII. Natural selection, dispersal and the diversity of molecular-functional variation patterns among butterfly species complexes (Colias: Lepidoptera, Pieridae)', *Molecular Ecology* 12: 1265–75.

Wheat, C.W. (2009). 'Phosphoglucose isomerase (Pgi) performance and fitness effects among Arthropods and its potential role as an adaptive marker in conservation genetics', *Conservation Genetics* 11: 387–97.

Wheat, C.W., Haag, C.R., Marden, J.H. *et al.* (2010). 'Nucleotide polymorphism at a gene (Pgi) under balancing selection in a butterfly metapopulation', *Molecular Biology and Evolution* 27: 267–81.

Wheat, C.W., Watt, W.B., Pollock, D.D. *et al.* (2006). 'From DNA to fitness differences: Sequences and structures of adaptive variants of Colias phosphoglucose isomerase (PGI)', *Molecular Biology and Evolution* 23: 499–512.

Zera, A.J. and Denno, R.F. (1997). 'Physiology and ecology of dispersal polymorphism in insects', *Annual Review of Entomology* 42: 207–30.

CHAPTER 9

Genetics of plant dispersal

Jocelyn C. Hall and Kathleen Donohue

9.1 Introduction

Remarkable natural variation exists in fruit structures that influence dispersal capabilities in flowering plants. Fruit morphology is extremely evolutionarily labile, exhibiting multiple origins of fleshy, dry dehiscent (fruit opens to release seeds), and dry indehiscent (fruit does not open) fruit types (Bolmgren and Eriksson 2005; Lorts *et al.* 2008). Ultimately, we want to identify and characterize the genetic changes responsible for such morphological variation. Major questions pertaining to the genetic basis of the evolution of diversity concern the contribution of novel genes derived through duplication, and the contribution of novel changes in gene expression of existing genes.

Experimental approaches to studying evolutionary developmental genetics of morphological variation typically include a combination of phylogenetics and molecular genetics of candidate genes in taxa with variable morphology (Cronk *et al.* 2002). Firstly, the phylogenetic distribution of morphological diversity across taxa provides the evolutionary context for interpreting variation in genetic pathways associated with those traits. Second, phylogenetic analysis is applied to candidate genes that have been identified in model organisms, which reveal occurrences of gene duplication. Third, characterization of gene expression in related taxa with different morphologies reveals conservation and deviation in gene-expression patterns associated with different morphologies. Fourth, genetic functional experiments test whether a particular candidate gene is causally responsible for morphological differences. This last step is challenging especially when working with non-model organisms. Here we review work that applies these different approaches to address the genetic bases of variation in dispersal traits.

9.2 Using phylogenetics to interpret evolutionary patterns in dispersal

Phylogenies are powerful tools for interpreting evolutionary patterns of variation, and they represent the starting point for many evolutionary developmental studies (Cronk *et al.* 2002). Robust organismal phylogenies provide information on the evolutionary lability of traits, co-evolution of traits with other traits or with ecological factors, and the sequence of trait transitions. Such information facilitates the identification of appropriate taxa for detailed genetic analysis, since closely related species with contrasting attributes offer the best chance of identifying genetic differences that cause those morphological differences. This information is invaluable for generating hypotheses regarding the ecological, developmental, and genetic basis of trait (co)variation.

Examining traits of species with known relationships enables estimates of the number and direction of changes in those traits. For example, phylogenetic and morphological analyses of the family Brassicaceae revealed that almost all fruit traits are homoplastic, having arisen multiple times independently and/or exhibiting evolutionary reversals of character states (Koch *et al.* 2003). For example, indehiscence has evolved multiple times within genera (e.g. *Lepidium*, Mummenhoff *et al.* 2009) and within tribes (e.g. Brassiceae, Hall *et al.* 2011).

Dispersal Ecology and Evolution. First Edition. Edited by Jean Clobert *et al.*

Consistent with this pattern of multiple transitions to indehiscence, genetic studies of knockout mutants have shown that dehiscence can be lost via the disruption of any one of several genes in Brassicaceae member *Arabidopsis thaliana* ('*Arabidopsis*' hereafter; reviewed in Dinneny *et al.* 2005). These genetic results suggest the testable hypothesis that the independent losses of dehiscence in *Lepidium* and in the Brassiceae are caused by different genetic changes.

Furthermore, analyses of Brassiceae showed that dehiscence is evolutionarily labile, including reversals from indehiscence to dehiscence (Hall *et al.* 2011). Such reversals indicate that the genetic pathways involved in the ancestral traits, in this case dehiscence, can remain functional even when they are not expressed in a given trait. Thus phylogenetic analysis can reveal evolutionary trajectories of character state changes and thereby propose genetic hypotheses regarding trait evolution.

9.3 Gene phylogenies: gene duplication and functional divergence

The above discussion pertains to organismal phylogenies. Phylogenetic inference is also applied to genes in order to determine how genes, gene copies, and alleles are related to one another. Gene phylogenies are used to identify lineage-specific gene duplications and losses as well as sequence divergence within and among gene copies.

Gene duplication and the retention of duplicated copies cause an increase in the number of genes in the genome. A major question is how new genes evolve new functions. Plants have higher rates of whole-genome duplication than animals, and the higher rate of polyploidy contributes to the higher prevalence of duplicated genes in plants. Interestingly, certain classes of genes, such as transcription factors, have an especially high copy number, suggesting that copy retention and/or additional duplication of such genes is also high (Shiu *et al.* 2005). Retention of functional gene duplicates in plant genomes may facilitate evolutionary divergence in gene products and gene regulation, and ultimately contribute to morphological variation (reviewed in Moore and Purugganan 2005). Three major fates await duplicated genes: (1) pseudogenization, when one copy becomes non-functional because of the accumulation of deleterious mutations, (2) neofunctionalization, when one duplicate acquires a novel function, or (3) subfunctionalization, when duplicated genes retain a subset of the ancestral repertoire. Redundancy, where genes maintain some overlap in function, is also predicted with subfunctionalization.

Phylogenetic analyses of gene families associated with carpel development reveal that duplication, redundancy, and subfunctionalization have occurred in these gene families (Kramer *et al.* 2004; Litt and Irish 2003). For example, *SHATTERPROOF1* (*SHP1*) and *SHATTERPROOF2* (*SHP2*) are paralogs resulting from a recent gene duplication that appears to be within the Brassicaceae (Vrebalov *et al.* 2009; Kramer *et al.* 2004), and they have redundant functions promoting dehiscence in *Arabidopsis* (Savidge *et al.* 1995). *SHP1*/*SHP2* also have functions during ovule development that overlap with *AGAMOUS* (*AG*) and *SEEDSTICK* (*STK*) (Pinyopich *et al.* 2003). All of these MADS-box genes are the result of an ancient duplication in flowering plants (Kramer *et al.* 2004). *FRUITFULL* (*FUL*) is another MADS-box gene involved in establishing dehiscence, and it also belongs to a gene family that experienced a duplication event around the same evolutionary time as the *AG* gene family (Litt and Irish 2003).

Another gene involved in the determining the dehiscence zone is *ALCATRAZ* (*ALC*), which encodes a bHLH protein (Rajani and Sundaresan 2001). Closely related to this gene is *SPATULA* (*SPT*) (Alvarez and Smyth 1999; Rajani and Sundaresan 2001). This gene appears to play a much broader role in carpel development than *ALC* (Alvarez and Smyth 1999), with the carpels of *spt* plants exhibiting defects in the development of marginal tissues. Interestingly, aspects of the dynamic expression pattern of *SPT* suggest that it may play a role in dehiscence, albeit redundantly with other factors (Heisler *et al.* 2001).

Thus in the more recent duplication of *SHP1* and *SHP2*, functional redundancy is apparent, whereas they overlap in function only partially with their more distant gene relatives, *AG* and *STK*. Likewise, in the more divergent gene pair of *ALC* and *SPT*, some functional redundancy may exist, but the two genes differ significantly in the timing and location of their gene expression. These examples show that duplicated genes are involved in dispersal processes. They also demonstrate that knowledge of the relationships among genes is valuable for inferring how genes diverge in their function and how novel functions can emerge with gene duplication. However, there is no experimental evidence as yet that duplication in the genes discussed above actually leads to variation in dehiscence among taxa.

9.4 Conservation of the molecular-genetic components of fruit development

Just as the function of related genes could be compared within a given organism, the function of the same gene or its ortholog can be compared across taxa. Comparing gene function and genetic pathways across taxa of known phylogenetic relatedness is a powerful method for investigating the evolution of gene function and its relationship to morphological diversity. One can characterize divergence in gene sequence and gene expression as well as identify pathways and genes that have been highly conserved through evolutionary time.

Two genetic model species with very different fruits have taken the foreground in fruit development: *Arabidopsis* and tomato (*Solanum lycopersicum*). *Arabidopsis* has a dry, dehiscent fruit (Figure 9.1) whereas the fruit of tomato is a fleshy berry. These two many-seeded fruits are dispersed quite differently; seeds of *Arabidopsis* are passively dispersed after the fruits open, whereas animals that eat the berries distribute seeds of tomato.

Despite significant differences in fruit characteristics between *Arabidopsis* and tomato, the molecular-genetic components of fruit development are conserved (Vrebalov *et al.* 2009), and they are closely associated with carpel development (Seymour *et al.* 2008). For example, *SHP1/SHP2* mentioned above are genes in *Arabidopsis* that redundantly contribute to proper fruit opening and ovule development (Liljegren *et al.* 2000; Pinyopich *et al.* 2003). The ortholog of *SHP1/2* in tomato, called Tomato *AGAMOUS-LIKE1* (*TAGL1*), is required for proper fruit expansion and ripening (Vrebalov *et al.* 2009). However, when *TAGL1* was transformed into a *shp1/shp2* double mutant, it failed to rescue the mutant phenotype, suggesting that the two proteins have diverged significantly in function (Vrebalov *et al.* 2009). *FUL* is another important gene involved in controlling dehiscence in *Arabidopsis*. *TDR4* is the putative ortholog of *FUL* in tomato (Litt and Irish 2003) and has been proposed to be involved in controlling fruit ripening (Eriksson *et al.* 2004).

Two hypotheses have been proposed to explain the conservation of the genetic components of the pathway of fruit development between these model species and other species (e.g. apple, Cevik *et al.* 2010; canola, Østergaard *et al.* 2006), even as their functions have diverged. The first is that MADS-box genes, such as *SHP1/2* and *FUL*, code for transcription factors and can evolve quickly by pairing with different partners for regulation. This flexibility in functional partnership has been proposed to contribute to angiosperm morphological evolution and explain the shared genetic components between *Arabidopsis* and tomato (Vrebalov *et al.* 2009). Second, fleshy fruits are hypothesized to be derived from dehiscent fruits (Bolmgren and Eriksson 2005). As a result, it is perhaps unsurprising that genes involved in carpel development and dehiscence have been recruited for a different maturation process in fleshy fruits (Eriksson *et al.* 2004; Seymour *et al.* 2008).

The observation that genetic components of developmental pathways are conserved even when fruit morphology itself differs suggests that orthologous genes can function differently across species, even as they contribute to the same general process of development. Either the gene products themselves may differ, or perhaps more likely, the timing and location of gene expression may differ. Distinguishing between these possibilities is a major goal of the field of evolutionary developmental genetics.

9.5 Changes in gene expression as potential causes of variation in dispersal

Variation in development is caused by (1) heterotopy, or the change in location of structures, (2) homeosis, or the replacement of one structure by another

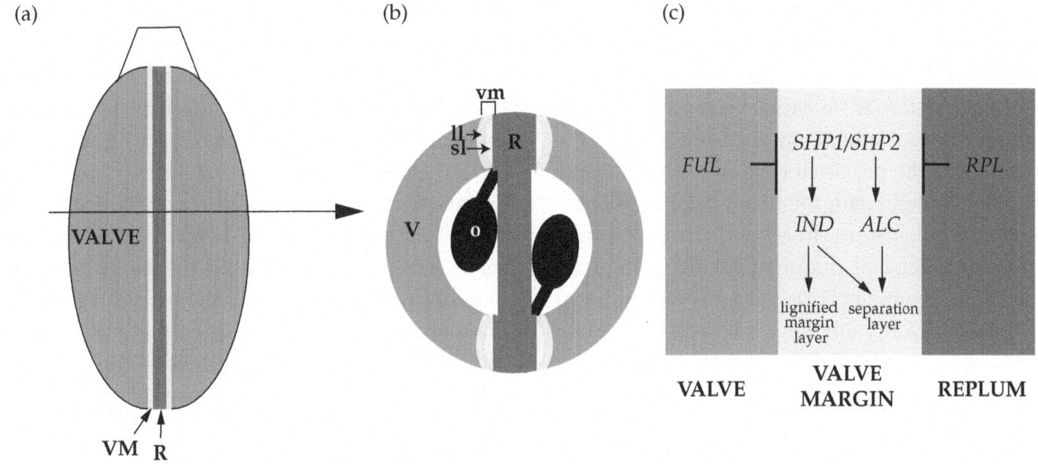

Figure 9.1 Fruit morphology and genetic pathway of dehiscence in *Arabidopsis thaliana*. (a) Fruit schematic, valves separate at maturity to release seeds (dehiscence). (b) Cross-section showing components important for dehiscence. (c) Genetic pathway. ll: lignified layer, sl: separation layer, o: ovule, r: replum, vm: valve margin.

structure, and (3) heterochrony, or the change in timing of developmental events. These three changes in development are caused by diversification of duplicated genes, temporal and spatial changes in gene expression, or other genetic or epigenetic changes (Li and Johnston 2000; Baum and Donoghue 2002).

The genetic basis of fruit morphology and dehiscence is well studied in *Arabidopsis* (reviewed in Dinneny *et al.* 2005). Moreover, relatives of this model species exhibit variation in dehiscence capabilities, representing potential systems to study the genetic mechanisms underlying differences in dehiscence across taxa (Hall *et al.* 2006; Mummenhoff *et al.* 2009).

Arabidopsis has the standard fruit morphology for Brassicaceae, hereafter referred to as a typical silique. At maturity, the two valves fall off the fruit and seeds are released. The valves detach from persistent placental tissue, called the replum, along the valve margin (Figure 9.1). The valve margin is composed of two layers: the separation layer consisting of the cells that degrade to release the valves, and the lignification layer, a band of lignified cells that provide mechanical tension required to break the valve from the replum.

The genes that control the differentiation of morphologically distinct cell types and their interactions are well characterized in *Arabidopsis* (Figure 9.1; reviewed in Dinneny *et al.* 2005). The valve margin is determined by the combined activities of *SHP1/SHP2*, *INDEHISCENT* (*IND*), and *ALC* (Dinneny *et al.* 2005; Liljegren *et al.* 2004). *SHP1/SHP2* are upstream of *IND* and *ALC*, but all are regulated by additional genes (Dinneny *et al.* 2005). *FUL* prevents expression of *SHP1/SHP2*, *IND*, and *ALC* in the valve, and *REPLUMLESS* (*RPL*) prevents their expression in the replum, thereby ensuring that they are expressed only in the valve margin (Ferrandiz *et al.* 2000; Roeder *et al.* 2003). Thus *FUL* and *RPL* position the valve margin. Aspects of this pathway are conserved across the family (Girin *et al.* 2010; Østergaard *et al.* 2006).

Heterotopy and homeosis—Heterotopy, the spatial change in an organ, structure, or developmental program, has been proposed to be an important process in plant developmental evolution (Kellogg 2000; Baum and Donoghue 2002). Heterotopy may be intimately connected with heterochrony because changes in the timing of genetic programs could feasibly alter the spatial placement of those programs (Kellogg 2000).

A prime example of a putative heterotopic change is the spatial change in the dehiscence zones of the heteroarthrocarpic fruits of the tribe Brassiceae (Figure 9.2). Heteroarthrocarpic fruits are laterally divided into two segments by a structure called the joint. Whereas the proximal (bottom) segment may or may not be dehiscent, the distal (top) segment is consistently indehiscent. In addition, the joint may or may not abscise at maturity, breaking the fruit into separate segments. This joint appears to be an evolutionary novelty enabling the dispersal of one to few seeds enclosed within protected propagules that are dispersed independently from each other (Hall *et al.* 2006).

Heteroarthrocarpic fruits have altered their dehiscence zone in two ways relative to that found in the typical siliques. First, the dehiscence zone is limited to the proximal segment instead of being completely congruent with the replum tissue as is observed in *Arabidopsis*. There is no dehiscence zone in the distal segment of the fruit because the ovary wall is not valvular in origin, and the dehiscence zone is established via the juxtaposition of the replum and valves (Hall *et al.* 2006). Second, the joint of heteroarthrocarpic fruits laterally bisects the fruits. Developmentally, the joint represents the distal portion of the valve margin with additional internal modifications (Hall *et al.* 2006). The joint may break the fruit into two independent segments, a process that typically occurs when the fruit is completely indehiscent. Thus the dehiscence function has moved from releasing the valves in a longitudinal orientation to completely bisecting the fruit. In this case of heteroarthrocarpy, heterotopy of the dehiscence zone appears to permit the dispersal of indehiscent propagules that are enclosed in a protective or dispersal-enhancing pericarp. This dispersal mode contrasts fundamentally with the dehiscence of valves, which releases independent, unprotected seeds into the environment.

Figure 9.2 Heteroarthrocarpic fruits. (a) *Erucaria erucarioides* with a dehiscent proximal segment. (b) *Cakile lanceolata* with a completely indehiscent fruit and a joint that breaks the fruit into two segments. Arrowhead = joint, scale bars = 0.5 cm (reprinted from Hall *et al.* 2006).

The distinction between heterotopy and homeosis is not always clear at the morphological level, and the two processes have been considered equivalent (Sattler 1988). Alternatively, homoesis has been considered as a special case of heterotopy (Baum and Donoghue 2002). For example, the distal segments of heteroarthrocarpic fruits may be interpreted as the replacement of valvular tissue by stylar-derived tissue or undifferentiated ovary wall. Likewise, the joint may be interpreted as the replacement of valvular tissue with a dehiscence zone.

The genetic basis of such morphological heterotopy is as yet unknown. A plausible hypothesis regarding the evolution of indehiscence is that some genes in the dehiscence pathway have been disrupted and lost their function in taxa with indehiscence (Avino *et al.* unpublished). However, given the evolutionary lability of indehiscence, with occasional reversions to dehiscence, it is also likely that the gene remains functional, but its expression may no longer occur in the dehiscence zone, even if it continues to occur and function somewhere else. Regarding heteroarthrocarpy, it is also possible that genes involved in valve dehiscence of typical siliques are expressed in the abscising joints. These hypotheses are all testable by comparing gene expression patterns in related taxa with contrasting morphologies.

Heterochrony—Heterochrony occurs when the timing of the development of a feature is altered, which can cause a change in the relative proportions among features, with some traits being exaggerated and others minimized. For example, the difference in size between wild and cultivated tomatoes is due, at least in part, to a heterochronic change in the locus *fw2.2* (Cong *et al.* 2002). As a direct regulator of cell division, a change in the timing of expression of *fw2.2* is sufficient to alter final fruit mass (Cong *et al.* 2002). Another example of heterochrony may be the development of pericarp features of dry fruits, such as spikes, wings, or other protuberances. Unfortunately, developmental and genetic studies are lacking that explicitly test these latter hypotheses.

9.6 Techniques to determine underlying genetic mechanisms

A number of methods are available to ascertain the genes responsible for particular morphological traits.

Firstly, changes in transcriptional regulation can lead to heterotopic or heterochronic changes associated with morphological evolution. Heterologous gene transformation is valuable for determining whether shifts in gene expression are due to cis-regulatory changes at a particular locus or trans-changes that affect localization of an upstream regulator. The basic premise can be explained by considering two orthologs, denoted A^x and A^y, which differ in expression between species X and Y. If the A^y locus and its regulatory regions are transformed into X, and reproduce the normal expression for A^y, this demonstrates that the expression shift is caused by changes in the cis-regulatory region of A^y itself. Alternatively, if the transgenic A^y is expressed in the pattern normally observed for A^x, this suggests that the expression shift of A^y is caused by changes in trans-acting factors. Experimental data has shown that both cis- and trans-regulation can occur in the same system (Yoon and Baum 2004).

Another suite of techniques to determine gene function includes methods that induce post-transcriptional down-regulation. For example, if valve margin genes are expressed in the joint of heteroarthrocarpic fruits that break apart at maturity, this would suggest, although not prove, that they have been recruited for this novel dehiscence zone. If fruits do not break into two segments when a candidate gene of interest is down regulated, this result would indicate that gene is essential for joint abscission. Virus-induced gene silencing (VIGS) has become a promising method for determining loss-of-function phenotypes (Burch-Smith *et al.* 2004). VIGS involves suppressing a target transcript by cloning a specific plant gene into a viral delivery system that is then injected or infiltrated into vegetative plant tissues (Burch-Smith *et al.* 2004) and has been applied to diverse species (Gould and Kramer 2007; Renner *et al.* 2009). VIGS and other methods of post-transcriptional down regulation (Pan *et al.* 2010) have not, to the best of our knowledge, been used to examine fruits of non-model species, although VIGS have been frequently employed in tomato (Liu *et al.* 2002; Fu *et al.* 2005).

Alteration of gene expression to infer function has successfully been applied to investigate dehiscence in relatives of canola (*Brassica*). Over-expression of *FUL* from *Arabidopsis* in transgenic *Brassica juncea* resulted in indehiscent fruits (Østergaard *et al.* 2006). Fruits also did not dehisce when *MADSB*, a putative ortholog of *FUL* from *Sinapis alba*, was over expressed in transgenic *Brassica* (Chandler *et al.* 2005). These studies reveal conservation in components of the valve margin genetic pathway, yet also demonstrate that changes in expression patterns can lead to impressive changes in dehiscence capabilities.

Experimental manipulation of gene expression has not been employed extensively to test hypotheses about the genetic basis regulating major dispersal traits. However, these methods are being more widely applied to non-model systems. Moreover, particular genetic pathways, initially characterized in genetic model organisms, are promising systems for testing whether particular candidate genes are causally implicated in variation in dispersal traits and whether the regulatory function of such genes is conserved in species with different or similar dispersal.

9.7 Conclusions and future research

In conclusion, the genetic basis of diversity in plant dispersal morphology is a promising research system for investigating how pathway structure evolves even while pathway components are conserved, and how evolutionarily labile regulatory networks are across taxa with contrasting or similar morphologies. In particular, it is an excellent context for investigating the role of gene duplication and subsequent functional overlap and diversification in the evolution of morphological novelty. Combining organismal phylogenies with analyses of genes, genetic pathways, and gene expression is a powerful approach for investigating how the evolution of genetic components and pathways translates to morphological evolution. Genetic transformations and manipulation of gene expression, in combination with these comparative techniques, can test the causal contributions of particular genetic changes to morphological change.

Such studies thus far have indicated that duplicated genes have important functions in dispersal, and that such genes may be associated with the diversity of dispersal in plants. While homologous genes are involved in dispersal in diverse taxa, the mode of their regulation, and the timing and spatial pattern of gene expression can be quite different. Comparing the timing and location of gene expression across taxa is an important focus for future studies on the genetic basis of dispersal variation. Using comparative approaches to characterize differences in genetic pathways of dispersal as well as their components is the most promising approach for elucidating evolutionary trajectories in the genetics of dispersal morphology.

References

Alvarez, J. and Smyth, D.R. (1999). '*CRABS CLAW* and *SPATULA*, two *Arabidopsis* genes that control carpel development in parallel with *AGAMOUS*', *Development* 126: 2377–86.

Baum, D. and Donoghue, M.J. (2002). 'Transference of function, heterotropy and the evolution of plant development. Pp. 52–69 In Cronk, Q.C.B., Bateman, R.M. and Hawkins, J.A. (Eds.). *Developmental Genetics and Plant Evolution*. Taylor and Francis, London.

Bolmgren, K. and Eriksson, O. (2005). 'Fleshy fruits—origins, niche shifts, and diversification', *Oikos* 109: 255–72.

Burch-Smith, T.M., Anderson, J.C., Martin, G.B. *et al.* (2004). 'Applications and advantages of virus-induced gene silencing for gene function studies in plants', *The Plant Journal* 39: 734–46.

Cevik, V., Ryder, C.D., Popovich, A. *et al.* (2010). 'A *FRUITFULL*-like gene is associated with genetic variation for fruit flesh firmness in apple (*Malus domestica* Borkh.)', *Tree Genetics and Genomes* 6: 271–9.

Chandler, J., Corbesier, L., Spielmann, P. *et al.* (2005). 'Modulating flowering time and prevention of pod shatter in oilseed rape', *Molecular Breeding* 15: 87–94.

Cong, B., Liu, J.P. and Tanksley, S.D. (2002). 'Natural alleles at a tomato fruit size quantitative trait locus differ by heterochronic regulatory mutations', *Proceedings of the National Academy of Sciences of the United States of America* 99: 13606–11.

Cronk, Q.C.B., Bateman, R.M. and Hawkins, J.A. (2002). *Developmental genetics and plant evolution*. Taylor and Francis, London.

Dinneny, J.R., Weigel, D. and Yanofsky, M.F. (2005). A genetic framework for fruit patterning in *Arabidopsis thaliana*', *Development* 132: 4687–96.

Eriksson, E.M., Bovy, A., Manning, K. *et al.* (2004). 'Effect of the *Colorless non-ripening* mutation on cell wall biochemistry and gene expression during tomato fruit development and ripening', *Plant Physiology* 136: 4184–97.

Ferrandiz, C., Liljegren, S.J. and Yanofsky, M.F. (2000). 'Negative regulation of the *SHATTERPROOF* genes by *FRUITFULL* during *Arabidopsis* fruit development', *Science* 289: 436–8.

Fu, D.Q., Zhu, B.Z., Zhu, H.L. *et al.* (2005). 'Virus-induced gene silencing in tomato fruit', *Plant Journal* 43: 299–308.

Girin, T., Stephenson, P., Goldsack, C.M.P. *et al.* (2010). 'Brassicaceae *INDEHISCENT* genes specify valve margin cell fate and repress replum formation', *Plant Journal* 63: 329–38.

Gould, B. and Kramer, E.M. (2007). 'Virus-induced gene silencing as a tool for functional analyses in the emerging model plant *Aquilegia* (columbine, Ranunculaceae)', *BMC Plant Methods* 3: 6.

Hall, J.C., Tisdale, T.E., Donohue, K. *et al.* (2006). 'Developmental basis of an anatomical novelty: heteroarthrocarpy in *Cakile lanceolata* and *Erucaria erucarioides* (Brassicaceae)', *International Journal of Plant Sciences* 167: 771–89.

Hall, J.C., Tisdale, T.E., Donohue, K. *et al.* (2011). 'Convergent evolution of a complex fruit structure in the tribe Brassiceae (Brassicaceae)', *American Journal of Botany* 98: 1989–2003.

Heisler, M.G.B., Atkinson, A., Bylstra, Y.H. *et al.* (2001). '*SPATULA*, a gene that controls development of carpel margin tissues in *Arabidopsis*, encodes a bHLH protein', *Development* 128: 1089–98.

Kellogg, E.A. (2000). 'The grasses: A case study in macroevolution'. *Annual Review of Ecology and Systematics* 31: 217–38.

Koch, M., Al-Shehbaz, I.A. and Mummenhoff, K. (2003). 'Molecular systematics, evolution, and population biology in the mustard family (Brassicaceae)', *Annals of the Missouri Botanical Garden* 90: 151–71.

Kramer, E.M., Jaramillo, M.A. and Di Stilio, V.S. (2004). 'Patterns of gene duplication and functional evolution during the diversification of the *AGAMOUS* subfamily of MADS box genes in angiosperms', *Genetics* 166: 1011–23.

Li, P. and Johnston, M.O. (2000). 'Heterochrony in plant evolutionary studies through the twentieth century', *Botanical Review*, 66, 57–88.

Liljegren, S.J., Ditta, G.S., Eshed, Y. *et al.* (2000). '*SHATTERPROOF* MADS-box genes control seed dispersal in *Arabidopsis*', *Nature* 404: 766–70.

Liljegren, S.J., Roeder, A.H.K., Kempin, S.A. *et al.* (2004). 'Control of fruit patterning in *Arabidopsis* by *INDEHISCENT*', *Cell* 116: 843–53.

Litt, A. and Irish, V.F. (2003). 'Duplication and diversification in the *APETALA1/FRUITFULL* floral homeotic gene lineage: Implications for the evolution of floral development', *Genetics* 165: 821–33.

Liu, Y.L., Schiff, M. and Dinesh-Kumar, S.P. (2002). 'Virus-induced gene silencing in tomato', *Plant Journal* 31: 777–86.

Lorts, C.M., Briggeman, T. and Sang, T. (2008). 'Evolution of fruit types and seed dispersal: A phylogenetic and ecological snapshot', *Journal of Systematics and Evolution* 46: 396–404.

Moore, R.C. and Purugganan, M.D. (2005). 'The evolutionary dynamics of plant duplicate genes', *Current Opinion in Plant Biology* 8: 122–8.

Mummenhoff, K., Polster, A., Muhlhausen, A. *et al.* (2009). '*Lepidium* as a model system for studying the evolution of fruit development in Brassicaceae', *Journal of Experimental Botany* 60: 1503–13.

Østergaard, L., Kempin, S.A., Bies, D. *et al.* (2006). 'Pod shatter-resistant *Brassica* fruit produced by ectopic expression of the *FRUITFULL* gene', *Plant Biotechnology Journal* 4: 45–51.

Pan, I.L., Mcquinn, R., Giovannoni, J.J. an *et al.* (2010). 'Functional diversification of *AGAMOUS* lineage genes in regulating tomato flower and fruit development', *Journal of Experimental Botany* 61: 1795–806.

Pinyopich, A., Ditta, G.S., Savidge, B. *et al.* (2003). 'Assessing the redundancy of MADS-box genes during carpel and ovule development', *Nature* 424: 85–8.

Rajani, S. and Sundaresan, V. (2001). 'The *Arabidopsis* myc/bHLH gene *ALCATRAZ* enables cell separation in fruit dehiscence', *Current Biology* 11: 1914–22.

Renner, T., Bragg, J., Driscoll, H.E. *et al.* (2009). 'Virus-induced gene silencing in the culinary ginger (*Zingiber officinale*): An effective mechanism for down-regulating gene expression in tropical monocots', *Molecular Plant* 2: 1084–94.

Roeder, A.H.K., Ferrandiz, C. and Yanofsky, M.F. (2003). 'The role of the *REPLUMLESS* homeodomain protein in patterning of *Arabidopsis* fruit', *Current Biology* 13: 1630–5.

Sattler, R. (1988). 'Homeosis in plants', *American Journal of Botany* 75: 1606–17.

Savidge, B., Rounsley, S.D. and Yanofsky, M.F. (1995). 'Temporal relationship between the transcription of two *Arabidopsis* MADS Box genes and the floral organ identity genes', *Plant Cell* 7: 721–33.

Seymour, G., Poole, M., Manning, K. (2008). 'Genetics and epigenetics of fruit development and ripening', *Current Opinion in Plant Biology* 11: 58–63.

Shiu, S.H., Shih, M.C. and Li, W.H. (2005). 'Transcription factor families have much higher expansion rates in plants than in animals', *Plant Physiology* 139: 18–26.

Vrebalov, J., Pan, I.L., Arroyo, A.J.M. *et al.* (2009). 'Fleshy fruit expansion and ripening are regulated by the tomato *SHATTERPROOF* Gene *TAGL1*', *Plant Cell* 21: 3041–62.

Yoon, H.S. and Baum, D.A. (2004). 'Transgenic study of parallelism in plant morphological evolution', *Proceedings of the National Academy of Sciences of the USA* 101: 6524–9.

PART III

The Association of Dispersal with Other Life-history Traits

CHAPTER 10

Dispersal syndromes

Ophélie Ronce and Jean Clobert

10.1 Introduction

10.1.1 Why should we study dispersal syndromes?

Dispersal syndromes describe patterns of covariation of morphological, behavioural, and/or life-history traits associated with dispersal (see Glossary). Covariation is a continuous measure of statistical association between traits composing a complex phenotype. Documenting such patterns of covariation can be seen of interest for four main reasons. Firstly, syndromes may help us predict *a priori*, from the observation of dispersal phenotypes, the intensity, nature, and modalities of movement. One may wish to predict the differential spread rates of species across space, which populations will produce more emigrants, or which particular individuals within a population are most likely to undertake dispersal. This is typically what is done when the possession of wings or the fraction of winged individuals are used to draw inferences about dispersal variation in wing dimorphic insects (e.g.Thomas *et al.* 2001). The term 'dispersal phenotype' rather than dispersal syndrome is often used in that context. In the plant literature, dispersal syndromes are predominantly used to predict the principal dispersal vectors of seeds based on fruit colour and morphology (Thomson *et al.* 2010), and, from there, draw some inference about the potential distribution of dispersal distances (e.g. Willson 1993; Portnoy and Willson 1993). Given that movement itself can often be difficult to measure, especially at large spatial scales, simple phenotypic predictors of dispersal are widely sought. Yet the success of such an enterprise relies on our ability to (i) identify the correct set of phenotypic traits explaining a large fraction of variation in effective movement, and (ii) extrapolate such syndromes to other systems and conditions. This may be problematic when patterns of covariation vary themselves continuously across environmental conditions or populations, as we will illustrate in this review.

The second motivation to study syndromes lies in their potential to inform us about the mechanistic determinants of dispersal and the constraints associated with movement. Which are the morphological and behavioural requisites for movement, and how does movement impact on the expression of other phenotypic traits? Syndromes may in particular inform us of the costs of dispersal and more generally of trade-offs affecting dispersal evolution (see review by Bonte *et al.* 2012). Many factors may, however, obscure the relationship between, on the one hand, patterns of phenotypic covariation and, on the other, mechanistic constraints and trade-offs (see detailed discussion following). This suggests the need for considerable caution when interpreting syndromes in a mechanistic framework, in the absence of additional data obtained through the experimental manipulation of phenotypes.

The third historically important reason to study dispersal syndromes is the hope that they would provide information about the proximate motivations and ultimate causes of dispersal. Under this view, syndromes are shaped by natural selection and may reflect major differences, among species or among habitats, in selection schemes affecting multiple traits. The 'colonizer syndrome' associating high dispersal ability, large fecundity, and the ability to self-fertilize as joint adaptations to the colonization of ephemeral habitats (Baker and Stebbins 1965) is a typical example. At a very different scale, patterns of association of body size or body

Dispersal Ecology and Evolution. First Edition. Edited by Jean Clobert *et al.*

condition with dispersal status have been proposed to be symptomatic of different kin competition avoidance strategies (Cote *et al.* 2007; see review in Clobert *et al.* 2009, and also theoretical considerations in Chapter 11). As previously noted, direct inferences from syndromes to patterns of selection are problematic. In particular, syndromes emerge in general from the interactions between mechanistic constraints (trade-offs) and the history of multivariate selection: disentangling the relative role of constraints and adaptation in shaping patterns of multi-trait covariation is often tricky. Yet this perspective on dispersal syndromes had the merit of drawing attention on the multivariate nature of adaptation, on the fact that different phenotypic traits evolve jointly along different environmental gradients, that predictions about single traits evolution could be modified by selective interactions with other traits, and finally helped integrate studies of dispersal within the general framework of life-history evolution (Ronce and Olivieri 2004).

The fourth, and probably most convincing reason to study dispersal syndromes is that patterns of covariation between traits associated with dispersal will critically affect both the demographic and genetic consequences of movement (Benard and McCauley 2008; Clobert *et al.* 2009). Whatever the precise proximate and ultimate causes creating syndromes, the fact that dispersing and non-dispersing individuals tend to differ in their life histories, mating tactics behaviour, and ecological niches will deeply affect spread rates, colonization patterns, the effective level of gene flow, and its fitness consequences. Such trait covariation is rarely taken into account in population genetics and population dynamics studies. More than disentangling mechanisms shaping syndromes for the sake of understanding their evolution, the critical issue here is to assess the stability of observed patterns of covariation and our ability to project into the future the consequences of such associations.

10.1.2 General issues about syndromes

As for any pattern of covariation among different phenotypic traits, the study of dispersal syndromes is complicated. This complexity may make inferences from the observation of syndromes difficult, or compromise our ability to extrapolate syndromes across different situations. The first very general issue concerns the relationship between mechanistic trade-offs and the mere statistical relationship described

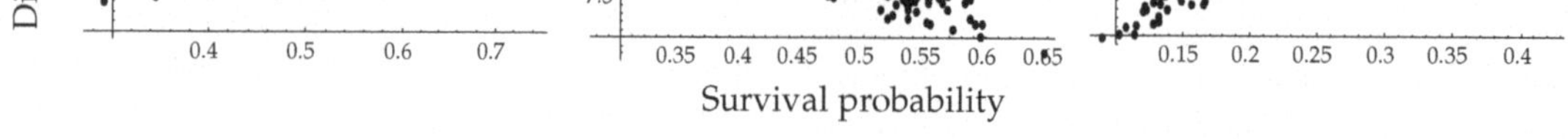

Figure 10.1 Dispersal syndromes and underlying sources of variation: a theoretical illustration. We have here used the classical Y model (see Roff and Fairbarn 2007) to describe covariation between dispersal distance and survival probability. It is here assumed that both dispersal distance and survival probability increase with individual condition. For a given individual condition, the survival probability decreases as the individual disperses further. More precisely dispersal distance is given by D = Dmax*P*T, where T is an index for individual condition, and P is the fraction of resources allocated to dispersal by this individual. The survival probability is given by S = Smin + (1-P)*T, where 1-P is the fraction of resource allocated to maintenance. For each individual, condition T and allocation P are drawn independently in centred Gaussian distributions, then transformed with an inverse-logit function to be distributed between 0 and 1. Panel A gives the pattern of covariation for a sample of 500 individuals where the variance in condition equals the variance in allocation ($\sigma_P^2 \approx \sigma_T^2 \approx 10^{-2}$); Panel B gives covariation when the variance in allocation is much greater than the variance in condition; Panel C gives covariation when the variance in condition among individuals is much greater than the variance in allocation ($\sigma_P^2 \approx 10^{-4}$; $\sigma_T^2 \approx 10^{-2}$).

by life-history syndromes (see a review in Roff and Fairbairn 2007). The sign and extent of covariation between phenotypic traits depends tightly on what drives phenotypic variation among individuals (Van Noordwijk and De Jong 1986), e.g. dispersal can be positively correlated to survival, despite being costly, if some hidden variable (e.g. condition), which increases both the probability to survive and the propensity to disperse, represents the dominant source of variation in the examined sample (see Figure 10.1). Conversely, negative relationships between dispersal and fitness components may not stem from direct functional trade-offs but from the hidden variation of a third variable.

The second issue regards the spatial scale over which syndromes are measured (Figure 10.2). When measuring covariation between different phenotypic traits, one may compare individuals with known dispersal status within the same family, from different families but from the same population, from different populations in the same metapopulation, from different metapopulations in different regions, or even from different species. Dispersal syndromes have been sought at all these scales; e.g. the 'colonizer syndrome' discussed by Baker and Stebbins (1965) is very much about understanding species differences, while discussion about dispersal phenotypes, involving body condition for instance, more often focus on inter-individual differences within the same population. Other types of comparisons are also possible, when contrasting phenotypes of individuals of known genotypes, or individuals confronted to different experimental treatments. There is no reason to expect patterns of variation shaping syndromes to be identical across scales; factors driving the structure of phenotypic variation within families are not expected to be necessarily the same as those explaining differences between species. In this review, we will illustrate how expected and observed syndromes can vary across different scales of analysis. In addition to matters of scale, syndromes are also sensitive to the precise gradients underlying variation; for instance, dispersal syndromes across gradients of disturbance may be different from that measured along gradients of fragmentation (e.g. Ronce *et al.* 2000). Different gradients at different scales (e.g. see the effect of body size variation within and between families on dispersal syndromes detailed in Chapter 11) can also be superimposed, blurring patterns of covariation. Scale- or gradient-dependency becomes an issue only because it is rarely taken explicitly into consideration, and information about potential source of variation in the sample used to infer the syndrome is often missing in empirical studies.

Covariation implies phenotypic variation. Both genetic and environmental sources of variation can shape dispersal syndromes. Distinguishing between the two is often difficult as it requires manipulating the environment or the genetic resemblance between individuals (see Chapter 2), or having large enough data sets to reconstruct deep pedigrees and sample enough environmental variation to discriminate between genetic and environmental effects (e.g. Ozgul *et al.* 2009). Artificial selection experiments also permit investigation of the genetic architecture of syndromes (e.g. Torriani *et al.* 2010). Once again, the sign and extent of genetic correlations may differ significantly from that of environmental correlations. For instance, short-distance dispersal correlates negatively with aggregation behaviour in *Tetrahymena* ciliates, but the phenotypic correlation has no genetic component despite significant genetic differences among clones for both dispersal and aggregation (Fjerdingstad *et al.* 2007; Schtickzelle

Figure 10.2 Scale-sensitivity of dispersal syndromes. In this putative example, larger wings allow larger dispersal distances. Within each patch, more dispersive individuals have smaller abdomen than less dispersive ones; yet variation in overall size between patches (due to variation of resource level for instance) results in positive covariation between abdomen size and dispersal distance. Syndromes thus differ when comparing individuals from the same patch or from different patches.

et al. 2009). Whether genetic or environmental correlations underlying phenotypic variation best reflects constraints, or the history of selection, is not clear. For instance, spurious genetic correlations between dispersal status and fitness components may emerge from inbreeding depression and variation in consanguinity in the sample (Mix *et al.* 2006; Bonte 2009). Conversely, plastic multivariate response to environmental cues affecting joint changes in dispersal and other traits (Meylan *et al.* 2009) may have been shaped by selection (see, for example, the theoretical study by King and Roff 2010). Genetic and environmental sources of variation can affect syndromes at various scales; for example, from differences between individuals within populations to regional differences. Both types of information (i.e. phenotypic and genetic patterns of covariation) are, however, necessary to predict short-term evolutionary trajectories, because selection 'sees' phenotypic variation but genetic variation determines how the effects of selection are transmitted across generations (Roff and Fairbairn 2007).

10.1.3 Specific issues about dispersal syndromes

The complexity of the dispersal phenotype leads to some specific issues. In principle, one could expect different combinations of behavioural, morphological, and life-history traits to be associated with different phases of the dispersal process; i.e. departure, transience, and settlement. For instance, the traits of emigrants may differ from that of philopatric individuals before departure, but a different combination of traits may distinguish successful immigrants and residents after settlement (see detailed discussion in Bonte *et al.* 2012). We could also expect different syndromes for different types of movements (short- and long-distance dispersal); there have been claims that the mechanisms, and thus the involved plant traits, promoting short-distance seed dispersal, differed from that allowing rare long-distance movements (Higgins *et al.* 2003; Nathan *et al.* 2008). In *Tetrahymena* ciliates, clones with the highest dispersal abilities at short distance differ from clones producing specialized long-distance dispersal morphs for a suite of morphological and life-history traits (Fjerdingstad *et al.* 2007). Different phenotypic traits may also be associated with different properties of the dispersal distance distribution; for instance, in *Arabidopsis thaliana*, different features of the plant architecture and fruit traits correlate with the mean seed distance and with the kurtosis of the seed distribution (Donohue *et al.* 2005). Finally, there have been claims that dispersal syndromes could differ depending on the exact ultimate causes of dispersal (Clobert *et al.* 2009). This suggests that when discussing dispersal syndromes, consideration should be given to what is exactly meant by 'dispersal' in each case.

In this review, we first examine data about observed syndromes and comment on the variability of patterns found in nature. We then list possible causes of covariation among phenotypic traits associated with dispersal, distinguishing proximate and ultimate explanation for syndromes. We discuss the difficulties in linking theoretical studies about syndromes to empirical patterns and more generally in inferring causes of covariation. Finally, we point to a number of potential genetic and demographic consequences of syndromes, many of which deserve further exploration.

10.2 Observed patterns of covariation

10.2.1 Morphology

Morphological adaptations to dispersal are diverse, such as airbone dispersal structures in plants (Cousens *et al.* 2008; Wilson and Traveset 2000, see Chapter 14), or wings in insects and vertebrates (Dingle 1996). However, many species have excellent colonizing abilities without having dedicated structure to enhance dispersal (Elton 2000). Variation in dispersal propensities is also found within species, in particular in the context of density dependence (Clobert *et al.* 2004). As density increases, dispersal is often enhanced by the production of winged individuals in insects (locusts, Williams 1957) or by the production of ballooning silk in spiders (Bell *et al.* 2005; De Meester and Bonte 2010). Similarly, deterioration in habitat quality triggers the production of a higher fraction of seeds equipped with a dispersal structure in the seed heteromorphic plant *Crepis sancta* (Imbert and Ronce 2001). In many species, however, morphological differences between resident and dispersing individuals are not overwhelming and are rather quantitative than discrete. For example, in vertebrates, residents are either smaller or bigger than immigrants (De Fraipont *et al.* 2000; Sinervo *et al.* 2006a; see other examples in Clobert *et al.* 2009), but nevertheless do not differ that much from one another. The only reported case of a real morphological differentiation between dispersing and non-dispersing morphs in vertebrates was found in the mole rat (O'Riain *et al.* 1996), where dispersers are much bigger and possess more vertebrae.

10.2.2 Physiology

In vertebrates, differences between residents and immigrants are more commonly reported on physiological traits. Higher titres of corticosterone and/or testosterone in particular were found to be associated with activity, aggression, and dispersal (e.g. corticosterone in the screech owl, Belthoff and Dufty 1998; and in the willow tit, Silverin 1997; testosterone in the spotted hyena, Holekamp and Smale 1998; and in the meadow vole, Nichols and Bondrupnielsen 1995). The relationship between hormones and dispersal is, however, complex: in the common lizard, dispersers have a lower titre of corticosterone (De Fraipont *et al.* 2000), dispersers in Belding's ground squirrels show no difference in testosterone (Nunes *et al.* 1999). These differences in action have often been ascribed to the organizational versus activational effects of hormones (Dufty and Belthoff 2001). Other hormones such as serotonin (Trefilov *et al.* 2000) have also been demonstrated to act on dispersal. In insects, the production of winged forms is often under the control of the juvenile hormone (Zera and Denno 1997; see a review in Zera *et al.* 2007), and an enhanced locomotor's activity is associated with a higher respiration rate (Crnokrak and Roff 2002). In the Glanville Fritillary butterflies, greater dispersal and mobility are not associated with some distinct morphology but with a different metabolism (Haag *et al.* 2005, see Box III and Chapter 13 for more details).

10.2.3 Behaviour

Hormones are not only acting on the metabolism and hence the locomotory activity of a species; they also influence individual behaviour. Although there are still a restricted number of examples, individuals who are resident seem indeed to differ in their personality from dispersers. Residents are less aggressive (Duckworth and Kruuk 2009), more social (Ims 1990; Székely *et al.* 2010; Cote and Clobert 2007b), more prone to collaboration (Sinervo and Clobert 2003), more neophobic (lizards, De Fraipont *et al.* 2000), less exploratory (fish, Haughland and Larsen 2004, mammals Krackow 2003), more risk-averse (birds, Dingemanse and de Goede 2004; Cote and Clobert 2010), and less bold (fish, Fraser *et al.* 2001) than dispersers. In primates, the timing and intensity of dispersal is often negatively associated to sociability (Mehlman *et al.* 1995). These differences are also found in single-cell species (ciliate, Schtickzelle *et al.* 2009). Boldness in particular seems to be always associated in the same way with dispersal (Cote *et al.* 2010). Other personality traits such as aggressive, social, or risk-taking behaviours are not always correlated with dispersal with the same sign (Cote *et al.* 2010).

10.2.4 Life-history traits

Empirical evidence for covariation between dispersal and other life-history traits (survival, age at maturity, fecundity) is abundant, with a wide taxonomic distribution (in plants, Moles and Westoby 2006; in invertebrates, Hanski *et al.* 2006; Roff and Bradford 1996; Chapter 4, and vertebrates Bélichon *et al.* 1996; see a review in Bonte *et al.* 2012). The sign of the association between dispersal and a particular life-history trait varies, however, widely both between and within species, which precludes any general assessment. Furthermore, many of these studies are observational and suffer from shortcomings, especially with respect to the measure of survival (Doligez and Part 2008). Dispersers can have a higher fecundity or survival (greater survival in the great tit, Clobert *et al.* 1988; greater fecundity for their second clutch in the parasitoid wasp, Innocent *et al.* 2010), or to the contrary, a smaller fecundity or survival (smaller survival in the damselfly *Calopteryx splendens*, Chaput-Bardy *et al.* 2010; smaller fecundity in the soil nematode *Caenorhabditis elegans*, Friedenberg 2003).

10.2.5 Some general properties

In summary, examples of traits associated to dispersal are now numerous and encompass all dimensions of the phenotype (morphology, e.g. Donohue *et al.* 2005; physiology, see the review in Dufty and Belthoff 2001; behaviour, review in Cote *et al.* 2010; life history, review in Bélichon *et al.* 1996; Clobert *et al.* 2009). Other dimensions of dispersal syndromes involving mating strategies (Givnish 1980; see discussion in Cheptou and Massol 2009), habitat preferences, niche breadth and plasticity (Frantz *et al.* 2009), or trophic level (Berg *et al.* 2010), still deserve further empirical exploration. It is too early to make any general conclusions about the nature of such associations (often involving multiple organizational levels, Sinervo and Clobert 2003; De Fraipont *et al.* 2000), their control (some are genetically determined, Trefilov *et al.* 2000; Innocent *et al.* 2010; others have at least some environmentally based determination, Dingemanse *et al.* 2003; Duckworth 2009), and the inducing factor (social, Gerlach 1998; parent-offspring competition, Cote and Clobert 2010; habitat, Mabry and Stamps 2008). Some general patterns emerge nevertheless. First, there is no systematic direction in the association of a phenotypic trait and dispersal, with the exception of dispersal-dedicated structure (i.e. winged individuals). The fluctuating direction of the association between a phenotypic trait and dispersal is even found at the within-species level. In two lizard species (*Uta stansburiana* and *Lacerta vivipara*), it has been experimentally demonstrated that, depending on the social habitat, the mating strategy, and hence the dispersal cause, either the smallest or the biggest individuals were dispersing (Sinervo and Clobert 2003; Cote and Clobert 2007b). The association between dispersal and life-history traits seem particularly inconsistent across case studies. Some syndromes, associating personality traits such boldness and exploration with dispersal, appear more robust.

Second, there is in general more than one phenotypic trait associated with dispersal, leading to the concept of dispersal syndromes. This term has been introduced by plant ecologists especially to explain species differences in seed dispersal either through wind or through zoochory (Thomson *et al.* 2010), involving complex sets of traits. Examples are now accumulating of multiple traits' association to dispersal in species ranging from unicellular (Fjerdingstad *et al.* 2007), to invertebrates (Hanski *et al.* 2006) to vertebrates (Sinervo *et al.* 2006b). These syndromes are labile, often with an important environmentally based determination. As we will develop in the next section, this lability of syndromes can be related to the proximate and ultimate causes at the origin of such associations.

10.3 Causes of covariation

10.3.1 Proximal causes of covariation

A first, proximal cause for dispersal syndromes is when dispersal itself affects the expression of other phenotypic traits (Figure 10.3). Dispersal alters phenotypic expression either because it diverts resources from other functions (e.g. trade-off between ovary mass and wing size in some insects as in Roff and Gelinas 2003), or because dispersers end up in different environments than philopatric individuals, which affects the expression of traits (see Figure 10.3). Such phenomena include the various costs of dispersal (see detailed discussion in Bonte *et al.* 2012) but should be broadened to more general phenotypic effects. Indeed, dispersing individual relocating in better habitats may exhibit better fitness traits than non-dispersers that remained in the original habitat (Lowe 2010). We may then expect different syndromes characterizing the phenotypes of emigrants (before departure) and immigrants (after settlement, see Figure 10.3). Experimental manipulation of movement can greatly help in disentangling the effect of dispersal on phenotypic expression from other sources of covariation (Boudjemadi *et al.* 1999). At a larger scale, syndromes associated with variation between metapopulations in dispersal intensity could also emerge because dispersal rates and dis-

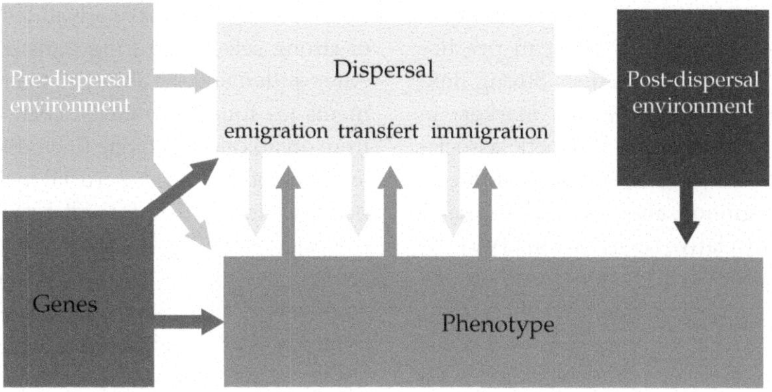

Figure 10.3 A schematic representation of the make-up of dispersal syndromes.

tances affect population dynamics, cycling and extinction probability (e.g. see the experimental evidence that intermediate dispersal minimizes extinction frequency in the plant *Cardamine hirsuta*, by Molofsky and Ferdy 2005). Conversely, syndromes may also emerge when an individual's phenotype influences habitat selection or directly affects dispersal (Figure 10.3). Seed or fruit traits enhance the probability of being taken over by dispersal vectors such as wind or animals (e.g. Thomson *et al.* 2010). Condition-dependence (association of dispersal with overall organismal condition) belongs in this respect to dispersal syndromes. Phenotype matching (see a review in Edelaar *et al.* 2008), whereby individuals with different phenotypes make different optimal emigration and immigration decisions, may also contribute to the emergence of syndromes. In the Glanville Fritillary butterfly, individual preference for different host plants determines emigration and immigration probability in habitat patches varying in plant composition (e.g. Hanski and Singer 2001). Heavier fledglings of great tits preferentially settle in the less crowded parts of Whytham wood, while lighter birds settle in denser areas (Garant *et al.* 2005). Note, however, that phenotype matching should produce phenotypic difference between emigrants and philopatric individuals, but not necessarily between immigrants and residents in the same locality if the latter share the same habitat preferences (see Figure 10.3).

Environmentally driven dispersal syndromes emerge when variation in the environment in the broad sense (including hormonal level, food, temperature, parasites, density, etc.) jointly affects the expression of dispersal and other phenotypic traits (Figure 10.3; see extensive review in Benard and McCauley 2008). Corticosterone manipulation in common lizard mothers affects their offspring size, behaviour, physiology, and dispersal strategies (Meylan *et al.* 2000; Belliure and Clobert 2004; Vercken *et al.* 2007). Similarly, testosterone manipulation in the Belding's ground squirrel (naturally increased during fights) causes the departure of young males (Holekamp *et al.* 1984). Warmer temperature experienced by spider *Erigone atra* juveniles during development both shortens the time till maturity and increases the motivation to engage into short distance dispersal movement by rappelling (Bonte *et al.* 2008). Note, however, that warm temperatures conversely inhibit ballooning, which is a long-distance dispersal behaviour (see Case Study III). Temperature variation thus results in different life-history phenotypes being associated with short- and long-distance movements in *E. atra* (Bonte *et al.* 2008).

Finally, syndromes emerge because of genetic correlations between dispersal and other traits (see Chapter 2). Genetic correlations can either be due to linkage disequilibrium between genes affecting dispersal and genes affecting other traits (such genetic associations being shaped by the history of selection, migration, and recombination), or because genes affecting dispersal have pleiotropic effects on other components of the behaviour or the life history. We expect different evolutionary stability of syndromes in the two situations, but in practice they might be difficult to distinguish. Strong linkage disequilibrium among molecular markers in side-blotched lizards suggests that genetic association between colour morph, reproductive strategy, life history, kin recognition, and dispersal are maintained in part by correlative selection (Sinervo *et al.* 2006b). As single genes with major effects on dispersal are discovered, we gain much insight on their pleiotropic effects (see Chapter 13, and Case Study III, for discussion of traits associated with molecular variation at the *Pgi* locus in butterflies). Allelic variation in the *npr-1* gene in *Caenorhabditis elegans* affects simultaneously dispersal strategies, gregariousness, and foraging tactics (Gloria-Soria and Azevedo 2008). Bearers of natural variants of the *foraging* (*for*) gene in *Drosophila melanogaster* ('rover' and 'sitter' alleles) exhibit differences in mobility, learning and memory (Mery *et al.* 2007).

10.3.2 Ultimate causes of covariation

Both environmental and genetic correlations between traits may have been shaped by selection. Syndromes may in particular emerge because of divergent selection on dispersers and non-dispersers. Dispersal syndromes would then represent integrated suites of traits optimized to maximize fitness of individuals with different life-history tactics. For example, in juvenile guanacos, more aggressive individuals leave earlier and reproduce better than submissive individuals (Ronald *et al.* 2003). Each stage of the dispersal process imposes very specific selection pressures on dispersers. Assessing the quality of local patch and of potential target patches is critical for making the right emigration decision, so it is for the choice of the immigration patch. Different cognitive abilities may then be advantageous during both emigration and immigration (Figure 10.3). In agreement with this expectation, the response to conspecific odours varies between dispersing and non-dispersing juveniles in the Common Lizard (Aragon *et al.* 2006). Locomotion ability (such as hind-leg length in invasive cane toad, Phillips *et al.* 2006), orientation skills, and metabolism sustaining movement (e.g. flight metabolic activity in insects, Haag *et al.* 2005) are all obviously the target of strong selection during transience (Figure 10.3). After settlement, the ability to tolerate new environments (as found in rabbits, Letty *et al.* 2008), both from an abiotic and biotic (including the social context) perspective, is also crucial to the success of dispersers (Figure 10.3). It is interesting to note in this respect that long-distance disperser clones in *Tetrahymena* ciliates show also the most frequent aggregation behaviour (Schtickzelle *et al.* 2009). Similarly, the best disperser genotypes in *C. elegans*, are the most gregarious (Gloria-Soria and Azevedo 2008). In these examples, enhanced group formation can be interpreted as a way to diminish the social cost of dispersal. Syndromes could thus be interpreted as adaptations to reduce the general cost of dispersal. However, this is not the only reason why selection pressures on dispersing individuals may differ from those on non-dispersing ones. Dispersers are for instance the founders of new populations (Hanski *et al.* 2004), where they experience very specific environmental conditions, such as relaxed competition, scarcity of potential mates, and rarity of specific pathogens (Phillips *et al.* 2010b). This may have profound consequences for the evolution of disperser life history and mating system, as suggested by the theoretically predicted change in reproductive effort, age at maturity, and inbreeding with increasing population age (see reviews in Ronce and Olivieri 2004; Phillips *et al.* 2010b). Consistent with this idea, females of the Glanville Fritillary butterfly originating from recently established populations are more dispersive (Hanski *et al.* 2004), have higher mobility, earlier mating, shorter lifespan, but same lifetime fecundity than females from older populations (Hanski and Saccheri 2006; Chapter 13). The phenotypic traits selected for in dispersing individuals may vary with the ultimate or proximate cause of dispersal (e.g. depending on whether colonization of empty space or kin competition is a driving force or not). There are very few theoretical studies that have addressed the question of the evolution of specific integrated dispersal phenotype (Bélichon *et al.* 1996). In Chapter 11, Kisdi and collaborators provide a nice theoretical example where diverging body size evolves in dispersed and philopatric offspring.

An ultimate cause of syndromes is when selection favours the evolution of condition-dependant dispersal strategies rather than fixed behaviours. Cost and benefits of dispersal vary with the phenotype, as explained above, which may promote the plastic expression of dispersal behaviour depending on phenotypic traits. For instance, experimental manipulation of offspring size during development in the side-blotched lizard, by either miniaturization or gigantization, affects offspring dispersal distance, consistently with the cost and benefits of dispersal in the different colour morphs (Sinervo *et al.* 2006a). The orange morph is the most dispersive (Sinervo and Clobert 2003): bigger orange males, with putatively smaller cost of dispersal and higher probability of usurping a territory, disperse further than small orange males (Sinervo *et al.* 2006a). Yellow males have different reproductive strategies and disperse less than other morphs (Sinervo and Clobert 2003): with no strong incentive to disperse, giant yellow males disperse less far than control yellow males. Similarly, local and regional crowding determines the cost and benefits of dispersal in the mongoose *Mungos mungo*, which decides whether aggressive or submissive individuals disperse (Cant *et al.* 2001). Phenotypes (e.g. condition) may also reflect habitat quality and motivate the dispersal decision (Figure 10.3). Information about habitat quality conveyed by internal state can be in particular mediated by hormones (Dufty and Belthoff 2001). Manipulation of hormonal level can mimic different types of chronic or acute stress, conveying different information about intrinsic habitat quality or extent of competition: common lizard juveniles have, for instance, different dispersal responses if their mother was exposed to brief or long corticosterone treatment (e.g. Vercken *et al.* 2007). Body size may both affect the probability to survive the dispersal event and reflect habitat quality in the natal patch. Kisdi *et al.* (Chapter 11) show how such conflicting pieces of information can theoretically lead to the evolution of complex reaction norms for size-dependant dispersal (Gyllenberg *et al.* 2008; Bonte and de la Pena 2009). Theory distinguishes cases when the phenotype evolves as a function of dispersal status versus when the dispersal propensity evolves as a function of phenotypic variation (both illustrated in Chapter 11). However, in natural populations, both phenotype and dispersal evolve together.

Syndromes may also emerge because of eco-evolutionary feedbacks: dispersal rate affects the genetic structure, social interactions, and population dynamics, which modifies the selection pressures on other traits. Table 10.1 lists a number of theoretical cases and some putative examples. The main difference with the syndromes discussed above is the scale at which covariation is expected. Most theoretical models deal with phenotypic syndromes emerging when comparing different species or different metapopulations (see Table 10.1). In these models, different phenotypic values evolve depending on the mean dispersal rate, without being necessarily distinct in dispersing and non-dispersing individuals. For instance, variation in gene flow mediated by dispersal in heterogeneous habitats, modifies the selection regime and affects the evolution of niche breadth (Ronce and Kirkpatrick 2001; Kisdi 2002). Consistent with such expectation, manipulation of dispersal in heterogeneous microcosms of the bacteria *Pseudomonas fluorescens* leads to the evolution of strains with varying abilities to exploit a diverse array of carbon sources (Venail *et al.* 2008). Conversely, variation in mating system or life history may affect the population and metapopulation characteristics and thus alter selection pressures on dispersal. In contrast to the classical colonizer syndrome associating selfing and high dispersal, Cheptou and Massol (2009), for instance, predicted that the ability or inability to self-fertilize modified the spatio-temporal variance in reproductive success, indirectly selecting for higher dispersal in outcrossing species, which suffered more common local extinction when pollinators become scarce. Dispersal and other life-history traits may also jointly evolve along various environmental gradients. For instance, Ronce *et al.* (2000) predicted that dispersal rates may covary positively with reproductive effort when both evolve along gradients of increasing disturbance (see Crowley and McLetchie 2002). Covariation patterns are, however, very sensitive to specific assumptions of the model (discussed in Ronce and Olivieri 2004). The only general finding of these models is that syndromes are tightly dependent on the nature of the underlying gradient of variation in life histories: there may be as many different syndromes as different environmental gradients (see Table 10.1)

10.3.3 Open questions

A first, open question concerns how syndromes may vary at different scales of analysis and variation (individual, population, species): do syndromes at higher scales reflect differences between migrants and non-migrants? Theoretical studies suggest that syndromes observed at the species scale may poorly reflect syndromes characterizing within population variation. For instance, higher frequency of dispersal events at the scale of the metapopulation may lead to more efficient space filling and higher average population density, which indirectly selects for decreased investment into reproduction and higher investment in maintenance (Ronce and Olivieri 1997; Ronce *et al.* 2000b). When comparing individuals originating from different networks, we might then expect negative covariation between average dispersal propensity and average reproductive effort (Ronce and Olivieri 2004). Yet within the same metapopulation, dispersers colonizing new sites are still expected to invest more into reproduction than philopatric individuals. The sign of covariation can therefore be reversed with a change in the scale of analysis. This is also found in empirical studies. For instance, in the Glanville Fritillary, butterflies bearing allele *Pgi-F* have higher mobility, fecundity, and lifespan (Saastamoinen 2007; Saastamoinen *et al.* 2009). At the scale of the patch, the relationship between the frequency of such genotypes and the population's growth rate is, however, complex, varying with the size of the patch and its connectivity (Hanski and Saccheri 2006).

Syndromes observed at different scales inform us about different evolutionary processes (i.e. global eco-evolutionary feedbacks versus selection of integrated phenotypes). Conflicting syndromes at different scales are not expected by default but syndrome variation can be seen itself as an interesting source of information.

A second open question concerns the relationship between proximal and ultimate causes of syndromes. Do proximal causes merely reflect constraints or have they been shaped by selection on dispersal phenotypes? Establishing that environmental correlations correspond to adaptive plasticity is a difficult task, as it ultimately requires comparing the fitness attached to different joint responses of phenotypic traits to some environmental cues. The pleiotropic effects of genes affecting dispersal can be seen as a very contingent constraint of extant natural genetic variation. Alternatively, these effects could have also been shaped by selection. For instance, the modified clutch size and lifespan associated with allelic variant of the *Pgi* polymorphism (Saastamoinen 2007; Saastamoinen *et al.* 2009) could be seen as side effects of strong selection on a specific genetic modifier of dispersal. An alternative view would be that these alleles have been selected precisely for their effects on whole syndromes among other dispersal variants lacking such pleiotropic effects. Answering this question would imply knowing how much genetic variation for pleiotropy is available for selection to act upon. Note that polymorphism at the *Pgi* locus is associated to strong genetic load, as the high dispersal allele is close to lethal in its homozygous form (only 10% of homozygotes survive, Orsini *et al.* 2009). This suggests that, at least, not all properties of this polymorphism have been optimized by selection and that selection on dispersal can be strong enough in this system to bypass serious costs. There is much empirical evidence that dispersal can be costly (see review in Bonte *et al.* 2012). A more puzzling question concerns how much of this cost can evolve. The same question emerges when discussing the evolution of any trade-off in the context of life-history evolution (see Roff and Fairbairn 2007, for a review) or that of ecological specialization (e.g. Magalhaes *et al.* 2009). Billiard and Lenormand (2005) provide a nice theoretical example where the cost of dispersal (due to loss of local adaptation and linkage disequilibrium between viability loci and dispersal locus) evolves together with the average dispersal rate: as higher dispersal rates are selected, the cost of dispersal evolves toward lower values (see Armsworth 2008). Evolution reducing the cost of resistance to insecticides has been documented in detail through

Table 10.1 Some theoretical predictions about dispersal syndromes.

Type of syndrome	Evolving trait	Scale of covariation	Underlying gradient	Effect found	Nature of covariation	Cause of dispersal	Reference	Putative examples
life history	offspring dispersal	among individuals	fecundity	positive covariation between maternal fecundity and offspring dispersal	environmental	kin competition	(Kisdi 2004)	Plants; Chapter 14
life history	dispersal	among individuals	fecundity	sign of covariation between fecundity and dispersal changes with distribution of variation of fecundity within and between patches	environmental	kin competition and escape from crowding	(Bonte and de la Pena 2009)	All taxonomic groups; Bélichon *et al.* 1996, Chapters 13, 14
life history	lifespan and dispersal	among metapopulations	isolation	positive covariation between dispersal and lifespan	genetic	kin competition	(Dytham and Travis 2006)	Vertebrates; Bélichon *et al.* 1996; insects Chapter 13
life history	age at maturity	among metapopulations	natal dispersal	positive covariation between dispersal and age at maturity	genetic	escape from crowding	(de Jong *et al.* 2000)	
life history	aging	among metapopulations	natal dispersal	positive covariation between dispersal and lifespan	genetic	kin competition	(Ronce and Promislow 2010)	
life history	allocation to flight	among patches	resource level	sign of covariation between fecundity and flight ability changes with predictability of resource level	environmental	escape from bad habitats	(King and Roff 2010)	Insects; Chapter 13
mating system	seed and pollen dispersal	among metapopulations	inbreeding depression	negative covariation between seed and pollen dispersal	genetic	kin competition	(Ravigne *et al.* 2006)	Plants; Chapter 14
mating system	seed dispersal and selfing	among metapopulations	pollinators availability	negative covariation between seed dispersal and selfing	genetic	escape from crowding	(Cheptou and Massol 2009)	
ecological niche	dispersal and niche breadth	among families	habitat heterogeneity	positive covariation between niche breadth and dispersal	genetic	kin competition and bet-hedging	(Kisdi 2002)	Insects; Chapter 13
morphology	dispersal and size	among individuals	resource level	dispersers are bigger	environmental	kin competition	Kisdi *et al.* This volume	Vertebrates; Chapter 12
morphology	natal dispersal	among patches	body condition	mostly positive covariation between body condition and dispersal	environmental	kin competition	(Gyllenberg *et al.* 2008)	

For more examples of syndromes involving life-history traits, see Table 10.2 in Ronce and Olivieri (2004), which all involved comparisons among metapopulations. Putative empirical examples are indicated.

the characterization of successive resistance alleles, replacing each other sequentially in regions treated with insecticides (Labbe *et al.* 2009). Similarly, when documenting dispersal syndromes, one may observe genetic modifiers of dispersal having swept in the population recently following a change in selection regime, or syndromes with diminished costs having been fine-tuned by selection during a long history of evolution. There have been few attempts to connect the cost of dispersal to the history of selection of the population. More generally, we need more empirical studies investigating the fitness consequences of dispersal syndromes to better understand their adaptive nature (e.g. Cote *et al.* 2008).

10.4 Consequences of covariation

10.4.1 Demographic consequences

Surprisingly we know little about the consequences of individual heterogeneity on population dynamics (see review by Benton et al. 2006 and Chapter 20). In human populations, individual heterogeneity is recognized as having a big impact on longevity and mortality (Carnes and Olshansky 2001), and hence on human demography. In non-human species, heterogeneity has been seen more like a noise parameter, the impact of which should be corrected in order to estimate survival or population size in the wild (Chao 1987). A theory of population dynamics under individual heterogeneity has not yet been really developed (Pelletier *et al.* 2007; Ozgul *et al.* 2009). Although the effect of dispersal on fragmented populations has generated much work from at least two decades (Hanski 1991), the study of the interplay between dispersal, individuals heterogeneity, and the dynamic of fragmented population is still in its infancy.

10.4.1.1 Colonization, invasion, range expansion

To-date, the best evidence we have concerning the impact of dispersal-dedicated phenotypes on the dynamics of fragmented populations comes from the study of species at their range limits or on a colonization front (Clobert *et al.* 2009). Colonization success is enhanced when bluebirds have an aggressive phenotype (Chapter 6). Populations in small isolated patches founded by butterflies with higher flight metabolic rate and larger clutch size (elements associated with a greater dispersal propensity) have higher population growth rate (Hanski and Saccheri 2006; Chapters 8 and 23; Case Study III). In the cane toad (*Bufo marinus*), individuals with longer limbs were found to be more frequent on the colonization front (Phillips *et al.* 2006). The success of this phenotype at invading new habitats in turn accelerates the rhythm of invasion (Phillips *et al.* 2010a). The connection between colonization and dispersal phenotypes has been experimentally demonstrated in the common lizard. In this species, dispersers arising from kin competition are more exploratory (De Fraipont *et al.* 2000), take more risk (Cote and Clobert 2010), and display a higher colonizing success than individuals with a different phenotype (Cote *et al.* 2007). In plants, at the species level, seed-size heterogeneity is assumed, at least theoretically, to have important impact on colonization rate through a trade-off between seed size and seed number (Coomes and Grubb 2003). Specific morphological structures involved in seed-dispersal syndromes were shown to affect colonization rates in surveys of vegetation dynamics (Fröborg and Eriksson 1997). Although the evidence is still scarce, the role of a dispersal phenotype in enhancing colonization propensity and success is probably important for many species.

10.4.1.2 Extinction/stability

Whenever the phenotype of a dispersing individual decreases dispersal costs, or increases settlement or reproductive success in the new environment, it is likely to result in some impact on the overall metapopulation dynamics (Chapter 20). Besides its role in colonizing empty habitats, heterogeneity in dispersal syndromes is likely to affect the dynamics of spatially structured populations. For example, house sparrow juveniles with low rank in the nest are more prone to disperse, and have a higher success after settlement than non-dispersing individuals in the same population, hence decreasing the overall extinction probability of the recipient population (Altwegg *et al.* 2000).

In the common lizard, global warming is associated with a decrease in movements between and within populations at the cold margin of the distribution area (Massot *et al.* 2008). Heat-tolerant morphs, which have a higher reproductive rate, earlier sexual maturity, but lower mobility, have increased in frequency (Lepetz *et al.* 2009, and Chapter 25). As temperature increases, the population has a higher potential growth rate, but the lower mobility leads to a decreased connection with other populations and therefore a higher extinction risk, as predicted by Massot *et al.* (2008), and empirically supported by Sinervo *et al.* (2010). The disperser phenotype might also constitute a source of public information to other individuals in the population, and may therefore influence their own dispersal behaviour. Indeed, the phenotype of a disperser carries information on its own history and/or on its population of origin. Dispersal decisions by individuals in the recipient population based on the immigrant phenotypes has been experimentally demonstrated in the common lizard (Cote and Clobert 2007a). If a dispersal phenotype acts as a source of information on the presence and state of other neighbouring populations, the rescuing effect of dispersal at the metapopulation level may increase, as is potentially the case in the bog fritillary (Baguette *et al.* 2011). Cote *et al.* (2010) and Clobert *et al.* (2009) have discussed the potential effects that heterogeneity in dispersal syndromes might have on metapopulation dynamic. To summarize, dispersal dedicated phenotypes coupled with heterogeneity in dispersal syndromes might either stabilize or destabilize spatially structured populations dynamic depending on their effect on transience and settlement. For example, syndromes that enhance habitat matching processes are likely to reduce stability. Conversely, heterogeneity in dispersal-associated syndromes produced by different dispersal causes should increase stability if different phenotypes are associated to different transience and settlement strategies. Empirical examples are, however, still scarce, and it is difficult to present a general picture at this stage.

10.4.1.3 Interactions with other species

Individual heterogeneity has been recently recognized to have marked effects on communities, trophic chains, and ecosystem functioning (Schmitz 2008). We therefore expect that dispersal syndromes will also have some impact on competition, predation and meta-ecosystem functioning. One of the best current examples of such an effect is the mountain bluebird (Chapter 7). In this species, dispersers have a higher level of aggressiveness than resident individuals, which enable them to displace successfully another bird species occupying similar habitats (Duckworth and Badyaev 2007; Duckworth 2008). At the species level, fish of the genus *Gambusia* that are less bold, have a higher rate of feeding, and better predator tactics avoidance are found to be more invasive (Rehage and Sih 2004). More generally, life-history syndromes, including dispersal syndromes, have been proposed as explanations for patterns of successional replacement (see discussion in Olivieri and Gouyon 1997). The fact that dispersal rates differ systematically among trophic levels (Berg *et al.* 2010) could strongly affect the response of whole communities to climate change, when different species in interaction track their optimal climate through space with different lags.

10.4.2 Genetic consequences

10.4.2.1 Maintenance of genetic polymorphism

The multivariate nature of phenotypic variation among dispersers and non-disperser may be critical for the maintenance of genetic polymorphism for dispersal. Dispersal cost, and more generally trade-offs involving dispersal propensity, are indeed central to understand cases of protected polymorphism. For instance, Zheng *et al.* (2009) modelling *Pgi* polymorphism in realistic metapopulations showed that different alleles were maintained at equilibrium because the increased mobility and higher growth rate of heterozygous bearers of the C allele were balanced by the much lower survival of homozygotes for this allele (Orsini *et al.* 2009). Intuitively, one could expect that co-existence of genotypes with varying dispersal strategies could be facilitated if such genotypes have divergent ecological niches. Differences in life history and regeneration strategies could reduce overlap and competition between them. Conversely, divergence of dispersal strategies may help the co-existence of genotypes with different resource uses: Kisdi (2002) showed theoretically that in heterogeneous habitats, generalists and specialists could co-exist if their dispersal rate jointly evolves with specialization, while such co-existence was precluded if both genotypes had the same dispersal strategy. Joint evolution of specialization with dispersal, however, greatly reduced the range of parameters where polymorphism for dispersal could be maintained (Kisdi 2002). More studies assessing how the multidimensionality of phenotype divergence contributes to the maintenance of polymorphism are needed.

10.4.2.2 Gene flow, adaptation, and divergence

The fact that the life histories of dispersers may differ systematically from those of residents is likely to modify strongly the patterns of gene flow. For instance, Ingvarson and Whitlock (2000) showed that the increased fecundity of immigrants due to heterosis (hybrid vigour) could theoretically increase effective gene flow (see an empirical example by Ebert *et al.* 2002). Conversely, any costs attached to dispersal might decrease gene flow (e.g. see Marr *et al.* 2002; Hansson *et al.* 2004 for empirical examples in birds). More generally, there has been little attention in theoretical population genetics to the consequences of phenotype or condition-dependent dispersal for the structuring of genetic diversity through space (see Chapter 21). Vuilleumier *et al.* (2010) explored how deviation from random dispersal movements in heterogeneous habitat (i.e. ideal free distribution or resident pre-emption models of dispersal) affects the probability of fixation for an advantageous mutation. How covariation of dispersal with phenotypic traits affects the process of adaptation remains largely unexplored (see Edelaar *et al.* 2008, for suggestions of how it might affect evolution of niche breadth). Phenotype-dependent dispersal strategies can in particular accentuate phenotypic divergence among populations rather than constrain it as is usually assumed. This was exemplified in the case of the differential spatial settlement of birds of different sizes in great tits (Garant *et al.* 2005), and is discussed in detail in Edelaar *et al.* (2008). Armsworth and Roughgarden (2008) in particular found that fitness-dependent dispersal behaviour could theoretically sharpen genetic clines along environmental gradients, but the precise cues motivating dispersal (e.g. the precise fitness components assessed in the departure and target patches) influenced the shape of these clines greatly. There has been recent interest about how evolution of dispersal affects speciation processes (Heinz *et al.* 2009). Phenotype-dependent dispersal or matching habitat choice could greatly facilitate divergence and specialization in heterogeneous habitats when mating occurs after dispersal (e.g. Ravigne *et al.* 2009; the review in Edelaar *et al.* 2008). Theory however suggests that the evolution of assortative mating can be slowed down or accelerated by phenotype-dependent dispersal, depending on disturbance dynamics (Armsworth and Roughgarden 2005).

10.5 Conclusions

Theoretical studies suggest that many different traits should covary with dispersal, at many different scales, and with diverse causes (see Table 10.1). Consistently, empirical studies have revealed the multidimensional nature of dispersal phenotypes, involving variation in morphology, personality, physiology, life history, mating strategy, and ecological niche. Both theory and data, however, reveal that such syndromes are highly idiosyncratic, varying widely both between and within species, and are quite sensitive to matters of scale (e.g. comparison within versus between families or populations). The search for a general dispersal syndrome, therefore, appears to be very unpromising. Variation in syndromes can instead be investigated as a source of information on the causes and consequences of such syndromes. Not all dimensions of the dispersal phenotype show the same degree of lability and idiosyncrasy. Despite the general agreement between data and theory on the lability of dispersal syndromes, theoretical studies and empirical studies have mostly explored syndromes at widely different scales (see Table 10.1), and thus addressed in essence quite different questions. There is an urgent need to increase convergence and overlap in questions addressed theoretically and experimentally. Experimental metapopulations (e.g. Venail *et al.* 2008) provide a fruitful way to address the question of syndromes evolution at a global scale when dispersal between populations can be manipulated. Conversely, there is no major technical impediment to addressing the theoretical evolution of phenotype divergence between dispersing and non-dispersing individuals as a specialized form of phenotypic plasticity (see Chapter 11 as example). Examples of theoretical studies investigating the evolution and consequences of dispersal syndromes at the individual level remain too rare (see Table 10.1). Finally, there is a vast field of research to explore concerning the demographic and genetic consequences of dispersal syndromes, which deserves much more investigation both from an experimental and theoretical point of view.

References

Altwegg, R., Ringsby, T.H. and Saether, B.E. (2000). 'Phenotypic correlates and consequences of dispersal in a metapopulation of house sparrows Passer domesticus', *Journal of Animal Ecology* 69: 762–70.

Aragon, P., Meylan, S. and Clobert, J. (2006). 'Dispersal status-dependent response to the social environment in the Common Lizard, Lacerta vivipara', *Functional Ecology* 20: 900–7.

Armsworth, P.R. (2008). 'The evolution of non-random movement along clines', *Evolutionary Ecology Research* 10: 967–85.

Armsworth, P.R., and Roughgarden, J.E. (2005). 'Disturbance induces the contrasting evolution of reinforcement and dispersiveness in directed and random movers', *Evolution* 59: 2083–96.

Armsworth, P.R., and Roughgarden, J.E. (2008). 'The Structure of Clines with Fitness-Dependent Dispersal', *American Naturalist* 172: 648–57.

Baguette, M., Clobert, J. and Schtickzelle, N. (2011). 'Metapopulation dynamics of the bog fritillary butterfly: experimental changes in habitat quality induced negative density-dependent dispersal', *Ecography* 34: 170–6.

Baker, H.G. and Stebbins, G.L. (1965). *The genetics of colonizing species*. Academic Press, New York.

Bélichon, S., Clobert, J. and Massot, M. (1996). 'Are there differences in fitness components between philopatric and dispersing individuals?', *Acta Oecologica-International Journal of Ecology* 17: 503–17.

Bell, J.R., Bohan, D.A., Shaw, E.M. *et al.* (2005). 'Ballooning dispersal using silk: world fauna, phylogenies, genetics and models', *Bulletin of Entomological Research* 95: 69–114.

Belliure, J. and Clobert, J. (2004). 'Behavioral sensitivity to corticosterone in juveniles of the wall lizard, Podarcis muralis', *Physiology and Behavior* 81:121–7.

Belthoff, J.R. and Dufty, A.M. (1998). 'Corticosterone, body condition and locomotor activity: a model for dispersal in screech-owls', *Animal Behaviour* 55: 405–15.

Benard, M.F. and McCauley, S.J. (2008). 'Integrating across life-history stages: Consequences of natal habitat effects on dispersal', *American Naturalist* 171: 553–67.

Benton, T.G., Plaistow, S.J., Coulson, T.N. (2006). 'Complex population dynamics and complex causation: devils, details and demography', *Proceedings of the Royal Society B* 273: 1173–81.

Berg, M.P., Kiers, E.T., Driessen, G. *et al.* (2010). 'Adapt or disperse: understanding species persistence in a changing world', *Global Change Biology* 16: 587–98.

Billiard, S. and Lenormand, T. (2005). 'Evolution of migration under kin selection and local adaptation', *Evolution* 59: 13–23.

Bonte, D. (2009). 'Inbreeding depresses short and long distance dispersal in three congeneric spiders', *Journal of Evolutionary Biology* 22: 1429–34.

Bonte, D. and de la Pena, E. (2009). 'Evolution of body condition-dependent dispersal in metapopulations', *Journal of Evolutionary Biology* 22: 1242–51.

Bonte, D., Travis, J.M.J., De Clercq, N. *et al.* (2008). 'Thermal conditions during juvenile development affect adult dispersal in a spider', *Proceedings of the National Academy of Sciences of the USA* 105:17000–5.

Bonte, D., Van Dyck, H., Bullock, J. *et al.* (2012). 'Costs of dispersal', *Biological Reviews* 87: 290–312.

Boudjemadi, K., Lecomte, J. and Clobert, J. (1999). 'Influence of connectivity on demography and dispersal in two contrasting habitats: an experimental approach', *Journal of Animal Ecology* 68: 1207–24.

Cant, M.A., Otali, E. and Mwanguhya, F. (2001). 'Eviction and dispersal in co-operatively breeding banded mongooses (Mungos mungo)', *Journal of Zoology* **254**: 155–62.

Carnes, B.A. and Olshansky, S.J. (2001). 'Heterogeneity and its biodemographic implications for longevity and mortality', *Experimental Gerontology* 36: 419–30.

Chao, A. (1987). 'Estimating the Population-Size for Capture Recapture Data with Unequal Catchability', *Biometrics* 43: 783–91.

Chaput-Bardy, A., Gregoire, A., Baguette, M. *et al.* (2010). 'Condition and Phenotype-Dependent Dispersal in a Damselfly, *Calopteryx splendens*', *PLoS One* 5: e10694.

Cheptou, P.O. and Massol, F. (2009). 'Pollination Fluctuations Drive Evolutionary Syndromes Linking Dispersal and Mating System', *American Naturalist* 174: 46–55.

Clobert, J., Ims, R.A. and Rousset, F. (2004). 'Causes, mechanisms and consequences of dispersal', in I. Hanski and O. Gaggiotti, eds, *Ecology, Genetics and Evolution of Metapopulations*. Academic Press, Amsterdam, 307–35.

Clobert, J., Le Galliard, J. F., Cote, J. *et al.* (2009). 'Informed dispersal, heterogeneity in animal dispersal syndromes and the dynamics of spatially structured populations', *Ecology Letters* 12: 197–209.

Clobert, J., Perrins, C.M., McCleery, R.H. *et al.* (1988). 'Survival Rate in the Great Tit Parus Major in Relation to Sex, Age, and Immigration Status', *Journal of Animal Ecology* 57: 287–306.

Coomes, D.A. and Grubb, P.J. (2003). 'Colonization, tolerance, competition and seed-size variation within functional groups', *Trends in Ecology and Evolution* 18: 283–91.

Cote, J. and Clobert, J. (2007a). 'Social information and emigration: lessons from immigrants', *Ecology Letters* 10: 411–17.

Cote, J. and Clobert, J. (2007b). 'Social personalities influence natal dispersal in a lizard', *Proceedings of the Royal Society B* 274: 383–90.

Cote, J. and Clobert, J. (2010). 'Risky dispersal: avoiding kin competition despite uncertainty', *Ecology* 91: 1485–93.

Cote, J., Clobert, J., Brodin, T. *et al.* (2010). 'Personality-dependent dispersal: characterization, ontogeny and consequences for spatially structured populations', *Philosophical Transactions of the Royal Society B* 365: 4065–76.

Cote, J., Clobert, J. and Fitze, P.S. (2007). 'Mother -offspring competition promotes colonization success', *Proceedings of the National Academy of Sciences of the USA* 104: 9703–8.

Cote, J., Dreiss, A. and Clobert, J. (2008). 'Social personality trait and fitness', *Proceedings of the Royal Society B - Biological Sciences* 275: 2851–8.

Cousens, R., Dytham, C. and Law, R. (2008). *Dispersal in Plants: A Population Perspective.* Oxford University Press, Oxford.

Crnokrak, P. and Roff, D.A. (2002). 'Trade-offs to flight capability in *Gryllus firmus*: the influence of whole-organism respiration rate on fitness', *Journal of Evolutionary Biology* 15: 388–98.

Crowley, P.H. and McLetchie, D.N. (2002). 'Trade-offs and spatial life-history strategies in classical metapopulations', *American Naturalist* 159: 190–208.

De Fraipont, M., Clobert, J., John, H. *et al.* (2000). 'Increased pre-natal maternal corticosterone promotes philopatry of offspring in common lizards *Lacerta vivipara*', *Journal of Animal Ecology* 69: 404–13.

de Jong, T.J., Klinkhamer, P.G.L. and de Heiden, J.L.H. (2000). 'The evolution of generation time in metapopulations of monocarpic perennial plants: some theoretical considerations and the example of the rare thistle *Carlina vulgaris*', *Evolutionary Ecology* 14: 213–31.

De Meester, N. and Bonte, D. (2010). 'Information use and density-dependent emigration in an agrobiont spider', *Behavioral Ecology* 21: 992–8.

Dingemanse, N.J., Both, C., vanNoordwijk, A.J. *et al.* (2003). Natal dispersal and personalities in great tits (*Parus major*)', *Proceedings of the Royal Society Series B* 270: 741–7.

Dingemanse, N.J. and de Goede, P. (2004). 'The relation between dominance and exploratory behavior is context-dependent in wild great tits', *Behavioral Ecology* 15:1023–30.

Dingle, H. (1996). *Migration: The biology of life on the move.* Oxford University Press, New York.

Doligez, B. and Part, T. (2008). 'Estimating fitness consequences of dispersal: a road to 'know-where'? Non-random dispersal and the underestimation of dispersers' fitness', *Journal of Animal Ecology* 77: 1199–211.

Donohue, K., Polisetty, C.R. and Wender, N.J. (2005). 'Genetic basis and consequences of niche construction: Plasticity-induced genetic constraints on the evolution of seed dispersal in *Arabidopsis thaliana*', *American Naturalist* 165: 537–50.

Duckworth, R.A. (2008). 'Adaptive dispersal strategies and the dynamics of a range expansion', *American Naturalist* 172: S4–17.

Duckworth, R.A. (2009). 'Maternal effects and range expansion: a key factor in a dynamic process?', *Philosophical Transactions of the Royal Society B* 364: 1075–86.

Duckworth, R.A. and Badyaev, A.V. (2007). 'Coupling of dispersal and aggression facilitates the rapid range expansion of a passerine bird', *Proceedings of the National Academy of Sciences of the USA* 104: 15017–22.

Duckworth, R.A. and Kruuk, L.E.B. (2009). 'Evolution of Genetic Integration between Dispersal and Colonization Ability in a Bird', *Evolution* 63: 968–77.

Dufty, A., and Belthoff, J.R. (2001). 'Proximate mechanisms of natal dispersal: the role of body condition and hormones',in J. Clobert, E. Danchin, A. A. Dhondt *et al.* eds, *Dispersal.* Oxford University Press, Oxford, 217–29.

Dytham, C., and Travis, J.M.J. (2006). 'Evolving dispersal and age at death', *Oikos* 113: 530–8.

Ebert, D., Haag, C., Kirkpatrick, M. *et al.* (2002). 'A selective advantage to immigrant genes in a Daphnia metapopulation', *Science* 295: 485–8.

Edelaar, P., Siepielski, A.M. and Clobert, J. (2008). 'Matching Habitat Choice Causes Directed Gene Flow: a

Neglected Dimension in Evolution and Ecology', *Evolution* 62: 2462–72.

Elton, C.S. (2000). *The Ecology of Invasions by Animals and Plants.* University of Chicago Press, Chicago.

Fjerdingstad, E.J., Schtickzelle, N., Manhes, P. *et al.* (2007). 'Evolution of dispersal and life history strategies -Tetrahymena ciliates', *BMC Evolutionary Biology* 7: 133.

Frantz, A., Calcagno, V., Mieuzet, L. *et al.* (2009). 'Complex trait differentiation between host-populations of the pea aphid Acyrthosiphon pisum (Harris): implications for the evolution of ecological specialisation', *Biological Journal of the Linnean Society* 97: 718–27.

Fraser, D.F., Gilliam, J.F., Daley, M.J. *et al.* (2001). 'Explaining leptokurtic movement distributions: Intrapopulation variation in boldness and exploration', *American Naturalist* 158: 124–35.

Friedenberg, N.A. (2003). 'Experimental evolution of dispersal in spatiotemporally variable microcosms', *Ecology Letters* 6: 953–9.

Fröborg, H. and Eriksson, O. (1997). 'Local colonization and extinction of field layer plants in a deciduous forest and their dependence upon life history features', *Journal of Vegetation Science* 8: 395–400.

Garant, D., Kruuk, L.E.B., Wilkin, T.A. *et al.* (2005). 'Evolution driven by differential dispersal within a wild bird population', *Nature* 433: 60–5.

Gerlach, G. (1998). 'Impact of social ties on dispersal, reproduction and dominance in feral house mice (Mus musculus domesticus)', *Ethology* 104: 487–99.

Givnish, T.J. (1980). 'Ecological Constraints on the Evolution of Breeding Systems in Seed Plants -Dioecy and Dispersal in Gymnosperms', *Evolution* 34: 959–72.

Gloria-Soria, A. and Azevedo, R.B.R. (2008). 'npr-1 Regulates Foraging and Dispersal Strategies in Caenorhabditis elegans', *Current Biology* 18: 1694–9.

Gyllenberg, M., Kisdi, E. and Utz, M. (2008). 'Evolution of condition-dependent dispersal under kin competition', *Journal of Mathematical Biology* 57: 285–307.

Haag, C.R., Saastamoinen, M., Marden, J.H. *et al.* (2005). 'A candidate locus for variation in dispersal rate in a butterfly metapopulation', *Proceedings of the Royal Society B* 272: 2449–56.

Hanski, I. (1991). 'Single-species metapopulation dynamics: concepts, models and observations', *Biological Journal of the Linnean Society* 42: 17–38.

Hanski, I., Eralahti, C., Kankare, M. *et al.* (2004). 'Variation in migration propensity among individuals maintained by landscape structure', *Ecology Letters* 7: 958–66.

Hanski, I., Saastamoinen, M. and Ovaskainen, O. (2006). 'Dispersal-related life-history trade offs in a butterfly metapopulation', *Journal of Animal Ecology* 75: 91–100.

Hanski, I., and Saccheri, I. (2006). 'Molecular-level variation affects population growth in a butterfly metapopulation', *PLoS Biology* 4: 719–26.

Hanski, I. and Singer, M.C. (2001). 'Extinction-colonization dynamics and host-plant choice in butterfly metapopulations', *American Naturalist* 158: 341–53.

Hansson, B., Bensch, S. and Hasselquist, D. (2004). 'Lifetime fitness of short-and long-distance dispersing great reed warblers', *Evolution* 58: 2546–57.

Haughland, D.L., and Larsen, K.W. (2004). 'Exploration correlates with settlement: red squirrel dispersal in contrasting habitats', *Journal of Animal Ecology* 73: 1024–34.

Heinz, S.K., Mazzucco, R. and Dieckmann, U. (2009). 'Speciation and the evolution of dispersal along environmental gradients', *Evolutionary Ecology* 23: 53–70.

Higgins, S. I., Nathan, R. and Cain, M.L. (2003). 'Are long-distance dispersal events in plants usually caused by nonstandard means of dispersal?', *Ecology* 84: 1945–56.

Holekamp, K.E. and Smale, L. (1998). 'Dispersal status influences hormones and behavior in the male spotted hyena', *Hormones and Behavior* 33: 205–16.

Holekamp, K.E., Smale, L., Simpson, H.B. *et al.* (1984). 'Hormonal influences on natal dispersal in free-living Belding's ground squirrels (*Spermophilus beldingi*)', *Hormones and Behavior* **18**: 465–83.

Imbert, E. and Ronce, O. (2001). 'Phenotypic plasticity for dispersal ability in the seed heteromorphic *Crepis sancta* (Asteraceae)', *Oikos* 93: 126–34.

Ims, R.A. (1990). 'Determinants of Natal Dispersal and Space Use in Gray-Sided Voles, *Clethrionomys Rufocanus* -a Combined Field and Laboratory Experiment', *Oikos* 57: 106–13.

Ingvarsson, P.K., and Whitlock, M.C. (2000). 'Heterosis increases the effective migration rate. *Proceedings of the Royal Society B* 267: 1321–6.

Innocent, T.M., Abe, J., West, S.A. *et al.* (2010). 'Competition between relatives and the evolution of dispersal in a parasitoid wasp', *Journal of Evolutionary Biology* 23: 1374–85.

King, E.G., and Roff, D.A. (2010). 'Modeling the Evolution of Phenotypic Plasticity in Resource Allocation in Wing-Dimorphic Insects', *American Naturalist* 175: 702–16.

Kisdi, E. (2002). 'Dispersal: Risk spreading versus local adaptation', *American Naturalist* 159: 579–96.

Kisdi, E. (2004). 'Conditional dispersal under kin competition: extension of the Hamilton-May model to brood size-dependent dispersal', *Theoretical Population Biology* 66: 369–80.

Krackow, S. (2003). 'Motivational and heritable determinants of dispersal latency in wild male house mice (*Mus musculus musculus*)', *Ethology* 109: 671–89.

Labbe, P., Sidos, N., Raymond, M. *et al.* (2009). 'Resistance Gene Replacement in the Mosquito Culex pipiens: Fitness Estimation From Long-Term Cline Series', *Genetics* 182: 303–12.

Lepetz, V., Massot, M., Chaine, A.S. *et al.* (2009). 'Climate warming and the evolution of morphotypes in a reptile', *Global Change Biology* 15: 454–66.

Letty J., J. Aubineau and S. Marchandeau (2008). 'Improving Rabbit Restocking Success: A Review of Field Experiments in France', *Lagomorph Biology* 5: 327–48.

Lowe, W.H. (2010). 'Explaining long-distance dispersal: effects of dispersal distance on survival and growth in a stream salamander', *Ecology* 91: 3008–15.

Mabry, K.E. and Stamps, J.A. (2008). 'Dispersing brush mice prefer habitat like home', *Proceedings of the Royal Society B* 275: 543–8.

Magalhaes, S., E. Blanchet, M. Egas, and I. Olivieri. (2009). 'Are adaptation costs necessary to build up a local adaptation pattern?', *BMC Evolutionary Biology* 9: 182.

Marr, A.B., Keller, L.F. and Arcese, P. (2002). 'Heterosis and outbreeding depression in descendants of natural immigrants to an inbred population of song sparrows (*Melospiza melodia*)', *Evolution* 56: 131–42.

Massot, M., Clobert, J. and Ferriere, R. (2008). 'Climate warming, dispersal inhibition and extinction risk', *Global Change Biology* 14: 461–9.

Mehlman, P.T., Higley, J.D., Faucher, I. *et al.* (1995). 'Correlation of CSF 5-HIAA concentration with sociality and the timing of emigration in free-ranging primates', *American Journal of Psychiatry* **152**: 907–13.

Mery, F., Belay, A.T., So, A.K.C. *et al.* (2007). 'Natural polymorphism affecting learning and memory in *Drosophila*', *Proceedings of the National Academy of Sciences of the USA* 104: 13051–5.

Meylan, S., Clobert, J. and De Fraipont, M. (2000). 'Maternal stress and juvenile dispersal in the common lizard', *American Zoologist* 40: 1132–1132.

Meylan, S., De Fraipont, M., Aragon, P. *et al.* (2009). 'Are Dispersal-Dependent Behavioral Traits Produced by Phenotypic Plasticity?', *Journal of Experimental Zoology Part A-Ecological Genetics and Physiology* 311A: 377–88.

Mix, C., Pico, F.X., van Groenendael, J.M. *et al.* (2006). 'Inbreeding and soil conditions affect dispersal and components of performance of two plant species in fragmented landscapes',. *Basic and Applied Ecology* 7: 59–69.

Moles, A. T. and Westoby, M. (2006). 'Seed size and plant strategy across the whole life cycle', *Oikos* 113: 91–105.

Molofsky, J. and Ferdy, J.B. (2005). 'Extinction dynamics in experimental metapopulations', *Proceedings of the National Academy of Sciences of the USA* 102: 3726–31.

Nathan, R., Schurr, F.M., Spiegel, O. *et al.* (2008). 'Mechanisms of long-distance seed dispersal', *Trends in Ecology and Evolution* 23: 638–47.

Nichols, R. and Bondrupnielsen, S. (1995). 'The Effect of a Single-Dose of Testosterone Propionate on Activity, and Natal Dispersal in the Meadow Vole', *Microtus-Pennsylvanicus*. *Annales Zoologici Fennici* 32: 209–15.

Nunes, S., Duniec, T.R., Schweppe, S.A. *et al.* (1999). 'Energetic and endocrine mediation of natal dispersal behavior in Belding's ground squirrels', *Hormones and Behavior* 35: 113–24.

O'Riain, M.J., Jarvis, J.U.M. and Faulkes, C.G. (1996). 'A dispersive morph in the naked mole-rat', *Nature* 380: 619–21.

Olivieri, I., and Gouyon, P.-H. (1997). 'Evolution of Migration Rate and Other traits: The Metapopulation Effect', in I. Hanski and M. E. Gilpin, eds, *Metapopulation Biology: Ecology, Genetics, and Evolution*. Academic Press, San Diego, 293–323.

Orsini, L., Wheat, C.W., Haag, C.R. *et al.* (2009). 'Fitness differences associated with Pgi SNP genotypes in the Glanville fritillary butterfly (*Melitaea cinxia*)', *Journal of Evolutionary Biology* 22: 367–75.

Ozgul, A., Tuljapurkar, S., Benton, T.G. *et al.* (2009). 'The Dynamics of Phenotypic Change and the Shrinking Sheep of St. Kilda', *Science* 325: 464–7.

Pelletier, F., Clutton-Brock, T. Pemberton, J. *et al.* (2007). 'The evolutionary demography of ecological change: Linking trait variation and population growth', *Science* 315: 1571–4.

Phillips, B.L., Brown, G.P. and Shine, R. (2010a). 'Evolutionarily accelerated invasions: the rate of dispersal evolves upwards during the range advance of cane toads', *Journal of Evolutionary Biology* 23: 2595–601.

Phillips, B.L., Brown, G.P. and Shine, R. (2010b). 'Life-history evolution in range-shifting populations', *Ecology* 91: 1617–27.

Phillips, B.L., Brown, G.P., Webb, J.K. *et al.* (2006). 'Invasion and the evolution of speed in toads', *Nature* 439: 803–803.

Portnoy, S. and Willson, M.F. (1993). 'Seed Dispersal Curves -Behavior of the Tail of the Distribution', *Evolutionary Ecology* 7: 25–44.

Ravigne, V., Dieckmann, U. and Olivieri, I. (2009). 'Live Where You Thrive: Joint Evolution of Habitat Choice and Local Adaptation Facilitates Specialization and Promotes Diversity', *American Naturalist* 174: E141–69.

Ravigne, V., Olivieri, I., Gonzalez-Martinez, S.C. *et al.* (2006). 'Selective interactions between short-distance pollen and seed dispersal in self-compatible species', *Evolution* 60: 2257–71.

Rehage, J.S. and Sih, A. (2004). 'Dispersal behavior, boldness, and the link to invasiveness: a comparison of four *Gambusia* species', *Biological Invasions* 6: 379–91.

Roff, D.A. and Bradford, M.J. 1996. 'Quantitative genetics of the trade-off between fecundity and wing dimorphism in the cricket *Allonemobius socius*', *Heredity* 76: 178–85.

Roff, D.A. and Fairbairn, D.J. (2007). 'The evolution of trade-offs: where are we?', *Journal of Evolutionary Biology* 20: 433–47.

Roff, D.A. and Gelinas, M.B. (2003). 'Phenotypic plasticity and the evolution of trade-offs: the quantitative genetics of resource allocation in the wing dimorphic cricket, *Gryllus firmus*', *Journal of Evolutionary Biology* 16: 55–63.

Ronald J. Sarno, Bank, R.J., Stern, M.S. *et al.* (2003). 'Forced dispersal of juvenile guanacos (*Lama guanicoe*): causes, variation, and fates of individuals dispersing at different times', *Behavioural Ecology and Sociobiology* 54: 22–9.

Ronce, O. and Kirkpatrick, M. (2001). 'When sources become sinks: migrational meltdown in heterogeneous habitats', *Evolution* 55: 1520–31.

Ronce, O. and Olivieri, I. (1997). 'Evolution of reproductive effort in a metapopulation with local extinctions and ecological succession', *American Naturalist* 150: 220–49.

Ronce, O. and Olivieri, I. (2004). 'Life history evolution in metapopulations', in I. Hanski and O.E. Gaggiotti, eds, *Ecology, Genetics, and Evolution of Metapopulations*. Academic Press, Amsterdam, 227–58.

Ronce, O., Perret, F. and Olivieri, I. (2000). 'Landscape dynamics and evolution of colonizer syndromes: interactions between reproductive effort and dispersal in a metapopulation', *Evolutionary Ecology* 14: 233–60.

Ronce, O. and Promislow, D. (2010). 'Kin competition, natal dispersal and the moulding of senescence by natural selection', *Proceedings of the Royal Society B* 277: 3659–67.

Saastamoinen, M. (2007). 'Life-history, genotypic, and environmental correlates of clutch size in the Glanville fritillary butterfly', *Ecological Entomology* 32: 235–42.

Saastamoinen, M., Ikonen, S. and Hanski, I. (2009). 'Significant effects of Pgi genotype and ',body reserves on lifespan in the Glanville fritillary butterfly. *Proceedings of the Royal Society B* 276: 1313–22.

Schmitz, O.J. (2008). 'Herbivory from Individuals to Ecosystems', *Annual Review of Ecology, Evolution and Systematics* 39: 133–52.

Schtickzelle, N., Fjerdingstad, E.J., Chaine, A. *et al.* (2009). 'Cooperative social clusters are not destroyed by dispersal in a ciliate', *BMC Evolutionary Biology* 9: 251.

Silverin, B. (1997). 'The stress response and autumn dispersal behaviour in willow tits', *Animal Behaviour* 53: 451–9.

Sinervo, B., Calsbeek, R., Comendant, T. *et al.* (2006a). 'Genetic and maternal determinants of effective dispersal: The effect of sire genotype and size at birth in side-blotched lizards', *American Naturalist* 168: 88–99.

Sinervo, B., Chaine, A., Clobert, J. *et al.* R. (2006b). 'Self-recognition, color signals, and cycles of greenbeard mutualism and altruism', *Proceedings of the National Academy of Sciences of the USA* 103: 7372–7.

Sinervo, B., and Clobert, J. (2003). 'Morphs, dispersal behavior, genetic similarity, and the evolution of cooperation', *Science* 300: 1949–51.

Sinervo, B., Mendez-de-la-Cruz, F., Miles, D.B. *et al.* (2010). 'Erosion of Lizard Diversity by Climate Change and Altered Thermal Niches', *Science* 328: 894–9.

Székely, T., A.J. Moore, and J. Komdeur. (2010). *Social Behaviour: Genes, Ecology and Evolution*. Cambridge University Press, Cambridge.

Thomas, C.D., Bodsworth, E.J., Wilson, R.J. *et al.* (2001). 'Ecological and evolutionary processes at expanding range margins', *Nature* 411: 577–81.

Thomson, F.J., Moles, A.T., Auld, T.D. *et al.* (2010). 'Chasing the unknown: predicting seed dispersal mechanisms from plant traits', *Journal of Ecology* 98: 1310–18.

Torriani, M.V.G., Mazzi, D., Hein, S. *et al.* (2010). 'Direct and correlated responses to artificial selection on flight activity in the oriental fruit moth (Lepidoptera: Tortricidae)', *Biological Journal of the Linnean Society* 100: 879–89.

Trefilov, A., J. Berard, M. Krawczak, and J. Schmidtke. (2000). 'Natal dispersal in rhesus macaques is related to serotonin transporter gene promoter variation', *Behavior Genetics* 30: 295–301.

Van Noordwijk, A.J. and De Jong, G. (1986). 'Acquisition and allocation of resources: their influence on variation in life history tactics', *American Naturalist* 128: 137–42.

Venail, P.A., MacLean, R.C., Bouvier, T. *et al.* (2008). 'Diversity and productivity peak at intermediate dispersal rate in evolving metacommunities', *Nature* 452: 210-U57.

Vercken, E., de Fraipont, M., Dufty, A.M. *et al.* (2007). 'Mother's timing and duration of corticosterone exposure modulate offspring size and natal dispersal in the common lizard (*Lacerta vivipara*)', *Hormones and Behavior* 51: 379–86.

Vuilleumier, S., Goudet, J. and Perrin, N. (2010). 'Evolution in heterogeneous populations From migration models to fixation probabilities', *Theoretical Population Biology* 78: 250–8.

Williams, C. B. (1957). 'Insect Migration', *Annual Review of Entomology* 2: 163–80.

Willson, M.F. (1993). 'Dispersal Mode, Seed Shadows, and Colonization Patterns', *Vegetatio* 108: 261–80.

Wilson, M., and Traveset, A. (2000). 'The Ecology of Seed Dispersal', in M. Fenner, ed, *Seeds: The ecology of regeneration in plant communities*. CAB International, Wallingford, 85–110.

Zera, A.J. and Denno, R.F. (1997). 'Physiology and ecology of dispersal polymorphism in insects', *Annual Review of Entomology* 42: 207–30.

Zera, A.J., Harshman, L.G. and Williams, T.D. (2007). 'Evolutionary endocrinology: The developing synthesis between endocrinology and evolutionary genetics', *Annual Review of Ecology Evolution and Systematics* 38: 793–817.

Zheng, C.Z., Ovaskainen, O. and Hanski, I. (2009). 'Modelling single nucleotide effects in phosphoglucose isomerase on dispersal in the Glanville fritillary butterfly: coupling of ecological and evolutionary dynamics', *Philosophical Transactions of the Royal Society B* - 364: 1519–32.

CHAPTER 11

Evolution of condition-dependent dispersal

Eva Kisdi, Margarete Utz, and Mats Gyllenberg

11.1 Introduction

There are a huge number of models exploring the evolution of dispersal, yet there is surprisingly little theoretical work addressing how dispersal should depend on the phenotype of individuals. The majority of models treat dispersal as a single number, the fraction of dispersers ('dispersal rate'). In reality, the decision to disperse depends on a number of factors including local population density, the presence and number of relatives, the quality of the natal habitat, sex, age, and properties relating to the competitive ability and survival of the focal individual or its mother (see Ims and Hjermann 2001; Bowler and Benton 2005; Benard and McCauley 2008; Clobert *et al.* 2009 for reviews; Chapters 4, 10, 12, and 13). While several models have investigated the evolution of density-dependent dispersal (e.g. Jánosi and Scheuring 1997; Travis *et al.* 1999; Gyllenberg and Metz 2001; Metz and Gyllenberg 2001; Poethke and Hovestadt 2002), dispersal in relation to the strength of kin competition (Ezoe and Iwasa 1997; Ronce *et al.* 1998; Kisdi 2004), and sex-related dispersal (Perrin and Mazalov 2000), models exploring how dispersal depends on body condition have just started to appear (Bonte and de la Peña 2009; Gyllenberg *et al.* 2008, 2011a, 2011b).

Empirical data show a variable relationship between dispersal propensity and measures of body condition such as body size or energy reserves. In some cases dispersers are weaker than those individuals who remain in the natal area, but more often the opposite holds so that the strong individuals are more prone to disperse (Ims and Hjermann 2001; Bowler and Benton 2005; Benard and McCauley 2008; Clobert *et al.* 2009). Inconsistent patterns can be expected because there are opposing selective forces acting on dispersal (Ims and Hjermann 2001). For example, having a strong body with large reserves may indicate that the natal site is in a high-quality habitat, where philopatry is advantageous. At the same time, strong individuals may have a greater capacity to disperse in terms of coping with the energetic costs of dispersal, avoiding hazards during transience, and successfully acquiring resources and mates in a new environment, which selects for dispersal. Yet having lower costs of dispersal cannot by itself explain why strong individuals disperse at all; they could as well choose to avoid any dispersal hazards, and instead use their strength to monopolize the resources of their natal site and force their weaker conspecifics to leave (Bonte and de la Peña 2009; the social dominance hypothesis or 'ideal despotic distribution' of Fretwell 1972). Integrating opposing selective forces in complex ecological systems is difficult via purely verbal arguments, partly because quantitative relationships matter and partly because less obvious selective forces or essential but hidden assumptions may be overlooked.

In this chapter, we summarize and extend our first attempts at modelling condition-dependent dispersal. Our focal question is whether competitively strong or weak individuals should disperse. First we assume that competitive ability varies due to some environmental variation that the organism cannot control, but it can adapt its dispersal strategy to the environment and to its own body condition. Alternatively, we assume that parents have a variable amount of resources that they can allocate to produce

Dispersal Ecology and Evolution. First Edition. Edited by Jean Clobert *et al.*

few strong or many weak offspring along the offspring size-number trade-off. Accordingly, we treat two families of models addressing (1) how the evolutionarily stable dispersal strategy depends on body condition, and (2) how offspring size co-evolves with dispersal. Our synopsis of model (1) is based on Gyllenberg *et al.* (2011a, 2011b). For model (2) we present new results and include the mathematical derivations in the Appendix.

11.2 Outline of models

Consider a population that inhabits small patches of habitat, which differ in their quality such that good patches tend to produce strong offspring in model (1) and provide much resource for offspring production in model (2). At the beginning of each generation, each patch is occupied by exactly one individual. The life cycle is as follows:

1. Survival till reproduction: here we assume the probability of survival, s, to be constant (see Gyllenberg *et al.* 2011b, for results where survival until maturity depends on the quality of the patch).
2. Reproduction: the parent produces a large number of clonal offspring and dies. The offspring differ in their body condition z within and between families. In model (1), the distribution of body condition within a family, $\beta(z, y)$, depends on patch quality y. In model (2), body condition follows a Gaussian distribution $\beta(z-m)$ with constant variance around the size of the offspring, m, which evolves subject to the offspring size-number trade-off.
3. Dispersal: in model (1), the probability of dispersal, $p(z, y)$, depends both on the body condition of the focal offspring and on the quality of its natal patch that determines the distribution of body condition among its sibs. In model (2), offspring equipped with a dispersal structure (wing, pappus, etc.) all disperse, and the parent controls dispersal via allocating the available resources between dispersing and non-dispersing offspring. Survival during dispersal, $\Pi(z)$, may increase or decrease with body condition z.
4. Competition: after dispersal, the offspring compete such that only one individual survives in each patch. Body condition determines competitive ability. Competition is modelled by a weighted lottery, where the competitive weight $g(z)$ of an offspring increases with its body condition z; the probability that an offspring wins the patch is proportional to its competitive weight $g(z)$ and is inversely proportional to the sum of the competitive weights of all offspring competing for the patch.

In these models, kin competition among the offspring selects for dispersal (Hamilton and May 1977; see Chapter 12), whereas the mortality cost of dispersal and, in model (2), the possible energetic cost of the dispersal structure select against it.

11.3 Model 1: evolutionarily stable dispersal strategies

In our model (1), the dispersal strategy is given by the probability that an offspring with condition z disperses from patch y, $p(z, y)$, and the shape of this function evolves. The ESS can be obtained only numerically, but qualitative properties of the ESS are found analytically (Gyllenberg *et al.* 2008, 2011a, 2011b).

Evolutionarily stable strategies need not be uniquely determined. If the probability of survival during dispersal does not depend on body condition (Π is constant), then many dispersal strategies are selectively neutral to each other. All that matters is how much competitive weight the family keeps in its natal patch y (formally given by $n(p,y)=\int[1-p(z,y)]\beta(z,y)g(z)dz$) and all dispersal strategies with the same competitive weight $n(p, y)$ are selectively indistinguishable. For example, a family may retain many weak offspring or fewer strong offspring in the natal patch to arrive at the same value of $n(p, y)$, or it may simply disperse each offspring from patch y with the same condition-independent probability $p(y)$. There is an evolutionarily stable value of the retained competitive weight, but this value corresponds to an entire class of many different strategies.

Because widely different strategies within the ESS class are selectively neutral, it is not possible to predict a definite relationship between body condition and dispersal. Populations of the same species living in similar habitats can thus exhibit different dispersal patterns, yet these all correspond to the same underlying ESS class. Strategies with contrasting patterns can also co-exist in one population, governed only by genetic drift. Inconsistent relationships between body condition and dispersal, found in empirical data (Ims and Hjermann 2001; Bowler and Benton 2005), might thus exist simply because they yield the same fitness.

Survival during dispersal determines which offspring of a family disperse. If the probability of survival during dispersal depends on body condition, then the above neutrality is broken and those offspring disperse that have the highest probability of surviving. Within each family, dispersal follows a 'bang-bang' strategy. If strong offspring have a higher chance of surviving dispersal (Benard and McCauley 2008; Clobert *et al.* 2009), then all offspring above a threshold body condition disperse and all below the threshold remain in the natal patch. If competitively weak individuals have higher survival during dispersal (e.g. because small individuals can hide more easily), then the pattern is the reverse. The threshold value of body condition depends on the quality of the patch, so that individuals in the same condition may or may not disperse depending on their patch.

Weak selection on weak offspring. Whatever dispersal strategy they follow, offspring of weak body condition are likely doomed. Therefore, dispersal strategies that only differ in how the weak offspring disperse have little differences in fitness, so that a small change in the model can substantially change the ESS dispersal strategy $p(z, y)$ at low values of z. This has been demonstrated by Gyllenberg *et al.* (2008) in a special case (see their example 3.4), but we expect it to be generally true.

Fixed versus random quality. Further details of the results depend on whether patches remain permanently good or bad (fixed quality), or the quality of a patch is changing from year to year (random quality, no autocorrelation assumed).

With fixed quality, bad patches are usually abandoned (Figure 11.1a). There is, however, an exception: if some individuals have very low probability of surviving dispersal *relative* to their siblings, then these may stay even in a bad patch, because some of their offspring will be substantially better dispersers than they themselves are (Gyllenberg *et al.* 2011b).

With random quality, every patch is equally valuable to retain because its future quality is independent of the present. If survival during dispersal does not depend on body condition, then each patch retains the same competitive weight (except families that have less weight in total; Figure 11.1b), which can be made up by many weak offspring in bad patches and fewer but stronger offspring in good patches. This result is analogous to Ezoe and Iwasa (1997) and Kisdi (2004), who assumed identical offspring and found that a constant number of them are retained in each patch. However, if strong offspring survive dispersal better, then families in very good patches retain

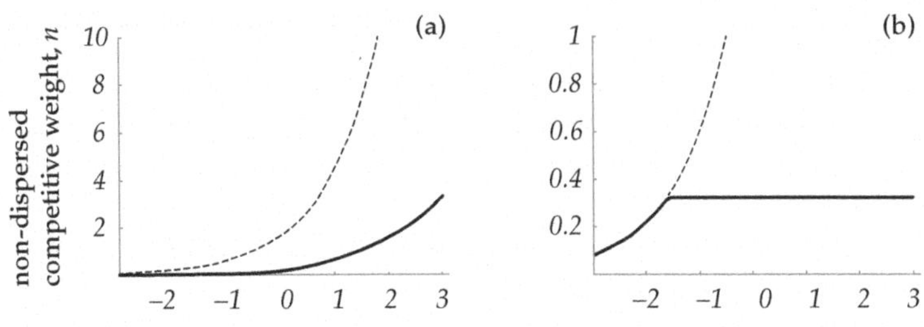

Figure 11.1 Non-dispersed competitive weight $n(p, y)$ at the ESS as a function of patch quality y (thick line) for constant Π and (a) fixed quality versus (b) random quality. The dashed lines show the total competitive weight produced in the patch $[\int\beta(z,y)g(z)dz]$, and the difference between the thick and dashed lines represents the dispersing competitive weight. In (a), $n(p, y) = 0$ for small y because all offspring disperse from permanently bad patches. In (b), every family retains the same competitive weight except in bad patches where the total weight is less than the target weight to retain; these families retain all. Assumptions and parameters: patch quality y follows a truncated normal distribution with mean standardized to 0 and variance 1; $\beta(z, y)$ is truncated normal with mean y and variance 1; $g(z) = e^z$; $s = 0.5$; $\Pi = 0.5$. (a) is from Gyllenberg *et al.* (2011b), reproduced with permission from Elsevier; (b) is modified from Gyllenberg et al. (2011a), reproduced with permission from Taylor and Francis.

less competitive weight in the natal patch (Gyllenberg *et al.* 2011a). This is because at the ESS, the reward from a dispersed offspring is equal to the reward from improving the chance to win the natal patch if one more offspring is kept at home (Gandon and Michalakis 1999; Kisdi 2004). The chance to win the natal patch saturates as the retained competitive weight increases. For families where the reward from dispersed offspring is higher (because they likely survive dispersal), the balance is reached at less competitive weight retained in the natal patch.

Dispersal as a function of body condition. Empirical studies measure the propensity of dispersing in relation to body condition (quantified, for example, by body size) among all individuals of a population. To predict this relationship, the left panels of Figure 11.2 show the ESS dispersal probability averaged across patches of origin. Whether strong or weak offspring disperse in a population depends on whether families in good or in bad patches disperse more offspring, and which offspring of each family disperse.

With random quality and increasing $\Pi(z)$, families in good patches disperse more of their strong offspring, and within each family, the stronger offspring disperse. As a consequence, strong offspring are dispersing in the population (Figure 11.2a). Conversely, with fixed quality and decreasing $\Pi(z)$, families in bad patches disperse all their offspring, and from other patches the weaker offspring disperse; as a consequence, all weak offspring of the population disperse and the probability of dispersal declines towards strong body condition (Figure 11.2d). In the remaining two cases, the within-family dispersal pattern is opposite to the pattern over different patches. With random quality and decreasing $\Pi(z)$, the weaker offspring of families living in good patches disperse most, hence the probability of dispersal peaks at an intermediate body condition (Figure 11.2b). With fixed quality and increasing $\Pi(z)$, the pattern is the opposite: Here the weaker offspring of good patches stay in the natal patch, resulting in a minimum of dispersal at intermediate conditions (Figure 11.2c). All weak offspring disperse because they are born in bad patches that they abandon; and all strong offspring disperse because they will survive dispersal better than their sibs. Benard and McCauley (2008) predicted a similar pattern by a verbal argument.

Note that the strongest offspring disperse if, and only if, they have a better chance of surviving dispersal, and this is independent of the spatio-temporal structure of the environment (compare Figure 11.2 (a)–(c) and (b)–(d) at high values of z). Random versus fixed quality however lead to different dispersal propensities of weak offspring, as their dispersal depends more on the particular circumstances (see *'weak selection on weak offspring'* earlier).

The distribution of body condition among dispersers and non-dispersers. Perhaps the most straightforward empirical test to use for the effect of body condition on dispersal is to compare the mean condition of dispersers with that of non-dispersers; a more complete analysis may compare the entire distributions of body condition rather than just the means. When the within- and between-family patterns of dispersal are similar, as in Figure 11.2a,d, then the distributions of dispersers and of non-dispersers are clearly different (compare the middle and right panels). However, when the within-family dispersal pattern is opposite to the pattern across families, as in Figure 11.2b,c, then there is little difference between the distributions of dispersing versus non-dispersing offspring, and the mean body conditions of dispersers and of non-dispersers are virtually the same. A simple empirical test may thus suggest no effect of body condition on dispersal, even though in reality there is a marked pattern both within and among families.

11.4 Model 2: co-evolution of dispersal and offspring size-number strategies

The competitive ability of offspring is likely to correlate with their body size at birth. To investigate the co-evolution of dispersal and offspring size, suppose that the patches differ in the amount of resources available for offspring production. In each patch, the parent must allocate the available resources between the number and the size of offspring, and thereby produce either many small offspring who tend to be weak, or a few large offspring who tend to be strong (with some residual variance in body condition not explained by birth size).

In order to disperse, the offspring may need to be equipped with special structures (wings, pappi, etc.) already at birth, when only their size

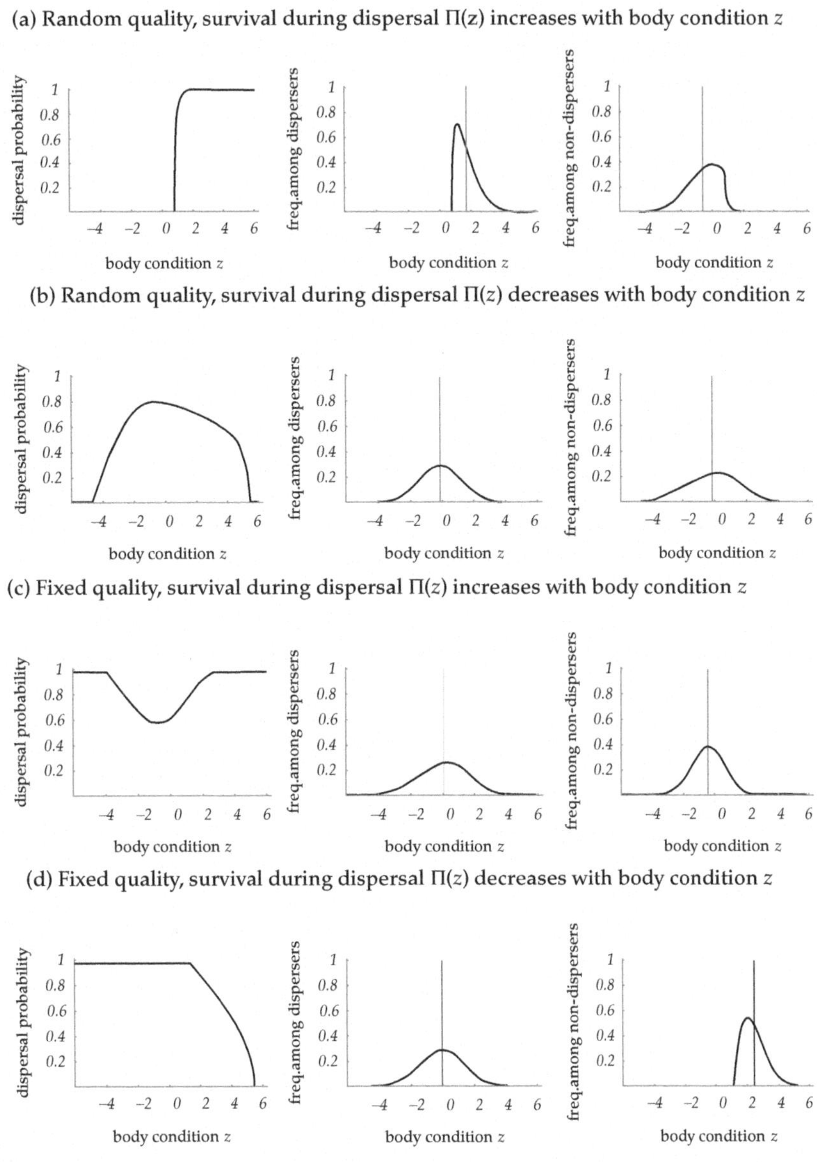

Figure 11.2 Left panels: the probability of dispersal for an individual with body condition z, averaged over the population. Middle and right panels: the distribution of body condition among the dispersers and among the non-dispersers, respectively; vertical lines indicate the mean. Assumptions and parameters as in Figure 11.1, except $\Pi(z)=0.2+0.6/(1+e^{-z})$ in (a,c) and $\Pi(z)=0.8-0.6/(1+e^{-z})$ in (b,d); $s=0.9$ in (a,b) and $s=0.5$ in (c,d). (a,b) is from Gyllenberg *et al.* (2011a), reproduced with permission from Taylor and Francis; (c,d) is from Gyllenberg *et al.* (2011b), reproduced with permission from Elsevier.

is known but their precise body condition is not; these structures may also be energetically costly. In other organisms, special dispersal structures are not necessary and the decision on dispersal can be delayed until after development, so that it can be based on the actual body condition of the offspring. Here we focus on the former case, assuming that all offspring equipped with dispersal structures will indeed disperse. The life cycle is as outlined at the beginning of this chapter, and the mathematical derivations are given in the Appendix.

Joint strategies of offspring size and dispersal. A parent living in patch y first has to split the amount of resources available in the patch, $E(y)$, into the amount used to produce non-dispersing offspring, $E_n(y)$, and the amount used to produce dispersing offspring, $E_d(y)$; naturally, $E_n(y)+E_d(y)=E(y)$ must hold. Next, the parent must decide on the size $M_n(y)$ of an offspring not equipped with dispersal structures; the number of non-dispersing offspring, $E_n(y)/m_n(y)$, follows from the size-number trade-off. Similarly, if each dispersing offspring is of size $m_d(y)$, then the number of dispersing offspring will be $E_d(y)/[m_d(y)+c]$, where c is the energetic cost of the dispersal structure. The life-history strategy is given by the choice of $E_n(y)$, $m_n(y)$, and $m_d(y)$, whereas the number and condition of dispersing and non-dispersing offspring follows respectively from the size-number trade-off and from the Gaussian distribution of body condition around offspring size, $\beta(z-m)$.

Optimal offspring size. The size of non-dispersing offspring must be chosen so that the total competitive weight remaining in the natal patch,

$$\frac{E_n(y)}{m_n(y)}\int \beta(z-m_n(y))g(z)dz, \qquad \text{(Equation 11.1)}$$

is maximal, whereas the size of dispersing offspring must maximize the successfully dispersed competitive weight

$$\frac{E_d(y)}{m_d(y)+c}\int \beta(z-m_d(y))\Pi(z)g(z)dz \qquad \text{(Equation 11.2)}$$

It is easy to see that the maximum points of expressions (11.1) and (11.2) are independent of $E_n(y)$ and $E_d(y)$. There is thus a single optimal size m_n of non-dispersing offspring and a single optimal size m_d of dispersing offspring that are the same in every patch and independent of how much resources are invested into the offspring. This result is analogous to classic life-history models addressing the size-number trade-off (Smith and Fretwell 1974). Note that m_n and m_d depend on the variance of the Gaussian distribution β, and if patches differed also in the variance of offspring condition (which we do not assume here), then m_n and m_d would vary with patch type y. For a given parent, the best choice of m_n and m_d are independent of what strategy other parents might follow, hence offspring size is optimized and is not merely an ESS (see Lloyd 1987; Metz *et al.* 2008).

Contrast between the body condition of dispersers and of non-dispersers. Whether dispersing offspring are on average stronger or weaker than non-dispersing offspring depends on whether their optimal size, which is also their expected body condition, is larger or smaller. There are two factors that influence m_d relative to m_n (compare expressions (11.1) and (11.2) above). First, increasing the cost of the dispersal structure (c) increases the optimal size m_d, so that offspring equipped with costly dispersal structures are larger and stronger than non-dispersers. This is a well-known effect of any fixed cost: parents should decrease the number of offspring (by increasing their size) to decrease the number of times the cost has to be paid (see e.g. de Jong and Klinkhamer 2005, p. 117).

The second difference between non-dispersing and dispersing offspring is that the latter should maximize the *successfully* dispersed competitive weight, which includes the body condition-dependent probability $\Pi(z)$ of survival during dispersal. The following proposition characterizes how changes of the shape of function (z) affect the optimal size of dispersing offspring:

> Let α be a parameter that controls the shape of $\Pi(z)$. Define $M=\int z\beta(z-m_d)g(z)\Pi(z)dz/\int\beta(z-m_d)g(z)\Pi(z)dz$; heuristically, M is the average of body condition weighted with both success in dispersal ($\Pi(z)$) and success in competition ($g(z)$). Increasing parameter α increases the optimal size of dispersing offspring if, and only if, it increases M. (In other words, $dm_d/d\alpha$ has the same sign as $\partial M / \partial \alpha$ at the optimal m_d.)

This proposition gives a straightforward numerical recipe to evaluate the effect of any small change in the shape of the survival function $\Pi(z)$ on the size and condition of dispersing offspring. Evaluate M using the 'original' and the 'new' function $\Pi(z)$ (keeping m_d at the original optimum); if the new function gives a higher value for M then it selects for larger offspring size. Further, we use this proposition to obtain our main result on the body condition of dispersing versus non-dispersing individuals (see Appendix):

> If survival during dispersal increases with body condition, then dispersing offspring are larger

and therefore, on average, stronger than non-dispersing offspring. If survival decreases with body condition, then it selects for dispersing offspring to be smaller than non-dispersing offspring; but the cost of dispersal structures (c) selects for the opposite, so that m_d may be larger or smaller than m_n depending on which mechanism has the stronger effect. Constant Π does not affect the optimal size (it factors out of the integral in (11.2)), so that dispersing offspring are larger if there is any cost to dispersal structures.

Example: sigmoid relationship between body condition and survival during dispersal. Since $\Pi(z)$ is a probability, it may be plausible to assume that it is a sigmoid function of body condition z. Let α be the parameter that controls the 'steepness' of the sigmoid function (see Figure 11.3a), such that increasing α makes survival increase more sharply with body condition. However, in this case increasing α does *not* always increase M, and therefore a more sharply increasing shape of $\Pi(z)$ will not increase the size of dispersing offspring beyond a threshold (Figure 11.3b). The optimal size of dispersing offspring starts to decrease when $\Pi(z)$ gets sharper than the thick line in Figure 11.3a. Because $\Pi(z)$ is increasing, dispersing offspring are always larger than the non-dispersing offspring (dotted line in Figure 11.3b).

The number of dispersing and non-dispersing offspring. Because the optimal offspring sizes m_n and m_d are constant across patches, the numbers of dispersing and non-dispersing offspring are directly proportional to the amount of resources invested into them. The last step is to determine the evolutionarily stable allocation of resources between non-dispersing and dispersing offspring. This is technically similar to finding the ESS in our model (1) and yields similar results.

If patch qualities $E(y)$ are fixed in time, then the evolutionarily stable strategy abandons patches with low resources ($E(y)$ below a threshold) by producing only dispersing offspring. The amount of resources allocated to non-dispersing offspring, and hence the number of non-dispersing offspring, increases with patch quality.

If patch qualities are changing randomly from year to year, then the ESS invests the same amount of resources into non-dispersing offspring in every patch (E_n is constant across patches), and therefore produces the same number of non-dispersing offspring of the same size m_n in every patch (except patches where the total amount of resources is less than E_n required by the ESS). This is because each patch is equally valuable to retain (cf. Ezoe and Iwasa 1997; Kisdi 2004). Any surplus resource is used to produce dispersing offspring of the same size m_d but in variable number.

Decoupling of dispersal and offspring size. In the special case when survival during dispersal does not depend on body condition (Π is constant) and there is no cost to any dispersal structure ($c = 0$), non-dispersing and dispersing offspring maximize the same quantity in expressions (11.1) and (11.2), so that every offspring has the same optimal size. Because

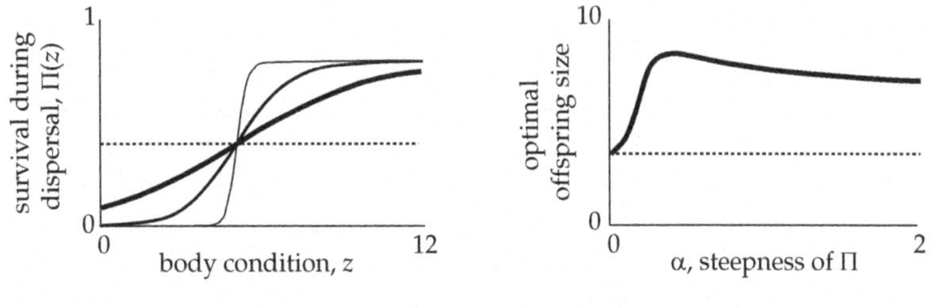

Figure 11.3 Optimal size of dispersing offspring assuming the sigmoid function $\Pi(z) = 0.8/(1+e^{-\alpha(z-5)})$. (a) The shape of $\Pi(z)$ with $\alpha = 0$ (constant, dotted line), $\alpha = 0.4$ (thick line), $\alpha = 1$ (intermediate line), and $\alpha = 5$ (thin line). (b) The optimal size of dispersing offspring as a function of α (thick line), assuming $c = 0$, $g(z) = max\ (0, 1-e^{0.3(1-z)})$, and $\beta(z - m_d)$ to be the standard normal distribution. The maximum of m_d at about $\alpha = 0.4$ corresponds to the thick line in panel (a). The dotted line shows the optimal size of non-dispersing offspring for comparison; because we assume no cost to any dispersal structure ($c = 0$), the size of non-dispersing offspring equals the size of dispersing offspring with constant Π.

Π is constant, there is an ESS class of dispersal strategies, and one ESS disperses every offspring with the same (patch-dependent) probability. In this case, the distributions of body condition in dispersing and in non-dispersing offspring are the same, and it does not matter whether dispersal is decided at birth or later.

11.5 Conclusions

Covariation between body condition and dispersal evolves in both OF our models, but the direction of causality is different. In model (1), body condition is determined by environmental factors, and dispersal responds to it; in model (2), body condition evolves (via body size) according to the dispersal status (Chapter 10). Our models assume that body condition can influence dispersal success (through the probability of survival during dispersal) and dispersal can be energetically costly (via a costly dispersal structure), but do not assume that dispersal is *a priori* constrained by insufficient body condition (Benard and McCauley 2008; Chapters 10 and 13).

As with a theoretical study of almost any complex system, the results may seem to say that many things matter in complicated ways. Our co-evolutionary model (2), however, does generate a simple and strong prediction: *dispersal syndromes are independent of the spatio-temporal structure of the environment.* The distributions of body condition among the dispersing and among the non-dispersing offspring are independent of the quality of their natal patch. Even more remarkably, these distributions are also independent of whether the qualities of patches are fixed, vary randomly, or vary but with some temporal autocorrelation. Because patch qualities are often difficult to measure, this independence facilitates testing the model against empirical data. The key assumption leading to this independence is that by equipping some of the offspring with dispersal structures, the parent decides which offspring will disperse already at birth, before knowing the precise body condition of the individual (recall the residual variation in condition independent of birth size). Offspring destined to disperse are then made of the same size and thus acquire the same distribution of body condition independently of the environment (and likewise for the non-dispersing offspring).

Both models highlight the role of survival during dispersal ($\Pi(z)$), i.e. whether competitively strong or weak individuals have better chances of surviving dispersal. In our co-evolutionary model (2), this is the main factor shaping condition-dependent dispersal (besides the trivial effect of the energetic cost of dispersal structures). In model (1), this alone determines whether strong or weak members of a family disperse (and this is combined with dispersal variation across families to obtain the population-level pattern). How survival during dispersal varies with body condition is therefore an essential ingredient to be measured in empirical studies. The role of possible condition-dependence in survival until reproduction (s) remains to be explored.

Model (1) highlights that pooling data across a population in a heterogeneous habitat can obscure the plastic strategies adapted to each individual's local environment. Take the example of Figure 11.2c, perhaps the most realistic scenario. In this case, families disperse their strongest offspring, but families in bad patches abandon their patches, and they therefore disperse many weak offspring. As a result, both the weakest and the strongest individuals of the population disperse (Figure 11.2c, left panel). Such a pattern cannot be captured by imposing a linear relationship between condition and dispersal in individual-based simulations (as in Bonte and de la Peña 2009), and a logistic regression analysis of data resembling Figure 11.2c (left panel) may erroneously find no condition-dependence in dispersal. The mean body conditions of dispersing and of non-dispersing individuals can be indistinguishable (Figure 11.2c, middle and right panels). Statistical tests on population data may therefore fail to demonstrate the condition-dependent dispersal strategies of individuals.

If the probability of survival during dispersal does not depend (or only weakly depends) on body condition, then many different dispersal strategies are (nearly) neutral to each other. The ESS class of strategies includes widely different patterns of condition-dependent dispersal; e.g. a strategy that disperses the weakest offspring of the family, a strategy that disperses the strongest offspring of the family, and many intermediates. This finding may help to explain why dispersal patterns appear to be inconsistent in nature (Ims and Hjermann 2001; Bowler and Benton 2005; Clobert *et al.* 2009).

When patches differ in the amount of resources available for reproduction, then parents in better patches can produce more offspring rather than stronger offspring. Unlike Ezoe (1998) and Levin and Muller-Landau (2000), our model (2) assumes that offspring size can evolve independently of dispersal (Sakai *et al.* 1998). Offspring destined to disperse are more likely to be larger and, on average, stronger than non-dispersing offspring; i.e., a dispersal syndrome evolves within the population and matches the common finding that individuals of better body condition disperse (Benard and McCauley 2008).

11.6 Further directions

Competition other than weighted lottery. The weighted lottery model has a special property, namely that many weak offspring are competitively equivalent to a few strong ones. This property is mathematically degenerate and constrains the results. A natural alternative assumption would be that the strongest competitor wins the patch; with infinitely many competitors, however, there is either no 'strongest' (if the support of body condition is not bounded), or the winner is always one with the best possible condition, rendering the strategy followed by all other body conditions irrelevant. A realistic model would then assume finite fecundity, which necessarily implies a stochastic number of offspring in each patch; but this is computationally very demanding. Gyllenberg *et al.* (2008) investigated some simple cartoon examples with finite numbers of competitors. Unless many patches received just one individual and hence there was no condition-dependent competition, the results were qualitatively similar to the case of weighted lottery, but the shape of the evolutionarily stable dispersal strategy at low body conditions was sensitive to model details (see '*weak selection on weak offspring*' under Section 11.3).

How much information is available? Our models make specific assumptions on the information available when the dispersal decision is made. In model (1), we assumed that the quality of the patch and therefore the distribution of body condition within the family is known; this is a natural assumption if dispersal is maternally controlled, but may require learning if dispersal is under offspring control. Evolving dispersal as a function solely of body condition ($p(z)$ rather than $p(z, y)$) would lead to different results. In model (2), the parent can optimize the number of eggs only if she knows the amount of available resources beforehand. We assumed that dispersal is decided already at birth; the mathematically more demanding problem of dispersal decided upon the actual body condition of the offspring at the time of dispersal needs future attention. In the latter case, body condition (via offspring size) and dispersal evolve jointly without unidirectional causality such that body condition would determine dispersal or vice versa (Chapter 10).

Immigration depending on the quality of the patch. In all of our models, we assumed that dispersers enter patches at random; but if patch quality can be assessed by the dispersers, immigration can be directed to good patches. Since this will cause intense competition for good patches, the evolutionarily stable immigration strategy is likely to be condition-dependent, as weak individuals may find it more profitable to avoid extreme competition.

Genetic conflicts in sexual populations. We assumed clonal inheritance, which implies that there is no conflict of interest within a family. This assumption is important for the prediction that strong offspring disperse whenever they survive dispersal with a higher probability than weak offspring. In sexual species, the interests of sibs do not coincide, and offspring have a selfish interest to avoid any dispersal hazards even if their own hazard is less than that of sibs (similarly to the social dominance hypothesis). Whether strong offspring force their weaker siblings to disperse depends on their degree of relatedness and on the relative costs of dispersal.

Acknowledgements

Our research was financially supported by the Academy of Finland and by the Finnish Graduate School in Computational Sciences.

Appendix

In this Appendix, we give a brief account of the mathematical analysis of the co-evolutionary Model (2). Since dispersal is decided already at birth, finding the ESS size and number of dispersing and non-dispersing offspring can be broken into two steps: (i) find the optimal size of dispersing and non-dispersing offspring given the amount of resource invested into them ($E_n(y)$ respectively $E_d(y)$); and (ii) find the amount of resource the ESS invests into dispersing versus non-dispersing offspring.

(i) Optimal size of offspring

Since $E_n(y)$ can be factored out in expression (11.1) of the main text, the optimal size of non-dispersing offspring, m_n, maximizes the non-dispersed competitive weight per unit investment,

$$F_n = \frac{1}{m_n}\int \beta(z - m_n)g(z)dz. \qquad \text{(Equation 11.3)}$$

Similarly, the optimal size of dispersing offspring, m_d, maximizes

$$F_d = \frac{1}{m_d + c}\int \beta(z - m_d)g(z)\Pi(z)dz. \qquad \text{(Equation 11.4)}$$

The optimal sizes are thus found from $\partial F_n/\partial m_n = 0$ and from $\partial F_d/\partial m_d = 0$, respectively, and they do not depend on the patch characteristics y as long as y does not affect the distribution of offspring condition, β. We assume throughout that the offspring condition is Gaussian with variance σ^2,

$$\beta(z - m) = \frac{1}{\sqrt{2\pi\sigma}}\exp\left[-\frac{(z-m)^2}{2\sigma^2}\right]. \qquad \text{(Equation 11.5)}$$

At the extrema, the second derivatives evaluate to

$$\frac{\partial^2 F_n}{\partial m_n^2} = \frac{1}{m_n\sigma^4}COV[(z - m_n)^2, g(z)] \qquad \text{(Equation 11.6)}$$

and

$$\frac{\partial^2 F_d}{\partial m_d^2} = \frac{1}{(m_d + c)\sigma^4}COV[(z - m_d)^2, g(z)\Pi(z)]$$

(Equation 11.7)

respectively, where the covariance is taken over the distribution of offspring condition, β. Assume that $g(z)$ and $g(z)\Pi(z)$ are analytic so that they can be substituted with their Taylor series around m_n and m_d, respectively, and recall that the odd central moments of the Gaussian distribution are zero whereas the $2n^{\text{th}}$ even moment is given by $(2n)!/(2^n n!)\sigma^{2n}$. A straightforward calculation shows that

$$\frac{\partial^2 F_n}{\partial m_n^2} = \frac{1}{m_n}\int \beta(z - m_d)g''(z)dz \qquad \text{(Equation 11.8)}$$

and

$$\frac{\partial^2 F_d}{\partial m_d^2} = \frac{1}{m_d + c}\int \beta(z - m_d)[g(z)\Pi(z)]''dz \qquad \text{(Equation 11.9)}$$

i.e. the extrema are optima if the second derivatives of $g(z)$ and of $g(z)\Pi(z)$ are negative *on average*. This generalizes the condition of optimality (e.g. Smith and Fretwell 1974; Lloyd 1987) to the case of distributed offspring. With similar techniques, we can show that the minimum viable offspring size, m_0, is *not* a local optimum for non-dispersing and for dispersing offspring if, respectively, $\int\beta(z-m_0)[g'(z)-g(z)/m_0]dz>0$ and $\int\beta(z-m_0)[(g(z)\Pi(z))'-g(z)\Pi(z)/(m_0+c)]dz>0$; analogous conditions hold for the maximum viable size. Note that because the number of offspring is assumed to be large, the maximum offspring size must be small compared to the amount of available resources.

Next, we investigate how m_d differs from m_n. To see the effect of c on the optimal size m_d, by the implicit function theorem we have $\frac{dm_d}{dc} = -\frac{\partial^2 F_d/\partial c\partial m_d}{\partial^2 F_d/\partial m_d^2}$, where the denominator is negative at the optimum so that dm_d/dc is sign-equivalent to $\partial^2 F_d/\partial c\partial m_d$. Using $\partial F_d/\partial m_d = 0$, this evaluates to $F_d/(m_d + c)^2 > 0$, which implies that increasing the cost of dispersal structures c always increases the optimal size of dispersing offspring. Because non-dispersing offspring have no costs associated with dispersal structures, $c > 0$ selects for $m_d > m_n$.

To see the effect of a parameter α that determines the shape of $\Pi(z)$, rewrite $\partial F_d/\partial m_d$ as $\gamma(M - m_d - \sigma^2/(m_d + c))$, where M is defined in the main text and $\gamma = \int\beta(z - m_d)g(z)\Pi(z)dz/((m_d + c)\sigma^2)$ is a positive factor. Using the implicit function theorem as above, $dm_d/d\alpha$ is sign-equivalent to $\partial^2 F_d/\partial\alpha\partial m_d = \gamma(\partial M/\partial\alpha)$, i.e., $dm_d/d\alpha$ has the same sign as $\partial M/\partial\alpha$ at the optimal m_d.

To compare the optimal sizes m_d and m_n, note that if the probability of survival during dispersal were independent of body condition, then it would factor out in (11.4) and would not cause any difference between m_d and m_n. Define $\Pi_\alpha(z) = \Pi_0 + (\Pi(z) - \Pi_0)$, where $0 < \Pi_0 < 1$ is an arbitrary constant, and let $\Phi(z)$ denote the probability density function $\Phi(z) = \beta(z - m_d)g(z) / \int \beta(z' - m_d)g(z')dz'$. $M(\Pi_\alpha)$ can be writtenas $M(\Pi_\alpha) = \int z\Phi(z)\Pi_\alpha(z)dz / \int \Phi(z)\Pi_\alpha(z)dz$, and a tedious but straightforward calculation shows that $\partial M(\Pi_\alpha)/\partial\alpha$ is sign-equivalent to the covariance of z and $\Pi(z)$ over the distribution Φ. Hence if $\Pi(z)$ is an increasing (decreasing) function, then $\partial M(\Pi_\alpha)/\partial\alpha$ is positive (negative) and therefore m_d strictly increases (decreases) with α in the interval [0,1]. If $c = 0$, then the optimal size at $\alpha = 0$ equals the optimal size of non-dispersing offspring m_n, and the optimal size at $\alpha = 1$ is the optimal size of dispersing offspring when the probability of survival during dispersal is given by $\Pi(z)$. Hence an increasing function $\Pi(z)$ selects for $m_d > m_n$, whereas a decreasing $\Pi(z)$ selects for $m_d < m_n$. Because $c > 0$ also selects for $m_d > m_n$, with increasing $\Pi(z)$ the dispersing offspring must always be larger than the non-dispersing offspring; but with decreasing $\Pi(z)$, the net difference depends on the balance of opposing selection from $c > 0$ and from decreasing $\Pi(z)$.

(ii) ESS investment into dispersing versus non-dispersing offspring

Consider a resident strategy that, when in patch y, invests $E_n(y)$ resources into non-dispersing offspring and invests the remaining $E_d(y) = E(y) - E_n(y)$ resources into dispersing offspring. Below we shall find the so-called best reply to $E_n(y)$, i.e. the mutant strategy $E_n^{mut}(y)$ that has the highest invasion fitness in the resident population of strategy $E_n(y)$. Throughout, we assume that all strategies produce the dispersing and non-dispersing offspring with their optimal sizes, m_d and m_n (otherwise the strategy cannot be a best reply, and hence cannot be an ESS). The ESS is the best reply to itself, and hence shares the qualitative properties of best reply strategies.

With randomly changing patch qualities, the remainder of the analysis is analogous to Gyllenberg *et al.* (2011a). The invasion fitness of the mutant is given by the functional

$$W^{mut} = s\int \phi(y)\left[\frac{E_n^{mut}(y)F_n}{E_n^{mut}(y)F_n + I} + (E(y) - E_n^{mut}(y))F_d V\right]dy$$

(Equation 11.10)

where s is the probability that the single occupant of a patch survives till reproduction, $\phi(y)$ is the probability density of patch y, F_n and F_d are as defined respectively in (11.3) and (11.4) with the optimal sizes m_n and m_d substituted, $I = sF_d\int \phi(y)(E(y) - E_n(y))dy$ is the competitive weight of immigrants and $V = \int \phi(y)\frac{s}{E_n(y)F_n + I}dy + \frac{1-s}{I}$ is the expected number of patches won by dispersed offspring per unit successfully dispersed competitive weight. The best reply $E_n^{mut}(y)$ maximizes the integrand of (11.10) (cf. Euler's equation) subject to the constraint $E_n^{mut}(y) \le E(y)$. A simple calculation gives

$$E_n^{mut}(y) = \begin{cases} \sqrt{\dfrac{I}{F_n F_d V}} - \dfrac{I}{F_n} & \text{if this is less than or equal to } E(y) \\ E(y) & \text{otherwise} \end{cases}$$

(Equation 11.11)

at almost every y. Note that $E_n^{mut}(y) \ge 0$ because $F_n \ge F_d$ and $IV \le 1$; I and V depend only on the resident strategy. Except in poor patches where all resources are used for non-dispersing offspring, the best reply strategy invests the same amount of resources into non-dispersing offspring in every patch; hence it produces the same number of non-dispersing offspring with the same size m_n in every patch that has sufficient resources to achieve this. The surplus resources are used to produce dispersing offspring, whose size m_d is constant across patches but their number increases with $E(y)$. The ESS value of I and V, and therefore the evolutionarily stable strategy $E_n(y)$ can be found numerically (similarly to Gyllenberg *et al.* 2011a) but this yields no further insight into the qualitative properties of the ESS.

If patch quality is fixed in time, then the analysis follows Gyllenberg *et al.* (2011b). The best reply maximizes the functional

$$R^{mut} = s\int \phi(y)v(y)(E(y) - E_n^{mut}(y))F_d T(E_n^{mut}(y))dy$$

(Equation 11.12)

where $v(y) = \frac{s}{E_n(y)F_n + I} + \frac{1-s}{I}$ is the probability that a unit competitive weight wins a randomly chosen patch y (which is occupied by a resident with probability s and is empty with probability 1-s), $T(E_n^{mut}(y)) = \left[1 - s\frac{E_n^{mut}(y)F_n}{E_n^{mut}(y)F_n + I}\right]^{-1}$ is the expected number of years a mutant family retains a patch of quality y, and $I = sF_d \int \phi(y)(E(y) - E_n(y))dy$ is the competitive weight of immigrants as above. The best reply

$$E_n^{mut}(y) = \begin{cases} \frac{-F_n I + \sqrt{F_n^2 I^2 - F_n^2 (1-s) I (I - sF_n E(y))}}{(1-s)F_n^2} & if\ E(y) \geq \frac{I}{sF_n} \\ 0 & \text{otherwise} \end{cases}$$

(Equation 11.13)

is found as above. The best reply abandons bad patches with $E(y)$ below the threshold I/sF_n, where only dispersing offspring are produced. Above this threshold, the amount of resources invested into non-dispersing offspring increases with $E(y)$, hence better patches produce more non-dispersing offspring; but the derivative $dE_n^{mut}(y)/dE(y)$ is less than 1, i.e., $E_d^{mut}(y)$ also increases with $E(y)$ so that better patches also disperse more offspring. As above, the ESS value of I and therefore the evolutionarily stable strategy $E_n(y)$ can be found only numerically (similarly to Gyllenberg *et al.* 2011b), but the qualitative properties of the ESS are the same as those of the best reply strategies $E_d^{mut}(y)$.

References

Benard M.F. and McCauley, S.J. (2008). 'Integrating across life-history stages: Consequences of natal habitat effects on dispersal', *American Naturalist* 171: 553–67.

Bonte, D. and de la Peña, E. (2009). 'Evolution of body condition-dependent dispersal in metapopulations', *Journal of Evolutionary Biology* 22: 1242–51.

Bowler, D.E. and Benton, T.G. (2005). 'Causes and consequences of animal dispersal strategies: Relating individual behaviour to spatial dynamics', *Biological Reviews* 80: 205–25.

Clobert, J., Le Galliard, J., Cote, J., Meylan, S. *et al.* (2009). 'Informed dispersial, heterogeneity in animal dispersal syndromes and the dynamics of spatially structured populations', *Ecology Letters* 12: 197–209.

De Jong, T. and Klinkhamer, P. (2005). *Evolutionary ecology of plant reproductive strategies*. Cambridge University Press, Cambridge.

Ezoe, H. (1998). 'Optimal dispersal range and seed size in a stable environment. *Journal of Theoretical Biology* 190: 287–93.

Ezoe, H. and Iwasa, Y. (1997). 'Evolution of condition-dependent dispersal: A genetic-algorithm search for the ESS reaction norm', *Researches in Population Ecology* 39: 127–38.

Fretwell, S.D. (1972). *Populations in a seasonal environment*. Princeton University Press, Princeton, NJ.

Gandon, S. and Michalakis, Y. (1999). 'Evolutionarily stable dispersal rate in a metapopulation with extinctions and kin competition', *Journal of Theoretical Biology* 199, 275–90.

Gyllenberg, M., Kisdi, E. and Utz, M. (2008). 'Evolution of condition-dependent dispersal under kin competition', *Journal of Mathematical Biology* 57: 285–307.

Gyllenberg, M., Kisdi, E. and Utz, M. (2011a). 'Variability within families and the evolution of body condition dependent dispersal', *Journal of Biological Dynamics* 5: 191–211.

Gyllenberg, M., Kisdi, E. and Utz, M. (2011b). 'Body condition dependent dispersal in a heterogeneous environment', *Theoretical Population Biology* 79: 139–54.

Gyllenberg, M. and Metz, J.A.J. (2001). 'On fitness in structured metapopulations', *Journal of Mathematical Biology* 43: 545–60.

Hamilton, W.D. and May, R.M. (1977). 'Dispersal in stable habitats', *Nature* 269: 578–81.

Ims, R.A. and Hjermann, D.O. (2001). 'Condition-dependent dispersal', in J. Clobert, E. Danchin, A.A. Dhondt *et al.* eds, *Dispersal.* Oxford University Press, Oxford, 203–16.

Jánosi, I.M. and Scheuring, I. (1997). 'On the evolution of density dependent dispersal in a spatially structured population model', *Journal of Theoretical Biology* 187: 397–408.

Kisdi, E. (2004). 'Conditional dispersal under kin competition: Extension of the Hamilton-May model to brood size-dependent dispersal', *Theoretical Population Biology* 66: 369–80.

Levin, S.A. and Muller-Landau, H.C. (2000). 'The evolution of dispersal and seed size in plant communities', *Evolutionary Ecology Research* 2: 409–35.

Lloyd, D.G. (1987). 'Selection of offspring size at independence and other size-versus-number strategies', *American Naturalist* 129: 800–17.

Metz, J.A.J. and Gyllenberg, M. (2001). 'How should we define fitness in structured metapopulation models?

Including an application to the calculation of evolutionarily stable dispersal strategies', *Proceedings of the Royal Society B* B 268: 499–508.

Metz, J.A.J., Mylius, S.D. and Diekmann, O. (2008). 'When does evolution optimize?', *Evolutionary Ecology Research* 10: 629–54.

Perrin, N. and Mazalov, V. (2000). 'Local competition, inbreeding, and the evolution of sex-biased dispersal', *American Naturalist* 155: 116–27.

Poethke, H.J. and Hovestadt, T. (2002). 'Evolution of density- and patch-size-dependent dispersal rates', *Proceedings of the Royal Society B* 269: 637–45.

Ronce, O., Clobert, J. and Massot, M. (1998). 'Natal dispersal and senescence', *Proceedings of the National Academy of Sciences USA* 95: 600–5.

Sakai, S., Kikuzawa, K. and Umeki, K. (1998). 'Evolutionarily stable resource allocation for production of wind-dispersed seeds', *Evolutionary Ecology* 12: 477–86.

Smith, C.C. and Fretwell, S.D. (1974). 'The optimal balance between size and number of offspring', *American Naturalist* 108: 499–506.

Travis, J.M.J., Murrell, D.J. and Dytham, C. (1999). 'The evolution of density-dependent dispersal', *Proceedings of the Royal Society B* 266: 1837–42.

CHAPTER 12

Dispersal syndromes in the common lizard: personality traits, information use, and context-dependent dispersal decisions

Julien Cote and Jean Clobert

12.1 Introduction

Dispersal is one of the fundamental life-history traits which affects processes as diverse as the dynamics and evolution of fragmented populations, gene flow, species' distributions, and species' ability to track favourable environmental conditions (Clobert *et al.* 2001; Bowler and Benton 2005; Clobert *et al.* 2009). Dispersal is a common response to many biotic and abiotic factors, therefore the multiple factors determining dispersal decisions have been studied often (see Chapter 1). A common idea is that, for a given environmental context, a species exhibits set dispersal behaviours depending on the dispersal phase (i.e. departure, transience, settlement). In recent years, we have shifted focus to recognize that variation between individuals is as important as their average behaviour, and this is as true with dispersal as in population dynamics. Faced with the same environmental conditions, different individuals will not make the same dispersal decisions (Cote and Clobert 2007b; Clobert *et al.* 2009). This idea suggests that dispersing individuals are not a random subset of the population; instead, dispersers often differ from non-dispersers in phenotype (Bowler and Benton 2005; Clobert *et al.* 2009). In general, phenotypic differences between dispersers and residents are likely to exist, for example, if some morphological, physiological, or behavioural specializations are needed to increase dispersal success (Bowler and Benton 2005; Clobert *et al.* 2009; Chapter 10). This 'dispersal syndrome' involves suites of phenotypic traits that may or may not be consistent over the lifetime of an individual, and that can be caused by reversible or irreversible ontogenetic changes (Chapters 6, 7, 10). Here, we would like to focus particularly on behavioural specializations of dispersers and illustrate those specializations with studies on common lizards (*Zootoca vivipara*). However, as behavioural specializations are often linked to individual condition, competitive ability, and/or health status (and thus to other phenotypic traits), we also briefly review condition-dependent dispersal (i.e. body size, body mass) in this species (Chapter 3). As previously stated, behavioural differences between dispersers and residents may be temporary shifts at the onset of dispersal decision, or may be intrinsic differences between individuals with these two dispersal strategies (Chapter 7). The latter possibility logically intrigued researchers working on consistent differences between individual in behavioural traits (also known as personality traits, behavioural syndrome, temperament), and it has been thought that at least in some species, dispersers display a specific set of personality traits driving their decisions through the dispersal process (Cote *et al.* 2010a).

Firstly, we review the evidence for a dispersal syndrome with a strong focus on long-lasting behavioural differences between dispersers and residents. We focus, in particular, on evidence on common lizards (*Zootoca vivipara*), a small lacertid inhabiting humid

Dispersal Ecology and Evolution. First Edition. Edited by Jean Clobert *et al.*

habitats in Eurasia (see Case Study I for more details on natural history of this species). Since the 1980s, a long-term study has investigated the effects of several abiotic and biotic factors on dispersal decisions in this species (Case Study I) using both long-term field monitoring of natural populations and a suite of experiments in semi-natural populations where environmental conditions were controlled or manipulated. For almost three decades, the research on common lizards aimed to integrate an individual-based approach into a (meta-) population approach. This is the best way to study phenotypic differences between dispersers and residents, as well as the ontogenetic processes leading to those differences. Moreover, the coupled individual- and population-based approach allows for the study of the consequences of dispersal syndromes for spatially structured populations. It is worth noting that most of these studies focused on the multiple factors determining natal dispersal decisions (Case study I). However, most of these studies also provided information on phenotypic differences between dispersers and residents (see Section 12.2.1). While at times dispersers and residents display diverging results for a specific trait across various studies, it is important to note that additional differences are only observed in a specific dispersal context (see Section 12.2.1). Hence, we review the evidence for dispersal syndromes in relationship to the environmental context of dispersal, and show the extent to which the environment itself contributes to dispersal syndromes in common lizards (Figure 12.1). In the second section, we integrate the idea of context-dependent dispersal syndrome in the dynamic of spatially structured populations and associate dispersal syndrome with informed dispersal through a heterogeneous landscape. Essentially, the phenotypes of dispersers provide cues about the surrounding landscapes to residents when the dispersers settle into a new patch.

12.2 Context-dependent dispersal syndrome

12.2.1 Condition-dependent dispersal varying with contexts

Before reviewing behavioural differences between dispersers and residents, we would like to give a brief overview of other phenotypic differences. Among phenotypic differences, body size, and body condition often differs between dispersers and residents (i.e. condition-dependent dispersal, Chapter 10). Classical theories about dispersal assume that dispersers are the least fit individuals or poorer competitors. Dispersal propensity should therefore be negatively correlated to body size and body condition as shown in several species (Bowler and Benton 2005; Chapter 10). On the contrary, the act of moving from one habitat to another entails major risks and costs to the disperser (Stamps 2001; Bonte *et al.* 2012) If this movement requires a certain level of energy to travel over long distances, or the crossing of unsuitable habitats, we can expect dispersers to be longer and/or heavier than residents. Moreover, during settlement, dispersers in good condition might compete better with residents of a new population or undergo the difficulties to find resources when colonizing an empty habitat. However, condition-dependent dispersal is expected to vary with the environmental context of dispersal. Indeed, this is the case more generally as many life-history decisions are context-dependent (Plaistow *et al.* 2006). According to this, the characteristics of dispersers are expected to vary with the environmental pressures on dispersal. Condition-dependent dispersal exists in common lizards but varies with the context. In most of the studies, dispersers are larger and in better condition than residents (Table 12.1). Ten studies clearly tested for the dependency of natal dispersal on body size and/or body condition at birth in several ecological contexts. Body condition at birth and dispersal probability were positively related in six studies, negatively related in three studies, and not related in three studies (note that in some studies, they can be positive and negative, or positive and not related, depending on treatments; see following). The relationship between body size at birth and dispersal probability is less clear, with two studies finding a positive relationship, and five studies finding no significant relationship. Because of the strong relationship between body mass and body size, the latter observation might be explained by statistical 'artifacts' (models testing for body size only or variance being soaked up by one of the two covariates). In this species, dispersers are thus often longer or heavier in many contexts, traits which indicate their future competitive ability and survival probability (Sorci and Clobert 1999; Le Galliard *et al.* 2004). In common lizards, bigger juveniles might be able to travel longer distances, spend more time searching for suitable habitats, or settle more easily in a novel habitat (e.g. dominance). After taking a closer look at those studies, the relationship between body condition/size and dispersal probability varies with the level of kin competition. Kin interactions, a factor driving social evolution (Hamilton 1964; Le Galliard *et al.* 2005b), are involved in the evolution of dispersal (see Chapter 1). Kin competition does not only affect the decision to stay or leave, but it also promotes the dispersal of morphologically and/or behaviourally different individuals (see Case Study I; Chapter 10). Empirical studies on common lizards support the hypothesis that kin-based dispersal decision covaries with phenotypic traits (Table 12.1). Kin competition may occur among parents and offspring, and among offspring. In the common lizard, however, mother-offspring competition was repeatedly found to be the most important component of kin competition in this species. In this species, there is no parental care and mothers' aggressive behaviour towards offspring is not common. However, offspring can discriminate between maternal odour and the odour of another female (Léna and de Fraipont 1998), and dispersal behaviour is related to the sensitivity to maternal odour (Léna *et al.* 2000). Among the ten studies, when juveniles are released in populations with their mother or in populations where the mothers of many juveniles are present (population level of kin competition), dispersal probability strongly depends on body condition or body size at birth (Table 12.1). Dispersers are generally bigger at birth than residents. However, this difference is reduced (Léna *et al.* 1998), cancelled (Cote *et al.* 2007; Meylan *et al.* 2007), or reversed (Le Galliard *et al.* 2003; Cote and Clobert 2010). Note that also, in Le Galliard *et al.* (2003), this is true for yearling dispersal when the mother is replaced by a surrogate female or when average kin competition within the population is low. Aragon *et al.* (2006b) also showed that for yearlings released without mothers into populations, there were no morphological differences between yearlings that did and those that did not previously disperse. Interestingly, these yearlings were released into populations without their mothers after their birth.

In Section 12.4, we discuss these results in a more general context and we explain how context dependence can affect the dynamic of spatially structured populations (Figure 12.1). This condition and context dependency strengthens the idea that dispersal is an individual strategy in common lizards. Individuals are not forced to disperse in many situations; they decide to disperse from a given environmental context and based on their own condition. Thus, this indicates that dispersal decisions are also probably linked to, or determined by, individual behavioural characteristics.

12.2.2 Behavioural syndrome of dispersers

In many taxa, behavioural traits partly or largely explain individual variation in dispersal (Bowler and Benton 2005; Cote *et al.* 2010a). For example, in mole rats (*Heterocephalus glaber*), dispersers are larger and fatter than residents, but also have higher locomotor and feeding activities and a reduced participation in cooperative activities in the colonies they join (O'Riain *et al.* 1996). This classical study and a few others strongly suggest the existence of disperser morphs with behavioural specializations. In addition, recent studies showed that dispersal tendencies are linked to individual differences in behavioural traits that can be stable over ontogeny or across situations (Cote *et al.* 2010a); these would be personality traits (Dall *et al.* 2004, Bell 2006; Sih and Bell 2008; Reale *et al.* 2010). While to-date, personality-dependent dispersal has been demonstrated in only a few species (Fraser *et al.* 2001; Dingemanse *et al.* 2003; Cote and Clobert 2007b; Duckworth and Badyaev 2007; Cote *et al.* 2010b), it has been thought to be a widespread phenomenon with important ecological consequences (Cote *et al.* 2010a).

In common lizards, we found behavioural differences between dispersers and residents (Table 12.1). These behavioural traits often affect dispersal patterns independently of the effect of body size/body condition at birth. Similarly to differences observed

Table 12.1 Context-dependent dispersal syndromes in common lizard

		Relationship with dispersal probability		
Trait	Dispersal context	Positive	Negative	None
Body condition at birth	High levels of kin competition	1, 2, 3, 4, 5		7
	Low levels of kin competition	1*, 2, 4*	1*, 2*	4*, 8
	Not clearly specified	6*	6*	
Body size at birth	High levels of kin competition	9, 10		1, 5, 9*
	Low levels of kin competition			1, 10
	Not clearly specified			6
Foraging activity	Not specified		11*	
Exploration novel environment	Not specified	12, 13, 14		
Sociability at birth	High-density populations		5	
	Low-density populations	5		

Here, we list examples of phenotypic traits linked to natal dispersal decision in common lizards. These dispersal syndromes sometimes depend on the context of dispersal. High level of kin competition exists when juveniles are released with their mother or, in Cote *et al*. 2007, when % juveniles released with their mothers is high. Low level of kin competition is when juveniles are not released with their mothers for the purpose of the experiment. 1*: negative relationship for low level of kin competition only in unconnected populations (positive in connected populations), 2*: true for yearling dispersal, 3*: significant interaction between kin competition level and body condition at birth, but no separate analysis for each level of kin competition, 6*: positive relationship for females, negative relationship for males, 9*: no relationship for only 1 of the 3 situations tested, 11* only significant for females.

1. J. Cote, J. Clobert (2010), *Ecology* **91:** 1485–1493; 2. J.-F. Le Galliard, R. Ferriere, J. Clobert (2003), *Proceedings of the Royal Society B* **270:** 1663–1669; 3. J.-F. Le Galliard, R. Ferriere, J. Clobert (2005b), *Journal of Animal Ecology* **74:** 241–249; 4. J.-P. Lena, J. Clobert, M. de Fraipont, J. Lecomte, G. Guyot (1998), *Behavioural Ecology* **9:** 500–507; 5. J. Cote, J. Clobert (2007b), *Proceedings of the Royal Society B* **274:** 383–390; 6. M. Massot, J. Clobert (2000), *Journal of Evolutionary Biology* **13:** 707–719; 7. K. Boudjemadi, J. Lecomte, J. Clobert (1999), *Journal of Animal Ecology* **68:** 1207–1224; 8. S. Meylan, J. Clobert, B. Sinervo (2007), *Oikos* **116:** 650–661; 9. J. Cote, J. Clobert (2007a), *Ecology Letters* **10:** 411–417; 10. J. Cote, J. Clobert, P. S. Fitze (2007), *Proceedings of the National Academy of Sciences USA* **104:** 9703–9708; 11. S. Meylan, M. de Fraipont, P. Aragon, E. Vercken, J. Clobert (2009), *Journal of Experimental Zoology A* **311A:** 377–388; 12. P. Aragon, S. Meylan, J. Clobert (2006b), *Functional Ecology* **20:** 900–907; 13. J. Clobert, M. Massot, J. Lecomte, G. Sorci, M. de Fraipont, R. Barbault (1994), in *Lizard Ecology: The third generation (J. Vitt and E.R. Pianka)*, Princeton University Press, 183–206; 14. M. de Fraipont, J. Clobert, H. John-Adler, S. Meylan (2000), *Journal of Animal Ecology* **69:** 404–413

in mole rats (O'Riain *et al.* 1996), Aragon *et al.* (2006b) and Meylan *et al.* (2009) suggested that dispersers and residents display consistent individual differences in activity/exploration and foraging activity. They showed that, in common lizards, dispersers were more active and took longer to catch a prey, 8–10 months after the dispersal event (Aragon *et al.* 2006b; Meylan *et al.* 2009). Meylan *et al.* (2009) provided a large set of behavioural tests at birth (before natal dispersal) and after ten months, which allows testing for consistency in dispersal syndrome over time. At birth, future dispersing females chased fewer prey and were thus less likely to eat than future non-dispersers, while, at ten months old, males and female dispersers took more time to catch prey than non-dispersers. In addition, Aragon *et al.* (2006) showed that in a novel environment, yearlings that previously dispersed spent more time walking than yearlings that stayed in their natal habitat, which is consistent with previous results on this species (Clobert *et al.* 1994; de Fraipont *et al.* 2000). These results suggest that dispersers display consistently lower foraging activity which can be explained by dispersers' time-allocation strategy to exploration more than to foraging. Increased exploratory activity would help dispersers to explore their novel habitats. However, these observed activity patterns were strongly dependent on social context. When faced with olfactory cues of non-interacting conspecifics, yearlings that dispersed decreased their time spent walking, while yearlings that did not disperse did not change their behaviour. However, when olfactory cues came from interacting conspecifics, dispersers tried to escape more from the experimental arena than when olfactory cues came from non-interacting conspecifcs, the opposite being true for individuals that did not disperse. Along with results from Lena *et al.* (1998), this shows that dispersers react differently, at birth and at older ages, to the presence of conspecifics and to their interactive behaviours. Although it was not clear whether these differences in social reaction were consistent over time or not, Cote and Clobert (2007b) tested to what extant dispersal decisions are linked to consistent individual differences in reaction to social context.

Bekoff (1977) predicted that individuals who socialized with, or are not repulsed by, others were less likely to disperse. However, individual variation in social behaviour (i.e. sociability) has only rarely been related to dispersal behaviour (Ims 1990; O'Riain *et al.* 1996; Cote and Clobert 2007b; Blumstein *et al.* 2009; Cote *et al.* 2010b; Cote *et al.* 2011), and only few studies defined the observed social behaviour as a personality trait: human, Jokela *et al.* 2008), mosquitofish, (Cote *et al.* 2010b) including the common lizard (Cote and Clobert 2007b). In the common lizard, individuals vary in the degree of their social tolerance (i.e. sociability). This behavioural trait is consistent throughout an individual's lifetime and independent of the social context in which it is observed (Cote and Clobert 2007b). Dispersal behaviour is also related to individual sociability levels but in interaction with population density. The results suggest that some juveniles leave their natal population in search of more socially attractive or dense populations, while other juveniles disperse in order to avoid crowded populations. Indeed, asocial individuals, who have increased fitness within a low density population and prefer to stay away from conspecifics, tend to disperse when densities get too high, while social individuals, who actively seek out conspecifics and have increased fitness at high densities, disperse when densities are too low (Cote and Clobert 2007b; Cote *et al.* 2008b). Proximate mechanisms explaining the link between sociability, density and fitness outcome are not known. However, the most likely explanation relates to sociability-dependent competitive strategies. Individual sociability/social tolerance should affect the intensity of individual social interactions, and therefore its space use. 'Asocial' individuals may avoid social interactions, and thus competition for suitable places (e.g. shelter, basking spot) and food with adults. This avoidance should decrease the direct (bites) and indirect (anxiety, stress) effects of competition but limit resource access. Competition for resources should be less intense in low-density populations. Therefore, asocial juveniles can access resources more easily in lower- than in higher- density populations. Although interactions among individuals should be more infrequent in lower- than in higher-density populations, 'social' individuals should still display more interactive behaviours with adults in a lower-density context than asocial individuals. Adult aggressiveness associated with a lower pay-off from competition might result in the lower survival of social individuals in low-density populations.

12.3 Dispersal syndromes, habitat preferences, and information use

Dispersing individuals often sample and compare environments and settle in those environments that best match their phenotype (habitat matching choices, Edelaar *et al.* 2008). As a consequence, different ecological conditions induce the departure of individuals with phenotypes dedicated to different settlement habitats. The production of context-dependent disperser types therefore link departure decisions with the future habitat selection strategies of dispersers (Clobert *et al.* 2009). In the common lizard, the interaction between sociability and density dependency of dispersal decisions and fitness is consistent with the habitat-matching theory. However, dispersal entails major risks to the disperser partly because of the uncertainty of the existence of suitable habitats. This uncertainty is known to be particularly high in the case of density-dependent habitat preferences, and uncertainty is known to constrain individuals' departure decisions. Indeed, candidate dispersers have to assess the 'likelihood' of finding a new population that better suits their density preferences than their present population. Common lizards use conspecifics to acquire information on other populations (i.e. social information). The context-dependent dispersal syndrome is involved in information use/transfer in two different ways. Firstly, Cote and Clobert (2007a) showed that candidate dispersers use the immigrants' dispersal syndrome to obtain information about the density in surrounding patches. Individuals can tell apart neighbours from 'foreign' lizards for two reasons. Firstly, this species has the necessary capacity for individual recognition through different cues (e.g. olfactory cues, Léna and de Fraipont 1998; Aragon *et al.* 2006a). Second, in natural populations, lizards rapidly explore an area of a similar size to the semi-natural populations in which we keep them (Lecomte *et al.* 2004). In the semi-natural populations, an individual has thus the possibility to encounter and/or interact rapidly with all lizards of its population and detects any change in the composition of its population. Moreover, as abiotic and biotic factors sculpt the phenotype of individuals, immigrants are subject to imprinting by their population of origin and can carry information on the conditions in those populations (e.g. density, sex ratio). For example, individuals from high-density populations are more active and more aggressive than individuals from low-density populations (Lecomte *et al.* 1994; Cote *et al.* 2008a). Thus, differences in dispersal syndrome between immigrants from high- and low-density populations may provide information about the origin of these individuals. This idea is also strengthened by other findings on social interactions between dispersers and residents (Aragon *et al.* 2006b; Meylan *et al.* 2009) and between individuals of different origins (Cote *et al.* 2008a). These studies have shown that individuals react differently to an interaction with a dispersing individual than to an interaction with a resident (individuals tried to escape more often when they were in the presence of a disperser; Meylan *et al.* 2009). In addition, the outcomes of social interactions with yearlings from low-density populations and with yearlings from high-density populations are different (Cote *et al.* 2008a). In conclusion, the context-dependent dispersal syndrome creates heterogeneity in information carried by dispersers, transferred to other populations, and used by candidate dispersers to make the right decisions.

The second way that information on dispersal syndromes is used depends on the personality of candidate dispersers, and with their habitat preferences. In other species, individuals of different personality types are known to use information differently to meet the needs linked to their strategy (Marchetti and Drent 2000; Kurvers *et al.* 2009; Kurvers *et al.* 2010). Likewise, individuals with different habitat preferences are believed to use different types of information to take decisions through dispersal phases. We think that variability in personality types is the mechanism involved in this process. For example, asocial individuals avoid interactions with conspecifics and are probably less sensitive to social information or have fewer encounters with transmitters of information than social individuals. Asocial individuals might then collect more information on habitat structure than on social context. The extreme example would be individuals that prefer to settle in an empty habitat than in already occupied habitats (i.e. colonizers).

In addition to individuals preferring low-density populations ('asocial' individuals) and ones preferring high-density populations ('social' individuals), we have circumstantial evidence that there may be a third type of individuals in common lizards: the colonizers (i.e. individuals preferring empty habitat, Le Galliard *et al.* 2005a) and that this preference may be driven by high-kin competition. Cote *et al.* (2007) provided an experimental demonstration that kin competition affects the success of colonization. As previously said, offspring leaving a population due to high overall levels of kin competition were bigger or larger than offspring leaving a population with low levels of kin competition. We also demonstrated that, when released in newly colonized habitats, dispersers from the high kin-competition populations displayed increased growth rate and survival, and newly established populations were less likely to go extinct than those founded by dispersers from low kin-competition populations. This demonstrates that an offspring's colonization ability depends on the level of kin competition experienced in its natal population. However, it does not show that kin-based dispersers prefer empty habitats. However, Cote and Clobert (2010) manipulated information about connection between habitat and the level of kin competition. By preventing any immigration and frustrating emigrants (by releasing them back into their populations), we simulated the absence of surrounding connected populations. While average emigration attempts were decreased in absence of connected populations, kin-based dispersers did not adjust their dispersal decisions on the basis of information on the existence of surrounding populations, and again kin-based dispersers were bigger. Kin competition may thus be a factor driving the departure of individuals with better colonizing ability (i.e. colonizers), but future studies should be directly testing for habitat preferences of kin-based dispersers.

12.4 Conclusion: dispersal syndrome and the dynamics of spatially structured population

In this chapter, we reviewed evidence for the existence of dispersal syndromes in common lizards (Figure 12.1). Although phenotypic differences between dispersers and residents have been found in many taxa for a long time, only recent review papers pointed out the importance of dispersal syndromes in population dynamic (Armsworth and Roughgarden 2005; Bowler and Benton 2005; Duckworth 2008; Clobert *et al.* 2009). In common lizards, many previous studies suggested a connection between phenotypic traits and dispersal decisions, and one that varies with context. For example, when kin competition is strong, dispersers are clearly longer and heavier than residents. This pattern disappears when kin competition is experimentally cancelled. Another clear example of context dependency of dispersal syndrome is the way dispersers and residents experience social interactions. An excess of social interactions has long been known to induce departure decision in some species, while intermediate to high densities of conspecifics is thought to attract individuals in some other species (Stamps 2001). Our results show that within-species variability of social attraction/repulsion exists and creates heterogeneity in dispersal syndromes. Some individuals search for conspecifics (social individuals), whilst others try to flee interactions with conspecifics. Dispersers then preferentially settle in habitats that best match their particular phenotype to increase their success in the selected habitat. Dispersers will use private and public information according to their needs/sensitivity to select a novel suitable habitat. As discussed in Cote and Clobert (2007b) and generalized in Clobert *et al.* (2009), this heterogeneity in dispersal syndromes, information use, and habitat preferences should drastically affect the dynamics of spatially structured populations. Given that individuals have different dispersal syndromes, this clearly affects the way individuals settle in patches that are empty or occupied; the ratio of dispersal syndromes will therefore affect metapopulation processes (Figure 12.1).

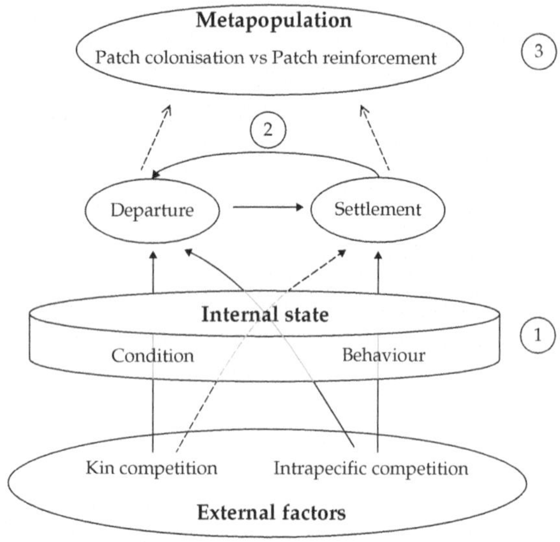

Figure 12.1 Relationship between phenotype-dependent, context-dependent dispersal, and metapopulation dynamic in common lizards. The diagram is based on the conceptual framework in Clobert *et al.* 2009. Three main effects are discussed: firstly, phenotypic differences (body condition/size and behavioural types) between dispersers and residents depend on external factors driving departure and settlement decisions (here kin and non-kin competition) (1); Information on surrounding populations is carried by immigrants through their phenotype which depends on dispersal contexts and individuals with different phenotypes might use different information in dispersal decisions (2); Altogether, phenotype-dependent, context-dependent, and informed dispersal may produce both colonizers and joiners. It will change the spatio-temporal dynamic of metapopulations (3). Dashed arrows show links that are still lacking strong evidence.

12.5 Final remark

We would like to end this review with a short list of future directions on dispersal syndromes. Phenotypic differences between dispersers and residents have been found in many species, leading to this new term of 'dispersal syndromes' (Clobert *et al.*

2009). However, few studies tested for the consistency of these differences over time and across life stages. This is particularly true for behavioural traits (Cote *et al.* 2010a). Second, in the future, studies should focus on proximate pathways connecting dispersal behaviour and phenotypic traits, especially personality traits. Dispersal syndrome can arise from genes (G), environment (E), the G × E interaction (Cote *et al.* 2010a), and epigenetic/maternal effects. For example, although several studies plausibly underlie the joint development of dispersal behaviour and personality traits, only a few studies have examined factors leading to the correlation between personality and dispersal behaviour. Finally, context-dependent dispersal syndromes should be integrated in the general notion of informed dispersal within a patchy environment; this would help to further understand the dynamics of metapopulation as well as biological invasions.

Acknowledgements

The authors want to thank Giovanna Montenegro for editing the paper and AXA research fund for a postdoctoral fellowship to Julien Cote.

References

Aragon, P., Massot, M., Clobert, J. *et al.* (2006a). 'Socially acquired information through chemical cues in the common lizard', *Animal Behaviour* **72**: 965–74.

Aragon, P., Meylan, S. and Clobert, J. (2006b). 'Dispersal status-dependent response to the social environment in the Common Lizard, Lacerta vivipara', *Functional Ecology* **20**: 900–7.

Armsworth, P.R. and Roughgarden, J.R. (2005). 'The Impact of Directed versus Random Movement on Population Dynamics and Biodiversity Patterns', *American Naturalist* **165**: 449–65.

Bekoff, M. (1977). 'Mammalian dispersal and the ontogeny of individual behavioral phenotypes', *American Naturalist* **111**: 715–32.

Bell, A. (2006). 'Future directions in behavioural syndromes research' *Proceedings of the Royal Society B* **274**: 755–61.

Blumstein, D.T., Wey, T.W. and Tang, K. (2009). 'A test of the social cohesion hypothesis: interactive female marmots remain at home', *Proceedings of the Royal Society B* **276**: 3007–12.

Bonte, D., Van Dyck, H., Bullock, J. (2012). 'Costs of dispersal', *Biological Reviews* (in press).

Boudjemadi, K., Lecomte, J., Clobert, J. (1999). Influence of connectivity on demography and dispersal in two contrasting habitats: an experimental approach. *Journal of Animal Ecology* **68**: 1207–24.

Bowler, D.E. and Benton, T.G. (2005). 'Causes and consequences of animal dispersal strategies: relating individual behaviour to spatial dynamics', *Biological Reviews* **80**: 205–25.

Clobert, J., Danchin, E., Dhondt, A.A. *et al.* (2001). *Dispersal.* Oxford University Press, New York.

Clobert, J., Le Galliard, J.F. Cote, J. *et al.* (2009). 'Informed dispersal, heterogeneity in animal dispersal syndromes and the dynamics of spatially structured populations', *Ecology Letters* **12**: 197–209.

Clobert, J., Massot, M., Lecomte, J. *et al.* (1994). 'Determinants of dispersal behavior: the common lizard as a case of study', in J. Vitt and E.R. Pianka, eds, *Lizard Ecology: The Third Generation.* Princeton University Press, Princeton NJ, 183–206.

Cote, J., Boudsocq, S. and Clobert, J. (2008a). 'Density, social information, and space use in the common lizard (*Lacerta vivipara*)', *Behavioral Ecology* **19**: 163–8.

Cote, J. and Clobert, J. (2007a). 'Social information and emigration: lessons from immigrants', *Ecology Letters* **10**: 411–17.

Cote, J. and Clobert, J. (2007b). 'Social personalities influence natal dispersal in a lizard', *Proceedings of the Royal Society B* **274**: 383–90.

Cote, J. and Clobert, J. (2010). 'Risky dispersal: avoiding kin competition despite uncertainty', *Ecology* **91**: 1485–93.

Cote, J., Clobert, J., Brodin, T. (2010a). 'Personality-dependent dispersal: characterization, ontogeny and consequences for spatially structured populations', *Philosophical Transactions of the Royal Society B* **365**: 4065–76.

Cote, J., Clobert, J. and Fitze, P.S. (2007). 'Mother-offspring competition promotes colonization success', *Proceedings of the National Academy of Sciences USA* **104**: 9703–8.

Cote, J., Dreiss, A. and Clobert, J. (2008b). 'Social personality trait and fitness', *Proceedings of the Royal Society B* **275**: 2851–8.

Cote, J., Fogarty, S.,Brodin, T. *et al.* (2011). 'Personality-dependent dispersal in the invasive mosquitofish: group composition matters', *Proceedings of the Royal Society B* **278**: 1670–8.

Cote, J., Fogarty, S., Weinersmith, K. *et al.* (2010b). 'Personality traits and dispersal tendency in the invasive mosquitofish (*Gambusia affinis*)', *Proceedings of the Royal Society B* **277**: 1571–9.

Dall, S.R.X., Houston, A.I. and McNamara, J. (2004). 'The behavioural ecology of personality: consistent individual differences from an adaptive perspective', *Ecology Letters* **7**: 734–9.

de Fraipont, M., Clobert, J., John-Adler, H. *et al.* (2000). 'Increased pre-natal maternal corticosterone promotes philopatry of offspring in common lizards *Lacerta vivipara*', *Journal of Animal Ecology* **69**: 404–13.

Dingemanse, N.J., Both, C., van Noordwijk, A. J. *et al.* (2003). 'Natal dispersal and personalities in great tits (*Parus major*)', *Proceedings of the Royal Society B* **270**: 741–7.

Duckworth, R.A. 2008. Adaptive dispersal strategies and the dynamics of a range expansion. *American Naturalist* **172**: S4–17.

Duckworth, R.A. and Badyaev, A.V. (2007). 'Coupling of dispersal and aggression facilitates the rapid range expansion of a passerine bird', *Proceedings of the National Academy of Sciences of the USA* **104**: 15017–22.

Edelaar, P., Siepielski, A.M. and Clobert, J. (2008). 'Matching Habitat Choice Causes Directed Gene Flow: A Neglected Dimension in Evolution and Ecology', *Evolution* **62**: 2462–72.

Fraser, D.F., Gilliam, J.F., Daley, M.J. *et al.* (2001). 'Explaining leptokurtic movement distributions: intrapopulation variation in boldness and exploration', *American Naturalist* **158**: 124–35.

Hamilton, W.D. (1964). 'The genetical evolution of social behaviour, I and II', *Journal of Theoretical Biology* **7**: 1–52.

Ims, R.A. (1990). 'Determinants of natal dispersal and space use in gray-sided voles, *Clethrionomys rufocanus*—a combined field and laboratory experiment', *Oikos* **75**: 106–13.

Jokela, M., Elovainio, M., Kivimäki, M. *et al.* (2008). 'Temperament and migration patterns in Finland', Psychological Sciences **19**: 831–7.

Kurvers, R.H.J.M., Prins, H.H.T., van Wieren, S.E. *et al.* (2010). 'The effect of personality on social foraging: shy barnacle geese scrounge more', *Proceedings of the Royal Society B* **277** (1681): 601–8.

Kurvers, R.H.J.M., Van Oers, K., Nolet, B.A. *et al.* (2010). 'Personality predicts the use of social information', *Ecology Letters* **13**: 829–37.

Le Galliard, J.F., Clobert, J. and Ferrière, R. (2004). 'Physical performance and darwinian fitness in lizards', *Nature* **432**: 502–4.

Le Galliard, J.F., Ferrière, R. and Clobert, J. (2003). 'Mother-offspring interactions affect natal dispersal in a lizard', *Proceedings of the Royal Society B* **270**: 1163–9.

Le Galliard, J.F., Ferrière, R. and Clobert, J. (2005a). 'Effect of patch occupancy on immigration on the common lizard', *Journal of Animal Ecology* **74**: 241–9.

Le Galliard, J.F., Ferrière, R. and Dieckmann, U. (2005b). 'Adaptive evolution of social traits: origin, trajectories, and correlations of altruism and mobility', *American Naturalist* **165**: 206–24.

Lecomte, J., Boudjemadi, K., Sarrazin, F. *et al.* (2004). 'Connectivity and homogenisation of population sizes: an experimental approach in *Lacerta vivipara*', *Journal of Animal Ecology* **73**: 179–89.

Lecomte, J., Clobert, J., Massot, M. *et al.* (1994). 'Spatial and behavioural consequences of a density manipulation in the common lizard', *Ecoscience* **1**: 300–10.

Léna, J.-P., Clobert, J., de Fraipont, M. *et al.* (1998). 'The relative influence of density and kinship on dispersal in the common lizard', *Behavioral Ecology* **9**: 500–7.

Léna, J.-P. and de Fraipont, M. (1998). 'Kin recognition in the common lizard', *Behavioral Ecology and Sociobiology* **42**: 341–7.

Léna, J.-P., de Fraipont, M. and Clobert, J. (2000). 'Affinity towards maternel odour and offspring dispersal in the common lizard', *Ecology Letters* **3**: 300–8.

Marchetti, C. and Drent, P.J. (2000). 'Individual differences in the use of social information in foraging by captive great tits', *Animal Behaviour* **60**: 131–40.

Massot, M., Clobert, J. (2000). 'Processes at the origin of similarities in dispersal behaviour among siblings', *Journal of Evolutionary Biology* **13**: 707–19.

Meylan, S., Clobert, J. and Sinervo, B. (2007). 'Adaptive significance of maternal induction of density-dependent phenotypes', *Oikos* **116**: 650–61.

Meylan, S., De Fraipont, M., Aragon, P. *et al.* (2009). 'Are dispersal-dependent behavioral traits produced by phenotypic plasticity?', *Journal of Experimental Zoology A* **311A**: 377–88.

O'Riain, M.J., Jarvis, J.U.M. and Faulkes, C. (1996). 'A dispersive morph in the naked mole-rat', *Nature* **380**: 619–21.

Plaistow, S.J., Lapsley, C.T. and Benton, T.G. (2006). 'Context-Dependent Intergenerational Effects: The Interaction between Past and Present Environments and Its Effect on Population Dynamics', *American Naturalist* **167**: 206–15.

Reale, D., Garant, D., Humphries, M.M. *et al.* (2010). 'Personality and the emergence of the pace-of-life syndrome concept at the population level', *Philosophical Transactions of the Royal Society B* **365**: 4051–63.

Sih, A. and Bell, A.M. (2008). 'Insights for Behavioral Ecology from Behavioral Syndromes', *Advances in the Study of Behavior* **38**: 227–81.

Sorci, G. and Clobert, J. (1999). 'Natural selection on hatchling body size and mass in two environments in the common lizard (*Lacerta vivipara*)', *Evolutionary Ecology Research* **1**: 303–16.

Stamps, J.A. (2001). 'Habitat selection by dispersers: integrating proximate and ultimate approaches', in J. Clobert, E. Danchin, A.A. Dhondt *et al.* eds, *Dispersal*. Oxford University Press, New York, 110–22.

CHAPTER 13

Dispersal syndromes in butterflies and spiders

Dries Bonte and Marjo Saastamoinen

13.1 Introduction

Dispersal is essential for the persistence of populations in fragmented landscapes (Chapters 20, 30, 31, 34), and can play a key role in determining the range expansion dynamics of invasive species and range-shifting in response to climate change (Chapter 26). Dispersal is an expensive life-history trait with costs imposed during dispersal itself, but often also during other stages of the life history (Bonte *et al.* 2012). The selection pressures on dispersal and related life-history traits are likely to vary across species and among individuals within a species. In addition to natural selection affecting dispersal phenotypes, environmental conditions during development and during the actual dispersal process may further shape individual phenotypes with covarying dispersal-related life-history traits (Chapter 10). Understanding the evolution of dispersal therefore requires an understanding of the way dispersal influences, and is influenced by, the behavioural and/or life-history traits that are associated with it. The covariation of life-history traits with dispersal defines a dispersal syndrome; i.e. dispersers may be similar in behaviour, morphology, and life history compared with non-dispersers. These dispersal-life history correlates have been traditionally studied in wing-dimorphic insects like crickets and bugs. In these species, polymorphism in wing and flight muscle development (pterygomorphism) is correlated with various life-history traits both within- and among species. In the herbivorous *Dysdericus* beetles, species from ephemeral habitats allocate resources into dispersal (wings and flight muscles) during development while delaying maturity and reproduction (oögenesis-flight syndrome). This trade-off is evident even though dispersive individuals are able to re-allocate resources from flight muscles (histolysis) back to reproduction after settlement (Derr *et al.* 1981; Harrison 1980). Similarly, within species, Roff and colleagues have demonstrated that life-history traits between winged and wingless cricket morphs mostly show negative covariation (Roff and DeRose 2001; Roff *et al.* 2002; Roff *et al.* 1999; Stirling *et al.* 2001). These negative correlations have been shown to be mediated through changes in hormone titres, feeding back on basal metabolism (Zera and Bottsford 2001; Zera and Mole 1994; Zera *et al.* 1994).

Alongside the large body of literature focusing on distinctive wing polymorphisms, we here aim to focus on arthropods with dispersal as a continuous, behaviourally mediated trait. Our aim is to discuss covariation of dispersal with other life-history traits in two groups (spiders and butterflies) which have no obvious morphological dispersal adaptations. As we will show here, dispersive phenotypes show strong phenotypic correlates with other life-history traits, and these correlations are often shaped by environmental conditions experienced during dispersal itself, during the life stages immediately preceding emigration, during the natal development, and even conditions experienced by previous generations. The correlated responses are therefore context-specific and highly depending on the scale of biological organization. Here, the focus will be on between- and within-population comparisons rather than on comparisons between species.

Dispersal Ecology and Evolution. First Edition. Edited by Jean Clobert *et al.*

13.2 Dispersal syndromes among populations: the Glanville fritillary system as an example of butterfly with a colonizer syndrome

In butterflies, dispersal relies on flight ability, which is also a fundamental prerequisite for other life-history functions, as individuals need to fly in order to forage, locate mating partners, and distribute eggs. Even though there are examples of both between (Stevens *et al.* 2010) and within-species (Hill *et al.* 1999; Thomas *et al.* 2001) that indicate how in some cases flight-related morphological traits are reliable proxies of dispersal, it is currently debatable how much interpatch movements depend on special behaviours rather than routine movements (Van Dyck and Baguette 2005). In the butterfly case study discussed here we have consistently showed that dispersal and flight-related components, such as within-patch mobility (Hanski *et al.* 2006) and flight metabolic rate (Niitepõld *et al.* 2009), are linked (for more details, see Case Study III and Chapter 23). In addition, variation in dispersal ability is tightly linked to variation in one gene, *Pgi*, encoding a glycolytic enzyme Phopshoglucose isomerase. Females with a particular genotype, *Pgi-f*, have higher peak flight metabolic rate, higher dispersal ability in the field, and this genotype is more frequent in the newly established populations (Haag *et al.* 2005). In contrast to females, males do not show similar positive correlation between dispersal and peak flight metabolic rate, presumably as high flight capacity in males is related to territorial behaviour and increases the number of matings, whereas inferior males may be those forced to disperse (Niitepõld *et al.* 2011). Thus flight capacity has opposite associations with dispersal rate in the two sexes, but as females are those that establish new populations, we will concentrate on them in Case Study III. Interestingly, however, there does not seem to be any significant differences in flight-related morphologies between the dispersive and non-dispersive individuals in either females or males.

As dispersal is energetically costly, correlations between dispersal and other life-history traits are often presumed to be negative simply because resources allocated to dispersal are no longer available for other energetically expensive traits. In accordance with this idea, in the speckled wood butterflies, for example, individuals from highly fragmented agricultural landscapes have reduced fecundity and lifespan relative to individuals from continuous woodland, due to the higher dispersal costs between suitable habitat fragments, and associated differences in flight morphology and behaviour (Gibbs and Van Dyck 2010). In this species, increased dispersal is associated with reduced investment in reproduction in populations that are expanding their range, and which possess higher dispersal rates than populations at the core of the range (e.g., Hughes *et al.* 2003). A contrasting dispersal syndrome is that of a 'colonizer syndrome', in which highly dispersive and/or mobile individuals show rapid development, early reproduction and high fecundity (Baker and Stebbins 1965). Individuals showing such positive correlations, between flight ability and other life-history traits, have also been called 'super dispersers' or 'fugitive species' (Hutchinson 1951). A positive correlation between dispersal and reproduction in dispersive species or populations has been explained by the selective advantage of such correlation due to increased success of individuals colonizing new habitats (Dingle *et al.* 1988; Ritte 1978; Chapter 10). Following, we will discuss in more detail an example of the colonizer syndrome in a butterfly case study.

The Glanville fritillary butterfly occurs in a highly fragmented landscape and has a classic metapopulation structure in the Åland Islands, south-west Finland (Hanski 1999), with a high rate of population turnover (extinctions and recolonizations; Nieminen *et al.* 2004). Hundreds of new populations are established each year on habitat patches that were unoccupied in the previous year, and these populations compensate for the loss of similar number of local population extinctions (Case Study III, and Chapter 23). Females establishing these new populations are more dispersive than females from already established old populations (populations that have existed for more than five years without going extinct; Hanski *et al.* 2002; Hanski *et al.* 2004; Chapter 23). This difference can simply be explained by females with high dispersal ability being more likely to successfully colonize new, habitat patches, especially isolated ones. As dispersal has a genetic bases (Klemme and Hanski 2009; Saastamoinen and Hanski 2008; Saastamoinen 2008), the offspring of these females also have higher dispersal ability.

Based on numerous different studies on the Glanville fritillary butterfly, in which we have compared within-patch mobility of females from newly established and old populations, or females with different *Pgi* genotypes in large outdoor enclosures, our general conclusion is that there is no evident trade-off between dispersal and reproduction. Instead numerous fecundity-related life-history traits show positive correlation with dispersal ability (Table 13.1 and Figure 13.2). More mobile females initiate reproduction (mating and/or oviposition) at younger age, and have higher mating frequency compared with the more sedentary females (Klemme and Hanski 2009; Saastamoinen 2007a; Saastamoinen 2007b, Figure 13.1). This may partially be due to the fact that females from newly established populations, which in general are also more mobile, mature eggs at faster rate than females from old populations (Wheat *et al.* in prep). Even though the potential fecundity, essentially the number of oocytes at eclosion, is lower in females from the newly established populations (Hanski *et al.* 2004), this trade-off does not translate to difference in lifetime egg production (Saastamoinen 2007b). It seems that under field conditions, females only rarely are able to lay all the eggs they have at eclosion. In addition, females with a particular genotype, *Pgi-f*, lay on average, 20–30% larger clutches compared with females with the *Pgi-d* genotype (Saastamoinen 2007a; Saastamoinen and Hanski 2008). Given that *Pgi-f* genotype is related to higher flight metabolic rate, higher dispersal, and is found in higher frequencies in newly established than old populations, this difference actually suggests a positive correlation between dispersal and clutch size. The higher clutch size of the *Pgi-f* females is at least partially explained by *Pgi-f* females being able to fly at lower ambient temperatures and hence being active earlier in the day, and thereby initiating oviposition during the most favourable conditions in the early afternoon, which in general increases clutch size (Saastamoinen and Hanski 2008).

Even though *Pgi-f* females lay larger and more clutches, the lifetime egg production seems to be approximately equal in the two types of population (Saastamoinen 2007b). Interestingly, however, in newly established populations within-patch mobility is positively correlated with lifetime reproduc-

Table 13.1 Comparisons of dispersal and related life-history traits between newly established and old populations in the Glanville fritillary butterfly.

	Newly established populations	Old populations
Dispersal-related traits:		
Dispersal	high	low
Within-patch mobility	high	low
Flight metabolic rate	high	low
Most common *Pgi*-genotype	*Pgi-f*	*Pgi-d*
Life-history traits:		
Development time*	equal	equal
Body size	equal	equal
Egg maturation	high	low
Age at first reproduction	low	high
Number of oocytes	low	high
Clutch size*	high	low
Lifetime egg production	equal	equal
Lifespan	low under stressful conditions[1]	high under stressful conditions[1]

* based solely on comparisons between the *Pgi*-genotypes.[1] under laboratory conditions individuals with *Pgi-f* genotype live longer.

tive success, whereas in old populations the opposite is true, and more mobile females have reduced lifetime egg production compared with sedentary females (Saastamoinen 2007b). These results are reminiscent of some previous results comparing dispersive versus sedentary species/populations (Lavie and Ritte 1978; Wu 1981), and have been suggested to indicate dissimilar selection on dispersal and related life-history traits between newly established and old populations. The positive correlation between dispersal and fitness (i.e. the colonizer syndrome) especially in the newly established isolated population is in general assumed to be characteristic for species frequently colonizing ephemeral habitats and hence living in highly unstable habitats (Rankin and Burchsted 1992; Ronce and Olivieri 2004). On the other hand, for individuals living in more permanent (i.e. old) populations, high dispersal rate is not so essential and in such cases selection is thought to favour individuals with other fecundity-related life-history traits, for instance increased longevity (Chapter 23).

Comparisons of individual life-history traits in *M. cinxia* between old and new populations in enclosures suggested that individuals from newly established populations have a shorter lifespan than females from established populations (Hanski *et al.* 2006; Zheng *et al.* 2007). This would imply that the new population females have a so-called 'fast' life-history strategy; they initiate oviposition at young age, produce larger clutches at faster rate but have a shorter lifespan. Equivalent lifetime egg production is achieved by females from old populations, but these females, who move less, seem to compensate for the slower reproduction by longer lifespan. Laboratory studies, conducted under constant optimal environmental conditions, have, however, shown that the story is more complicated, as females with *Pgi-f* genotype, which is more frequent in the new populations, actually live longer than females without this genotype (Klemme and Hanski 2009; Saastamoinen *et al.* 2009). Taken together, the often significant but conflicting results on differences in lifespan under different experimental conditions suggest that the effect of genotype and time since the population was established are highly dependent on the environmental conditions. Possibly under unlimited food conditions in the laboratory, the more mobile individuals can increase food intake sufficiently to compensate for the energetic cost of higher mobility, whereas the trade-off between mobility and lifespan may be apparent under more restricted food availability under semi-natural conditions in the population enclosure. This hypothesis is also supported by the fact that the effect of the rate of reproduction on abdomen weight loss (indication of depletion of resources) is much greater in the *Pgi-f* than in the other genotypes (Saastamoinen *et al.* 2009). This suggests a trade-off between fecundity and maintenance, since if the *Pgi-non-f* females are able to reproduce with a smaller reduction of resources in the abdomen, it leaves more resource left for maintenance, potentially prolonging the lifespan under some conditions (e.g. mild food stress). As concluded by Reznick *et al.* (2000), 'super phenotypes' may be real, but they can only be 'super' under particular environmental conditions.

13.3 Dispersal syndromes within populations: dispersal reaction norms and syndromes in spiders

In spiders, aerial dispersal is mediated by distinct behaviours preceding transfer. By means of tiptoe behaviour (i.e. stretching of legs, raising abdomen, and producing silk threads) spiders initiate either long-distance dispersal when silk threads are transported in the air (*ballooning*) or short-distance dispersal when silk threads are used as bridges (*rappelling*) (Case Study II). Due to aerodynamic constraints, ballooning is restricted to smaller individuals. In temperate regions, these comprise predominantly the adults of smaller money spiders (Linyphiidae) or the juveniles of larger wolfspiders (Lycosidae). In the latter, routine cursorial movements (Bonte *et al.* 2004) or maternally assisted spreading of offspring (Bonte *et al.* 2007b) exist as alternative adult and natal dispersal mechanisms, respectively. Obviously, the different dispersal modes largely differ in efficiency and underlying costs. In case of ballooning, costs are directly related to the availability of suitable habitat within the dispersal perimeter (Englund and Hamback 2007),

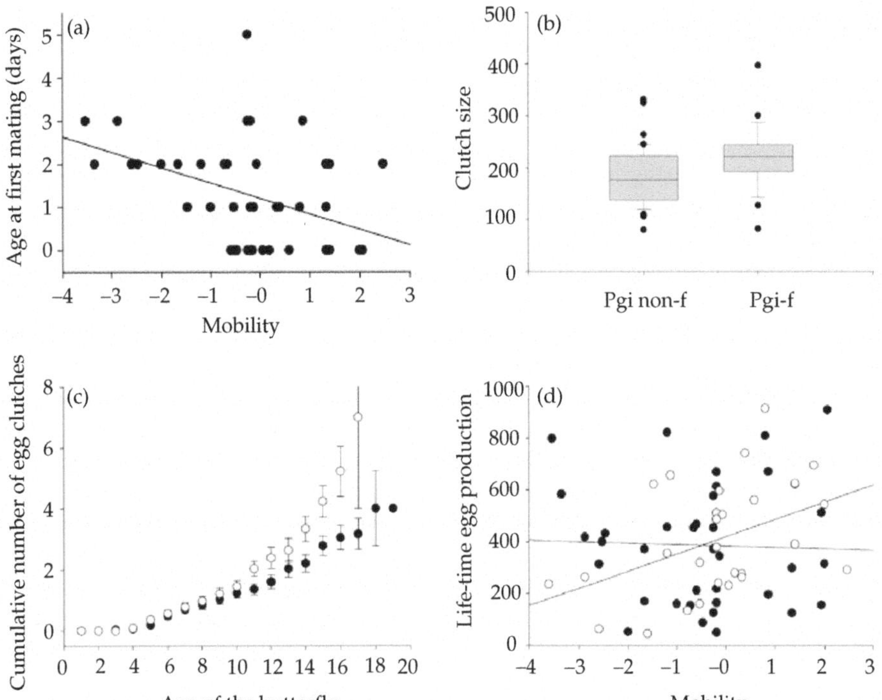

Figure 13.1 Dispersal syndromes in *Melitaea cinxia*: (a) more mobile females reproduce at younger age, (b) females with a *Pgi* genotype that is related with high dispersal rate (*Pgi-f*) lay larger egg clutches, (c) females from newly established populations (open symbols), in which females have higher dispersal rate, lay eggs at faster rate, (d) and show a positive correlation between mobility and lifetime egg production, whereas females from old populations (closed symbols) show no such correlation. Data are adopted from Hanski *et al.* 2006 and Saastamoinen 2007a.

while short-distance dispersal success is more likely correlated with the spatial configuration of resources and habitat (Englund and Hamback 2007).

To investigate dispersal syndromes in spiders, we mainly applied a behavioural reaction norm approach (i.e. assessing the set of behavioural phenotypes that a single individual produces in a given set of environments; Dingemanse *et al.* 2010). We focused on species belonging to the genus *Erigone* which are predominantly associated with disturbed habitats. In these species, a strong female-biased dispersal is observed. In contrast to aerial dispersal in wolfspiders (Bonte *et al.* 2007a; Bonte *et al.* 2006), food shortage starvation induce a reduction of dispersal initiation in adult *Erigone* individuals (Bonte *et al.* 2008a). Dispersal was also strongly reduced by inbreeding (Bonte 2009) and the prevalence of endosymbiontic bacteria (Goodacre *et al.* 2009), suggesting a positive correlation between body condition and dispersal. This pattern has been confirmed in a study by varying body condition through variable developmental temperature (from 15–30 °C). When the different dispersal modes (so *ballooning* versus *rappelling*) were analysed separately, high developmental temperatures induced proportionally higher rates of rappelling relative to ballooning, while individuals raised at lowest temperatures dispersed more by ballooning compared to by rappelling (see Figures 13.2a,b). This shift in dispersal behaviour according to the developmental temperature is assumed to be adaptive since it reflects seasonal changes in habitat configuration related to the harvesting of crops (Bonte *et al.* 2008b). It thus suggests that temperature during development is perceived as a cue to adopt specific dispersal behaviours. Patterns were slightly different for males and females, with a positive correlation between the two dispersal modes among temperature treatments for females, and a slight negative one for males (Figure 13.2c). Interestingly, the positive correlations with proxies for body condition were also positive within temperature treatment (Bonte *et al.* 2008b).

Different temperatures during development subsequently induce different dispersal phenotypes. In females, these phenotypes are additionally linked to distinct settlement strategies. Lab experiments demonstrated that ballooning phenotypes of female spiders built webs above suitable microhabitat much faster than philopatric individuals (Figure 13.3a) and that they were much more eager to take over webs that were already occupied (Figure 13.3b; from Bonte *et al.* 2011). Long-distance dispersal phenotypes of the spider *E. atra* possess in consequence settlement advantages compared to residential phenotypes. This faster and more competitive settlement behaviour of ballooning compared to residential phenotypes can be the result of environmental constraints, enabling only those individuals in best conditions to emigrate. However, because ballooning dispersal has been shown to be a consistent trait and partly genetically determined (Bonte *et al.* 2009; Bonte and Lens 2007), they are likely to be adaptive dispersal strategies. Indeed, the evolution of positive covariation between individual condition and dispersal may evolve in landscapes with high levels of environmental stochasticity and either vacant or crowded habitat (Bonte and de la Pena 2009). These are the conditions experienced by *Erigone* spiders when emigrating in spring towards freshly sawn crops or emigrating en masse in autumn towards litter-rich vegetation, subsequently facing overcrowding due to concentration effects (Bonte *et al.* 2008b). The evolution of long-distance dispersal associated syndrome shows large parallels with the bold-aggression-activity syndrome (a syndrome in which levels of aggression, boldness, and general activity are positively correlated; Sih *et al.* 2004; Chapter 10). Indeed, competitive behaviour during settlement

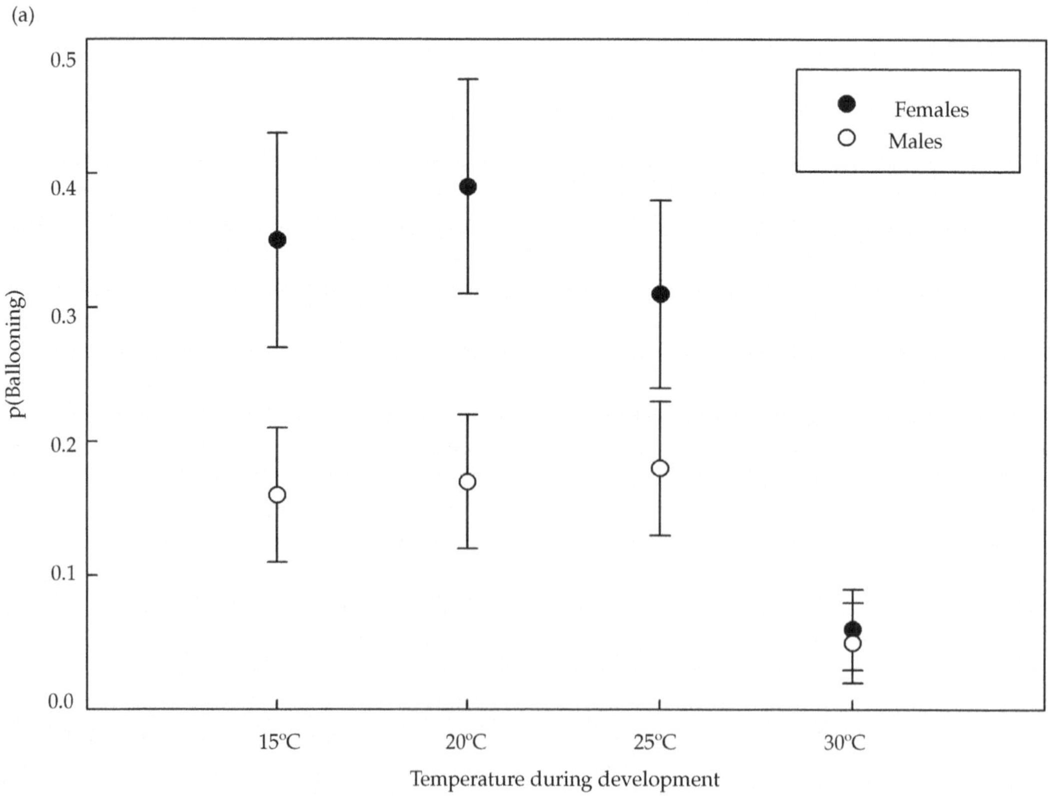

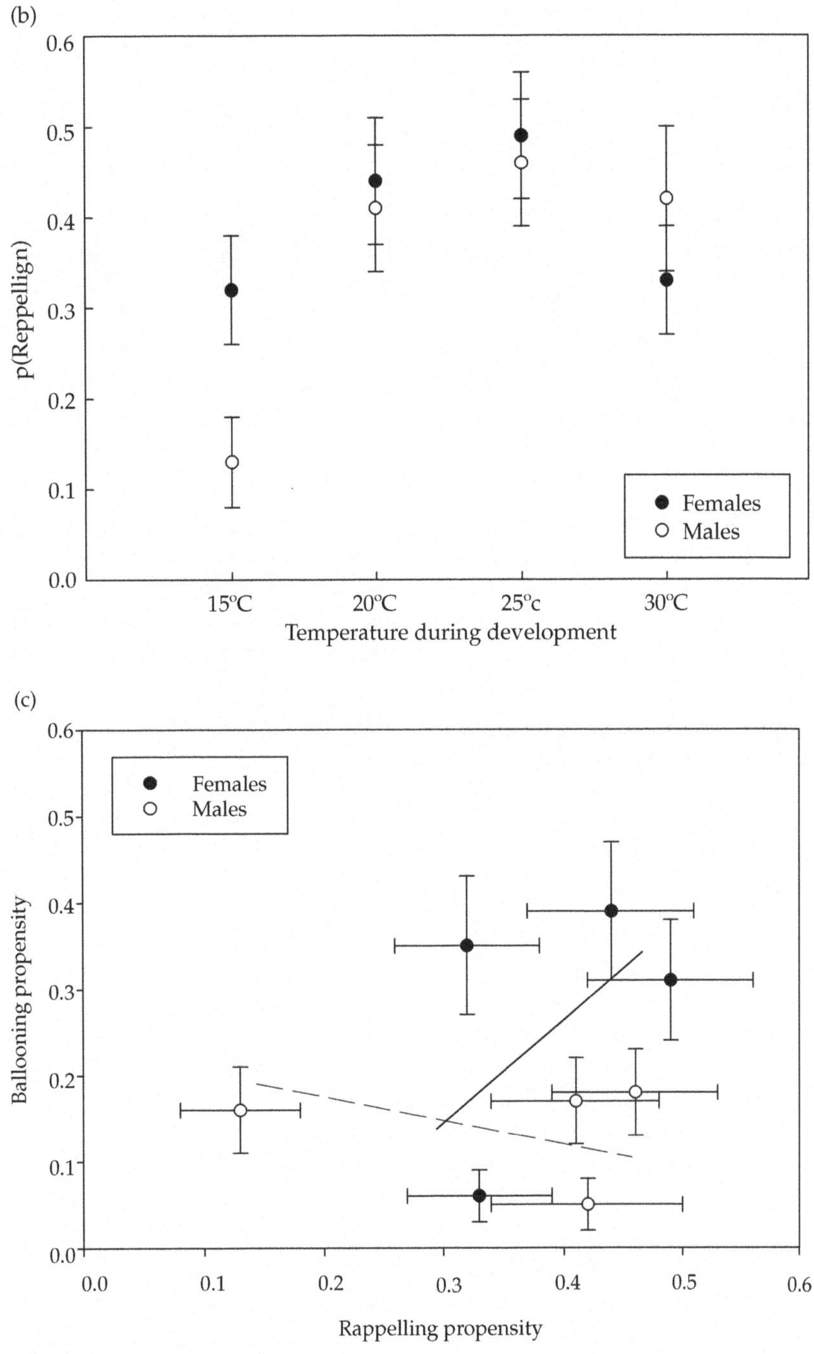

Figure 13.2 Dispersal syndromes in *Erigone atra*: temperature during development induces different dispersal reaction norms for ballooning (a) and rappelling (b) propensity. When combining both, females show a positive correlation (full line) between long and short distance dispersal propensity (c - filled symbols), while males not (open symbols, dashed line). Error bars represent standard errors. Data are adapted from Bonte *et al*. 2008b.

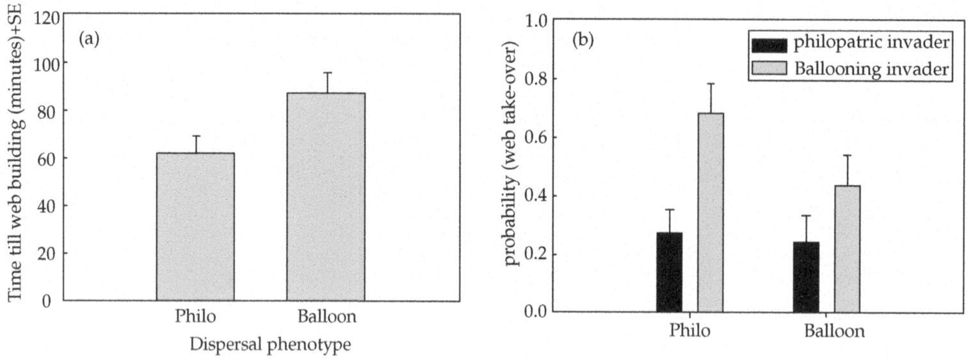

Figure 13.3 Correlated responses of long-distance dispersal behaviour (ballooning) in *E. atra* females with web-building related behaviours: (a) latency to web building, and (b) web-take-over behaviour. From Bonte *et al.* (2011), used with permission from Elsevier.

implicates the presence of individual differences in boldness (i.e. taking decision to occupy new microhabitat sites) and aggression (i.e. the willingness to go into confrontation with conspecifics to take over web sites).

13.4 Context dependence of dispersal and the organization of syndromes

In spiders, dispersal strategies are strongly sex- and context-dependent. Temperature during development (Bonte *et al.* 2008b), local densities of males and females (De Meester and Bonte 2010), but probably also internal physiological changes due to the presence of endosymbionts (Goodacre *et al.* 2009), provide information to which spiders adjust their dispersal behaviour. These contexts also induce differences in individual condition that impact (constrain) the eventual dispersal strategy and are likely to shape the direction of dispersal syndromes. The *M. cinxia* system additionally shows that dispersal syndromes can have a genetic background when dispersal or genetically correlated traits have a heritable background (example of *Pgi-genotype* affecting dispersal and fecundity via effects on flight metabolic rate). The strong selection for dispersal genotypes and associated syndromes as colonizers of vacant habitat in true metapopulations is then likely to scale up syndromes among populations (i.e. new versus old ones). The fact that dispersive individuals are not a random subsample from the (meta)population has profound consequences for population genetic structure and dynamics. In both *E. atra* and *M. cinxia* the positive relationship between body condition, dispersal, and settlement strategies can be expected to have a disproportionally larger effect on colonization potential and gene flow compared to species lacking this positive correlation. Of course, these conclusions are drawn from experiments under favourable conditions in which variation in body condition was prominent. Given the prominent importance of other proximate stress-inducing factors under natural conditions such as local density and the evolution of informed dispersal strategies (Chapter 10), it will be interesting to study reaction norms of these syndromes in relation to variable environmental contexts that impact dispersal-related life-history traits (Reznick *et al.* 2000).

References

Baker, H.G. and Stebbins, G.L. (1965). *The Genetics of Colonising Species*. Academic Press, New York.

Bonte, D. (2009). 'Inbreeding depresses short and long distance dispersal in three congeneric spiders', *Journal of Evolutionary Biology* **22**: 1429–34.

Bonte, D., Bossuyt, B. and Lens, L. (2007a). 'Aerial dispersal plasticity under different wind velocities in a salt marsh wolf spider', *Behavioral Ecology* **18**: 438–43.

Bonte, D., De Clercq, N., Zwertvaegher, I. *et al.* (2009). 'Repeatability of dispersal behaviour in a common dwarf spider: evidence for different mechanisms behind short- and long-distance dispersal', *Ecological Entomology* **34**: 271–6.

Bonte, D. and de la Pena, E. (2009). 'Evolution of body condition-dependent dispersal in metapopulations', *Journal of Evolutionary Biology* **22**: 1242–51.

Bonte, D., De Meester, N. and Matthysen, E. (2011). 'Selective integration advantages when transience is costly: immigration behaviour in an agrobiont spider', *Animal Behaviour* **81**: 837–41.

Bonte, D. and Lens, L. (2007). 'Heritability of spider ballooning motivation under different wind velocities', *Evolutionary Ecology Research* **9**: 817–27.

Bonte, D., Lens, L. and Maelfait, J.P. (2004). 'Lack of homeward orientation and increased mobility result in high emigration rates from low-quality fragments in a dune wolf spider', *Journal of Animal Ecology* **73**: 643–50.

Bonte, D., Lukac, M. and Lens, L. (2008a). 'Starvation affects pre-dispersal behaviour of Erigone spiders', *Basic and Applied Ecology* **9**: 308–15.

Bonte, D., Travis, J.M.J., De Clercq, N. *et al.* (2008b). 'Thermal conditions during juvenile development affect adult dispersal in a spider', *Proceedings of the National Academy of Sciences of the USA* **105**: 17000–5.

Bonte, D., Van Belle, S. and Maelfait, J.P. (2007b). 'Maternal care and reproductive state-dependent mobility determine natal dispersal in a wolf spider', *Animal Behaviour* **74**: 63–9.

Bonte D., Van Dyck H., Bullock J.M. *et al.* (2012). 'Costs of dispersal', *Biological Reviews* **87**: 290–312.

Bonte, D., Vanden Borre, J., Lens, L. *et al.* (2006). 'Geographical variation in wolf spider dispersal behaviour is related to landscape structure. *Animal Behaviour* **72**: 655–62.

De Meester, N. and Bonte, D. (2010). 'Information use and density-dependent emigration in an agrobiont spider', *Behavioral Ecology* **21**: 992–8.

Derr, J.A., Alden, B. and Dingle, H. (1981). 'Insect life histories in relation to migration, body size, and host plant array—a comparative-study of Dysdercus', *Journal of Animal Ecology* **50**: 181–93.

Dingemanse, N.J., Kazem, A.J.N., Réale, D. *et al.* (2010). 'Behavioural reaction norms: animal personality meets individual plasticity', *Trends in ecology* and *evolution (Personal edition)* **25**: 81–9.

Dingle, H., Evans, K.E. and Palmer, J.O. (1988). 'Responses to selection among life-history traits in a nonmigratory population of milkweed bugs (*Oncopeltus fasciatus*)', *Evolution* **42**: 79–92.

Englund, G. and Hamback, P.A. (2007). 'Scale dependence of immigration rates: models, metrics and data', *Journal of Animal Ecology* **76**: 30–5.

Goodacre, S.L., Martin, O.Y., Bonte, D. *et al.* (2009). 'Microbial modification of host long-distance dispersal capacity', *BMC Biology* **7**: 32.

Hanski, I. (1999). *Metapopulation ecology*. Oxford University Press, New York..

Hanski, I., Breuker, C.J., Schops, K. *et al.* (2002). 'Population history and life history influence the migration rate of female Glanville fritillary butterflies', *Oikos* **98**: 87–97.

Hanski, I., Erälahti, C., Kankare, M. *et al.*, Ovaskainen, O. (2004). 'Variation in migration propensity among individuals maintained by the landscape structure', *Ecology Letters* **7**: 958–66.

Hanski, I., Saastamoinen, M. and Ovaskainen, O. (2006). 'Dispersal-related life history trade-offs in a butterfly metapopulation', *Journal of Animal Ecology* **75**: 91–100.

Harrison, R.G. 1980. 'Dispersal polymorphisms in insects', *Annual Review of Ecology and Systematics* **11**: 95–118.

Hill, J.K., Thomas, C.D. and Blakeley, D.S. (1999). 'Evolution of flight morphology in a butterfly that has recently expanded its geographic range', *Oecologia* **121**: 165–70.

Hughes, C.L., Hill, J.K. and Dytham, C. (2003). 'Evolutionary trade-offs between reproduction and dispersal in populations at expanding range boundaries', *Proceedings of the Royal Society B* **270**: S147–50.

Hutchinson, G.E. (1951). 'Copepodology for the ornithologist', *Ecology* **32**: 571–7.

Klemme, I. and Hanski, I. (2009). 'Heritability of and strong single gene (Pgi) effects on life-history traits in the Glanville fritillary butterfly', *Journal of Evolutionary Biology* **22**: 1944–53

Kuussaari, M., van Nouhuys, S., Hellmann, J. *et al.* (2004). 'Larval biology', in P. R. Ehrlich and I. Hanski, eds, *On The Wings of Checkerspots: A Model System for Population Biology*. Oxford University Press, Oxford, 138–60.

Lavie, B. and Ritte, U. (1978). 'The relation between dispersal behaviour and reproductive fitness in the flour beetle *Tribolium castaneum*', *Canadian Journal of Genetics and Cytology* **20**: 589–95.

Niitepold, K., Mattila, A.L.K, Harrison, P.J. *et al.* (2011). 'Flight metabolic rate has contrasting effects on dispersal in the two sexes of the Glanville fritillary butterfly more options', *Oecologia* **165**: 847–54.

Niitepõld, K., Smith, A.D., Osborne, J.L. *et al.* (2009). 'Flight metabolic rate and Pgi genotype influence butterfly dispersal rate in the field', *Ecology* **90**: 2223–32.

Rankin, M.A. and Burchsted, J.C.A. (1992). 'The cost of migration in insects', *Annual Review of Entomology* **37**: 533–59.

Reznick, D., Nunney, L. and Tessier, A. (2000). 'Big houses, big cars, superfleas and the costs of reproduction', *Trends in Ecology and Evolution* **15**: 421–5.

Roff, D.A. and DeRose, M.A. (2001). 'The evolution of trade-offs: Effects of inbreeding on fecundity relationships in the cricket *Gryllus firmus*', *Evolution* **55**: 111–21.

Roff, D.A., Mostowy, S. and Fairbairn, D. J. (2002). 'The evolution of trade-offs: Testing predictions on response to selection and environmental variation', *Evolution* **56**: 84–95.

Roff, D.A., Tucker, J., Stirling, G. *et al.* (1999). 'The evolution of threshold traits: effects of selection on fecundity and correlated response in wing dimorphism in the sand cricket', *Journal of Evolutionary Biology* **12**: 535–46.

Ronce, O. and Olivieri, I. (2004). 'Life history evolution in metapopulations', in I. Hanski and O.E. Gaggiotti, eds, *Ecology, Genetics, and Evolution of Metapopulations*. Elsevier Academic Press, San Diego, 227–57.

Saastamoinen, M. (2007a). 'Life-history, genotypic, and environmental correlates of clutch size in the Glanville fritillary butterfly', *Ecological Entomology* **32**: 235–42.

Saastamoinen, M. (2007b). 'Mobility and lifetime fecundity in new versus old populations of the Glanville fritillary butterfly', *Oecologia* **153**: 569–78.

Saastamoinen, M. (2008). 'Heritability of dispersal rate and other life history traits in the Glanville fritillary butterfly', *Heredity* **100**: 39–46.

Saastamoinen, M. and Hanski, I. (2008). 'Genotypic effects on flight activity and oviposition in the Glanville fritillary butterfly', *American Naturalist* **171**: 701–12.

Saastamoinen, M., Ikonen, S. and Hanski, I. (2009). 'Significant effects of *Pgi* genotype and body reserves on lifespan in the Glanville fritillary butterfly', *Proceedings of the Royal Society of London Series B* **276**: 1313–22.

Sih, A., Bell, A. and Johnson, C. (2004). 'Behavioural syndromes: an ecological and evolutionary overview', *Trends in Ecology* and *Evolution* **19**: 372–8.

Stirling, G., Fairbairn, D.J., Jensen, S. *et al.* (2001). 'Does a negative genetic correlation between wing morph and early fecundity imply a functional constraint in *Gryllus firmus*?', *Evolutionary Ecology Research* **3**: 157–77.

Thomas, C.D., Bodsworth, E.J., Wilson, R.J. *et al.* (2001). 'Ecological and evolutionary processes at expanding range margins', *Nature* **411**: 577–81.

Van Dyck, H. and Baguette, M. (2005). 'Dispersal behaviour in fragmented landscapes: Routine or special movements?', *Basic and Applied Ecology* **6**: 535–45.

Wu, A.C. (1981). *Life history traits correlated with emigration in flour beetle populations*. Ph.D. Thesis, University of Illinois, Chicago.

Zera, A.J. and Bottsford, J. (2001). 'The endocrine-genetic basis of life-history variation: The relationship between the ecdysteroid titer and morph-specific reproduction in the wing-polymorphic cricket *Gryllus firmus*', *Evolution* **55**: 538–49.

Zera, A.J. and Mole, S. (1994). 'The physiological costs of flight capability in wing-dimorphic crickets. *Researches on Population Ecology* **36**: 151–6.

Zera, A.J., Mole, S. and Rokke, K. (1994). 'Lipid, carbohydrate and nitrogen content of long-winged and short-winged *Gryllus firmus*—Implications for the physiological cost of flight capability', *Journal of Insect Physiology* **40**: 1037–44.

Zheng, C., Ovaskainen, O., Saastamoinen, M. and *et al.* (2007). 'Age-dependent survival analyzed with Bayesian models of mark-recapture data', *Ecology* **88**: 1970–6.

CHAPTER 14

Plant dispersal phenotypes: a seed perspective of maternal habitat selection

Rafael Rubio de Casas, Charles G. Willis, and Kathleen Donohue

14.1 Introduction

An individual's fitness depends on its ability to reach an appropriate habitat for development and reproduction. Fitness therefore depends on dispersal. In plants, unlike many animals, most of the traits that influence dispersal (i.e. propagule morphology, nutrition, detachment position, mechanics, shape, and size) are primarily maternal characteristics (or paternal in the case of pollen grains). Thus plant dispersal is typically not under the control of the dispersed individual, but instead it results from the interaction between a maternally determined propagule and environmental factors (including dispersal vectors). Consequently, dispersal simultaneously affects both maternal and progeny fitness. As a result, the evolutionary dynamics of dispersal phenotypes derive from correlations among traits within and across generations. In this chapter we discuss how interactions between maternally determined dispersal traits and environmental conditions influence the ecology and evolution of dispersal in plants.

14.2 Plant dispersal syndromes

Plants disperse via pollen or seed, and in some cases, via vegetative structures (e.g. stolons or rhizomes). Vegetative and sexual (i.e. via pollen and seed) dispersal are not mutually exclusive, and both often contribute to population maintenance and expansion (Weppler *et al.* 2006).

The most common form of dispersal and range expansion in clonal plants is through vegetative growth or the dispersal of vegetative fragments, as in a number of aquatic species (Rottenberg and Parker 2004). However, vegetative dispersal cannot be considered the product of selection solely on dispersal itself because it is usually the direct product of vegetative growth (e.g. a stolon is a secondary stem).

Both pollen and seeds play a significant role in gene flow in plant populations. Their relative contribution is still a matter of debate, partly because it is often difficult to assign parental genotypes with precision under natural conditions (De-Lucas *et al.* 2008; Grivet *et al.* 2009). However, pollen alone cannot colonize new patches. For this reason, in this chapter we focus on seed-mediated dispersal, although many of the ideas presented can be extended to pollen and even to vegetative dispersal.

The variability in plant dispersal phenotypes depends on the anatomical structures that contribute to dispersal: maternal plant architecture, structures surrounding the propagules, and properties of the propagules themselves. Some of these characteristics are shared across ecological and taxonomic categories and tend to be associated with specific dispersal strategies, or 'syndromes' (Cousens *et al.* 2008), and many have long been recognized in plants (Ridley 1930). Most of these syndromes are defined based on the putative dispersal vector (the transience phase; Figure 14.1); for example, winged propagules and wind dispersal (i.e. anemochory; Augspurger 1986, Augspurger and Franson 1987), fleshy fruits and endozoochory (Howe and Smallwood 1982; Herrera 1987), or spongy coats and hydrochory (van den Broek *et al.* 2005). The different syndromes often entail specific settlement characteristics. In endozoochorous dispersal, for

Dispersal Ecology and Evolution. First Edition. Edited by Jean Clobert *et al.*

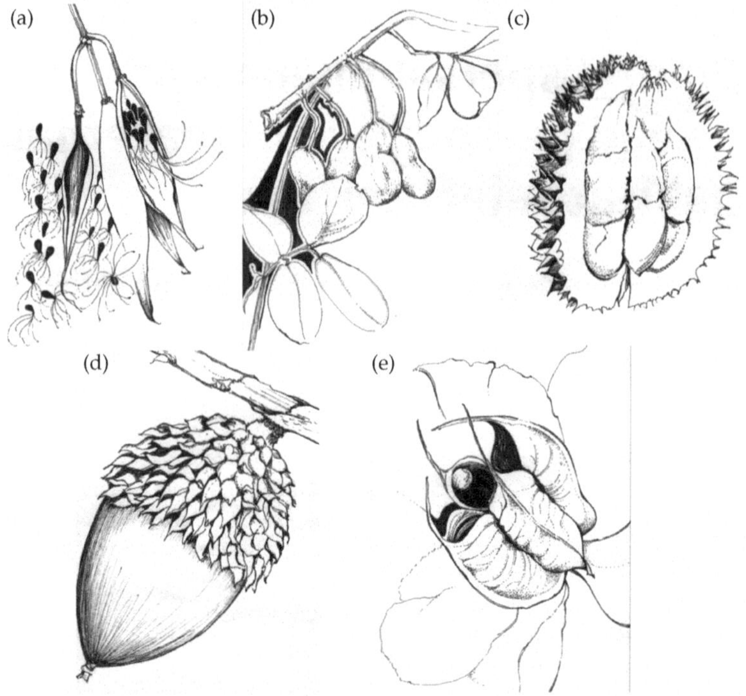

Figure 14.1 Examples of dispersal syndromes in seed plants. (a) *Asclepias syriaca* seeds have a silky coma that facilitates their dispersal by wind (anemochory). (b) *Arachis hypogaea* seeds are buried underground by continued stem growth following pollination (geocarpy) (c) Seeds of the genus *Durio* are encased in a fleshy aril that is attractive to large mammals, such as orang-outangs, that can disperse the seeds over long distances in their gut (endozoochory). (d) *Quercus coccifera* fruits also attract animal vectors, but instead of passing through the gut, they are typically cached (synzoochory). (e) *Helleborus foetidus* seeds have an elaiosome, a protein, and lipid rich appendage, that attracts ants and facilitates dispersal (anemochory). Illustrations by Ann Willis.

example, seeds are deposited in nutrient-rich patches following gut passage. Some structures, such as the bristles of *Avena* and *Erodium* fruits, promote epizoochory—transport on the exterior of animals—and also contribute to the placement of the seeds in the substrate by mechanically pushing them into the soil.

While the adaptive significance of particular dispersal syndromes is often not well established experimentally, some general trends can be described: hydrocory is more prevalent and is likely to be more efficient in communities growing in or near water bodies, anemochory in sparsely vegetated grasslands, epizoochory in open communities, and bird endozoochory in forests and shrub communities (Ozinga *et al.* 2004). However, it is hard to establish *a priori* what mode of dispersal will be favoured in a particular case without information concerning micro-habitats, the local variation in dispersal agents, and causes of natural selection. There are many exceptions to the above patterns, and every kind of syndrome can usually be found in every community.

Secondary dispersal, or diplochory, occurs when more than one vector contributes to the propagule's dispersal. Diplochory is frequent, and may even be the norm (Levin *et al.* 2003; Ozinga *et al.* 2004; Van der Wall and Longland 2004; Chapter 5). For example, in some species, diplochory occurs when ants take propagules that have before been dispersed ballistically (i.e. by explosion of the fruit or seed capsule, Ohkawara and Higashi 1994), or by birds (Christianini and Oliveira 2010).

Historically, the apparent complementarity of propagule morphology and individual vectors led authors to assume a close association between them

(Van der Pijl 1972). However, the exclusivity of associations has been challenged in recent years (see Higgins *et al.* 2003 and Nathan *et al.* 2008, for detailed discussions). Dispersal is a stochastic process and undoubtedly, not all winged propagules will be dispersed by wind (Tackenberg *et al.* 2003), nor all fleshy fruits eaten by vertebrates (Hampe 2004). Nevertheless, the morphological structure of a propagule is often indicative of its most likely vector. For example, Tackenberg *et al.* (2003) showed that the proportion of seeds with high potential for wind dispersal was highest among taxa that had plumed or winged propagules. Although a plumed fruit can attach itself to animal fur or float on the water, a fruit that has no specialized structure will not be efficiently dispersed by wind under normal circumstances.

Arguments against specialized dispersal stem in part from its insufficiency to explain long-distance dispersal (LDD; Higgins *et al.* 2003; Nathan 2006). There is evidence that LDD events are fairly rare and made possible by unusual circumstances, such as extreme atmospheric events or the participation of non-standard vectors (Nathan *et al.* 2008). Despite this inherent unpredictability of LDD, overwhelming evidence suggests that certain attributes of plants and dispersing propagules do influence the distance as well as the probability of dispersal, and as reviewed below, sometimes the environment in which seeds germinate and establish. Additionally, some predictable vectors, such as migratory birds, appear to be fairly efficient LDD agents (Sanchez *et al.* 2006), and at least some plants mature fruits in synchrony with bird activity (Noma and Yumoto 1997; Poulin *et al.* 1999). More research is needed to quantify the relative importance of different vectors for LDD (Nathan *et al.* 2008; Chapter 15).

Because environmental factors appear to have a strong influence on seed dispersal, there has been a tendency to consider plant dispersal to be primarily influenced by stochasticity in dispersal vectors (Wheelwright and Orians 1982; Nathan *et al.* 2008). However, plant dispersal is not a neutral process (Clark 2009), and examples of active dispersal, such as ballistic autochory, can be regarded as maternally controlled (Van der Pijl 1972). Moreover, maternal growth conditions significantly influence dispersal distance and distribution, providing at least the opportunity for maternally mediated plasticity in dispersal (Figure 14.2; Donohue 1998, 1999).

The effect of selection on dispersal syndromes can be quite fast, with propagule morphology evolving over only a few generations (Chapter 24; CaseStudy IV). For instance, *Crepis sancta* varies in the relative production of winged fruits (long-distance dispersal), and non-winged achenes (short-dispersal distance, i.e. barochorous). In fewer than 12 generations, *C. sancta* plants growing in sites with heterogeneous habitat quality, where maternal home sites were predictably of better quality than random sites elsewhere, had proportionally fewer plumed fruits than those growing in locations with more homogeneous habitat quality (Figure 14.3; Cheptou *et al.* 2008; Case Study IV). As another potential example, *Acer rubrum* samaras from early successional populations have a lower wing loading (mass/area) and are thus more likely to disperse farther than samaras from late successional populations (Peroni 1994). In addition, fruit morphology and dispersability were found to differ between island versus mainland populations of three arctic species (Fresnillo and Ehlers 2008). Therefore, while dispersal events may be somewhat stochastic for individuals, it is likely that deterministic adaptive patterns may be discernible at an evolutionary scale, and that dispersal syndromes can significantly affect ecological and evolutionary trajectories (Chapters 15 and 24).

14.3 Dispersal as habitat selection

Dispersal provides the opportunity for an individual to reach an environment where it will develop under conditions that will increase its fitness. In most plants, seed dispersal is the only mechanism to reach new localities, and is therefore the primary means of habitat selection (Bazzaz 1991). The first question is whether propagules should remain at the maternal site or change location (Donohue 2003). Staying at home might be beneficial when the home site is advantageous relative to other sites, or when the cost of dispersal is high (Levin *et al.* 2003; Kisdi 2002), as in the case of locally adapted genotypes, or when maternal plants positively condition the home site relative to the surrounding landscape. For instance, maternal plants can protect seeds and seedlings and increase nutrient and water availability (Flores and Jurado 2003; Gomez-Aparicio *et al.* 2008). Conversely, limited spatial dispersion of propagules can have deleterious consequences if sibling competition is high near the maternal plant (Cheplick and Kane 2004), if resource depletion or pathogen load increases with density (Petermann 2008; Martin and Canham 2010), or if the probability of inbreeding is higher. Thus depending on the quality of the home site, the probability and distance of dispersal can determine the relative quality of the environment experienced after dispersal.

14.3.1 Directed dispersal

Beyond the basic contrast of remaining in the maternal site or dispersing away from it, some propagules have attributes that facilitate their ability to reach particular habitats (Table 14.1). This preferential dispersal to sites where offspring survival is most likely is termed directed dispersal (Howe and Smallwood 1982; Wenny 2001). Directed dispersal has traditionally been described in the context of animal-mediated dispersal. Conventional examples include vertebrate endozoochory and ant removal facilitating settlement in nutrient-rich spots, such as dung piles, latrines, or ant nests (Chapter 5). Strictly speaking, all instances in which dispersal to a certain habitat occurs more frequently than expected at random should be considered directed dispersal (Wenny 2001). For example, in willows (*Populus* sp.), cottony hairs facilitate settlement in favourable riverine sandbanks by adhering to sandy substrates with a specific level of moisture (Seiwa *et al.* 2008), while attachment to waterfowl enables *Eichhornia paniculata* to reach ephemeral ponds that constitute its preferred habitat (Purves and Dushoff 2005). Even syndromes such as hydrochory, which have not traditionally been associated with directed dispersal have recently been shown to be (Merrit and Wohl 2002; Nilsson *et al.* 2002). It can be further argued that mechanisms that strongly limit dispersal are a form of directed dispersal that ensure the occupation by the progeny of a site that has already proved suitable for the maternal plant.

Some potentially adverse effects of directed dispersal have been identified. In particular, if seeds are concentrated in specific patches, then density-dependent dynamics would be enhanced (Spiegel and Nathan 2010), and competition could offset the benefit of otherwise favourable environments. However, the competition intensity depends on the patchiness and abundance of favourable sites.

14.3.2 Phenotypes that lead to enhanced dispersal variance: the case of heterocarpy

Dispersal within the maternal site or at very short distances has traditionally been associated with heterogeneous environments: if suitable patches are rare it pays to stay in one (Venable and Brown 1993; Levin *et al.* 2003). Indeed, most seed dispersal kernels are leptokurtic (Willson 1993), and some plants have highly specialized mechanisms, such as the growth of fruits underground (geocarpy), to ensure that dispersal distance is limited. However, a number of plants have seeds or fruits that differ in their propensity to remain at the home site. This plural dispersal strategy is thought to be an adaptation to temporally variable environments, in which the maternal home site may be of high quality in the maternal generation but of unpredictable quality in progeny (or subsequent) generations. Mathematical models predict that more than one dispersal morph can co-exist within a plant when the likelihood of encountering an appropriate site away from the maternal location balances the costs of dispersal (Hiebeler 2007; Snyder 2011). The production by the same maternal plant of two or more fruit morphs that differ in their size, shape, and associated structures (e.g. 'wings', 'plumes') is termed heterocarpy. Usually one morph has limited dispersal, while the other(s) has longer dispersal distances. A particularly dramatic case of heterocarpy is amphicarpy, in which a single plant produces aerial and underground fruits (Cheplick 1987). The different morphs tend to derive from flowers that have contrasting mating systems and produce seeds that vary in their dormancy and size (Brandel 2007; Ma *et al.* 2010). Often, the locally dispersing morphs derive from self-pollinated flowers. This association of limited or no dispersal with self-fertilization is expected to maximize the retention of locally adapted alleles, while the combination of 'dispersing' phenotype and outcrossing reduces the risk of inbreeding. A similar association is apparent in plants that produce vegetative (local dispersal only) and sexual (longer distance dispersal) propagules (Rottenberg and Parker 2004).

14.4 Seed dispersal is maternally determined

The dispersal attributes of a propagule are primarily maternally determined (Figure 14.1). The maternal plant influences the dispersal morphology of the propagule, the resources contained within the propagule, and the position and time of detachment. Furthermore, these maternal traits can be plastic, varying with environmental conditions experienced by the maternal plant (Figure 14.2).

14.4.1 Maternal determination of dispersal

While seed size may be partially under the control of embryonic genotypes, propagule mass and morphology are largely determined by the maternal genome, as is the architectural platform from which those propagules are released (Roach and Wulf 1987). In gymnosperms, the most obvious maternal influence on dispersal structures is in the morphology of female cones, which vary widely in shape and size. Dispersal structures in gymnosperms may also develop from the integument, the external cell layer of the ovule, as in the winged fruits of *Pinus*, the fleshy *Gingko* 'nut', or the aril of genus *Taxus*.

In angiosperms, dispersal structures typically derive from the ovary wall of the maternal plant, and form during the development of the pericarp. The pericarp can be further differentiated into three layers called exo, meso, and endocarp that can independently contribute to dispersal morphology. The modular nature of fruit development has resulted in a wide diversity of pericarp-derived dispersal structures, ranging from the wing of an *Acer* samara to the sugary pulp of a *Mangifera indica* fruit. Dispersal structures in angiosperms may also develop from other maternal tissue within the ovary such as the aril and seed coat. For instance, in the family Bignoniaceae, seeds are encased in a thin, papery wing that derived from the seed coat. Additional dispersal structures may develop from floral structures other than ovary. In Rosaceae, the pome, the dispersal structure of apples and pears is derived partly from the receptacle tissue, while the dispersal structure of the Dipterocarpaceae are two modified sepals that act as wings. Finally, dispersal structures may derive from the fusion, aggregation, or combined effect of multiple floral and vegetative tissues. The fig is an aggregate fruit derived from the enclosure of the inflorescence into a single dispersal unit. At the extreme, the entire plant may function as the dispersal propagule as in the case of species that form 'tumbleweeds' (e.g. *Kochia scoparia*, made famous by movie westerns).

Maternal plant architecture itself strongly influences seed dispersal: the height, fruit placement,

Table 14.1 Examples of directed dispersal in plants, with the corresponding syndrome, the trait favouring the directionality, the dispersal vector and the targeted site.

Syndrome	Trait favouring dispersal	Vector	Site of settlement	Taxon	Reference
Endozoochory	Fleshy pulp	Vertebrates	Perches/Latrines	*Prunus* sp. *Olea* sp.	Howe and Smallwood 1982; Herrera 1987
Endozoochory	Non-dehiscent cones	Corvids	Underground caches	*Pinus* sp.	Tomback and Linhart 1990
Epizoochory	Small seeds	Waterfowl	Ephemeral ponds	*Eichhornia paniculata*	Purves and Dushoff 2005
Myrmecochory	Elaiosomes	Ants	Ant hills/nests	*Elleborus* sp.	Giladi 2006
Anemochory	Cottony hairs	Wind	Dry sandbanks	*Populus* sp.	Seiwa *et al.* 2008
Hydrochory	Buoyant and angular seeds	Water	Emergent substrates	*Taxodium distichum*	Schneider and Sharitz 1988
Blastochory	Underground fruit	Stem growth	Maternal site	*Arachis* sp.	Cheplick and Quinn, 1982

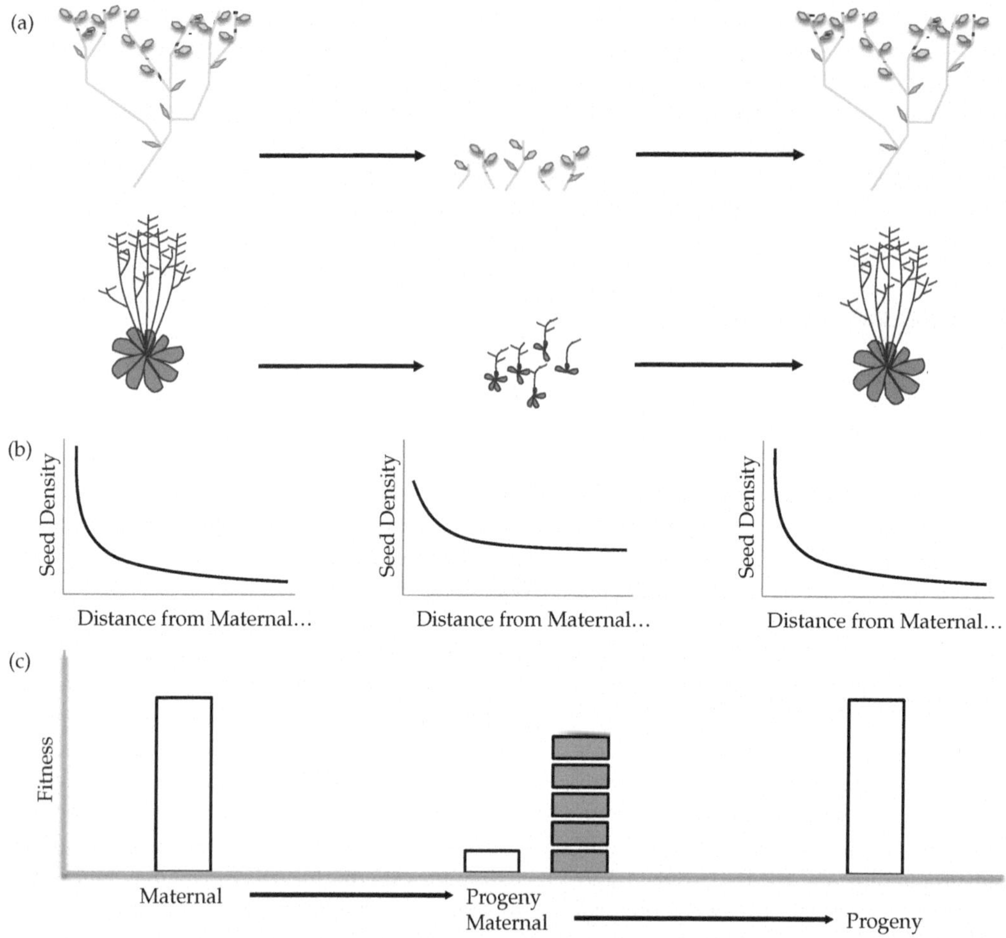

Figure 14.2 Illustration of environmentally mediated maternal effects on dispersal. (a) Three generations of two plant species, *Cakile edentula* (top) and *Arabidopsis thaliana* (bottom). (b) Seed dispersal of each generation. (c) Fitness of each generation, with white bars indicating individual fitness and grey bars indicating total fitness of the population. In the first generation, a low-density environment results in a maternal phenotype (high branch number) that limits long-distance dispersal. In the second generation, a high-density environment results in a maternal phenotype (low branch number) that permits long-distance dispersal. While long-distance dispersal is generally adaptive, the evolution of a purely long-distance dispersal strategy is constrained by adaptive value of high branch number under low-density conditions. Note that under high density, relative fitness of each individual plant is reduced; i.e. the number of fruits per plant is lower. However, the total fitness can be high (grey bars) because the production of an elevated number of fruits by the high-branching, low-density maternal plant leads to a numerous population.

infructescence architecture, branch number, and branch placement have all been shown to influence dispersal distance and post-dispersal seed density (McEvoy and Cox 1987; McCanny and Cavers 1989; Donohue 1998, 1999; Cousens and Rawlinson 2001). Genetic variation in architectural traits such as height and branching is common, and the influence of such variation on dispersal has been documented (Sinha and Davidar 1992; Donohue 1997; Gomez 2007).

In summary, maternal plant architecture and maternal tissue within dispersal propagules largely determine patterns of seed dispersal. As a consequence, traits that influence dispersal influence the fitness not only of the dispersed progeny but also that of the maternal plants.

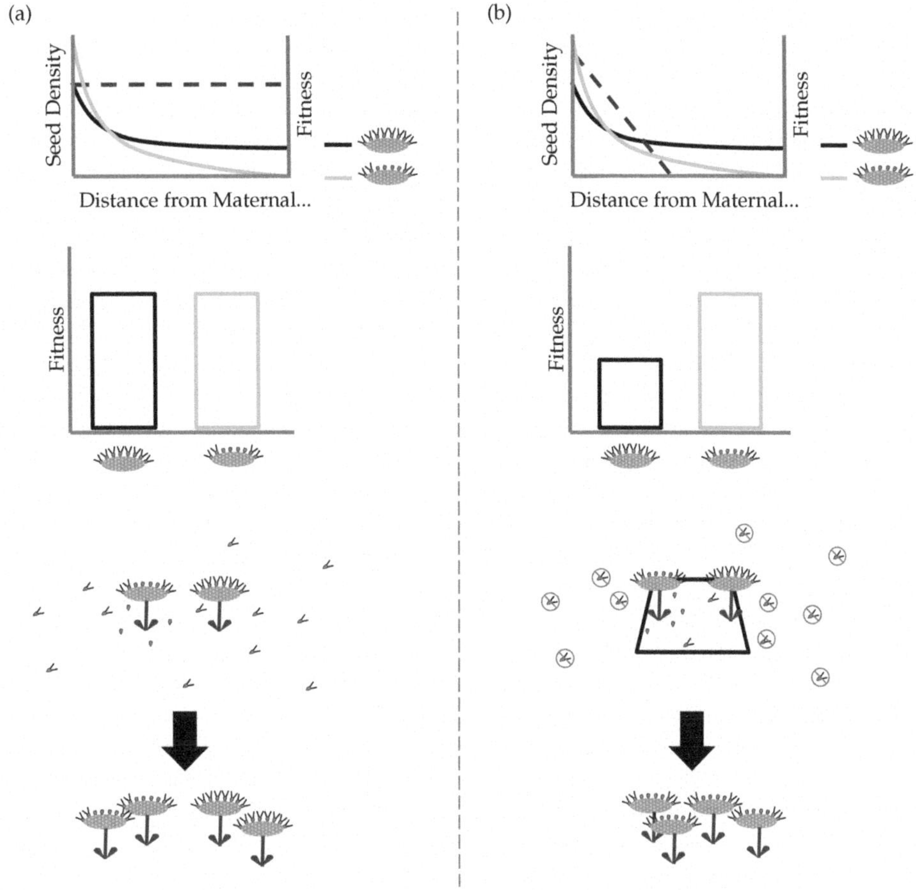

Figure 14.3 Evolution of dispersal phenotypes. *Crepis sancta* is heterocarpic and can produce fruits with either long-distance or short-distance dispersal morphologies. Long-distance dispersal fruits have a pappus that facilitates wind dispersal, while the short-distance fruit do not have a pappus. The ratio of long-distance fruits to short-distance fruits is heritable. Furthermore, different ratios are adaptive under specific conditions (see Cheptou *et al.* 2008). Here we illustrate the fate of two genotypes with different ratios of the two fruit morphs under a homogeneous environment (a), and a heterogeneous environment (b). (a) Maternal fitness (dashed line) does not depend on the dispersal distance from the maternal plant. Thus, there is no adaptive difference between genotypes with high-ratio (black line) and low-ratio (grey line) of dispersing fruits. Both genotypes persist in the population. (b) Maternal fitness declines sharply with dispersal distance from the maternal plant. The suitable environment is localized (out-lined by black square). High-ratio genotypes, consequently, have lower fitness. The low-ratio genotypes eventually rise to fixation.

14.4.2 Plasticity of dispersal phenotypes

The biotic or abiotic environment of the maternal plant can influence dispersal by altering propagule morphology, the ratio of heteromorphic fruit types, or maternal plant architecture. For example, *Raphanus raphanistrum*, a heterocarpic species, exhibits a higher frequency of fruit detachment when exposed to drought stress (Taghizadeh *et al.* 2009). In addition, the maternal environment commonly influences seed size (Roach and Wulf 1987; Agrawal 2001; Hereford and Moriuchi 2005), which may in turn influence dispersibility, particularly in species in which the propagule is a naked seed (Burrows 1975).

In heteromorphic species, the maternal environment can alter the ratio of dispersing versus non-dispersing morphs (Imbert 2002). This variation has been considered to be a form of adaptive plasticity, where a shift toward producing more dispersing morphs is predicted in response to local environmental stress. Alternatively, stress has sometimes been considered to be a cue of environmental unpredictability, which could favour variance in dispersal as a bet-hedging strategy, as discussed earlier. This phenomenon has been observed in numerous Asteraceae species that have both disc achenes with a pappus (dispersal morph), and ray (non-dispersal) achenes (Baker and O'Dowd 1982; Imbert 2002; Pico *et al.* 2003; Brandel 2007). Similarly, Mandak and Pysek (1999) found that increased density and nutrient stress resulted in a higher proportion of dispersing morphs in the heteromorphic species *Atriplex sagittata*. Thus dispersal exhibits plasticity to environmental conditions through changes in the proportions of different propagule types.

Plasticity in architectural traits is common (Sultan 2000) and affects dispersal (Thiede and Augspurger 1996). Height and branching patterns are known to vary in response to light (Weinig 2000), density (Figure 14.2; Donohue *et al.* 2000), herbivory (Weinig *et al.* 2003), CO_2 (Pritchard *et al.* 1999), and resource availability. Some plastic responses in architectural traits that affect dispersal have fitness consequences to progeny. In *Cakile edentula* and *Arabidopsis thaliana* (Figure 14.2), high density resulted in relatively sparser branching that increased dispersal distance and thus facilitated the 'escape' of seeds from intraspecific competition (Donohue 1998; Wender *et al.* 2005).

Maternal effects have been considered to be cross-generational plasticity in which progeny traits (in this case dispersal) respond to environmental conditions experienced by (or created by) the maternal plant during seed development. Such plasticity can actually alter correlations between parent and offspring traits and thereby alter evolutionary trajectories, as discussed next.

14.5 Evolutionary consequences of the maternal determination of dispersal

The maternal determination of seed dispersal has important consequences for its evolution. First, with maternal determination of offspring traits, natural selection on those traits occurs through fitness effects on both parents and offspring. As a consequence, the optimal dispersal strategy is not solely determined by progeny fitness. Second, maternal determination of traits can influence the correlation between parents and offspring and thereby alter evolutionary responses to selection and evolutionary trajectories.

The optimum dispersal strategy is likely to differ for parents and offspring. From the parental perspective, fitness is maximized when the combined fitness of all progeny is maximized, whereas from the progeny perspective, its fitness is optimized when it alone has high fitness, regardless of the fitness of its siblings (Trivers 1974). Thus, even though a single seed may have high fitness by remaining at home and avoiding the risks of dispersal, maternal fitness may be maximized by dispersing some offspring to sub-optimal sites in order to decrease kin competition and hedge its evolutionary bets (Motro 1983; Starrfelt and Kokko 2010; see Figure 14.2).

Maternal determination of progeny characters can impose evolutionary constraints when a trait under selection in the offspring also influences the fitness of the maternal parent (Kirkpatrick and Lande 1989; Wade 1998; Wolf and Wade 2001). In many theoretical and empirical treatments of the evolution of dispersal, the fitness consequences of dispersal to the dispersing propagule are considered to determine fully the adaptive value of dispersal. If dispersal reduces propagule fitness, it is assumed to be maladaptive, whereas if it increases propagule fitness, it is considered adaptive. However, because dispersal is maternally determined, fitness consequences for the maternal parent must also be considered in order to assess accurately the true adaptive value of dispersal. For example, in *Cakile edentula*, plant architecture was a strong determinant of dispersal, such that plants with many branches had impeded dispersal (Donohue 1998). Increased dispersal distance was clearly advantageous to the dispersed progeny, because of decreased sibling competition (Donohue 1997). However, a reduction in branch number, required for efficient dispersal, was associated with a severe reduction of maternal fitness (Donohue 1999). The evolution of this dispersal enhancing trait would therefore be constrained by correlated selection acting through the maternal parent. Thus the fitness consequences to maternal plants of dispersal of their progeny can be a strong constraint on the evolution of dispersal, and they are likely to be a major cause of the ubiquity of inefficient dispersal that appears to be sub optimal from the perspective of dispersed progeny alone (Figure 14.2). However, these conflicts between maternal and progeny traits are not always manifest. For instance, in *Erysimum mediohispanicum*, increased plant height was found to be associated both with higher maternal fitness and with greater dispersal distance (Gomez 2007). Such traits that increase fitness in both maternal parents (by increasing plant size) and progeny (by enhancing progeny dispersal) would favour increased dispersal without constraint.

The maternal determination of progeny characters can also alter evolutionary trajectories because of non-Mendelian contributions to correlations between maternal and offspring characters (Falconer 1965; Kirkpatrick and Lande 1989; Wade 1998; Rasanen and Kruuk 2007). While Mendelian inheritance is expected to cause positive correlations between maternal parents and offspring due to shared nuclear genes, maternal inheritance can alter inter-generational correlations between parents and offspring. For example, a negative maternal effect was found for seed dispersion patterns in *Cakile edentula* (Brassicaceae). Large plants with many branches produced many seeds that had impeded dispersal and consequently experienced high post-dispersal densities (Figure 14.2). Those seeds then produce small plants with few branches which dispersed their seeds to low densities (Donohue 1999). In this case, the maternal effect was negative, such that adults originating from well-dispersed seeds tended to disperse their seeds poorly, while adults originating from poorly dispersed seeds tended to disperse their seeds more efficiently. A similar dynamic was seen in *Arabidopsis thaliana* (Wender *et al.* 2005; Donohue *et al.* 2005, Figure 14.2). Such negative maternal effects can actually cause negative responses to selection in the short term.

In short, the maternal determination of progeny traits, such as dispersal, can influence the overall adaptive value of that trait, as well as the trajectory of evolutionary response to selection on those traits. To predict the evolutionary dynamics of seed dispersal, explicit functional studies are necessary to identify the traits that determine seed dispersion patterns. Second, the fitness consequences of these traits need to be measured—not only with respect to how they influence progeny fitness via their effects on dispersal, but also how they influence maternal fitness. Third, the responses of these traits to environmental factors (Donohue 1999; Peroni 1994; Imbert and Ronce 2001; Mazer and Lowry; 2003) need to be characterized, since this plasticity determines the nature of the correlations between maternal and offspring traits via maternal effects. This is important because the correlations among these traits, both within and across generations, are what determine the evolutionary constraints and the dynamics of maternally determined characters (Kirkpatrick and Lande 1989; Rasanen and Kruuk 2007). Combined, this information enables predictions of the evolutionary dynamics of dispersal by providing estimates of the across-generation correlations for dispersal, and by revealing potential selective constraints on its evolution.

14.6 Consequences of habitat selection via dispersal

As discussed earlier, attributes of maternal plants and dispersing propagules influence the environmental conditions that plants experience after dispersal. This ability of dispersal to influence in a predictable fashion the environment experienced by progeny has some important evolutionary consequences. These consequences arise from the influence of the environment directly on fitness, its influence on natural selection, and its influence on the expression of genetic variation.

First, when dispersal predictably transports progeny to locations in which they experience higher fitness, selection will favour greater dispersal. Thus, while bet-hedging may be favoured in unpredictable environments, directed dispersal to favourable microhabitats (habitat matching), or dispersal as an escape from poor conditions, will predictably cause an increase in progeny fitness.

While dispersal may direct progeny to sites of higher than average quality, that 'quality' may be relative to the genotype that is being dispersed; that is, local adaptation may determine the adaptive value of dispersal. To the extent that progeny end up in locations to which they are locally adapted, dispersal is more likely to be favourable. Such habitat selection is theoretically advantageous because it matches genotypes with environments in which they have higher fitness (Holt 1987; Whitlock 1996).

Likewise, when dispersal predictably determines the environment that is experienced by progeny, it can influence natural selection on other progeny phenotypes. In particular, it can increase exposure to some environments and reduce exposure to others. As such, it can accelerate adaptive evolution, allowing more rapid specialization (Holt 1987; Whitlock 1996). Under some circumstances, when variation in both dispersal and loci under selection are available, habitat selection through dispersal can maintain genetic variation (e.g. Levins 1968; Holt 1987; reviewed in Donohue 2003). However, very little information exists in plants on whether habitat selection and environment dependent performance are variable and whether they are positively correlated. This lack of information is likely because the various methods by which plants determine the environment they experience are not typically recognized.

When dispersal influences the environment experienced by progeny, it also influences the expression of plastic progeny phenotypes. Such plasticity can alter not only selection on phenotypes, but it can also alter correlations between parent and offspring traits and thereby alter evolutionary trajectories, as discussed above.

Moreover, when different genotypes exhibit different levels of plasticity to post-dispersal conditions, the expression of genetic variation can be environment-dependent. For example in many plants, increased dispersal results in lower post-dispersal density (Figure 14.2; Wender *et al.* 2005; Donohue *et al.* 2005; Janzen 1978; Baker and O'Dowd 1982; Augspurger, 1983; Rees and Brown 1991; Augspurger and Kitajima 1992). In *Arabidopsis thaliana* (Wender *et al.* 2005; Donohue *et al.* 2005), genetic variation for traits that influence dispersal (e.g. height) and genetic variation for dispersal itself (measured as post-dispersal density) was density-dependent, such that significant genetic variation for dispersal was detected only at high density. While seldom documented, such environment-dependent genetic expression for dispersal-related phenotypes is likely to be a common occurrence; dispersal frequently alters the density that plants experience, and density has major effects on plant size and architecture as well as the expression of variation for architectural traits that are likely to influence dispersal.

This act implies that genetic variation for dispersal can depend on dispersal itself. In the above example of *A. thaliana*, if natural selection favours dispersal to lower post-dispersal density, a population that is presently at high density can evolve more efficient dispersal to lower density because genetic variation for dispersal is expressed at high density. However, once low post-dispersal density is achieved, genetic variation for dispersal will cease to be expressed. Thus further evolution of dispersal would be constrained. Note that this lack of genetic variation is not caused by the depletion of variation through selection, but rather it occurs because less genetic variation is expressed, even when the genotypes are identical.

In this particular example, the dynamics operated so as to reduce the expression of genetic variation and thereby constrain its continued evolution. However, if genetic variation for dispersal were higher at low density than at high density, genetic variation may actually increase with further evolution. This phenomenon could facilitate sustained evolutionary responses to selection by enabling more genetic variance to be expressed (Donohue 2009). Therefore the ability of dispersal to determine the environment that organisms experience can cause unexpected evolutionary dynamics that can either constrain or facilitate the evolution of such characters.

14.7 Conclusions

Dispersal phenotypes are determined primarily by the maternal plant. This fact has important evolutionary and ecological consequences. First, the true adaptive value of dispersal depends on the fitness consequences of these phenotypes not only to progeny but also to the maternal plants. Because the dispersal structures and detachment of the propagule are conditioned predominantly by maternal architecture and maternal tissue, some maternal traits are indissolubly linked to dispersal and will be affected by selection on dispersal. Moreover, plasticity of these traits can alter correlations between parent and offspring dispersal traits, and thereby alter evolutionary responses to selection and other evolutionary dynamics. Thus identifying particular plant traits that determine dispersal is necessary for making evolutionary predictions about dispersal.

Characterizing the plasticity of those traits provides information on plasticity in dispersal as well as on the correlations between parents and offspring.

Just as the environment can alter dispersal, so can dispersal determine the environmental conditions experienced by progeny. The ability of dispersal to determine post-dispersal environments also has profound evolutionary consequences, including altering the expression of phenotypes, the selection on phenotypes, and the expression of genetic variation for those phenotypes. Thus characterizing the ability of particular traits to produce habitat selection through dispersal is relevant for predicting evolutionary dynamics of dispersal as well as other plant traits.

Moreover, other traits that have a strong influence on habitat selection are likely correlated with dispersal. For instance, the adaptive value of seed dormancy (the state in which a viable seed does not germinate when exposed to conditions favourable for germination) is expected to be influenced by dispersal, since dormancy is in effect dispersal through time and thereby subject to similar types of natural selection as dispersal through space. The evolution of dispersal phenotypes is likely contingent on dormancy phenotypes and *vice versa*.

So far, dispersal phenotypes have been catalogued extensively and numerous theoretical models of dispersal have been developed. Conversely, empirical tests of the hypotheses concerning causes and consequences of variation in dispersal and dispersal-related phenotypes are scarce. Recent research indicates that long-held assumptions on plant dispersal, such as the correspondence between propagule morphology and vector, or the rarity of directed dispersal, might not be true. Also, the maternal determination of dispersal and the capacity of dispersal to determine the environment experienced by the progeny have evolutionary ramifications that remain to be tested empirically. Questions such as the preponderance of directed dispersal, the relative frequency of specialized dispersal, the influence of plasticity of dispersal traits on the evolution of dispersal, and the correlated evolution of dispersal and other plant traits remain to be elucidated. Further experimental research is needed to untangle how plant phenotypes determine what habitat plants are exposed to and what the evolutionary consequences of this habitat selection might be.

References

Agrawal, A. (2001). 'Maternally induced plant resistance: Genetic variation, signals, and consequences', *Ecological Society of America Annual Meeting Abstracts* **86**: 41.

Augspurger, C.K. (1983). 'Offspring recruitment around tropical trees—Changes in cohort distance with time', *Oikos* **40**: 189–96.

Augsperger, C.K. (1986). 'Morphology and dispersal potential of wind-dispersed diaspores of Neotropical trees', *American Journal of Botany* **73**: 353–63.

Augsperger, C.K. and Franson, S.E. (1987). 'Wind dispersal of artificial fruits varying in mass, area and morphology', *Ecology* **68**: 27–42.

Augspurger, C.K. and Kitajima, K. (1992). 'Experimental studies of seedling recruitment from contrasting seed distributions', *Ecology* **73**: 1270–84.

Baker, G.A. and O'Dowd, D.J. (1982). 'Effects of parent plant-density on the production of achene types in the annual *Hypochaeris glabra*', *Journal of Ecology* **70**: 201–15.

Bazzaz, F. A. (1991). 'Habitat selection in plants', *American Naturalist* **137**: S116–30.

Brandel, M. (2007). 'Ecology of achene dimorphism in *Leontodon saxatilis*', *Annals of Botany* **100**: 1189–97.

Burrows, F.M. (1975). 'Wind-born seed and fruit movement', *New Phytologist* **75**: 405–18.

Cheplick, G.P. (1987). 'The ecology of amphicarpic plants', *Trends in Ecology and Evolution* **2**: 97–101.

Cheplick, G.P., Kane, K.H. (2004). 'Genetic relatedness and competition in *Triplasis purpurea* (Poaceae): Resource partitioning or kin selection?', *International Journal of Plant Sciences* **165**: 623–30.

Cheplick, G.P., Quinn, J.A. (1982). '*Amphicarpum purshii* and the pessimistic strategy in amphicarpic annuals', *Oecologia* **52**: 327–32.

Cheptou, P.O., Carrue, O., Rouifed, S. *et al.* (2008). 'Rapid evolution of seed dispersal in an urban environment in the weed *Crepis sancta*', '*Proceedings of the National Academy of Sciences of the USA* **105**: 3796–9.

Christianini, A.V. and Oliveira, P.S. (2010). 'Birds and ants provide complementary seed dispersal in a neotropical savanna', *Journal of Ecology* **98**: 573–82.

Clark, J.S. (2009). 'Beyond neutral science', *Trends in Ecology and Evolution* **28**: 8–15.

Cousens, R., Dytham, C., Law, R. (2008). *Dispersal in plants: a population perspective*. Oxford University Press, New York.

Cousens, R.D., Rawlinson, A.A. (2001). 'When will plant morphology affect the shape of a seed dispersal 'kernel'?', *Journal of Theoretical Biology* **211**: 229–38.

Cruz-Mazo, G., Buide, M.L., Samuel, R. *et al.* (2009). 'Molecular phylogeny of *Scorzoneroides* (Asteraceae): Evolution of heterocarpy and annual habit in unpredictable environments'. *Molecular Phylogenetics and Evolution* **53**: 835–47.

De-Lucas, A.I., Robledo-Arnuncio, J.J., Hi-Dalgo, E. *et al.* (2008). 'Mating system and pollen gene flow in Mediterranean maritime pine', *Heredity* **100**: 390–9.

Donohue, K, (1997). 'Seed dispersal in *Cakile edentula* var. *lacustris*: Decoupling the fitness effects of density and distance from the home site', *Oecologia* **110**: 520–7.

Donohue, K. (1998). 'Maternal determinants of seed dispersal in *Cakile edentula*: Fruit, plant, and site traits', *Ecology* **79**: 2771–88.

Donohue, K. (1999). 'Seed dispersal as a maternally influenced character: Mechanistic basis of maternal effects and selection on maternal characters in an annual plant', *American Naturalist* **154**: 674–89.

Donohue, K. (2003). 'Setting the stage: phenotypic plasticity as habitat selection', *International Journal of Plant Sciences* **164**: S79–92.

Donohue, K. (2009). 'Some Evolutionary Consequences of Niche Construction with Genotype-Environment Interaction', in J.Vanderwerf, H.U. Graser, R. Frankham *et al.* eds, *Adaptation and Fitness in Animal Populations—Evolutionary and Breeding Perspectives on Genetic Resource Management.* Springer, Berlin, Germany. pp. 131–150.

Donohue, K., Messiqua, D., Pyle, E. H. *et al.* (2000). Evidence of adaptive divergence in plasticity: Density- and site-dependent selection on shade-avoidance responses in *Impatiens capensis*. *Evolution*, **54**: 1956–68.

Donohue, K., Polisetty, C.R., Wender, N.J. (2005). 'Genetic consequences of niche construction: Plasticity-mediated genetic constraints on the evolution of seed dispersal in *Arabidopsis thaliana*', *American Naturalist* **165**: 537–50.

Falconer, D.S. (1965). 'Maternal effects and selection response', in S.J. Geerts (ed), *Genetics Today, Proceedings of the VI International Congress on Genetics.* Pergamon, Oxford, UK.

Flores, J. and Jurado, E. (2003). 'Are nurse-protege interactions more common among plants from arid environments?', *Journal of Vegetation Science* **14**: 911–16.

Fresnillo, B. and Ehlers, B.K. (2008). 'Variation in dispersability among mainland and island populations of three wind dispersed plant species', *Plant Systematics and Evolution* **270**: 243–55.

Galloway, L.F., Etterson, J.R., McGlothin, J.W. (2009). 'Contribution of direct and maternal genetic effects to life-history evolution', *New Phytologist* **183**: 826–38.

Giladi, I. (2006). 'Choosing benefits or partners: a review of the evidence for the evolution of myrmecochory', *Oikos* **112**: 481–92.

Gomez, J.M. (2007). 'Dispersal-mediated selection on plant height in an autochorously dispersed herb', *Plant Systematics and Evolution* **268**: 119–30.

Gomez-Aparicio, L., Zamora, A., Castro, J. *et al.* (2008). 'Facilitation of tree saplings by nurse plants: Microhabitat amelioration or protection against herbivores?'. *Journal of Vegetation Science* **19**: 161–72.

Grivet.D., Robledo-Arnuncio, J.J., Smouse, P.E. *et al.* (2009). 'Relative contribution of contemporary pollen and seed dispersal to the effective parental size of seedling population of California valley oak (*Quercus lobata*, Nee)', *Molecular Ecology* **18**: 3967–79.

Hall, J.C., Tisdale, T.E., Donohue, K. *et al.* (2006). 'Developmental basis of an anatomical novelty: Heteroarthrocarpy in *Cakile lanceolata* and *Erucaria erucarioides* (Brassicaceae).', *International Journal of Plant Sciences* **167**: 771–89.

Hampe, A. (2004). 'Extensive hydrochory uncouples spatiotemporal patterns of seedfall and seedling recruitment in a 'bird-dispersed' riparian tree', *Journal of Ecology* **92**: 797–807.

Hereford, J. and Moriuchi, K.S. (2005). 'Variation among populations of *Diodia teres* (Rubiaceae) in environmental maternal effects', *Journal of Evolutionary Biology* **18**: 124–31.

Herrera, C.M. (1987). 'Vertebrate-dispersed plants of the Iberian Peninsula- A study of fruit characteristics', *Ecological Monographs* **57**: 305–31.

Hiebeler, D.E. (2007). 'Competing populations on fragmented landscapes with spatially structured heterogeneities: improved landscape generation and mixed dispersal strategies', *Journal of Mathematical Biology* **54**: 337–56.

Higgins, S.I., Nathan, R., Cain, M.L. (2003). 'Are long-distance dispersal events in plants usually caused by nonstandard means of dispersal?', *Ecology* **84**: 1945–56.

Holt, R.D. (1987). 'Population dynamics and evolutionary processes: the manifold roles of habitat selection', *Evolutionary Ecology* **1**: 331–47.

Howe, H.F. and Smallwood, J. (1982). 'Ecology of seed dispersal', *Annual Review of Ecology and Systematics* **13**: 201–28.

Imbert, E. (2002). 'Ecological consequences and ontogeny of seed heteromorphism', *Perspectives in Plant Ecology Evolution and Systematics*, **5**: 13–36.

Imbert, E. and Ronce, O. (2001). 'Phenotypic plasticity for dispersal ability in the seed heteromorphic *Crepis sancta* (Asteraceae).', *Oikos* **93**: 126–34.

Janzen, D.H. (1978). 'The size of a local peak in a seed shadow', *Biotropica* **10**: 78.

Kirkpatrick, M. and Lande, R. (1989). 'The evolution of maternal characters', *Evolution* **43**: 485–503.

Kisdi, E. (2002). 'Dispersal: Risk spreading versus local adaptation', *American Naturalist* **159**: 579–96.

Levin, S.A., Muller-Landau, H.C., Nathan, R. *et al.* (2003). 'The ecology and evolution of seed dispersal: A theoretical perspective', *Annual Review of Ecology Evolution and Systematics* **34**: 575–604.

Levins, R. (1968). *Evolution in changing environments.* Princeton University Press, Princeton, New Jersey.

Ma, W.B., Zhao, X.J., Tan, D.Y. *et al.* (2010). 'Nutlet dimorphism in individual flowers of two cold desert annual *Lappula* species (Boraginaceae): implications for escape by offspring in time and space', *Plant Ecology* **209**: 361–74.

Mandak, B. and Pysek, P. (1999). 'Effects of plant density and nutrient levels on fruit polymorphism in *Atriplex sagittata*', *Oecologia* **119**: 63–72.

Martin, P.H., Canham, C.D. (2010). 'Dispersal and recruitment limitation in native versus exotic tree species: life-history strategies and Janzen-Connell effects', *Oikos* **119**: 807–24.

Mazer, S.J. and Lowry, D.E. (2003). 'Environmental, genetic, and seed mass effects on winged seed production in the heteromorphic *Spergularia marina* (Caryophyllaceae)', *Functional Ecology* **17**: 637–50.

McCanny, S.J. and Cavers, P.B. (1989). 'Parental effects on spatial patterns of plants—A contingency table approach', *Ecology* **70**: 368–78.

McEvoy, P.B. and Cox, C.S. (1987). 'Wind dispersal distances in dimorphic achenes of ragwort, *Senecio jacobaea*', *Ecology* **68**: 2006–15.

Merrit, D.M. and Wohl, E.E. (2002). 'Processes governing hydrochory along rivers: Hydraulics, hydrology, and dispersal phenology', *Ecological Applications* **12**: 1071–87.

Motro, U. (1983). 'Optimal rates of dispersal 3: Parent—offspring conflict', *Theoretical Population Biology* **23**: 159–68.

Nathan, R. (2006). 'Long-distance dispersal of plants', *Science* **313**: 786–8.

Nathan, R., Schurr, F.M., Siegel, O. *et al.* (2008). 'Mechanisms of long-distance seed dispersal', *Trends in Ecology and Evolution* **23**: 638–47.

Nilsson, C., Andersson, E., Merritt, D.M. *et al.* (2002). 'Differences in riparian flora between riverbanks and river lakeshores explained by dispersal traits.', *Ecology* **83**: 2878–87.

Noma, N. and Yumoto, T. (1997). 'Fruiting phenology of animal-dispersed plants in response to winter migration of frugivores in a warm temperate forest on Yakushima Island, Japan', *Ecological Research* **12**: 119–29.

Ohkawara, K. and Higashi, S. (1994). 'Relative importance of ballistic and ant dispersal in two diplochorous *Viola* species (Violaceae)', *Oecologia* **100**: 135–40.

Ozinga, W.A., Bekker, R.M., Schaminee, J. *et al.* (2004). 'Dispersal potential in plant communities depends on environmental conditions', *Journal of Ecology* **92**: 767–77.

Peroni, P.A. (1994). 'Seed size and dispersal potential of *Acer rubrum* (Aceraceae) samaras produced by populations in early and late successional stages', *American Journal of Botany* **81**: 1428–34.

Petermann, J.S., Fergus, A.J.F., Turnbull, L.A. *et al.* (2008). 'Janzen-Connell effects are widespread and strong enough to maintain diversity in grasslands', *Ecology* **89**: 2399–406.

Pico, F.X., Ouborg, N.J., Van Groenen-Dael, J.M. (2003). 'Fitness traits and dispersal ability in the herb *Tragopogon pratensis* (Asteraceae): Decoupling the role of inbreeding depression and maternal effects', *Plant Biology* **5**: 522–30.

Poulin, B., Wright, S.J., Lefebvre, G. *et al.* (1999). 'Interspecific synchrony and asynchrony in the fruiting phenologies of congeneric bird-dispersed plants in Panama', *Journal of Tropical Ecology* **15**: 213–27.

Pritchard, S.G., Rogers, H.H., Prior, S.A. *et al.* (1999). 'Elevated CO2 and plant structure: a review', *Global Change Biology* **5**: 807–37.

Purves, D.W. and Dushoff, J. (2005). 'Directed seed dispersal and metapopulation response to habitat loss and disturbance: application to *Eichhornia paniculata*', *Journal of Ecology* **93**: 658–69.

Rasanen, K. and Kruuk, L.E.B. (2007). 'Maternal effects and evolution at ecological time-scales', *Functional Ecology* **21**: 408–21.

Rees, M. and Brown, V.K. (1991). 'The effect of established plants on recruitment in the annual forb *Sinapis arvensis*', *Oecologia* **87**: 58–62.

Ridley, H.N. (1930). The dispersal of plants around the world, Ashford, Reeve.

Roach, D.A. and Wulf, R.D. (1987). 'Maternal effects in plants', *Annual Review of Ecology and Systematics* **18**: 209–35.

Rottenber, A. and Parker, J.S. (2004). 'Asexual populations of the invasive weed Oxalis pes-caprae are genetically variable', *Proceedings of the Royal Society B* **271**: S206–8.

Sanchez, M.I., Green, A.J., Castellanos, E.M. (2006). 'Internal transport of seeds by migratory waders in the Odiel marshes, south-west Spain: consequences for long-distance dispersal', *Journal of Avian Biology* **37**: 201–6.

Schneider, R.L. and Sharitz, R.R. (1988). 'Hydrochory and regeneration in a bald cypress water Tupelo swamp', *Ecology* **69**: 1055–63.

Seiwa, K., Tozawa, M., Ueno, N. *et al.* (2008). 'Roles of cottony hairs in directed seed dispersal in riparian willows', *Plant Ecology* **198**: 27–35.

Sinha, A. and Davidar, P. (1992). 'Seed dispersal ecology of a wind dispersed rain-forest tree in the Western Ghats, India', *Biotropica* **24**: 519–26.

Snyder, R.E. (2006). 'Multiple risk reduction mechanisms: can dormancy substitute for dispersal?', *Ecology Letters* **9**: 1106–14.

Snyder, R.E. (2011). 'Leaving home ain't easy: non-local seed dispersal is only evolutionarily stable in highly unpredictable environments', *Proceedings of the Royal Society B* **278**: 739–44.

Spiegel, O., Nathan, R.Spiegel, O. and Nathan, R. (2010). 'Incorporating density dependence into the directed-dispersal hypothesis', *Ecology* **91**: 1538–48.

Starrfelt, J., Kokko, H. (2010). 'Parent-Offspring Conflict and the Evolution of Dispersal Distance', *American Naturalist* **175**: 38–49.

Sultan, S.E. (2000). 'Phenotypic plasticity for plant development, function and life history', *Trends in Plant Science* **5**: 537–42.

Tackenberg, O., Poschlod, P., Bonn, S. (2003). 'Assessment of wind dispersal potential in plant species', *Ecological Monographs* **73**: 191–205.

Taghizadeh, M.S., Crawford, S., Nicolas, M.E. *et al.* (2009). 'Water deficit changes the anatomy of the fruit abscission zone in *Raphanus raphanistrum* (Brassicaceae)', *Australian Journal of Botany* **57**: 708–14.

Theide, D.A., Augspurger, C.K. (1996). 'Intraspecific variation in seed dispersion of *Lepidium campestre* (Brassicaceae)', *American Journal of Botany* **83**: 856–66.

Tomback, D.F., Linhart, Y.B. (1990). 'The evolution of bird—dispersed pines', *Evolutionary Ecology* **4**: 185–219.

Trivers, R.L. (1974). 'Parent—offspring conflict', *American Zoologist* **14**: 249–64.

Van Den Broek, T., Van Diggelen, R., Bobbink, R. (2005). 'Variation in seed buoyancy of species in wetland ecosystems with different flooding dynamics', *Journal of Vegetation Science* **16**: 579–86.

Van Der Pijl, L. (1972). *Principles of dispersal in higher plants.* Springer-Verlag, New York.

Van der Wall, S.B., Longland, W.S. (2004). 'Diplochory: are two seed dispersers better than one?', *Trends in Ecology and Evolution* **19**: 155–61.

Venable, D.L. (1985). 'The evolutionary ecology of seed heteromorphism', *American Naturalist* **126**: 577–95.

Venable, D.L., Brown, J.S. (1993). 'The population-dynamic functions of seed dispersal', *Vegetatio* **108**: 31–55.

Wade, M.J. (1998). *The evolutionary genetics of maternal effects.* Oxford University Press New York.

Weinig, C. (2000). 'Plasticity versus canalization: population differences in the timing of shade-avoidance responses', *Evolution* **54**: 441–51.

Weinig, C., Stinchcombe, J.R., Schmitt, J. (2003). 'Evolutionary genetics of resistance and tolerance to natural herbivory in *Arabidopsis thaliana*', *Evolution* **57**: 1270–80.

Wender, N.J., Polisetty, C.R., Donohue, K. (2005). 'Density-dependent processes influencing the evolutionary dynamics of dispersal: A functional analysis of seed dispersal in *Arabidopsis thaliana* (Brassicaceae)', *American Journal of Botany* **92**: 960–71.

Wenny, D.G. (2001). 'Advantages of seed dispersal: A re-evaluation of directed dispersal', *Evolutionary Ecology Research* **3**: 51–74.

Weppler, T., Stoll, P., Stocklin, J. (2006). 'The relative importance of sexual and clonal reproduction for population growth in the long-lived alpine plant *Geum reptans*', *Journal of Ecology* **94**: 869–79.

Wheelwright, N.T. Orians, G.H. (1982). 'Seed dispersal by animals—Contrasts with pollen dispersal, problems of terminology and constraints on coevolution', *American Naturalist* **119**: 402–13.

Whitlock, M.C. (1996). 'The red queen beats the Jack-of-all-trades: the limitations on the evolution of phenotypic plasticity and niche breadth', *American Naturalist* **148**: S65–77.

Willson, M.F. (1993). 'Dispersal mode, seed shadows and colonization patterns', *Vegetatio* **108**: 261–80.

Wolf, J. B. and Wade, M.J. (2001). 'On the assignment of fitness to parents and offspring: whose fitness is it and when does it matter?', *Journal of Evolutionary Biology* 14: 347–56.

PART IV

Distribution of Dispersal Distances: Dispersal Kernels

CHAPTER 15

Dispersal kernels: review

Ran Nathan, Etienne Klein, Juan J. Robledo-Arnuncio, and Eloy Revilla

15.1 Basic concepts and definitions

Dispersal distance, the Euclidian distance between 'start' and 'end' points of a dispersal event, is recognized as a fundamental characteristic of the dispersal process, defined here as the movement of dispersers—called diaspores, propagules, gametes, or dispersing individuals in the literature—from their natal site or their breeding home range to a new place of potential establishment. Distance constitutes the most basic spatial descriptor of dispersal. For sexually reproducing plants, pollen and seed dispersal distances outline, respectively, the net displacement of the entire lifetime track of the male gamete or the individual; for migratory animals, in contrast, the dispersal distance often constitutes a tiny proportion of the lifetime track (Nathan *et al.* 2008). Although the dispersal distance cannot capture all aspects of how dispersal shapes post-dispersal survival and subsequent population and community dynamics, it has been used extensively in this context because it represents some key features of these relationships (Janzen 1970; Connell 1971; Kot *et al.* 1996; Nathan and Muller-Landau 2000). In this chapter, we discuss the basic concepts, questions, tools, and future directions in using dispersal kernels to study the behavior and ecology of animals and plants, and refer the readers to Chapter 16 for a discussion on the evolution of dispersal and dispersal kernels (Ronce 2007).

The statistical distribution of dispersal distances in a population is termed the 'dispersal kernel'. It is a probability density function (pdf) describing the distribution of the post-dispersal locations relatively to the source point. The term 'dispersal kernel' is equivalent to various terms previously used in the literature such as 'contact distribution' (Mollison 1977) and 'dispersal distribution' (Tufto *et al.* 1997). These terms encompass the probabilistic special case of the 'dispersal curve', a general popular term used for any graphical representation of the distribution of dispersal distances or the density of dispersing individuals at different distances from the source. The term 'dispersal kernel' emerged from mathematical studies of integro-differential equations for population spread (Thieme 1977; Diekmann 1978; Van den Bosch *et al.* 1990; see Box 15.1). Indeed, 'kernel' was used in integration theory since Hilbert (1904) and etymologically means the 'core', 'nucleus', or 'inner part'; hence, a dispersal kernel represents the contribution of a specific core (source point) to the re-organization of certain units (e.g. individuals) in a larger entity (population). Three influential studies analysing the relationship between the shape of the dispersal kernel and population spread (Mollison 1991; Kot *et al.* 1996; Clark *et al.* 1999) popularized the term in ecology. It was then integrated successively in various fields of dispersal ecology, mostly in studies of plant seeds and pollen (e.g. Nathan and Muller-Landau 2000; Klein *et al.* 2003; Robledo-Arnuncio *et al.* 2004) and of larvae of marine organisms (e.g. Largier 2003; Siegel *et al.* 2003; Kinlan *et al.* 2005), but also in studies of various other taxa including pathogens (Skelsey *et al.* 2005), water fleas (Havel *et al.* 2002), insects (Baguette 2003), birds (van Houtan *et al.* 2007), and mammals (Revilla and Wiegand 2008). The rich ensemble of species of various taxonomic groups for which dispersal kernels were estimated is illustrated in Table 15.1.

Cousens *et al.* (2008) argued that 'dispersal kernel' has been used inconsistently and confusingly to denote either the distribution of the distances travelled or the distribution of the post-dispersal locations relative to the source location. Various terms were suggested to distinguish these two

Dispersal Ecology and Evolution. First Edition. Edited by Jean Clobert *et al.*

Table 15.1 Various functions used for the dispersal kernel, mostly in phenomenological approaches. Expressions for the dispersal location kernel $k_L(r)$ have been re-parameterized to include a scale parameter (*a*) homogeneous to a distance and a shape parameter (*b*) determining the shape of the curve, and particularly the relative weight of long-distance events.

Kernel name	$k_L(r)$ expression*	Parameter values	Mean dispersal distance	Tail fatness	Taxonomic groups applied	Model comparison	Recommended use
Gaussian	$\frac{1}{\pi a^2}\exp\left(-\frac{r^2}{a^2}\right)$	$a > 0$	$\frac{a\sqrt{\pi}}{2}$	Thin-tailed	Pollen (Austerlitz *et al.* 2004; Goto *et al.* 2006), Seeds (Sagnard *et al.* 2007), Seedlings (Goto *et al.* 2006), Beetles (Chapman *et al.* 2007), Moths (Bianchi *et al.* 2009), Fish (Coombs and Rodriguez 2007), Birds (Van Houten *et al.* 2007—named Rayleigh therein), Mammals (Krkosek *et al.* 2007)	Poorer fits than more leptokurtic kernels in general, except occasionally for heavy seeds (Sagnard *et al.* 2007)	In general, should be used only as a reference against more leptokurtic kernels. Adequate to represent the result of dispersal through diffusion or completely random walk during a constant time.
(Negative) Exponential	$\frac{1}{2\pi a^2}\exp\left(-\frac{r}{a}\right)$	$a > 0$	$2a$	Exponential	Pollen (Austerlitz *et al.* 2004; Goto *et al.* 2006), Seeds (Willson 1993; Clark *et al.* 1999; Nathan *et al.* 2000; Sagnard *et al.* 2007; Jones and Muller-Landau 2008), Seedlings (Goto *et al.* 2006), Butterflies (Baguette 2003; Fric and Konvicka 2007), Mosquitos (Estep *et al.* 2010), Beetles (Chapman *et al.* 2007; Carrasco *et al.* 2010)	Generally outperforms the Gaussian but is outperformed by more leptokurtic kernels, except occasionally for seed dispersal and butterflies (Fric and Konvicka 2007)	In general, should be used only as a reference against more fat-tailed kernels. Represents well a travel at constant speed in a random direction with a constant stopping rate. Also obtained from a correlated random walk with settlement (Hovestadt *et al.* 2011)
Exponential Power (Gaussian for $b = 2$; Exponential for $b = 1$)	$\frac{b}{2\pi a^2\Gamma(2/b)}\exp\left(-\frac{r^b}{a^b}\right)$	$a, b > 0$	$a\frac{\Gamma(3/b)}{\Gamma(2/b)}$	$b > 1$: thin-tailed, $b < 1$: fat-tailed. Always thinner than power laws	Pollen (Hardy *et al.* 2004; Oddou-Muratorio *et al.* 2005, Robledo-Arnuncio and Gil 2005, Goto *et al.* 2006), Seeds (Clark *et al.* 1998, González-Martínez *et al.* 2006, Schurr *et al.* 2008), Seedlings (Ribbens *et al.* 1994 with b=3, Goto *et al.* 2006), Beetles (Chapman *et al.* 2007), Moths (Bianchi *et al.* 2009), Mammals (Krkosek *et al.* 2007)	Outperforms the Gaussian, Exponential, 2Dt and Geometric functions for pollen (Austerlitz *et al.* 2004) Values for b generally found from 0.3 to 0.6	Well suited for pollen dispersal. Encompasses the Gaussian and Exponential as special cases and therefore adequate for kernel's shape comparisons.

2Dt	$\frac{(b-1)}{\pi a^2}\left(1+\frac{r^2}{a^2}\right)^{-b}$	$a > 0; b > 1$	$a\frac{\sqrt{\pi}}{2}\frac{\Gamma\left(b-\frac{3}{2}\right)}{\Gamma(b-1)}$ ∞ for $b < 3/2$	Fat-tailed Power-law tail	Pollen (Austerlitz *et al.* 2004; Goto *et al.* 2006), Seeds (Clark *et al.* 1999; Greene *et al.* 2004; Canham and Uriarte 2006; Schurr *et al.* 2008; Venable *et al.* 2008), Seedlings (Greene *et al.* 2004; Goto *et al.* 2006) Beetles (Carrasco *et al.* 2010, with $b = 1$)	Outperforms many other kernels in seed dispersal studies, except occasionally the lognormal and WALD (Clark *et al.* 1999; Greene *et al.* 2004; Schurr *et al.* 2008; Jones and Muller-Landau 2008). Compares more poorly to other kernels for pollen (Austerlitz *et al.* 2004).	Well suited for seed dispersal studies. Obtained as a continuous mixture of Gaussian kernels with variance parameters distributed as the inverse of a Gamma distribution. The particular case $b = 3/2$ was obtained as the deposition location on the ground of a 3D drift Brownian motion from an elevated source when gravity is negligible (Stockmarr 2002)
(Inverse) Power-law	$\frac{(b-2)(b-1)}{2\pi a^2}\left(1+\frac{r}{a}\right)^{-b}$	$a > 0; b > 2$	$\frac{2a}{b-3}$ ∞ for $b < 3$	Fat-tailed Power-law tail	Pollen (Austerlitz *et al.* 2004; Devaux *et al.* 2007)	Outperformed the exponential-power family when fitting long-distance pollen dispersal data (Devaux *et al.* 2007)	Adequate for describing very fat tails.
(Inverse) Power-law (undefined)	$\left(\frac{r}{a}\right)^{-b}$	Not a PDF (infinite integral)	Not a PDF (infinite integral)	Fat-tailed Power-law Not a PDF (infinite integral)	Seeds (Willson 1993; Nathan *et al.* 2000), Butterflies (Baguette 2003; Fric and Konvicka 2007)	Less sensitive than the negative exponential to the spatial scale of the dispersal observations (Fric and Konvicka 2007)	Have been used traditionally in numerous studies to fit dispersal curves, but does not fit the definition of a dispersal kernel since it is not a PDF. For dispersal kernel representations, should be replaced by the power-law PDF given above.
Logistic	$\frac{b}{2\pi a^2\Gamma\left(\frac{2}{b}\right)\Gamma\left(1-\frac{2}{b}\right)}\left(1+\frac{r^b}{a^b}\right)^{-1}$	$a > 0; b > 2$	$a\frac{\Gamma\left(\frac{3}{b}\right)\Gamma\left(1-\frac{3}{b}\right)}{\Gamma\left(\frac{2}{b}\right)\Gamma\left(1-\frac{2}{b}\right)}$	Fat-tailed Power-law tail	Pollen (Klein *et al.* 2006b) Beetles (Chapman *et al.* 2007)	Like generally the power-law functions, it outperformed exponential families in Chapman *et al.* (2007)	Suitable for frequent long-distance dispersal events together with weak effect of distance close to the source

(continued)

Table 15.1 *Continued*

Kernel name	$k_L(r)$ expression*	Parameter values	Mean dispersal distance	Tail fatness	Taxonomic groups applied	Model comparison	Recommended use
Lognormal	$\frac{1}{(2\pi)^{3/2} br^2}\exp\left(-\frac{\log(r/a)^2}{2b^2}\right)$	a>0; b>0	$a\exp\left(\frac{b^2}{2}\right)$	Fat-tailed	Seeds (Greene *et al.* 2004; Stoyan and Wagner 2001; Canham and Uriarte 2006; Schurr *et al.* 2008; Jones and Muller-Landau 2008), Seedlings (Greene *et al.* 2004)	Outperforms many other kernels, except occasionally the 2Dt and WALD, in seed dispersal studies (Greene *et al.* 2004; Schurr *et al* 2008; Jones and Muller-Landau 2008).	Well suited for seed dispersal studies, especially when the peak of the distribution is presumably not at zero distance from the source (e.g. seedlings after establishment under Janzen-Connell effects)
Gaussian Mixture (particular case of the general mixture below)	$\frac{p}{\pi a_1^2}\exp\left(-\frac{r^2}{a_1^2}\right)+\frac{1-p}{\pi a_2^2}\exp\left(-\frac{r^2}{a_2^2}\right)$	$a_1, a_2 > 0$ $0 < p < 1$	$\frac{\sqrt{\pi}}{2}(pa_1+(1-p)a_2)$	Leptokurtic Never fat-tailed	Kernel integrating several vectors	Mostly used in theoretical studies (e.g. Bialozyt *et al.* 2006; Fayard *et al.* 2009)	
General Mixture	$pK_1(r)+(1-p)K_2(r)$				Exponential + inverse power and a variety of other mixtures used for Pollen (Goto *et al.* 2006; Slavov *et al.* 2009), Seeds (Bullock and Clarke 2000), Fish (Coombs and Rodriguez 2007), Butterflies (Hovestadt *et al.* 2011)	Outperformed non-mixture kernels when tested	Very versatile, though at the cost of increased dimensionality. Well-suited if short- and long-distance dispersal are suspected to be governed by different processes or vectors.
Inverse Gaussian (Wald)	$\frac{\sqrt{b}}{\sqrt{8\pi^3 r^5}}\exp\left(-\frac{b(r-a)^2}{2a^2 r}\right)$	$a > 0; b > 0$	a	Leptokurtic Exponential tail Mode at $r > 0$	Wind dispersed propagules (Katul *et al.* 2005; Schurr *et al.* 2008)	Outperformed exponential, lognormal and 2Dt kernels in a seed dispersal study including source effects (Schurr *et al.* 2008)	Derived from a mechanistic wind dispersal model, hence well suitable for wind dispersal studies, especially when independent information on wind dispersal parameters is available.
Weibull (Gaussian for b = 2)	$\frac{b}{2\pi a^2}r^{b-2}\exp\left(-\frac{r^b}{a^b}\right)$	$a > 0; b > 0$	$a\frac{\Gamma(1/b)}{b}$	Fat-tailed for $b < 1$ Mode at $r > 0$ for $b > 2$ $k_L(0) = \infty$ for $b < 2$	Pollen (Austerlitz *et al.* 2004; Goto *et al.* 2006), Seeds (Greene *et al.* 2004; Canham and Uriarte 2006), Seedlings (Greene *et al.* 2004; Goto *et al.* 2006), Flies (Taylor *et al.* 2010, with $b = 1.5$), Birds (Paradis *et al.* 2002; Kesler *et al.* 2010)	Generally outperformed by other kernels, such as the 2Dt and lognormal (Greene *et al.* 2004; Jones and Muller-Landau 2008), but has sometimes provided the best fits in pollen dispersal studies (Austerlitz *et al.* 2004)	Can fit well fat tails, but should be examined against other fat-tailed kernels in general.

Gamma (Exponential for $b = 2$)	$\frac{1}{2\pi a^2\Gamma(b)}\left(\frac{r}{a}\right)^{b-2}\exp\left(-\frac{r}{a}\right)$	$a > 0; b > 0$	ab	Peaks at 0 for $b \leq 2$ $k_L(0) = \infty$ for $b < 2$ Mode at $r > 0$ for $b > 2$ Exponential tail	Birds (van Houtan *et al.* 2007)	Was outperformed by a power-law tail (log-sech) in van Houten *et al.* 2007	Can be used alternatively to the related Weibull family function
Log-sech distribution (Cauchy for $b = 1$)	$\frac{1/(\pi^2 br^2)}{(r/a)^{1/b}+(r/a)^{-1/b}}$	$a > 0; b > 0$	∞ for $b \geq 1$	Fat-tailed Power-law tail Flat shape close to 0 for $b < 1$ Sharp decrease in 0 for $b, \geq 1$	Birds (Paradis *et al.* 2002, van Houtan *et al.* 2007), Mammals (Krkosek *et al.* 2007 used a Cauchy in a 1D space)	Outperformed the Gaussian and the gamma (van Houtan *et al.* 2007). Outperformed the non power-law tails (Weibull, Exponential) in Paradis *et al.* 2002	Suitable for frequent long-distance dispersal events together with weak effect of distance close to the source

* For sake of consistency, for all models, we provide expressions for the dispersal location kernel [$k_L(r)$]. The expression can thus appear different to that originally published in the cited reference if the authors had provided the dispersal distance kernel [$k_D(r)$]. The relations between these two are detailed in Box 15.1.

Box 15.1 Basic formulations of dispersal kernels

Consider a single point source is the origin site of n dispersers drawn from a large population in a continuous space (Figure 15.1). The dispersal distance of each disperser is the net displacement from the source, or the Euclidian distance between the 'start' (the source) and the 'end' point of the dispersal event. The histogram (Figure 15.2) shows a hypothetical example for $n = 100$ dispersers. Assume this histogram represents a continuous parametric distribution k characteristic of this population. Function k is a probability density function (pdf) termed the 'dispersal kernel', which can also be a discrete pdf in lattice space (Chesson and Lee 2005).

The dispersal kernel k can be defined as the pdf of the distribution of the end location of a disperser relative to the source point. It can be defined on a line or one-dimension (1D) space, a plane or two-dimensional (2D) space, and even on a three-dimensional (3D) space for dispersal in water or canopy. This type of dispersal kernel was termed 'dispersal kernel' (Nathan and Muller-Landau 2000), 'two-dimensional (2D) dispersal kernel' (Cousens and Rawlinson 2001) or 'dispersal density kernel' (Nathan *et al.* 2008b). Here we propose the alternative term 'dispersal location kernel' to avoid confusion with 2D space and with the word 'density' of the pdf. In a 2D space, either a polar

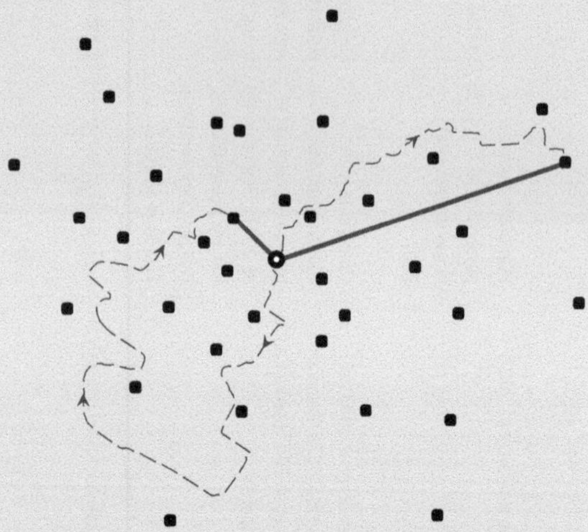

Figure 15.1 A schematic representation of dispersal from a single source (open circle at the center), showing the end points of dispersal events (black squares). The movement path (thin dashed line with arrows) and the dispersal distance (thick solid line) are shown for two dispersers.

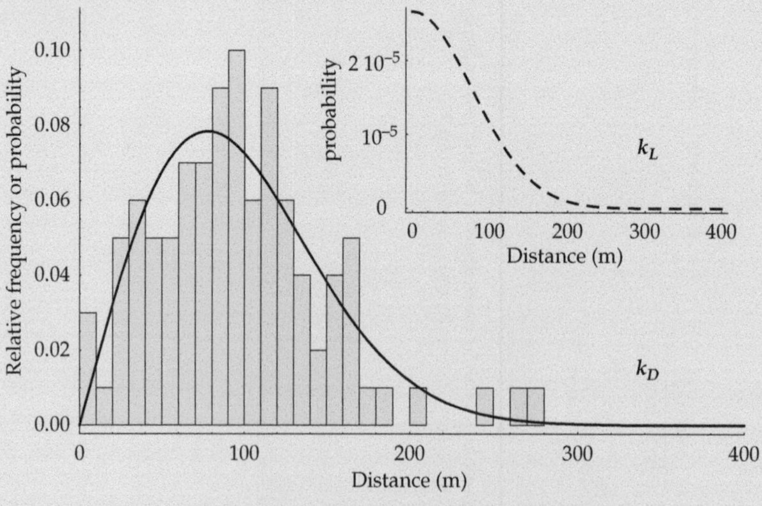

Figure 15.2 A histogram of 100 dispersal distances and the fitted dispersal distance kernel ($k_D(r)$, solid line); i.e. the probability density function of the distribution of the dispersal distance travelled by a disperser. We also plotted the dispersal location kernel ($k_L(r)$, dashed line, encapsulated figure); i.e. the probability density function of the distribution of the final location of a disperser (see Figure 5.3 in Cousens *et al.* 2008 for similar representations).

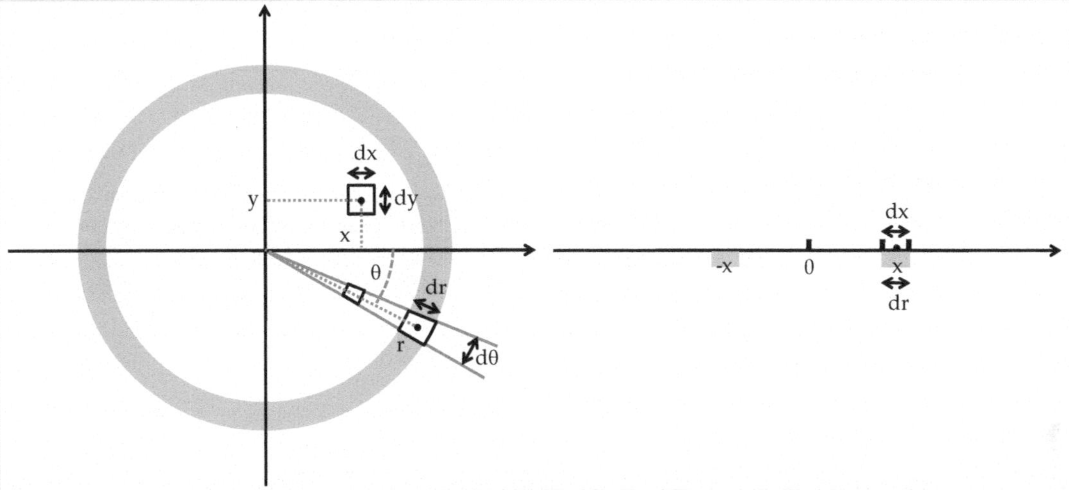

Figure 15.3 For a source at the origin 0, the dispersal location kernel denoted as $k_L(r)$ provides the density of the probability of the dispersal end point in the 1D, 2D, or 3D space according to the spatial dimension of the study system. For instance, in 2D space, $k_L(r)dA$ is the probability of a dispersal end point to be within a small 2D area dA around the location (x,y) in cartesian coordinates (left picture, top half) or (r,θ) in polar coordinates (left figure, bottom half). The surface dA depicted by the empty polygons is given by dxdy or rdrdθ. Since a probability is unitless and dA is an area, $k_L(r)$ is expressed in per unit area in a 2D space (see Table 15.2). The dispersal distance kernel denoted as $k_D(r)$ is the probability density function of the dispersal distance and is 1D regardless of the space dimension (Table 15.2). $k_D(r)dr$ equals the probability of a dispersal end point to be at a distance between r and r + dr from the origin regardless of the direction, i.e. within the grey area of the circular ring depicted in the left figure. For instance, in 1D space (right picture), $k_L(r)dr$ is the probability to end in the white segment around *x* whereas $k_D(r)dr$ is the probability to end in any of the grey segments around *x* and –*x*.

(r, θ) or a cartesian (x,y) coordinate system can be used to specify the locations of the source point $r = 0$, or $(x, y) = (0,0)$ and of the final point at distance and direction (r, θ) with coordinates (x, y). Formally, given that $x = r\cos(\theta)$ and $y = r\sin(\theta)$, the probability of having a dispersal end point in an infinitesimally small area $dA = dxdy = rdrd\theta$ around a single point (x,y) or (r, θ) is:

$$k_{X,Y}(x,y)dA = k_{R,\theta}(r,\theta)dA, \quad \text{(Equation 15.1)}$$

where x,y vary in $(-\infty,+\infty)$, r in $(0, +\infty)$ and θ in $(0,2\pi)$ (Figure 15.3).

Dispersal kernels are simpler to present in polar coordinates because distance *r* is a meaningful variable, and because the moments of the pdf are simpler to calculate (Clark *et al.* 1999). Also, under the common assumption of radial symmetry, $k_{R,\theta}(r, \theta)$ depends only on *r* and can be denoted as $k_L(r)$. This kernel integrates to 1 over the whole 2D space:

$$\int_{-\infty}^{\infty}\int_{-\infty}^{\infty} k_{X,Y}(x,y)dxdy = \int_0^{\infty}\int_0^{2\pi} k_{R,\theta}(r,\theta)rdrd\theta = \int_0^{\infty}\int_0^{2\pi} k_L(r)rdrd\theta = 1.$$

In 1D space, the cartesian and polar systems of coordinates can also be used alternatively and provide

$$k_X(x)dx = k_{R,\theta}(r,\theta)dr, \quad \text{(Equation 15.2)}$$

with x varying in $(-\infty,+\infty)$, r in $(0, +\infty)$ and θ being either 0 or π. Being pdfs, these kernels integrate to unity:

$$\int_{-\infty}^{\infty} k_X(x)dx = \int_0^{\infty} k_{R,\theta}(r,0)dr + \int_0^{\infty} k_{R,\theta}(r,\pi)dr = 1.$$

An alternative formulation only provides the density of probability $k_D(r)$ that a dispersal event ends at a distance *r* away from the source, regardless of the direction. The dispersal kernel $k_D(r)$ always gives a probability per unit distance (Figure 15.3). It was termed 'distance distribution' (Nathan and Muller-Landau 2000), 'one-dimensional (1D) dispersal kernel' (Cousens and Rawlinson 2001), and 'dispersal distance kernel' (Nathan *et al.* 2008b), the latter was adopted in this review. It is defined for *r* in $(0,+\infty)$ and integrates to 1 on this interval.

Location and distance kernels are related through the relations

$$k_D(r) = \int_0^{2\pi} k_{R,\theta}(r,\theta)rd\theta = 2\pi r k_L(r) \quad \text{(Equation 15.3)}$$

in 2D space, and

Box 15.1 *Continued*

$$k_D(r) = k_{R,\theta}(r,0) + k_{R,\theta}(r,\pi) = 2k_L(r) \quad \text{(Equation 15.4)}$$

in 1D space. In both (15.3) and (15.4) the last equality is restricted to the radial symmetric case.

Table 15.2

Space dimension	Dispersal distance kernel	Dispersal location kernel
1D	$k_D(r)$ in per unit distance	$k_x(x) = k_L(r)$ in per unit distance
2D	$k_D(r)$ in per unit distance	$k_{x,y}(x, y) = k_L(r)$ in per unit area
3D	$k_D(r)$ in per unit distance	$k_{x,y}(x, y, z) = k_L(r)$ in per unit volume

Computing the 'seed shadow' around a source, i.e. the number of dispersers at different places, requires multiplying the dispersal kernel by the total number of dispersers (Q) originated from that source (Clark *et al.* 1999; Nathan and Muller-Landau 2000). The units are numbers per unit of distance when the $k_D(r)$ formulation is used, and densities (individuals per unit distance, area, or volume, depending on space dimension) when the $k_L(r)$ formulation is used. When several individual seed shadows overlap, the locations of all sources and the number of dispersers each produces interact with the dispersal kernel in a convolution product to predict the numbers or densities of dispersers. In a 2D space and assuming a single kernel for all sources, for instance, the density of dispersers arriving at location (*x*,*y*) equals

$$\int_{-\infty}^{+\infty}\int_{-\infty}^{+\infty} Q(x_i,y_i)\,k_{X,Y}(x-x_i,y-y_i)\,dx_i dy_i, \quad \text{(Equation 15.5)}$$

where (x_i,y_i) denotes a potential source location and $Q(x_i,y_i)$ is the number of dispersers originated from the source per unit area. When the sources can be considered as points, these expressions of the density of dispersers arriving at location (*x*,*y*) expresses through discrete sums

$$\sum_{i:\,all\,sources} Q(x_i,y_i)k_{X,Y}(x-x_i,y-y_i), \quad \text{(Equation 15.6)}$$

or under isotropic dispersal

$$\sum_{i:\,all\,sources} Q(x_i,y_i)k_L(r_{xy}), \quad \text{(Equation 15.7)}$$

where $r_{i,xy}$ is the distance between (x_i,y_i) and (x,y). Equation 15.7 is at the core of in inverse modelling techniques, which employ expected disperser obtained in this way to estimate the most likely parameter values of $k_L(r)$, given observed densities at a set of locations (Ribbens *et al.* 1994). If the different sources are distinguishable, dispersal location kernels also allow predicting the respective proportion (rather than density) of dispersers from each individual source *j* arriving at a given location (*x*,*y*):

$$\frac{Q(x_j,y_j)k_L(r_{j,xy})}{\sum_{i:\,all\,sources} Q(x_i,y_i)k_L(r_{i,xy})}, \quad \text{(Equation 15.8)}$$

assuming again isotropic dispersal in 2D space and invariant kernel across sources.

alternative representations of a dispersal kernel (Nathan and Muller-Landau 2000; Cousens *et al.* 2008; Nathan *et al.* 2008b; see Box 15.1), yet none of them has been sweepingly adopted. Evaluating the shortcomings of previous suggestions, we propose the use of 'dispersal distance kernel' and 'dispersal location kernel', denoted as k_D and k_L, respectively (for details, see Box 15.1). The term dispersal kernel should not be used indiscriminately. Wrong conclusions could result from confusing k_D and k_L while comparing, for example, dispersal kernels and spread rates (Lewis *et al.* 2006). Therefore, authors should always explicitly mention what type of kernel is presented. In addition, the term 'dispersal kernel' should be reserved only for probabilistic representations of 'dispersal curves', because it clearly disentangles fecundity and dispersal (Ribbens *et al.* 1994).

Another important confusion associated with dispersal kernels comes from ambiguous definitions of 'dispersal'. Strictly, a dispersal event ends when the movement of the disperser ends; for instance, when an animal- or wind-dispersed seed arrives to its potential establishment site, or a bird reaches a potential new breeding site. The resulting 'basic dispersal kernel' encompasses the movement and the survival of the disperser during dispersive movements. More broadly, dispersal encompasses also the 'post-arrival' survival of dispersers to include only effective dispersal events that lead to establishment and/or successful reproduction. The resulting 'effective dispersal kernel' can be defined using different levels of establishment and/or reproduction. In plant studies, for example, effective dispersal can refer to seedlings (e.g. González-Martínez *et al.* 2006), saplings, and even to adults of the next generation (Steinitz *et al.* 2011). Because post-arrival survival varies in time and space and with distance from the source, effective dispersal kernels can differ substantially from the basic dispersal kernels, as found when comparing kernels estimated for seeds, seedlings, and adult plants (e.g. González-Martínez *et al.* 2006; Steinitz *et al.* 2011). Effective pollen dispersal requires viable pollen transport, fertilization, and the development of a viable seed. The spatial arrangement of individual plants and the scale and shape of the pollen dispersal kernel jointly determine the structure of the mating network (Meagher and Vassiliadis 2003; Fortuna *et al.* 2008).

Similar terminological and conceptual problems also appear in dispersal studies of active movers. The dispersal concept can encompass different processes (e.g. natal dispersal and breeding dispersal) that are likely to produce different kernels. Philopatry, the settlement at the natal site, is often neglected when defining the range of dispersal events. This practice can greatly affect the estimation and usefulness of dispersal kernels and therefore the way philopatry was treated should be specified. Capture–recapture studies of active movers must cope with the challenge of identifying the starting natal site, the establishment end point or the fate of censored animals (Turchin 1998). Such difficulties can be resolved by following the complete trajectory of a dispersing animal. Nevertheless, nearly all studies based on capture–recapture or individual tracking actually estimate effective dispersal kernels, since dispersers that did not survive or establish after reaching the dispersal end point site are either unknown or deliberately excluded. In a few studies (e.g. Larsen and Boutin 1994; Anders *et al.* 1998), data on post-dispersal mortality of animals is available, and in principle, the basic and effective dispersal kernels can be compared. On the other hand, mortality during dispersal movements (before settlement) is more commonly estimated for actively dispersed animals (e.g. Larsen and Boutin 1994; Clobert *et al.* 2004; Revilla and Wiegand 2008; Johnson *et al.* 2009) than for passively dispersed animals or plants. In any case, any consideration of effective dispersal kernels necessitates clear explicit criteria to delimit the starting and establishment sites and their temporal resolution, for both passive and active movers.

15.2 The questions

The dispersal kernel concept is central to fundamental research on dispersal processes and spatial patterns. Here we review several key questions covering the patterns observed in nature for dispersal kernels and their variations, the underlying mechanisms shaping them, and the large-scale consequences of dispersal kernels shapes, especially for long-distances.

15.2.1 What are the most common characteristics of dispersal kernels found in nature?

The history of dispersal research has experienced a drastic shift from Gaussian and Exponential dispersal kernels that dominated until the late 1990s, to various fat-tailed dispersal kernels that prevail nowadays. Early studies on actively dispersing animals, mostly insects, had already noted that dispersal location kernels are leptokurtic and thus deviate from the Gaussian case (Bateman 1950). Yet Gaussian and other exponentially-bounded kernels constitute, for example, the core mathematical feature of the classical theories of reaction-diffusion (Skellam 1951), genetic isolation by distance (Wright 1943), island biogeography (MacArthur and Wilson 1967), and spatial population ecology (Hanski 1994; Moilanen and Hanski 2001). The major drivers of the shift to fat-tailed kernels were studies demonstrating that observed population spread rates cannot be explained by simple exponentially-bounded kernels, but necessitate mixture (Lewis 1997) or fat-tailed ones (Mollison 1991; Kot *et al.* 1996; Clark 1998; Clark *et al.* 1999). In parallel, simple random walks associated to normal diffusion and Gaussian dispersal kernels have been questioned by studies showing that Lévy flights associated with anomalous diffusion and fat-tailed distributions fit animal movement data better (Viswanathan *et al.* 1996; Ramos-Fernández *et al.* 2004; Sims *et al.* 2008; but see Edwards *et al.* 2007). Nearly all studies comparing the fit of different dispersal kernels to empirical dispersal data have concluded that fat-tailed kernels outperform thin-tailed ones. Examples span from pollen (Robledo-Arnuncio and Gil 2005; Klein *et al.* 2006b) and seeds (Bullock and Clarke 2000; Clark *et al.* 2005; González-Martínez *et al.* 2006; Robledo-Arnuncio and García 2007; Martínez and González-Taboada 2009), to fish (Fujiwara *et al.* 2006) and birds (Paradis *et al.* 2002; van Houtan *et al.* 2007). However, dispersal kernels are typically estimated from data confined to relatively small spatial scales, and the statistical properties of the far tails have rarely been tested (Niggemann *et al.* 2012).

Dispersal kernels are based on the assumption that distance is the primary factor determining the probability of connecting two sites: the farther away, the lower the probability, with the exception of the opposing trend at short distances in the case of non-zero mode. Yet dispersal data suggest that for many active movers (or passive dispersal mediated by active movers), distance-decaying kernels are insufficient. Habitat heterogeneity greatly affects the probability of arriving at any site (Thomas and Kunin 1999). Spatially structured populations comprise complex spatial distributions of breeding habitats embedded in a non-breeding habitat matrix. Here, distance-decaying kernels might suit dispersal movements within the available breeding habitat, yet dispersers moving in the matrix often follow different rules and thus, in the simplest case, require an additional kernel. The matrix itself can be highly heterogeneous, and possibly each matrix class can yield a matrix-specific kernel (Vandermeer and Carvajal 2001; Revilla *et al.* 2004). More generally, active movers make decisions affecting their trajectories as a function of their interaction with the environment and a set of trade-offs that the organisms need to take into account (Nathan *et al.* 2008a), giving rise to complex multi-peaked or spatially restricted dispersal kernels (Haas 1995; Wiens *et al.* 1997; Haddad *et al.* 2003; Fisher *et al.* 2009; Lees and Peres 2009).

15.2.2 Uncertainty and variability of dispersal kernels

Dispersal is a complex process affected by multiple factors, some of which are highly variable in nature, others are difficult to estimate empirically, representing unpredictable contingencies. Three sources of uncertainties were identified in predictive studies of population rate of spread (Clark *et al.* 2003; Higgins *et al.* 2003a). The first, model uncertainty, is due to over-simplifications of the processes and sources of variability included in the model (e.g. stage structure in populations, spatial variations of the environment, variability of dispersal vectors). More detailed investigation of the movement ecology of dispersers should help to improve this aspect. The second, parameter uncertainty, is due to an imperfect knowledge/estimation of the parameters governing the processes. More powerful experimental designs and statistical analyses of large-scale datasets aim at reducing this one. Even if the dispersal kernel is known perfectly, the third source, inherent uncertainty, results from the stochastic nature of dispersal, the disproportionate effect or rare discrete events concerning dispersers or features of the landscape. Schtickzelle *et al.* (Chapter 18), for butterfly examples, found significant inter-annual variation in dispersal kernels not explained by meteorological variation. Revilla and Wiegand (2008) demonstrated that mortality patterns of resident lynx generated highly dynamic changes in the availability of settlement sites for dispersers. Consequently, the stochastic realization of the same dispersing behavior and the same survival risk deeply affected the effective dispersal kernels between years. Thus, although observed patterns might appear consistent across space, they might not be replicable along time even if all else is equal. Long-distance dispersal events typically contribute to inherent uncertainty with fatter tailed dispersal kernels and higher net reproduction rates, providing more unpredictable extreme events (Clark *et al.* 2001). Given these uncertainties, understanding to what extent dispersal kernels are repeatable and/or determined by measurable characteristics of the environment is challenging.

15.2.3 What determines the properties of dispersal kernels?

Dispersal, like other types of movement, depends on four different types of proximate factors: the internal state determining the motivation to move, the motion and the navigation capacities of the individual, and external factors such as landscape properties and biotic interactions (Nathan *et al.* 2008a). Mechanistic models developed to tailor specific types of dispersal, have striven to incorporate the key processes affecting the studied dispersal system, and the system-specific parameters to investigate their role in shaping the dispersal kernels, and to enable predictions. Examples span across many taxonomic groups, growth forms, and biomes, yet perhaps the most basic distinction is between passively and actively dispersed organisms.

There are three major stages in the dispersal of organisms passively dispersed by either an abiotic or biotic vector: initiation, transport, and termination (Isard and Gage 2001). Each of these stages is associated with a key parameter: the load of the vector, its displacement velocity, and the passage time of the disperser within the vector, respectively (Nathan *et al.* 2008b). In wind-driven passive dispersal, the vector load is the number of dispersal units originated from the source, the vector displacement velocity is determined by the horizontal wind speed during transport, and the passage time is determined by attributes of the source location, the terminal velocity of the dispersal unit, and the vertical wind speed (see, for example, Thomas *et al.* 2003 for ballooning spiders, and Nathan *et al.* 2011b for wind-dispersed seeds). For animal-mediated dispersal, the vector load is the number of dispersal units taken by the animal, the vector displacement velocity is determined by the animal movement and the passage time is determined, for example, by the retention time in the gut or the fur of the animal (Tsoar *et al.* 2011). Note that the dispersal kernel itself is determined only by two basic parameters (vector displacement velocity and passage time), whereas the vector load is the multiplier, or source strength, that transforms the kernel's probabilities to numbers or densities (see Box 15.1).

Sensitivity analyses of mechanistic models have elucidated, qualitatively and quantitatively, the relative role of different biological and environmental factors in determining the properties of dispersal kernels. It has been shown, for example, that wind speed parameters have greater impact on intraspecific variation in median dispersal distances and long-distance dispersal (LDD) compared to biological parameters (Soons *et al.* 2004; Nathan and Katul 2005). Of particular interest are atmospheric conditions favouring turbulent updrafts originating LDD (Nathan *et al.* 2002; Kuparinen *et al.* 2007a). Air movement may in turn be locally affected by physiography and canopy structure, and experimental evidence suggests that canopy opening and fragmentation tend to enhance airborne pollen dispersal distances (e.g. Bacles and Ennos 2008), a trend also noticed in insect-pollinated species (e.g. White *et al.* 2002). Effective pollen dispersal kernels are also strongly affected by pollen viability, which can drastically decrease in only a few hours (e.g. maize in hot conditions, Aylor 2004; but see Bohrerova *et al.* 2009). For animal-dispersed seeds, the movement of the animal vector, greatly affected by landscape features and complex behaviors, has the strongest effects on dispersal patterns (Russo *et al.* 2006; Spiegel and Nathan 2007; Levey *et al.* 2008). Both small- and large-scale spatial heterogeneity in the landscape can also influence the dispersal of seeds by wind (Bullock and Moy 2004; Bohrer *et al.* 2008; Pounden *et al.* 2008; Schurr *et al.* 2008).

All the discussion above assumes that there is a single major dispersal vector. Yet a problem of model uncertainty appears when important vectors are neglected or missed. In plants, for example, it has long been assumed that the 'standard' dispersal vector is the one inferred from seed morphology (van der Pijl 1982). Yet many plants, if not most, are actually dispersed by additional 'non-standard' vectors, which play a key role especially in LDD (Higgins *et al.* 2003b). The existence of multiple vectors operating at different spatial scales has been compellingly demonstrated for various plant species (Dennis and Westcott 2007; Jordano *et al.* 2007; Spiegel and Nathan 2007). A given species might be passively dispersed by many types of dispersal vectors, each potentially operating at a different scale and/or generating a different dispersal kernel. The combination of these different kernels, weighted by the relative load dispersed by each vector, constitutes the 'total dispersal kernel' (Nathan *et al.* 2008b), the complete descriptor of dispersal distances in the studied dispersal system. Overall, for many organisms, quantifying the dispersal kernel generated by the common dispersal vector might be insufficient, because it could deviate substantially from the total dispersal kernel.

In actively dispersed organisms individuals search for a place to reproduce. However, their movement trajectory is not only affected by this aim, since they have to consider simultaneously other factors such as their perceived mortality risk and the availability of food (Nathan *et al.* 2008a), integrating the effects of several processes normally acting at different spatiotemporal scales (Fryxell *et al.* 2008; Delgado *et al.* 2010). This is presumably the most fundamental difference between passive and active dispersers. The kernels obtained for active dispersers do not purely describe the dispersal process alone, but integrate different behaviors such as food, water, or refuge search, and predator avoidance, while searching for a new place to settle. The fact that there are several simultaneous processes behind empirical movement data hinders the identification of the onset and end of dispersal *per se*. The only way to identify a dispersal event is to detect deviations from a 'normal' behaviour by defining rules based on thresholds in movement parameters estimated using moving time windows, such as changes in short-term home ranges, distance to the origin, velocities, and turning angles (e. g. Palomares *et al.* 2000; Fryxell *et al.* 2008). The existence of strong behavioural differences among individuals, in many cases due to varying personalities (e.g. McDougall *et al.* 2006; Cote and Clobert 2007), make these approaches inherently subjective, and the interpretation of who is dispersing under those rules qualitative. A typical example is the difficulty in classifying the movements of young animals moving away but eventually returning to their natal site, mimicking actual dispersal patterns (i.e. when they finally settle and reproduce). Whether those excursions are part of the normal behaviour, or an initial training of the individual for later dispersal, is very relevant because in the end dispersal is a searching process during which individuals can have lots of a priori information (e.g. dispersing badgers between adjacent groups, Macdonald *et al.* 2008), or be naïve, with a broad range of intermediate cases (McConnell *et al.* 2002; Stamps and Davis 2006; Brown *et al.* 2008).

Nevertheless, we have a strong theoretical basis to predict movement properties of searching individuals, as a function of the spatio-temporal distribution of targets and the available prior information. Trajectories can be strongly linear when the cost of moving is high and the individual has poor information on target location (Zollner and Lima 1999), leading to leptokurtic kernels. However, if the spatial autocorrelation of targets is known, the movements of dispersers should initially be restricted locally to find a nearby high quality area, given that the natal site has proved favourable. When targets are clustered, the search can be Lévy-like with long straight moves until detecting some cue on a potential target, and then performing a restricted area search (Bartumeus *et al.* 2005), leading to fat-tailed kernels. Animals can even use systematic strategies such as foray search by which individuals sample the landscape, systematically obtaining information on the absence and/or relative quality on the surrounding areas before making the dispersal decision (Conradt *et al.* 2003). Search movements are additionally modified by factors such as cueing strategies, in which individuals trade off the onset of dispersal as a function of their relative quality against conspecifics (Strevens and Bonsall 2011) or conspecific attraction (Gamble *et al.* 2007), which can overrun all the previous information available to the individual (Serrano *et al.* 2001).

Estimating multi-process kernels for active dispersers is not necessarily more complex than for passive dispersers. However, building mechanistic models remains very challenging since it requires an integrated understanding of all the processes involved, including the movements and habitat selection before dispersal. The predicted kernels depend not only on model structure, but also on the actual parameterization, the initial conditions, the temporal and spatial heterogeneity, and resolution used.

15.2.4 Long-distance dispersal and large-scale dynamics

Although dispersal kernels are typically fitted based on dispersal data collected at relatively short distances from the source, increasing interest in dispersal kernels comes largely from the recognition of LDD importance (Petit *et al.* 1997; Clark *et al.* 1999; Higgins and Richardson 1999; Cowen *et al.* 2000; Cain *et al.* 2000; Kinlan and Gaines 2003; Bowler and Benton 2005; Trakhtenbrot *et al.* 2005; Nathan 2006). The magnitude and frequency of LDD events—derived from the properties of the dispersal kernel and especially the shape parameter determining the 'fatness' of the distribution tail—have been found to play an important role in a variety of ecological and genetic processes: the rate of spread of an expanding population (Kot *et al.* 1996; Neubert and Caswell 2000; Clark *et al.* 2001), the response to climate changes (Nathan *et al.* 2011a), the connectivity among isolated populations (Muñoz *et al.* 2004), the survival and genetic variation in metapopulations (Sork *et al.* 1999; Bohrer *et al.* 2005), the effects of forest fragmentation (van Houtan *et al.* 2007), the spatial distribution of genetic diversity (Ibrahim *et al.* 1996; Petit 2004; Bialozyt *et al.* 2006; Wingen *et al.* 2007; Fayard *et al.* 2009), and the transfer of genes between locally adapted populations (Kuparinen *et al.* 2010). LDD can be defined in a relative way, characterizing the distance above which a given fraction of dispersal events occurs (say 1% or 1‰), or in absolute terms, characterizing the fraction of dispersal events that occurs above a given threshold distance associated with the biology of the species and the environment (Nathan *et al.* 2008b). Both definitions are intimately related to the dispersal kernel and especially to the 'fatness' of its tail, whose characterization remains a challenge both for mechanistic and empirical approaches.

15.3 The tools

Dispersal kernels can be estimated by various experimental and mechanistic modelling approaches. Section 15.2 outlines the principal components of mechanistic models of passive and active dispersal; this section focuses on experimental approaches.

Experimentally, basic dispersal kernels can be estimated from the trajectories of particular dispersers (Lagrangian approach) or from the amount and/or diversity of dispersers at particular sampling points (Eulerian approach) (Bullock *et al.* 2006). The Lagrangian approach has been applied mostly to animals or animal-dispersed seeds or pollen, using mark/recapture (e.g., van Houtan *et al.* 2007), or tracking designs such as radio telemetry or GPS tags (e.g. Larsen and Boutin 1994; Anders *et al.* 1998, Ovaskainen *et al.* 2008; Revilla and Wiegand 2008; Johnson *et al.* 2009). It was also applied to smaller passive dispersers, through visual tracking of seeds dispersed by wind (Andersen 1991; Soons *et al.* 2004), radio tracking of tagged acorns dispersed by birds (Pons and Pausas 2007), or searching for radio-labeled gamma emitting seeds dispersed by ants (Kalisz *et al.* 1999). This approach provides a more straightforward distribution fitting compared to Eulerian methods, since dispersal end points are not confined to the sampling sites. Another advantage is the possibility of correlating disperser traits with dispersal capacities. Yet the number of dispersal units sampled is limited because tagging and tracking individual dispersers is costly. Finally, observations are concentrated on a short time period, which may be a problem if variable environmental conditions result in temporally fluctuating dispersal patterns (Skarpaas *et al.* 2011).

The Eulerian approach relies on physical or biological 'traps' at some particular points where dispersers are sampled. Live trapping of moving animals allows the mark and posterior recapture of individuals, including the non-invasive camera trapping (e.g. Gardner *et al.* 2010). Traps can also be containers or nets to catch seeds (e.g. Bullock and Clarke 2000 on the ground; Dauer *et al.* 2009 in the air; Middleton 1995 in water), ground plots where seeds or seedlings are collected (e.g. García *et al.* 2007), pheromone attractive systems (Qureshi *et al.* 2005), light traps or sticky traps for insects (Collier and Smith 1998), nets for catching animals (Chifflet *et al.* 2011), sticky slides or rotorods to collect pollen (e.g. Tufto *et al.* 1997), or mother-plants to sample viable pollen (e.g. Oddou-Muratorio *et al.* 2005). Some combinations of trapping systems can inform about the post-dispersal survival of dispersers (e.g. mother-plant versus sticky slides for pollen) enabling to distinguish effective dispersal from the basic dispersal kernel.

Most studies have measured the number of dispersers collected per trap. Some also measured a morphological characteristic associated to different sources (e.g. Klein *et al.* 2003). An increasing number of studies have genotyped the dispersers at several neutral markers to assign them to their original source, more or less categorically (e.g. Jordano *et al.* 2007; Jones and Muller-Landau 2008; see following). Even when the trapping design is not explicit (e.g. 'some seedlings chosen randomly throughout a forest plot'), studies based on traps must account for the positions of the sampling sites relatively to the sources to retrieve the dispersal kernel (e.g. Figure 1 in Nathan and Muller-Landau 2000). Robledo-Arnuncio and García (2007) indeed showed that simply estimating a dispersal kernel from the effective dispersal distances without accounting for the sampling design results in biases (Jones and Muller-Landau 2008).

A first set of methods to fit the dispersal kernel relies on the information carried by the number of dispersers in the traps (Equations 15.5–15.7; Box 15.1). Observations from a single point source are the easiest to analyse (Bullock and Clarke 2000; Skarpaas *et al.* 2004). Experimental designs where several undiscernible sources are located all around the traps are analysed through inverse modelling (Ribbens *et al.* 1994; Clark *et al.* 1998; Schurr *et al.* 2008), which can be extended to consider independent seed shadows as replicates if dispersers can be assigned categorically to different sources (e.g. by means of genetic methods; Jones and Muller-Landau 2008).

Because the numbers of dispersers in the traps might be unknown (e.g. mother tree as a pollen trap) or strongly determined by external factors independent of dispersal (e.g. germination and survival rates producing seedling number variation across sampling sites), the second set of methods to estimate dispersal kernels uses the proportions from different discernible sources in the traps (Equation 15.8; Box 15.1). This solution was first used for few sources with discernible traits (e.g. different colours of kernels in corn, Klein *et al.* 2003). Next, the use of highly polymorphic neutral markers was used to conduct a categorical parentage analysis and then analyse the retrieved dispersal events (e.g. Hardy *et al.* 2004 for pollen dispersal; Robledo-Arnuncio and García 2007 for seeds). More advanced methods integrate into a single likelihood function the Mendelian rules used in parentage analysis and the dispersal kernel and mass-action law used in spatial analysis (Adams *et al.* 1992; Burczyk *et al.* 2002; Oddou-Muratorio *et al.* 2005; Hadfield *et al.* 2006; Klein *et al.* 2008). These methods can estimate the basic dispersal kernel even if survival varies spatially, as long as the survival rate is source-independent. They rely on the probability of originating from the different sources independently of the factors affecting the probability of presence of the disperser at the sampling point. This is a clear benefit in studies sampling individuals a long time after dispersal, like settled seedlings of trees (Moran and Clark 2011).

A variety of other uses of molecular information to estimate dispersal are found in the literature and are not detailed here; these include assignments to geo-referenced populations with different allelic frequencies (e.g. Gaggiotti *et al.* 2002), patterns of isolation by distance (Hardy *et al.* 2006), and other methods from landscape genetics (Guillot *et al.* 2009), or genetic differentiation among pollen pools (TwoGener method, e.g. Austerlitz *et al.* 2004).

From historical least square estimation (e.g. Klein *et al.* 2003), the toolbox of statistical approaches has been expanded with non-parametric estimations (Chesson and Lee 2005), maximum likelihood (Tufto *et al.* 1997; Clark *et al.* 1999; Burczyk *et al.* 2002; Canham and Uriarte 2006; Schurr *et al.* 2008), and Bayesian analyses for hierarchical models (Clark *et al.* 2004; Hadfield *et al.* 2006; Snäll *et al.* 2007; Klein *et al.* 2008; Jones and Muller-Landau 2008; Moran and Clark 2011). The latter enables accounting for additional sources of variability (e.g. individual fecundity obtained as a function of basal area plus a random component in Clark *et al.* 2004; Chapter 19) and to consider jointly various types of observations (Clark and Gelfand 2006).

Optimizing the observation design can also improve the estimation of the dispersal kernel. The general objective is to cover the widest range of environments and trapping distances to sample dispersal events independently over the actual range, including rare LDD events. But the optimal sampling design is a compromise between statistical power and sampling time and costs (see Bullock *et al.* 2006 for a thorough discussion). Mathematical and computational techniques can determine the optimal sampling distances and sampling effort at each distance (Stoyan and Wagner 2001; Skarpaas *et al.* 2005). But this requires a good *a priori* knowledge about the dispersal kernel from a pilot study or mechanistic models.

Estimating the dispersal kernel empirically requires using appropriate families of parameterized functions (e.g. Austerlitz *et al.* 2004; Greene *et al.* 2004; Klein *et al.* 2006b; Jongejans *et al.* 2008). In Table 15.1 we summarize a wide array of kernel families used in the literature, providing statistical formulations, properties, performance, and actual and recommended applications.

Dispersal kernels are first characterized by a scale parameter, providing the spatial extent of most dispersal events (e.g. mean dispersal distance). The importance of LDD is further characterized by the overall shape of the kernel indicated by the kurtosis, or alternatively by the fatness of its tail. Kurtosis is the fourth moment of the kernel and equals 0 for a Gaussian, typically mesokurtic. As noted above, the Gaussian kernel has been extensively used in dispersal research, yet this function does not model LDD realistically. The use of leptokurtic distributions to characterize dispersal is almost always necessary (Table 15.1), yet not sufficient: the exponential distribution and the mixture of two Gaussians, two leptokurtic alternatives to the Gaussian, often provide worse fits than distributions with fatter tails (Table 15.1). Thus, the shape of the tail, not only the kurtosis, should be characterized to elucidate how fast the function decreases towards zero at relatively large distances. Fat-tailed dispersal kernels decrease more slowly than any exponential (Kot *et al.* 1996), while power-law kernels (e.g. 2Dt) decrease more slowly than any exponential-power function (Klein *et al.* 2006a). Recent studies of LDD generalized the use of fat-tailed kernels, not merely leptokurtic kernels (Table 15.1). The key challenge here is to obtain data on the frequency and spatial extent of the rare LDD events (Nathan 2006).

The shape of the dispersal kernel close to the origin also varies among functions: some are strongly peaked (power-exponential with $b < 1$) or even infinite at 0 (some Weibull), some show a convex shape close to the origin (2Dt, Clark *et al.* 1999) or even increase from 0 at short distances (log-normal, Stoyan and Wagner, 2001).

When the family of functions is not adequate both near and far, the fit reaches a compromise to catch the shape of the whole function. This can result in a bad characterization of the tail. It is strongly advisable to fit a variety of dispersal families and select the best-fitting one (Bullock *et al.* 2006), when possible the more mechanistic one, yet remembering that model selection is sensitive to the study site area and that extrapolations beyond the observational scale may be problematic (Fric and Konvicka 2007; Kuparinen *et al.* 2007b). The examples given in Table 15.1 show that the Gaussian and the Exponential are (still) widely used kernels, despite their overall poor performance. The simple inverse power-law function $\left(\frac{r}{a}\right)^{-b}$ (Table 15.1) is a fat-tailed distribution that has long been used to fit 'dispersal curves', yet it cannot be integrated close to zero unlike other formulations of the power-law that provide true dispersal kernels (Levin *et al.* 2003; 2Dt, power-law or logistic in Table 15.1). The use of various fat-tailed distributions is steadily becoming mainstream practice, exhibiting a notable upsurge in the last decade. Different fat-tailed functions have been adopted for different taxonomic groups (Table 15.1), reflecting more field-specific research histories than biological differences among taxa. The exponential power function, for example, has been used intensively in plant dispersal research since Clark *et al.* (1998), but is rarely applied to animals.

The need for a better mechanistic interpretation of dispersal kernel parameters has led to the proposition of more sophisticated expressions, mostly for wind dispersal of seeds, in the lineage of Greene and Johnson (1989) and Okubo and Levin (1989) and recently reviewed in Nathan *et al.* (2011b). Models of animal-mediated dispersal have recently gone beyond simply multiplying animal velocity by seed passage time, to incorporate, for example, landscape heterogeneity as well (e.g. Russo *et al.* 2006; Levey *et al.* 2008). To model mechanistically how self-propelled animals move during their own dispersal, all different components of the movement ecology framework need to be incorporated

(Nathan *et al.* 2008b; Revilla and Weigand 2008; Tsoar *et al.* 2011). These components have equivalents among passively dispersing organisms (Nathan *et al.* 2008a; Wright *et al.* 2008; Bacles and Jump 2011; Tsoar *et al.* 2011), and these two basic dispersal systems, previously considered incomparable, are likely to converge.

Even when mechanistic and phenomenological models fit comparably well to a particular experimental dataset (e.g. Skarpaas *et al.* 2004), a clear benefit of mechanistic models is that they can be used both to investigate the underlying mechanisms (Section 15.2) and to provide predictions based on *a priori* knowledge of environmental conditions and biological parameters.

15.4 Synthesis and future directions

By providing an individual-based description of dispersal probabilities at different distances and directions, dispersal kernels are statistical tools most suited for (i) characterizing dispersal scale and direction *per se*, (ii) investigating the relation between dispersal patterns, individual traits and ecological and demographic factors, or (iii) correcting for distance effects in studies of other processes (e.g. mating systems, selection gradients, or community assemblages). Dispersal kernels are key components of biological models that need to consider individual movement in explicit space, such as reproduction, recruitment, competition, range expansion, metapopulation dynamics, population resilience to habitat disturbance, gene flow, and transgene escape.

The dispersal kernel, though, may not be ideal to address all questions. For instance, metapopulation models may not require explicit individual information and may be more efficiently parameterized using pairwise migration rates among subpopulations (e.g. Bullock *et al.* 2006). Further, commonly employed dispersal kernels (and especially phenomenological functions) provide a simple quantitative description of the average dispersal process across individuals and environments. If they are well suited for comparative and modelling purposes, they are potentially too coarse-grained for detailed descriptions of the actual dispersal pattern when it involves complex behavioural mechanisms, intricate phenotypic correlations, and strong interactions with environmental factors, resulting in substantial inter-individual variation in dispersal ability and distance-independent fluctuations of dispersal probability across heterogeneous landscapes. In the latter case, the observed distribution of dispersal events will contain valuable information that may be lost upon oversimplified kernel fitting. The challenge will be to develop versatile enough kernels to unravel the actual variability in the dispersal process, for which mechanistic approaches are likely to excel as explicative and predictive tools. Researchers should not be discouraged by the apparent complexity of mechanistic models; we advocate efforts to acquire the helpful analytical skills and attitude required for modelling and understanding the fundamental mechanisms driving plant and animal movement.

Dispersal kernels should not be considered as goals in themselves or as fixed within species. Dispersal kernels are individually and temporally variable laws subject to many influential factors of interest. Much could be gained when modelling movement patterns using dispersal kernels by including data on morphological, physiological, behavioural, genetic, and developmental attributes of focal individuals, as well as on environmental variables, such as resource availability, landscape configuration, weather conditions, or relative abundance of intervening dispersal vectors (Nathan *et al.* 2008b). Comparative observational studies or controlled manipulation of dispersal-related traits and suspected environmental determinants are still scarce (Chapter 19), while empirical evidence is crucially needed. A careful assessment of the statistical uncertainty associated with kernel estimates, a frequently neglected issue, will be required for such kind of comparative inference. On the other hand, longitudinal studies should add the informative temporal dimension often missing in dispersal kernel evaluations. For instance, animal lifetime tracks can reveal spatio-temporal scaling properties of movement and their association with internal and external factors, to which separate dispersal kernels could be fitted in order to dissect total movement into specific processes of interest, such as foraging, natal, or breeding dispersal.

The development of dispersal kernels reflecting LDD events accurately is a remaining issue concerning both phenomenological and mechanistic models. The limited spatial scale of observations is a key drawback of phenomenological methods, since dispersal functions with very different tails may yield similarly good fits (Kuparinen *et al.* 2007b). Mechanistic kernels are more promising since they provide a basis for generalization beyond the scale of empirical analysis, but they still confront spatial scale constraints when it comes to experimental validation of their long-distance predictions. How to progress then? For species distributed in discrete populations, among-population pairwise migration rates might prove pivotal quantities for long-distance validation of kernels. Migration rates could be independently estimated in the field, using for instance capture–recapture or individual genetic assignment procedures (Manel *et al.* 2005), and compared against the expected proportion of migrants predicted by integrating the candidate kernel across individuals in the recipient and source populations (e.g. Revilla *et al.* 2004; Klein *et al.* 2006b). Alternatively, migration rates could be combined with local dispersal data to produce better kernel fits across different spatial scales (e.g. Goto *et al.* 2006). The estimation of migration rates can however be problematic if there are many large candidate source populations or for continuously distributed species. Indeed, there might often be insurmountable difficulties in obtaining the detailed individual information required for estimating and validating dispersal kernels over both short and long distances. Whether the dispersal kernel concept remains practical and testable over very broad spatial scales is questionable. Alternative approaches not based on individual dispersers deserve further exploration, such as connectivity maps, which have proved valuable to predict long-term continental-scale movements of both passive dispersers (Muñoz *et al.* 2004) and strong flyers (Felicísimo *et al.* 2008).

Finally, dispersal kernels can be regarded as complex phenotypic traits; assessing their heritability, plasticity, and among-population genetic divergence could inform on the selection shaping them and the ultimate evolutionary causes of dispersal (Chapter 16). For most plants and many animals, the dispersal kernel or the morphological traits affecting the dispersal kernel could be measured in provenance or progeny trials raised in common gardens, to infer genetically determined phenotypic differences among populations and their environmental and demographic correlates (Donohue *et al.* 2005; Ovaskainen *et al.* 2008). Additionally, genomic advances are providing whole genome and transcriptome sequence information (e.g. Vera *et al.* 2008), enabling functional genomic and association studies that could reveal the genetic architecture of dispersal. On this point, because short- and long-distance dispersal may rely on different mechanisms and evolve independently (Ronce 2007), dispersal kernel families used should not enforce artefactual shape constraints: the parametrization of near-the-origin and tail behaviors should be clearly separated.

Overall, the dispersal kernel concept has proven essential for dispersal research. The simplest models capture the fundamental distance-decay principle of ecology and geography (Nekola and White 1999) and were used to develop classical theories on biogeography and spatial dynamics. More complex leptokurtic and fat-tailed dispersal kernels usually better fit empirical data, at least over the necessarily limited scale of analysis, implying that the spatial extent of dispersal is much larger, possibly by several orders of magnitude (Nathan *et al.* 2011b), compared to the simple exponentially bounded functions. Recent advances in statistical analyses of the shape of dispersal kernels elucidated important phenomena such as rapid population spread. More complex formulations of mixed kernels account for higher degrees of variability in the dispersal process, yet this promising research avenue remains rather dormant thus far. Nevertheless, all these simple and advanced dispersal models are inherently limited by the basic assumption that the end point location of a disperser depends first and foremost on the distance from its source. Although partially true in all dispersal systems, distance might be only a minor determinant of the dispersal probability in many dispersal systems, particularly for actively moving organisms, and for passive movers dispersed by active movers. For these many cases, the dispersal kernel concept should be elaborated to include additional sources of variation, accounting for landscape heterogeneity concerning the major factors determining dispersal movements. The rarely questioned assumption about independent moves of the different dispersers that jointly construct the empirical dispersal kernel should be relaxed too. Because apparently many plants and animals move, passively or actively, in a correlated manner, the scope of this phenomenon needs to be assessed, the underlying mechanisms should be unveiled, and statistical tools to cope with lack of independence should be developed. Addressing these challenges would require further refinement of methods to quantify the paths and end points of dispersers, the key features of the landscape they travel through, as well as enhancement of data analysis tools and theoretical frameworks.

References

Adams, W.T., Griffin, A.R., Moran, G.F. (1992). 'Using paternity analysis to measure effective pollen dispersal in plant populations', *American Naturalist* **140**: 762–80.

Anders, A.D., Faaborg, J., Thompson, F.R. (1998). 'Postfledging dispersal, habitat use, and home-range size of juvenile Wood Thrushes', *Auk* **115**: 349–58.

Andersen, M. (1991). 'Mechanistic models for the seed shadows of wind-dispersed plants', *American Naturalist* **137**: 476–97.

Austerlitz, F., Dick, C.W., Dutech, C. *et al.* (2004). 'Using genetic markers to estimate the pollen dispersal curve', *Molecular Ecology* **13**: 937–54.

Aylor, D.E. (2004). 'Survival of maize (*Zea mays*) pollen exposed in the atmosphere', *Agricultural and Forest Meteorology* **123**: 125–33.

Bacles, C.F.E. and Ennos, R.A. (2008). 'Paternity analysis of pollen-mediated gene flow for *Fraxinus excelsior* L. in a chronically fragmented landscape', *Heredity* **101**: 368–80.

Bacles C.F.E. and Jump, A.S. (2011). 'Taking a tree's perspective on forest fragmentation genetics', *Trends in Plant Science* **16**: 13–18.

Baguette, M. (2003). 'Long distance dispersal and landscape occupancy in a metapopulation of the cranberry fritillary butterfly', *Ecography* **26**: 153–60.

Bartumeus, F., Da Luz, M.G.E., Viswanathan G.M. *et al.* (2005). 'Animal search strategies: a quantitative random-walk analysis', *Ecology* **86**: 3078–87.

Bateman, A.J. (1950). 'Is gene dispersion normal?', *Heredity* **4**: 353–63.

Bialozyt, R., Ziegenhagen, B., Petit, R.J. (2006). 'Contrasting effects of long distance seed dispersal on genetic diversity during range expansion', *Journal of Evolutionary Biology* **19**: 12–20.

Bianchi, F., Schellhorn, N. A., van der Werf, W. (2009). 'Predicting the time to colonization of the parasitoid *Diadegma semiclausum*: The importance of the shape of spatial dispersal kernels for biological control', *Biological Control* **50**: 267–74.

Bohrer, G., Katul, G.G., Nathan, R. *et al.* (2008). 'Effects of canopy heterogeneity, seed abscission and inertia on wind-driven dispersal kernels of tree seeds', *Journal of Ecology* **96**: 569–80.

Bohrer, G., Nathan, R., Volis, S. (2005). 'Effects of long-distance dispersal for metapopulation survival and genetic structure at ecological time and spatial scales', *Journal of Ecology* **93**: 1029–40.

Bohrerova, Z., Bohrer, G., Cho, K.D. *et al.* (2009). 'Determining the viability response of pine pollen to atmospheric conditions during long-distance dispersal', *Ecological Applications* **19**: 656–67.

Bowler, D.E. and Benton, T.G. (2005). 'Causes and consequences of animal dispersal strategies: relating individual behaviour to spatial dynamics', *Biological Reviews* **80**: 205–25.

Brown, C.R., Brown, M.B., Brazeal, K.R. (2008). 'Familiarity with breeding habitat improves daily survival in colonial cliff swallows', *Animal Behaviour* **76**: 1201–10.

Bullock, J. M. and Clarke, R.T. (2000). 'Long distance seed dispersal: measuring and modelling the tail of the curve', *Oecologia* **124**: 506–21.

Bullock, J.A. and Moy, I.L. (2004). 'Plants as seed traps: inter-specific interference with dispersal', *Acta Oecologica* **25**: 35–41.

Bullock, J.M., Shea, K., Skarpaas, O. (2006). 'Measuring plant dispersal: an introduction to field methods and experimental design', *Plant Ecology* **186**: 217–34.

Burczyk, J., Adams, W.T., Moran, G.F. *et al.* (2002). 'Complex patterns of mating revealed in a *Eucalyptus regnans* seed orchard using allozyme markers and the neighbourhood model', *Molecular Ecology* **11**: 2379–91.

Cain, M.L., Milligan, B.G., Strand, A.E. (2000). 'Long-distance seed dispersal in plant populations', *American Journal of Botany* **87**: 1217–27.

Canham, C.D. and Uriarte, M. (2006). 'Analysis of neighborhood dynamics of forest ecosystems using likelihood methods and modelling', *Ecological Applications* **16**: 62–73.

Carrasco, L.R., Harwood, T.D., Toepfer, S. *et al.* (2010). 'Dispersal kernels of the invasive alien western corn rootworm and the effectiveness of buffer zones in eradication programmes in Europe', *Annals of Applied Biology* **156**: 63–77.

Chapman, D.S., Dytham, C., Oxford, G.S. (2007). 'Modelling population redistribution in a leaf beetle: an evaluation of alternative dispersal functions', *Journal of Animal Ecology* **76**: 36–44.

Chesson, P. and Lee, C.T. (2005). 'Families of discrete kernels for modeling dispersal', *Theoretical Population Biology* **67**: 241–56.

Chifflet, R., Klein, E.K., Lavigne, C. *et al.* (2011). 'Spatial scale of insect-mediated pollen dispersal in oilseed rape in an open agricultural landscape', *Journal of Applied Ecology* **48**: 689–96.

Clark, C.J., Poulsen, J.R., Bolker, B.M. *et al.* (2005). 'Comparative seed shadows of bird, monkey, and wind-dispersed trees. *Ecology* **86**: 2684–94.

Clark, J.S. (1998). 'Why trees migrate so fast: confronting theory with dispersal biology and paleo-record', *American Naturalist* **152**: 204–24.

Clark, J.S. and Gelfand, A.E. (2006). 'A future for models and data in environmental science', *Trends in Ecology and Evolution* **21**: 375–80.

Clark, J.S., LaDeau, S., Ibanez, I. (2004). 'Fecundity of trees and the colonization-competition hypothesis', *Ecological Monographs* **74**: 415–42.

Clark, J.S., Lewis, M., Horvath, L. (2001). 'Invasion by extremes: Population spread with variation in dispersal and reproduction', *American Naturalist* **157**: 537–54.

Clark, J.S., Lewis, M., McLachlan, J.S. *et al.* (2003). 'Estimating population spread: What can we forecast and how well?', *Ecology* **84**: 1979–88.

Clark, J.S., Macklin, E., Wood, L. (1998). 'Stages and spatial scales of recruitment limitation in southern Appalachian forests', *Ecological Monographs* **68**: 213–35.

Clark, J.S., Silman, M., Kern, R. *et al.* (1999). 'Seed dispersal near and far: patterns across temperate and tropical forests', *Ecology* **80**: 1475–94.

Clobert, J., Ims, R.A., Rousset, F. (2004). 'Causes, mechanisms and consequences of dispersal', in I. Hanski, O.E. Gaggiotti (eds), *Ecology, Genetics and Evolution of Metapopulations*. Elsevier, Academic Press, London, 307–335.

Collier, K.J. and Smith, B.J. (1998). 'Dispersal of adult caddisflies (Trichoptera) into forests alongside three New Zealand streams', *Hydrobiologia* **361**: 53–65.

Connell J.H. (1971). 'On the role of natural enemies in preventing competitive exclusion in some marine animals and in forest trees', in, P.J. den Boer, G.R. Gradwell (eds), *Dynamics of populations*. Centre for Agricultural Publishing and Documentation, Wageningen. 298–312.

Conradt, L., Zollner, P., Roper, T. *et al.* (2003). 'Foray search: an effective systematic dispersal strategy in fragmented landscapes', *American Naturalist* **161**: 905–15.

Coombs, M.F. and Rodriguez, M.A. (2007). 'A field test of simple dispersal models as predictors of movement in a cohort of lake-dwelling brook charr', *Journal of Animal Ecology* **76**: 45–57.

Cote, J. and Clobert, J. (2007). 'Social personalities influence natal dispersal in a lizard', *Proceeding of the Royal Society of London Series B* **274**: 383–90.

Cousens, R., Dytham, C., Law, R. (2008). *Dispersal in Plants: A Population Perspective*. Oxford University Press, Oxford.

Cousens, R.D. and Rawlinson, A.A. (2001). 'When will plant morphology affect the shape of a seed dispersal 'kernel'?', *Journal of Theoretical Biology* **211**: 229–38.

Cowen, R.K., Lwiza, K.M.M., Sponaugle, S. *et al.* (2000). 'Connectivity of marine populations: open or closed?', *Science* **287**: 857–9.

Dauer, J.T., Mortensen, D.A., Luschei, E.C. *et al.* (2009). '*Conyza canadensis* seed ascent in the lower atmosphere', *Agricultural and Forest Meteorology* **149**: 526–34.

Delgado, M.D., Penteriani, V., Revilla E. *et al.* (2010). 'The effect of phenotypic traits and external cues on natal dispersal movements', *Journal of Animal Ecology* **79**: 620–32.

Dennis, A.J. and Westcott, D.A. (2007). 'Estimating dispersal kernels produced by a diverse community of vertebrates', in, A.J. Dennis, R.J. Green, E.W. Schupp, D.A. Westcott (eds), *Seed Dispersal: Theory and its Application in a Changing World*. CAB International, Wallingford, UK, 201–28.

Devaux, C., Lavigne, C., Austerlitz, F *et al.* (2007). 'Modelling and estimating pollen movement in oilseed rape (*Brassica napus*) at the landscape scale using genetic markers', *Molecular Ecology* **16**: 487–99.

Diekmann, O. (1978). 'Thresholds and traveling waves for the geographical spread of infection', *Journal of Mathematical Biology* **6**: 109–30.

Donohue, K., Polisetty, C.R., Wender, N.J. (2005). 'Genetic basis and consequences of niche construction: plasticity-induced genetic constraints on the evolution of seed dispersal in *Arabidopsis thaliana*', *American Naturalist* **165**: 537–50.

Edwards, A.M., Phillips, R.A., Watkins, N.W. (2007). 'Revisiting Lévy flight search patterns of wandering albatrosses, bumblebees and deer', *Nature* **449**: 1044–8.

Estep, L.K., Burkett-Cadena, N.D., Hill, G.E. *et al.* (2010). 'Estimation of dispersal distances of *Culex erraticus* in a focus of eastern equine encephalitis virus in the southeastern United States', *Journal of Medical Entomology* **47**: 977–86.

Fayard, J., Klein, E.K., Lefevre, F. (2009). Long distance dispersal and the fate of a gene from the colonization front', *Journal of Evolutionary Biology* **22**: 2171–82.

Felicísimo, A.M., Muñoz, J., González-Solis, J. (2008). 'Ocean surface winds drive dynamics of transoceanic aerial movements', *PLoS ONE* **3**: e2928.

Fisher, D.O., Lambin, X., Yletyinen, S.M. (2009). 'Experimental translocation of juvenile water voles in a Scottish lowland metapopulation', *Population Ecology* **51**: 289–95.

Fortuna, M.A., García, C., Guimaraes, P.R. *et al.* (2008). 'Spatial mating networks in insect-pollinated plants', *Ecology Letters* **11**: 490–8.

Fric, Z., Konvicka, M. (2007). 'Dispersal kernels of butterflies: power-law functions are invariant to marking frequency', *Basic and Applied Ecology* **8**: 377–86.

Fryxell, J.M., Hazell, M., Borger, L. *et al.* (2008). 'Multiple movement modes by large herbivores at multiple spatiotemporal scales', *Proceedings of the National Academy of Sciences of the USA* **105**: 19114–9.

Fujiwara, M., Anderson, K.E., Neubert, M.G. *et al.* (2006). 'On the estimation of dispersal kernels from individual mark-recapture data', *Environmental and Ecological Statistics* **13**: 183–97.

Gaggiotti, O.E., Jones, F., Lee, W.M. *et al.* (2002). 'Patterns of colonization in a metapopulation of grey seals', *Nature* **416**: 424–7.

Gamble L.R., McGarigal, K., Compton B.W. (2007). 'Fidelity and dispersal in the pond-breeding amphibian, *Ambystoma opacum*: implications for spatio-temporal population dynamics and conservation', *Biological Conservation* **139**: 247–57.

García, C., Jordano, P., Godoy, J.A. (2007). 'Contemporary pollen and seed dispersal in a *Prunus mahaleb* population: patterns in distance and direction', *Molecular Ecology* **16**: 1947–55.

Gardner B., Reppucci, J., Lucherini, M. *et al.* (2010). 'Spatially explicit inference for open populations: estimating demographic parameters from camera-trap studies', *Ecology* **91**: 3376–83.

González-Martínez, S.C., Burczyk, J., Nathan, R. *et al.* (2006). 'Effective gene dispersal and female reproductive success in Mediterranean maritime pine (*Pinus pinaster* Aiton)', *Molecular Ecology* **15**: 4577–88.

Goto, S., Shimatani, K., Yoshimaru, H *et al.* (2006). 'Fat-tailed gene flow in the dioecious canopy tree species *Fraxinus mandshurica* var. *japonica* revealed by microsatellites', *Molecular Ecology* **15**: 2985–96.

Greene, D.F., Canham, C.D., Coates, K.D. *et al.* (2004). 'An evaluation of alternative dispersal functions for trees', *Journal of Ecology* **92**: 758–66.

Greene D.F. and Johnson E.A. (1989). 'A model of wind dispersal of winged or plumed seeds', *Ecology* **70**: 339–47.

Guillot, G., Leblois, R., Coulon, A. *et al.* (2009). 'Statistical methods in spatial genetics', *Molecular Ecology* **18**: 4734–56.

Haas, D.A (1995). 'Dispersal and use of corridors by birds in wooded patches on an agricultural landscape', *Conservation Biology* **9**: 845–54.

Haddad, N.M., Bowne, D.R., Cunningham, A. *et al.* (2003). 'Corridor use by diverse taxa', *Ecology* **84**: 609–15.

Hadfield, J.D., Richardson, D.S., Burke, T. (2006). 'Towards unbiased parentage assignment: combining genetic, behavioural and spatial data in a Bayesian framework', *Molecular Ecology* **15**: 3715–30.

Hanski, I. (1994). 'A practical model of metapopulation dynamics', *Journal of Animal Ecology* **63**: 151–62.

Hardy, O.J., González-Martínez, S.C., Freville, H. *et al.* (2004). 'Fine-scale genetic structure and gene dispersal in *Centaurea corymbosa* (Asteraceae) I. Pattern of pollen dispersal', *Journal of Evolutionary Biology* **17**: 795–806.

Hardy, O.J., Maggia, L., Bandou, E. *et al.* (2006). 'Fine-scale genetic structure and gene dispersal inferences in 10 Neotropical tree species', *Molecular Ecology* **15**: 559–71.

Havel, J.E., Shurin, J.B., Jones J.R. (2002). 'Estimating dispersal from patterns of spread: spatial and local control of lake invasions', *Ecology* **83**: 3306–18.

Higgins, S.I., Clark, J.S., Nathan, R. *et al.* (2003a). 'Forecasting plant migration rates: managing uncertainty for risk assessment', *Journal of Ecology* **91**: 341–7.

Higgins, S.I., Nathan, R., Cain, M.L. (2003b). 'Are long-distance dispersal events in plants usually caused by nonstandard means of dispersal?', *Ecology* **84**: 1945–56.

Higgins, S.I. and Richardson, D.M. (1999). 'Predicting plant migration rates in a changing world: the role of long-distance dispersal', *American Naturalist* **153**: 464–75.

Hilbert, D. (1904). 'Grundzüge einer allgemeinen Theorie der linearen Integralgleichungen. (Erste Mitteilung.)', *Nachrichten von der Königl. Gesellschaft der Wissenschaften zu Göttingen, Mathematisch-physikalische Klasse* **1**: 49–91.

Hovestadt, T., Binzenhöfer, B., Nowicki, P. *et al.* (2011). 'Do all inter-patch movements represent dispersal? A mixed kernel study of butterfly mobility in fragmented landscapes', *Journal of Animal Ecology* **80**: 1070–77.

Ibrahim, K.M., Nichols, R.A., Hewitt, G.M. (1996). 'Spatial patterns of genetic variation generated by different forms of dispersal during range expansion', *Heredity* **77**: 282–91.

Isard, S.A. and Gage, S.H. (2001). *Flow of Life in the Atmosphere*. Michigan State University Press, East Lancing.

Janzen, D.H. (1970). 'Herbivores and the number of tree species in tropical forests' *American Naturalist* **104**: 501–28.

Johnson, C.A., Fryxell, J.M., Thompson, I.D. *et al.* (2009). 'Mortality risk increases with natal dispersal distance in American martens', *Proceedings of the Royal Society B* **276**: 3361–7.

Jones, F.A. and Muller-Landau, H.C. (2008). 'Measuring long-distance seed dispersal in complex natural environments: an evaluation and integration of classical and genetic methods', *Journal of Ecology* **96**: 642–52.

Jongejans, E., Skarpaas, O., Shea, K. (2008). 'Dispersal, demography and spatial population models for conservation and control management', *Perspectives in Plant Ecology Evolution and Systematics* **9**: 153–70.

Jordano, P., García, C., Godoy, J.A. *et al.* (2007). 'Differential contribution of frugivores to complex seed dispersal patterns', *Proceedings of The National Academy of Sciences of the USA* **104**: 3278–82.

Kalisz, S., Hanzawa, F.M., Tonsor, S.J. *et al.* (1999). 'Ant-mediated seed dispersal alters pattern of relatedness in a population of *Trillium grandiflorum*', *Ecology* **80**: 2620–34.

Katul, G.G., Porporato, A., Nathan, R. *et al.* (2005). 'Mechanistic analytical models for long-distance seed dispersal by wind', *American Naturalist* **166**: 368–81.

Kesler, D.C., Walters, J.R., Kappes, J.J. (2010). 'Social influences on dispersal and the fat-tailed dispersal distribution in red-cockaded woodpeckers', *Behavioural Ecology* **21**: 1337–43.

Kinlan, B.P., Gaines, S.D. (2003). 'Propagule dispersal in marine and terrestrial environments: a community perspective', *Ecology* **84**: 2007–20.

Kinlan, B.P., Gaines S.D., Lester, S.E. (2005). 'Propagule dispersal and the scales of marine community process', *Diversity and Distributions* **11**: 139–48.

Klein, E.K., Desassis, N., Oddou-Muratorio, S. (2008). 'Pollen flow in the wildservice tree, *Sorbus torminalis* (L.) Crantz. IV. Whole interindividual variance of male fecundity estimated jointly with the dispersal kernel', *Molecular Ecology* **17**: 3323–36.

Klein, E.K., Lavigne, C., Foueillassar, X. *et al.* (2003). 'Corn pollen dispersal: quasi-mechanistic models and field experiments', *Ecological Monographs* **73**: 131–50.

Klein, E.K., Lavigne, C. Gouyon, P.H. (2006a). 'Mixing of propagules from discrete sources at long distance: comparing a dispersal tail to an exponential', *BMC Ecology* **6**: 3.

Klein, E.K., Lavigne, C., Picault, H. *et al.* (2006b). 'Pollen dispersal of oilseed rape: estimation of the dispersal function and effects of field dimension', *Journal of Applied Ecology* **43**: 141–51.

Kot, M., Lewis, M.A. Van Den Driessche, P. (1996). 'Dispersal data and the spread of invading organisms', *Ecology* **77**: 2027–42.

Krkosek, M., Lauzon-Guay, J.S., Lewis, M.A. (2007). 'Relating dispersal and range expansion of California sea otters', *Theoretical Population Biology* **71**: 401–7.

Kuparinen, A., Markkanen, T., Riikonen, H. *et al.* (2007). 'Modeling air-mediated dispersal of spores, pollen and seeds in forested areas', *Ecological Modelling* **208**: 177–88.

Kuparinen, A., Savolainen, O., Schurr, F.M. (2010). 'Increased mortality can promote evolutionary adaptation of forest trees to climate change', *Forest Ecology and Management* **259**: 1003–8.

Kuparinen, A., Snall, T., Vanska, S. *et al.* (2007). 'The role of model selection in describing stochastic ecological processes', *Oikos* **116**: 966–74.

Largier, J.L. (2003). 'Considerations in estimating larval dispersal distances from oceanographic data', *Ecological Applications* **13**: S71–89.

Larsen, K.W. and Boutin, S. (1994). 'Movements, survival, and settlement of red squirrel (*Tamiasciurus hudsonicus*) offspring', *Ecology* **75**: 214–23.

Lees, A.C. and Peres, C.A. (2009). 'Gap-crossing movements predict species occupancy in Amazonian forest fragments', *Oikos* **118**: 280–90.

Levey, D.J., Tewksbury, J.J., Bolker, B.M. (2008). 'Modelling long-distance seed dispersal in heterogeneous landscapes', *Journal of Ecology* **96**: 599–608.

Levin, S.A., Muller-Landau, H.C., Nathan, R. *et al.* (2003). 'The ecology and evolution of seed dispersal: a theoretical perspective', *Annual Review of Ecology Evolution and Systematics* **34**: 575–604.

Lewis, M.A. (1997). 'Variability, patchiness and jump dispersal in the spread of an invading population', in, D.Tilman, P.M. Kareiva (eds), *Spatial Ecology*. Princeton University Press, New York.

Lewis, M.A., Neubert, M.G., Caswell, H. *et al.* (2006). 'A guide to calculating discrete-time invasion rates from data', in, M.W. Cadotte, S.M. McMahon, T. Fukami (eds), *Conceptual Ecology and Invasion Biology: Reciprocal Approaches to Nature*. Springer, Dordrecht, 169–92.

MacArthur, R.H. and Wilson, E.O. (1967). *The theory of island biogeography*. Princeton University Press, Princeton.

Macdonald, D.W., Newman, C., Buesching, C.D. *et al.* (2008). 'Male-biased movement in a high-density population of the eurasian badger (*Meles meles*)', *Journal of Mammalogy* **89**: 1077–86.

Manel, S., Gaggiotti, O.E., Waples, R.S. (2005). 'Assignment methods: matching biological questions with appropriate techniques', *Trends in Ecology and Evolution* **20**: 136–42.

Martínez, I. and González-Taboada, F. (2009). 'Seed dispersal patterns in a temperate forest during a mast event: performance of alternative dispersal kernels', *Oecologia* **159**: 389–400.

McConnell, B., Fedak, M., Burton, H.R. *et al.* (2002). 'Movements and foraging areas of naive, recently weaned southern elephant seal pups', *Journal of Animal Ecology* **71**: 65–78.

McDougall, P.T., Reale, D., Sol, D. *et al.* (2006). 'Wildlife conservation and animal temperament: causes and consequences of evolutionary change for captive, reintroduced, and wild populations', *Animal Conservation* **9**: 39–48.

Meagher, T.R. and Vassiliadis, C. (2003). 'Spatial geometry determines gene flow in plant populations', in, R.Hails, J.Beringer, H.C.J. Halifax (eds), *Genes in the Environment*. Blackwell Science Ltd, Oxford, 76–90.

Middleton, B.A. (1995). 'Sampling devices for the measurement of seed rain and hydrochory in rivers', *Bulletin of the Torrey Botanical Club* **122**: 152–5.

Moilanen, A. and Hanski, I. (2001). 'On the use of connectivity measures in spatial ecology', *Oikos* **95**: 147–52.

Mollison, D. (1977). 'Spatial contact models for ecological and epidemic spread', *Journal of the Royal Statistical Society Series B* **39**: 283–326.

Mollison, D. (1991). 'Dependence of epidemic and population velocities on basic parameters', *Mathematical Biosciences* **107**: 255–87.

Moran, E.V. and Clark, J.S. (2011). 'Estimating seed and pollen movement in a monoecious plant: a hierarchical Bayesian integrating genetic and ecological data'. *Molecular Ecology* **20**: 1248–62.

Muñoz, J., Felicísimo, A.M., Cabezas, F. *et al.* (2004). 'Wind as a long-distance dispersal vehicle in the Southern Hemisphere', *Science* **304**: 1144–7.

Nathan, R. (2006). 'Long-distance dispersal of plants', *Science* **313**: 786–8.

Nathan R., Getz W.M., Revilla E. *et al.* (2008a). 'A movement ecology paradigm for unifying organismal movement research', *Proceedings of the National Academy of Sciences of the USA* **105**: 19052–9.

Nathan, R., Horvitz, N., He, Y. *et al.* (2011a). 'Spread of North-American wind-dispersed trees in future environments'. *Ecology Letters* **14**: 211–19.

Nathan, R. and Katul, G.G. (2005). 'Foliage shedding in deciduous forests lifts up long-distance seed dispersal by wind', *Proceedings of the National Academy of Sciences of the USA* **102**: 8251–6.

Nathan, R., Katul, GG., Bohrer, G. *et al.* (2011b). 'Mechanistic models of seed dispersal by wind', *Theoretical Ecology* **4**: 113–32.

Nathan, R., Katul, G.G., Horn, H.S. *et al.* (2002). 'Mechanisms of long-distance dispersal of seeds by wind', *Nature* **418**: 409–13.

Nathan, R. and Muller-Landau, H.C. (2000). 'Spatial patterns of seed dispersal, their determinants and consequences for recruitment', *Trends in Ecology and Evolution* **15**: 278–85.

Nathan, R., Safriel, U.N., Noy-Meir, I. *et al.* (2000). 'Spatiotemporal variation in seed dispersal and recruitment near and far from *Pinus halepensis* trees', *Ecology* **81**: 2156–69.

Nathan R., Schurr, F.M., Spiegel, O. *et al.* (2008b). 'Mechanisms of long-distance seed dispersal', *Trends in Ecology and Evolution* **23**: 638–47.

Nekola, J.C. and White, P.S. (1999). 'The distance decay of similarity in biogeography and ecology', *Journal of Biogeography* **26** 867–78.

Neubert, M.G. and Caswell, H. (2000). 'Demography and dispersal: calculation and sensitivity analysis of invasion speed for structured populations', *Ecology* **81**: 1613–28.

Niggemann, M., Wiegand, T., Robledo-Arnuncio, J.J. *et al.* (2012). 'Marked point pattern analysis on genetic paternity data for uncertainty assessment of pollen dispersal kernels', *Journal of Ecology* **100**: 264–76.

Oddou-Muratorio, S., Klein, E.K., Austerlitz, F. (2005). 'Pollen flow in the wildservice tree, *Sorbus torminalis* (L.) Crantz. II. Pollen dispersal and heterogeneity in mating success inferred from parent-offspring analysis', *Molecular Ecology* **14**: 4441–52.

Okubo, A. and Levin, S.A. (1989). 'A theoretical framework for data analysis of wind dispersal of seeds and pollen', *Ecology* **70**: 329–38.

Ovaskainen, O., Smith, A.D., Osborne, J.L. *et al.* (2008). 'Tracking butterfly movements with harmonic radar reveals an effect of population age on movement distance', *Proceedings of the National Academy of Sciences of the USA* **105**: 19090–5.

Paradis, E., Baillie, S. R., Sutherland, W.J. (2002). 'Modeling large-scale dispersal distances', *Ecological Modelling* **151**: 279–92.

Palomares F., Delibes, M., Ferreras P. *et al.* (2000). 'Iberian lynx in a fragmented landscape: pre-dispersal, dispersal and post-dispersal habitats', *Conservation Biology* **14**: 809–18.

Petit, R.J. (2004). 'Biological invasions at the gene level', *Diversity and Distributions* **10**: 159–65.

Petit, R.J., Pineau, E., Demesure, B. *et al.* (1997). 'Chloroplast DNA footprints of postglacial recolonization by oaks', *Proceedings of the National Academy of Sciences of the USA* **94**: 9996–10001.

Pons, J. and Pausas, J.G. (2007). 'Acorn dispersal estimated by radio-tracking', *Oecologia* **153**: 903–11.

Pounden, E., Greene, D. F., Quesada, M. *et al.* (2008). 'The effect of collisions with vegetation elements on the dispersal of winged and plumed seeds', *Journal of Ecology* **96**: 591–8.

Qureshi, J.A., Buschman, L.L., Throne, J.E. *et al.* (2005). 'Adult dispersal of *Ostrinia nubilalis* Hubner (Lepidoptera: Crambidae) and its implications for resistance management in B.T.-maize', *Journal of Applied Entomology* **129**: 281–92.

Ramos-Fernández, G., Mateos, J.L., Miramontes, O. *et al.* (2004). 'Lévy walk patterns in the foraging movements of spider monkeys (*Ateles geoffroyi*)', *Behavioural Ecology and Sociobiology* **55**: 223–30.

Revilla, E. and Wiegand, T. (2008). 'Individual movement behavior, matrix heterogeneity and the dynamics of spatially structured populations', *Proceedings of the National Academy of Sciences of the USA* **105**: 19120–5.

Revilla, E., Wiegand, T., Palomares, F. *et al.* (2004). 'Effects of matrix heterogeneity on animal dispersal: from individual behaviour to metapopulation-level parameters', *American Naturalist* **164**: E130–53.

Ribbens, E., Silander, J.A., Pacala, S.W. (1994). 'Seedling recruitment in forests: calibrating models to predict patterns of tree seedling dispersion', *Ecology* **75**: 1794–806.

Robledo-Arnuncio, J.J., Alia, R., Gil, L. (2004). 'Increased selfing and correlated paternity in a small population of a predominantly outcrossing conifer, *Pinus sylvestris*', *Molecular Ecology* **13**: 2567–77.

Robledo-Arnuncio, J.J. and García, C. (2007). 'Estimation of the seed dispersal kernel from exact identification of source plants', *Molecular Ecology* **16**: 5098–109.

Robledo-Arnuncio, J.J. and Gil, L. (2005). 'Patterns of pollen dispersal in a small population of *Pinus sylvestris* L. revealed by total-exclusion paternity analysis', *Heredity* **94**: 13–22.

Ronce, O. (2007). 'How does it feel to be like a rolling stone? Ten questions about dispersal evolution', *Annual Review of Ecology, Evolution, and Systematics* **38**: 231–53.

Russo, S.E., Portnoy, S., Augspurger, C.K. (2006). 'Incorporating animal behaviour into seed dispersal models: Implications for seed shadows', *Ecology* **87**: 3160–74.

Sagnard, F., Pichot, C., Dreyfus, P. *et al.* (2007). 'Modelling seed dispersal to predict seedling recruitment: Recolonization dynamics in a plantation forest', *Ecological Modelling* **203**: 464–74.

Schurr, F.M., Steinitz, O., Nathan, R. (2008). 'Plant fecundity and seed dispersal in spatially heterogeneous environments: models, mechanisms and estimation', *Journal of Ecology* **96**: 628–41.

Serrano, D., Tella, J.L., Forero, M.G. *et al.* (2001). 'Factors affecting breeding dispersal in the facultatively colonial lesser kestrel: individual experience versus conspecific cues', *Journal of Animal Ecology* **70**: 568–78.

Siegel, D.A., Kinlan, B.P., Gaylord, B. *et al.* (2003). 'Lagrangian descriptions of marine larval dispersion', *Marine Ecology Progress Series* **260**: 83–96.

Sims, D.W., Southall, E.J., Humphries, N.E. *et al.* (2008). 'Scaling laws of marine predator search behaviour', *Nature* **451**: 1098–102.

Skarpaas, O., Shea, K., and Bullock, J. M. (2005). 'Optimizing dispersal study design by Monte Carlo simulation'. *Journal of Applied Ecology* **42**: 731–9.

Skarpaas, O., Shea, K., and Jongejans, E. (2011). 'Watch your time step: trapping and tracking dispersal in autocorrelated environments', *Methods in Ecology and Evolution*, **2**: 407–415.

Skarpaas, O., Stabbetorp, O.E., Ronning, I. *et al.* (2004). 'How far can a hawk's beard fly? Measuring and modelling the dispersal of *Crepis praemorsa*', *Journal of Ecology* **92**: 747–57.

Skellam, J.G. (1951). 'Random dispersal in theoretical populations', *Biometrika* **38**: 196–218.

Skelsey, P., Rossing, W.A.H., Kessel, G.J.T. *et al.* (2005). 'Influence of host diversity on development of epidemics: An evaluation and elaboration of mixture theory', *Phytopathology* **95**: 328–38.

Slavov, G.T., Leonardi, S., Burczyk, J. *et al.* (2009). 'Extensive pollen flow in two ecologically contrasting populations of *Populus trichocarpa*', *Molecular Ecology* **18**: 357–73.

Snäll, T., O'Hara, R.B., Arjas, E. (2007). 'A mathematical and statistical framework for modelling dispersal', *Oikos* **116**: 1037–50.

Soons, M.B., Heil, G.W., Nathan, R. *et al.* (2004). 'Determinants of long-distance seed dispersal by wind in grasslands', *Ecology* **85**: 3056–68.

Sork, V.L., Nason, J., Campbell, D.R. *et al.* (1999). 'Landscape approaches to historical and contemporary gene flow in plants', *Trends in Ecology and Evolution* **14**: 219–24.

Spiegel, O. and Nathan, R. (2007). 'Incorporating dispersal distance into the disperser effectiveness framework: frugivorous birds provide complementary dispersal to plants in a patchy environment', *Ecology Letters* **10**: 718–28.

Stamps, J.A. and Davis, J.M. (2006). 'Adaptive effects of natal experience on habitat selection by dispersers', *Animal Behaviour* **72**: 1279–89.

Steinitz, O., Troupin, D., Vendramin, G.G. *et al.* (2011). 'Genetic evidence for a Janzen-Connell recruitment pattern in reproductive offspring of *Pinus halepensis* trees.', *Molecular Ecology* **20**: 4152–64.

Stockmarr, A. (2002). 'The distribution of particles in the plane dispersed by a simple 3-dimensional diffusion process', *Journal of Mathematical Biology* **45**: 461–9.

Stoyan, D. and Wagner, S. (2001). 'Estimating the fruit dispersion of anemochorous forest trees', *Ecological Modelling* **145**: 35.

Strevens, C.M.J. and Bonsall, M.B. (2011). 'Density-dependent population dynamics and dispersal in heterogeneous metapopulations', *Journal of Animal Ecology* **80**: 282–93.

Taylor, D.B., Moon, R.D., Campbell, J.B. *et al.* (2010). 'Dispersal of stable flies (Diptera: Muscidae) from larval development sites in a nebraska landscape', *Environmental Entomology* **39**: 1101–10.

Thieme, H.R. (1977). 'Model for spatial spread of an epidemic', *Journal of Mathematical Biology* **4**: 337–51.

Thomas, C.D. and Kunin, W.E. (1999). 'The spatial structure of populations', *Journal of Animal Ecology* **68**: 647–57.

Thomas, C.F.G., Brain, P., Jepson, P.C. (2003). 'Aerial activity of linyphiid spiders: modelling dispersal distances from meteorology and behaviour', *Journal of Applied Ecology* **40**: 912–27.

Trakhtenbrot, A., Nathan, R., Perry, G. *et al.* (2005). 'The importance of long-distance dispersal in biodiversity conservation', *Diversity and Distributions* **11**: 173–81.

Tsoar, A., Shohami, D., Nathan, R. (2011). 'A movement ecology approach to study seed dispersal and plant invasion: an overview and application of seed dispersal by fruit bats', in, D.M. Richardson (ed), *Fifty Years of Invasion Ecology: the Legacy of Charles Elton.* Wiley-Blackwell, London, 103–19.

Tufto, J., Engen, S., Hindar, K. (1997). 'Stochastic dispersal processes in plant populations', *Theoretical Population Biology* **52**: 16–26.

Turchin, P. (1998). *Quantitative Analysis of Movement.* Sinauer, Sunderland.

Van den Bosch, F., Metz, J.A.J., Diekmann, O. (1990). 'The velocity of spatial population expansion', *Journal of Mathematical Biology* **28**: 529–65.

van der Pijl L. (1982). *Principles of Dispersal in Higher Plants.* Springer, Berlin. Third Edition.

van Houtan, K.S., Pimm, S.L., Halley, J.M. *et al.* (2007). 'Dispersal of Amazonian birds in continuous and fragmented forest', *Ecology Letters* **10**: 219–29.

Vandermeer, J. and Carvajal, R. (2001). 'Metapopulation dynamics and the quality of the matrix', *American Naturalist* **158**: 211–20.

Venable, D.L., Flores-Martinez, A., Muller-Landau, H.C. *et al.*, (2008). 'Seed dispersal of desert annuals', *Ecology* **89**: 2218–27.

Vera, J.C., Wheat, C.W., Fescemyer, H.W. *et al.* (2008). 'Rapid transcriptome characterization for a non-model organism using 454 pyrosequencing', *Molecular Ecology* **17**: 1636–47.

Viswanathan, G.M., Afanasyev, V., Buldyrev, S.V. *et al.* (1996). 'Lévy flight search patterns of wandering albatrosses', *Nature* **381**: 413–15.

White, G.M., Boshier, D.H., Powell, W. (2002). 'Increased pollen flow counteracts fragmentation in a tropical dry forest: an example from *Swietenia humilis* Zuccarini', *Proceedings of the National Academy of Sciences of the USA* **99**: 2038–42.

Wiens, J.A., Schooley R.L., Weeks, R.D. (1997). 'Patchy landscapes and animal movements: do beetles percolate?', *Oikos* **78**: 257–64.

Willson, M.F. (1993). 'Dispersal mode, seed shadows, and colonization patterns', *Vegetatio* **107, 108**: 261–80.

Wingen, L.U., Brown, J.K.M., Shaw, M.W. (2007). 'The population genetic structure of clonal organisms generated by exponentially bounded and fat-tailed dispersal', *Genetics* **177**: 435–48.

Wright, S. (1943). Isolation by distance. *Genetics* **28**, 114–38.

Wright, S.J., Trakhtenbrot, A., Bohrer, G. *et al.* (2008). 'Understanding strategies for seed dispersal by wind under contrasting atmospheric conditions', *Proceedings of the National Academy of Sciences of the USA* **105**: 19084–9.

Zollner, P.A., Lima, S.L. (1999). 'Search strategies for landscape-level interpatch movements', *Ecology* **80**: 1019–30.

CHAPTER 16

Evolution and emergence of dispersal kernels—a brief theoretical evaluation

Thomas Hovestadt, Dries Bonte, Calvin Dytham, and Hans Joachim Poethke

16.1 Introduction

We want to start with a definition of the term 'dispersal kernel' we use, as it is not consistent throughout the literature (see Cousens *et al.* 2008). Ambiguity exists in several respects. (1) It has been used to describe the **density** probability density function (pdf); i.e. the density (per area) of seeds deposited in a certain distance from the source, or the **distance** pdf, i.e. the probability to disperse a certain distance. (2) It has been used as the summary description of dispersal events for a whole population or just of a single individual. (3) It can be used as the statistical description of real dispersal data or of a 'hypothetical' kernel from which a single dispersal event is drawn as a random realization. In this chapter, we typically talk about hypothetical kernels for single individuals as this kernel defines the long-term average success of a dispersal strategy, and is the ultimate target of selection; consequently it is the appropriate definition in the context of (evolutionary) models of dispersal. If other meanings are intended, we will express this clearly. We also always take the perspective of the individual that has control over movement; e.g. in the case of seed dispersal the perspective of the tree (and not the seed) that sheds its one million seeds, but the perspective of the moving individual in case of, for example, dispersing foxes.

A detailed knowledge of dispersal kernels is of paramount importance when, for example, predicting the expansion of invasive species (Chapter 26), or the recolonization of sites in fragmented landscapes (Chapter 31). Kernels are, however, not phenotypic attributes in their own right. Instead, dispersal events we document are the consequence of some rule × environment interaction (Cousens *et al.* 2008), where rules can be coded into the genome but could also be acquired from personal or cultural experience (Nathan *et al.* 2008; Schick *et al.* 2008). We use the term 'rule' in a very broad sense here, including not only behavioural rules but also morphological and physiological attributes that influence movement. Consequently, while the dispersal kernel is the ultimate target of selection, selection will act upon the morphological, physiological, or behavioural attributes that influence dispersal (ability); e.g. the wing loading of seeds, the readiness to emigrate, or the timing of dispersal. Especially where the rule × environment interaction component is strong—and we are inclined to believe that this is usually the case—resolving the interaction term will be a prerequisite for generating landscape-specific predictions about the dispersal of individuals or whole populations.

Studying dispersal kernels thus raises three major issues, all of considerable complexity. Empiricists are confronted with the problem of fitting a given statistical kernel model to dispersal data; we do not consider these problems in this chapter (Turchin 1998; Chapters 15, 19). In the following, we address two aspects of modelling dispersal kernels. In the first section, we consider models on kernel evolution—what would the optimal distribution (of offspring) be? In the second, we consider the generation of dispersal kernels based on modelling some dispersal mechanisms.

Dispersal Ecology and Evolution. First Edition. Edited by Jean Clobert *et al.*

16.2 Evolution of dispersal kernels

In parallel with the accelerating interest in dispersal data, dispersal theory has prospered (e.g. Bowler and Benton 2005; Ronce 2007) (Chapters 1, 11). However, most theory considers the evolution of emigration probabilities or rates from clearly defined local populations, with individuals dispersing through a (hostile) matrix. This approach is rooted in the concept of metapopulations and appears well suited for cases where habitat and matrix are clearly distinct and individuals can control whether they leave a patch of suitable habitat or not. Yet theory on the evolution of dispersal kernels as described by some scale and shape parameters is more limited.

The seminal work of Hamilton and May (1977) demonstrated the importance of kin-competition in maintaining dispersal, even if dispersal were extremely costly. Hamilton and May did not directly address the evolution of dispersal distance, but their underlying argumentation makes it clear that in an infinite and homogeneous world without costs to dispersal, a uniform density pdf of offspring would be the optimal strategy, as it best minimizes kin-competition. In the real world, this is not attainable though, as the world is not homogeneous over infinite scales, dispersal cannot be free of costs or risks, and there are generally no perceivable mechanisms that could truly generate such a distribution (see Section 16.3). However, the minuscule fungal or microbial spores are dispersed extremely widely and may at least approximate uniform density distributions (de Wit and Bouvier 2006).

A few papers have considered the evolution of mean dispersal distance but have assumed a fixed shape (function) of the kernel; fewer have allowed the evolution of kernel shape (see e.g. Starrfelt and Kokko 2010 for both approaches). Muller-Landau *et al.* (2003) specifically focus on the evolution of long-distance dispersal but do not consider the evolution of a dispersal kernel as a whole.

Rousset and Gandon (2002) investigated the influence of explicit distance-dependent costs on the evolution of dispersal kernels in spatially homogenous systems exposed to kin competition. Such costs could be due to investment in structures or abilities that enhance dispersal at the expense of other fitness-relevant attributes, such as fecundity or survival. They showed that depending on the shape of the cost function, different evolutionary responses might emerge. For linearly or asymptotically increasing costs, they predicted the evolution of uni-modal distributions for the distance pdf. Indeed, the specific shape of an (evolutionary-stable) dispersal kernel should always balance the cost of dispersing a certain distance and the benefits achieved by reducing kin competition in that distance so that net benefits become identical across distance.

Hovestadt *et al.* (2001) simulated the evolution of seed dispersal kernels in fractal landscapes—artificial landscapes that resemble natural landscapes. They did not assume explicit costs for dispersal, but an implicit penalty was paid due to seed deposition in unsuitable habitat. In such landscapes, selection would favour the evolution of 'fat-tailed kernels', with a large fraction of seeds deposited near mother trees and a small fraction of seeds spread out over large distances (Figure 16.1). The argument underlying this prediction is the balance between kin competition (that always favours long-distance dispersal) and the fact that the probability of arriving in suitable habitat is largest in the neighbourhood of an established tree. However, the risk of ending up in unsuitable habitat increases asymptotically over distance—results thus agree well with those derived by Rousset and Gandon (2002). Empirical data indeed suggest that fat-tailed dispersal kernels are frequent (Chapter 15).

Some animal species can choose between distinctly different modes of movement. For example, some spider species either disperse 'on foot' over short distances, or they let themselves be carried away by wind ('ballooning'). These distinct modes of dispersal would not be well captured by a single kernel model. Bonte *et al.* (2010) thus simulated the evolution of a 'mixed dispersal kernel' in such a way that one gene determines whether a spider disperses either locally or over long distances (implemented as uniform density pdf), while a second gene defines the mean dispersal distance of a local Gaussian dispersal kernel. Increasing overall habitat availability selected for an increased probability of dispersing long-distance (p)—but never beyond

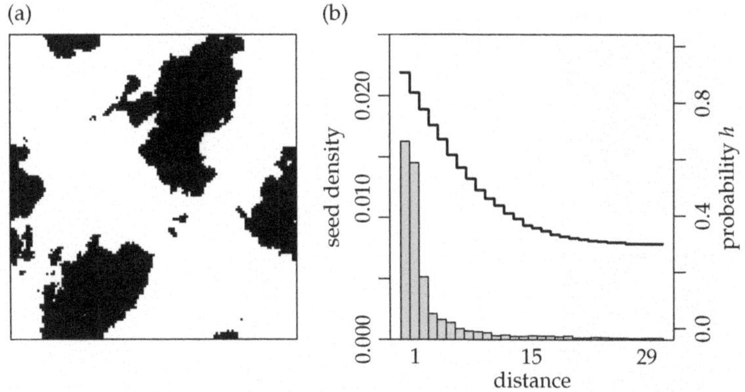

Figure 16.1 (adjusted from Hovestadt *et al.* 2001). (a) Exemplarily snapshot of a fractal landscape, and (b) of the evolved density pdf (histogram). Seeds are located in larger densities near the mother tree but at larger distance, seeds are evenly distributed. This implies a linearly increasing **distance** pdf. The line indicates the distance dependent probability h to encounter suitable habitat for a seed released in suitable habitat itself (at distance zero $h = 1$).

p=0.1 even in the most favourable conditions simulated. Autocorrelation in the distribution of habitat selected for reduced long-distance dispersal but an increased mean distance of the local dispersal kernel (Figure 16.2a). The predictions are supported by data describing the regional variance in dispersal behaviour within spiders (Bonte *et al.* 2006). An especially interesting finding is that under conditions of high habitat clumping but low habitat availability, local adaptation in the dispersal strategy may occur (Figure 16.2b); i.e. different mean dispersal distances evolve in large or small habitat patches. This is remarkable as dispersal is the very process that undermines the evolution of locally adapted strategies.

Murrell *et al.* (2002) simulated the evolution of dispersal distance in homogeneous landscapes with costs of dispersal increasing linearly and the shape of the kernel fixed to a negative exponential distance pdf. They showed that mean dispersal distance would increase with increasing fecundity and also be affected by the type of competition. The fecundity effect can be explained through the increase in kin competition that comes with an increase in the number of offspring—and intensity of competition becomes stronger (at high densities) under scramble competition. Parent-offspring competition would also select for increased dispersal distance if life expectancy increases (Dytham and Travis 2006). However, Dytham and Travis (2006) showed that the argument works the other way round too—low dispersal distance, possibly imposed by high costs of dispersal, would select for earlier death if fecundity declines with age. In this case, it is at some age in the interest of a parent to allow its own replacement by a more fecund offspring. The mean distance evolving will also depend on whether offspring or mothers have control over dispersal (Starrfelt and Kokko 2010); if distance-dependent costs exist, an individual offspring in control of its movement prefers shorter dispersal as it is more closely related to itself than to its mother. Both examples make clear that theory should invest more into understanding the joint evolution of dispersal related traits and other life-history attributes.

An area of theoretical research that has recently attracted interest is selection on dispersal in gradients (e.g. Gaston 2009). A study looking at the evolution of mean dispersal distance in constant gradients clearly suggests that the selection of dispersal distance –with the distance pdf fixed as a negative exponential—would depend on the type of gradient (Dytham 2009; Gros *et al.* 2006; Chapter 31). However, a gradient that affects the potential benefits of dispersal, for example, an increase in habitat turnover, would favour elevated dispersal distance towards the range margin. More on this topic—especially for the case of range expansion in non-equilibrium conditions can be found in Chapter 26.

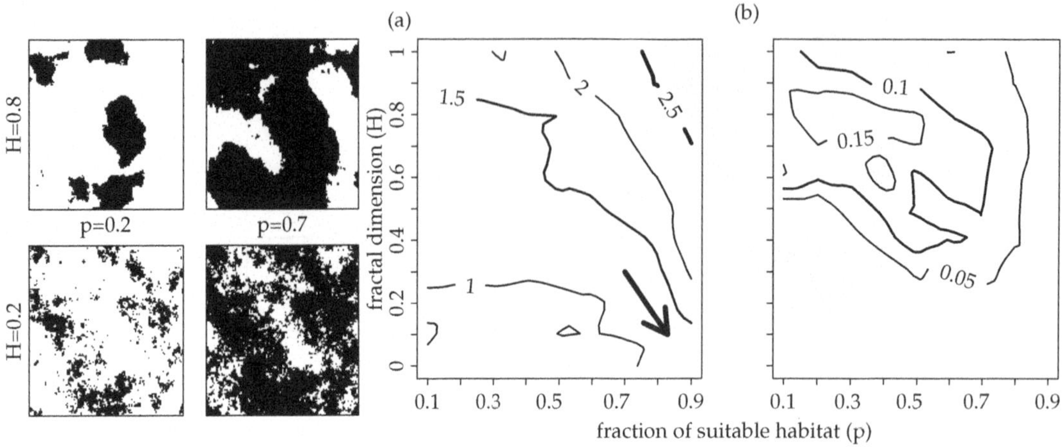

Figure 16.2 (adjusted from Bonte *et al.* 2010) (a) Effect of habitat availability *p* and fractal dimension *H* on the evolution of the mean dispersal distance of a local Gaussian distance pdf (contour lines). The arrow indicates the parameter region where the fraction of offspring dispersing globally was maximized. (b) Emergence of local adaptation (measured as strength of correlation between dispersal traits of individuals and fraction of suitable habitat in their surrounding in the *p*-*H* parameter space. Each habitat cell contains a population with density-dependent (logistic) growth. The panels on left show four exemplary landscapes where fraction *p* of suitable habitat increase from left to right and the Hurst index *H* (analogue to autocorrelation) from bottom to top.

16.3 The genesis of dispersal kernels (from first principles)

Predictions generated by evolutionary models are typically either not constrained by assumptions concerning the dispersal mechanism at all, or they make arbitrary assumptions concerning allowable shapes of kernels. Models of the type outlined in the previous section are thus not likely to quantitatively conform to empirical data. However, such strategic models are valuable as they indicate in which way natural selection is likely to operate on dispersal mechanisms. They may thus allow us to interpret differences in dispersal-related traits and rules when comparing species or populations within species. Yet clearly, the design of strategic evolutionary models would benefit from a better understanding of the mechanisms responsible for the emergence of dispersal kernels.

In the last two decades we have seen a massive increase in our understanding of movement mechanisms and rules that are the basis for the emergence of dispersal events. This interest has been inspired by insights that successfully predicting dispersal under altered conditions, e.g. increased habitat fragmentation or the effects of climate change, will require a better knowledge of the mechanisms underlying dispersal. As we will outline, this is indeed prerequisite for understanding how landscape (context) and movement mechanisms and rules **interact** to generate dispersal kernels. Nonetheless, we should not forget that an assumed mechanistic basis underlies earlier approaches to describing the spread of populations as a diffusion process or as random walk (see Turchin 1998). At least under certain conditions a diffusion parameter can be directly estimated from step length and time needed to take a step, given that the number of steps is not too small (e.g. Kareiva and Shigesada 1983; Turchin 1998).

Diffusion models have occasionally failed in predicting the expansion of a species range—typically range expansions were faster (and even accelerating) than predicted by simple diffusion models (Cain *et al.* 2000; Kot *et al.* 1996). There are several possibilities for adjusting the diffusion framework. These include assuming that movement behaviour, and thus the diffusion parameter, is variable between individuals and over time (e.g. Petrovskii and Morozov 2009), that movement occurs by different vectors (Higgins *et al.* 2003), or that it is variable between habitat types with rules regulating behaviour at borders between habitat types (e.g.

Zheng *et al.* 2009). Diffusion theory may have its limits when considering the movement of (smart) individuals in complex landscapes, and may be an inadequate framework when considering the evolution of movement strategies. Space is too limited to go further into the details of diffusion theory: we recommend, for example, Turchin (1998) or Levin *et al.* (2003) for further study.

One of the fields in which studying the mechanistic basis of individuals' dispersal kernels has advanced especially is dispersal of seeds by wind (see Levin *et al.* 2003). First attempts to model dispersal distance in dependence of seed-specific terminal velocity and wind speed date back 40 years. Since then much effort has been invested into refinement of such models with a strong focus on documenting the influence of a plant's surrounding, i.e. the height and density of the vegetation, on the wind profile and turbulences (e.g. Tackenberg 2003; Chapters 15, 19), and thus on the shape of an emergent dispersal kernel. Nathan *et al.* (2002) showed that a small fraction of seeds becomes airborne above the crown layer and is dispersed over much larger distances than those seeds that remain under the crown layer where wind velocities are much lower. The effect is enhanced by existing autocorrelation in the direction of airflows that may transport some seeds to considerable heights. This can lead to distinctly bimodal dispersal kernels for individual trees with a small fraction of seeds dispersed over much larger distances than the mean dispersal distance. Other studies have shown that plants may control under which wind conditions seeds are released (Soons and Bullock 2008); releasing seeds only at higher wind speed may lead to more long-distance dispersal events but possibly at the cost that more seeds will not be removed at all (Horn *et al.* 2001). We refer to Chapter 19 in this book for further details.

Another area where models have proved useful is the dispersal of seeds by animals. Whenever animals pick up seeds, the movement of the animals and the retention time of seeds, i.e. the distribution of the time seeds remain associated with their animal dispersal vector, will define the resulting seed dispersal pattern (e.g. Wehncke *et al.* 2003). This holds for seeds ingested or seeds carried externally. If animals, for example, move according to a random walk they will at any time $t > 0$ be distributed according to a Gaussian density pdf. However, seed-retention time itself is distributed according to another probability distribution. In the case of seeds attached to fur, this could be a simple negative exponential distribution (cf. Römermann *et al.* 2005) or a 'stretched exponential' where the detachment rate of seeds declines over time (Bullock *et al.* 2011). For gut passage, the distribution is typically unimodal (Cousens *et al.* 2008; Vellend *et al.* 2003). The time seeds remain in or on the animal vector depends on attributes of both animals and seeds (e.g. Levey and Del Rio 2001; Römermann *et al.* 2005)—selection could thus influence the seed attributes to modify the resulting kernel. The two distributions need to be combined to generate the kernel for the distribution of seeds (Cousens *et al.* 2008 p. 60; Levin *et al.* 2003). With a simple random walk (diffusion) model for the animal, and a negative exponential distribution for retention times, an analytical tractable solution for the resulting kernel does exist (Levin *et al.* 2003, p. 584).

Analytical approaches may fail or become too complex, however, for animal vectors that move within a home range, within a heterogeneous landscapes, in groups, or by directed movement towards favourable sites (Spiegel and Nathan 2010) and equally if the distribution of retention times does not follow a simple distribution. Matters may further be complicated if seeds are picked up by secondary dispersal agents, which requires the operation of kernel convolution to generate the overall dispersal kernel (e.g. Bullock *et al.* 2006). It should also be noted that the seeds of most plant species are not dispersed by one animal species only. The 'total dispersal kernel' may thus consist of a mixture of many different kernels provided by animals with very different movement rules, e.g. by small birds and elephants, where different vectors play very different roles (e.g. Clark *et al.* 2005; Spiegel and Nathan 2007). Indeed, 'smart' animals have considerable perceptual abilities allowing collection of information and corresponding adjustment of movement (Getz and Saltz 2008). They possess not only inherited but also culturally and personally acquired experience, and possibly can—especially when residing in a home range—build cognitive maps of their environment (Sutherland and Gass 1995). At the extreme, we could imagine individuals with accurate cognitive maps of their home range that choose least-cost paths (in terms of risk, energy) connecting desired locations (e.g. Janmaat *et al.* 2006). Movement of such organisms might better be described as 'traplining' resources (e.g. Ohashi and Thomson 2005) or as a solution to a travelling salesman problem (e.g. Noser and Byrne 2010)—and so would be the movement of e.g. seeds transported by them (Russo *et al.* 2006).

In such circumstances, we may need to turn to (individual-based) simulations to generate the dispersal kernel provided by animal vectors (Russo *et al.* 2006; Will and Tackenberg 2008). Simulations are increasingly utilized to generate dispersal kernels based on underlying (complex) movement rules, and can be powerful and flexible predictive tools. The flexibility may, however, also be a disadvantage of the approach as results often apply to very specific circumstances and do not necessarily lead to conclusions of general insight and applicability. Nonetheless, simulations are especially useful to account for the interaction between movement (morphological, physiological, behavioural) and landscape attributes—when matters become truly complex, they may indeed be the only approach after analytical or numerical approaches fail. Below, we will present two examples showing how dramatic this interaction between landscape and rules may become, and that standard kernels may not adequately describe kernels emerging from such interactions.

In the first example we simulate the dispersal of propagules, e.g. 'tumble seeds' blown over the surface by wind. We assume that the movement of individual propagules is linear with constant speed and that wind blows from a constant direction. Movement is implemented as a completely correlated walk composed of small movement steps but no change in the movement direction; the latter is chosen randomly at initialization. The landscape is represented as a grid of cells that are either suitable or non-suitable habitat. Suitable habitat is either distributed randomly (Figure 16.3a) or as a fractal (Figure 16.3d). Dispersal kernels are generated by simulating the fate of 10 000 seeds released at random locations within suitable habitat.

We apply two different movement rules for the dispersal of seeds. (1) Seeds only settle in suitable habitat where movement is terminated with a constant stepwise probability. In a completely homogeneous landscape with 100% habitat, this set of rules would clearly result in a negative exponential distance pdf. (2) Alternatively, we assume that non-suitable habitat (white areas in the panels of Figure 16.3) constitutes a barrier for movement—movement is terminated as soon as the moving seed encounters a border between habitat and non-habitat.

Figure 16.3 demonstrates that the property of a negative exponential kernel is maintained in the random landscape (b and c) but that the two different movement rules would lead to vastly different mean dispersal distances. In the fractal landscape, however, the shape of the dispersal kernel clearly deviates from the negative exponential, the interaction between rule and landscape is obvious, and the probability for long-distance dispersal is hugely increased in the fractal compared to the random landscape.

This is a simple example that could, however, easily be modified for greater realism. Movement direction could be less correlated as wind is not blowing constantly from the same direction. Settlement probability could be adjusted to decrease or increase over time or to account for variable 'stickiness' of different habitat types. Also, landscapes could be modified to include more than two types of habitat, to include specific obstacles, or include mortality traps. The example, however, clearly demonstrates that landscape can have a considerable influence on emerging dispersal kernel, but equally that for a given landscape, different movement rules lead to very different results.

In this example, we have generated a distance pdf for a whole population. Such a kernel is possibly useful for predicting population level consequences but it would not be the primary focus of selection. This requires identification of the interaction between the specific conditions under which movement is initiated and the rules applied by an individual.

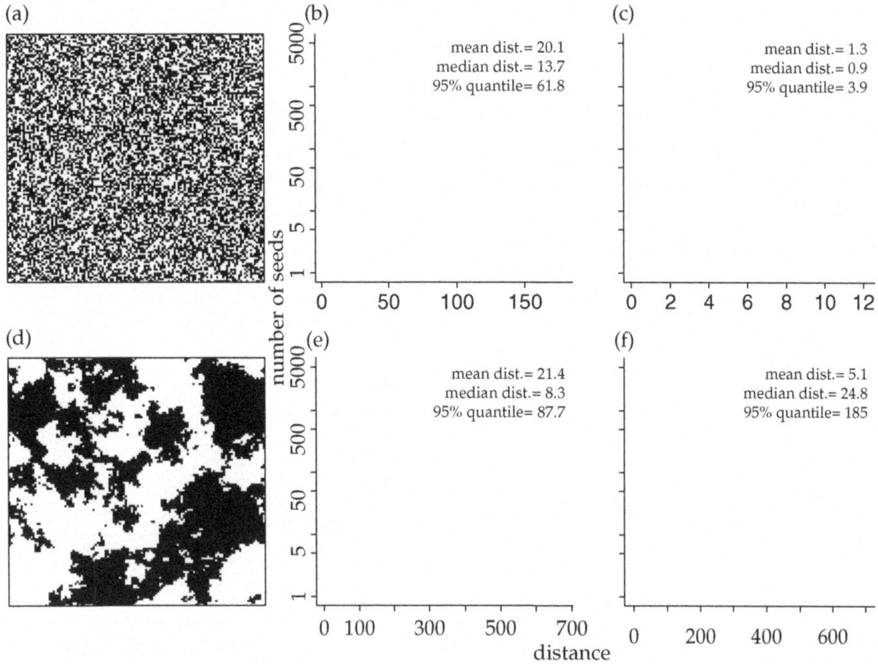

Figure 16.3 Effect of two landscapes and two movement rules on emergent population level non-normalized distance pdfs. To generate these kernels we simulated the movement of 10 000 virtual tumble seeds blown over the surface by wind. We assume that seeds are randomly released within suitable habitat (black; habitat fraction is 0.5) and move on a linear trajectory. In (b) and (c), habitat is distributed randomly (a); in (e) and (f), the landscape is fractal (d); with Hurst index $H = 0.5$). In (b) and (e) seeds settle with constant stepwise probability (0.1 per distance unit), but only in suitable habitat. In (c) and (f), we instead assume that a seed stops movement as soon as it hits non-suitable habitat; i.e. the latter constitutes an obstacle for movement. In random landscapes, both movement rules lead to a distance pdf that can be approximated by a negative exponential (linear decline in semi-logarithmic scale). Note, however, that mean dispersal distance would be much shorter with the second movement rule (stopping when encountering non-habitat); also note different scales for x-axes.

We explore this with another example. We simulate the egg-deposition behaviour of, for example, butterflies, which search for suitable host plants; their detection radius is small compared to the typical distance between such plants. We assume that an insect only deposits single eggs, e.g. to avoid kin competition, and needs a certain recovery time before it can lay the next egg. Movement is implemented as a correlated random walk with a certain stepwise mortality risk. More specifically, we compare three different movement rules: (1) a weakly correlated random walk where directional persistence from step to step is low, (2) a more strongly correlated random walk, and (3) a correlated walk where correlation increases the longer the time past since the last encounter with a host plant. This 'area-restricted' search is a simple approximation of a Bayesian forager, as the time since the last encounter is indicative of the plant density in the area in which the insect is searching (van Gils 2010). Host plants are assumed to be either randomly distributed (Figure 16.4a) or to occur in clusters (Figure 16.4d). The simulation is continuous in space. We record the distance between point of release for the insect and the location where eggs are deposited. Again, the resulting population level kernels become 'distorted' in the clustered landscape (Figure 16.4e) and cannot readily be described by any of the mathematical standard kernels. Bimodality (if not trimodality) emerges with many eggs deposited in the natal patch, but some in more distant patches; we stress, however, that the concept of habitat patches and emigration from such patches are not part of the movement rules. More importantly, in Figure 16.4f we show that the exact location where the insect starts its journey has a considerable effect on the resulting (expected) individual dispersal kernel.

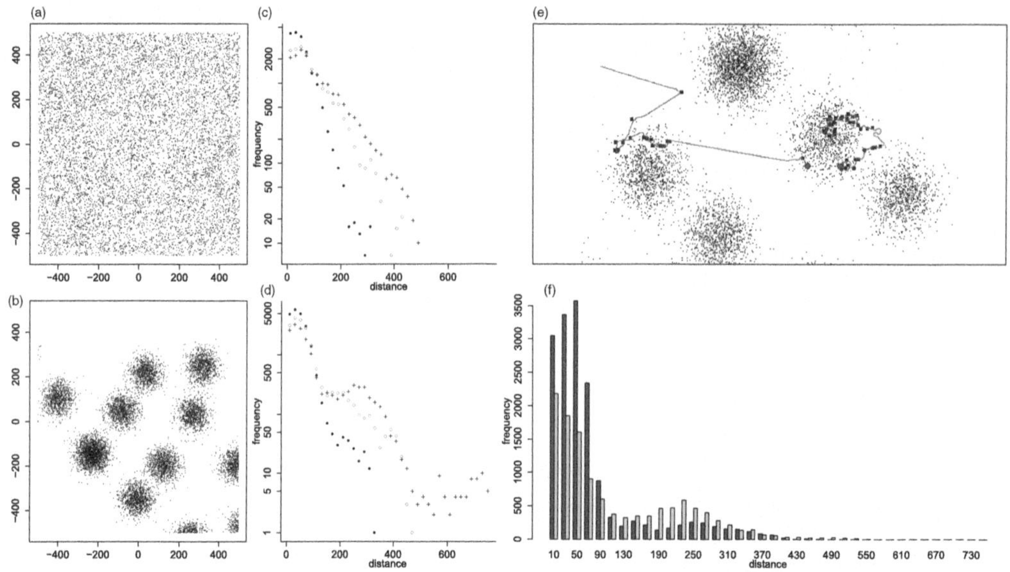

Figure 16.4 Emergence of distance frequency distributions for egg dispersal distances (distance from mother's natal site) generated by virtual insects depositing single eggs on suitable host plants. Hosts are either distributed randomly (a), or in clusters (d). Insects move according to three different rules: (i) a correlated random walk (CRW) with weak (black dots) or stronger correlation (diamonds; right trajectory in (c), or a 'area-restricted search' (e.g. Dias *et al.* 2009) where correlation increases as time since the last encounter with a host plant increases (crosses; left trajectory in (c). In all scenarios insects perish with a constant stepwise probability. Panels (b) and (e) show resulting distance pdf for a population of insects released randomly in the central section in (a) respectively in the central cluster (grey circle in (d)) of the landscape. Panel (f) shows the expected distance pdf for an individual either starting on a specific central (light-grey bars) or a specific peripheral plant (dark grey) of the central cluster in (d). All data are based on 500 replicate insects released, for generating kernels shown in (f), movement started always at the identical position.

It is thus inadequate to equate the population level kernel for a given set of movement rules with the expected kernel for an individual starting at a specific location, nor can we base estimates of (inclusive) fitness consequences for a certain set of movement rules (a strategy) directly on the population-level kernel. Instead, for a certain strategy, we would need to simulate kernels for each possible starting condition (at least for a number of reasonable categories of starting conditions), estimate the fitness consequences of each of these kernels, and then add these fitness values weighted according to the probability that an individual (with that strategy) will find itself in the corresponding starting condition. Alternatively, we can utilize evolutionary simulations where movement rules and attributes may evolve to find set(s) of rules that achieve the highest fitness in certain landscapes.

In a recent step forward in model development, Travis *et al.* (2010) have linked selection on dispersal distance with that of the mechanisms of dispersal itself. They simulate selection on tree height that influences the travel distance of seeds released into the wind—yet in the model, increasing tree height is associated with a trade-off in fecundity. Travis *et al.* (2010) then explore how landscape attributes affect selection on tree height. We think that this is a promising new direction to link evolutionary models with mechanistic models.

With this example, we leave the brief excursion into simulation of mechanisms generating dispersal kernels and the methods we can use to model them. More examples and a more thorough overview can be found in the recent book by Cousens *et al.* (2008) and Chapters 15, 17, 18, and 19.

16.4 Conclusions

We have seen great advances in both the ability to collect empirical data on the mechanisms of dispersal and in developing mechanistically based models that allow the generation of more accurate predictions about dispersal. Recent advances in 'kernel research' may be summarized as progress from a phenomenological (statistical) description of dispersal kernels to a mechanistic understanding of their genesis. Development could also be described as a shift from a Eulerian perspective, that focuses on the (statistical description of the) spatial distribution of the endpoints of dispersal to a Lagrangian approach focusing on the mechanisms and rules underlying individual dispersal events (Nathan *et al.* 2008; Turchin 1998). Despite the welcome shift towards a Lagrangian approach, a Eulerian perspective is nonetheless needed to evaluate the long-term fitness consequences of certain dispersal strategies, as it is the whole dispersal kernel that is the unit of selection.

A unifying result from studying and modelling the rules of movement is that emerging dispersal kernels are typically not species-specific constants. They are strongly affected by the interaction between movement rules—that themselves underlie landscape specific selection—and the environment. Thus the 'average kernel' may change considerably when environmental conditions change.

Studying dispersal in plants and animals will continue to benefit from this more mechanistic approach. Our understanding will broaden if not only ecologists and evolutionary biologists participate in the endeavour, but also sensory physiologists, animal psychologists, or physicists, to name just a few. In parallel, models that are based on mechanisms beyond simple diffusion or random walks will gradually augment or replace models that just take a phenomenological approach. Future models should also become more explicit about the possible trade-offs associated with certain dispersal strategies, and separating investment costs from costs associated with the very process of dispersal itself (e.g. Fronhofer *et al.* 2011).

Probably the most important danger to avoid in this process is getting lost in an infinite universe of particular models. Ideally, we will be able to identify a number of important and unifying principles governing the rules of movement and dispersal across the diversity of species. If so we may at some time be able to compile a tool-box of movement rules, from which at least reasonably accurate dispersal models can be built for most systems.

References

Bonte, D., Hovestadt, T., Poethke H.J. (2010). 'Evolution of dispersal polymorphism and local adaptation of dispersal distance in spatially structured landscapes', *Oikos* **119**: 560–6.

Bonte, D., Van den Borre, J., Lens L. *et al.* (2006). 'Geographical variation in wolf spider dispersal behaviour is related to landscape structure', *Animal Behaviour* **72**: 655–62.

Bowler, D.E., Benton, T.G. (2005). 'Causes and consequences of animal dispersal strategies: relating individual behaviour to spatial dynamics', *Biological Reviews* **80**: 205–25.

Bullock J.M., Galsworthy S.J., Manzano P. *et al.* (2011). 'Process-based functions for seed retention on animals: a test of improved descriptions of dispersal using multiple data sets', *Oikos* **120**: 1201–08.

Bullock, J.M., Shea, K., Skarpaas, O. (2006). 'Measuring plant dispersal: an introduction to field methods and experimental design', *Plant Ecology* **186**: 217–34.

Cain, M.L., Milligan, B.G., Strand, A.E. (2000). 'Long-distance seed dispersal in plant populations', *American Journal of Botany* **87**: 1217–27.

Clark, C.J., Poulsen, J.R., Bolker, B.M. *et al.* (2005). 'Comparative seed shadows of bird, monkey, and wind dispersed trees', *Ecology* **86**: 2684–94.

Cousens, R., Dytham, C., Law, R. (2008). *Dispersal in plants. A population perspective.* Oxford University Press, Oxford.

de Wit, R., Bouvier, T. (2006). 'Everything is everywhere, but, the environment selects'; what did Baas Becking and Beijerinck really say?', *Environmental Microbiology* **8**: 755–8.

Dias, M.P., Granadeiro J.P., Palmeirim, J.M. (2009). 'Searching behaviour of foraging waders: does feeding success influence their walking?', *Animal Behaviour* **77**: 1203–9.

Dytham, C. (2009). 'Evolved dispersal strategies at range margins', *Proceedings of the Royal Society of London Series B-Biological Sciences* **276**: 1407–13.

Dytham, C., Travis J.M.J. (2006). 'Evolving dispersal and age at death', *Oikos* **113**: 530–8.

Fronhofer E., Kubisch A., Hovestadt T. & Poethke H.J. (2011). Assortative mating counteracts the evolution of dispersal polymorphisms. *Evolution* **65**: 2461–2469.

Gaston, K.J. (2009). 'Geographic range limits of species', *Proceedings of the Royal Society of London Series B-Biological Sciences* **276**: 1391–3.

Getz, W.M., Saltz D. (2008). A framework for generating and analyzing movement paths on ecological landscapes. *Proceedings of the National Academy of Sciences*, **105**, 19066–71.

Gros, A., Poethke, H.J., Hovestadt, T. (2006). 'The evolution of spatial patterns in dispersal strategies', *Oikos* **114**: 544–52.

Hamilton, W.D., May, R.M. (1977). 'Dispersal in stable habitats', *Nature* **269**: 578–81.

Higgins, S.I., Nathan, R., Cain, M.L. (2003). 'Are long-distance dispersal events in plants usually caused by non-standard means of dispersal?', *Ecology* **84**: 1945–56.

Horn, H.S., Nathan, R., Kaplan, S.R. (2001). 'Long-distance dispersal of tree seeds by wind', *Ecological Research* **16**: 877–85.

Hovestadt, T., Messner, S., Poethke, H.J. (2001). 'Evolution of reduced dispersal mortality and 'fat-tailed' dispersal kernels in auto-correlated landscapes', *Proceedings of the Royal Society B* **268**: 385–91.

Janmaat, K.R.L., Byrne, R.W., Zuberbuhler, K. (2006). 'Evidence for a spatial memory of fruiting states of rainforest trees in wild mangabeys', *Animal Behaviour* **72**: 797–807.

Kareiva, P.M., Shigesada, N. (1983). 'Analyzing insect movement as a correlated random walk', *Oecologia* **56**: 234–8.

Kot, M., Lewis, M.A., van den Driessche, P. (1996). 'Dispersal data and the spread of invading organisms', *Ecology* **77**: 2027–42.

Levey, D.J., Del Rio, C.M. (2001). 'It takes guts (and more) to eat fruit: Lessons from avian nutritional ecology', *Auk* **118**: 819–31.

Levin, S.A., Muller-Landau, H.C., Nathan R. *et al.* (2003). 'The ecology and evolution of seed dispersal: A theoretical perspective', *Annual Review of Ecology Evolution and Systematics* **34**: 575–604.

Muller-Landau, H.C., Levin, S.A., Keymer, J.E. (2003). 'Theoretical perspectives on evolution of long-distance dispersal and the example of specialized pests', *Ecology* **84**: 1957–67.

Murrell, D.J., Travis J.M.J., Dytham, C. (2002). 'The evolution of dispersal distance in spatially-structured populations', *Oikos* **97**: 229–36.

Nathan, R., Getz, W.M., Revilla, E. *et al.* (2008). 'A movement ecology paradigm for unifying organismal movement research', *Proceedings of the National Academy of Sciences of the USA* **105**: 19052–9.

Nathan, R., Katul, G.G., Horn H.S. *et al.* (2002). 'Mechanisms of long-distance dispersal of seeds by wind', *Nature* **418**: 409–13.

Noser, R., Byrne, R.W. (2010). 'How do wild baboons (Papio ursinus) plan their routes? Travel among multiple high-quality food sources with inter-group competition', *Animal Cognition* **13**: 145–55.

Ohashi, K., Thomson, J.D. (2005). 'Efficient harvesting of renewing resources', *Behavioral Ecology* **16**: 592–605.

Petrovskii, S., Morozov, A. (2009). 'Dispersal in a Statistically Structured Population: Fat Tails Revisited', *American Naturalist* **173**: 278–89.

Römermann, C., Tackenberg, O., Poschlod, P. (2005). 'How to predict attachment potential of seeds to sheep and cattle coat from simple morphological seed traits', *Oikos* **110**: 219–30.

Ronce, O. (2007). 'How does it feel to be like a rolling stone? Ten questions about dispersal evolution', *Annual Review of Ecology, Evolution, and Systematics* **38**: 231–53.

Rousset, F., Gandon, S. (2002). 'Evolution of the distribution of dispersal distance under distance-dependent cost of dispersal', *Journal of Evolutionary Biology* **15**: 515–23.

Russo, S.E., Portnoy, S. Augspurger, C.K. (2006). 'Incorporating animal behavior into seed dispersal models: Implications for seed shadows', *Ecology* **87**: 3160–74.

Schick, R.S., Loarie, S.R., Colchero F. *et al.* (2008). 'Understanding movement data and movement processes: current and emerging directions', *Ecology Letters* **11**: 1338–50.

Soons, M.B., Bullock, J.M. (2008). 'Non-random seed abscission, long-distance wind dispersal and plant migration rates', *Journal of Ecology* **96**: 581–90.

Spiegel, O., Nathan, R. (2007). 'Incorporating dispersal distance into the disperser effectiveness framework: frugivorous birds provide complementary dispersal to plants in a patchy environment', *Ecology Letters* **10**: 718–28.

Spiegel, O., Nathan, R. (2010). 'Incorporating density dependence into the directed-dispersal hypothesis', *Ecology* **91**: 1538–48.

Starrfelt, J., Kokko, H. (2010). 'Parent-offspring conflict and the evolution of dispersal distance', *American Naturalist* **175**: 38–49.

Sutherland, G.D., Gass, C.L. (1995). 'Learning and Remembering of Spatial Patterns by Hummingbirds', *Animal Behaviour* **50**: 1273–86.

Tackenberg, O. (2003). 'Modelling long-distance dispersal of plant diaspores by wind', *Ecological Monographs* **73**: 173–89.

Travis, J.M.J., Smith H.S., Ranwala, S.M.W. (2010). 'Towards a mechanistic understanding of dispersal evolution in plants: conservation implications', *Diversity and Distributions* **16**: 690–702.

Turchin, P. (1998). *Quantitative analysis of movement. Measuring and modelling population redistribution in animals and plants*. Sinauer Associates, Sunderland, MA.

van Gils, J.A. (2010). 'State-dependent Bayesian foraging on spatially autocorrelated food distributions', *Oikos* **119**: 237–44.

Vellend, M., Myers, J.A., Gardescu, S. *et al.* (2003). 'Dispersal of *Trillium* seeds by deer: Implications for long-distance migration of forest herbs', *Ecology* **84**: 1067–72.

Wehncke, E.V., Hubbell S.P., Foster R.B. *et al.* (2003). 'Seed dispersal patterns produced by white-faced monkeys: implications for the dispersal limitation of neotropical tree species', *Journal of Ecology* **91**: 677–85.

Will, H., Tackenberg, O. (2008). 'A mechanistic simulation model of seed dispersal by animals', *Journal of Ecology* **96**: 1011–22.

Zheng, C.Z., Pennanen, J., Ovaskainen, O. (2009). 'Modelling dispersal with diffusion and habitat selection: Analytical results for highly fragmented landscapes', *Ecological Modelling* **220**: 1495–505.

CHAPTER 17

Quantifying individual differences in dispersal using net squared displacement

Luca Börger and John Fryxell

17.1 Introduction

Quantifying individual differences and the phenotypic correlates of dispersal are of considerable interest for obtaining a better understanding of the mechanisms of dispersal (Bowler and Benton 2005; Clobert *et al.* 2009; Morales *et al.* 2010). The aim of this chapter is to present a new approach for modelling animal dispersal, based on net squared displacement statistics combined with a non-linear hierarchical modelling framework. It allows efficient construction of accurate population redistribution kernels, quantification of individual differences in dispersal, and testing hypothesized correlates of the latter (Börger *et al.* unpublished manuscript).

The chapter is organized as follows. Firstly, we explore the theoretical basis for using net squared displacement as a statistical modelling approach. This is followed by a simulation study to investigate the data requirements and power of the proposed method, ending with general conclusions.

17.2 Theoretical considerations

17.2.1 Net squared displacement as a synthetic measure of animal movement rate

Consider an organism, or in general any object, moving for a certain interval of time t. The straight-line, or Euclidian, distance from the start to the end point is called the net displacement and the square of this value is the net squared displacement (NSD; Turchin 1998). The NSD is a single time-dependent distance statistic which is fundamental for quantifying movements of organisms or particles (Codling *et al.* 2008; Nouvellet *et al.* 2009; Skellam 1951; Turchin 1998), and in general random processes (Uhlenbeck and Ornstein 1930).

The mean (i.e. expected) value of the NSD over time is called the mean net squared displacement (MSD). MSD estimates can be obtained as the average over multiple recording sessions, or as the average over a given time unit for single-trajectory data (Elliott *et al.* 2011; Turchin 1998). In the simplest case of a freely moving Brownian particle over longer timescales (the so-called diffusive regime) the MSD will increase linearly with time (t):

$$MSD = 4Dt \qquad \text{(Equation 17.1)}$$

where D is the diffusion constant (multiplied by four for two-dimensional movements, by two for movements in one dimension) and t denotes the time since start, whereas over very short timescales the value increases as a quadratic function of time due to inertial forces, called ballistic motion (Einstein 1905; Langevin 1908; Li *et al.* 2010; Moorcroft and Lewis 2006; Turchin 1998; Uhlenbeck and Ornstein 1930). In general, the form of the MSD over time may deviate from a linear pattern due to movement constraints or external forces; thus a more general equation for the MSD over time is the power-law formulation (Codling *et al.* 2008; Metzler and Klafter 2000; Turchin 1998):

$$MSD = Dt^{\alpha} \qquad \text{(Equation 17.2)}$$

When $\alpha > 1$ the motion is called super-diffusive, sub-diffusive if $\alpha < 1$, and normal diffusion for $\alpha = 1$.

Dispersal Ecology and Evolution. First Edition. Edited by Jean Clobert *et al.*

The inferior limit to sub-diffusion occurs when movements are constrained by an external boundary or when animals remain always in the same area, called the home range (reviewed in Börger *et al.* 2008). In this case, it can be shown (Turchin 1998) that the MSD initially will increase linearly over time as for diffusion (Equation 17.1), but the rate of increase will decrease over time and reach a steady-state constant value:

$$MSD = \frac{3D^2}{c^2} \qquad \text{(Equation 17.3)}$$

where c is an advection coefficient quantifying the strength of the tendency to return to the home range centre (the localizing tendency, see Moorcroft and Lewis 2006; Chapter 10).

Given these results, we can also predict the broad-scale pattern of the MSD over time for dispersal (see Fryxell *et al.* 2008 for a data-driven approach using GAM models). Specifically, dispersal is a three-stage multi-phase process (*sensu* Fryxell *et al.* 2008; Nathan *et al.* 2008); i.e. two stationary space use phases (the departure and settlement phases) separated by an exploratory transience phase (Bowler and Benton 2005; Clobert *et al.* 2004; Van Dyck and Baguette 2005). Thus, the MSD value will be close to zero during the pre-dispersal (or departure) phase, increase linearly over time during the transience phase, and converge to a stationary value at the settlement phase, hence leading to a sigmoid shape for the expected MSD. Note that we consider here only large-scale patterns—for example, short-scale exploratory movements are often observed during the departure or the transience phases (Clobert *et al.* 2009), yet are considered here as temporary deviations from the long-term pattern; see Turchin (1998) for the importance of focusing on the timescale of interest, whilst ignoring shorter time patterns. This approach can be extended to predict the MSD patterns of migratory animals (Kölzsch and Blasius 2008) by considering migration as a combination of two dispersal events, where the second 'dispersal' event returns the animal to the point of origin (and thus the MSD decreases again to zero, which leads to an expected double-sigmoid shape), as we consider in detail elsewhere (Börger *et al.* 2011; Bunnefeld *et al.* 2011).

The NSD is thus a single time-dependent movement statistic which encapsulates key statistical properties of movement processes, allowing to link empirical measures directly with theoretical expectations of its mean (expected) value (the MSD). It circumvents several of the difficulties of traditional random walk analysis based on modelling distributions of step length and turning angle (Nouvellet *et al.* 2009). Importantly, it does not rely on equally spaced, regularly sampled locations, as displacement is modelled as a continuous function of time. Thus, NSD constitutes an ideal probe for assessing the support from the data for alternative movement models (Moorcroft and Lewis 2006; Nouvellet *et al.* 2009; Turchin 1998). In the following section we describe an efficient inferential approach to link models to data, based on non-linear models.

17.2.2 Capturing the functional form of displacement patterns with non-linear models

Animal movements are best understood as lying on a continuum between the extremes of sedentarism and nomadism (Börger *et al.* 2011). In the previous section, we showed that the NSD efficiently captures key statistical properties of movement processes and that the expected value of the displacement over time (MSD) can be derived from theory for different movement processes (e.g. Equations 17.1–17.3). Here we show how a system of non-linear models can be used to link the theoretical expectations to movement data.

A key characteristic of nomadic animals is that they never remain in the same area for a long time. Over larger scales this will lead to a diffusive space use pattern, with a linear increase of the MSD over time (Equation 17.1) or, more generally, a power model (Equation 17.2). Obviously, at even larger scales, the movements may deviate from a linear pattern due to, for example, regional constraints to movements, but we do not consider such complications here. At the other extreme, for sedentary animals restricting their movements within a home range area the MSD value will initially increase linearly but eventually approach a steady-state value (Equation 17.3). Consequently, the change over time of the MSD can be modelled using a simple equation for asymptotic regression through the origin:

$$MSD = \phi_1[1 - \exp(\phi_2 t)] \qquad \text{(Equation 17.4)}$$

where Φ_1 is the asymptote at the steady-state equilibrium, and Φ_2 is the logarithm of the rate constant.

Finally, the expected form of the MSD will be sigmoid for dispersers (see previous section) and thus can be modelled using the equation for the logistic curve:

$$MSD = \frac{\phi_3}{1+\exp\left[\frac{(\phi_4 - t)}{\phi_5}\right]} \quad \text{(Equation 17.5)}$$

where Φ_3 is the asymptote at the steady-state equilibrium (settlement) and thus quantifies dispersal distance; Φ_4 is a parameter for the inflexion point, in units of time since start (t), and thus models the timing of the transience phase (note that MSD = $\Phi_3/2$ at time t = Φ_4; i.e. half the dispersal distance has been reached); Φ_5 is a scale parameter on the time axis governing the shape of the curve—given Φ_3 and Φ_4 it determines the time elapsed between reaching half and approximately 3/4 of the asymptote—and thus predicts how long dispersers stay in the transience phase. Specifically, 95% of the dispersal distance will be reached at the following settlement time:

$$t_{settle} = \phi_4 + 3\phi_5. \quad \text{(Equation 17.6)}$$

Thus, t_{settle} can be taken as a reasonable estimate of the start of the settlement phase (similarly for the start of the transience phase).

Having derived a set of non-linear MSD equations, maximum likelihood methods within a model selection framework can be used to identify the movement type and estimate model parameters from observational data (for an example applied to migration, see Bunnefeld *et al.* 2011; but see following). In order to quantify also variation between individuals, an efficient approach is to use non-linear mixed effects models, as we detail in the next section.

17.3 The modelling approach: hierarchical non-linear models

17.3.1 Non-linear mixed effects models

Non-linear mixed effects models, a form of hierarchical non-linear model, are an efficient approach to estimate parameters and make inferences on features underlying profiles of continuous, repeated measurements from a sample of individuals from a given population of interest (Davidian and Giltinan 2003; Pinheiro and Bates 2000). Of particular interest for observational data, such as most movement data, is that mixed effects models are robust to missing or unbalanced data, and that subject-specific estimates for individuals with limited data are improved by 'borrowing strength' from information available for other individuals (Pinheiro and Bates 2000).

Non-linear mixed effects models are an extension of linear mixed models and are usually viewed as two-stage (Davidian and Giltinan 2003; Pinheiro and Bates 2000):

Stage 1: Individual-level model.

$$y_{ij} = f(\Phi_{ij}, x_{ij}) + \varepsilon_{ij}, i = 1,\ldots,M, j = 1,\ldots,n_i \quad \text{(Equation 17.7)}$$

where f is a non-linear function of an individual-specific parameter vector Φ_{ij} and of the covariate vector $\mathbf{x}_{ij}$, ε_{ij} is a normally distributed noise term, Equations 17.7 and 17.8: M is the number of individuals, and n_i is the number of records for the ith individual. Note that the noise term may include the effects of measurement error as well as of deviations caused by other, short-scale, movement processes. Specifically, the function f is non-linear in one or more components of the individual-specific parameter vector Φ_{ij}, where the latter is modelled in the second stage as:

Stage 2: Population model.

$$\Phi_{ij} = A_{ij}\beta + B_{ij}b_i, b_i \sim N(0, \sigma^2 D) \quad \text{(Equation 17.8)}$$

where β is a vector of fixed population parameters (*fixed effects*), $\mathbf{b}_i$ is a *random effects* vector associated with the ith individual (and invariant with j), $\mathbf{A}_{ij}$ and $\mathbf{B}_{ij}$ are design matrices for the fixed and random effects, and $\sigma^2\mathbf{D}$ is a variance-covariance matrix ($\mathbf{D}$ characterizes the magnitude of *natural* inter-individual variation; i.e. variation in the elements of Φ_{ij} not accounted for by systematic association with the individual attributes in β, the fixed effects vector). Both the matrices $\mathbf{A}_{ij}$ and $\mathbf{B}_{ij}$ depend on the individual, and also on the values of some covariates at the jth record (thus also time-varying covariates can be included in the fixed and random effects). Observations on different individuals are assumed to be independent and the residuals are assumed to be normally distributed and independent of the random effects (but these assumptions can be relaxed, see Pinheiro and Bates 2000; Chapter 8). Two types of predictions are obtained: population predictions, corresponding to random effects equal to zero, and individual-specific predictions, obtained by adding the estimated random effects (which model subject-specific deviations from the population patterns).

Two important issues to consider are determining which effects should be modelled with an associated random component and which should be entirely a fixed effect, and how to use covariates to explain inter-individual parameter variability (Pinheiro and Bates 2000). Regarding the random effects, the preferred approach is to start with all parameters as mixed effects (i.e. allowing for individual variation for all model parameters), then using AIC to compare alternative models. This may not always be possible, due to non-convergence of the full model, or one of the random effects may be very small compared to the others. In such cases, a common approximation is to choose a model without an associated random effect for the component of concern, in order to achieve parsimony and numerical stability in fitting (Davidian and Giltinan 2003). For example, we found that in many instances the dispersal model (Equation 17.5) may be best fitted without including a random effect for the scale parameter (ϕ_5), in order to avoid overfitting. Once the most appropriate random effect structure has been determined, fixed effects covariates are fitted as linear models of the model parameters.

Model parameters can be estimated using several methods (Pinheiro and Bates 2000). We used the likelihood-based methods implemented in the nlme package (Pinheiro *et al.* 2011) for R (R Development Core Team 2011, version 2.12.2), based on maximum log-likelihood.

Given the availability of theoretical models for the expected NSD relationships (Section 17.2), and an efficient modelling approach to estimate time-related variation for each individual, an important consideration can be raised. Traditional approaches focus on MSD patterns, i.e. values averaged over time or between repeated recording sessions in order to obtain the broad-scale patterns, while averaging out shorter-scale variation due to other sources. For many animal movement studies, however, sufficient data to calculate MSD values are not available, due to logistical constraints. Yet once a model for a theoretical expectation is available, non-linear mixed models offer an efficient approach to identifying the relevant signal even with highly noisy and unbalanced data. This suggests that similar inferences may be obtained by applying the approach to the observed, non-averaged NSD data directly derived from each individual trajectory. In Section 17.4 we investigate this suggestion using a simulation study.

17.3.2 The model selection approach

Given the set of *a priori* models derived from theory linking different movement behaviours to the expected squared displacement patterns (Section 17.2) it would appear that information-theoretic criteria, such as AIC (Burnham and Anderson 2002), could be used to identify the most likely movement type used by the animals under study. However, given that non-linear mixed models make predictions at the population and individual levels, the situation is considerably more complex, and global measures of goodness-of-fit (GOF) at the population level, as AIC or also R^2, may be insufficient or even grossly misleading in determining the GOF of fitted models (Huang *et al.* 2009). For example, often a sample of individual location data may contain both sedentary and dispersing individuals and thus it is crucial to assess GOF for individual subjects. Huang *et al.* (2009) developed formulae for non-linear mixed models, highlighting especially the advantages of using the concordance correlation (CC) coefficient calculated at the subject-specific level (see equation 7 in Huang *et al.* 2009, p.2421):

$$CC_i = 1 - \frac{\sum_{j=1}^{n_i}\left(y_{ij} - \hat{y}_{ij}\right)^2}{\sum_{j=1}^{n_i}\left(y_{ij} - \bar{y}\right)^2 + \sum_{j=1}^{n_i}\left(\hat{y}_{ij} - \bar{\hat{y}}\right)^2 + n_i\left(\bar{y} - \bar{\hat{y}}\right)^2}.$$

(Equation 17.9)

where $\bar{y}$ and $\bar{\hat{y}}$ are the means of the observed (y_{ij}) and predicted ($\hat{y}_{ij}$) values for individual *i* and n_i is the subject-specific number of locations. The concordance criterion measures the level of agreement between the observed and predicted values, and is a combined measure of the degree of accuracy and precision of pairs of values of y_i and $\hat{y}_i$ on the identity line (the concordance line or the 45° line through the origin) and any $CC \leq 0$ indicates lack of fit (Huang *et al.* 2009). To evaluate the GOF of the model, it is important to investigate individual-specific GOF measures and their frequency distribution (Huang *et al.* 2009). Using simulations, we evaluate the adequacy of the different approaches in Section 17.4.

17.3.3 Deriving dispersal kernels from net squared displacement models

The estimated NSD distances are a prediction of the average location of an individual over time. Thus, an efficient way to obtain an equation for dispersal kernels that takes into account the observed individual variability in movements is to estimate the distribution of parameter values for Equation 17.5 that captures the population distribution. Given that in most populations the distribution of dispersal distances is typically characterized by a long tail of large distance values, a robust approach is to calculate the quartiles of the predicted individual displacement distances for each time step and to fit separate non-linear models to each. The obtained parameter estimates are the values predicting where 50% and 100% of the population will be found at each time step.

17.4 Simulation study: evaluating data requirements and power

We used simulations to test whether individual-level dispersal parameters can be estimated using the proposed NSD approach, even in the presence of substantial individual variability. To do so, we simulated daily movements of 140 dispersers over three years, using the adehabitat library in R (Calenge 2006, version 1.8.4). Sedentary space-use patterns during the pre-dispersal and settlement phases were modelled using a Ornstein-Uhlenbeck process (Blackwell 1997), as implemented in the *sim.mou* function, which simulates the movements of an individual moving randomly but constantly attracted to a central location (Börger *et al.* 2008). Movements during the transience phase were simulated using a correlated random walk as implemented in the *sim.crw* function. The start of the transience phase was randomly selected for each individual between 30 and 515 days after start. The transience phase lasted between 30 and 90 days, selected randomly for each individual and independently from the timing of the transience phase. The last location of the transience phase became the attraction point of the settlement phase Ornstein-Uhlenbeck function and the distance of this location from the start point was taken as the true dispersal distance.

For each simulated movement path we calculated the NSD after start for each location. To investigate the sensitivity of our approach to different measures of MSD, we calculated MSDs over intervals of three, seven, 14, 30, and 90 days (with each value attributed to the mid-time point) and repeated all analyses for each. Similarly, to test if NSD values can be used directly (see Section 3.1) and evaluate the impact of sampling frequency, we also simulated five different sampling regimes (one location every three, seven, 14, 30, and 90 days). We fitted the dispersal model (Equation 17.5) and extracted the estimated individual-level dispersal parameters (using Equation 17.6 for obtaining the estimates of the end and start of dispersal). We repeated this procedure 100 times, for a total 14 000 simulated movement paths and 140 000 models fitted to the ten different MSD and NSD measures.

True simulated dispersal distances varied over two orders of magnitude and the samples contained both individuals dispersing over the first as well the second year after start. Results were clear (Figure 17.1)—the true values of distance and timing of dispersal were always recovered with high efficiency ($r > 0.95$), with no difference between using NSD or MSD values (e.g. for estimating dispersal distances, it made no difference to use one location every 90 days or one location every three days). Only estimates of the departure and settlement times become inaccurate when the sampling interval is coarser then the true process—see the estimates for the 30- and 90-days sampling regime in Figure 17.1. For

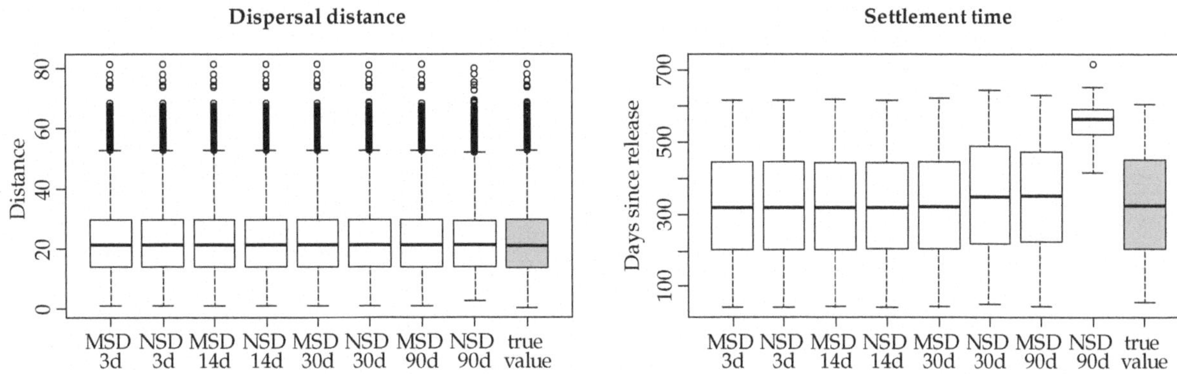

Figure 17.1 Distribution of estimated and true simulated individual-level dispersal parameters (100 simulations of a set of 140 movement paths). Models were fitted to simulated data using five different types of mean net squared displacement statistics (MSD; calculated by averaging the simulated daily net squared displacement values over five different time periods, ranging from three ('MSD3d)' to 90 days ('MSD90d') or non-averaged net squared displacement values (NSD) obtained from five different simulated sampling regimes (from one location every three days ('NSD3d') to one locations every 90 days ('NSD90d'). See text for further details.

similar problems associated with too infrequently sampled location data, see Mysterud *et al.* (2011); the impact of these problems will, however, be markedly reduced by using the hierarchical modelling approach (combined with the CC statistic) that we present here (see results following), as compared to the individual-fit approach we presented elsewhere for migration (Bunnefeld *et al.* 2011) and adopted by Mysterud *et al.* (2011).

In order to test the ability to identify the correct movement process, we fitted to the data also the models for sedentary and nomadic animals (Equations 17.1–17.4) and compared all models using both global measures of goodness-of-fit (AIC/BIC, R^2, and global CC statistics) as well as individual-level measures (individual-level CC_i). In all cases, the dispersal model unambiguously received the highest support, but only the individual-level CC statistics highlighted poor model fit ($CC <= 0$) for a small set of simulated data. An inspection of the associated movement paths indicated that the simulated movement paths were peculiar; e.g. the correlated random walk during the transience phase had initially moved far from the start location, but by chance had returned to the departure location before the start of the settlement phase.

To investigate the ability of the approach to separate dispersal from other large-scale movement patterns, we simulated datasets with individuals moving according to three different movement types: 120 were modelled as dispersers as described, ten always remained within a home range (modelled using an Ornstein-Uhlenbeck process), and ten were nomadic over the whole time period (modelled with a correlated random walk). We also evaluated an alternative model selection approach, by separately fitting the alternative models to each individual location set and evaluating the adequacy of the competing models using the CC, R2, and AIC weight statistics calculated for each individual movement path. We repeated this procedure 100 times. Results were clear—classification success using the hierarchical approach combined with the individual-level CC statistic was markedly higher compared to fitting alternative models separately to each single movement path (overall 81% correct space use modes identified; 88% for dispersers, with up to 100% accuracy for single simulations consisting of a set of 140 simulated movement paths).

We lastly evaluated the power of the approach to correctly estimate population-level differences such as sex differences: we simulated a sample of 140 individuals coming from two groups with different mean dispersal distances, timing of dispersal, or both, by modelling individual model parameters for the logistic curve (Equation 17.5) using normal distributions with different means. We simulated three scenarios, with groups differing in mean

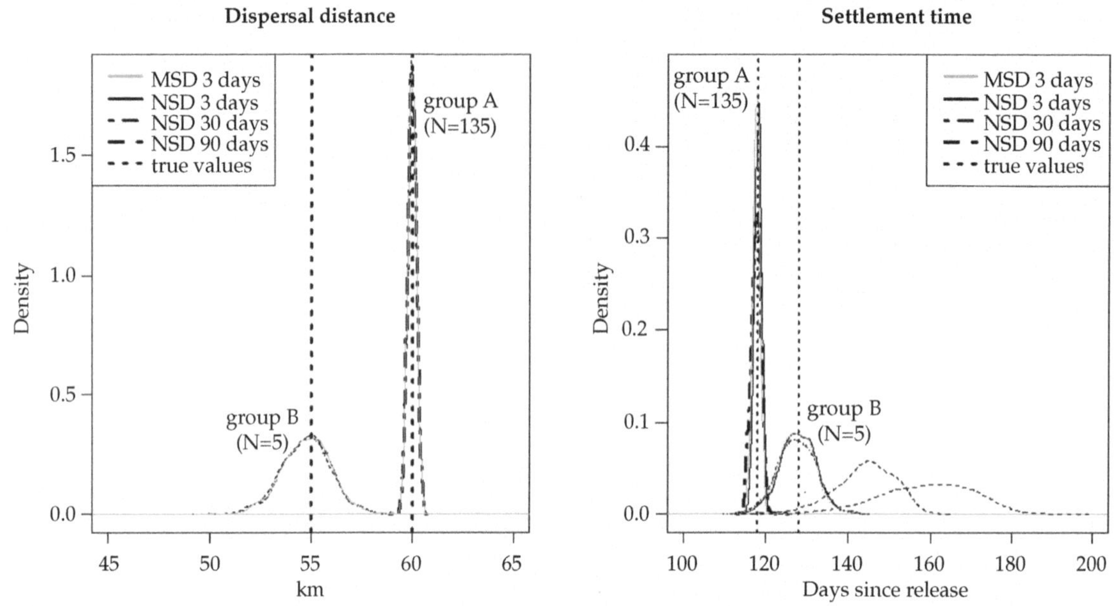

Figure 17.2 Distribution of estimated and true simulated population-level dispersal parameters (N = 1000 simulations of a set of 140 dispersers). The aim was to evaluate the power of the approach to estimate true dispersal parameters of two groups of individuals from a same population characterized by fixed mean differences. Results are shown for 1000 simulations of the case of unbalanced sample size (five versus 135 individuals), using four different squared displacement metrics. See text for further details.

dispersal distance (mean = 60 km versus 39 km or versus 55 km, variance = 300 km^2) or/and also timing of dispersal (mean = 100 versus 110 days after start, variance = 10, with 60 km versus 55 km mean dispersal distance, respectively). The scale parameter was always sampled from a normal distribution with mean = 6 and variance = 1. Each simulation consisted of 140 movement paths with equal or unequal sample size between the groups (in the latter case, 130 versus 10 or 135 versus five individuals). We used the individual model parameters to generate the true NSD curves over three years, to which we added two error sources, taken from normal distributions with mean zero and standard deviations of 14 km or 1.7 km, respectively, to simulate stochasticity caused by other movement types and by location error. We calculated NSD/MSD values with different sampling regimes, fitted the dispersal model and extracted the fixed effects population predictions and associated p-values. For each scenario we repeated the simulations 1000 times, for a total of 140 000 simulated NSD patterns.

Results were clear (Figure 17.2). The difference in true mean dispersal distance could be recovered with near certainty even in the most restrictive scenario (five versus 135 individuals) with no appreciable difference even when using only one location every 90 days (and $p < 0.05$ in 99% of the 1000 simulations). Similarly, differences in the timing of dispersal could be detected, except when using too sparsely sampled locations (one location every 30 and, especially, 90 days; Figure 17.2), in accordance with previous simulations. The statistical power, however, was lower, as a significant p-value was obtained only in 58% of the simulations (27% for the 90-days sampling regime). The power increased with better sample sizes between the two groups.

17.5 Conclusions

A problem highlighted by current research is to develop methods to account for the typically high degree of individual variability in dispersal behaviour, in order to develop dispersal kernels of improved predictive ability. We demonstrated that by modelling NSD patterns within a non-linear mixed effects modelling framework, highly accurate dispersal kernels can be estimated and determinants of individual differences can be tested within the same framework. By including a general theoretical framework for movement studies, we demonstrated how to objectively separate dispersers from non-dispersing individuals. The method works also on sparsely sampled movement data, without requiring equal sampling intervals between locations, and thus might prove applicable for a wide range of taxa.

References

Blackwell, P. G. (1997). 'Random diffusion models for animal movement', *Ecological Modelling* **100**: 87–102.

Börger, L., Dalziel, B.D., Fryxell, J.M. (2008). 'Are there general mechanisms of animal home range behaviour? A review and prospects for future research.', *Ecology Letters* **11**: 637–50.

Börger, L., Matthiopoulos, J., Holdo, R.M. *et al.* (2011). 'Migration quantified: Constructing models and linking them with data', in E.J. Milner-Gulland, J.M. Fryxell, A.R.E. Sinclair (eds), *Animal Migration-A Synthesis.* Oxford, Oxford University Press. 111–128.

Börger, L., McIntosh, T., Ryckman, M. *et al.* (unpublished manuscript). 'Where will they go? Predicting space use behavior and survival of dispersing animals', *Journal of Animal Ecology.*

Bowler, D.E., Benton, T.G. (2005). 'Causes and consequences of animal dispersal strategies: relating individual behaviour to spatial dynamics', *Biological Reviews* **80**: 205–25.

Bunnefeld, N., Börger, L., Van Moorter, B. *et al.* (2011). 'A model-driven approach to quantify migration patterns: individual, regional and yearly differences', *Journal of Animal Ecology* **80**: 466–76.

Burnham, K.P., Anderson, D.R. (2002). *Model selection and multi-model inference. A practical information-theoretical approach.* Springer Verlag, Berlin, Germany.

Calenge, C. (2006). The package 'adehabitat' for the R software: A tool for the analysis of space and habitat use by animals', *Ecological Modelling* **197**: 516–19.

Clobert, J., Ims, R.A., Rousset, F. (2004). 'Causes, mechanisms and consequences of dispersal', in I. Hanski, O.E. Gaggiotti (eds), *Ecology, Genetics and Evolution of Metapopulations.* Academic Press, Elsevier, London.

Clobert, J., Le Galliard, J.F., Cote, J. *et al.* (2009). 'Informed dispersal, heterogeneity in animal dispersal syndromes and the dynamics of spatially structured populations', *Ecology Letters* **12**: 197–209.

Codling, E.A., Plank, M.J., Benhamou, S. (2008). 'Random walk models in biology', *Journal of the Royal Society Interface* **5**: 813–34.

Davidian, M., Giltinan, D.M. (2003). Non-linear Models for Repeated Measurement Data: An Overview and Update.', *Journal of Agricultural, Biological and Environmental Statistics* **8**: 387–419.

Einstein, A. (1905). 'Über die von der molekularkinetischen Theorie der Wärme geforderte Bewegung von in ruhenden Flüssigkeiten suspendierten Teilchen', *Annalen der Physik* **322**: 549–60.

Elliott, L.C.C., Barhoum,M., Harris, J.M. (2011). Trajectory analysis of single molecules exhibiting non-Brownian motion', *Physical Chemistry Chemical Physical* **13**:

Fryxell, J.M., Hazell, M., Börger, L. *et al.* (2008). 'Multiple movement modes by large herbivores at multiple spatiotemporal scales', *Proceedings of the National Academy of the USA* **105**: 19114–19.

Huang, S., S. Meng, S.X., Yang, Y. (2009). 'Assessing the goodness of fit of forest models estimated by non-linear mixed-models methods', *Canadian Journal of Forest Research, Revue Canadienne De Recherche Forestiere* **39**: 2418–36.

Kölzsch, A., Blasius, B. (2008). 'Theoretical approaches to bird migration', *European Physical Journal-Special Topics* **157**: 191–208.

Langevin, P. (1908). Sur la théorie du mouvement brownien', *Comptes-Rendus de l'Academie des Sciences (Paris)* **146**: 530–3.

Li, T., Kheifets, S., Medellin, D. *et al.* (2010). 'Measurement of the Instantaneous Velocity of a Brownian Particle', *Science* **328**: 1673–5.

Metzler, R., Klafter, J. (2000). 'The random walk's guide to anomalous diffusion: a fractional dynamics approach', *Physics Reports* **339**: 1–77.

Moorcroft, P.R., Lewis, M.A. (2006). *Mechanistic Home Range Analysis. Monographs in Population Biology.* Princeton University Press, Princeton.

Morales, J.M., Moorcroft, P.R., Matthiopoulos, J. *et al.* (2010). 'Building the bridge between animal movement and population dynamics', *Philosophical Transactions of the Royal Society B* **365**: 2289–301.

Mysterud, A., Loe, L.E., Zimmermann, B. *et al.* (2011). 'Partial migration in expanding red deer populations at northern latitudes—a role for density dependence?', *Oikos* 120 (12): 1817–25.

Nathan, R., Getz W.M. Revilla, E. *et al.* (2008). 'A movement ecology paradigm for unifying organismal movement

research.', *Proceedings of the National Academy of Sciences of the USA* **105**: 19052–9.
Nouvellet, P., Bacon, J.P., Waxman, D. (2009). 'Fundamental Insights into the Random Movement of Animals from a Single Distance-Related Statistic', *American Naturalist* **174**: 506–14.
Pinheiro, J., D. Bates, S. Deb Roy, D. *et al.* (2011). nlme: Linear and Non-linear Mixed Effects Models, version 3.1–98. http://cran.r-project.org/web/packages/nlme/index.html
Pinheiro, J.C., Bates, D.M. (2000). *Mixed-effects models in S and S-Plus: Statistics and Computing.* Springer-Verlag, New York.
R Development Core Team. (2011). *R: A Language and Environment for Statistical Computing.* R Foundation for Statistical Computing, Vienna, Austria.
Skellam, J.G. (1951). 'Random dispersal in theoretical populations'. *Biometrika* **38**: 196–218.
Turchin, P. (1998). *Quantative analysis of movement.* Sinauer Associates, Inc., Sunderland, MA.
Uhlenbeck, G.E. and Ornstein, L.S. (1930). 'On the Theory of the Brownian Motion', *Physical Review* **36**: 823.
Van Dyck, H. and Baguette, M. (2005). Dispersal behaviour in fragmented landscapes: Routine or special movements?', *Basic and Applied Ecology* **6**: 535–45.

CHAPTER 18

Temporal variation in dispersal kernels in a metapopulation of the bog fritillary butterfly (*Boloria eunomia*)

Nicolas Schtickzelle, Camille Turlure, and Michel Baguette

18.1 Introduction

Dispersal kernels are frequently used to model dispersal (Chapter 15), especially among local populations in metapopulation models (e.g. Hanski 1999). Variation of dispersal kernels in space has been thoroughly documented among different populations of the same species (e.g. Schtickzelle, Mennechez and Baguette 2006; Wang *et al.* 2011). The ultimate driver of such geographical patterns is the modification of individual movements according to variations in landscape settings (habitat geometry and quality, including inter- and intra-specific interactions) (e.g. Baguette and Van Dyck 2007; Hanski *et al.* 2002; Hanski *et al.* 2004; Phillips *et al.* 2008), which may have a genetic basis (Phillips, Brown and Shine 2010).

Variation of dispersal kernels in time has received less attention, even if temporal change in dispersal rates among local populations has been repeatedly documented in the metapopulation literature (e.g. Baguette, Clobert and Schtickzelle 2011; Hanski *et al.* 2004; Schtickzelle *et al.* 2006). Changes in individual mobility that generate temporal shifts in dispersal kernels would obviously be context- and phenotypic- dependent. Especially in ectothermic organisms, moderate modifications of mean temperature, periods of sun or rain will entail context-dependent variation in mobility (e.g. Clobert *et al.* 2009; Hill, Griffiths and Thomas 2011; Parmesan 2006), which is potentially dependent on individual phenotypes. For example, the possession of different alleles of the Pgi (Phosphogluco-isomerase) enzyme, which plays a central role in the Krebs cycle, confers contrasting movement abilities to individual arthropods according to climatic conditions (Watt *et al.* 2003; Wheat *et al.* 2011; Wheat *et al.* 2006; Case Study III; Chapters 8 and 23). Variation in conspecific density will also entail context-dependent responses in terms of individual motivation to move (Clobert *et al.* 2009; Hovestadt and Nieminen 2009; Matthysen 2005; De Meester and Bonte 2010), which will be in turn modulated by individual phenotypes.

Both environmental conditions (such as climatic variables) and conspecific density are thus expected to play a central role in temporal variation of dispersal kernels. Here we used standardized capture-mark-recapture (CMR) data from long-term monitoring (19 yr) of the bog fritillary butterfly, *Boloria eunomia*, metapopulation dynamics in a single landscape. For each generation of *B. eunomia*, we computed male and female dispersal kernels from all recorded butterfly movements, and investigated: (1) the amount of temporal (i.e. between years) variability, and (2) the amount of this temporal variability that was explained by climatic variables and conspecific density. The analysis of the dispersal kernels was centred on both the overall magnitude of dispersal in the patch network and the decay of dispersal probability with the distance. The main aim of this study is to illustrate that dispersal varies between

Dispersal Ecology and Evolution. First Edition. Edited by Jean Clobert *et al.*

years, and hence to challenge the validity of dispersal snapshot measures (i.e. dispersal kernel estimated using one or a few datasets for a given species).

18.2 Methods

The bog fritillary, *Boloria eunomia*, is a specialist butterfly restricted, in Belgium, to peat bogs and unfertilized wet meadows. Adults fly in one generation in May and June in our study area. Both adults and caterpillars feed on a single plant, the bistort, *Persicaria bistorta*.

The 'Prés de la Lienne' nature reserve (50 °18′ N, 5 °49′ E) hosts one patchy population of *B. eunomia* (Figure 18.1). This population has been intensively studied using CMR over 19 generations (1992–2010) (e.g. Baguette *et al.* 2011; Baguette and Nève 1994; Schtickzelle and Baguette 2004). During the flight period of adults, the 11 patches of habitat (totalling 2.5 ha) were visited each day when weather conditions allowed butterfly activity. Every adult was caught with a net, individually marked on the underside of the left hind wing and immediately released at the spot of capture. For each (re)capture, we recorded the marking code, sex, date, hour, and patch location.

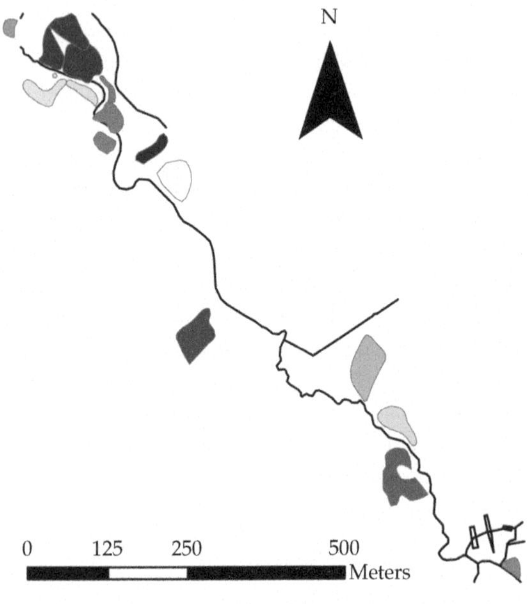

Figure 18.1 Map of the study area: the 'Prés de la Lienne' nature reserve comprises 11 habitat patches for *B. eunomia* (some consist in several contiguous host plant patches), along the Lienne river (black line). Each patch is depicted with a different grey value.

We extracted dispersal data out of these 19 CMR datasets, using each pair of successive captures of any given marked individual as a dispersal event. Pairs of successive captures may indeed be considered as dispersal events because *B. eunomia* butterflies can disperse all along their life, females being able to lay eggs and males being able to mate in each visited patch. All distances were computed between centres of habitat patches.

Total population size (number of different *B. eunomia* adults present at least during part of their life in the population over the generation) and the sex ratio (defined as female/male total population sizes) were estimated from CMR data for each year with Popan models implemented in Mark software (Schtickzelle, Le Boulengé and Baguette 2002; N. Schtickzelle, unpublished data).

Weather data were obtained from a Belgian Royal Meteorological Institute weather station at Nadrin (ca. 17 km from Prés de la Lienne): maximum temperature, number of hours of sunshine, and amount of rainfall were computed for each day of the flight season for each year. These three variables were obviously correlated, and accordingly were summarized by the first axis of a Principal Component Analysis, which explained 61% of the total variation. This axis expressed the quality of weather for butterfly activity, being positively correlated with temperature and sunshine, and negatively with rainfall.

We defined the dispersal kernel as the probability to disperse farther than a certain distance. Accordingly, observed data to estimate the kernel were, for each observed dispersal distance, 1-S, with S the cumulative distribution of the proportion of dispersal movements up to that distance. In this way, dispersal probability was zero for the maximum observed distance (no butterflies went farther than the maximum), while at distance zero it represented the emigration rate (proportion of pairs of successive (re)capture implying a movement > 0 m). We modelled the dispersal kernel by an inverse-power function: $P = a \cdot D^{-b}$ with P the probability to disperse farther than the distance D (expressed in m); the a parameter expressed the emigration rate, and the *b* parameter scaled the decrease in dispersal probability with distance. This function was fitted to the data in its linearized form ln $(P + 1) = \ln(a) - b\cdot \ln(D + 1)$ using a generalized linear model with normal error distribution and identity link function.

The model included additional explanatory variables, both as main effects (affecting the probability of dispersal across the whole distance range) and as interactions with ln(*D*) (affecting the decay of dispersal with distance). In a first step, we included the following explanatory variables: the sex of the individual, the sampling year and their interaction terms, to test a model containing the following effects: intercept, sex, year, sex*year, ln(*D*), sex*ln(*D*), year*ln(*D*), sex*year*ln(*D*). In a second step, we replaced the year effect by the combination of total population size (Ntot), sex ratio (SR) and weather (W) conditions during the flight period, with a model containing: intercept, sex, Ntot, SR, W, sex*Ntot, sex*SR, sex*W, ln(*D*), sex*ln(*D*), Ntot*ln(*D*), SR*ln(*D*), W*ln(*D*), sex*Ntot*ln(*D*), sex*SR*ln(*D*), sex*W*ln(*D*).

To determine factors affecting the dispersal kernel, we used AICc model selection (Anderson 2008; Burnham and Anderson 2002). We fitted the global model and all its submodels (consisting of all the possible combinations of the explanatory variables) and computed for each model its AICc value and its AICc weight. In the case where no single model was clearly more supported than others, we used the multimodel averaging approach (Anderson 2008; Burnham and Anderson 2002; note under Table 18.2).

18.3 Results

Over the 19 generations, we marked 1940 females and 2815 males. Data for 1998 and 2005 to 2008 were discarded from the analysis because of the very low number of observed movements (Table 18.1). Using such datasets might bias the estimation of the dispersal kernels; some spurious variation between years might appear due to estimation error and hence not reflecting real variation due to weather conditions or conspecific density. The analysis thus pertains to 5877 movements (1370 for females and 4507 for males), of 4224 marked butterflies (1749 females and 2475 males). Longest movements recorded in a given year ranged from 473 to 1046 m for females and 551 to 1136 m for males, for possible inter-patch distances in this system ranging from 57 to 1136 m.

The observed distribution of the dispersal distances fitted an inverse power function, with a very strong decrease with distance (Figure 18.2). The inverse power function did not fit well the early part of the kernel (0–100 m) partly because very short distances cannot be observed with CMR data as implemented here (positions were recorded as 'within a patch', and patch centroid coordinates used). Being

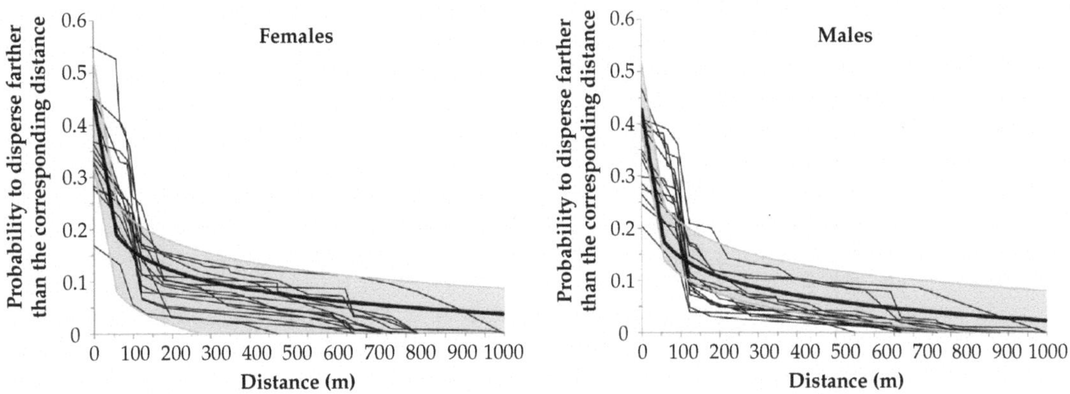

Figure 18.2 Dispersal kernels. Thin lines represent observed kernels for each year from 1992 to 2010 (except 1998, and 2005 to 2008, see text). Thick black line represents the average kernel. Grey area is delimited by the lowest and highest fitted yearly kernels (1999 and 2010 for females, 1997 and 1995 for males, respectively).

Table 18.1 Data summary. See text for details on computation methods. Years shown in italics were discarded from the analysis because of the very low number of observed movements.

	Marked butterflies		Pairs of successive captures		Longest observed movement (m)		Total population size			Sex ratio (F/M)	Index of good weather
Year	Females	Males	Females	Males	Females	Males	Females	Males	Pooled		
1992	343	295	255	330	1046	668	576	470	1046	1.22	−0.461
1993	113	247	60	260	668	928	231	361	592	0.64	−0.065
1994	123	139	178	287	831	668	162	186	348	0.87	0.730
1995	64	65	64	105	1046	1046	92	96	188	0.96	−0.422
1996	136	180	132	563	831	1046	176	200	376	0.88	0.214
1997	172	228	125	588	734	1046	259	251	510	1.03	−0.100
1998	*44*	*155*	*8*	*49*	*122*	*551*	*61*	*340*	*401*	*0.18*	*−0.487*
1999	85	195	53	431	473	668	166	237	403	0.70	0.117
2000	90	129	57	191	763	763	146	184	330	0.79	0.016
2001	42	36	54	61	668	551	67	52	119	1.27	−0.413
2002	63	106	62	248	831	928	104	129	233	0.80	0.051
2003	105	222	117	674	831	831	165	260	425	0.63	0.697
2004	132	181	87	349	831	1136	259	235	494	1.10	−0.201
2005	*52*	*53*	*30*	*92*	*668*	*668*	*93*	*77*	*170*	*1.21*	*0.363*
2006	*32*	*41*	*18*	*53*	*178*	*473*	*99*	*56*	*155*	*1.78*	*0.843*
2007	*41*	*46*	*22*	*80*	*197*	*831*	*79*	*53*	*132*	*1.48*	*−0.267*
2008	*22*	*45*	*10*	*39*	*92*	*668*	*43*	*89*	*132*	*0.48*	*−0.505*
2009	68	130	33	122	831	831	116	192	308	0.60	0.013
2010	213	322	93	298	763	763	545	470	1015	1.16	0.357

constrained by the minimum distance between two patch centroids (57 m), a recorded movement is necessary above that minimum distance.

When assessing variations of the kernel between males and females and across years, one model appeared more supported than others: Sex + Year + Sex*Year + ln(*D*), with an AICc value of –1101.18 and an AICc weight of 69%. This result indicated that the overall dispersal frequency (dispersal probability across the whole distance range): (1) was higher for females than for males, (2) varied according to the year (generation), and (3) in a different way for each sex. On the contrary, the decay of dispersal probability with distance did not significantly differ between sexes or years. Figure 18.2 illustrates these variations in the dispersal kernel. In comparison, the next model in the list sorted according to the AICc value was less supported (Sex + Year + Sex*Year + ln(*D*) + Sex*ln(*D*) term: AICc = –1098.82 and weight = 21%). AICc values of the ln(*D*)-only model (a common dispersal kernel for both sexes and all years) and the intercept only model (no decay of dispersal with distance) were much higher (AICc = –1052.36 and weight = 0% and AICc = –642.07 and weight = 0%, respectively).

When trying to explain the yearly variation in dispersal kernel by a combination of descriptors (total population size, sex ratio, and index of weather quality for butterfly activity), a relatively large number of models had similar AICc support from the data. We then quantified the effect of each explanatory variable by its AICc weight, using a multimodel averaging approach. Besides the effects already observed in the previous analysis (higher dispersal probability of females across the whole distance range compared to males, and clear decay of dispersal probability with distance), the following effects were significant (Table 18.2): (1) during years of higher butterfly abundance, females increased their dispersal probability across the whole distance range while males decreased it; (2) the decay with distance was 1% smaller for females than for males, although this effect is likely to be too small to be biologically important; (3) the decay with distance was also sharper in years in which butterflies were more abundant, this effect again being more pronounced for females than for males (a change in one standard deviation of Ntot induced a decay of dispersal with distance 13% higher for females, and 4% higher for males).

Only a fraction of the yearly variation in the dispersal kernel could be explained by effects of total population size, sex ratio, and weather conditions (the best model with these factors was: Sex + Ntot + Sex-ratio + Sex*Ntot + ln(*D*) + Sex*ln(*D*) + Ntot*ln(*D*) + Sex*Ntot*ln(*D*), with an AICc value of –1068.61; i.e. 33 AICc units above the best model with yearly variation).

18.4 Discussion

We investigated here the temporal variation in dispersal kernel within a metapopulation over 14 generations, based on one of the largest and most detailed set of CMR data available for a butterfly metapopulation. Estimating kernels from CMR data is not trivial and biases exist. Discussing these biases in detail is outside the scope of this chapter, but the most obvious are that observed movements i) are influenced by the likelihood to be caught and the structure of the habitat network, ii) can be aggregated into different statistics used to fit the kernel (e.g. individual movements as net displacement vs total distance travelled by an individual in its lifetime), and iii) can be modelled using different functions (such as inverse power and negative exponential functions commonly used in the butterfly literature); each approach having advantages and disadvantages (Chapter 15). The method we chose can obviously be criticized on some of these aspects. However, our chapter aims to quantify the temporal variation existing in the dispersal kernel, and the major point was to make the kernels comparable. The CMR protocol had been standardized at the beginning of the study and performed in the same way during all 19 years. Furthermore, habitat geometry was kept largely constant from one generation to the next. All these precautions allow these dispersal kernels to be compared reliably between years, and then prevent biases/errors while quantifying the temporal variation in dispersal.

Our results indicated an important temporal (between generations) variation in the dispersal kernel. This variation between years resided far more in the overall dispersal intensity than in its

Table 18.2 Multimodel averaging of explanatory variables involved in modelling the dispersal kernel. See text and note for details on the methodology used to obtain these estimates. Effects significant at the 0.05 level are in bold. Sex effect is given for females, males being fixed at 0. Continuous explanatory variables have been standardized to make the comparison of magnitude of effects comparable. Main effects relate to the probability of dispersal across the whole distance range, while ln(*D*) and its interactions with other terms relate to the decay of dispersal probability with the distance.

Explanatory variable	Level	AICc weight	P value	Parameter estimate	SD	Lower 95% IC	Upper 95% IC
Intercept		.	.	0.1106	0.0040	0.1028	0.1185
Sex	**F**	**100%**	**0.000**	**0.0124**	**0.0061**	**0.0004**	**0.0245**
Ntot		**99%**	**0.001**	**−0.0100**	**0.0049**	**−0.0197**	**−0.0003**
Sex-ratio		75%	0.117	0.0037	0.0034	−0.0029	0.0104
Weather		43%	0.639	−0.0001	0.0017	−0.0034	0.0033
Sex*Ntot	**F**	**98%**	**0.000**	**0.0212**	**0.0063**	**0.0088**	**0.0336**
Sex*Sex-ratio	F	26%	0.112	0.0016	0.0023	−0.0029	0.0060
Sex*Weather	F	13%	0.380	0.0005	0.0010	−0.0014	0.0024
ln(*D*)		**100%**	**0.000**	**−0.0816**	**0.0038**	**−0.0891**	**−0.0741**
Sex*ln(*D*)	**F**	**65%**	**0.024**	**0.0008**	**0.0040**	**−0.0070**	**0.0086**
ln(*D*)*Ntot		**92%**	**0.002**	**−0.0033**	**0.0051**	**−0.0134**	**0.0067**
ln(*D*)*Sex-ratio		28%	0.096	−0.0008	0.0012	−0.0031	0.0015
ln(*D*)*Weather		12%	0.454	0.0000	0.0004	−0.0009	0.0009
Sex*ln(*D*)*Ntot	**F**	**51%**	**0.001**	**−0.0071**	**0.0047**	**−0.0164**	**0.0022**
Sex*ln(*D*)*Sex-ratio	F	2%	0.109	−0.0001	0.0002	−0.0006	0.0003
Sex*ln(*D*)*Weather	F	1%	0.228	−0.0001	0.0001	−0.0004	0.0002

Note: The *AICc weight* of each variable was computed as the sum of the weight of all the models where this variable appears. Because models with interactions but not the corresponding main effects were not considered, the interaction terms appeared in fewer models than the main effects; this is likely to bias the final weight of the explanatory variable, the weight of a rarer variable being the sum of a smaller number of terms (weight of individual models containing the variable). We therefore used a permutation test (N. Schtickzelle and G. San Martin y Gomez, unpublished) to assess whether the observed weight of each explanatory variable is significantly higher than expected under the null hypothesis (no relation between the response and the explanatory variables). This test comprises randomly shuffling the response variable column, disrupting any relationship that could exist with explanatory variable, and running the full set of models to compute the weight of all explanatory variables; 1000 random permutations were done, giving the distribution of variable weights under the null hypothesis. Then the *P value* was computed for each explanatory variable as the proportion of the 1000 random permutations where the variable weight was higher than its value in the observed dataset. Variables with a P value < 0.05 were considered as significant. The *parameter estimate* (and associated *SD*) for each variable was computed as its average over all models, weighted by the AICc weight of each model, considering a value of 0 for the parameter estimate (and missing value for its SD) when the variable was not present in a model; this is the shrinkage approach to model averaging (Burnham and Anderson 2002, Anderson 2008).

decay with distance, and could be partly explained by butterfly density, whereas it seems not linked to sex ratio and weather conditions. Furthermore, there was a clear difference between sexes both in the kernel but also in butterfly reaction to density.

The lack of weather influence on dispersal kernel variation was surprising, as there is a direct relation between weather condition and butterfly activity. One possible explanation of this discrepancy between behavioural observations and dispersal patterns may be in the meteorological data we used. These data originated from a weather station 17 km away from the study site, and only a single daily measure was available. The response of butterflies to changes in weather conditions is often immediate, so these data might simply not reflect the actual weather conditions encountered by the butterflies, and hence have limited power to explain observed variations in dispersal kernels. For example, during hot days, bog fritillary butterflies are on the wing even during short periods of sunshine, and they are able to cover long dispersal distances in a very short time (2.5 km in two hours is the current recorded maximum: N. Schtickzelle, unpublished data). In contrast, after days of continuous rain they need a long time to recover normal activity levels. In both cases, dispersal and weather conditions would not correlate. The quantitative relationship between weather conditions, butterfly activity, and dispersal patterns certainly deserves further attention.

Observed changes in dispersal kernel according to butterfly density corresponded to insights from behavioural observations. Females of the bog fritillary butterfly mate only once, soon after their emergence (Baguette and Nève 1994). Mated females spent most of their time hidden, thermoregulating in the vegetation. When they fly, mated females are constantly harassed by courting males over their lifetime, which may trigger their dispersal out of habitat patches at high male density (Baguette, Convié and Nève 1996; Baguette *et al.* 1998). In contrast, males constantly patrol within habitats in search for females. They are attracted by small, orange objects, from dry leaves to other butterfly species such as the map butterfly (*Araschnia levana*, spring generation). Such visual contacts seem to motivate the males to stay in the same patch. If they do not meet such contacts, they leave the habitat after having prospected it for a while. Such emigration events are more frequent in small habitat patches in fragmented landscapes (Baguette *et al.* 1996; Baguette *et al.* 1998; Schtickzelle and Baguette 2003). At high butterfly density, females will thus emigrate and hence disperse more, whereas males will emigrate and disperse less. These behavioural differences between males and females provide a reliable mechanistic explanation to the relation between dispersal kernels and butterfly density. Such sex-biased density-dependent dispersal in butterflies has also been documented in other butterfly species (e.g. Hovestadt and Nieminen 2009).

The relation between butterfly density and a steeper decay of dispersal kernels with distance can also be interpreted in this framework. This decay with distance might be attributed to a higher likelihood to be caught of individuals at higher densities that inflates the number of small movements. This explanation can be discarded, as the mean likelihood to be caught is lower when butterfly density is high (N. Schtickzelle, unpublished results). Another possibility is that females hide more when males are abundant, whereas males are more motivated to stay in habitat patches by frequent conspecific encounters.

Considering the variation in dispersal kernels among years, but also between sexes, and the opposite reactions of males and females to butterfly local density (a parameter known to vary greatly in space and time), blind adherence to a given equation to describe the dispersal kernel of a species would have dramatic consequences on the predictions of dispersal and more generally, of any process depending on dispersal, such as metapopulation functioning or viability. Using a single dataset acquired under specific environmental and population conditions, to estimate a dispersal kernel that is then viewed as largely applicable to the species is likely not appropriate for many species. As an example, the predictions of dispersal probability for the mean distance between pairs of patches (421 m) would range from 0% to 13% for females, and 3.5%

to 13% for males (Figure 18.2), which are not trivial differences in terms of impact on metapopulation functioning. Even if there are cases where no other solution might exist (e.g. Schtickzelle, Wallis De Vries and Baguette 2005), we therefore caution scientists to be careful when generalizing dispersal kernels, whatever the generalization is: between years, sexes, landscapes, or even species. The possibility of comparing dispersal for various landscapes with only one or very few kernel estimates for each seems particularly difficult; several kernel estimates seem necessary to disentangle temporal variation from the effect of interest (e.g. Schtickzelle *et al.* 2006).

Acknowledgements

Nicolas Schtickzelle is Research Associate and Camille Turlure is Postdoctoral Researcher of the Fund for Scientific Research-FNRS and they acknowledge its financial support. Their work in this chapter is referenced as BRC245 at the Biodiversity Research Centre (UCL). Michel Baguette is funded by the ANR (Agence Nationale de la Recherche) through the EU FP6 Biodiversa ERANET TenLamas project and by the EU FP7 SCALES project (project no. 226852).

References

Anderson, D.R. (2008). *Model based inference in the life science: a primer on evidence.* Springer Verlag, Berlin.

Baguette, M., Clobert, J., Schtickzelle, N. (2011). 'Metapopulation dynamics of the bog fritillary butterfly: experimental changes in habitat quality induced negative density-dependent dispersal' *Ecography* **34**: 170–6.

Baguette, M., Convié, I., Nève, G. (1996). 'Male density affects female spatial behaviour in the butterfly *Proclossiana eunomia*', *Acta Oecologica-International Journal of Ecology* **17**: 225–32.

Baguette, M. and Nève, G. (1994). 'Adult movements between populations in the specialist butterfly *Proclossiana eunomia* (Lepidoptera, Nymphalidae). *Ecological Entomology* **19**: 1–5.

Baguette, M. and Van Dyck, H. (2007). 'Landscape connectivity and animal behaviour: functional grain as a key determinant for dispersal', *Landscape Ecology* **22**: 1117–29.

Baguette, M., Vansteenwegen, C., Convié, I. *et al.* (1998). 'Sex-biased density-dependent migration in a metapopulation of the butterfly *Proclossiana eunomia*', *Acta Oecologica-International Journal of Ecology* **19**: 17–24.

Burnham, K.P. and Anderson, D.R. (2002). *Model selection and multimodel inference: a practical information-theoretic approach.* Springer-Verlag, New York.

Clobert, J., Le Galliard, J.F., Cote, J. *et al.* (2009). 'Informed dispersal, heterogeneity in animal dispersal syndromes and the dynamics of spatially structured populations', *Ecology letters* **12**: 197–209.

De Meester, N. and Bonte, D. (2010). 'Information use and density-dependent emigration in an agrobiont spider', *Behavioral Ecology* **21**: 992–8.

Hanski, I. (1999). *Metapopulation ecology*. Oxford University Press, Oxford.

Hanski, I., Breuker, C.J., Schöps, K. *et al.* (2002). 'Population history and life history influence the migration rate of female Glanville fritillary butterflies', *Oikos* **98**: 87–97.

Hanski, I., Eralahti, C., Kankare, M. *et al.* (2004). 'Variation in migration propensity among individuals maintained by landscape structure', *Ecology letters* **7**: 958–66.

Hill, J.K., Griffiths, H.M., Thomas, C.D. (2011). 'Climate Change and Evolutionary Adaptations at Species' Range Margins', *Annual Review of Entomolgy* **56**: 143–59.

Hovestadt, T. and Nieminen, M. (2009). 'Costs and benefits of dispersal in butterflies', in J. Settele, M. Konvicka, T.G. Shreeve *et al.* (eds), *Ecology of butterflies in Europe.* Cambridge University Press, Cambridge. **97**–106.

Matthysen, E. (2005) 'Density-dependent dispersal in birds and mammals', *Ecography* **28**: 403–16.

Parmesan, C. (2006). 'Ecological and evolutionary responses to recent climate change'. *Annual Review of Ecology Evolution and Systematics* **37**: 637–69.

Phillips, B.L., Brown, G.P., Shine, R. (2010). 'Evolutionarily accelerated invasions: the rate of dispersal evolves upwards during the range advance of cane toads', *Journal of Evolutionary Biology* **23**: 2595–601.

Phillips, B.L., Brown, G.P., Travis, J.M.J. *et al.* (2008). 'Reid's paradox revisited: The evolution of dispersal kernels during range expansion', *American Naturalist* **172**: S34–48.

Schtickzelle, N. and Baguette, M. (2003). 'Behavioural responses to habitat patch boundaries restrict dispersal and generate emigration-patch area relationship in fragmented landscapes with low quality matrix', *Journal of Animal Ecology* **72**: 533–45.

Schtickzelle, N. and Baguette, M. (2004). 'Metapopulation viability analysis of the bog fritillary butterfly using RAMAS/GIS', *Oikos* **104**: 277–90.

Schtickzelle, N., Le Boulengé, E., Baguette, M. (2002). 'Metapopulation dynamics of the bog fritillary butterfly: demographic processes in a patchy population', *Oikos* **97**: 349–60.

Schtickzelle, N., Mennechez, G., Baguette, M. (2006). 'Dispersal depression with habitat fragmentation in the bog fritillary butterfly', *Ecology* **87**: 1057–65.

Schtickzelle, N., Wallis De Vries, M.F., Baguette, M. (2005). 'Using surrogate data in population viability analysis: the case of the critically endangered cranberry fritillary butterfly', *Oikos* **109**: 89–100.

Wang, R.J., Ovaskainen, O., Cao, Y.D. *et al.* (2011). 'Dispersal in the Glanville fritillary butterfly in fragmented versus continuous landscapes: comparison between three methods', *Ecological Entomology* **36**: 251–60.

Watt, W.B., Wheat, C.W., Meyer, E.H. *et al.* (2003). 'Adaptation at specific loci. VII. Natural selection, dispersal and the diversity of molecular-functional variation patterns among butterfly species complexes (Colias: Lepidoptera, Pieridae)', *Molecular Ecology* **12**: 1265–75.

Wheat, C.W., Fescemyer, H.W., Kvist, J. *et al.* (2011) Functional genomics of life history variation in a butterfly metapopulation', *Molecular Ecology* **20**: 1813–28.

Wheat, C.W., Watt, W.B., Pollock, D.D. *et al.* (2006). 'From DNA to fitness differences: Sequences and structures of adaptive variants of Colias phosphoglucose isomerase (PGI)', *Molecular Biology and Evolution* **23**: 499–512.

CHAPTER 19

How random is dispersal? From stochasticity to process in the description of seed movement

Frank M. Schurr

19.1 Process and stochasticity in the modelling of plant dispersal

Distance and direction of seed dispersal are typically highly variable (e.g. Levin *et al.* 2003; Ghazoul 2005). This variation arises because seeds are dispersed by multiple vectors, because one and the same vector behaves differently under different circumstances, and because of variation in plant and seed traits relevant for dispersal. In particular, many plant species show leptokurtic distributions of dispersal distance since the conditions under which their seeds are dispersed over long distances are rare compared to the conditions generating short-distance dispersal (Nathan *et al.* 2008). Despite this variation, it should in principle be possible to predict almost exactly the dispersal of a given seed (within fundamental physical limits, e.g. Binder 2008). Yet this would require perfect knowledge about how traits and environmental conditions determine dispersal, and about the dynamics and initial conditions of these determinants.

In practice, we are far from having perfect knowledge of dispersal determinants. Hence, dispersal models only describe some processes deterministically and summarize effects of the remaining processes in stochastic terms. Following Clark (2009), one can thus write a general dispersal model as

dispersal distance = f (traits, environment) + error

The first term is deterministic and explanatory, whereas the second term is stochastic and has no explanatory power. The simplest possible dispersal models have no deterministic component that describes how traits and environment cause variation in dispersal. These purely stochastic models thus only consist of a probability distribution of dispersal distance (a simple 'dispersal kernel'; Chapter 15). In order to describe leptokurtic distributions of dispersal distance, several widely used simple dispersal kernels have fat tails with high stochastic variance (Chapter 15). Yet the replacement of process descriptions through high stochastic variance comes at a cost. For instance, fat-tailed dispersal kernels can cause large uncertainty about the future spread dynamics of plant populations (Clark *et al.* 2003). More generally, improved understanding and prediction of dispersal requires the reshuffling of variation from the second, stochastic term to the first, deterministic term of the above equation (Clark 2009).

Hence it is desirable to replace some of the stochasticity in simple dispersal kernels with a deterministic description of mechanisms. In the following example, I show how simple kernels can be expanded to describe more processes. This not only helps to understand dispersal and the selective forces acting on it, but also improves dispersal predictions.

19.2 Effects of release height and environmental variability on seed dispersal by wind

19.2.1 A field experiment

To illustrate the benefits of moving from purely stochastic to more deterministic descriptions of dispersal, I use data from an experiment investigating how a key dispersal trait (the height of seed release, h_r) affects the wind-driven seed dispersal of Common

Dispersal Ecology and Evolution. First Edition. Edited by Jean Clobert *et al.*

Cat's-ear (*Hypochaeris radicata*). This species is a herbaceous, self-incompatible perennial that frequently occurs in nutrient-poor grasslands (Turkington and Aarssen 1983; Pico *et al.* 2004). Its plumed seeds are borne in capitula and are typically released from heights between 0.15 m and 0.6 m above the ground surface (Turkington and Aarssen 1983).

In the experiment, students of a plant ecology course followed seeds released from predefined heights, and measured their dispersal distance as the distance between the release point and the location where the seed first hit the vegetation surface. In each of five years, the experiment was conducted during ca. 1.5 h on a 2 ha section of a mown meadow (vegetation height < 10 cm) at the University of Potsdam's Ecological Station Gülpe. The experiment comprised multiple blocks of four seeds that were released at the same location within 5 min. The four seeds per block started from heights of 0.15 m, 0.25 m, 0.5 m, and 1 m, respectively. In total, 436 seeds were released in 109 experimental blocks (Figure 19.1).

19.2.2 The mechanistic Wald kernel

I statistically fitted the Wald (or inverse Gaussian) kernel (Katul *et al.* 2005) to the experimentally observed distributions of dispersal distance. The Wald kernel is a prime example for a mathematically simple dispersal kernel that has a mechanistic interpretation (Chapter 15). It is a simplification of a coupled Eulerian-Lagrangian closure (CELC) model that mechanistically predicts seed transport by turbulent winds (Nathan *et al.* 2002; Soons *et al.* 2004a; Chapter 15). Given dispersal traits and summary statistics of wind and vegetation properties, the Wald kernel describes variation in dispersal distance that results from turbulent fluctuations in the

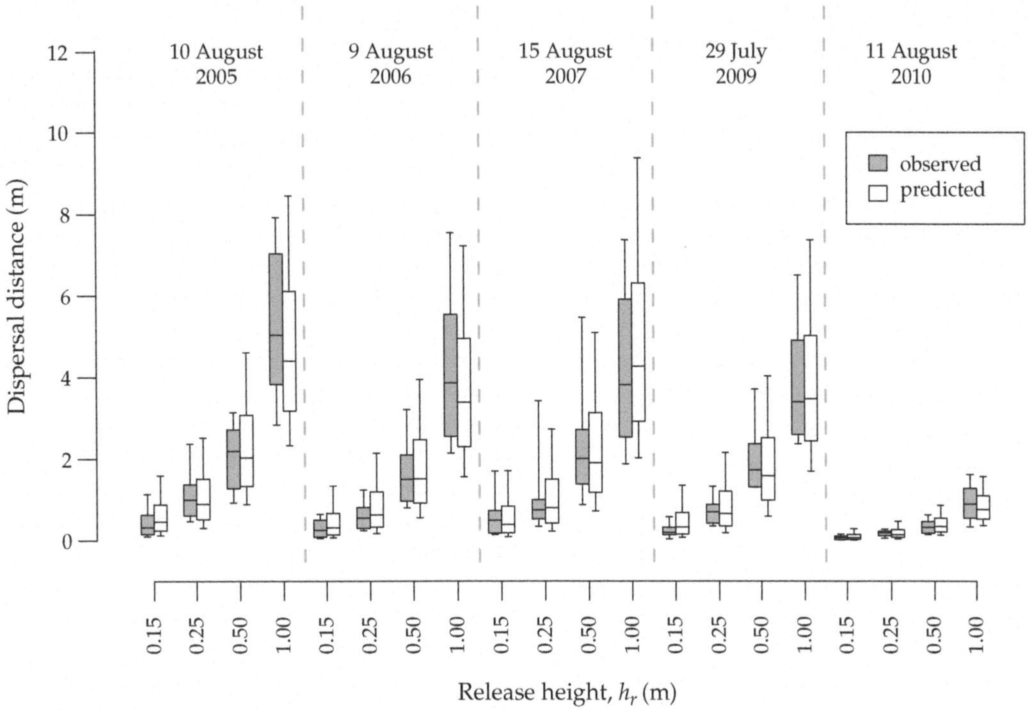

Figure 19.1 Dispersal distances of *Hypochaeris radicata* seeds that were released at five dates from four different heights (grey) and predictions of the mechanistic Wald model fitted to these data (white). Boxplots show variation between seeds released in different experimental blocks (bold line: median, box: interquartile range, whiskers: 10%–90% interquantile range).

wind conditions encountered by individual seeds. The stochastic terms of the Wald kernel (and the stochastic elements of the underlying CELC model) subsume the effect of these turbulent fluctuations. According to the Wald model, the probability density of dispersal over distance x is

$$p(x) = (\frac{\lambda}{2\pi x^3})^{1/2} \exp[-\frac{\lambda(x-\mu)^2}{2\mu^2 x},$$

where μ is mean dispersal distance and λ is a shape parameter. The mechanistic nature of Wald is evident from the fact that these two parameters can be derived from independent information on environmental conditions and dispersal traits (Katul *et al.* 2005). In particular,

$$\mu = h_r \bar{U} / v_t = h_r \alpha, \qquad \text{(Equation 19.1)}$$

where h_r is seed release height, v_t is the terminal falling velocity of seeds in still air, and $\bar{U}$ *is* the mean horizontal wind speed at a reference height above the ground surface. Since $\bar{U}$ and v_t are not known for each experimentally released seed, I subsume these two quantities in a dimensionless parameter $\alpha = \bar{U} / v_t$.

The shape parameter of the Wald kernel is

$$\lambda = h_r^2 \bar{U} / 2kH\sigma_w = h_r^2 \beta, \qquad \text{(Equation 19.2)}$$

where H is mean vegetation height and σ_w the standard deviation of the vertical wind speed. The coefficient k is close to 0.4 for very sparse vegetation like the one examined here (Katul *et al.* 2005). Again, I subsume quantities that were not measured in the experiment in an aggregate parameter $\beta = \bar{U} / 2kH\sigma_w$ (which has units of m^{-1}).

19.2.3 Fitting the Wald kernel to experimental data

Analysis of the seed release experiment proceeds in three steps. First, I fit the Wald kernel to a subset of the data (the dispersal distance distributions of seeds released from a single height in a single year). Secondly, I evaluate the performance of this simple—and purely stochastic—Wald kernel for the full dataset (the dispersal distance distribution of seeds released from all heights in all years). In the third step, I develop an extended model which deterministically describes how Wald's parameters change with release height, and which additionally describes environmental variation between years and blocks (see Equations 19.1 and 19.2). This extended Wald model is then fitted to the full dataset. Since the extended model is difficult to fit with frequentist (Maximum Likelihood) methods (Chapter 15), I conducted all analyses in a Hierarchical Bayesian framework using WinBUGS 1.4 (Lunn *et al.* 2000) with uninformative prior distributions for all parameters. To compare alternative models fitted to the same data, I used the Deviance Information Criterion (DIC, Spiegelhalter *et al.* 2002). As the Akaike Information Criterion (AIC) used in frequentist statistics, DIC trades off model fit against model complexity, with lower DIC values indicating better model performance.

The first step of analysis only considers the dispersal distances of seeds that were released from 1m height in 2006. As shown in Figure 19.2a, the Wald kernel fits well to this distribution of dispersal distances. In fact, with a DIC of 51.3 the Wald kernel describes this subset of the data substantially better than the more commonly used lognormal (Stoyan and Wagner 2001) and 2Dt (Clark *et al.* 1999) kernels, which have DICs of 134.3 and 139.8, respectively (see Chapter 15 for a description of these two alternative kernels).

The superior performance of Wald for this subset of the data (and for ten of the 19 other subsets) might suggest that Wald is also suitable for describing the entire dataset. This is, however, not the case: when fitted to data for all years and release heights, the Wald kernel (DIC: 1686.6) performs markedly worse than the lognormal and 2Dt kernels (DIC: 1431.6 and 1515.2, respectively). Specifically, Wald overestimates the shortest and longest dispersal distances but underestimates dispersal distances close to the median (Figure 19.2b).

This poor performance is in fact not surprising if we consider Wald's mechanistic derivation: Wald describes the distribution of dispersal distances for seeds released from a single height in a given environment that is characterized by summary statistics of environmental conditions. However, the full dataset comprises dispersal distances of seeds released (1) from various heights, and (2) under various wind conditions that cannot adequately be described by a single set of summary statistics (from very windy and turbulent conditions in 2007, to stable conditions with very little wind in 2010). Since the simple Wald kernel cannot describe these two sources of variation in μ and λ, it poorly describes the entire dataset. However, the first source of variation can be described deterministically using the mechanistic knowledge summarized in Equations 19.1 and 19.2: while μ should be proportional to release height h_r, λ should be proportional to h_r^2.

To integrate this mechanistic knowledge, I formulate an extended Wald model which has a deterministic backbone describing how μ and λ vary with h_r. This extended model subsumes unmeasured environmental conditions in two parameters α and β (see Equations 19.1 and 19.2). The unmeasured environmental conditions experienced by two seeds are expected to be more similar if the seeds were released in the same year, and even more similar if the seeds were released in the same experimental block. The extended model describes this second source of variation through nested random effects assuming that α and β vary lognormally between years and between blocks within the same year.

The extended Wald model closely fits the full dataset (Figure 19.2c). With a DIC of 984.6, it performs much better than the lognormal as the best simple dispersal kernel (DIC: 1431.6). This superior performance arises because the extended model describes systematic variation between release heights, years, and experimental blocks (Figure 19.1).

Stochastic terms play two different roles in the extended Wald model: firstly, stochastic Wald-distributed residuals describe effects of turbulent fluctuations on seed dispersal distance in an environment that is characterized by summary statistics of environmental conditions, and secondly, stochastic random effects describe how these summary statistics vary between years and blocks. These two types of stochastic elements differ in how easily they can be replaced by deterministic terms. The residual variation could

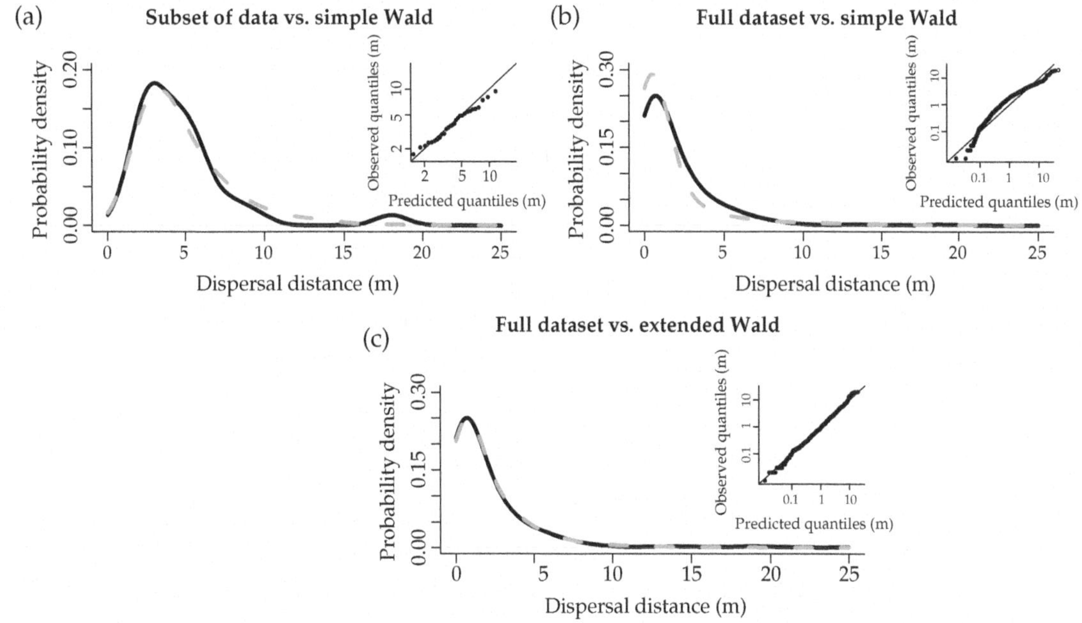

Figure 19.2 Fit of mechanistic Wald kernels to dispersal distance distributions of experimentally released *Hypochaeris radicata* seeds. Subplots compare (a) a simple Wald kernel with data for seeds released in 2006 from 1 m height, (b) a simple Wald kernel with the full dataset, and (c) an extended Wald model with the full dataset. For each comparison, the large plot shows density estimates of observed (black) and predicted (grey) distance distributions, whereas the insert shows a quantile–quantile plot of observed and predicted quantiles.

be reduced somewhat by using more elaborate turbulence models (such as the CELC model, Soons *et al.* 2004a). Yet, a strong reduction of residual variation seems essentially impossible as this would require both highly resolved data on the wind conditions experienced by each individual seed, and a highly resolved deterministic description of turbulence. In contrast, the second type of stochasticity can readily be reduced given that we have data on how summary statistics of wind conditions and vegetation structure vary in time and space.

19.3 The benefits of replacing stochasticity by process

Thanks to its deterministic backbone, the extended Wald model yields insights into seed dispersal that are impossible from a simple, purely stochastic kernel. In the following, I illustrate these benefits by showing how the extended Wald model (1) helps to understand how fat-tailed dispersal kernels are generated, (2) reduces uncertainty in the prediction of long-distance dispersal, and (3) predicts how kin selection should affect release height.

19.3.1 The making of a fat-tailed dispersal kernel

Simple Wald kernels are thin-tailed (or exponentially bounded): on the semi-log plot in Figure 19.3, their tails form straight lines. In this respect, simple Wald kernels differ qualitatively from lognormal or 2Dt kernels that have fat tails which curve away from the x-axis of a semi-log plot (Clark *et al.* 1999; Stoyan and Wagner 2001). Consequently, simple Wald kernels predict lower rates of long-distance dispersal than lognormal or 2Dt kernels with the same mean and variance.

If simple Wald kernels are thin-tailed, is the extended Wald kernel then also thin-tailed? To examine this, we have to know how the parameters of the extended Wald model (α, β and h_r) vary between seeds. For environmental conditions (α and β this is straightforward: by fitting the extended Wald model to the experimental data, we have already estimated the variation of α and β between years and experimental blocks (Figure 19.3). For release height h_r, the case is slightly different: since h_r was experimentally manipulated, its natural distribution in *Hypochaeris* cannot be estimated from the experimental data. To realistically describe variation in h_r, I thus chose a lognormal distribution with a mean of 0.3 m and a standard deviation of 0.06 m (Figure 19.3). More than 99.9% of this release height distribution falls within the empirically reported range from 0.15–0.6 m (Turkington and Aarssen 1983). The dispersal kernel predicted by the extended Wald model now arises as a mixture of simple Wald kernels for all possible values of α, β and h_r. More precisely, it arises as a weighted mixture because the contribution of each simple Wald kernel is weighted by how likely the respective values of α, β and h_r are.

It turns out that this extended Wald kernel has a very fat tail (Figure 19.3 shows the kernel calculated from posterior mean estimates of the extended Wald model). Thus, even though simple Wald kernels are thin-tailed for any combination of environmental conditions (α, β) and release heights (h_r), a fat-tailed kernel arises when these thin-tailed kernels are combined according to realistic distributions of α, β and h_r. Similarly, Clark *et al.* (1999) showed that the fat-tailed 2Dt kernel arises as a continuous mixture of thin-tailed Gaussian kernels. Yet the extended Wald model goes further in enabling us to examine how variation in measurable dispersal traits and environmental conditions generates fat tails. Since the extended Wald model explicitly describes the effect of release height, we can use it to understand the role of this trait for the ecology and evolution of wind-driven dispersal. This is illustrated in the following two examples.

19.3.2 Reducing uncertainty about long-distance dispersal

Long-distance seed dispersal is of prime importance for many ecological and evolutionary processes (e.g. Trakhtenbrot *et al.* 2005; Nathan 2006; Nathan *et al.* 2008; Schurr *et al.* 2009). However, simple fat-tailed kernels result in great uncertainty about long-distance seed dispersal which can severely limit our ability to predict the spread of plant populations (Clark *et al.* 2003). This uncertainty arises

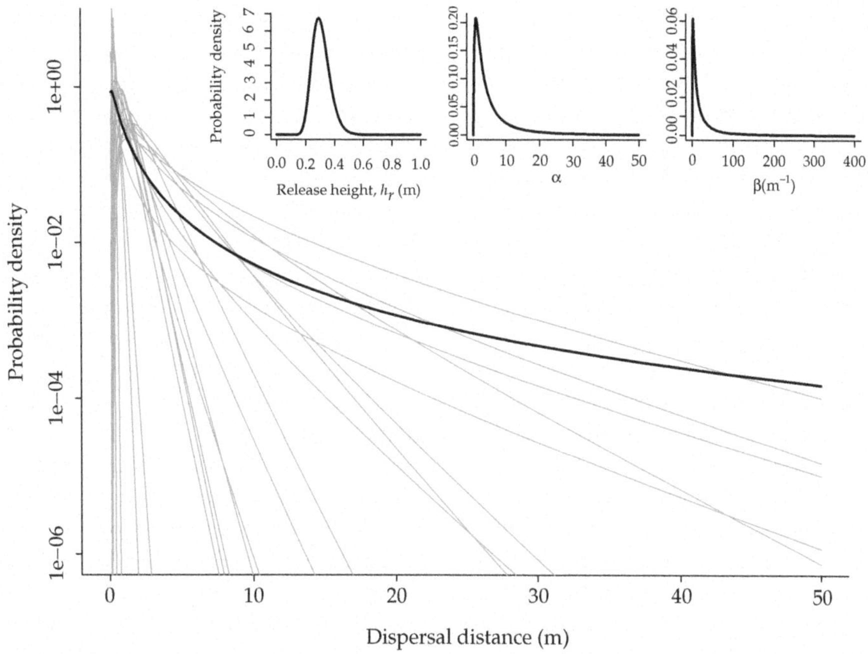

Figure 19.3 The making of a fat-tailed dispersal kernel. Thin lines show Wald kernels for 20 values of release height and environmental conditions (α and β), that were drawn at random from the distributions shown in the inserts. Distributions of α and β were estimated with the extended Wald model, whereas the height distribution was chosen to reflect realistic variation in the release height of *Hypochaeris radicata* seeds. The extended Wald kernel (bold line) arises as a weighted mixture of simple Wald kernels for all possible release heights and environmental conditions. The semi-log plot shows that all simple Wald kernels have thin tails (which form a straight line), whereas the extended Wald kernel is fat-tailed (curves away from the x-axis).

because simple kernels are purely stochastic and do not represent any knowledge about which seeds are likely to be dispersed far. For example, the simple lognormal kernel fitted to the experimental data predicts that each seed has a 1.5% probability of moving more than 20 m, irrespective of seed release height.

In contrast, the deterministic backbone of the extended Wald model enables one to identify which seeds are likely to be dispersed far: the extended model predicts that—across all environments—a seed starting from 0.15 m height only has a 0.6% probability of dispersing beyond 20 m. Yet, for a seed starting from 0.6 m height, this probability is fivefold higher (3.0%). When combined with knowledge about the conditions under which *Hypochaeris* plants grow tall (Soons and Heil 2002), the extended model thus helps to reduce uncertainty about long-distance dispersal.

19.3.3 Seed release height and the intensity of sibling competition

Kin competition can have major effects on the evolution of dispersal (e.g. Hamilton and May 1977; Rousset and Gandon 2002; Levin *et al.* 2003; Ronce 2007). In particular, kin selection should act to reduce the probability that sibling seeds are dispersed to neighbouring sites as this reduces competition among siblings. In *Hypochaeris* one might expect kin competition to be particularly intense between seeds borne in the same capitulum. This is because (1) bees, bumblebees, and hoverflies as the main pollinators (Becker *et al.* 2008) may visit multiple flowers on the same capitulum, so that seeds from the same capitulum are likely to share the same father and thus be full sibs, and (2) seeds from the same capitulum are likely to be dispersed at approximately the same time and thus under similar environmental conditions.

From the fitted kernels one can calculate the probability that two seeds dispersed from the same capitulum land less than 0.5 m apart and may thus compete after establishment. This probability is thus a proxy for the intensity of sibling competition. For simplicity, I assume that both seeds are dispersed in the same direction (so that seeds dispersed over similar distances end up close to each other). The simple lognormal kernel predicts this probability of sibling competition to be 29.0%. Note that this probability is the same for all possible pairs of seeds, since simple kernels do not represent any processes that cause the dispersal distances of individual seeds to be correlated. In contrast, the extended Wald model describes the fact that two seeds released from the same capitulum under similar conditions are likely to be dispersed over similar distances. Consequently, this model predicts a far higher probability of sibling competition (63.7%) for two seeds dispersed from 0.15 m height in the same environment (comparable to the same experimental block). Yet the extended Wald model also describes the fact that the variance in dispersal distance increases with release height (Figure 19.1). This counteracts the positive correlation induced by shared environmental conditions. Hence, the extended Wald model predicts that seeds starting from 0.6 m height have a co-occurrence probability of only 22.2%. It thus suggests that kin selection should act to increase seed release height.

19.4 How much process can and should we include in dispersal models?

The examples in the previous section demonstrate some of the benefits of replacing stochasticity by process in the description of dispersal. But how much process can we include in dispersal models? Clearly, the extended Wald model introduced in Section 19.2.3 constitutes only a first step towards more deterministic descriptions of seed dispersal. In the *Hypochaeris* example, further process and determinism could be introduced by including explicit measurements of the terminal velocity of individual seeds, and of spatio-temporal variation in wind and vegetation properties. This would not only reduce uncertainty about dispersal, but would also enable one to assess how directed changes in environmental conditions and dispersal traits affect dispersal. For instance, one could examine how global warming induced changes in wind conditions affect dispersal (Soons *et al.* 2004b; Kuparinen *et al.* 2009; Nathan *et al.* 2011).

More process-based descriptions of dispersal are not limited to seed dispersal by wind. In general, one can extend simple statistical models of plant dispersal through deterministic descriptions of source and path effects (Schurr *et al.* 2008). Source effects arise from the fact that dispersal depends on properties of the mother plant (be they traits or environmental conditions at the mother plant location). Path effects arise from the fact that dispersal depends on all environments a seed or pollen grain encounters along its dispersal path. Amongst others, this framework has potential for describing seed and pollen dispersal by animals: source effects can describe attraction of animals to certain plants, whereas path effects can describe how environmental heterogeneity affects animal movement and the deposition of dispersal units (Schurr *et al.* 2008).

Our ability to deterministically describe plant dispersal will ultimately be limited: for instance, it seems virtually impossible to replace stochastic descriptions of turbulence (see Section 19.2.2) with deterministic terms. In most systems, we are, however, far from reaching these ultimate limits. Hence, purely stochastic kernels should be regarded as the start rather than the end point of dispersal studies; they should increasingly be replaced by extended dispersal models that have deterministic components describing how dispersal depends on traits and environmental conditions. Statistical methods of model selection and uncertainty analysis can then be used to identify how much deterministic process description we need to reliably predict dispersal. This integration of mechanistic models with statistical techniques may substantially advance our ability to understand the ecological consequences and the evolutionary causes of variation in dispersal. Such an enhanced mechanistic understanding will also improve our ability to predict dispersal under the novel conditions that plants are likely to experience in the future.

Acknowledgements

Many thanks to all participants and tutors of the 'Plant Population Ecology' field course that conducted the seed release experiment. For knowledgeable comments on an earlier manuscript version, I thank Merel Soons, James Bullock, and an anonymous reviewer. Furthermore, I am grateful to the staff of the University of Potsdam's Ecological Field Station Gülpe and acknowledge funding from the European Union through Marie Curie Transfer of Knowledge Project FEMMES (MTKD-CT-2006-042261).

References

Becker, U., Dostal, P., Jorritsma-Wienk, L.D. *et al.* (2008). 'The spatial scale of adaptive population differentiation in a wide-spread, well-dispersed plant species', *Oikos* 117: 1865–73.

Binder, P.M. (2008). 'Theories of almost everything', *Nature* 455: 884–5.

Clark, J.S. (2009). 'Beyond neutral science', *Trends in Ecology and Evolution* 24: 8–15.

Clark, J.S., Lewis, M., McLachlan, J.S. *et al.* (2003). 'Estimating population spread: What can we forecast and how well?', *Ecology* 84: 1979–88.

Clark, J.S., Silman, M., Kern, R. *et al.* (1999). 'Seed dispersal near and far: Patterns across temperate and tropical forests', *Ecology* 80: 1475–94.

Ghazoul, J. (2005). 'Pollen and seed dispersal among dispersed plants', *Biological Reviews* 80: 413–43.

Hamilton, W.D. and May, R.M. (1977). 'Dispersal in stable habitats', *Nature* 269: 578–81.

Katul, G.G., Porporato, A., Nathan, R. *et al.* (2005). 'Mechanistic analytical models for long-distance seed dispersal by wind', *American Naturalist* 166: 368–81.

Kuparinen, A., Katul, G., Nathan, R. *et al.* (2009). 'Increases in air temperature can promote wind-driven dispersal and spread of plants', *Proceedings of the Royal Society B* 276: 3081–7.

Levin, S.A., MullerLandau, H.C., Nathan, R. *et al.* (2003). 'The ecology and evolution of seed dispersal: a theoretical perspective', *Annual Review of Ecology, Evolution, and Systematics* 34: 575–604.

Lunn, D.J., Thomas, A., Best, N. *et al.* (2000). 'WinBUGS—A Bayesian modelling framework: Concepts, structure, and extensibility', *Statistics and Computing* 10: 325–37.

Nathan, R. (2006). 'Long-distance dispersal of plants', *Science* 313: 786–8.

Nathan, R., Horvitz, N., He, Y.P. *et al.* (2011). 'Spread of North American wind-dispersed trees in future environments', *Ecology Letters* 14: 211–19.

Nathan, R., Katul, G.G., Horn, H.S. *et al.* (2002). 'Mechanisms of long-distance dispersal of seeds by wind', *Nature* 418: 409–13.

Nathan, R., Schurr, F.M., Spiegel, O. *et al.* (2008). 'Mechanisms of long-distance seed dispersal', *Trends in Ecology and Evolution* 23: 638–47.

Pico, F.X., Ouborg, N.J., Van Groenendael, J. (2004). 'Influence of selfing and maternal effects on life-cycle traits and dispersal ability in the herb Hypochaeris radicata (Asteraceae)', *Botanical Journal of the Linnean Society* 146: 163–70.

Ronce, O. (2007). 'How does it feel to be like a rolling stone? Ten questions about dispersal evolution', *Annual Review of Ecology Evolution and Systematics* 38: 231–53.

Rousset, F. and Gandon, S. (2002). 'Evolution of the distribution of dispersal distance under distance-dependent cost of dispersal', *Journal of Evolutionary Biology* 15: 515–23.

Schurr, F.M., Spiegel, O., Steinitz, O. *et al.* (2009). 'Long-distance seed dispersal', in L. Østergaard (ed), *Fruit Development and Seed Dispersal*. John Wiley and Sons, Chichester, 204–37.

Schurr, F.M., Steinitz, O. and Nathan, R. (2008). 'Plant fecundity and seed dispersal in spatially heterogeneous environments: models, mechanisms and estimation', *Journal of Ecology* 96: 628–41.

Soons, M.B. and Heil, G.W. (2002). 'Reduced colonization capacity in fragmented populations of wind-dispersed grassland forbs', *Journal of Ecology* 90: 1033–43.

Soons, M.B., Heil, G.W., Nathan, R. *et al.* (2004a). 'Determinants of long-distance seed dispersal by wind in grasslands', *Ecology* 85: 3056–68.

Soons, M.B., Nathan, R., Katul, G.G. (2004b). 'Human effects on long-distance wind dispersal and colonization by grassland plants', *Ecology* 85: 3069–79.

Spiegelhalter, D.J., Best, N.G., Carlin, B.R. *et al.* (2002). 'Bayesian measures of model complexity and fit', *Journal of the Royal Statistical Society Series B-Statistical Methodology* 64: 583–616.

Stoyan, D. and Wagner, S. (2001). 'Estimating the fruit dispersion of anemochorous forest trees', *Ecological Modelling* 145: 35–47.

Trakhtenbrot, A., Nathan, R., Perry, G. and Richardson, D.M. (2005). 'The importance of long-distance dispersal in biodiversity conservation', *Diversity and Distributions* 11: 173–81.

Turkington, R. and Aarssen, L.W. (1983). 'Biological Flora of the British Isles: *Hypochoeris radicata* L', *Journal of Ecology* 71: 999–1022.

PART V

Dispersal and Population Spatial Dynamics

CHAPTER 20

Linking dispersal to spatial dynamics

Tim G. Benton and Diana E. Bowler

20.1 Introduction

What determines a species' spatial dynamics? It is both profound and trivial to highlight that the spatial dynamics of a population arise from the sum of individual behaviour and life histories (births, deaths, movements). For some individuals, movement only happens once in their life: a plant moves as a seed, a mussel as a planktonic larva. Others move continually throughout life: a bluefin tuna can forage across the whole Atlantic over its lifetime (Block *et al.* 2005). Some individuals migrate between two centres of activity, such as swallows summering in Europe and wintering in Africa. For many species, individuals move around relatively small localities for most of the time (their territory, home range, or 'patch') occasionally moving longer distances. Given this continuum of movements, it is clear that there is no simple definition of dispersal as a behaviour that is universally applicable; most definitions have clear applicability to specific taxa or systems. The clearest conceptual definition is 'any movement of individuals or propagules with potential consequences for gene flow across space' (Ronce 2007) and so can be defined by the result rather than the behaviour *per se*. Movements in space will create spatial gene flow if breeding occurs away from the natal area, whatever the proximal cause of movement (to find food, shelter, mates, or escape predators), and whether the movement was 'dispersal behaviour' *per se* or 'routine movements' (Van Dyck and Baguette 2005).

For simplicity, it is useful to think of habitat being in 'patches' because spatio-temporal variation in the environment means that there is almost always some inherent structure imposed on populations as some areas provide more suitable habitat than others. Clearly, however, this 'patch-centred view' is a simplification. Firstly, patches are not likely to be discrete, may vary in suitability, and may have 'fuzzy boundaries'. Second, thinking in terms of habitat being in isolated patches leads to the overly simplistic notion that dispersal is defined as movement between patches, whereas in many real systems, organisms may also disperse within a patch as well as between patches (e.g. woodland birds inhabiting a patch of wood may disperse between suitable territories within it). Many of the dynamical consequences of movement arise from patches being connected, but not all dispersal will contribute to connectivity if some dispersal is within a patch. Third, patches may be spatially clustered at many hierarchical levels. For example, trees clump into woods, woods into valleys, valleys into regions, and so movement between any two patches may be very different in character depending on their overall spatial configuration (hence for some systems short and long distance dispersal behaviours exist; Bonte *et al.* 2010). The patch, and therefore dispersal between them, can be defined differently depending on the level in the hierarchy. Although many investigations of spatial dynamics, for simplicity, define discrete patches within a modelling framework, researchers are increasingly defining space continuously to make the models more realistic (e.g. North and Ovaskainen 2007; Cornell and Ovaskainen 2008; Travis *et al.* 2010). In continuous space, habitat variability is defined by spatial autocorrelation between points, and 'patches' are not explicitly delineated, but they will functionally occur when nearby points are good habitat:

Notwithstanding that thinking in terms of patches is a simplification, it remains conceptually useful because many of the ecological consequences of

Dispersal Ecology and Evolution. First Edition. Edited by Jean Clobert *et al.*

spatial connectivity depend on three simple concepts: patch quality, size, and inter-patch distance. Patch quality and size determine the local population size, which in turn determines the likelihood of persistence (as smaller populations are more likely to go extinct due to both demographic and environmental stochasticity); the spacing between patches determines the connectivity or fragmentation of the habitat.

All environments vary over time, and for any isolated population, where resources are finite, stochastic population theory predicts that in the long run it will go extinct just by chance (Renshaw 1991). Thus, if there was no movement between patches, the inevitable fate of the overall population is extinction. In this sense, dispersal can be regarded as the 'glue' that holds the population together: without dispersal extinct patches would not be re-colonized, and populations with low density would not be 'rescued' by immigration (demographically or genetically).

We often tend to think in terms of spatial dynamics being about changes in the number of individuals across space (i.e. population dynamics). However, population dynamics are necessarily related to phenotypic, genetic, and evolutionary dynamics as well. If individuals move, they affect (a) phenotypic dynamics: if the dispersing individuals are non-random with respect to phenotype, it will change the average phenotype in a patch when individuals join or leave; there will also be indirect effects as movement will typically alter local population densities, stage-structure, and competition and therefore affect phenotypically plastic traits of individuals, (b) genetic dynamics via altering gene flow (by definition) and hence the genetic composition within and between patches, and (c) evolutionary dynamics via affecting genetic and phenotypic

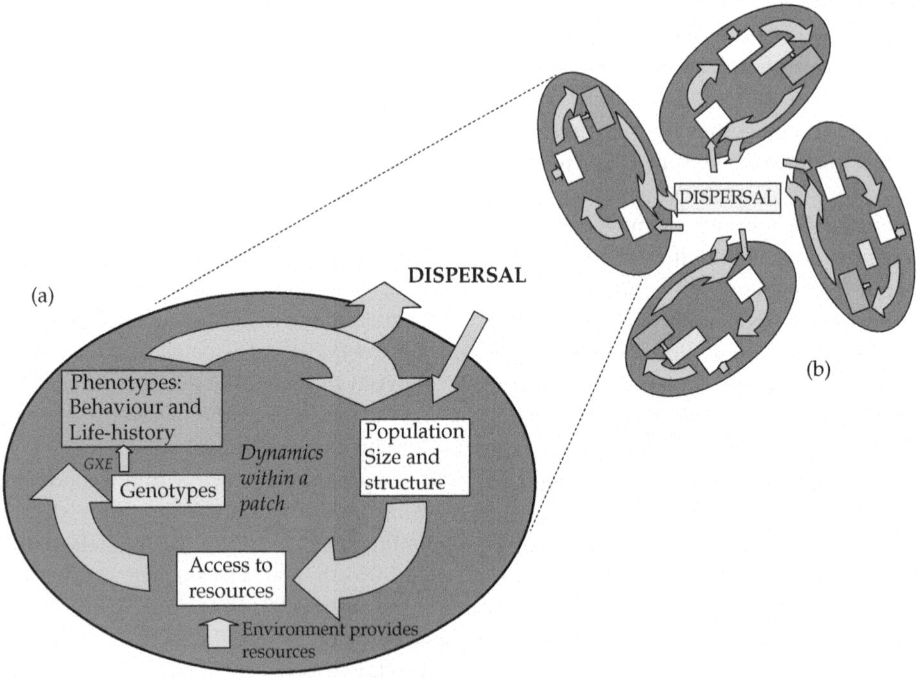

Figure 20.1 A metapopulation consists of smaller populations (dark grey patches) connected by dispersal. Whether an individual disperses from a patch is unlikely to be a fixed probability, but is more likely to depend on conditions within the patch. For example, the resources in the patch will, in combination with population size and structure, determine the resources available for individuals. The phenotype (whether life history, or dispersal phenotype) thus emerges from gene × environment interactions. Dispersal is thus likely to be both context- and condition-dependent. The rate of dispersal out of a patch will further affect the dynamics in other patches due to their effects on gene flow and via perturbing population processes.

dynamics and potential effects on inbreeding/outbreeding, local adaptation, and range-wide adaptation. Therefore, spatial population dynamics arise as an emergent property of a system of spatially connected patches, where each patch is itself a system of linked genetic, phenotypic and population dynamics (Figure 20.1).

In this chapter, we focus on population dynamics, and how the demographic processes underpinning them are affected by dispersal and its evolution. For the sake of conciseness, we do not cover the rich literature on the role of space in population genetic structure and diversity (but also see Part II), inbreeding, the invasion dynamics of rare alleles, and the evolution of specific traits and behaviours (such as movement behaviour, social systems, and life history). We structure this chapter by examining the dynamical consequences of dispersal on (a) changes in local population density, (b) changes in non-local population density, (c) colonization of empty patches, and (d) spatial gene flow.

20.2 Dispersal changes local population density, which has many consequences

The movement of individuals into or out of a patch clearly has a direct effect on the population size within the patch, which can be predicted to be most important for small populations already at increased risk of extinction. For any small population, extinction risk is inversely proportional to population size (Lande 1993) due to the effects of demographic stochasticity and Allee effects. Demographic stochasticity is the randomness associated with individuality (Fox 2005) (such as a mother giving birth to same-sex broods), and its impact is most prevalent in small populations because such stochastic variation averages out in large ones. As demographic stochasticity increases extinction risk, it acts as a selection pressure for dispersal (Cadet *et al.* 2003; Parvinen *et al.* 2003). In addition to demographic stochasticity, small populations may suffer from Allee effects, which occur when, below a threshold population size, density dependence on population growth goes from negative to positive (e.g. when mates become rare, reproduction cannot happen) (Courchamp *et al.* 1999). Dispersal away from a small population can therefore increase the risk of a local extinction (Menendez *et al.* 2002) while immigration can rescue a population by raising its size above the extinction threshold (Brown and Kodric-Brown 1977; Kendall and Fox 1998).

Access to resources is typically density-dependent, so emigration of individuals from a high density patch may also free up resources leading to changes in resource acquisition of the philopatric individuals (e.g. Gundersen *et al.* 2001b; Keeley 2001). This may potentially lead to a compensatory increase in population growth. In an experimental study on root voles, *Microtus*, the loss of dispersers from a population was shown to lead to an increase in the per capita recruitment rate of the remaining individuals (Gundersen *et al.* 2001b). This effect is particularly important perhaps among related individuals, where dispersal away from the natal patch reduces kin competition for resources, as well as reducing the potential for inbreeding (see Section 20.5 below) (Hamilton and May 1977; Ridley and Sutherland 2002; Cote *et al.* 2007); dispersal away from kin can therefore be considered an altruistic trait (Rousset and Ronce 2004; Jansen and Vitalis 2007). The freeing up of resources as individuals emigrate may also influence the emigration of others as the dispersal of many species depends on their condition, which may therefore increase as emigrants leave (Bowler and Benton 2005; see Part I, and Bonte and de la Pena 2009). Conversely, immigration increases local density and therefore adds to local density-dependence, which in turn can impact on any density-dependent phenotype (growth rate, fecundity, survival, territorial acquisition, dispersal, etc.) that may result in a decrease in population growth. Thus, the longer-term demographic effects of the immediate changes in density from emigration and immigration depend on the density-dependence of processes affecting local dynamics (Ives *et al.* 2004). Immigrants may have a smaller effect when moving into a population that already has a high density compared to when moving into a small population due to density-dependent competition (Gundersen *et al.* 2002).

The realized effect of dispersal depends on net movement into a patch, determined by the balance between immigration and emigration. Thus in turn critically depends on the presence and pattern (e.g. autocorrelation) of variation in the environment, as this causes differences in the population size of different patches and differences in the number of emigrants that they produce (Ives *et al.* 2004; Bowler and Benton 2011). For some populations, demographic stochasticity, or more complex intrinsic population dynamics, may also be responsible for generating variation in density among patches, affecting net movements.

20.3 Dispersal has non-local impacts on dynamics

Dispersal clearly has a range of effects on spatial dynamics beyond the immediate changes in local density created by immigration/emigration. For instance, the total population density will be affected by dispersal mortality. In mites the total population size decreased with increasing distance between patches as a greater proportion of individuals suffered dispersal mortality (Bowler and Benton 2009). Dispersal can also have additive effects on total population size. Ives *et al.* (2004) used a fungus laboratory system to demonstrate how dispersal could increase average density when populations are exposed to local environmental variation and density-dependence on population growth within patches is weak. Average density increases as long as the net immigration rate is positive when populations are growing and negative when they are declining (Ives *et al.* 2004).

The concept of source-sink population dynamics emphasizes that dispersal allows some populations to persist that could not in the absence of dispersal: populations with negative population growth rates (where mortality exceeds reproduction; so-called 'sinks') are sustained via emigration from populations with positive population growth rates ('sources'). There is considerable evidence for this in empirical systems (Gundersen *et al.* 2001a; e.g. Bonte *et al.* 2004; Tittler *et al.* 2006; Milot *et al.* 2008). In fact, in spatially stochastic situations, theory suggests that one can have the counterintuitive result that metapopulations can persist even if on average every population is a sink, as long as at any one time, some populations are at an above-average state and performing as sources (Schreiber 2010). An important factor affecting the dynamics of source-sink systems is whether dispersing individuals discriminate between source and sink patches during dispersal; i.e. whether individuals only move into sink habitat when there is no space in the source patch, or whether individuals passively disperse from the source to the sink without discrimination (Pulliam and Danielson, 1991). If dispersing individuals are not able to recognize a sink patch with any associated cues, the net loss of individuals due to dispersal may act as a drain on the source patch. This may occur if anthropogenic environmental changes results in mismatches between cues used by dispersing individuals and actual patch quality, dispersal choices may be maladaptive and lead to negative population growth (Kokko and Sutherland 2001).

If two populations are linked by dispersal at a fixed rate, their densities will tend to become more similar as the patch with the higher density produces more emigrants, leading to higher immigration and higher density in the coupled patch. When populations show oscillations in density that are driven by endogenous factors (e.g. delayed density-dependent cycles such as those driven by maternal effects or predator-prey cycles), their dynamics can become phase-locked such that they fluctuate exactly in synchrony (Figure 20.2) (Tanaka *et al.* 1997; Jansen 1999).

Phase locking is a purely physical property of linked oscillators (whether they are pendulums or population dynamical oscillators). If two oscillators are connected, they will tend to become entrained and oscillate in synchrony; this process is governed by the coupling strength (Bennett *et al.* 2002), which in ecological systems will be the dispersal rate. If oscillatory populations are all connected (e.g. by 'global' dispersal), all populations will oscillate together. However, if oscillators are coupled only to neighbours, it is possible to get complex patterns of spatial synchrony. A common one, with empirical support, is that of a travelling wave. Imagine populations in a row. If the population at the start has a high density, it sends out many dispersers. These immigrate into the population next door. This increases the population size there, but the peak

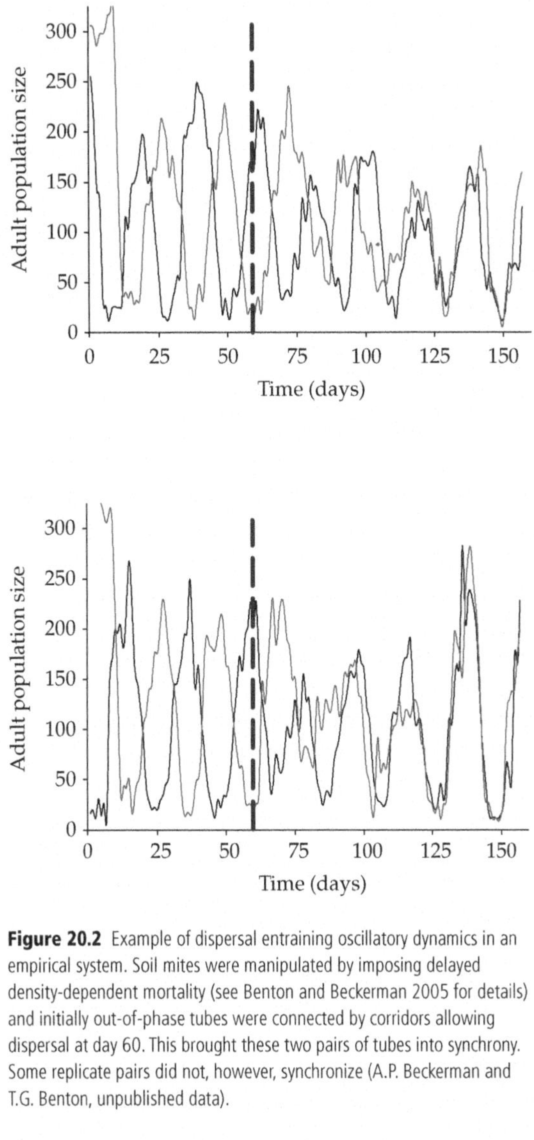

Figure 20.2 Example of dispersal entraining oscillatory dynamics in an empirical system. Soil mites were manipulated by imposing delayed density-dependent mortality (see Benton and Beckerman 2005 for details) and initially out-of-phase tubes were connected by corridors allowing dispersal at day 60. This brought these two pairs of tubes into synchrony. Some replicate pairs did not, however, synchronize (A.P. Beckerman and T.G. Benton, unpublished data).

size follows the peak size at the first population with a time lag. As a result, the peak population passes, like a Mexican wave, along the line of patches (Ranta and Kaitala 1997; Sherratt *et al.* 2000; Sherratt 2001; Takahashi *et al.* 2005). Travelling waves are not just a theoretical phenomenon, having been observed in field studies of vole populations (Lambin *et al.* 1998). In realistic landscapes, where there is interaction with landscape characteristics, travelling waves can be highly complex in form (Blasius *et al.* 1999; Sherratt *et al.* 2002; Smith *et al.* 2008). If the population growth rates within any given patch are high, dispersal linkage can lead to spatial chaos in many situations (Doebeli and Ruxton 1997; Blasius and Stone 2000; Petrovskii and Malchow 2001).

Many of the studies on spatial patterns of population dynamics have considered populations that are cyclic, perhaps because this predisposes them to be noticed. However, although many species do show some degree of periodicity (Kendall *et al.* 1999) and theory indicates that cycling is an evolutionarily stable dynamical pattern (Greenman *et al.* 2005), the last decade has brought consensus that ubiquitous environmental noise such as spatio-temporal variation in weather can interact with density dependence to create population dynamics of great complexity. This means that understanding dynamics requires understanding both density-dependent and density-independent processes (Bjørnstad and Grenfell 2001). For example, noise can interact with deterministically stable (equilibrium) dynamics to create population cycles via a process akin to resonance (Greenman and Benton 2003, 2005). The recognition that spatially coupled systems exposed to environmental noise display a range of dynamics has led to a considerable literature exploring this interaction in a spatial setting (Lundberg *et al.* 2000; Turchin and Ellner 2000; Greenman and Benton 2001; Haydon *et al.* 2003; Liebhold *et al.* 2004; Tobin and Bjørnstad 2005; Ovaskainen and Cornell 2006a, 2006b; Ross 2010).

Population synchrony has been of particular interest as it has important implications for population persistence (Heino *et al.* 1997; Cazelles *et al.* 2001; Koenig 2001; Greenman and Benton 2005). Persistence of a spatially structured metapopulation relies on asynchronous patch dynamics: for there to be re-colonization of empty patches, or rescue of populations at low density, other patches have to be occupied, or have a higher density to ensure a source of dispersers. For populations to be in synchrony, they need to be dynamically connected. Dispersal is one way populations can be connected (Figure 20.2). There is some empirical information to support a synchronizing effect of dispersal (Paradis *et al.* 1999; Post and Forchhammer 2002; Schwartz *et al.* 2002; Sundell *et al.* 2004; Fontaine and Gonzalez 2005; Ruetz *et al.* 2005; Liebhold *et al.* 2006). High rates of dispersal in experimental metapopulations of *Drosophila* led to population synchrony, while lower rates of dispersal did not (Dey and Joshi 2006).

Dynamical connection, and therefore synchrony, can also arise without dispersal, due to weather, which is spatially autocorrelated over large regions, or via populations sharing predators, such that changes in predator pressure simultaneously impact many prey populations (Ims and Andreassen 2000; Petty *et al.* 2000). Synchrony arising through shared environmental fluctuations is termed the Moran effect after a famous population biologist (Greenman and Benton 2001; Royama 2005). In realistic situations, the Moran effect is not a strong synchronizer because noise tends to amplify any differences in structure between different populations (Grenfell *et al.* 1998; Benton *et al.* 2001; Greenman and Benton 2001; Royama 2005). In nature, it is likely that all three causes—dispersal, spatially autocorrelated noise and predators—interact in producing patterns of synchrony (Kendall *et al.* 2000; Liebhold *et al.* 2006). The spatial scale at which populations are brought into synchrony will depend on details of the interaction (Snyder 2007). Dispersal rates generally decline as the distances between patches increase, according to the movement ability of the organism, and any costs, such as any energetic and mortality costs that it experiences when moving through the landscape. Thus, dispersal can act to synchronize population dynamics over a spatial scale that depends on dispersal distance (Ranta *et al.* 1998; Lande *et al.* 1999; Paradis *et al.* 1999), and, in fact, the spatial scale of population synchrony is often taken as evidence for or against dispersal as a causative agent (e.g. Post and Forchhammer 2002; Sundell *et al.* 2004), and, even to infer dispersal rates (Schneeberger and Jansen 2006).

The effect of distance on dispersal and, in turn, synchrony, also means that the spatial configuration of patches is important for persistence. In a study of experimental annual plant metapopulations, there was a non-linear relationship between the inter-patch distance and extinction risk so that those separated by an intermediate distance persisted the longest (Molofsky and Ferdy 2005). The most connected populations were at increased risk of extinction because of the synchronizing effect of dispersal.

The distance dispersed by individuals (i.e. the dispersal kernel, Chapter 15) is clearly very important for spatial dynamics as it determines rates of gene flow, how perturbations in density spread across the landscape, and spatial synchrony. The importance of the dispersal kernel lies not just in the mean distance dispersed, but also the variance in dispersal distances between individuals. This variance is important for two main reasons. First, rare, long-distance dispersal events, described by the tail of the distribution, can be very important dynamically by creating jumps across the landscape (Lindstrom *et al.* 2008). A 'fat-tail' of the dispersal kernel can arise both through randomness in distance travelled but also heterogeneities between individuals (Petrovskii *et al.* 2008) such as in their physical condition (see Section 20.1). Second, the variance between individuals affects the immigration rate into any given patch, and therefore the magnitude of the perturbation in its population density. Imagine living next door to a source patch: if all the sources' emigrants immigrate into your patch, density dependence is strengthened compared to the situation if individuals disperse across the landscape. The shape of the dispersal kernel clearly can also evolve in response to habitat structure (Hovestadt *et al.* 2001; Murrell *et al.* 2002; Bonte *et al.* 2010; Starrfelt and Kokko 2010). As a result, populations of the same species from different types of landscape may disperse in different ways (Schtickzelle *et al.* 2006; Wang *et al.* 2011).

Thus far, we have focused on a single species interacting with its environment. However, species clearly exist in interaction networks with other species (competitors, predators, parasites, mutualists, etc.). Dispersal of the interacting species can obviously impact on the population dynamics of any focal species. This is most apparent if one thinks of a parasite or disease's transmission rate as its dispersal (Cook *et al.* 2007), because a disease's transmission patently affects the host-population dynamics. There are many complex interactions between the spatial scales of dispersal of interacting species and their population growth rates (Murrell 2005). A recent empirical example of the importance of dispersal in structuring meta-communities is given by Bergerot *et al.* (2010). They showed that a herbivorous butterfly was unaffected by habitat fragmentation along an urban gradient, but its parasitoid wasp was, leading to a decrease in parasitism rates with increased fragmentation.

20.4 Dispersal allows colonization of empty patches

The sections above focus on the way that dispersal affects population sizes by changing population densities and perhaps creating patterns of spatial synchrony. A particularly important form of changing population density occurs when an empty patch is colonized. This can be the colonization of a patch that was previously occupied and has gone extinct, or the colonization of new patches during range changes.

Perhaps the most powerful, and simple, conceptualization of spatial dynamics was articulated by Levins in 1969, and is based around the dynamics of colonization and extinction. The Levins' metapopulation model consists of a collection of patches that are either occupied or vacant. Occupied patches go extinct at a given rate, and empty patches can be colonized at a rate that depends on the number of occupied patches and a dispersal rate. The key insight of this work has been that regional persistence is possible despite local extinctions if there is sufficient dispersal among patches. A central assumption here is that within-patch dynamics are fast relative to between-patch dynamics—so if a patch is 'occupied' it instantaneously reaches dynamical equilibrium (i.e. dispersal does not affect, or is affected by, patch dynamics). Levins-type models have been very powerful despite, and perhaps because of, their simplicity, and have been extended in various ways to make them biologically more realistic such as by making space explicit (Roy *et al.* 2008), adding dynamics within patches (Jansen and Vitalis 2007), or non-random dispersal, e.g. density-dependent dispersal (Saether *et al.* 1999). There is a whole literature based around Levins-type models that focus on the dynamics of the patches in terms of their extinctions and colonizations and make predictions on patch occupancy. For some empirical systems, Levins-type models are very appropriate because patches are small and dynamic (e.g. ephemeral or extinction-prone due to environmental fluctuation), and the overall dynamic is governed by colonizations and extinctions happening at an appreciable rate (e.g. butterflies, Saccheri *et al.* 1998; Schtickzelle *et al.* 2005; Wilson *et al.* 2010), watervoles (Telfer *et al.* 2001), aquatic plants (Dahlgren and Ehrlen 2005), or plants living in ephemeral/successional habitats (Jacquemyn *et al.* 2006; Milden *et al.* 2006). Adding more biological realism to patch-occupancy models alters the predictions and can make them more applicable to other systems (Higgins and Cain 2002; Verheyen *et al.* 2004).

A priori, we would expect edges of species 'ranges to be shaped by extinction-colonization dynamics. A species' range can be defined as the spatial extent over which its population persists (i.e. the population growth rate within a patch is on average ≥ 1). The implication is that, in an equilibrium situation, patches on the very edge of a species' range are sinks. Therefore, at the edge of a range, there is an expectation that local populations are more vulnerable to extinction (and as a result are often the focus of conservation effort) (Yakimowski and Eckert 2007). Dispersal determines how quickly a species moves across a landscape, and therefore the rate of change in the range (Buckley *et al.* 2010). Clearly, there are two situations where the interaction between dispersal and a species' range is particularly important: where the range is changing in response to environmental change or in the case of a biological invasion. There is an important difference between these two. In a biological invasion scenario, the assumption is that habitat is static and unchanging (so distant patches are as habitable as near patches). The shape of the dispersal kernel, in particular the number of rare long-distance dispersers, determines the rate of invasion; a rare long-distance disperser can 'spark' a new invasion in advance of the range front (Keeling *et al.* 2001). In an environmental-change scenario, the assumption is that the range is constrained by biotic or abiotic factors that typically prevent an organism inhabiting distant patches (Polechova *et al.* 2009), thus long-distance dispersal outside the range do not generally lead to colonization. Climate change creates new habitat on the leading edge and removes it on the trailing edge. Depending on the rate of change of the suitable habitat, the range will shift, typically, with some degree of a time lag (Mustin *et al.* 2009). The movement of the range is also associated with a change in shape as the leading edge becomes more abrupt (Mustin *et al.* 2009).

Dispersal behaviour can be an important factor shaping the range or changes in range (see Part VI); an organism with passive dispersal potentially settles in any patch that can be reached regardless of quality, while a more selective disperser may only settle in higher quality habitat and therefore occupy a more restricted distribution (Armsworth and Roughgarden 2005). Dispersal rates can often be different at range margins, reflecting differences in the costs and benefits of dispersal compared to the population core (Simmons and Thomas 2004; Llewelyn *et al.* 2010). This difference can arise from a number of causes such as evolution (see Section 20.5), change in expression of plastic traits, and parental effects on dispersal (Benard and McCauley 2008; Duckworth 2009; Polechova *et al.* 2009; Berg and Ellers 2010; Chevin *et al.* 2010). The change in dispersal behaviour clearly interacts with the speed at which the range change can occur: typically where range change is occurring, range-edge populations are more dispersive as they can reap the benefits of colonizing an empty patch, leading to an acceleration of range change (Llewelyn *et al.* 2010).

20.5 Dispersal, spatial gene flow, and evolutionary dynamics

Implicit in the above discussion is that if spatial population dynamics are being affected, genetic and evolutionary dynamics will be as well. Changes in population density due to dispersal affect competition within the patch, and hence selection, and strength of selection, for particular traits. Extinction dynamics also affects selection pressures, and may create selection via differential survival of groups (demic or group selection) as well as through natural selection on genes. Dispersal is also under selection, shaped by the benefits and costs of dispersal, and the way they trade off against other investments. For example, a recent evolutionary model (Travis *et al.* 2012) suggests that when dispersal costs are high, there will be selection for increased perceptual range to allow more direct, and therefore less costly, movement paths.

By definition, dispersal results in spatial gene flow; i.e. the movement of genes through space. An individual can indirectly affect gene flow by moving, even without reproducing, due to the impacts on local dynamics affecting genetic processes. However, direct influences on spatial gene flow require movement followed by reproduction. It is worth highlighting that it is possible for the demographic and genetic consequences of dispersal to become de-coupled. A single immigrant *can* have a dramatic impact on the genetic diversity of a population (Madsen *et al.* 1999; Vila *et al.* 2003), depending on age, sex, and life-history traits such as fecundity, as well as properties of the patch that the immigrant enters (e.g. mate availability, resources, population size, etc.). For instance, immigrants into an inbred population that mate with residents may result in offspring with hybrid vigour, which increases the realized gene flow (Ebert *et al.* 2002). In other cases, maternal or early environmental effects can radically affect adult phenotype, which may create phenotypic mismatches that reduce the establishment of immigrants into new patches, and therefore their genetic contribution to them (Marshall *et al.* 2010).

Spatial gene flow arising from dispersal has many consequences for genetic diversity and evolutionary processes. Movement away from the natal territory reduces the potential for inbreeding and its deleterious effects, while immigration into a distant patch is likely to promote outbreeding. Spatially separated populations can become locally adapted (Kawecki and Ebert 2004; Hedrick 2006), and this is perhaps most likely at range margins where environmental conditions can differ from those in the core (Kunin *et al.* 2009). Gene flow is typically thought to act as a brake on local adaptation, but this is not always the case as it depends on the demographic benefit of local adaptation versus the costs of maladaptation (Hanski *et al.* 2011). Furthermore, if there is life-history variation associated with dispersal choices, dispersal can result in assortment of like with like, and contribute to genetic differentiation ('directed gene flow', Edelaar *et al.* 2008). Similarly, at dynamic range margins, where populations are small and genetic drift can be strong, dispersal-related gene flow can cause iterated founders' effects that can drive mutants to fixation very rapidly ('mutation surfing', McInerny *et al.* 2009).

Gene flow resulting from dispersal clearly influences evolutionary dynamics. There is a very broad literature about the way spatial processes affect the evolution of different traits (social behaviour, mating systems, dispersal behaviour itself, or even phylogenetic branching patterns). We here pick three areas to highlight those particularly topical.

Firstly, environmental change, such as climate change, is causing geographic shifts in the locations to which species are climatically adapted (the 'climate envelope'). Species can move with the shift in the envelope, adapt, or die (Chevin *et al.* 2010). There is a growing literature that suggests some organisms have increased in dispersal propensity at the range margin (Thomas *et al.* 2001), and for slow-moving species, this may well mitigate against climate change by helping them track the envelope. Similarly, during invasions, there is often an evolutionary increase in dispersal as dispersers achieve a higher fitness. This then accelerates the rate of invasion in the new landscape (Phillips *et al.* 2008; Travis *et al.* 2009), potentially increasing the ecological problem of the invasion. Evolutionary changes in dispersal may also lead to changes in any correlated traits (e.g. lower reproduction), so that predicting rates of spread requires detailed understanding of the life history and how traits trade off against dispersal at the range margin (Hughes *et al.* 2003; Burton *et al.* 2010).

Second, climate change is only one aspect of contemporary environmental change. Humanity is rapidly changing the natural environment, affecting habitat quality, its fragmentation, and the costs of moving between patches. Changes in any of these will change the costs and benefits, and hence selection pressures, of dispersal. Numerous studies have shown that habitat configuration influences aspects of dispersal including, for example, dispersal propensity, the ability to select habitat, and the average and variance of distance moved (Murrell *et al.* 2002; Schtickzelle *et al.* 2006; Barton *et al.* 2009). Both short- and long-distance dispersal modes can be selected when patches are themselves clumped (Bonte *et al.* 2010). Furthermore, if the environment also varies over time, the way it changes, in terms of temporal autocorrelation, or in other words, predictability, affects the evolution of dispersal (Travis 2001). As discussed above, any evolutionary change in dispersal behaviour will affect the overall properties of the spatial system from genes, to individuals, to species interactions. For example, if habitat becomes fragmented and the costs of dispersal (e.g. mortality moving between patches) increases, there can be selection against dispersal. This may result in lowered gene flow and therefore greater isolation of habitat patches, which in turn can lead to an increase in extinction risk and 'evolutionary suicide' (Gyllenberg *et al.* 2002; Travis *et al.* 2010). Alternatively, if habitat fragmentation increases the benefits of dispersal—for instance by creating more empty patches—to a greater extent than the costs, there may be increases in dispersal and 'evolutionary rescue' of the population (Heino and Hanski 2001).

Third, a prime reason for individuals clumping in space is because they are related. Kin selection is therefore an important determinant of dispersal behaviour and *vice versa*. An interesting example is that if dispersal is local, such as trees shedding seeds nearby, death of the parent can be selected because it frees up space for the offspring—leading to the evolution of senescence (Travis 2004; Dytham and Travis 2006). Similarly, rates of dispersal can affect selection on offspring sex ratio via affecting the intensity of local mate competition (Foster and Benton 1992; Wild *et al.* 2006).

20.6 Conclusions

A species rarely, if ever, lives in a uniform environment where the population is well mixed. Instead, populations are spatially structured, connected by the movement of individuals between patches. The rates of movement then become crucial for defining the way the dynamical system (from genes, to phenotypes, to population and community dynamics) behaves, and how it responds to any changes in the environment—whether they be biotic ones (such as invasion of an alien species) or abiotic changes (such as habitat degradation, loss, or fragmentation, or climate change).

Spatial dynamics arise as an emergent property of a complex system comprising many interacting individuals (Figure 20.1). The complexity of the system, and the feedbacks within it, mean that we should not treat spatial population dynamics as a phenomenon and study it in isolation from the system that creates it. Instead, we should study the system in which the spatial population dynamics emerge. For example, an individual leaving a patch will free up resources for those that remain, which may alter their dispersal propensity. The individual who leaves will have developed in a given environment and this may influence the way they move across the landscape, the distance travelled, and the way they integrate into a new patch. Thinking of spatial dynamics within the concept of a simple dynamical system (i.e. as populations of size N connected by dispersal at a fixed rate) is unlikely to capture the way the system behaves.

Ecology is undergoing a change of mindset. In recent decades, we have used a range of more-or-less simple models designed to clarify and elucidate simple principles. The 'biology-as-physics' approach (Levins 1966) aims to develop simple explanations that are general, but may do so at the expense of realism. Although these simple models can be useful to isolate and allow understanding of the effects of specific parts of systems, it is increasingly important to capture the dynamics of real, and inherently complex, systems to predict how they will respond to anthropogenic change (Evans *et al.* 2012). Incorporating complex biological details leads to process-based rather than phenomenological models (Buckley *et al.* 2010). This move towards more process-based modelling has added a richness of detail to studies of spatial dynamics, by incorporating, for example, phenotypic plasticity (Berg and Ellers 2010; Jongejans *et al.* 2010), within-patch dynamics (Jansen and Vitalis 2007), and explicit space (Roy *et al.* 2008). In particular, individual-based simulation models, though lacking the generality of an elegant analytical model, are increasingly being used to generate new insights and predictions on the way that complex spatial systems respond to changes in the environment (e.g. Travis 2003; Hovestadt and Poethke 2006; Mustin *et al.* 2009; Bonte *et al.* 2010; Travis *et al.* 2010).

References

Armsworth, P.R. and Roughgarden, J.E. (2005). 'The impact of directed versus random movement on population dynamics and biodiversity patterns', *American Naturalist* **165**: 449–65.

Barton, K.A., Phillips, B.L., Morales, J.M. *et al.* (2009). 'The evolution of an 'intelligent' dispersal strategy: biased, correlated random walks in patchy landscapes', *Oikos* **118**: 309–19.

Benard, M.F. and McCauley, S.J. (2008). 'Integrating across life-history stages: Consequences of natal habitat effects on dispersal', *American Naturalist* **171**: 553–67.

Bennett, M., Schatz, M.F., Rockwood, H. *et al.* (2002). 'Huygens's clocks', *Proceedings of the Royal Society A* **458**: 563–79.

Benton T.G, Beckerman A.P. (2005) 'Population dynamics in a noisy world: lessons from a mite experimental system' in, R.A. Desharnis, ed., *Advances in ecological research: population dynamics and laboratory ecology. Vol. 37*, Elsevier, London, 143–181.

Benton, T.G., Lapsley, C.T., Beckerman, A.P. *et al.* (2001). 'Population synchrony and environmental variation: an experimental demonstration', *Ecology Letters* **4**: 236–43.

Berg, M.P. and Ellers, J. (2010). 'Trait plasticity in species interactions: a driving force of community dynamics', *Evolutionary Ecology* **24**: 617–29.

Bergerot, B., Julliard, R., Baguette, M. (2010). 'Metacommunity Dynamics: Decline of Functional Relationship along a Habitat Fragmentation Gradient', *PLoS One* **5**(6): e11294 http://www.plosone.org/article/info%3Adoi%2F10.1371%2Fjournal.pone.0011294

Bjrnstad, O.N. and Grenfell, B.T. (2001). 'Noisy clockwork: Time series analysis of population fluctuations in animals', *Science* **293**: 638–43.

Blasius, B., Huppert, A., Stone, L. *et al.* (1999). 'Complex dynamics and phase synchronization in spatially extended ecological systems', *Nature* **399**: 354–9.

Blasius, B. and Stone, L. (2000). 'Chaos and phase synchronization in ecological systems', *International Journal of Bifurcation and Chaos* **10**: 2361–80.

Block, B.A., Teo, S.L.H., Walli, A. *et al.* (2005). 'Electronic tagging and population structure of Atlantic bluefin tuna', *Nature* **434**: 1121–7.

Bonte, D., Baert, L., Lens, L. *et al.* (2004). 'Effects of aerial dispersal, habitat specialisation, and landscape structure on spider distribution across fragmented grey dunes', *Ecography* **27**: 343.

Bonte, D. and de la Pena, E. (2009). 'Evolution of body condition-dependent dispersal in metapopulations', *Journal of Evolutionary Biology* **22**: 1242.

Bonte, D., Hovestadt, T., Poethke, H.J. (2010). 'Evolution of dispersal polymorphism and local adaptation of dispersal distance in spatially structured landscapes', *Oikos* **119**: 560–6.

Bowler, D.E. and Benton, T.G. (2005). 'Causes and consequences of animal dispersal strategies: relating individual behaviour to spatial dynamics', *Biological Reviews* **80**: 205–25.

Bowler, D.E. and Benton, T.G. (2009). 'Impact of dispersal on population growth: the role of inter-patch distance', *Oikos* **118**: 403–12.

Bowler, D.E. and Benton, T.G. (2011). 'Testing the interaction between environmental variation and dispersal strategy on population dynamics using a soil mite experimental system', *Oecologia* **166**: 111–19.

Brown, J.H. and Kodric-Brown, A. (1977). 'Turnover rates in insular biogeography: effect of immigration on extinction', *Ecology* **58**: 445–9.

Buckley, L.B., Urban, M.C., Angilletta, M.J. *et al.* (2010). 'Can mechanism inform species' distribution models?', *Ecology Letters* **13**: 1041–54.

Burton, O.J., Phillips, B.L., Travis, J.M.J. (2010). 'Trade-offs and the evolution of life-histories during range expansion', *Ecology Letters* **13**: 1210–20.

Cadet, C., Ferriere, R., Metz, J.A.J. *et al.* (2003). 'The evolution of dispersal under demographic stochasticity', *American Nauralist* **162**: 427–41.

Cazelles, B., Bottani, S., Stone, L. (2001). 'Unexpected coherence and conservation', *Proceeding of the Royal Society B* **268**: 2595–602.

Chevin, L.M., Lande, R., Mace, G.M. (2010). 'Adaptation, Plasticity, and Extinction in a Changing Environment: Towards a Predictive Theory', *PLoS Biology* **8** (4): e1000357.

Cook, A.R., Otten, W., Marion, G. *et al.* (2007). 'Estimation of multiple transmission rates for epidemics in heterogeneous populations', *Proceedings of the National Academy Of Sciences of the USA* **104**: 20392.

Cornell, S.J. and Ovaskainen, O. (2008). 'Exact asymptotic analysis for metapopulation dynamics on correlated dynamic landscapes', *Theoretical Population Biology* **74**: 209–25.

Cote, J., Clobert, J., Fitze, P.S. (2007). 'Mother—offspring competition promotes colonization success', *Proceedings of theNationalAcademy of Sciences of the USA* **104**: 9703.

Courchamp, F., Clutton-Brock, T., Grenfell, B. (1999). 'Inverse density dependence and the Allee effect', *Trends in Ecology and Evolution* **14**: and 405–10.

Dahlgren, J.P. and Ehrlen, J. (2005). 'Distribution patterns of vascular plants in lakes—the role of metapopulation dynamics', *Ecography* **28**: 49–58.

Dey, S. and Joshi, A. (2006). 'Stability via asynchrony in Drosophila metapopulations with low migration rates', *Science* **312**: 434–6.

Doebeli, M. and Ruxton, G.D. (1997). 'Controlling spatial chaos in metapopulations with long-range dispersal', *Bulletin of Mathematical Biology* **59**: 497–515.

Duckworth, R.A. (2009). 'Maternal effects and range expansion: a key factor in a dynamic process?', *Philosophical Transactions of the Royal Society B* **364**: 1075–86.

Dytham, C. and Travis, J.M.J. (2006). 'Evolving dispersal and age at death', *Oikos* **113**: 530–8.

Ebert, D., Haag, C., Kirkpatrick, M. *et al.* (2002). 'A selective advantage to immigrant genes in a Daphnia metapopulation', *Science* **295**: 485–8.

Edelaar, P., Siepielski, A.M., Clobert, J. (2008). 'Matching Habitat Choice Causes Directed Gene Flow: A Neglected Dimension in Evolution and Ecology', *Evolution* **62**: 2462.

Evans, M.R., Norris, K. and Benton, T.G. (2012) Predictive Ecology: systems approaches. *Philosophical Transactions of the Royal Society* B **367**: 163–169.

Fontaine, C. and Gonzalez, A. (2005). 'Population synchrony induced by resource fluctuations and dispersal in an aquatic microcosm', *Ecology* **86**: 1463–71.

Foster, W.A. and Benton, T.G. (1992). 'Sex-Ratio, Local Mate Competition and Mating-Behavior in the Aphid Pemphigus-Spyrothecae', *Behavioural Ecology and Sociobiology* **30**: 297–307.

Fox, G.A. (2005). 'Extinction risk of heterogeneous populations *Ecology* **86**: 1191–8.

Greenman, J.V. and Benton, T.G. (2001). 'The impact of stochasticity on the behaviour of non-linear population models: synchrony and the Moran effect', *Oikos* **93**: 343–51.

Greenman, J.V. and Benton, T.G. (2003). 'The amplification of environmental noise in population models: Causes and consequences', *American Naturalist* **161**: 225–39.

Greenman, J.V. and Benton, T.G. (2005). 'The impact of environmental fluctuations on structured discrete time population models: Resonance, synchrony and threshold behaviour', *Theoretical Population Biology* **68**: 217–35.

Greenman, J.V., Benton, T.G., Boots, M. *et al.* (2005). 'The evolution of oscillatory behavior in age-structured species', *American Naturalist* **166**: 68–78.

Grenfell, B.T., Wilson, K., Finkenstadt, B.F. *et al.* (1998). 'Noise and determinism in synchronized sheep dynamics', *Nature* **394**: 674–7.

Gundersen, G., Andreassen, H.P., Ims, R.A. (2002). 'Individual and population level determinants of immigration success on local habitat patches: an experimental approach', *Ecology Letters* **5**: 294–301.

Gundersen, G., Johannesen, E., Andreassen, H.P. *et al.* (2001). 'Source-sink dynamics: how sinks affect demography of sources', *Ecology Letters* **14**: 14.

Gundersen, G., Johannesen, E., Andreassen, H.P. *et al.* (2001). 'Source-sink dynamics: how sinks affect demography of sources', *Ecology Letters* **4**: 14–21.

Gyllenberg, M., Parvinen, K., Dieckmann, U. (2002). 'Evolutionary suicide and evolution of dispersal in structured metapopulations', *Journal of Mathematical Biology* **45**: 79–105.

Hamilton, W. and May, R. (1977). 'Dispersal in stable habitats', *Nature* **269**: 578–81.

Hanski, I., Mononen, T., Ovaskainen, O. (2011). 'Eco-Evolutionary Metapopulation Dynamics and the Spatial Scale of Adaptation', *American Naturalist* **177**: 29–43.

Haydon, D.T., Greenwood, P.E., Stenseth, N.C. *et al.* (2003). 'Spatio-temporal dynamics of the grey-sided vole in Hokkaido: identifying coupling using state-based Markov-chain modelling', *Proceedings of the Royal Society B* **270**: 435–45.

Hedrick, P.W. (2006). 'Genetic polymorphism in heterogeneous environments: The age of genomics *Annual Review of Ecology, Evolution and Systematics* **37**: 67–93.

Heino, M. and Hanski, I. (2001). 'Evolution of migration rate in a spatially realistic metapopulation model', *American Naturalist* **157**: 495–511.

Heino, M., Kaitala, V., Ranta, E. *et al.* (1997). 'Synchronous dynamics and rates of extinction in spatially structured populations', *Proceedings of the Royal Society B* **264**: 481–6.

Higgins, S.I. and Cain, M.L. (2002). 'Spatially realistic plant metapopulation models and the colonization-competition trade-off', *Journal of Ecology* **90**: 616–26.

Hovestadt, T., Messner, S., Poethke, H.J. (2001). 'Evolution of reduced dispersal mortality and 'fat-tailed' dispersal kernels in autocorrelated landscapes', *Proceedings of the Royal Society B* **268**: 385.

Hovestadt, T. and Poethke, H. (2006). 'The control of emigration and its consequences for the survival of populations', *Ecological Model* **190**: 443–53.

Hughes, C.L., Hill, J.K., Dytham, C. (2003). 'Evolutionary trade-offs between reproduction and dispersal in populations at expanding range boundaries', *Proceeding of the Royal Society B* **270**: S147–50.

Ims, R.A. and Andreassen, H.P. (2000). 'Spatial synchronization of vole population dynamics by predatory birds', *Nature* **408**: 194–6.

Ives, A.R., Woody, S.T., Nordheim, E.V. *et al.* (2004). 'The synergistic effects of stochasticity and dispersal on population densities', *American Naturalist* **163**: 375–87.

Jacquemyn, H., Honnay, O., Van Looy, K. *et al.* (2006). 'Spatiotemporal structure of genetic variation of a spreading plant metapopulation on dynamic riverbanks along the Meuse River', *Heredity* **96**: 471–8.

Jansen, V.A.A. (1999). 'Phase locking: another cause of synchronicity in predator-prey systems', *Trends in Ecology and Evolution* **14**: 278–9.

Jansen, V.A.A. and Vitalis, R. (2007). 'The evolution of dispersal in a Levins' type metapopulation model', *Evolution* **61**: 2386–97.

Jongejans, E., Huber, H., de Kroon, H. (2010). 'Scaling up phenotypic plasticity with hierarchical population models', *Evolutionary Ecology* **24**: 585–99.

Kawecki, T.J. and Ebert, D. (2004). 'Conceptual issues in local adaptation', *Ecology Letters* **7**: 1225–41.

Keeley, E.R. (2001). 'Demographic responses to food and space competition by juvenile steelhead trout', *Ecology* **82**: 1247–59.

Keeling, M.J., Woolhouse, M.E.J., Shaw, D.J. *et al.* (2001). 'Dynamics of the 2001 UK foot and mouth epidemic: Stochastic dispersal in a heterogeneous landscape', *Science* **294**: 813–17.

Kendall, B.E., Bjornstad, O.N., Bascompte, J. *et al.* (2000). 'Dispersal, environmental correlation, and spatial synchrony in population dynamics', *American Naturalist* **155**: 628–36.

Kendall, B.E., Briggs, C.J., Murdoch, W.W. *et al.* (1999). 'Why do populations cycle? A synthesis of statistical and mechanistic modeling approaches', *Ecology* **80**: Ecology 80, 1789–805.

Kendall, B.E. and Fox, G.A. (1998). 'Spatial structure, environmental heterogeneity, and population dynamics: Analysis of the coupled logistic map', *Theoretical Population Biology* **54**: 11–37.

Koenig, W.D. (2001). 'Spatial autocorrelation and local disappearances in wintering North American birds', *Ecology* **82**: 2636–44.

Kokko, H. and Sutherland, W.J. (2001). 'Ecological traps in changing environments: Ecological and evolutionary consequences of a behaviourally mediated Allee effect', *Evolutionary Ecology* **3**: 537–51.

Kunin, W.E., Vergeer, P., Kenta, T. *et al.* (2009). 'Variation at range margins across multiple spatial scales: environmental temperature, population genetics and metabolomic phenotype', *Proceedings of the Royal Society B* **276**: 1495–506.

Lambin, X., Elston, D.A., Petty, S.J. *et al.* (1998). 'Spatial asynchrony and periodic travelling waves in cyclic populations of field voles', *Proceedings of the Royal Society B* **265**: 1491–6.

Lande, R. (1993). 'Risks of population extinction from demographic and environmental stochasticity and random catastrophies', *American Naturalist* **142**:American Naturalist 142, 911–27.

Lande, R., Engen, S., Saether, B.E. (1999). 'Spatial scale of population synchrony: Environmental correlation versus dispersal and density regulation', *American Naturalist* **154**: 271–81.

Levins, R. (1966). 'The strategy of model building in population ecology', *American Science* **54**: 421–31.

Levins, R. (1969). 'Some demographic and genetic consequences of environmental heterogeneity for biological control', *Bulletin of the Entomological Society of America* **15**: 237–40.

Liebhold, A., Koenig, W.D., Bjornstad, O.N. (2004). 'Spatial synchrony in population dynamics', *Annual Review of Ecology, Evolution and Systematics* **35**: 467–90.

Liebhold, A.M., Johnson, D.N., Bjornstad, O.N. (2006). 'Geographic variation in density-dependent dynamics impacts the synchronizing effect of dispersal and regional stochasticity', *Population Ecology* **48**: 131–8.

Lindstrom, T., Hakansson, N., Westerberg, L. *et al.* (2008). 'Splitting the tail of the displacement kernel shows the unimportance of kurtosis', *Ecology* **89**: 1784–90.

Llewelyn, J., Phillips, B.L., Alford, R.A. *et al.* (2010). Locomotor performance in an invasive species: cane toads from the invasion front have greater endurance, but not speed, compared to conspecifics from a long-colonised area', *Oecologia* **162**: 343–8.

Lundberg, P., Ranta, E., Ripa, J. *et al.* (2000). 'Population variability in space and time', *Trends in Ecology and Evolution* **15**: 460–4.

Madsen, T., Shine, R., Olsson, M. *et al.* (1999). 'Conservation biology—Restoration of an inbred adder population', *Nature* **402**: 34–5.

Marshall, D.J., Monro, K., Bode, M. *et al.* (2010). 'Phenotype-environment mismatches reduce connectivity in the sea', *Ecology Letters* **13**: 128–40.

McInerny, G.J., Turner, J.R.G., Wong, H.Y. *et al.* (2009). 'How range shifts induced by climate change affect neutral evolution', *Proceedings of the Royal Society B* **27**: 1527–34.

Menendez, R., Gutierrez, D., Thomas, C.D. (2002). 'Migration and allee effects in the six-spot burnet moth Zygaena filipendulae', *Ecological Entomology* **27**: 317–25.

Milden, M., Munzbergova, Z., Herben, T. *et al.* (2006). 'Metapopulation dynamics of a perennial plant, Succisa pratensis, in an agricultural landscape', *Ecological Modelling* **199**: 464–75.

Milot, E., Weimerskirch, H., Bernatchez, L. (2008). The seabird paradox: dispersal, genetic structure and population dynamics in a highly mobile, but philopatric albatross species', *Molecular Ecology* **17**: 1658–73.

Molofsky, J. and Ferdy, J.B. (2005). 'Extinction dynamics in experimental metapopulations', *Proceedings of the National Academy of Sciences of the USA* **102**: 3726–31.

Murrell, D.J. (2005). 'Local spatial structure and predator-prey dynamics: Counterintuitive effects of prey enrichment', *American Naturalist* **166**: 354–67.

Murrell, D.J., Travis, J.M.J., Dytham, C. (2002). 'The evolution of dispersal distance in spatially-structured populations', *Oikos* **97**: 229–36.

Mustin, K., Benton, T.G., Dytham, C. *et al.* (2009). 'The dynamics of climate-induced range shifting; perspectives from simulation modelling', *Oikos* **118**: 131–7.

North, A. and Ovaskainen, O. (2007). 'Interactions between dispersal, competition, and landscape heterogeneity', *Oikos* **116**: 1106–19.

Ovaskainen, O. and Cornell, S.J. (2006a). 'Asymptotically exact analysis of stochastic metapopulation dynamics with explicit spatial structure', *Theoretical Population Biology* **69**:Theoretical Population Biology 69, 13–33.

Ovaskainen, O. and Cornell, S.J. (2006b). 'Space and stochasticity in population dynamics', *Proceedings of the National Academy of Sciences of the USA* **103**: 12781–6.

Paradis, E., Baillie, S.R., Sutherland, W.J. *et al.* (1999). 'Dispersal and spatial scale affect synchrony in spatial population dynamics', *Ecology Letters* **2**: 114–20.

Parvinen, K., Dieckmann, U., Gyllenberg, M. *et al.* (2003). 'Evolution of dispersal in metapopulations with local density dependence and demographic stochasticity', *Journal of Biology* **16**: 143–53.

Petrovskii, S., Morozov, A., Li, B.L. (2008). 'On a possible origin of the fat-tailed dispersal in population dynamics', *Ecological Complexity* **5**: 146–50.

Petrovskii, S.V. and Malchow, H. (2001). 'Wave of chaos: New mechanism of pattern formation in spatio- temporal population dynamics', *Theoretical Population Biology* **59**: 157–74.

Petty, S.J., Lambin, X., Sherratt, T.N. *et al.* (2000). 'Spatial synchrony in field vole Microtus agrestis abundance in a coniferous forest in northern England: The role of vole-eating raptors', *Journal of Applied Ecology* **37**: 136–47.

Phillips, B.L., Brown, G.P., Travis, J.M.J. *et al.* (2008). 'Reid's paradox revisited: The evolution of dispersal kernels during range expansion', Symposium of the American Society of Naturalists, Christchurch, New Zealand, S34–*48*.

Polechova, J., Barton, N., Marion, G. (2009). 'Species' Range: Adaptation in Space and Time', *American Naturalist* **174**: E186–204.

Post, E. and Forchhammer, M.C. (2002). 'Synchronization of animal population dynamics by large-scale climate', *Nature* **420**: 168–71.

Pulliam, H.R. and Danielson, B.J. (1991). Sources, Sinks and Habitat Selection—A landscape perspective on population dynamics', *American Naturalist* **137**: S50–S66.

Ranta, E. and Kaitala, V. (1997). 'Travelling waves in vole population dynamics', *Nature* **390**: 456–456.

Ranta, E., Kaitala, V., Lundberg, P. (1998). 'Population variability in space and time: the dynamics of synchronous population fluctuations', *Oikos* **83**: 376–82.

Renshaw, E. (1991). *Modelling Biological Populations in Space and Time.* Cambridge University Press, Cambridge.

Ridley, J. and Sutherland, W.J. (2002). 'Kin competition within groups: the offspring depreciation hypothesis', *Proceedings of the Royal Society B* **269**: 2559–64.

Ronce, O. (2007). 'How does it feel to be like a rolling stone? Ten questions about dispersal evolution', *Annual Review of Ecology, Evolution and Systematics* **38**: 231.

Ross, J.V. (2010). 'Computationally exact methods for stochastic periodic dynamics: Spatiotemporal dispersal and temporally forced transmission', *Journal of Theoretical Biology* **262**: 14–22.

Rousset, F. and Ronce, O. (2004). 'Inclusive fitness for traits affecting metapopulation demography', *Theoretical Population Biology* **65**: 127.

Roy, M., Harding, K., Holt, R.D. (2008). 'Generalizing Levins metapopulation model in explicit space: Models of intermediate complexity', *Journal of Theoretical Biology* **255**: 152–61.

Royama, T. (2005). 'Moran effect on non-linear population processes', *Ecological Monographs* **75**: 277–93.

Ruetz, C.R., Trexler, J.C., Jordan, F. *et al.* (2005). 'Population dynamics of wetland fishes: spatio-temporal patterns synchronized by hydrological disturbance?', *Journal of Animal Ecology* **74**: 322–32.

Saccheri, I., Kuussaari, M., Kankare, M. *et al.* (1998). 'Inbreeding and extinction in a butterfly metapopulation', *Nature* **39**: 491–4.

Saether, B.E., Engen, S., Lande, R. (1999). 'Finite metapopulation models with density-dependent migration and stochastic local dynamics', *Proceedings of the Royal Society B* **266**: 113–18.

Schneeberger, A. and Jansen, V.A.A. (2006). 'The estimation of dispersal rates using the covariance of local populations', *Ecological Modelling* **196**: 434–46.

Schreiber, S.J. (2010). 'Interactive effects of temporal correlations, spatial heterogeneity and dispersal on population persistence', *Proceeding of the Royal Society B* **277**: 1907–14.

Schtickzelle, N., Choutt, J., Goffart, P. *et al.* (2005). 'Metapopulation dynamics and conservation of the marsh fritillary butterfly: Population viability analysis and management options for a critically endangered species in Western Europe', *Biological Conservation* **126**: 569–81.

Schtickzelle, N., Mennechez, G., Baguette, M. (2006). 'Dispersal depression with habitat fragmentation in the bog fritillary butterfly', *Ecology* **87**: 1057–65.

Schwartz, M.K., Mills, L.S., McKelvey, K.S. *et al.* (2002). 'DNA reveals high dispersal synchronizing the population dynamics of Canada lynx', *Nature* **415**: 520–2.

Sherratt, J.A. (2001). 'Periodic travelling waves in cyclic predator-prey systems', *Ecology* **4**: 30–7.

Sherratt, J.A., Lambin, X., Thomas, C.J. *et al.* (2002). 'Generation of periodic waves by landscape features in cyclic predator-prey systems', *Proceedings of the Royal Society B* **269**:Proceedings of the Royal Society of London Series B-Biological Sciences **269**, 327–34.

Sherratt, T.N., Lambin, X., Petty, S.J. *et al.* (2000). 'Use of coupled oscillator models to understand synchrony and travelling waves in populations of the field vole Microtus agrestis in northern England', *Journal of Applied Ecology* **37**: 148–58.

Simmons, A.D. and Thomas, C.D. (2004). 'Changes in dispersal during species' range expansions', *American Naturalist* **164**: 378–95.

Smith, M.J., Sherratt, J.A., Lambin, X. (2008). 'The effects of density-dependent dispersal on the spatiotemporal dynamics of cyclic populations', *Journal of Theoretical Biology* **254**: 264–74.

Snyder, R.E. (2007). 'Spatiotemporal population distributions and their implications for species coexistence in a variable environment', *Theoretical Population Biology* **72**: 7–20.

Starrfelt, J. and Kokko, H. (2010). 'Parent-Offspring Conflict and the Evolution of Dispersal Distance', *American Naturalist* **175**: 38–49.

Sundell, J., Huitu, O., Henttonen, H. *et al.* (2004). 'Large-scale spatial dynamics of vole populations in Finland revealed by the breeding success of vole-eating avian predators', *Journal of Animal Ecology* **73**: 167–78.

Takahashi, L.T., Maidana, N.A., Ferreira, W.C. *et al.* (2005). 'Mathematical models for the Aedes aegypti dispersal dynamics: travelling waves by wing and wind', *Bulletin of Mathematical Biology* **67**: 509–28.

Tanaka, H.A., Lichtenberg, A.J., Oishi, S. (1997). 'Self-synchronization of coupled oscillators with hysteretic responses', *Physica D* **100**: 100, 279–300.

Telfer, S., Holt, A., Donaldson, R. *et al.* (2001). 'Metapopulation processes and persistence in remnant water vole populations', *Oikos* **95**: 31–42.

Thomas, C.D., Bodsworth, E.J., Wilson, R.J. *et al.* (2001). 'Ecological and evolutionary processes at expanding range margins', *Nature* **411**: 577–81.

Tittler, R., Fahrig, L., Villard, M.A. (2006). 'Evidence of large-scale source-sink dynamics and long-distance dispersal among wood thrush populations', *Ecology* **87**: 3029–36.

Tobin, P.C. and Bjornstad, O.N. (2005). 'Roles of dispersal, stochasticity, and non-linear dynamics in the spatial structuring of seasonal natural enemy-victim populations', *Population Ecology* **47**: 221–7.

Travis, J.M.J. (2001). The colour of noise and the evolution of dispersal', *Ecological Research* **16**: 157.

Travis, J.M.J. (2003). 'Climate change and habitat destruction: a deadly anthropogenic cocktail', *Proceedings of the Royal Society B* **270**: 467–73.

Travis, J.M.J. (2004). 'The evolution of programmed death in a spatially structured population', *Journals of Gerontology Series A-Biological Sciences and Medical Sciences* **59**: 301–5.

Travis, J.M.J., Mustin, K., Benton, T.G. *et al.* (2009). 'Accelerating invasion rates result from the evolution of density-dependent dispersal', *Journal of Theoretical Biology* **259**: 151–8.

Travis, J.M.J., Smith, H.S., Ranwala, S.M.W. (2010). 'Towards a mechanistic understanding of dispersal evolution in plants: conservation implications', *Diversity and Distributions* **16**: 690–702.

Travis, J.M.J., Mustin, K., Bartoń, K.A. *et al.* (2012). 'Modelling dispersal: an eco-evolutionary framework incorporating emigration, movement and settlement behaviour and the multiple costs involved', *Methods in Ecology and Evolution* doi: 10.1111/j.2041-210X.2012.00193.x.

Turchin, P. and Ellner, S.P. (2000). 'Living on the edge of chaos: Population dynamics of Fennoscandian voles', *Ecology* **81**: 3099–116.

Van Dyck, H. and Baguette, M. (2005). 'Dispersal behaviour in fragmented landscapes: Routine or special movements?', *Basic and Applied Ecology* **6**: 535–45.

Verheyen, K., Vellend, M., Van Calster, H. *et al.* (2004). 'Metapopulation dynamics in changing landscapes: A new spatially realistic model for forest plants', *Ecology* **85**: 3302–12.

Vila, C., Sundqvist, A.K., Flagstad, O., Seddon, J. *et al.* (2003). 'Rescue of a severely bottlenecked wolf (Canis lupus) population by a single immigrant', *Proceedings of the Royal Society B* **270**: 91–7.

Wang, R.J., Ovaskainen, O., Cao, Y.D. *et al.* (2011). 'Dispersal in the Glanville fritillary butterfly in fragmented versus continuous landscapes: comparison between three methods', *Ecological Entomology* **36**: 251–60.

Wild, G., Greenwood-Lee, J., Taylor, P.D. (2006). 'Sex allocation and dispersal in a heterogeneous two-patch environment', *Theoretical Population Ecology* **70**: 225–35.

Wilson, R.J., Davies, Z.G., Thomas, C.D. (2010). 'Linking habitat use to range expansion rates in fragmented landscapes: a metapopulation approach', *Ecography* **33**: 73–82.

Yakimowski, S.B. and Eckert, C.G. (2007). 'Threatened peripheral populations in context: Geographical variation in population frequency and size and sexual reproduction in a clonal woody shrub', *Conservation Biology* **21**: 811–22.

CHAPTER 21

Demographic consequences of the selective forces controlling density-dependent dispersal

François Rousset

21.1 Introduction

The main forces that select for dispersal are local extinction that always recreates free space to colonize (e.g. van Valen 1971; Olivieri *et al.* 1995) and, more generally, local fluctuations in habitat quality (Gadgil 1971), kin competition (e.g. Hamilton and May 1977; Taylor 1988), and inbreeding avoidance (e.g. Greenwood 1980; Motro 1991; Roze and Rousset 2009). Forces acting against dispersal include physiological costs of dispersal structures, the costs of dispersal in general, and to unsuitable habitat in particular, and benefits received from interacting with relatives (Chapter 11). These different forces have generally been analysed in simple models aiming at demonstrating their impact.

However, with increased concerns about population persistence, there is increasing interest in how dispersal and its evolution can facilitate or impede the persistence of populations (e.g. Olivieri and Gouyon 1997; del Mar Delgado *et al.* 2011). Ideally, one could expect that when dispersal evolves so as to reduce resource competition, this will favour population persistence. However, natural selection does not always lead to such an outcome. Evolution of dispersal will not necessarily maximize population size (e.g. Roff 1975; Hamilton and May 1977; Motro 1982; Olivieri and Gouyon 1997) and can even lead to selection-driven extinction or 'evolutionary suicide' (Dieckmann and Ferrière 2004). Such selection-driven extinction particularly, though not only, occurs in metapopulation models with Allee effects (Gyllenberg *et al.* 2002; Rousset and Ronce 2004).

There is thus a need to understand how the different forces that affect the evolution of dispersal balance each other in their effects on population persistence. However, there is no systematic study of this issue under realistic demographic scenarios, partly because analytical decompositions of the selection process into different components are not systematically considered, and the more so as such decompositions may be uneasy to formulate.

The response to different levels of resource competition is a form of context-dependent dispersal. Such context-dependence has been considered in some of the earliest studies of dispersal (Gadgil 1971), and is the subject of renewed theoretical efforts (Bowler and Benton 2005; Chapter 11). This chapter considers the example of departure decisions conditional on current local population size. This appears to be the scenario most actively investigated, partly because of its perceived empirical relevance, although local population age since colonization (e.g. Olivieri *et al.* 1995; Hanski *et al.* 2004; Ronce *et al.* 2005), and information about potential destination populations certainly matters.

This chapter will first summarize the building blocks of the analysis: the force that makes dispersal evolve to reduce resource competition in the absence of a dispersal cost, the equalization of marginal gains in reproductive value in the presence of a dispersal cost, and the effects of kin competition. It will next examine a metapopulation model where an analytical partition of the different forces, such a reproductive value imbalance and kin competition is possible. The effects on population size and on

Dispersal Ecology and Evolution. First Edition. Edited by Jean Clobert *et al.*

recolonization as measured by the average number of immigrants, of context-dependent dispersal, and of individual selection versus maximization of population size will then be compared.

21.2 Minimization of resource competition

This section briefly introduces some models that have been used to formalize the idea that evolution can lead to a minimization of resource competition.

The simplest of these models considers patches of constant carrying capacity, where individuals can settle in a fixed number of breeding areas ('territories') with the same number of juveniles produced. For example, consider that there are two patches, one with 100 territories and the other with 20 territories. Intuitively, in the absence of dispersal cost, all individuals may be expected to send five out six children to the larger patch; i.e. panmixia is expected. The earliest attempt to formalize these ideas is Fretwell and Lucas' (1970) concept of the ideal free distribution (IFD). In this perspective, individuals are expected to distribute themselves in proportion to available resources, so that they maximize their potential contribution to the future gene pool of the species.

21.2.1 Reproductive value equilibration

In evolutionary ecology this potential contribution is usually described in terms of reproductive value (e.g. Charlesworth 1994; Bulmer 1994). The reproductive value of an individual is its expected ultimate genetic contribution to a distant future population, measured as the probability of being the ancestor of a given descendant in the future population. This definition is readily extended to the deme; that is, the set of individuals inhabiting a patch. The reproductive value of a deme is the sum of the reproductive values of the deme members, and the reproductive value of any deme-size class is the sum of the deme reproductive values for all demes of a given size class. Reproductive value and local population growth rate are distinct, insofar as 'local growth rates' are indeed defined from local demographic rates, while reproductive values depend on the fate of emigrants. A discrepancy may occur, for example, in extreme scenarios where local growth rate is so that the next generation cannot reproduce. On the other hand, approximations for local populations of infinite size (Metz and Gyllenberg 2001), or casual inferences from more complex models (Cadet *et al.* 2003, p. 435), suggest that the two concepts (reproductive value and local population growth rate) will often be qualitatively congruent. Indeed, if emigrants disperse randomly over the landscape, the reproductive value of demes of a given size should vary in proportion to the total production over successive generations of successful emigrants from such demes, which can then be related to emigration rates and to some measure of local growth rate.

Intuitively, if there is a class of offspring with higher individual reproductive value (say, dispersers to a good habitat), it pays to produce more of this type of offspring. However, as this will result in more competition for the good habitat, this reduces the individual reproductive value of this class of offspring, and the end point is a state of equal reproductive value for all types of offspring. This is the classical argument for sex ratio equilibration (Fisher 1958), where the two 'habitats' are the female and male share of the genetic pool.

Elementary computations show that, in a two-patch model, equilibration of adult individual reproductive value occurs when the number of successful emigrants equals the number of successful immigrants in each patch (Rousset, 1999), where *successful* individuals are those that manage to settle after the competition step. The latter pattern was observed in simulations of dispersal evolution in the absence of dispersal cost, and dubbed balanced dispersal by McPeek and Holt (1992). McPeek and Holt (1992) actually claimed that the evolutionarily stable level of dispersal would not be uniquely determined, but would rather move randomly along a line of balanced dispersal; that is, along all strategies that yield equal number of emigrants and immigrants. Panmixia would not be the unique ESS.

With several patches, performing the IFD or balanced dispersal strategies generally requires information about potential destination patches. In natural populations, such information is imperfect, so that equalization of reproductive values could only be reached on average over a distribution of demographic situations encountered by offspring, more specifically over the distribution of deme sizes encountered by immigrants, rather than conditionally on each such deme size. In the metapopulation model analysed in the following, decisions are conditional on parental deme size but dispersal is random with respect to the size of the destination patch, at least below a size threshold.

21.2.2 Which reproductive value equilibration, and when?

Reproductive value equilibration is an attractive concept, but attempts to formalize it actually show that what is actually expected is equilibration of marginal reproductive value returns, where the marginal returns are the small changes in reproductive value of either emigrant or resident offspring resulting from a small change in dispersal behaviour. That is, if each emigrant offspring trades against one resident offspring, the ESS should be the point where the reproductive value return of an additional resident offspring matches that of a successful emigrant. Further, the reproductive value of an emigrant is reduced by the dispersal cost, and then one needs to distinguish *emigrant* reproductive value (measured before any dispersal cost has been paid) from *immigrant* reproductive value (that of juveniles that have survived the dispersal stage), which is itself distinct from the reproductive value of *successful* offspring (those who survive competition among juveniles). When dispersal is costly, an adult must produce more than one emigrant offspring for each fewer resident offspring, if both types have equal reproductive value after dispersal.

In a model without spatial heterogeneity, the reproductive value of a settled immigrant is the same as that of a settled resident, and thus no equilibration of reproductive value returns can occur. Conversely, since each emigrant trades against one resident juvenile, equilibration of reproductive value returns can occur only when the reproductive value of immigrant juveniles is higher than that of resident juveniles after the dispersal step (but that of emigrant juveniles is lower than that of resident ones at the same stage), and one could expect dispersal rates to evolve so as to achieve such an imbalance in spatially heterogeneous landscapes. An analogous result holds in sex ratio evolution: when (say) male juvenile survival is lower, so that adult sons are more costly to produce than adult daughters, birth sex ratios must be biased toward the most costly sex in order to attain equal reproductive value returns at the adult offspring stage. At the latter stage, sex ratio is biased toward the less costly sex (females) and thus adult sons have higher individual reproductive value.

With spatial heterogeneity, the equalization of marginal reproductive value benefits can lead to the evolution of dispersal. Greenwood-Lee and Taylor (2001) show that this happens even if this spatial heterogeneity is fixed in time, and illustrate this fact through a simple two-patch model with two life stages, juveniles and adults. Juveniles survive better in one patch and adults reproduce better in the other patch. With suitably chosen survival and fecundity values, at the ESS, juveniles disperse partially one way and adults disperse partially the other way (a limit case with complete dispersal would represent bird migration rather than natal dispersal). Greenwood-Lee and Taylor's model offers a remarkable example where dispersal can evolve in the absence of both kin competition and temporal fluctuations in demographic properties of the different habitats. Their example still illustrates the idea that evolution of dispersal leads to equalization of marginal reproductive value benefits.

21.2.3 Accounting for kin competition

Attempts to formalize equilibration of marginal reproductive value benefits or balanced dispersal tend to ignore kin competition. An allele that does not disperse cannot invade a population, and therefore there is always a selective pressure in favour of dispersal, which has been neglected in the analysis. In particular, any scenario with finite deme size implies some form of kin competition, which would select for some degree of dispersal.

Selection measures taking both reproductive value variation and kin competition into account can be computed (e.g. Taylor 1990; Taylor and Frank 1996). When a finite number of patches is considered, a finite population formulation of kin selection (Rousset and Billiard 2000) is required. In the end, though, much can be understood from models assuming an infinite number of patches. The analysis of the finite population case then mainly serves to sort out which of the conclusions from two-patch models make sense in a many-patches context.

We re-analysed McPeek and Holt's model without demographic stochasticity, but now taking finite deme size into account (Leturque and Rousset 2002). We found that under the assumptions of no variation in fecundity across habitats in space and time, and with costly dispersal, non-zero dispersal evolves at the ESS, and there is a deviation from balanced dispersal (and thus from equalization of reproductive values of juveniles among the patches after dispersal) of order proportional to the dispersal cost. Intuitively, since relatedness is higher in smaller demes for a given dispersal rate, the deviation due to kin competition should be in the direction of higher migration rate from smaller demes. This is indeed so for emigration probability, but not generally so for number of successful emigrants. With large cost (relative to deme size, i.e. large value of the product Nc of departure deme size and dispersal cost), there may be more successful emigrants from than successful immigrants to the larger demes, where individual adult reproductive value is thus higher. A unique ESS which is IFD or more simply panmixia is also found under the same assumptions than above but with a cost-free dispersal. This is a form of balanced dispersal, but all balanced dispersal states are not neutrally equivalent. Yet, as it is related to deme size, the strength of selection along the line of balanced dispersal strategies is weak in models with two large demes, and then confounded with neutrality. This difference persists as deme size is increased, but it matters most in scenarios of many small demes. Here the strength of selection against deviations from the ESS will also be inversely related to deme size, because in small demes local competition will be most strongly influenced by fluctuations in dispersal strategies.

Finally, context-dependent dispersal allows for the evolution of higher levels of dispersal than unconditional dispersal, a conclusion already reached by McPeek and Holt (1992). The reason is that a high emigration rate from a large habitat to a small one is more strongly deleterious than a small emigration rate from the small habitat to the large one.

A completely different discrete deme size generalization of McPeek and Holt's scenario has been built with an infinite number of demes and a given distribution of deme sizes K constant in time. With a continuous time formulation and a dispersal cost c, a remarkably simple result is that the ES dispersal rate from a deme of size K is $1/(cK)$ if all deme sizes K are $> 1/c$ (Massol *et al.* 2011).

21.2.4 Optimal resource use under demographic stochasticity

In the context of stochastic population dynamics, a strategy that can intuitively be expected to minimize resource competition is to send away all individuals beyond a given threshold T. If there are fewer individuals than T, none leave, and if there are more, the emigration probability is such as to leave only when T individuals are competing in the patch. If only one individual takes a decision at a time (as in continuous-time models), this 'T-stay' strategy leads to an all-or-none decision rule, where the individual always leaves as soon as there are already T individuals in the patch, but always stays otherwise. If decisions to leave are taken simultaneously (as in discrete-time models), one form of the T-stay strategy is that the dispersal rates m_n conditional on the number of individuals present, are such that $n(1 - m_n) = T$ for $n \geq T$; that is, $m_n = 1 - T/n$.

Several studies have obtained results that fit with the above expectations. Metz and Gyllenberg (2001) consider evolution of the whole array of conditional dispersal rates, under some infinite deme size approximation. In a model with negative density dependence, they found that dispersal should follow an all-or-none pattern, where there should be no dispersal from demes small enough to have positive growth rate, and complete dispersal above this threshold. Alizon and Taylor (2008) were more concerned with the effect of demographic stochasticity on the evolution of some helping behaviour, but first let all conditional dispersal rates evolving. They assumed no cost of dispersal and found an all-or-none strategy analogous to the one obtained by Metz and Gyllenberg. The *T*-stay strategy also results in models considering departure decisions based on the number of juveniles in competition, where all juveniles are the offspring of a single parent per deme (Ozaki 1995; Ezoe and Iwaza 1997; Kisdi 2004).

More models have considered the evolution of dispersal conditional on parental deme size where at least part of the variation in deme size is due to demographic stochasticity (Travis *et al.* 1999; Poethke and Hovestadt 2002; Cadet *et al.* 2003; Hovestadt *et al.* 2010; for studies of context-dependent dispersal under other sources of variation in deme size, see Bowler and Benton, 2005, Table 21.2), but have not derived dispersal rates conditional on the whole array of deme sizes. Indeed, the level of kin competition may vary with the number of parents in the deme which complicates both the mathematical analysis and the task of numerically assessing the stability of all conditional dispersal rates. Rather, these studies have assumed more constrained phenomenological relationship between dispersal and deme size (reviewed by Hovestadt *et al.* 2010). Cadet *et al.* (2003) assumed that no more than a certain number *T* of individuals could reproduce in a deme, and that individuals in excess of *T* always had to disperse. They then investigated the evolution of a single dispersal rate common to all demes of low size. The dispersal rate could be either zero or non-zero depending on the strength of competition for resource below *T*. The other studies have used individual-based simulations. Hovestadt *et al.* (2010) further put the pre-existing phenomenological models in pairwise competition in a discrete-time setting. In this set-up, the best strategy was what they described as a 'threshold-asymptotic' pattern, in which there is still a minimal threshold size for dispersal above which dispersal is an increasing, but concave (decelerating), function of the deme size. The more specific threshold-asymptotic model they considered is actually a *T*-stay strategy.

21.3 Departures from ideal resource use: a metapopulation model

Although the distribution of offspring according to local resources seems the best way to exploit the habitat, selected dispersal strategies will depart from such patterns. In the presence of dispersal costs, one can expect selection for lower levels of dispersal than those leading to the best use of resource, and ultimately this should have negative consequences on population size. Conversely, kin competition, by favouring dispersal, could ultimately have positive effects on population persistence. In the following, the impact of these forces is examined in a metapopulation model, where context-dependent and context-independent dispersal are further contrasted. There is both demographic stochasticity (resulting from variation in number of juveniles produced according to a Poisson distribution) and environmental stochasticity (in the form of a deme extinction rate independent of deme size). All demes are equivalent in terms of available resource, and dispersal follows an island model.

The life cycle is as follows (Rousset and Ronce 2004). The population is asexual and generations are discrete. Each parent produces a finite, Poisson-distributed, number of juveniles, each of which behave independently, with dispersal probability determined by the mother's genotype and the parental deme size. There is a reduction *c* in survival probability during dispersal (the dispersal cost). The survival of juveniles that have passed the dispersal stage is then affected by competition within each deme: each juvenile survives independently with probability $s(j) = sj^{a-1}e{-}kj$ where *j* is the number of juveniles in the deme after dispersal. This form allows in particular to consider Allee effects on juvenile survival when $a > 1$ and $k > 1$, as illustrated below, and yields a Ricker model when $a = 1$ and $k > 0$. For computation, a maximum adult deme size *N*max is imposed. Ideally, other demographic parameters are set so that a higher *N*max would little affect the results. As the computation of the joint strategy of *N*max rates is time consuming (in part because the equilibrium demography must be recomputed for each set of dispersal rates), *N*max values of up to 20 were only considered in numerical computations, which is realistic, for example, in the case of the urban populations of *Crepis sancta* (Cheptou and Avendaño 2006; Case Study IV and Chapter 24).

Selection gradients (S) will be evaluated such that the change in allele frequency p of a dispersal modifier, with effect δ on one of the conditional dispersal rates, is $\Delta p = \delta Sp(1 - p) + O(\delta^2)$. Rousset and Ronce (2004) provide computational formulae for selection gradients on unconditional behaviours in metapopulation models (Rousset and Ronce 2004, their eqs. 26 and 27), which are the sums of the gradients for parental deme-size dependent behaviour (sums over n in that paper's notation). The gradients for the context-dependent dispersal rates are thus given as the term for each n in those expressions. The gradients depend on reproductive values and on relatedness coefficients, computed according to eqs. (28) and (29) in Rousset and Ronce (2004). A candidate ESS (cESS) is identified as an intermediate (neither 0 or 1) resident dispersal rate where the gradient vanishes, or as non-intermediate (either 0 or 1) one if the gradient is negative or positive. A set of conditional dispersal rates that jointly satisfy these constraints for all deme sizes is obtained numerically (using Mathematica). The possibility of finding coexisting stable strategies is not excluded, though it has not been reported in the works reviewed above. Additional computations would be necessary to check joint convergence stability (Leimar 2009) and the absence of disruptive selection (Eshel 1996; Geritz *et al.* 1998; Ajar 2003). However, context-dependent dispersal rates are not conducive to disruptive selection, compared to the context-independent case (Massol *et al.* 2011).

21.3.1 Numerical patterns

With moderate fecundity ($r = 3$), small Nmax = 10, and simple negative density dependence ($s(j) = e^{-j/20}$ i.e. $a = 1$, $k = 1/20$, and $s = 1$), the selected dispersal rates were computed for extinction rate $e = 0$ to 0.2 and dispersal cost $c = 0.1$ to 0.5. We can compare the results for extreme e and c to the following cases. First, similar context-dependent dispersal patterns were obtained for $r = 6$, although higher dispersal was favoured overall. We will not detail them here. Second, weaker density-dependence ($k = 1/40$) were explored, for which a higher maximal deme size is allowed (Nmax = 20 versus Nmax = 10; these two cases will be conveniently labelled according to their Nmax). Third, we will compare context-dependent and -independent cESSs. Finally, dispersal and demographic properties at the cESSs will be compared to those observed under dispersal rates that maximize average population size, providing a reference point against which the demographic effects of the other factors can be assessed.

Figure 21.1 shows that there is little difference overall between average deme size attained at the candidate ESSs and the maximum possible size, under either context-dependent or context-independent dispersal. Variances in deme size were similarly marginally affected (details not shown). Figure 21.1 also shows that there is little difference between average deme size attained at the candidate ESS under context-dependent versus -independent dispersal, the largest differences being observed for high dispersal cost and high extinction rate. In the same conditions, we find that the maximum possible average deme size differs most from the cESSs one, and that the maximum average deme size differs most between context-dependent versus context-independent dispersal. These results suggest that context-dependent dispersal has the most important demographic consequences when demographic conditions are bad.

The recolonization potential of an empty deme can be measured by the average number of immigrants that reach it. Although equilibrium deme size at the candidate ESSs is not far below its potential maximum, the dispersal rates are decreased so that the average number of immigrants is substantially reduced at the candidate ESSs (Figure 21.2). The number of immigrants at the context-dependent cESSs is slightly lower that at the context-independent cESSs for $c = 0.1$, and marginally higher for $c = 0.5$. As expected, the number of immigrants is increased by selection in the presence of extinctions.

Context-dependent dispersal patterns (right-hand panels of Figures 21.3 and 21.4) do not exactly fit a 'threshold asymptotic' pattern, while being hard to distinguish from it in practice. Rather, the

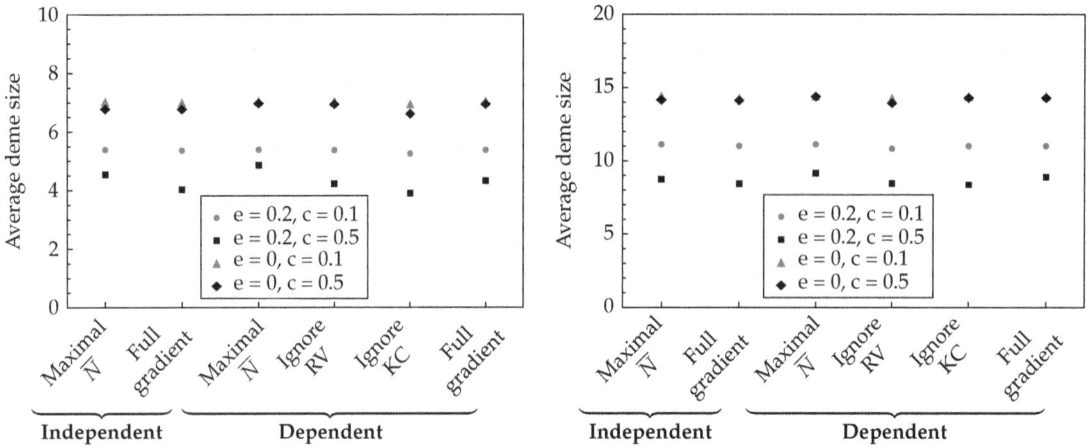

Figure 21.1 Average deme size under different selective regimes. Left: *N*max = 10; right: *N*max = 20. Results for context-independent dispersal are shown in the left two columns of each panel, for maximized average deme size *N* and for candidate ESSs computed from the full selection gradient. For context-dependent dispersal, the results when either reproductive value ('Ignore RV') or kin competition ('Ignore KC') are ignored are also shown.

number of juveniles staying in their natal deme still slightly increases in large-sized demes.

21.3.2 Analysing the selective forces

As previously discussed, the average reproductive values of immigrants should be higher than those of resident juveniles (the consequence of selection on marginal reproductive value gains). Although also increasing with dispersal cost, this difference is very strong when there are local extinctions. Similar trends are observed at the adult stage. With $c = 0.5$, $e = 0.2$ and Nmax = 10, the juvenile relative reproductive values are 1.24 and 0.93, and the one of adults are 1.1 and 0.96. With Nmax = 20, these values become 1.37, 0.92, 1.16, and 0.96, respectively. By contrast, when dispersal is such as to maximize population size, these values become 1.18, 0.93, 1.07, and 0.97 respectively, illustrating that reproductive values become more balanced when population size is maximized, and that selection on marginal gains in reproductive value leads to a reduction in population size. The variance of standardized individual reproductive value (with mean 1) in demes of different sizes follows similar trends at the candidate ESSs, growing from 0.001 when $e = 0$ and $c = 0.1$ to 0.034 when $e = 0.2$ and $c = 0.5$ (Nmax = 20 in both cases).

The selection gradients represent the total effects of individual selection on allele frequency change, including kin (or group) effects as well as variation in reproductive value of offspring (Rousset and Ronce 2004). One can investigate the effect of variation in reproductive value on cESSs by taking this variation out of the formulas for selection gradients. Individual reproductive value is forced to be independent of deme size, by replacing the class reproductive value of demes of a given size by the product of the frequency of such demes and deme

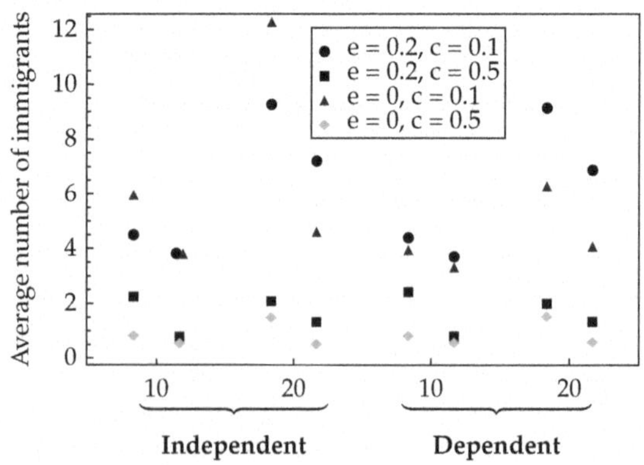

Figure 21.2 Average number of immigrants per deme. Results are shown for *N*max = 10 and 20 and context-independent and -dependent dispersal. For each of the four pairs of columns, the results for maximization of deme size are shown on the left and for the candidate ESS on the right.

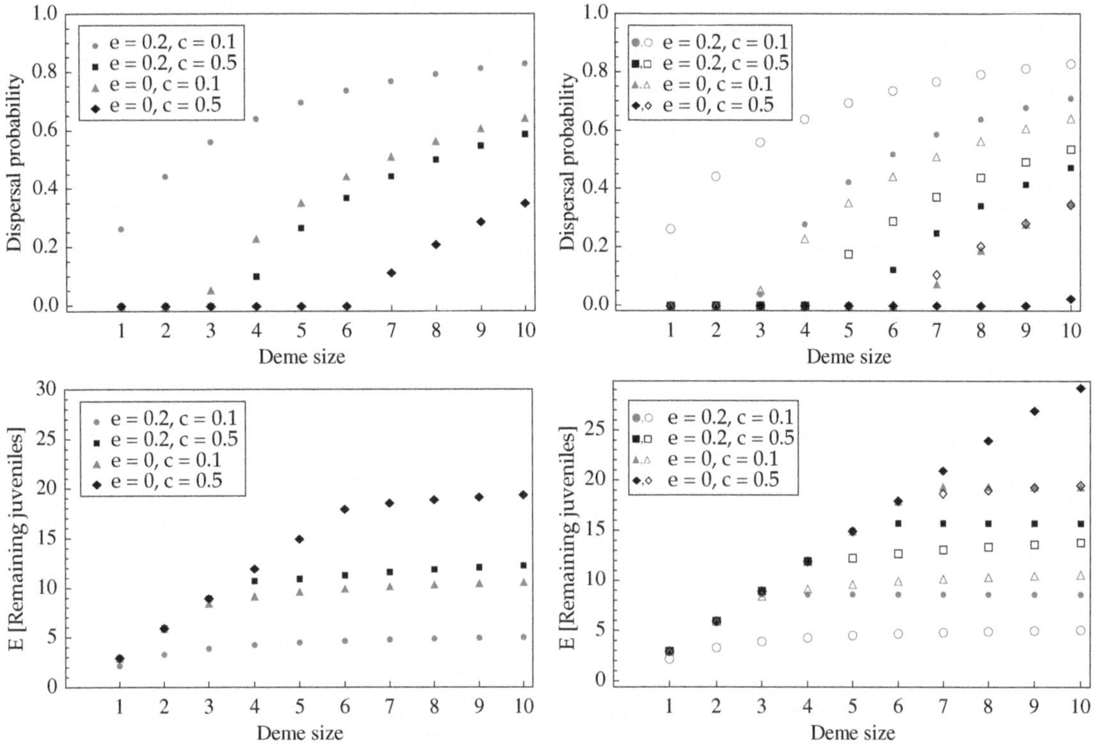

Figure 21.3 Candidate ESSs for *N*max = 10, and outcomes that would be predicted if different components of the gradient were ignored. Top: context-dependent dispersal probabilities; bottom: corresponding expected numbers of juveniles staying in their natal deme. Left: full gradient computation; right: predicted outcome if either reproductive value variation (filled symbols) or kin competition (empty symbols) were ignored. See chapter for further details.

size. In this computation, the fitness measure used reduces to the average number of adult offspring of an individual, irrespective of offfspring reproductive value. Likewise, since the selection gradients are expressed as an inclusive fitness partition of individual fitness effects, one may ignore all kin competition terms by setting all relatedness coefficients to zero. All these modifications had weak impact, but in the case of both high extinction rate and high environmental extinction rate. In particular, a simple counting of offspring number is sufficient to obtain good approximations for the population sizes predicted from the full gradients. Conversely, if only equilibration of marginal reproductive value benefits was considered as a force of dispersal evolution, but not kin competition, the predicted equilibrium deme sizes will be lower when *N*max = 10, showing that the selective component in relation to marginal reproductive value reduces population size, but that kin competition increases it. For *N*max = 20, the reverse pattern is observed, as the average population sizes predicted when reproductive value is ignored are close to those predicted from the full gradients, and higher than when kin competition is ignored, except in the conditions leading to the lowest average population sizes (e = 0.2, c = 0.5). Thus, as expected, the effect of kin competition is less important when population size is greater.

Similar comparisons can be made in order to understand the evolution of dispersal rates themselves. Unsurprisingly, given the range of deme sizes considered, kin competition has an impact, especially in absence of environmental stochasticity. The dispersal rates predicted by taking kin competition into account, but ignoring reproductive value variation, are close to the candidate ESSs (compare open symbols in right-hand panels to

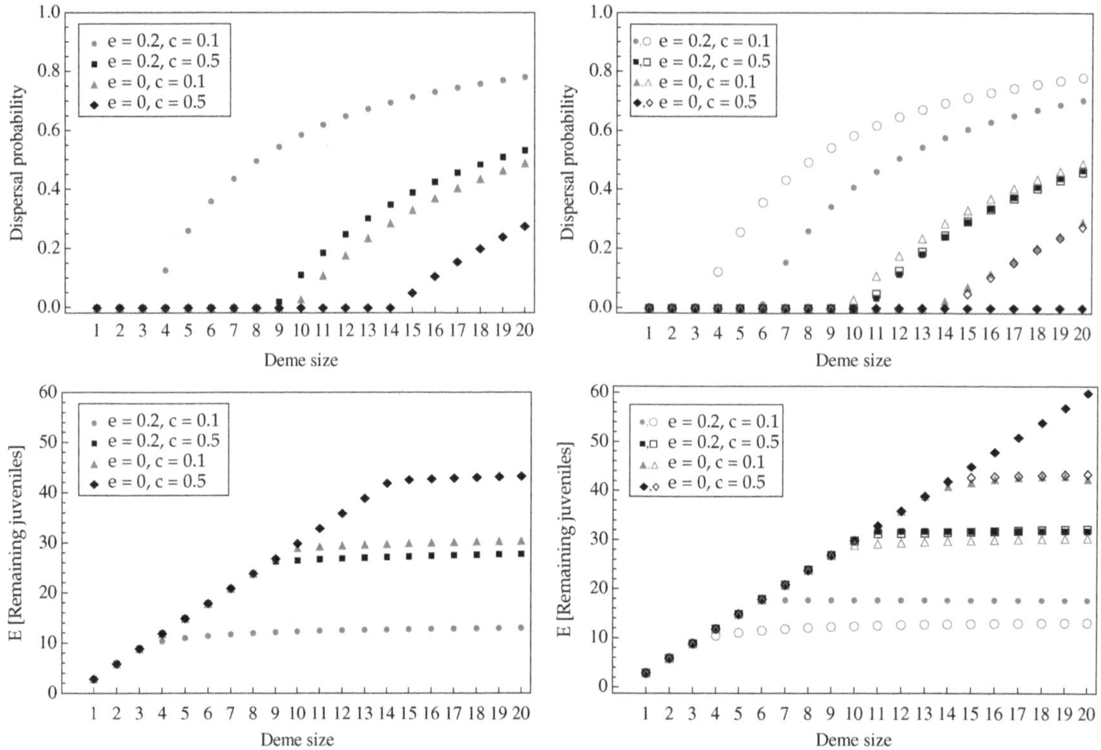

Figure 21.4 Candidate ESSs for *N*max = 20. As in Figure 21.3, filled and empty symbols stand for dispersal rates predicted from the full gradient, and for those that would be predicted if kin competition was ignored, respectively. See chapter for further details.

left-hand panels in each of Figures 21.3 and 21.4). The deviations from the *T*-stay pattern (in particular, the fact that the number of juveniles staying in demes keeps increasing at larger deme sizes) may partly be a result of decreasing kin competition with increasing deme size. However, dispersal rates computed when kin competition is ignored still do not follow the *T*-stay pattern.

21.3.3 Allee effects

These numerical results suggest that context-dependent dispersal has generally small demographic consequences. However, opposite results are obtained in the presence of Allee effects. In this case, evolution of context-independent dispersal can lead to 'evolutionary suicide' (Gyllenberg *et al.* 2002). This process is observed, for example, under the same metapopulation model considered earlier, but now with an Allee effect on juvenile survival $s(j)$. For example, when $r = 4$, $s(j) = 0.05j^{1.8}e^{-0.13j}$, *N*max = 20 and $c = 0.7$, evolutionary suicide occurs for $e > 0.11$ (Rousset and Ronce 2004; Figure 21.4). With conditional dispersal, evolutionary suicide no longer occurs at the cESS for e up to at least 0.2115.

Why not disperse out of the smallest demes? Although resident offspring from a single parent do not survive maximally, their position may not be so bad (Figure 21.5). In particular, an Allee effect on juvenile survival was considered, so that if the parental fecundity is high enough, or if the number of immigrants is high enough, the survival of resident juveniles survival is not so low, and those who will survive may have high reproductive value as their own juveniles may find themselves in close to ideal density conditions. High dispersal out of the smallest demes would therefore appear more favoured by Allee effects on both survival and fecundity, and low immigrant flux (by increased dispersal cost, for example). But such conditions are

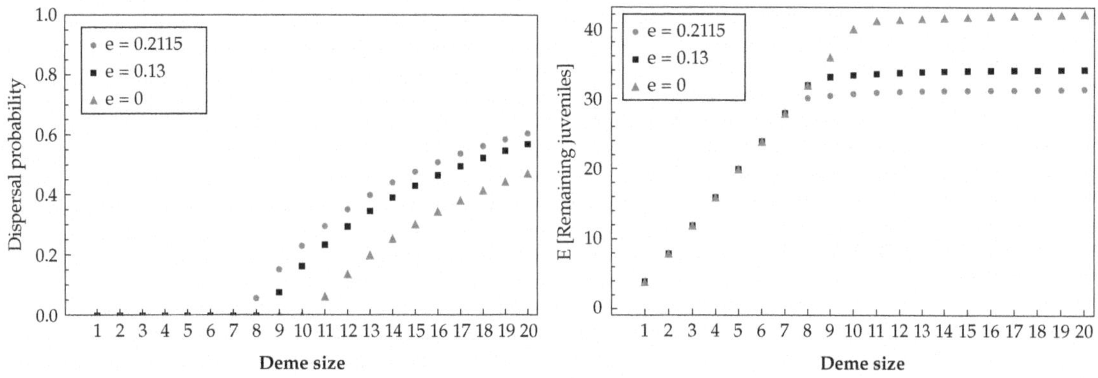

Figure 21.5 Candidate ESSs under Allee effects. Left: context-dependent dispersal probabilities; right: corresponding expected numbers of juveniles staying in their natal deme.

very unfavourable to population persistence, and it may be expected that extant populations in such hard conditions have evolved alternative dispersal strategies, as further discussed below.

21.4 Discussion

Context-dependent dispersal may intuitively be expected to favour population persistence (Hovestadt and Poethke 2006). The escape from evolutionary suicide under Allee effects most clearly illustrate this intuitive prediction. Likewise, Ruxton and Rohani (1998) have shown that the *T*-stay dispersal strategy can stabilize population dynamics, although their model differs in many ways from the ones considered here. When an absence of Allee effects is considered, the effects of individual selection on equilibrium population size are small, except with a strong dispersal costs, a result congruent with simulations by del Mar Delgado *et al.* (2011). However, the effects on the recolonization potential are stronger. The dispersal rate maximizing population size is higher than the candidate ESSs for dispersal. Previous works have found that this was so when environmental fluctuations were considered (Roff 1975; Olivieri and Gouyon 1997) but that the opposite result held when only demographic stochasticity was acting (Hamilton and May 1977; Motro 1982). However, the present result holds in both cases, showing that finer details of the demographic process matter.

The present model has numerous shortcomings, and the list following only considers some of the most important and easily overlooked ones.

21.4.1 Demographic regime

The metapopulation model assumes that dispersal is dependent on parental deme population size. This could represent passive dispersal in a plant where a parent can feel the competition with its neighbours and equip some seeds with a dispersal structure. In many cases it would seem better for juveniles to take dispersal decisions based on the number of competing juveniles. A basic reason for not considering this here is the computation load, which increases with the number of traits considered (up to 20 conditional dispersal rates in the present case). Decisions based on number of parents and on number of juveniles will also be equivalent if fecundity is very large and the dispersal trait is determined before population regulation, so that juvenile numbers accurately reflect adult numbers. Such an assumption may not always be wrong, as, for example, wing dimorphisms in insects are controlled by population density at various developmental stages (Zera and Denno 1997). The dynamics of the model itself are unchanged when fecundity is increased by some factor *F*, and when density-independent mortality has probability $1 - 1/F$ after dispersal but before density-dependent competition, so that dispersal dependent on parent or on offspring numbers evolves in convergent ways as *F* increases. However, this still ignores the potential for parent-offspring conflicts on dispersal in sexual populations (Taylor 1988). This also does not represent a good model of active dispersal under juvenile control, as, in the latter case, one could expect dispersal to be also dependent on destination-deme size. Such an increase in discriminating behaviour can be expected to have distinct effects on population demography.

It was also assumed that all variation in habitat quality, as perceived by a breeding individual, follows from stochastic fluctuations in the number of adults. This generally favours an increase of dispersal with deme size. Alternatively, simple models considering fixed differences in deme size predict the opposite pattern (e.g. McPeek and Holt 1992). In general, patterns of context-dependent dispersal must depend on how much variation in deme size depends on stochastic fluctuations versus temporally constant spatial variation, and the two have been considered separately (for some simulations that consider both together, see Enfjäll and Leimar 2009). Natural populations may be anywhere between these two extremes, but information about this is hard to find in empirical studies of metapopulation model systems. Such information should be critical for further understanding of patterns of dispersal in natural populations.

Finally, it was assumed that all juveniles that have survived the dispersal stage are equally competitive. Conversely, when dispersal is dependent on body condition, fitter individuals may disperse more or less than less fit ones (Clobert *et al.* 2009; Gyllenberg *et al.* 2011 and references therein; Bonte and de la Peña 2009; Chapter 11). Asymmetries in competitive abilities may not only cause deviations from the above conclusions but more importantly generate behavioural syndromes that may be the general patterns rather than epiphenomena (Clobert *et al.* 2009).

21.4.2 Alternative behavioural responses

In many instances, individuals are able to identify close kin by familiarity. Then it could be expected that dispersal evolves not simply in response to average levels of relatedness generated by the demographic processes, but rather as a plastic response to the realized dispersal events of relatives. Inbreeding avoidance based on familiarity should also operate in response to inbreeding depression.

We have seen that context-dependent dispersal can rescue a population otherwise susceptible to evolutionary suicide. There are other behavioural responses through which such evolutionary suicide can be avoided. One is the colonization of habitats (or the invasion of already-occupied habitat) by groups of individuals, thereby escaping the costs of solitary emigration implied by Allee effects. This mode of dispersal is indeed common in social insects (colony fission or budding in ants and bees, for example; Peeters and Ito 2001), vertebrates (e.g. lions), and collective swarming in bacteria (Fraser and Hughes 1999). Although the evolution of dispersal rates under group dispersal has been considered (Gandon and Michalakis 1999), the evolution of dispersal in groups versus solitary dispersal has been disregarded.

Another strategy that can retain the benefits of the presence of neighbours, yet provides access to new resource, is the evolution of very short dispersal distances. In arid environments, for example, plant patches benefit from a better retention of water than solitary individuals. Short-distance dispersal can then lead to the striking reticulate patterns of habitat occupation that are observed in such habitats (Pueyo *et al.* 2008). Similar dispersal responses and distribution patterns may occur in bacterial biofilms (Xavier and Foster 2007; Xavier *et al.* 2009). This shows that the distribution of dispersal distance matters, even in an abiotic environment without intrinsic heterogeneity.

21.5 Conclusion

This chapter has considered the demographic consequences of the evolution of dispersal and their analysis in terms of a decomposition of the selective forces involved. The two forces considered here were equilibration of marginal reproductive value returns and kin competition. Their effects were quantified in a model with negative-frequency dependence following a Ricker equation. The former force leads to a decrease in population size relative to the maximum feasible size, and kin competition leads to a closely compensating increase, so that population size at the candidate ESS is close to the maximum. These results suggest that comparatively large selection-induced reductions of population size may occur when deme sizes are large (so that kin selection is weaker). Further, all demographic effects were weak, except for when there is a large dispersal cost and large local extinction rate. Although context-dependent dispersal may allow individuals to better exploit habitat resources and lead to an increase in population size, the demographic effects were again weak under the Ricker model. The number of immigrants, which measures the colonization potential, was comparatively more strongly reduced by individual selection than average population size was. Still, it differed little under context-dependent versus context-independent dispersal. Thus, the demographic importance of context-dependent dispersal is not obvious. These results clearly call for further numerical investigations to assess their generality. Context-dependent dispersal can still have drastic consequences, as shown in the case of Allee effects, but this might occur only in the harshest environmental conditions.

Acknowledgements

This work was supported by the Agence Nationale de la Recherche, programme '6th extinction', through the project EVORANGE (ANR-09-PEXT-011). J. Clobert and T. Benton provided much advice, and a reviewer made very useful suggestions.

References

Ajar, É. (2003). 'Analysis of disruptive selection in subdivided populations', *BMC Evolutionary Biology* **3**: 22.

Alizon, S. and Taylor, P. (2008). 'Empty sites can promote altruistic behaviour', *Evolution* **62**: 1335–44.

Billiard, S. and Lenormand, T. (2005). 'Evolution of migration under kin selection and local adaptation', *Evolution* **59**: 13–23.

Bonte, D. and de la Peña, E. (2009). 'Evolution of body condition-dependent dispersal in metapopulations', *Journal of Evolutionary Biology* **22**: 1242–51.

Bowler, D.E. and Benton, T.G. (2005). 'Causes and consequences of animal dispersal strategies: relating individual behaviour to spatial dynamics', *Biological Review* **80**: 205–25.

Bulmer, M.G. (1994). *Theoretical evolutionary ecology*. Sinauer, Sunderland, Mass.

Cadet, C., Ferrière, R., Metz, J.A.J. *et al*. (2003). 'The evolution of dispersal under demographic stochasticity', *American Naturalist* **162**: 427–41.

Charlesworth, B. (1994). *Evolution in age-structured populations*, 2nd edition. Cambridge University Press, Cambridge.

Cheptou, P.-O. and Avendaño, L.G. (2006). 'Pollination processes and the Allee effect in highly fragmented populations: consequences for the mating system in urban environments', *New Phytology* **172**: 774–83.

Clobert, J., le Gailliard, J.-F., Cote, J. *et al*. (2009). 'Informed dispersal, heterogeneity in animal dispersal syndromes, and the dynamics of spatially structured populations', *Ecological Letters* **12**: 197–209.

del Mar Delgado, M., Ratikainen, I.I., Kokko, H. (2011). 'Inertia: the discrepancy between individual and common good in dispersal and prospecting behaviour', *Biological Review* **86**: 717–32.

Dieckmann, U., and Ferrière, R. (2004). 'Adaptive dynamics and evolving biodiversity', in, R. Ferrière, U. Dieckmann, and D. Couvet (eds), *Evolutionary Conservation Biology*. Cambridge University Press, Cambridge, 188–224.

Enfjäll, K. and Leimar, O. (2009). 'The evolution of dispersal—the importance of information about population density and habitat characteristics', *Oikos* **118**: 291–9.

Eshel, I. (1996). 'On the changing concept of evolutionary population stability as a reflection of a changing point of view in the quantitative theory of evolution', *Journal of Mathematical Biology*. **34**: 485–510.

Ezoe, I. and Iwaza, Y. (1997). 'Evolution of condition-dependent dispersal: a genetic-algorithm search for the ESS reaction norm', *Researches in Population Ecology* **39**: 127–37.

Fisher, R.A. (1958). *The genetical theory of natural selection*, 2nd edition. Dover, New York.

Fraser, G.M. and Hughes, C. (1999). 'Swarming motility', *Current Opinion in Microbiology* **2**: 630–5.

Fretwell, S.D. and Lucas, Jr., H.L. (1970). 'On territorial behaviour and other factors influencing habitat distribution in birds', *Acta Biotheoretica* **19**: 16–36.

Gadgil, M. (1971). 'Dispersal: population consequences and evolution', *Ecology* **52**: 253–60.

Gandon, S. and Michalakis, Y. (1999). 'Evolutionarily stable dispersal rate in a metapopulation with extinctions and kin competition', *Journal of Theoretical Biology* **199**: 275–90.

Geritz, S.A.H., Kisdi, É., Meszéna, G. *et al.* (1998). 'Evolutionarily singular strategies and the adaptive growth and branching of the evolutionary tree', *Evolutionary Ecology* **12**: 35–57.

Greenwood, P.J. (1980). 'Mating systems, philopatry and dispersal in birds and mammals', *Animal Behaviour* **28**: 1140–62.

Greenwood-Lee, J.M. and Taylor, P.D. (2001). 'The evolution of dispersal in spatially varying environments', *Evolutionary Ecology Research* **3**: 649–65.

Gyllenberg, M., Kisdi, É., Utz, M. (2011). 'Body condition dependent dispersal in a heterogeneous environment' *Theoretical Population Biology* **79**: 139–54.

Gyllenberg, M., Parvinen, K. Dieckmann, U. (2002). Evolutionary suicide and evolution of dispersal in structured metapopulations', *Journal of Mathematical Biology* **45**: 79–105.

Hamilton, W.D. and May, R. (1977). 'Dispersal in stable habitats', *Nature* **269**: 578–81.

Hanski, I., Erälahti, C., Kankare, M. *et al.* and (2004). Variation in migration propensity among individuals maintained by landscape structure', *Ecology Letters* **7**: 958–66.

Hovestadt, T., Kubisch, A., Poethke, H.-J. (2010). 'Information processing in models for density-dependent emigration: a comparison', *Ecological Modelling* **221**: 405–10.

Hovestadt, T. and Poethke, H.J. (2006). 'The control of emigration and its consequences for the survival of populations', *Ecological Modelling* **190**: 443–53.

Kisdi, É. (2004). 'Conditional dispersal under kin competition: extension of the Hamilton-May model to brood size-dependent dispersal', *Theoretical Population Biology* **66**: 369–80.

Leimar, O. (2009). 'Multidimensional convergence stability', *Evolutionary. Ecology Res.* **11**: 191–208.

Leturque, H. and Rousset, F. (2002). 'Convergence stability, probabilities of fixation, and the ideal free distribution in a class-structured population', *Theoretical Population Biology* **62**: 169–80.

Massol, F., Duputié, A., David, P. *et al.* (2011). 'Asymmetric patch size distribution leads to disruptive selection on dispersal', *Evolution* **65**: 490–500.

McPeek, M.A. and Holt, R.D. (1992). 'The evolution of dispersal in spatially and temporally varying environments', *American Naturalist* **140**: 1010–27.

Metz, J.A.J. and Gyllenberg, M. (2001). 'How should we define fitness in structured metapopulation models? Including an application to the calculation of evolutionarily stable dispersal strategies', *Proceedings of the Royal Society B* **268**: 499–508.

Motro, U. (1982). 'Optimal rates of dispersal. I. Haploid populations', *Theoretical Popular Biology* **21**: 394–411.

Motro, U. (1991). 'Avoiding inbreeding and sibling competition: the evolution of sexual dimorphism for dispersal', *American Naturalist* **137**: 108–15.

Olivieri, I. and Gouyon, P.-H. (1997). 'Evolution of migration rate and other traits: the metapopulation effect', in I.A. Hanski, and M.E. Gilpin, *Metapopulation biology: ecology, genetics and evolution*. Academic Press, San Diego, California, 293—323.

Olivieri, I., Michalakis, Y., Gouyon, P.-H. (1995). 'Metapopulation genetics and the evolution of dispersal', *American Naturalist* **146**: 202–228.

Ozaki, K. (1995). 'Intergall migration in aphids; a model and a test of ESS dispersal rate', *Evolutionary Ecology* **9**: 542–9.

Peeters, C. and Ito, F. (2001). 'Colony dispersal and the evolution of queen morphology in social Hymenoptera', *Annual Review of Entomology* **46**: 601–30.

Poethke, H.J. and Hovestadt, T. (2002). 'Evolution of density- and patch-size-dependent dispersal rates', *Proceedings of the Royal Society B* **269**: 637–45.

Pueyo, Y., Kéfi, S., Alados, C.L. *et al.* (2008). 'Dispersal strategies and spatial organization of vegetation in arid ecosystems', *Oikos* **117**: 1522–1532.

Roff, D.A. (1975). 'Population stability and the evolution of dispersal in a heterogeneous environment', *Oecologia* **19**: 217–237.

Ronce, O., Brachet, S., Olivieri, I. *et al.* (2005). 'Plastic changes in seed dispersal along ecological succession: theoretical predictions from an evolutionary model', *Journal of Ecology* **93**: 431–40.

Rousset, F. (1999). 'Reproductive value versus sources and sinks', *Oikos* **86**: 591–6.

Rousset, F. and Billiard, S. (2000). 'A theoretical basis for measures of kin selection in subdivided populations: finite populations and localized dispersal', *Journal of Evolutionary Biology* **13**: 814–25.

Rousset, F. and Ronce, O. (2004). 'Inclusive fitness for traits affecting metapopulation demography', *Theoretical Population Biology* **65**: 127–41.

Roze, D. and Rousset, F. (2009). 'Strong effects of heterosis on the evolution of dispersal rates', *Journal of Evolutionary Biology* **22**: 1221–1233.

Ruxton, G.D. and Rohani, P. (1998). 'Fitness-dependent dispersal in metapopulations and its consequences for persistence and synchrony', *Journal of Animal Ecology* **67**: 530–539.

Taylor, P.D. (1990). 'Allele-frequency change in a class-structured population', *American Naturalist* **135**: 95–106.

Taylor, P.D. (1988). 'An inclusive fitness model for dispersal of offspring', *Journal of Theoretical Biology* **130**: 363–78.

Taylor, P.D. and Frank, S.A. (1996). 'How to make a kin selection model', *Journal of Theoretical Biology* **180**: 27–37.

Travis, J.M.J., Murrell, D.J., and Dytham, C. (1999). 'The evolution of density-dependent dispersal', *Proceedings of the Royal Society B* **266**: 1837–42.

van Valen, L. (1971). 'Group selection and the evolution of dispersal', *Evolution* **25**: 591–8.

Xavier, J.B. and Foster, K.R. (2007). 'Cooperation and conflict in microbial biofilms', *Proceedings of the National Academy of Sciences of the USA* **104**: 876–81.

Xavier, J.B., Martinez-Garcia, E., Foster, K.R. (2009). 'Social evolution of spatial patterns in bacterial biofilms: when conflict drives disorder', *American Naturalist* **174**: 1–12.

Zera, A.J. and Denno, R.F. (1997). 'Physiology and ecology of dispersal polymorphism in insects', *Annual Review Entomology* **42**: 207–30.

CHAPTER 22

Landscape effects on spatial dynamics: the natterjack toad as a case study

Virginie M. Stevens and Aurélie Coulon

22.1 Introduction

Most organisms live in heterogeneous landscapes. Different components of the landscape have specific biotic and abiotic properties that may interact with organisms' biology and influence their ability to move across the landscape (With *et al.* 1999; Goodwin and Fahrig 2002): the permeability of a landscape differs depending on the organisms considered. This heterogeneity goes beyond the binary habitat/uniform matrix representation of the landscape that once predominated in landscape ecology. Indeed, even within the 'matrix' (the landscape outside suitable habitat patches, which usually is more or less inhospitable to the species of interest though it may in some cases offer some, but not all, resources), there are different types of environments with specific properties. As a result, the nature of all the elements composing a landscape and their spatial organisation will affect the way individuals move within this landscape (e.g. Jonsen and Taylor 2000; Rondinini and Doncaster 2002; Dickson *et al.* 2005; Bakker 2006; Loarie *et al.* 2009), some facilitating dispersal ('corridors'), others impeding or reducing the passage of dispersers ('barriers'). The differential permeability of landscape features may affect the movements within each, but also the willingness of the organisms to enter or leave them. Landscape characteristics hence have the potential to affect the different aspects of dispersal processes: emigration, movement trajectories, and immigration.

Two reasons make dispersal a key process in population dynamics (Chapter 20). First, movement of individuals in itself impacts population dynamics. Second, dispersal movement also has important consequences for the degree of genetic mixing that we expect to find among populations, and hence on the genetic variation found within populations. Both genetic mixing and genetic variation may affect population dynamics, through their consequences on individual performances and the adaptive potential of populations.

The effects of landscape on dispersal processes hence have the potential to affect spatial dynamics significantly. Understanding which elements within a landscape significantly influence movements and predicting how landscape changes may affect spatial dynamics are crucial in the current context of global changes. Here we illustrate how the question can be addressed through the example of a species with limited mobility and which has been extensively studied in this regard: the Natterjack toad (*Epidalea calamita*). Firstly, empirical studies were used to measure the effects of landscape elements on movement patterns. Second, the results from these empirical studies were used to model the effects of landscape on movement patterns. Third, the model was validated through the comparison of the movements predicted by the models with those estimated by the analysis of gene flow among populations. Finally, the model was used to predict natterjack population dynamics under different landscape management scenarios. We then discuss how the effect of dispersal goes beyond a simple numerical redistribution of individuals across landscapes and give some examples of the effect of dispersal on population dynamics attributable to gene flow.

Dispersal Ecology and Evolution. First Edition. Edited by Jean Clobert *et al.*

22.2 Measuring landscape effects on movement patterns: functional connectivity

The effect of landscapes on movement patterns is referred to as functional connectivity: the ability of a landscape to be crossed by an organism (With *et al.* 1999). Functional connectivity includes two components: i) patch resistance or viscosity, the degree to which a landscape component impedes or encourages individuals to move (With 1994; Wiens *et al.* 1997), and ii) boundary permeability, the probability of crossing a boundary between two landscape components (Stamps *et al.* 1987; Wiens *et al.* 1997; Jonsen and Taylor 2000). Measuring functional connectivity hence requires assessing patch resistance and boundary permeability. These were assessed experimentally in the Natterjack toad (Stevens *et al.* 2004, 2006a).

A first experiment was designed to assess patch resistance through the comparison of movement patterns in 3.5 m^2 indoor arenas mimicking the five most frequent types of land uses encountered by natterjacks (sand, cement, forest, grass, and field) (Stevens *et al.* 2004). Toadlets were individually released in the arenas, and their movement trajectories were recorded over 2 min by placing sequentially numbered marks at each stopping and turning angle. Toadlets exhibited significant differences in their performances in the different arenas (Figure 22.1). Overall, the toadlet trajectories had lowest bout lengths, were slowest, and most tortuous in simulated forests, and the reverse was observed in the sand arena. Those differences revealed differential resistances of the corresponding landscape features to toadlet movements. Based on the values of effective speed (i.e. the linear distance of the trajectory divided by trial duration), the relative resistance (unitless) of each substrate to toadlet movement was ca. 1, 1.5, 2, 2.5, and 5 for sand, cement, field, grass, and forest respectively.

Boundary permeability was then assessed in a second experiment in which toadlets were placed in (1 meter long) Y-shaped experimental devices which mimicked the conditions in two environments; the line where the two environments met corresponded to a boundary (Stevens *et al.* 2006a). The same five landscape components as in the patch resistance experiment were tested. The permeability of boundaries was assessed as the proportion of individuals crossing the boundary within 5 min. The crossing frequencies varied according to the boundary configurations (Figure 22.1): the toadlets left the simulated agricultural environments ('field' and 'grass') more often than the bare environments ('cement' and 'sand'), and entered the latter more often than the former. The authors tested whether the permeability of a boundary was determined by the relative resistance of the environments (as inferred in the first experiment) on the two sides of the boundary (the 'resistance-based permeability (RBP) hypothesis'). According to this hypothesis, the crossing frequencies should be proportional to the ratio of the resistance for the environments on both sides of the boundary. Stevens *et al.* found that crossing frequencies generally did not differ from expectations under the RBP hypothesis, except for the boundary configurations involving forest environments: those boundaries showed a higher permeability than expected under the RBP hypothesis, as high as that of the sand environment (Figure 22.1). This unexpected pattern could be due to the fact that forests and hedgerows, although not favourable to movements, usually offer hiding and resting places, abundant food, and favourable temperature and moisture conditions, and hence can be attractive to naive juvenile toads even though they offer resistance to movements.

This type of experimental approach to assess functional connectivity has the advantage of allowing precise estimates based on numerous replicates in controlled environments. An alternative for larger species, for which it is difficult to perform this type of experiments, is the monitoring of dispersal movements in the wild through the use of tracking devices such as GPS (Global Positioning System) collars. Statistical analyses then allow inference of which type of landscape elements are preferred or avoided during movements (e.g. O'Brien *et al.* 2006; Driezen *et al.* 2007; Arellano *et al.* 2008; Coulon *et al.* 2008; Ovaskainen *et al.* 2008; Chietkiewicz and Boyce 2009; Cushman and Lewis 2010). Tracking however tackles larger time and spatial scales, and thus may be more appropriate than micro-landscape experiments to investigate disperser–landscape interaction for

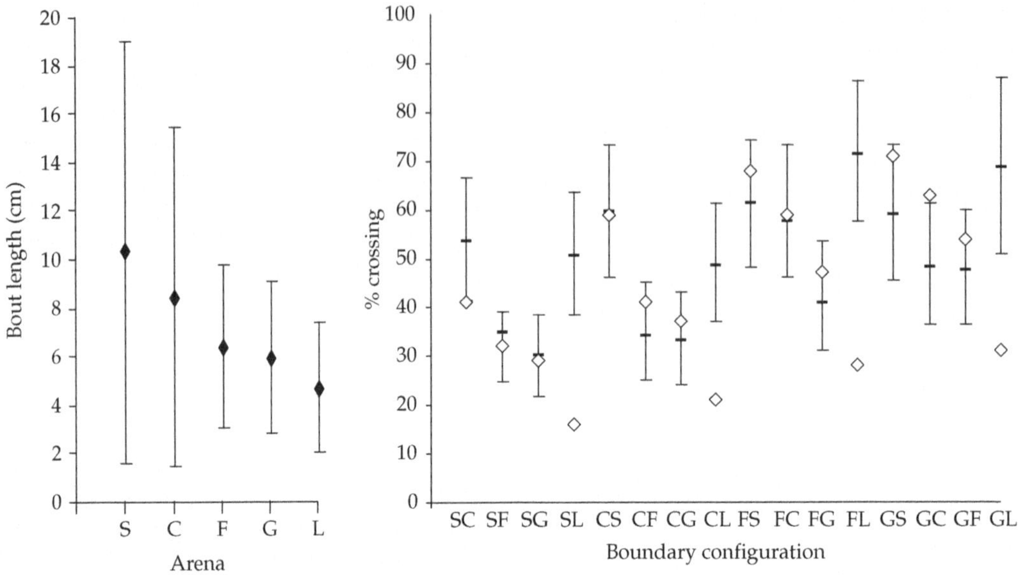

Figure 22.1 Experimental assessment of functional connectivity in the natterjack toad. Left: patch resistance: mean bout length (and s.e.) of paths realized in five arenas mimicking soil conditions prevailing in most common landscape features. Right: boundary permeability (realized in a Y-shaped device: black symbols with 95% CI) compared to the expected crossing frequency if the choice of the toadlets was based on the relative resistance of the two elements proposed in the device (white diamonds). The first letter refers to the release environment; the second to the environment behind the boundary. For both parts of the figure: S: sand for natterjack habitats; C: cement for roads; F: cereal fields; G: grasslands; L: deciduous forests litter.

species with a protracted dispersal phase (Chapter 32). Alternatively, several variables describing the landscape can be used to build models for presence-absence data, which also give some information about the relative resistance of landscape elements. This approach however requires large datasets that can be split into a 'training' part used to fit the model, and a 'validation' part used to verify the adequacy of the model (see, for instance, this approach in Laiola and Tella 2006; Wang *et al.* 2008). The method also needs the habitat requirement to be precisely described, and that species presence and absence can reliably be recorded to avoid false absences in the data. Moreover, presence-absence data reflect not only landscape effects on movements, but also metapopulation dynamics or the probability of settlement and survival in the different habitat patches.

22.3 Modelling landscape effects on movement patterns

Once the relative functional connectivities of the different components of a landscape have been assessed, it is then possible to model movement patterns across heterogeneous landscapes. The approach that has been most widely used in the recent past is least-cost path modelling: a 'cost-grid' of the study area is built, in which each element of the landscape is assigned a hypothetical cost value representing the degree to which this type of element facilitates or impedes dispersal movements for the studied organism. Movements are then modelled as the 'least-cost paths' (LCP) between pairs of locations in this system: the LCP between two locations is the route that has the lowest cumulative cost, estimated as the sum of the costs of the cells crossed by that route. In other words, the least-cost path is the optimum route—cost-wise—between two points. A critical parameter of LCP modelling is the cost assigned to the different landscape components. These cost values should reflect the response of the studied organism to landscape and hence require accurate data on habitat selection and movement ability in the various landscape components.

The LCP approach was used to model natterjack toad functional connectivity (Stevens *et al.* 2006b).

Different scenarios were simulated, all assuming that long-distance and long-term gene flows can be extrapolated from short-term and small-scale behaviours observed in the micro-landscape experiments, and they only differed in the hypothesis made about dispersal route selection. In the first scenario, called 'resistance', toads were supposed to select the route that minimized the resistance to movements, by minimizing the use of high resistance components. The cost values under this scenario were the time needed to perform a net displacement of 10 m in the landscape components, as determined by the patch resistance experiment described in the previous section. In the second scenario, called 'preference', it was hypothesized that toadlets select the route that maximized the use of preferred habitats. Cost values were hence assessed from the boundary permeability experiment (cost = 100—mean % of toadlets entering an environment when starting in another one). In the last scenario, called 'mixed', the hypothesis was that both resistance and boundary permeability determined the selection of the dispersal route. A range of their relative influence was screened by using various combinations of the costs used in the first two scenarios (cost mixed = cost resistance^a × cost preference^b; five combinations of a and b were tested). This modelling was applied to a study area in southern Belgium which encompasses four natterjack populations.

The experimental estimates of resistance and boundary permeability of the landscape elements predicted rather different patterns of dispersal through the study landscape, mainly due to differences in the cost attributed to forest relatively to agricultural elements (see Figure 22.1). To assess which of the proposed models best match real dispersal movements, Stevens *et al.* (2006b) compared the results of their LCP simulations to gene-flow pattern within the study area. To do so, they sampled and genotyped at neutral loci (microsatellites) individuals from the four populations and estimated from those genotypes the number of migrants per generation between each population. They then tested the correlation between the estimated migration rates and the cost of the LCPs linking the different populations, under each scenario. The correlation was significant with the LCPs based on the preference scenario (partial Mantel coeff. = 0.75), while it was not significant with the LCPs based on the resistance or the mixed scenarios (partial Mantel coefficients were all < 0). This result showed that LCPs based on the choice at habitat boundaries best predicted gene-flow patterns in the study area. Such a process of validation of LCP modelling is crucial, because this type of approach is sensitive to the relative cost values attributed to each element. A major hurdle to this validation, however, is that the output of alternative scenarios may be somewhat correlated (which was not the case here), which may strongly reduce the power to discriminate their fit to genetic data.

The LCP modelling approach is appealing because it requires a moderate number of parameters, namely the relative cost of the different landscape features—although accurate measures of those costs are difficult to get for most organisms. One of its main limitations is its implicit assumption of a perfect knowledge of the entire landscape by the moving organism. This is not realistic for most dispersal movements that go beyond the range previously explored by an individual, especially for species with limited perceptual ranges. This assumption of omniscience however might be valid over evolutionary timescales as even if there is no real choice of the dispersal route, the toadlets that follow (by chance) low-cost paths have a higher probability of success in dispersal than those following high-cost paths, and eventually only these will be selected. The LCP modelling approach can be improved by integrating this limited perception of the landscape: for example, Palmer *et al.* (2011) produced a movement simulator in which the trajectories follow the least-cost path according to the cost of the cells within the perceptual range of the simulated organisms. Vuilleumier and Metzger (2006) also integrated this aspect of limited perceptual range in their object-oriented model of dispersal movements. Animal movement modelling has made substantial progress in the last decade, mostly as stochastic models incorporating correlation and bias in the movement rules (e.g. Morales *et al.* 2004, 2005; Heinz and Strand 2006; Vuilleumier and Perrin 2006; Bartón *et al.* 2009). However, those models do not integrate landscape effects on movements. Adding a landscape component into these new simulators should allow modelling movements more accurately. But these more complex approaches come at the cost of a higher demand on empirical parameters, often difficult to get. Studies comparing the performance of the LCP approach and of its alternatives are now needed in order to know whether this extra data demand is necessary to get satisfying models.

22.4 The effect of dispersal on population dynamics

A general consequence of habitat fragmentation is the functional isolation of small populations, and as a consequence, dispersal, by maintaining gene flow among these demographic units, has a prominent role in shaping both genetic diversity and genetic structure of populations in fragmented landscapes. Besides this genetic effect, dispersal also may have important demographic consequences, like the rescue of declining populations and the increase in synchrony among local populations (Chapter 20). In the natterjack toad, both movement rates and gene flow were shown to be important for population persistence. Hence landscape management may affect local dynamics through these two kinds of effects (genetic and numerical).

A population viability analysis (PVA) illustrated the effect of movement rates, without consideration for the indirect effects of gene flows. In their PVA, Stevens and Baguette (2008) simulated the fate of isolated natterjack populations, as contrasted to that of otherwise similar populations connected via dispersal to larger populations. They showed that even quite rare dispersal events (≤ 1 immigrant per generation) positively affected the dynamics of local populations, efficiently rescuing small, declining, local populations. In this analysis, the possible fitness differences between philopatric and dispersing individuals were totally ignored, as was the indirect impact of dispersal that may arise through the genetic composition of local populations, or from inbreeding and outbreeding effects.

The genetic output of dispersal may also impact population dynamics. Inbreeding depression was evidenced in several isolated populations where drift had decreased the genetic diversity, with potential negative impact on population persistence and on individual fitness through genetic load (Nieminen *et al.* 2001; Keller and Waller 2002; O'Grady *et al.* 2006). A common-garden experiment showed that larval growth rate (a critical fitness attribute, Werner 1986) of natterjack tadpoles was positively correlated to the degree of heterozygosity in the population of origin (Rowe *et al.* 1999; Rowe and Beebee 2005). This effect of isolation on inbreeding thus probably has a negative impact on individuals' fitness, and consequently on population growth rate.

Theory predicts that adaptive divergence between populations results from a balance between the diversifying effect of selection and the homogenizing effect of gene flow (Endler 1977, 1986; Slatkin 1987). In the current context of habitat fragmentation, however, dispersal has the potential to affect local adaptation positively because it buffers populations against genetic drift and helps maintain a certain level of genetic variation within small local populations, needed for adaptive selection to take place. In the frog *Rana temporaria*, F_{ST} (a measure of neutral genetic population differentiation) was more pronounced in fragmented than in continuous habitats, while the opposite was found for Q_{ST} (a measure of population differentiation for quantitative traits), suggesting that fragmented populations had less adaptive potential than connected populations (Johansson *et al.* 2007). The relationship between the genetic variation within populations and the adaptive response of populations to environmental changes has been only partly investigated so far in the natterjack toad. Rogell *et al.* (2010) found positive, but not significant correlation between neutral and quantitative genetic variation in island natterjack populations. In their study, the Q_{ST}-F_{ST} comparison suggested that selection on quantitative traits cannot be distinguished from genetic drift, and hence that genetic drift is another possible explanation to the small divergence observed in quantitative traits.

Predicting the effects of gene flow on population dynamics requires that we know the effect of individual genotypes on phenotypes, and then that we take into account the contribution of the various genotypes present in the population to get the population's growth rate. Unfortunately, the situation is more complex because the demographic effects of individual genotypes are not additive. In a pond experiment, eggs originating from either inbred or outbred natterjack populations were mixed and larval development was allowed to proceed naturally until tadpoles approached metamorphosis. Rowe and Beebee (2005) then measured larval survival and larval growth rate and showed that competition reduced the survival of inbred larvae in shared ponds. This unbalanced outcome of the competition has the potential to purge populations from inbred individuals after immigration of outbred toads, even when the two classes of individuals do not mate together. This might also occur as a consequence of landscape management: when connectivity is restored or improved, the immigration into isolated populations of individuals from genetically healthy populations might purge inbreeding more rapidly than expected from only the 'dilution' consecutive to mixed mating.

22.5 Discussion

The approach presented here on the natterjack toad was recognized as one of the most thorough ways to assess the effect of landscape structure on dispersal and thus gene flow (Knowlton and Graham 2010; Spear *et al.* 2010) and it was applied with slight variations to other study cases (noticeably by Michels *et al.* 2001). The methodological advance is that cost grids in the least-cost model are derived from behavioural rules obtained experimentally, and gene flow, assessed independently from behavioural rules, is then used to validate the model. Most landscape genetic studies adopt another strategy, inferring costs from observed patterns of gene flow by model optimization, without access to the processes at the origin of the observed genetic patterns. The bottom-up approach adopted for the natterjack is of course more demanding than an iterative search for the combination of costs that best match gene flows. However, it has the advantage of accounting for two independent types of data: movement rules, coming from experiments or from individual tracking, and dispersal flows, assessed directly by mark-recapture studies or indirectly by population genetics. The method should thus allow testing of specific hypotheses on the processes that govern functional connectivity, or address temporal changes in landscape connectivity, for instance by contrasting historical gene flows and contemporary exchanges among populations with the same set of behavioural rules. The solution adopted for the natterjack, combining experimental and genetic data, of course is not the single best solution, and other sources of data can be used for both the assessment of costs and the validation step. Others studies have, for instance, inferred costs from expert advice or from ecologically derived best guess, (like prey biomass: Zalewski *et al.* 2009), and other non-genetic informative data could be tracking data or presence-absence data (see a review in Spear *et al* 2010). The choice of both datasets will often be dictated practically by technical limitations, although other elements should be considered. For instance, Lambin *et al.* (Chapter 32) show how short-term behaviour, measured at small spatial and temporal scales, cannot be extrapolated to infer connectivity at larger scales.

To complete the picture of the landscape effects on spatial dynamics, we nevertheless still lack crucial information (Figure 22.2). Firstly, in most studies, dispersing individuals are supposed to be a random sample of local populations, and phenotypic differences among individuals are largely ignored. However, dispersal patterns (frequency of emigration, and dispersal-associated costs) were shown to vary among populations (e.g. Schtickzelle *et al.* 2006; Fellous *et al.* 2011) and mounting evidence shows that dispersal success also varies within populations, depending on a variety of traits, including behavioural personality, morphology or physiological state (reviewed in Clobert *et al.* 2009), which definitely shatters the myth of the 'average' disperser. Similarly, there is no reason to believe that transfer through the landscape matrix would be phenotype-independent: movement trajectories (and associated costs and probability of success) indeed might depend on a variety of traits as well as on the disperser's own experience. We therefore expect functional connectivity to be phenotype-dependent, the landscape being variously passable for individuals with different phenotypes.

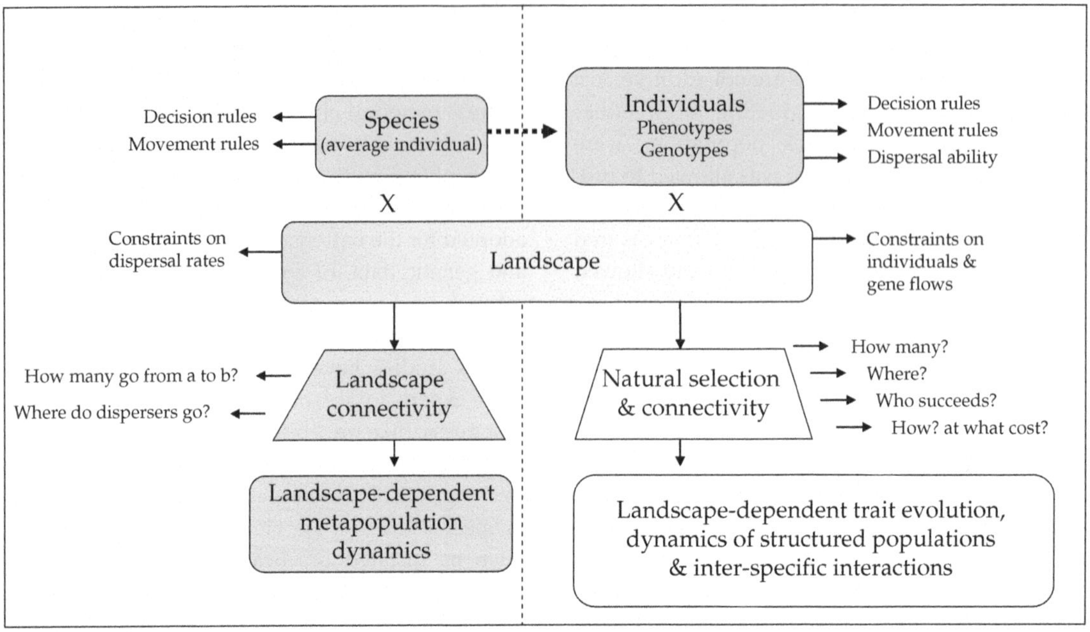

Figure 22.2 The gradual changes in the way we consider dispersal processes, triggered by our accounting for the inter-individual heterogeneity in dispersal-associated traits: from its numerical effect in dynamics of structured populations to its integrated function (both numerical and genetic effects) at the crossing point of evolutionary ecology, metapopulation dynamics, and community ecology. White bubbles are proposed future directions in the study of landscape effects on dispersal processes.

Landscape effects on dispersal are increasingly assessed through the analysis of genotypes at neutral molecular markers with an array of analytical methods commonly referred to as 'landscape genetics' (Manel *et al.* 2003). However, the use of neutral markers limits our understanding of the landscape processes acting on population dynamics to the patterns of gene flow (as in Stevens *et al.* 2006b; and in papers reviewed by Storfer *et al.* 2010), and both the dispersers' genotype *per se* and the genetic composition of local populations are still largely ignored, as are the effects of genetic mixing. Most studies make the simplification that the dispersers are a random genetic sample of populations. However, this should obviously not be the case as dispersal costs and benefits might be genotype-dependent. For instance, Wheat *et al.* (2011) detected significant effects of several loci under selection on dispersal-associated traits. Genotype-dependent dispersal could have strong effects on dynamics and evolutionary processes in wild populations (see, for instance, Garant *et al.* 2005). To our knowledge, no attempt has been made so far to investigate if landscape connectivity varies among genotypes, but we expect that the genetic mixing resulting from dispersal will not be a random mixing of two or more local genetic compositions if dispersal success is genotype-dependent.

Investigating which combinations of phenotypic and genetic traits predispose the dispersers to succeed in crossing a particular landscape matrix, locate, and settle into new habitat patches in this landscape, and finally produce offspring there, can be addressed with similar experimental setups as those used to assess functional connectivity in the natterjack, coupled with controlled breeding experiments. Another, complementary option is to extend the landscape genetic to landscape genomics, to study the spatial patterns of adaptive variation incorporating the process of selection along with spatially explicit behavioural ecology. This conceptual shift is already occurring, moving from describing patterns of neutral genetic variation to exploring the spatial variation in genes with known functional roles (Eckert *et al.* 2010; Manel *et al.* 2010; Sork and Waits 2010).

Following dispersal, immigrants' genes undergo genetic mixing with the resident gene pool. This may have indirect consequences for population dynamics, that may also be difficult to assess. For example, genetic mixing may negatively affect individual performances through outbreeding depression due to the disruption of co-adapted gene complexes or the loss of local adaptation (Templeton 1986; Brown 1991; Marr *et al.* 2002; Sagvikl *et al.* 2005). Consequently, the evolutionary trajectory of populations will depend on the selection operated by the landscape both during emigration and transfer, and also on the interaction between genotypes after immigration. The study of indirect (genetic) effects of dispersal on population dynamics thus requires deeper understanding of the genetics, and of the genomics, of dispersal and mixing.

We have so far mostly considered that the spatial ecology of one species is independent of that of other species. This is often a necessary simplification of the complexity of biological interactions. The resource/consumer interaction has often been the only interaction considered (implicitly) in the studies addressing landscape connectivity, but in addition, each organism will be part of a more or less complex network of inter-specific interactions, such as facilitation, competition, etc. All these interactions potentially modulate the way a landscape will affect dispersal and metapopulation dynamics. For instance, Bergerot *et al.* (2010) experimentally showed the decoupled dispersal dynamics of a butterfly and one of its parasitoid in urbanized landscapes. Such decoupling of a crucial inter-specific interaction is another way the landscape can affect spatially structured population dynamics, even in the case where dispersal of the species of interest is not affected.

Dispersal has the potential to affect spatially structured dynamics both directly (numerical effect) and indirectly (genetic effects). There are now many studies addressing how the landscape structure affects the number of individuals dispersing, but to our knowledge, little progress has been made to disentangle a potential effect of landscape structure on the genetic effect of dispersal on population dynamics. Investigating how direct (numerical) and indirect (genetic) effects of dispersal combine, and how these are affected by the composition and the structure of the landscape to determine the net output of dispersal on population dynamics, inter-specific relationships, and on trait evolution is a challenging and exciting venue for future research in landscape ecology.

References

Arellano, L., León-Cortés, J.L., Ovaskainen, O. (2008). 'Patterns of abundance and movement in relation to landscape structure: a study of a common scarab (*Canthon cyanellus cyanellus*) in Southern Mexico', *Landscape Ecology* **23**: 69–78.

Bakker, V.J. (2006). 'Microhabitat features influence the movements of red squirrels (*Tamiasciurus hudsonicus*) on unfamiliar grounds', *Journal of Mammalogy* **87**: 124–30.

Bartón, K.A., Phillips, B.L., Morales, J.M. *et al.* (2009). 'The evolution of an 'intelligent' dispersal strategy: biased, correlated random walks in patchy landscapes', *Oikos* **118**: 309–19.

Bergerot, B., Julliard, R., Baguette, M. (2010). 'Metacommunity Dynamics: Decline of Functional Relationship along a Habitat Fragmentation Gradient', *PLoS ONE* **5**: (6), Article e11294.

Brown, A.F. (1991). 'Outbreeding depression as a cost of dispersal in the harpacticoid copepod, *Tigriopus californicus*', *Biological Bulletin* **181**: 123–6.

Chietkiewicz, C.B. and Boyce, M.S. (2009). 'Use of resource selection functions to identify conservation corridors', *Journal of Applied Ecology* **46**: 1036–47.

Clobert, J., Le Gailliard, J.F., Cote, J. *et al.* (2009). 'Informed dispersal, heterogeneity in animal dispersal syndromes and the dynamics of spatially structured populations', *Ecology Letters* **12**: 197–209.

Coulon, A., Morellet, N., Goulard, M. *et al.* (2008). 'Inferring the effects of landscape structure on roe deer (*Capreolus capreolus*) movements using a step selection function', *Landscape Ecology* **23**: 603–14.

Cushman, S.A. and Lewis, J.S. (2010). 'Movement behaviour explains genetic differentiation in American black bears', *Landscape Ecology* **25**: 1613–25.

Dickson, B.G., Jenness, J.S. and Beier, P. (2005). 'Influence of vegetation, topography, and roads on cougar movement in Southern California', *Journal of Wildlife Management* **69**: 264–76.

Driezen, K., Adriaensen, F., Rondinini, C. *et al.* (2007). 'Evaluating least-cost model predictions with empirical dispersal data: a case-study using radiotracking data of hedgehogs (*Erinaceus europaeus*)', *Ecological Modelling* **209**: 314–22.

Eckert, A.J., Bower, A.D., González-Martínez, S.C. *et al.* (2010). 'Back to nature: ecological genomics of loblolly pine (*Pinus taeda*, Pinaceae)', *Molecular Ecology* **19**: 3789–805.

Endler, J.A. (1977). *Geographic Variation, Speciation, and Clines*. Princeton University Press, Princeton.

Endler, J.A. (1986). *Natural Selection in the Wild*. Princeton University Press, Princeton.

Fellous, S., Quillery, E., Duncan, A.B. *et al.* (2011). 'Parasitic infection reduces dispersal of cilliate host', *Biology Letters* **7**: 327–9.

Garant, D., Kruuk, L.E.B., Wilkin, T.A. *et al.* (2005). 'Evolution driven by differential dispersal within a wild bird population', *Nature* **433**: 60–5.

Goodwin, B.J. and Fahrig, L. (2002). 'How does landscape structure influence landscape connectivity?', *Oikos* **99**: 552–70.

Johansson, M., Primmer, C.R., Merilä, J. (2007). 'Does habitat fragmentation reduce fitness and adaptability? A case study of the common frog (*Rana temporaria*);, *Molecular Ecology* **16**: 2693–700.

Jonsen, I.D. and Taylor, P.D. (2000). 'Fine-scale movement behaviours of calopterygid damselflies are influenced by landscape structure: an experimental manipulation', *Oikos* **88**: 553–62.

Keller, L.F. and Waller, D.M. (2002). 'Inbreeding effects in wild populations', *Trends in Ecology and Evolution* **17**: 230–41.

Knowlton, J.L. and Graham, C.H. (2010). 'Using behavioural landscape ecology to predict species' responses to land-use and climate change', *Biological Conservation* **143**: 1342–54.

Laiola, P. and Tella, J.L. (2006). 'Landscape bioacoustics allow detection of the effects of habitat patchiness on population structure', *Ecology* **87**: 1203–14.

Loarie, S.R., Aarde, R.J.V. and Pimm, S.L. (2009). 'Fences and artificial water affect African savannah elephant movement patterns', *Biological Conservation* **142**: 3086–98.

Manel, S., Joost, S., Epperson, B.K. *et al.* (2010). 'Perspectives on the use of landscape genetics to detect genetic adaptive variation in the field', *Molecular Ecology* **19**: 3760–72.

Manel, S., Schwartz, M.K., Luikart, G *et al.* (2003). Landscape genetics: combining landscape ecology and population genetics', *Trends in Ecology and Evolution* **18**: 189–97.

Marr, A.B., Keller, L.F. and Arcese, P. (2002). 'Heterosis and outbreeding depression in descendants of natural immigrants to an inbred population of song sparrows (*Melospiza melodia*).', *Evolution* **56**: 131–42.

Michels, E., Cottenie, K., Neys, L. *et al.* (2001). 'Geographical and genetic distances among zooplankton populations in a set of interconnected ponds: a plea for using GIS modelling of the effective geographical distance', *Molecular Ecology* **10**: 1929–38.

Morales, J.M., Fortin, D., Frair, J.L. *et al.* (2005). 'Adaptive models for large herbivore movements in heterogeneous landscapes', *Landscape Ecology* **20**: 301–16.

Morales, J. M., Haydon, D. T., Frair, J. *et al.* (2004). 'Extracting more out of relocation data: building movement models as mixtures of random walks', *Ecology* **85**: 2436–45.

Nieminen, M., Singer, M.C., Fortelius, W. *et al.* and (2001). 'Experimental confirmation that inbreeding depression increases extinction risk in butterfly populations', *American Naturalist* **157**: 237–44.

O'Brien, D., Manseau, M., Fall, A. *et al.* (2006). 'Testing the importance of spatial configuration of winter habitat for woodland caribou: an application of graph theory', *Biological Conservation* **130**: 70–83.

O'Grady, J.J., Brook, B.W., Reed, D.H. *et al.* (2006). 'Realistic levels of inbreeding depression strongly affect extinction risk in wild populations', *Biological Conservation* **133**: 42–51.

Ovaskainen, O., Rekola, H., Meyke, E. *et al.* (2008). 'Bayesian methods for analyzing movements in hetereogeneous landscapes from mark-recapture data', *Ecology* **89**: 542–54.

Palmer, S.C.F., Coulon, A., Travis, J.M.J. (2011). 'Introducing a 'stochastic movement simulator' for estimating habitat connectivity', *Methods in Ecology and Evolution* **2**: 258–68.

Rogell, B., Eklund, M., Thorngren, H. *et al.* (2010). 'The effects of selection, drift and genetic variation on life-history trait divergence among insular populations of natterjack toad, *Bufo calamita*', *Molecular Ecology* **19**: 2229–40.

Rondinini, C. and Doncaster, P.C. (2002). 'Roads as barriers to movements for hedgehogs', *Functional Ecology* **16**: 504–9.

Rowe, G., Beebee, T.J.C., Burke, T. (1999). 'Microsatellite heterozygosity, fitness and demography in natterjack toads *Bufo calamita*', *Animal Conservation* **2**: 85–92.

Rowe, G. and Beebee, T.J.C. (2005). 'Intraspecific competition disadvantages inbred natterjack toad (*Bufo calamita*) genotypes over outbred ones in a shared pond environment', *Journal of Animal Ecology* **74**: 71–6.

Sagvikl, J., Uller, T., Olsson, M. (2005). 'Outbreeding depression in the common frog, *Rana temporaria*', *Conservation Genetics* **6**: 205–11.

Schtickzelle, N., Mennechez, G., Baguette, M. (2006). 'Dispersal depression with habitat fragmentation in the bog fritillary butterfly', *Ecology* **87**: 1057–65.

Slatkin, M. (1987). 'Gene flow and the geographic structure of natural populations', *Science* **236**: 787–92.

Sork, V.L. and Waits, L. (2010). 'Introduction: Contributions of landscape genetics—approaches, insights, and future potential', *Molecular Ecology* **19**: 3489–95.

Spear, S.F., Balkenhol, N., Fortin, M.-J. *et al.* (2010). 'Use of resistance surfaces for landscape genetic studies: considerations for parameterization and analysis', *Molecular Ecology* **19**: 3576–91

Stamps, J.A., Buechner, M., Krishnan, V.V. (1987). 'The effects of edge permeability and habitat geometry on emigration from patches of habitat', *American Naturalist* **129**: 533–52.

Stevens, V.M. and Baguette, M. (2008). 'Importance of Habitat Quality and Landscape Connectivity for the Persistence of Endangered Natterjack Toads', *Conservation Biology* **22**: 1194–204.

Stevens, V.M., Leboulengé, E., Wesselingh, R.A. *et al.* (2006a). 'Quantifying functional connectivity: experimental assessment of boundary permeability for the natterjack toad (*Bufo calamita*)', *Oecologia* **150**: 161–71.

Stevens, V.M., Polus, E., Wesselingh, R.A. *et al.* (2004). 'Quantifying functional connectivity: experimental evidence for patch-specific resistance in the Natterjack toad (*Bufo calamita*)', *Landscape Ecology* **19**: 829–42.

Stevens, V.M., Verkenne, C., Vandewoestijne, S. *et al.* (2006b). 'Gene flow and functional connectivity in the natterjack toad', *Molecular Ecology* **15**: 2333–44.

Storfer, A., Murphy, M.A., Spear, S.F. *et al.* (2010). 'Landscape genetics: where are we now?', *Molecular Ecology* **19**: 3496–514.

Templeton, A.R. (1986). 'Coadaptation and outbreeding depression', in M.E. Soulé (ed), *Conservation Biology: The Science of Scarcity and Diversity*. Sinauer Associates, Sunderland, MA, 105–116.

Vuilleumier, S. and Metzger, R. (2006). 'Animal dispersal modelling: handling landscape features and related animal choices', *Ecological Modelling* **190**: 159–70.

Vuilleumier, S. and Perrin, N. (2006). 'Effects of cognitive abilities on metapopulation connectivity', *Oikos* **113**: 139–47.

Wang, Y.-H., Yang, K.-C., Bridgman, C.L. *et al.* (2008). 'Habitat suitability modeling to correlate gene flow with landscape connectivity', *Landscape Ecology* **23**: 989–1000.

Werner, E.E. (1986). 'Amphibian metamorphosis—Growth-rate, predation risk, and the optimal size at transformation', *American Naturalist* **128**: 319–41.

Wheat, C.W., Fescemyer, H.W., Kvist, J. *et al.* (2011). 'Functional genomics of life history variation in a butterfly metapopulation', *Molecular Ecology* **20**: 1913–828.

Wiens, J.A., Schooley, R.L., Weeks, R.D.J. (1997). 'Patchy landscapes and animal movements: do beetles percolate?', *Oikos* **78**: 257–64.

With, K.A. (1994). 'Ontogenic shifts in how grasshoppers interact with landscape structure—an analysis of movement patterns', *Oecologia* **150**: 161–71.

With, K.A., Cadaret, S.J., Davis, C. (1999). 'Movement responses to patch structures in experimental fractal landscapes', *Ecology* **80**: 1340–53.

Zalewski, A., Piertney, S.B., Zalewska, H. *et al.* (2009). 'Landscape barriers reduce gene flow in an invasive carnivore: geographical and local genetic structure of American mink in Scotland', *Molecular Ecology* **18**: 1601–15.

CHAPTER 23

Dispersal and eco-evolutionary dynamics in the Glanville fritillary butterfly

Ilkka Hanski

23.1 Butterfly dispersal in highly fragmented landscapes

Dispersal is a ubiquitous feature of all organisms, but different species of animals and plants differ greatly in the traits that make dispersal possible, and dispersal plays dissimilar roles in their life cycles. The ultimate goal of dispersal research is to develop a unified cross-species framework for the ecology and evolution of dispersal (Nathan *et al.* 2008), but this first requires empirical knowledge a range of 'model' species, each with different ecology, such as migratory colonial birds, solitary territorial mammals, flowering plants, freshwater protists, and so forth. This chapter has been written in this spirit with one well-studied group of animals in mind, butterflies, with special consideration for one particular species, the Glanville fritillary butterfly, which we have studied for 20 years (Ehrlich and Hanski 2004; Hanski 1999). Many of the processes that influence, and are influenced by, dispersal in the Glanville fritillary are likely to be equally significant in other species of insects and invertebrates.

Butterflies are ectothermic, they have short generation times, small body sizes and high intrinsic rates of population increase, and they often exhibit high host plant and habitat specificities (Murphy *et al.* 1990). In many landscapes, the host plants occur in habitats that have a highly fragmented distribution, and hence the respective butterfly species have spatially structured populations, often in the form of a network of local populations called a metapopulation (Hanski 1999). Being small bodied and ectothermic, the growth and development of caterpillars and the performance of adults are strongly influenced by the prevailing environmental, and especially thermal, conditions, which increases the impact of environmental stochasticity on their population dynamics. For this and other reasons, the often small populations have a high risk of local extinction. On the other hand, the potentially high rate of population increase facilitates the establishment of new populations, and thereby many species occur in a balance between stochastic local extinctions and recolonizations in highly fragmented landscapes. In this situation, dispersal is obviously a key life-history trait with critical consequences for population dynamics.

Different organisms differ in their cognitive capacity for sophisticated dispersal decisions (Clobert *et al.* 2009; Chapter 12). Birds may extensively use cues about the state of the environment, plants do not; butterflies are closer to plants than to birds. Perhaps the most appropriate conceptual framework for dispersal research in butterflies is the correlated random walk (CRW), supported by tracking (Kareiva and Shigesada 1983; Ovaskainen *et al.* 2008c; Root and Kareiva 1984; Schtickzelle *et al.* 2007) and mark-release-recapture studies (Ovaskainen *et al.* 2008a; Ovaskainen *et al.* 2008b). An important advantage of the CRW framework is that it can be formulated in mathematical models that facilitate both theoretical and empirical studies. A series of studies by Ovaskainen and his collaborators has developed the CRW approach beyond the classic models of diffusion (Turchin, 1998, and

Dispersal Ecology and Evolution. First Edition. Edited by Jean Clobert *et al.*

references therein) for natural populations living in heterogeneous environments (Ovaskainen 2004; Ovaskainen and Crone 2008). These models include habitat selection and habitat-specific parameter values: as an individual disperses in a heterogeneous landscape, the random walk movement behaviour is continuously modified depending on which kind of habitats the individual happens to visit, and which kind of habitat boundaries it happens to encounter. Dispersal behaviour is thus context-dependent (Chapter 1). Even in butterflies, however, the CRW framework is not applicable to all species. For example, long-distance migrants such as the monarch butterfly (Oberhauser and Solensky 2004) and the red admiral (Stefanescu 2001) fly in a directed way, and some large tropical butterflies, such as *Heliconius* (Brown 1972), have the capacity to forage systematically within large areas of their habitat. There is also some work on temperate butterflies suggesting that dispersal behaviour becomes more complex than envisioned by CRW near habitat boundaries (Conradt *et al.* 2000; Conradt *et al.* 2003; but see Crone and Schultz 2008; Schtickzelle and Baguette 2003).

The aim of this chapter is to explore the links between dispersal and spatial dynamics. I start by outlining in Section 23.2 the biology of the Glanville fritillary, the focal species of this chapter, and the general features of its dispersal. Section 23.3 examines variation in dispersal rate, both among individuals within local populations as well as among local populations in metapopulations. A dispersal syndrome exists in the Glanville fritillary that is associated with a single nucleotide polymorphism in the gene phosphoglucose isomerase (*Pgi*), but dispersal is further influenced by a range of environmental, phenotypic, and genotypic factors that are discussed in Section 23.4. Section 23.5 is devoted to the population dynamic consequences of dispersal and variation in dispersal rate among individuals. Recent studies have shown that dispersal is under ongoing natural selection in the Glanville fritillary, and that metapopulation dynamics influence the evolutionary dynamics of dispersal and vice versa. Section 23.6 describes modeling studies addressing such eco-evolutionary dynamics of dispersal in the Glanville fritillary.

23.2 The Glanville fritillary butterfly and the general features of its dispersal

The Glanville fritillary butterfly (*Melitaea cinxia*), a relative of the American checkerspot butterflies, is widely distributed in Europe and temperate Asia. In the Åland Islands in Finland, it inhabits a large network of around 4000 dry meadows with at least one of the two larval host plant species, *Plantago lanceolata* and *Veronica spicata* (Hanski 1999; Nieminen *et al.* 2004; Case Study III). The meadows are small, with an average area of 0.17 ha, and local populations inhabiting these meadows are correspondingly small. For instance, in the autumn 2004 there were 727 local populations detected in the entire network, of which 196 consisted of just a single larval group, and only 136 populations had more than 10 larval groups; the very largest one had 143 larval groups. Larval groups consists of around 100 larvae in the autumn, but as the larvae in each group are mostly full sibs (Boggs and Nieminen 2004), the genetic effective sizes of the populations are small. The small local populations are prone to go extinct for many reasons (Hanski 1998). Thus of the 464 meadows that had a local population in 1993, only five meadows have been continuously occupied until 2010, and they all have gone through small bottlenecks. On the other hand, local extinctions are compensated for by the establishment of new populations by dispersing butterflies, and the metapopulation has persisted in a balance between stochastic local extinctions and recolonizations. The size of the entire metapopulation has remained relatively stable over the past 20 years despite a high rate of population turnover (Case Study III).

In northern Europe, the Glanville fritillary has one generation per year. Females lay their eggs in clutches of 150–200 eggs and the larvae remain gregarious until the final instar in the following spring, having diapaused as a group in a dense silken web spun around the host plant. The 'winter webs' are conspicuous in the field in late summer and allow large-scale population surveys (Nieminen *et al.* 2004; Case Study III). Adult butterflies have a lifespan of one to three weeks, depending on weather, and they typically spend only a few days in a particular habitat patch before moving on, especially if the habitat patch is small. Based on the results of an individual-based model parameterized with empirical data (Zheng *et al.* 2009), the total number of habitat patches visited during the lifetime of a butterfly varies from one to more than ten, depending on the density of meadows around the natal meadow, but also on the mobility of a particular butterfly (Section 23.3.1). Generally, females have a higher rate of emigration than males, and the pattern of emigration is somewhat different: females appear to have a relatively constant probability of emigration, whereas a fraction of males remain in the natal population permanently (Hanski *et al.* 1994; Kuussaari *et al.*, 1996), apparently having established a territory (Wahlberg, 1995; Wiklund *et al.* 2003). In summary, there is substantial but sex-biased dispersal between nearby populations, up to 2–3 km from the natal population. However, the average figures hide a large amount of variation among individuals within and between local populations in the metapopulation. Such variation, if heritable, is essential for evolutionary dynamics but has also consequences for population dynamics.

23.3 Variation in dispersal rate

23.3.1 Variation among individuals in dispersal capacity

Individuals differ in their capacity to disperse for both genetic and non-genetic reasons. There is no evidence for inherited morphological differences in the Glanville fritillary that would make an obvious difference to dispersal, but the metabolic capacity to sustain active flight varies greatly among individuals. Flight metabolic rate has been measured as the output of CO_2 during 10 min continuous flight experiments (Niitepõld, 2010), and it has great significance for dispersal. In free-flying females tracked by harmonic radar in the field, flight metabolic rate explained 32% of the variation in the distance moved in 1 hr (Figure 23.1; Niitepõld *et al.* 2009). The peak flight metabolic rate has significant broad-sense heritability (Mattila and Hanski, unpublished data), and Saastamoinen (2008) found, based on parent-offspring regression, significant heritability for mobility in female butterflies in large outdoor population cages.

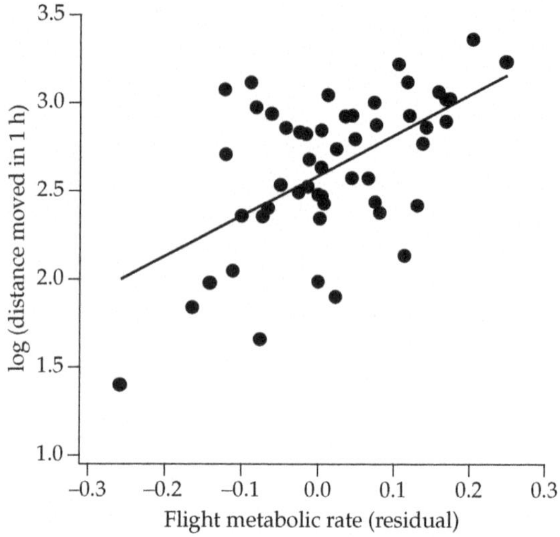

Figure 23.1 Distance dispersed in one hour against the flight metabolic rate in female Glanville fritillaries. Dispersal distance was measured by tracking free-flying butterflies with harmonic radar (adapted from Niitepõld *et al.* 2009).

One cause of variation in flight metabolic rate in the Glanville fritillary is genetic variation in the gene phosphoglucose isomerase (*Pgi*), which encodes the glucolytic enzyme with the same name (PGI). Orsini *et al.* (2009) have identified a single nucleotide polymorphism (*Pgi_111*) in the coding region of *Pgi* with an equally strong association with flight metabolic rate as the fast allozyme allele *PgiI-f* previously reported by Haag *et al.* (2005). Over a broad range of measurement temperatures, the AC heterozygotes in *Pgi_111* have roughly 40% higher flight metabolic rate than the AA homozygotes, which translates to an even greater difference in dispersal rate (Niitepõld, 2010). One intriguing feature of *Pgi_111* in the Glanville fritillary is the scarcity of CC homozygotes in the metapopulation in the Åland Islands, though CC homozygotes are common elsewhere. In Åland, a large fraction of the CC homozygotes dies at an early stage of development, possibly due to linkage with a lethal mutation on a common haplotype (Orsini *et al.* 2009; Wheat 2010). Whatever the cause of the poor performance of the CC homozygotes, it prevents the fixation of the C allele. The frequency of the AC heterozygotes is high because they have superior performance to the AA homozygotes in most fitness components (Table 23.1).

Table 23.1 Effects of the *Pgi* genotype on life-history traits in the Glanville fritillary. The second column indicates the difference in the performance of the AC heterozygotes in SNP *Pgi_111* in comparison with the performance of the AA homozygotes. The performances of the two genotypes are typically affected by other factors, and hence the figures given in the table are approximate and refer to commonly observed situations in the field or conditions used in laboratory experiments (from Hanski 2011).

Trait	AC performance	Interactions with	Ref.
Flight metabolic rate	+ 50%	temperature, body size	1,2
Dispersal rate in the field	+ 70%	ambient air temperature	2
Body temperature at flight	+ 10%	ambient air temperature	3
Clutch size	+ 20%	complex interactions	4
Lifespan	+ 15%	reproduction, sex, resource availability	5,6
Pupal weight	− 15%	temperature	7
Population growth rate	+ or −	depending on population parameters	8

References: 1 Niitepõld (2010); 2 Niitepõld *et al.* (2009); 3 Saastamoinen and Hanski (2008); 4 Saastamoinen (2007); 5 Klemme and Hanski (2009); 6 Saastamoinen *et al.* (2009); 7 Kallioniemi and Hanski (2011); 8 Hanski and Saccheri (2006)

23.3.2 Variation among populations in mean dispersal rate

A metapopulation with a high rate of population turnover presents an invaluable opportunity to address questions about dispersal and colonization in an entirely natural setting. Given that there is substantial variation in dispersal rate among individuals, one could expect that new populations are typically established by more dispersive females than the average female in the metapopulation, and even more so if the habitat patch to be colonized is highly isolated from the source populations, making it more difficult to reach.

These expectations are strongly supported by the empirical results. In a field experiment in which females from newly established and old (inhabited for > 5 years) populations were simultaneously released in the same environment and tracked with harmonic radar, the new-population females moved significantly longer distances in 2 hrs than females from old populations (mean 1850 m versus 460 m) (Ovaskainen *et al.* 2008c). A detailed analysis of the radar-tracked flight paths indicated that the difference was primarily in the frequency of flight bouts. The Glanville fritillary typically flies in short bouts, with an average length of 32 m in the above experiment (n = 354). The new-population females had roughly twice as high probability of initiating a new flight bout than the old-population females (Ovaskainen *et al.* 2008c). The same difference in dispersal rate between butterflies from new versus old populations has been observed in experiments conducted in a large population cage (Hanski *et al.* 2006), though not in all years when the comparison has been made (Saastamoinen 2007), indicating an interaction in the effects of population age and the prevailing environmental conditions on dispersal (Section 23.4). From these results, we may conclude that there is biologically and statistically significant variation in the mean dispersal rate of butterflies in local populations with dissimilar ages. This variation is partly due to the effects of *Pgi*, but there are also other genetic differences between newly-established and old local populations (Section 23.4).

23.3.3 The dispersal syndrome and trade-offs

The contrast in the rate of dispersal of female butterflies in new versus old populations comprises a dispersal syndrome, covariation among a range of life-history traits and dispersal (Part III), which here also involves the physiological performance of individuals as well as known genetic differences. The clustering of individuals into two types is not exact, of course, but the strong association between dispersal rate and *Pgi* polymorphism implies that there may be a tendency towards two broad classes of individuals.

The *Pgi* genotype has significant effects not only on dispersal but also on fecundity and lifespan

(Table 23.1). Given that *Pgi* is involved in central metabolism, it is not surprising that the dispersal syndrome in the Glanville fritillary involves a contrast between 'fast' and 'slow' butterflies. The fast butterflies have higher flight metabolic rate and dispersal rate, and are hence good colonizers, but they also start ovipositing sooner after eclosion than the slow butterflies and, surprisingly, they have longer intrinsic lifespan at least under some conditions (Table 23.1). Indeed, there are no simple trade-offs between the key life-history traits (Chapter 13), though there are more complex trade-offs due to interactions with environmental conditions (Section 30.4). One likely cost is that the fast females appear to run out of resources faster than the slow butterflies when resource use is high and availability is low. This is demonstrated by the observation that both thorax and abdomen masses are smaller in AC than AA females at the end of their lives, suggesting that the former are generally using body reserves at a faster rate and may become more easily resource limited especially under sub-optimal conditions (Saastamoinen *et al.* 2009). These results lead to the hypothesis that, in general, the fast butterflies have higher fitness under favourable environmental conditions but are more affected by resource limitation and possibly by other environmental challenges.

Finally, there is an often overlooked trade-off that contributes to the maintenance of variation in dispersal rate in insects (Hanski *et al.* 2006). As we have seen, individuals that are particularly dispersive contribute disproportionately to immigration, gene flow, and colonization of empty habitat patches, but at the same time they and their dispersive offspring are likelier than the others to emigrate away from existing populations. In an extreme case, an individual that has just arrived at an unoccupied meadow with plentiful resources may emigrate away in the same or during the following day, simply because the most dispersive butterflies easily fly away, accidentally, from small habitat patches. This happens in the correlated random walk model that has been used to model dispersal in the Glanville fritillary (Zheng *et al.* 2009), but also in reality based on direct observations. In this respect, vertebrates are likely to behave in a very different manner.

23.4 Environmental, phenotypic, and genotypic effects on dispersal

Insect activity is strongly temperature-dependent at molecular, organismal, and population levels, and hence dispersal is much affected by the prevailing environmental conditions. Low temperatures constrain activity and thereby restrict dispersal with consequences for population and metapopulation dynamics. These effects are, however, species- and population-specific because selection may have improved the performance of some species, populations, and genotypes under generally adverse conditions, presumably with a cost to some other component of fitness.

Butterflies require resources (nectar) to fuel active flight and dispersal. Examining the factors that influence the rates of emigration from and immigration to particular meadows, Kuusaari *et al.* (1996) found that abundant nectar plants reduced emigration and increased immigration in the Glanville fritillary. Apart from the nectar plants, ovipositing females search for host plants, and larval host plant density could be expected to influence dispersal. Indeed, Austin *et al.* (2011) found that the numbers of females colonizing a previously unoccupied habitat patch increased with the density of larval host plants.

When adult butterflies are limited by the availability of nectar, for instance due to low temperatures limiting foraging, they can utilize their body reserves that were accumulated during the larval stage. Saastamoinen *et al.* (2009) found very significant decreases in the abdomen and thorax weights of the Glanville fritillary during the adult life, up to 30–40%, and they found that body reserves were used for both maintenance and reproduction. Furthermore, the relative use of resources in the thorax and abdomen was different under different environmental conditions, with an interaction with the *Pgi* genotype. It is particularly intriguing that thorax mass declined by up to 40% under some conditions—one could expect such a large reduction to impair dispersal. However, even a large reduction in flight muscle mass does not necessarily have proportional impact on flight capacity for two reasons. First, relative flight muscle mass (flight muscle mass/body mass) does not decrease with age in most species of butterflies that have been investigated (Ahman and Karlsson 2009), indicating that the overall body mass declines at the same rate than the flight muscle mass. Second, a study by Marden *et al.* (2008) on the alternatively spliced forms of the flight muscle protein *troponin-t* (*Tnt*) in the Glanville fritillary points to a possible mechanism that may allow butterflies to retain adequate flight capacity in spite of loss of flight muscle mass. In simple terms, selective loss of muscle cells maintains the large *Tnt* isoform, which promotes force and power output (Marden *et al.* 2001).

Dispersal is likely to be condition-dependent, though how significant this might be for population dynamics and evolution is not clear. Saastamoinen *et al.* (2010) working on the butterfly *Bicyclus anynana* showed that limited food availability at the larval stage reduced adult body size but nonetheless improved flight performance, perhaps due to the observed increase in relative allocation to thorax mass. A few other studies on butterflies have shown dispersal-related morphological changes in response to the quality or quantity of larval resources (Boggs and Freeman 2005; Pellegroms *et al.* 2009). Changes in dispersal behaviour in response to food shortage are well documented in, for example, aphids (Muller *et al.* 2001; Sutherland 1969), in which they clearly represent an adaptive mechanism to avoid crowded populations.

Turning to genetic effects, there is a strong *Pgi* genotype × temperature interaction affecting both the peak flight metabolic rate and the dispersal rate in the field (Figure 23.2). The AA homozygotes perform much worse than the AC heterozygotes in low ambient temperatures, whereas at high temperatures there is no difference or the difference is even reversed. The genotype × temperature interaction is consistent with a molecular trade-off between kinetic efficiency versus thermal stability of the different PGI genotypes, a hypothesis first put forward by Watt (1983; Watt *et al.* 2003; Watt 2003) in the context of his studies on *Colias* butterflies. The consequences of the genotype × temperature interaction for dispersal are clear: in summers when temperatures are generally high, the *Pgi* genotype does not constrain dispersal in the same way as it does in cool summers. Thus we may predict that in warm summers the newly colonized populations are established by a more random set of butterflies than in cool summers.

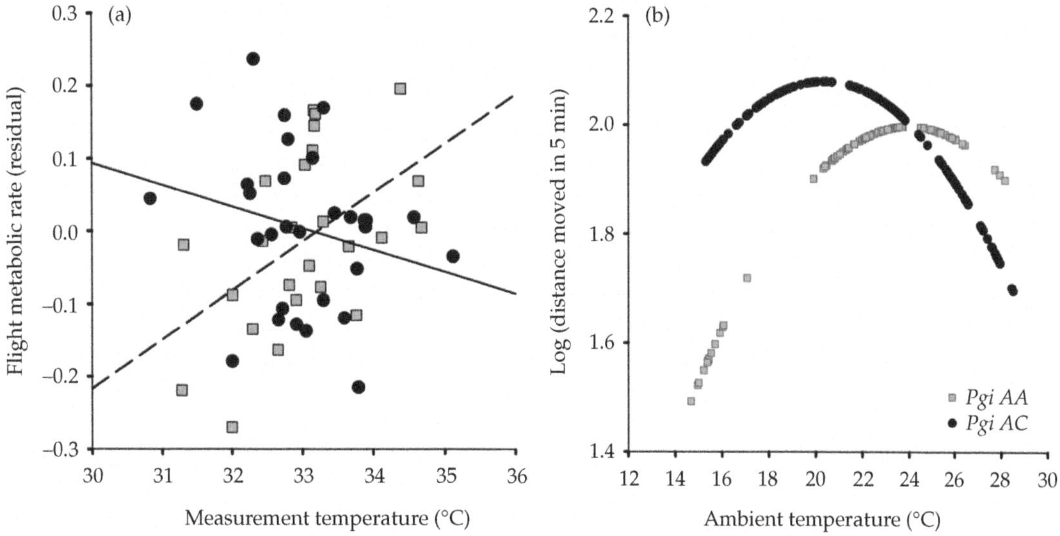

Figure 23.2 (a) Peak flight metabolic rate against measurement temperature in female Glanville fritillaries, comparing the AC heterozygotes and AA homozygotes in *Pgi_111*. The interaction with measurement temperature is statistically significant (P = 0.005; Niitepõld *et al.* 2009). (b) Distance moved in a period of 5 min against the ambient air temperature in female Glanville fritillaries. The points are the model-fitted values. The temperature x genotype interaction is statistically significant (P = 0.02; Niitepõld *et al.* 2009).

The strong association between *Pgi_111* and flight capacity in the Glanville fritillary is a striking example of single-gene effects on dispersal, though the actual molecular mechanisms have not yet been conclusively pinned down. Furthermore, *Pgi* is hardly the only gene that is related to differences among individuals in dispersal and to variation in the dynamics of local populations. A micro-array study comparing individuals with different peak flight metabolic rates and individuals originating from new versus old local populations found many other genes with significant expression differences (Wheat *et al.* 2011). In particular, females originating from newly established populations had higher expression of genes involved in egg provisioning and maintenance of flight-muscle proteins than females originating from old populations. These differences are consistent with the general dispersal syndrome described in Section 23.3.3.

23.5 Population dynamics

Immigration and emigration are the two processes, in addition to births and deaths, that lead to changes in population size. It is often assumed that within local populations, immigration and emigration balance each other out, with no net change in local population size, but this is unlikely to happen in heterogeneous environments. Emigration is more or less independent of the spatial position of the focal population in the landscape, but immigration is not: populations that are located close to other populations will receive more immigrants than populations located far away from the others. In the case of butterflies, we cannot directly count the actual numbers of immigrants arriving at a particular population, but the following measure of connectivity (Hanski, 1994; Hanski, 1999) provides a helpful proxy

$$S_i = \sum_{j \neq i} N_j \frac{\alpha^2}{2\pi} e^{-\alpha d_{ij}}. \tag{23.1}$$

S_i is the connectivity of habitat patch i, measuring the expected numbers of immigrants, N_j is the size of population j, d_{ij} is the distance between patches i and j and $1/\alpha$ is the average dispersal distance. The factor $\alpha^2/2\pi$ ensures that the exponential dispersal kernel integrates to one over the two-dimensional space. Often there are no data on population sizes, in which case one may replace N_j with p_jA_j, where A_j is the area of patch j and p_j is the probability of patch j being occupied. For other refinements of Equation 23.1 see Hanski (1999) and Moilanen and Nieminen (2002).

To assess the impact of immigration on population dynamics, I calculated the change in the sizes of local populations from year t to year $t + 1$ using the long-term data for hundreds of local populations of the Glanville fritillary for the years 1993 to 2010. I excluded populations with fewer than five larval groups in year t to reduce the effect of demographic stochasticity. Four factors had a highly significant ($P<0.0001$) effect on the value of $\log(N_{t+1}/N_t)$ in a multiple regression model, which explained 33% of the variation ($n = 1077$): log population size in year t (negative density dependence, t–= –8.52), log patch area ($t = 6.94$), regional trend in population sizes (reflecting the impact of spatially correlated environmental stochasticity, $t = 16.32$; see Hanski and Meyke 2005; Ovaskainen and Hanski 2003), and log connectivity ($t = 4.19$). Therefore, the higher the rate of immigration as measured by connectivity, the greater the increase in population size from one year to another.

Many models assume that emigration is positively density dependent and thereby contributes to population regulation (Chapter 20). Empirical studies on various groups of animals have reported both positively and negatively density-dependent emigration (Hansson 1991), and the same applies to butterflies: positively density-dependent emigration has been found in, for example, *Pieris protadice* females (Shapiro 1970) and in *Euphydryas editha* (Murphy and White 1984), while negatively density-dependent emigration has been reported for *Euphydryas chalcedona* (Brown and Ehrlich 1980). Kuussaari *et al.* (1996), working on field populations that had been experimentally established to control for habitat quality and population density, found negatively density-dependent emigration in the Glanville fritillary, which was interpreted as indicating that butterflies use the presence of conspecifics as a cue of habitat quality.

23.5.1 Dispersal and colonization

Connectivity increases the growth rate of existing populations but also the rate of establishment of new populations at currently unoccupied habitat patches. In the pooled data for the years from 1993 to 2010 (see earlier), there were 8728 opportunities for colonization (an unoccupied meadow in year *t*), of which 1069 resulted in the establishment of a new population (12%). In a logistic regression model, the probability of colonization increased with connectivity and the area of the habitat patch (both $P<0.0001$). The effect of connectivity on colonizations is observed very generally in metapopulations inhabiting fragmented habitats (Hanski 1999) and, together with the effect of patch area on extinctions, comprises the hard core of the spatially realistic metapopulation theory (Hanski 1994; Hanski 2001a).

The effect of connectivity (immigration rate) on colonizations represents what we might call a first-order effect, which is modified by variation in dispersal and colonization capacities of individuals. In the Glanville fritillary, there are differences in the dispersal rate of individuals associated with their *Pgi* genotype (Section 23.3.1), and we could hence expect that the rate of establishment of new populations depends on the *Pgi* genotypic composition of the nearby populations that function as sources of colonists. That this is indeed so is demonstrated by systematic differences between new and old populations in their *Pgi* allele frequencies. The AC heterozygotes, which are superior dispersers (Section 23.3.1), are more common than the AA homozygotes in new populations, and especially in isolated new populations (Hanski and Saccheri 2006; Hanski and Mononen 2011), demonstrating a genetic effect on colonizations no underlying.

Do the colonists to a particular unoccupied habitat patch originate from one or many source populations? This makes a big difference for the maintenance and distribution of genetic diversity in general among local populations in metapopulations (Pannell and Charlesworth 1999; Wade and McCauley 1988; Whitlock and McCauley 1990). Austin *et al.* (2011) used microsatellite markers to determine the numbers and genetic relatednesses of individuals that established new local populations in a network of 277 habitat patches, of which 107 were occupied by a local population of the Glanville fritillary in 2002. There were 24 colonizations, half of which had been established by a single mated female, as only a single larval group was present in 2002. In the remaining populations, several females had successfully reproduced following their dispersal. The two parents of a single larval group were on average much more related to each other than two randomly chosen individuals from the metapopulation as a whole (relatedness $r = 0.39 \pm 0.28$ sd). In contrast, individuals that were the parents of different larval groups in the same newly established population were on average not more related than randomly selected individuals from the metapopulation (0.04 ± 0.26). These results imply that females had typically mated in their natal populations prior to dispersal, whereas two or more females colonizing the same habitat patch typically originated from different local populations in the metapopulation.

23.5.2 Dispersal and rescue effects

Dispersal is necessary for the establishment of new populations, but dispersal may also play a role in local extinctions. A high rate of emigration reduces the growth rate of a local population and may thereby increase the risk of local extinction. On the other hand, a high rate of immigration may rescue a local population from extinction by increasing the growth rate as shown above. Hanski (1999: Table 8.2) indeed found that the annual risk of extinction was significantly lower in well-connected than in similar-sized but poorly connected populations.

Small populations of the Glanville fritillary suffer inbreeding depression, which reduces egg hatching rate and other fitness components (Haikola *et al.* 2001) and increases the risk of population extinction in the field (Nieminen *et al.* 2001; Saccheri *et al.* 1998). Saccheri and Brakefield (2002) demonstrated experimentally that a small number of unrelated immigrants can significantly reduce inbreeding depression in laboratory populations of *Bicyclus anynana*. A similar genetic rescue effect is likely to occur in the Glanville fritillary, though distinguishing between the demographic and genetic rescue effects is difficult without experimental studies.

23.5.3 Metapopulation viability

Long-term persistence of classic metapopulations consisting of extinction-prone local populations requires that the rate of dispersal is sufficiently high to generate a sufficiently high rate of colonizations to compensate for extinctions. To specify, for real metapopulations, what is 'sufficiently high' calls for models that include a description of the spatial structure of the landscape and a description of how that structure influences the extinction and colonization processes (Hanski 2001b).

Figure 23.3 illustrates, for many sub-networks of the 4000-meadow network in the Åland Islands, the relationship between the empirically measured metapopulation size and the metapopulation capacity, showing a good fit to a spatially realistic model (Hanski and Ovaskainen 2000). Metapopulation capacity, the 'carrying capacity' of a fragmented landscape, is a measure that captures the impact of the pooled area of habitat and its spatial configuration on metapopulation persistence (Hanski and Ovaskainen 2000). Notice that the response of metapopulation size to improving network quality (moving to the right in Figure 23.3) is non-linear: below the extinction threshold, the metapopulation is extinct though there is still suitable habitat left in the network. The important message here is that metapopulation persistence is possible, in a particular fragmented landscape, only if the metapopulation capacity of that landscape exceeds a critical value that is determined by the traits of the species, its extinction-proneness and colonization capacity (Hanski and Ovaskainen 2000). The model thus quantifies what is a 'sufficiently high' rate of dispersal for long-term persistence in a particular landscape. Figure 23.3 gives a clear-cut example of an extinction threshold. In many cases, the picture may be somewhat blurred by stochastic extinctions of entire metapopulations that are located above but close to the extinction threshold (Ovaskainen and Hanski 2003; the model prediction in Figure 23.3 is based on a deterministic approximation), and by temporary presence of the species, due to dispersal from outside, in networks that are located below the extinction threshold.

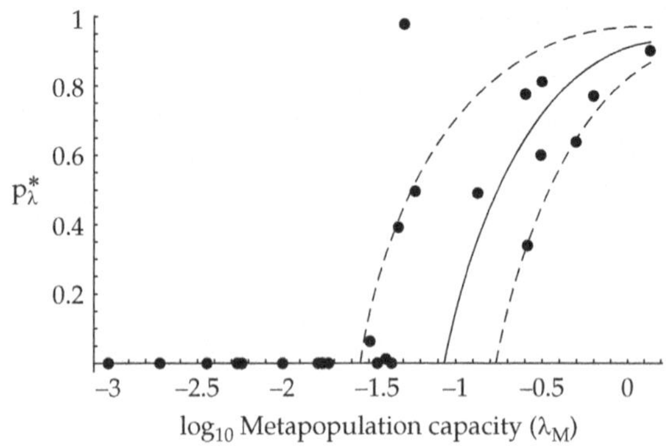

Figure 23.3 The size of the Glanville fritillary metapopulation as a function of the metapopulation capacity in 25 habitat patch networks in the Åland Islands (different sub-networks of the entire 4000 meadow network). The vertical axis shows the size of the metapopulation based on a survey of habitat patch occupancy in one year. The empirical data have been fitted by a spatially realistic model (from Hanski and Ovaskainen (2000), reprinted with permission from Macmillan Publishers Ltd).

The model predicts that the number of local extinctions per year increases, and the number of colonizations per year decreases, with an increasing fraction of habitat patches occupied (Hanski 1994). The empirical results for the Glanville fritillary for the years from 1993 to 2010 bear out this prediction beautifully (Case Study III), and thus explain the long-term stability of the metapopulation.

23.5.4 Interactions with other species

Models of predator-prey and host-parasitoid dynamics have elucidated the role of dispersal in these interactions (Hassell 1978). If a natural enemy increases the risk of local extinction of the prey, the two species may still persist at the metapopulation level if the prey has a sufficiently high rate of establishment of new populations, which initially lack the natural enemy and thereby provide a temporary refuge. The interaction between the Glanville fritillary and its specialist braconid parasitoid *Cotesia melitaearum* provides an excellent example. The wasp has a high intrinsic rate of population increase and may cause local extinction of the host butterfly (Lei and Hanski 1998; Lei and Camara 1999), but the wasp is a poor disperser with maximum per year colonization distance of 1.5 km, much less than 5 km in the host (van Nouhuys and

Hanski 2002). The wasp only persists in regions where the density of habitat patches and host populations is high. In contrast, the second specialist parasitoid attacking the Glanville fritillary in the Åland Islands, the ichneumonid *Hyposoter horticola*, is an excellent disperser and occurs throughout the host metapopulation. This wasp is, however, not as efficient in parasitizing the host as *C. melitaearum* (van Nouhuys and Hanski 2004), which facilitates local co-existence with the host.

23.6 Eco-evolutionary dynamics of dispersal

In the Glanville fritillary, the *Pgi* genotype influences the rate of dispersal, and hence the individuals that represent the fast-dispersing genotype, the AC heterozygotes in the SNP *Pgi_111*, contribute disproportionately to gene flow and the establishment of new populations. At the same time, the ecological dynamics may influence the micro-evolutionary dynamics at the *Pgi* locus. For instance, any ecological processes that increase the rate of local extinctions generate more opportunities for recolonization, which will benefit, at least locally, the more dispersive *Pgi* genotype. Such reciprocal effects between ecological (demographic) and micro-evolutionary dynamics have been called eco-evolutionary dynamics for short (Hanski 2011; Pelletier *et al.* 2009). In the case of dispersal, these effects have to be placed in a broader context, however, as there is a multitude of other processes affecting ecological spatial dynamics and the evolution of dispersal. Thus dispersal is selected against by the cost of dispersal, including dispersal mortality but also time lost for reproduction (Hanski *et al.* 2006). Locally, dispersal is selected against by dispersal itself: more dispersive individuals emigrate at a higher rate than the less dispersive individuals, and hence the mean dispersal rate in the local population tends to decline (Section 23.3.3). Kin selection apart from temporal variation in fitness among local populations is expected to select for dispersal especially in small populations (Ronce 2007).

Given the many processes that influence spatial population dynamics and the evolution of dispersal, it is difficult to construct and analyse realistic models, and it is especially difficult to develop testable models for particular species living in heterogeneous environments. The Glanville fritillary metapopulation described in this chapter has the advantage that much information is available to parameterize detailed individual-based models. Heino and Hanski (2001) and Zheng *et al.* (2009) have constructed two somewhat dissimilar models, which were parameterized with data on life histories, dispersal, and metapopulation dynamics for the Glanville fritillary and its close relative, *Melitaea diamina*. The models include all the main processes that could be expected to affect the evolution of dispersal, namely resource and kin competition, dispersal mortality (influenced by the spatial configuration of the habitat), and spatio-temporal population dynamics. Heino and Hanski (2001) assumed haploid asexual reproduction and that dispersal is determined by many loci with small additive effects, whereas Zheng *et al.* (2009) assumed sexual reproduction and that dispersal is determined by one locus with two alleles, corresponding to *Pgi_111*.

The models can be run on any kind of patch network, which has the advantage that one may generate quantitative predictions for real metapopulations. Heino and Hanski (2001) predicted the value of emigration propensity in the *M. diamina* metapopulation by first fixing all other model parameters at their estimated values while allowing the emigration propensity to evolve (emigration rate was assumed to depend on patch area as $\varepsilon = \eta A^{\zeta}$, where η is emigration propensity, the daily rate of emigration from a patch of 1 ha). The predicted value was 0.104 (sd 0.005), which did not differ significantly from the value estimated empirically in a mark-release-recapture study, 0.130 (95% confidence limits of 0.103 and 0.176). Zheng *et al.* (2009) predicted that the C allele in *Pgi_111* has highest frequency in newly-established isolated populations, which has been observed empirically (Section 23.5.1). At the network level, the frequency of the C allele was predicted to be highest in small metapopulations with low incidence of patch occupancy, which has been observed empirically (Hanski 2011; Zheng *et al.* 2009). Zheng *et al.* (2009) demonstrated with their model, by perturbing population sizes and/or *Pgi* allele frequencies, that the demographic dynamics influence the dynamics of *Pgi* allele frequency and *vice versa*, making the dynamics of dispersal in the Glanville fritillary a prime example of reciprocal eco-evolutionary dynamics (Hanski 2011).

Complex simulation models have the advantage that one may include any number of processes to make the model more realistic, but this comes with the cost of many parameters and the results are often difficult to interpret. To overcome these limitations, Hanski and Mononen (2011) have constructed a simpler model consisting of fast-dispersing and slow-dispersing individuals by combining the stochastic patch-occupancy metapopulation model with a model for changes in the fraction of fast-dispersing individuals in local populations. The model predicts two things, the long-term probability of each patch in a network being occupied, and the corresponding mean dispersal phenotype (fraction of fast-dispersing individuals) conditional on occupancy. The model predicts that the expected mean dispersal phenotype in a particular habitat patch increases with increasing immigration rate (because fast-dispersing individuals are overrepresented among the immigrants), with increasing extinction rate (which increases the influence of founder events), and with increasing frequency of fast-dispersing individuals among the immigrants. Identifying the AC heterozygotes in *Pgi-111* as fast-dispersing and the AA homozygotes as slow-dispersing individuals, the results for the Glanville fritillary supported these predictions, and the model explained 40% of spatial variation in *Pgi* allele frequency among local populations (Hanski and Mononen 2011). The model also clarifies why habitat loss and fragmentation is expected to select for increasing dispersal in the Glanville fritillary, but may select for decreasing dispersal in other species. The outcome depends on how much habitat loss and fragmentation increase the extinction rate of local populations (which selects for increasing dispersal) and how much it reduces immigration to existing populations (which selects for decreasing dispersal). There is no single answer; what happens depends on the biology of the species. This is a sobering conclusion but also highlights the need for predictive quantitative models in the study of the ecology and evolution of dispersal.

References

Ahman, M. and Karlsson, B. (2009). 'Flight endurance in relation to adult age in the green-veined white butterfly *Pieris napi*', *Ecological Entomology* **34**: 783–7.

Austin, A., Ovaskainen, O., Hanski, I. (2011). 'Size and genetic composition of the colonizing propagules in a butterfly metapopulation', *Oikos* **120**: 1357–65.

Boggs, C. and Nieminen, M. (2004). 'Checkerspot reproductive biology', in P.R. Ehrlich, I. Hanski (eds), *On the wings of Checkerspots: A model system for population biology*. Oxford University Press, New York, 92–112.

Boggs, C.L. and Freeman, K.D. (2005). 'Larval food limitation in butterflies: effects on adult resource allocation and fitness', *Oecologia* **144**: 353–61.

Brown, I.L. and Ehrlich P.R. (1980). 'Population biology of the checkerspot butterfly, *Euphydryas chalcedona*: structure of the Jasper Ridge colony', *Oecologia* **47**: 239–51.

Brown, K.S. (1972). 'Heliconians of Brazil (Lepidoptera: Nymphalidae). 3. Ecology and biology of *Heliconius nattereri*, a key primitive species near extinction, and comments on evolutionary development of *Heliconius* and *Eueides*', *Zoologica* **57**: 41–69.

Conradt, L., Bodsworth, E.J., Roper, T.J. *et al.* (2000). 'Nonrandom dispersal in the butterfly *Maniola jurtina*: implications for metapopulation models', *Proceedings of the Royal Society B* **267**: 1505–10.

Conradt, L., Zollner, P.A., Roper, T.J *et al.* (2003). 'Foray search: An effective systematic dispersal strategy in fragmented landscapes', *American Naturalist* **161**: 905–15.

Crone, E.E. and Schultz, C.B. (2008). 'Old models explain new observations of butterfly movement at patch edges', *Ecology* **89**: 2061–7.

Ehrlich, P. and Hanski, I. (eds), (2004). *On the wings of Checkerspots: A model system for population biology*. Oxford University Press, New York.

Haag, C., Saastamoinen, M., Marden J. *et al.* (2005). 'A candidate locus for variation in dispersal rate in a butterfly metapopulation', *Proceedings of the Royal Society B* **272**: 2449–56.

Haikola, S., Fortelius, W., O'Hara, R.B. *et al.* (2001). 'Inbreeding depression and the maintenance of genetic load in *Melitaea cinxia* metapopulations', *Conservation Genetics* **2**: 325–35.

Hanski, I. (1994). 'A practical model of metapopulation dynamics', *Journal of Animal Ecology* **63**: 151–62.

Hanski, I. (1998). 'Metapopulation dynamics', *Nature* **396**: 41–9.

Hanski, I. (1999). *Metapopulation ecology*. Oxford University Press, New York.

Hanski, I. (2001a). 'Spatially realistic models of metapopulation dynamics and their implications for ecological, genetic and evolutionary processes', in J. Silvertown, J. Antonovics (eds), *Integrating ecology and evolution in a spatial context*. Blackwell Science, Oxford, 139–56.

Hanski I. (2001b) 'Spatially realistic theory of metapopulation ecology'. *Naturwissenschaften* **88**: 372–81.

Hanski, I. (2011). 'Eco-evolutionary spatial dynamics in the Glanville fritillary butterfly', *Proceedings of the National Academy of Sciences of the USA* **108**: 14397–404.

Hanski, I., Kuussaari, M., Nieminen, M. (1994). 'Metapopulation structure and migration in the butterfly *Melitaea cinxia*', *Ecology* **75**: 747–62.

Hanski, I. and Meyke, E. (2005). 'Large-scale dynamics of the Glanville fritillary butterfly: Landscape structure, population processes, and weather', *Annales Zoologici Fennici* **42**: 379–95.

Hanski, I. and Mononen, T. (2011). 'Eco-evolutionary dynamics of dispersal in spatially heterogeneous environments', *Ecology Letters* **19**: 1025–1034.

Hanski, I. and Ovaskainen, O. (2000). 'The metapopulation capacity of a fragmented landscape', *Nature* **404**: 755–8.

Hanski, I., Saastamoinen, M., Ovaskainen, O. (2006). 'Dispersal-related life-history trade-offs in a butterfly metapopulation', *Journal of Animal Ecology* **75**: 91–100.

Hanski, I. and Saccheri, I. (2006). 'Molecular-level variation affects population growth in a butterfly metapopulation', *PLoS Biology* **4**: 719–26.

Hansson, L. (1991). 'Dispersal and connectivity in metapopulations', *Biological Journal of the Linnean Society* **42**: 89–103.

Hassell, M.P. (1978). *The Dynamics of arthropod predator-prey systems*. Princeton University Press, Princeton.

Heino, M. and Hanski, I. (2001). 'Evolution of migration rate in a spatially realistic metapopulation model', *American Naturalist* **157**: 495–511.

Kallioniemi, E. and Hanski I. (2011). 'Interactive effects of *Pgi* genotype and temperature on larval growth and survival in the Glanville fritillary butterfly', *Functional Ecology* **25**: 1032–39.

Kareiva, P.M. and Shigesada, N. (1983). 'Analyzing insect movement as a correlated random walk', *Oecologia* **56**: 234–8.

Klemme, I. and Hanski, I. (2009). 'Heritability of and strong single gene (*Pgi*) effects on life-history traits in the Glanville fritillary butterfly', *Journal of Evolutionary Biology* **22**: 1944–53.

Kuussaari, M., Nieminen, M., Hanski, I. (1996). 'An experimental study of migration in the Glanville fritillary butterfly *Melitaea cinxia*', *Journal of Animal Ecology* **65**: 791–801.

Lei, G. and Camara, M.D. (1999). 'Behaviour of a specialist parasitoid, *Cotesia melitaearum*: From individual behaviour to metapopulation processes', *Ecological Entomology* **24**: 59–72.

Lei, G. and Hanski, I. (1998). 'Spatial dynamics of two competing specialist parasitoids in a host metapopulation', *Journal of Animal Ecology* **67**: 422–33.

Marden, J.H., Fescemyer H.W., Saastamoinen M. *et al.* (2008). 'Weight and nutrition affect pre-mRNA splicing of a muscle gene associated with performance, energetics and life history', *Journal of Experimental Biology* **211**: 3653–60.

Marden, J.H., Fitzhugh, G.H., Girgenrath M. *et al.* (2001). 'Alternative splicing, muscle contraction and intraspecific variation: Associations between troponin T transcripts, Ca2+ sensitivity and the force and power output of dragonfly flight muscles during oscillatory contraction', *Journal of Experimental Biology* **204:** 3457–70.

Moilanen, A. and Nieminen, M. (2002). 'Simple connectivity measures for spatial ecology', *Ecology* **83**: 1131–45.

Muller, C.B., Williams, I.S., Hardie, J. (2001). 'The role of nutrition, crowding and interspecific interactions in the development of winged aphids', *Ecological Entomology* **26**: 330–40.

Murphy, D.D., Freas, K.E., Weiss, S.B. (1990). 'An environment-metapopulation approach to population viability analysis for a threatened invertebrate', *Conservation Biology* **4**: 41–51.

Murphy, D.D. and White, R.R. (1984). 'Rainfall, resources and dispersal in southern populations of *Euphydryas editha* (Lepidoptera, Nymphalidae)', *The Pan-Pacific Entomologist* **60**: 350–54.

Nathan, R., Getz, W.M., Revilla E. *et al.* (2008). 'A movement ecology paradigm for unifying organismal movement research', *Proceedings of the National Academy of Sciences of the USA* **105:** 19052–59.

Nieminen, M., Siljander, M., Hanski, I. (2004). 'Structure and dynamics of *Melitaea cinxia* metapopulations', in P.R. Ehrlich, I. Hanski (eds), *On the wings of Checkerspots: A model system for population biology*. Oxford University Press, New York, 63–91.

Nieminen, M., Singer, M.C., Fortelius, W. *et al.* (2001). 'Experimental confirmation that inbreeding depression increases extinction risk in butterfly populations', *American Naturalist* **157**: 237–44.

Niitepõld, K. (2010). 'Genotype by temperature interactions in the metabolic rate of the Glanville fritillary butterfly', *Journal of Experimental Biology* **213**: 1042–8.

Niitepõld, K., Smith, A.D., Osborne J.L. *et al.* (2009). 'Flight metabolic rate and *Pgi* genotype influence butterfly dispersal rate in the field', *Ecology* **90:** 2223–32.

Oberhauser. K., Solensky, M.E. (2004). *The monarch butterfly. Biology and conservation.* Cornell University Press, Ithaca, New York.

Orsini, L., Wheat, C.W., Haag, C.R. *et al.* (2009). 'Fitness differences associated with *Pgi* SNP genotypes in the Glanville fritillary butterfly (*Melitaea cinxia*)', *Journal of Evolutionary Biology* **22:** 367–75.

Ovaskainen, O. (2004). 'Habitat-specific movement parameters estimated using mark–recapture data and a diffusion model', *Ecology* **85:** 242–57.

Ovaskainen, O. and Crone, E. (2008). 'Modeling movement behavior with diffusion' in S. Cantrell *et al.* (eds), *et al Emergent challenges in spatial ecology.* Princeton University Press, Princeton.

Ovaskainen, O. and Hanski, I. (2003). 'Extinction threshold in metapopulation models', *Annales Zoologici Fennici* **40:** 81–97.

Ovaskainen, O. and Luoto, M., Ikonen, I. *et al.* (2008) 'An empirical test of a diffusion model: Predicting clouded apollo movements in a novel environment', *American Naturalist* **171:** 686–98.

Ovaskainen, O., Rekola, H., Meyke E. *et al.* (2008). 'Bayesian methods for analyzing movements in heterogeneous landscapes from mark-recapture data', *Ecology* **89:** 542–54.

Ovaskainen, O., Smith, A.D., Osborne J.L. *et al.* (2008). 'Tracking butterfly movements with harmonic radar reveals an effect of population age on movement distance', *Proceedings of the National Academy of Sciences of the USA* **105:** 19090–5.

Pannell, J.R., Charlesworth, B. (1999). 'Neutral genetic diversity in a metapopulation with recurrent local extinction and recolonization', *Evolution* **53:** 664–76.

Pellegroms, B., Van Dongen, S., Van Dyck H. *et al.* (2009). 'Larval food stress differentially affects flight morphology in male and female speckled woods (*Pararge aegeria*)', *Ecological Entomology* **34:** 387–93.

Pelletier, F., Garant D., Hendry, A.P. (2009). 'Eco-evolutionary dynamics. Introduction', *Philosophical Transactions of the Royal Society B* **364:** 1483–9.

Ronce, O. (2007). 'How does it feel to be like a rolling stone? Ten questions about dispersal evolution',: 231–53.

Root, R.B. and Kareiva, P.M. (1984). 'The search for resources by cabbage butterflies (*Pieris rapae*): Ecological consequences and adaptive significance of markovian movements in a patchy environment', *Ecology* **65:** 147–65.

Saastamoinen, M. (2007). 'Life-history, genotypic, and environmental correlates of clutch size in the Glanville fritillary butterfly', *Ecological Entomology* **32:** 235–42.

Saastamoinen, M. (2008). 'Heritability of dispersal rate and other life history traits in the Glanville fritillary butterfly', *Heredity* **100:** 39–46.

Saastamoinen, M. and Hanski, I. (2008). 'Genotypic and environmental effects on flight activity and oviposition in the Glanville fritillary butterfly', *American Naturalist* **171:** 701–12.

Saastamoinen, M., Ikonen, S., Hanski, I. (2009). 'Significant effects of *Pgi* genotype and body reserves on lifespan in the Glanville fritillary butterfly', *Proceedings in Biological Sciences* 1313–22.

Saastamoinen, M., van der Sterren, D., Vastenhout N. *et al.* (2010). 'Predictive adaptive responses: Condition-dependent impact of adult nutrition and flight in the tropical butterfly *Bicyclus anynana*', *American Naturalist* **176:** 686–98.

Saccheri, I.J. and Brakefield, P.M. (2002), 'Rapid spread of immigrant genomes into inbred populations', *Proceedings of the Royal Society B* **269:** 1073–8.

Saccheri, I.J., Kuussaari, M., Kankare M. *et al.* (1998). 'Inbreeding and extinction in a butterfly metapopulation', *Nature* **392:** 491–4.

Schtickzelle, N. and and Baguette, M. (2003). 'Behavioural responses to habitat patch boundaries restrict dispersal and generate emigration-patch area relationships in fragmented landscapes', *Journal of Animal Ecology* **72:** 533–45.

Schtickzelle, N., Joiris A., Van Dyck H. *et al.* (2007). 'Quantitative analysis of changes in movement behaviour within and outside habitat in a specialist butterfly', *BMC Evolutionary Biology* **7.**

Shapiro, A. (1970). 'The role of sexual behaviour in density-related dispersal of pierid butterflies', *American Naturalist* **104:** 367–72.

Stefanescu, C. (2001). 'The nature of migration in the red admiral butterfly *Vanessa atalanta*: Evidence from the population ecology in its southern range', *Ecological Entomology* **26:** 525–36.

Sutherland, O.R.W. (1969). 'Role of crowding in production of winged forms by 2 strains of aphid *Acyrthosiphon pisum*', *Journal of Insect Physiology* **15:** 1385–1410.

Turchin, P. (1998). *Quantative analysis of movement: Measuring and modelling population redistribution in animals and plants.* Sinaeur Associates, Sunderland, MA.

van Nouhuys. S. and Hanski, I. (2002). 'Colonization rates and distances of a host butterfly and two specific parasitoids in a fragmented landscape', *Journal of Animal Ecology* **71:** 639–50.

van Nouhuys, S. and Hanski, I. (2004). 'Natural enemies of checkerspot butterflies', in P.R. Ehrlich, I. Hanski (eds), *On the wings of Checkerspots: A model system for population biology.* Oxford University Press, New York, 161–80.

Wade, M.J. and McCauley, D.E. (1988). 'Extinction and recolonization—their effects on the genetic differentiation of local populations', *Evolution* **42**: 995–1005.

Wahlberg, N. (1995). *One day in the life of a butterfly; A study of the biology of the Glanville fritillary Melitaea cinxia.* University of Helsinki, Helsinki.

Watt, W.B. (1983). 'Adaptation at specific loci .2. Demographic and biochemical elements in the maintenance of the *Colias Pgi* Polymorphism', *Genetics* **103**: 691–724.

Watt, W.B. *et al.* (2003). 'Adaptation at specific loci. VII. Natural selection, dispersal and the diversity of molecular-functional variation patterns among butterfly species complexes (*Colias*: Lepidoptera, Pieridae)', *Molecular Ecology* **12**: 1265–75.

Wheat, C.W. (2010). 'Phosphoglucose isomerase (*Pgi*) performance and fitness effects among Arthropods and its potential role as an adaptive marker in conservation genetics', *Conservation Genetics* **11**: 387–97.

Wheat, C.W., Fescemyer, H.W., Kvist J. *et al.* (2011). 'Functional genomics of life history variation in a butterfly metapopulation', *Molecular Ecology* **20**: 1813–28.

Whitlock, M.C. and McCauley, D.E. (1990). 'Some population genetic consequences of colony formation and extinction—genetic correlations within founding groups', *Evolution* **44**: 1717–24.

Wiklund, C., Gotthard, K., Nylin, S. (2003). 'Mating system and the evolution of sex-specific mortality rates in two nymphalid butterflies', *Proceedings of the Royal Society B* **270**: 1823–8.

Zheng, C., Ovaskainen, O., Hanski, I. (2009). 'Modelling single nucleotide effects in phosphoglucose isomerase on dispersal in the Glanville fritillary butterfly: Coupling of ecological and evolutionary dynamics', *Philosophical Transactions of the Royal Society B* **364**: 1519–32.

CHAPTER 24

Urban metapopulation dynamics, and evolution of dispersal traits in the weed *Crepis sancta*

Pierre-Olivier Cheptou and Antoine Dornier

24.1 Introduction

A species' habitat is often unstable at a local scale but the maintenance of a population at a regional scale is usually possible due to a species' dispersal capacity (Cousens *et al.* 2008). For sessile organisms such as higher plants, major dispersal episodes are restricted to specific stages in the life cycle such as pollen dispersal during reproduction and seed dispersal after reproduction. Morphological structures that aid dispersal are easily identifiable and involve both biotic (e.g. birds) and abiotic (e.g. wind, water) vectors (Cousens *et al.* 2008). As seed development requires favourable habitat, and reaching favourable habitat requires dispersal, dispersal is closely related to a species' habitat preference. Thus, rather than a single trait, plant dispersal encompasses a syndrome of traits (Baker and Stebbins 1965). As a consequence, ecological and evolutionary processes of dispersal in plants are intimately linked to the spatial habitat structure and how it changes over time.

Naturalists have, for a long time, observed associations between habitat structure and dispersal traits. In the 1960s, Carlquist (1966) reported reduced dispersal structures in several species of Asteraceae on Pacific islands (see Roff 1990, for flightlessness in insects). The loss of diaspores in 'island' environments surrounded by unfavourable habitat has given us an intuitive interpretation of the pattern. However, determining causal relationships between habitat characteristics and dispersal traits has turned out to be challenging (Roff 1990). On the other hand, theoretical models have identified factors favouring dispersal such as competition among kin and heterogeneity of habitat (Gandon and Michalakis 2001). Selection against dispersal arises from the associated costs of dispersal, such as the cost of producing a dispersal structure or that diaspores can land in unsuitable habitat.

While we have made much theoretical progress or in experimental populations (Molofsky and Ferdy 2005), analysing dispersal in the field remains challenging (Ronce 2007). In terms of dynamics, dispersal is particularly important at a regional level, where populations are linked by dispersal processes. In this context, the metapopulation model *sensu* Levins (Levins 1969) which assumes constant extinction rates and constant colonization rates appears as the simplest way to formalize regional dynamics. Not all spatially structured populations are, however, metapopulations. According to Hanski (1998), four conditions must be fulfilled for a species to persist as a metapopulation. First, suitable habitats have the potential to be occupied by a reproducing population. Second, every metapopulation has a measurable risk of extinction, even the largest. Third, isolation must be in a way that recolonization after extinction is possible and fourth, local population dynamics are not completely synchronized. While plants have sometimes been viewed as an example of patchy distribution fitting with metapopulation assumptions (Husband and Barrett 1996), some authors have pointed out specific limitations in plants (Freckleton and Watkinson 2002). The difficulty of defining binary habitat (suitable versus unsuitable), the question of seed dormancy, or the low population turnover

Dispersal Ecology and Evolution. First Edition. Edited by Jean Clobert *et al.*

(Eriksson 1996; Ouborg and Eriksson 2004) may obscure the signature of a colonization/extinction dynamics. Recent progress has been made for the statistical inference of spatially realistic metapopulation models (e.g. Stochastic Patch Occupancy Models, hereafter SPOMs; Moilanen, 1999; Etienne *et al.* 2004) but this theory has been seldom applied to plants. Hence, the degree to which plants fit the metapopulation model remains unknown.

Human influences tend to simplify ecosystems, and studying such simplified systems may therefore render metapopulation analysis easier and allow inference of evolutionary processes. In this context, we used the urban environment as a model system to study metapopulation dynamics and the contemporary evolution of dispersal traits (Cheptou *et al.* 2008). Specifically, we focused on the highly fragmented populations of the heterocarpic weed *Crepis sancta* occupying small patches on pavements, around trees planted within the city of Montpellier (south of France). We modelled patch dynamics using metapopulation models. In addition, comparing highly fragmented urban populations and more continuous rural populations allowed us to study micro-evolutionary processes of dispersal within the urban metapopulation.

24.2 A plant metapopulation model

24.2.1 Urban fragmented populations

Urban environments represent an unsuitable habitat matrix, extending up to many square kilometres, covered in asphalt and buildings punctuated by isolated patches of suitable plant habitat (Figure 24.1a). Consequently, we can assume a binary classification of plant habitat (suitable versus unsuitable) which can be inferred *a priori*. There are several thousand trees planted in the city, and each is surrounded by patches of soil. This suitable habitat represents less than 1% of the total surface, and constitutes a highly fragmented landscape in a regular arrangement. About 120 weed species have colonized these patches and field observations have revealed that *C. sancta* has a high annual turnover.

24.2.2 The model species: *Crepis sancta* (Case Study IV)

In the French Mediterranean region, the heterocarpic and allogamous species *Crepis sancta* is present in wasteland, vineyards, and on roadsides (Cheptou *et al.* 2002) and has colonized the centre of cities. *C. sancta* produces both non-dispersing achenes without a pappus (about 10–15 per head)

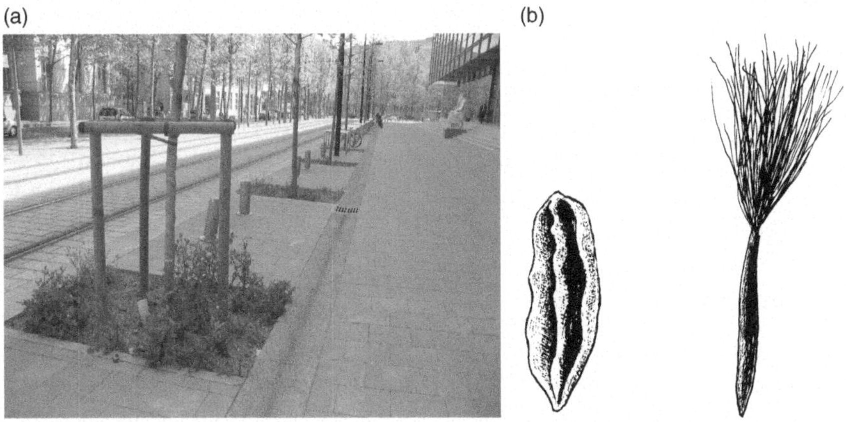

Figure 24.1 The heterocarpic plant *Crepis sancta* in urban environment. (a) urban patchy environment in Montpellier, southern France (here, Antigone district; photo courtesy G. Przetak). (b) Heterocarpy in *Crepis sancta*. On the left, heavy achenes deprived of pappus are produced by outer florets in the capitulum. On the right, light achenes with pappus are produced by inner florets. Drawings by Rene Ferris.

at the periphery of the capitulum, and wind-dispersing achenes with a pappus (about 80–100 per head) at the centre of the capitulum (Figure 24.1b). The two types of achenes exhibit contrasting wind dispersal ability: non-dispersing achenes are heavy (0.27 ± 0.02 mg) and have a high rate of falling (0.23 ± 0.01 m sec-1) while dispersing achenes are light (0.1 ± 0.01 mg) and have a low rate of falling (1.48 ± 0.08 m sec-1) (Imbert *et al.* 1996). Such heterocarpy is widespread in Asteraceae and found in more than 50 genera (Imbert 2002). While achene heteromorphism in Asteraceae is sometimes associated with differential seed dormancy (Venable and Brown 1988), neither type of achenes in *C. sancta* exhibit dormancy (Imbert 1999). Soil samples from urban patches did not provide evidence for the existence of a seed bank in *C. sancta* (Dornier *et al.* 2011).

Thanks to the two distinct classes of achenes, individual investment in dispersal in heterocarpic plants can be estimated as the ratio:

$$R_d = \frac{\textit{Number of non-dispersing achenes}}{\textit{Total number of achenes}}$$

In urban patches, *Crepis sancta* forms small populations (from 0–40 plants) linked by pollen transfer by bees (Cheptou and Avendano 2006). In contrast, rural populations (20–30kms from the city) are large continuous populations consisting of several hundred thousand plants per hectare.

24.3 Metapopulation dynamics

As a case study, *Crepis sancta* in urban patches exhibits a high annual turn-over (e.g. 1/3 of the patches go extinct each year). It thus provides an ideal situation for metapopulation inferences.

24.3.1 The SPOM framework

SPOMs model extinction and colonization processes using multi-year occupancy data (Moilanen 1999). It assumes a set of suitable patches within an unsuitable matrix and summarizes the demographical states of patches as binary: occupied or empty. We assume that $X_i(t)$ represents the state of a patch i at time t, $\overline{X}(t)$ represents the vector of the total state of the metapopulation at time t and the dependence of $\overline{X}(t+1)$ on $\overline{X}(t)$ follows a Markov chain process. An occupied patch goes extinct with the probability E_i or remains occupied with probability $1 - E_i$. Following Moilanen (1999), the probability of observing a metapopulation state change from $\overline{X}(t+1)$ to $\overline{X}(t+1)$ is:

$$P[\overline{X}(t+1) \mid \overline{X}(t)] = \Pi \begin{cases} E_i & \textit{if } X_i(t) = 1 \textit{ and } X_i(t+1) = 0 \\ 1 - E_i & \textit{if } X_i(t) = 1 \textit{ and } X_i(t+1) = 1 \\ C_i & \textit{if } X_i(t) = 0 \textit{ and } X_i(t+1) = 1 \\ 1 - C_i & \textit{if } X_i(t) = 0 \textit{ and } X_i(t+1) = 0 \end{cases}$$

(See Dornier *et al.* 2011, for methodological details)

In annual plants, occupancy in patches is assessed during flowering time and so the time step is generally considered on a yearly basis as a consequence.

24.3.2 Metapopulation scenarios and model selection

The SPOM framework and estimation procedure allows flexibility in the model assumptions that relate to different metapopulation dynamics. SPOMs can be understood as variations of the classical Levins model (Gotelli 1991). Levins' model made the assumption that each occupied patch contributes equally to colonization of empty patch (so that colonization results from the sum of all occupied patches) and that extinction is homogeneous and constant. Thus, the dynamics of patch occupancy are as follows:

$$\frac{dp}{dt} = c.p.(1-p) - e.p$$

Where p is the proportion of occupied patch, c and e are colonization and extinction rates respectively. Gotelli (1991) showed that the dependency of colonization and extinction on the proportion of occupied patch (p) allows the derivation of various ecologically meaningful metapopulation scenarios. If colonization is constant and does not depend on patch occupancy, it suggests a constant rain of propagules across the metapopulation, hence this scenario is called the Propagule Rain Model (PRM) by Gotelli (1991). If the colonization rate is not constant but increases with patch occupancy, this corresponds to an Allee-like effect, a model similar to the

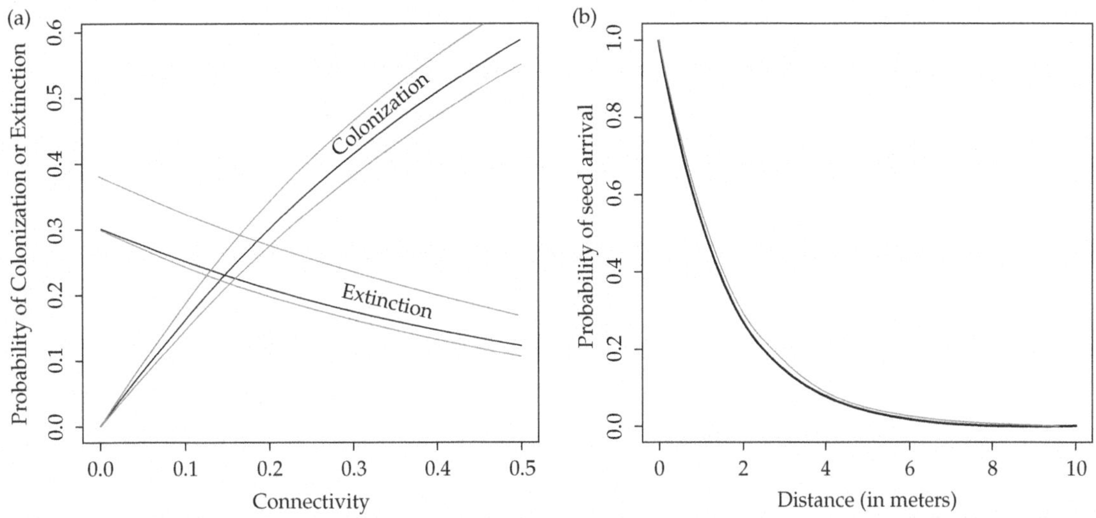

Figure 24.2 Estimated extinction and colonization probabilities as a function of connectivity for the JC sub-network (a), and seed dispersal kernel for the same cluster of patches (b). Black lines represent the estimate, and grey lines represent the upper and lower bounds of 95% confidence interval.

Incidence Function Model (IFM) (Hanski 1994). Finally, if extinction is not constant but positively related to occupancy rates, this signals a rescue effect; i.e. the fact that extinction rate decreases in occupied patches. Thus the combination of colonization and extinction functions allows defining various scenarios to be defined (detailed in Table 24.1).

While the Levins model is spatially implicit and assumes that patches are equally connected, spatially realistic models account for limited dispersal. Limited dispersal is captured by the connectivity metric *Si* (*t*) based on a negative exponential kernel (Moilanen and Nieminen 2002):

$$S_i(t) = \sum_{j \neq i} \exp(-\alpha d_{ij}) X_j(t)$$

Where d_{ij} is the distance between patch *i* and patch *j* and α the amplitude of exponential decrease. The metapopulation scenarios (Table 24.1) can be defined in a spatially realistic way by combining colonization and extinction functions (Table 24.1).

24.3.3 Application to urban metapopulations

For our purpose, we constructed five metapopulation scenarios including the possibility of an Allee-like effect, a rescue effect, and external colonization (see Table 24.1). Parameters were estimated using SPOMSIM software (Moilanen 2004), and the best metapopulation scenario was chosen using the AICc informative criterion. Metapopulation inferences were derived from two networks in the city of Montpellier at a distance of four kilometers, using a total of 289 patches. The 'HM network' (Henri Marès street) is at the city's outskirts and consists of a single line of 96 patches of about 2 m² spaced from 7–10 m apart. The 'JC network' (Jacques Cartier street in Antigone district; see Figure 24.1) is in the city centre and consists of a set of four lines for a total of 193 patches of 1.5 m² and spaced from 5–10 m apart within and between lines (see Dornier *et al.* 2011 for details).

Presence/absence data were collected over four years (2005–2008) for the HM network and over six years (2005–2010) for the JC network. Because observations indicated spatial heterogeneity in occupancy rates that may impede precise estimation of statistical parameters (Etienne *et al.* 2004), a 'western' sub-network of 81 similar patches was isolated in JC based on field observations. Two sub-networks (36 patches [HMpharma] and 60 patches [HMinra]) were considered in HM due to slight differences in the patch areas and inter-patch distance. The statistical analysis has revealed that treating the HM network as a single entity was parsimonious, which proved that the division *a priori* into two sub-networks was not relevant.

Table 24.1 In spatially realistic model, metapopulation scenarios are the combinations of colonization and extinction functions potentially depending on the connectivity metric *Si* (see text). We consider five classical scenarios in the urban system based on whether (and how) extinction and colonization depend on connectivity.

Colonisation	Extinction	Metapopulation scenarios
$C_i(t)=1-\exp(-y\,S_i(t))$	$C_i=y$	Spatially Realistic Levins Model (SRLM)
$C_i=y$	$C_i=y$	Propagule Rain Model (PRM)
$C_i(t)=\frac{S_i(t)^2}{S_i(t)^2+y^2}$	$C_i=y$	Incidence Function Model (IFM)
$C_i(t)=\frac{S_i(t)^2}{S_i(t)^2+y^2}$	$E_i(t)=E_i(1-C_i(t))^R$	IFM + rescue effect
$C_i(t)=1-\exp(-yS_i(t))$	$E_i(t)=E_i(1-C_i(t))^R$	SRLM + rescue effect

The best metapopulation scenario (Table 24.2) was the PRM (at least apart from JC western). The estimated colonization probability was 0.37 and the estimated extinction probability was 0.20. The absence of connectivity effect is the signature of an external colonization, and it is likely that a large peripheral population contributes to HM colonization. The PRM model was also the best model for the whole JC network, but in this case, the division into two sub-networks was statistically significant. Interestingly, the Spatially Realistic Levins Model with rescue effect was the best model for the 'western JC network'. This model provides an estimate of the colonization kernel ($\alpha = 0.64$) for a mean colonization distance of 1.66 m (Figure 24.2a), i.e. about one-third of the among-patch distance, and shows the dependency of colonization and extinction to connectivity (Figure 24.2b). The presence of a rescue effect is most likely to be due to the positive presence of neighbouring patches that may enhance pollination process (Cheptou and Avendano 2006).

In spite of similar habitat in urban environment and similar dispersal traits for the species *Crepis sancta* (Cheptou *et al.* 2008), metapopulation dynamics vary over time and space due to the potential demographic role of populations external to those studied.

24.4 Reduction of seed dispersal in the fragmented metapopulation

The extreme fragmentation in urban habitat provides the opportunity to test evolutionary predictions regarding the evolution of dispersal. Specifically, the urban system allowed us to measure the actual cost of dispersal endured by dispersing seeds and the heterocarpy of *Crepis sancta* allowed us to estimate recent shift in dispersal traits, R_d in urban environments. These empirical data allowed us to characterize selection processes and the speed of evolutionary processes.

24.4.1 How costly is dispersal in fragmented habitat?

While fragmentation may favour the reduction of dispersal (Carlquist 1966; Cody and Overton 1996; Cordeiro and Howe 2003; Riba *et al.* 2009), the effect of fragmentation on dispersal is currently unknown because of the difficulty of quantifying causal processes in natural systems. *C. sancta* in urban environments provides a good opportunity to investigate adaptive processes on dispersal traits on a short timescale. We can also take the advantage of two contrasting habitats at the regional scale: large and unfragmented rural habitat and small and highly fragmented patches in urban habitat. While urban patch colonization from rural populations should initially favour phenotypes with a high ratio of dispersing seeds (low R_d), these phenotypes should be at disadvantage in an urban environment because of the high cost of dispersal in a fragmented habitat (as most achenes will be lost on the concrete matrix). Using artificial patches with sticky surfaces, Cheptou *et al.* (2008) measured the probability

Table 24.2 AICc, number of parameters and weights (wi) for the best metapopulation models applied HM network (four years data) and JC network (six years data) for the species *Crepis sancta*.

	Model	Number of parameters	AICc	wi
HM entire	PRM	2	486.96	0.99
HM Pharma	PRM	2	153.54	0.87
HM INRA	PRM	2	358.38	0.56
	SRLM	3	360.27	0.22
JC entire	PRM	2	2342.8	0.92
JC 'western' sub-network	SRLM + rescue	3	1026.78	0.88

of dispersing and non-dispersing achenes falling out of the patch. Cheptou *et al.* (2008) found that dispersing akenes had a 55% greater chance of falling out of the patch than non-dispersing akenes, which indicates that dispersal is costly in urban patches.

24.4.2 Population differentiation for dispersal traits

In heterocarpic species, shifts in dispersal traits can be simply assessed by comparing mean population R_d. To detect potential shifts in R_d and to avoid environmental effects (Imbert and Ronce 2001), eleven populations of *Crepis sancta* were studied in an ordinary garden (actually a greenhouse: see Cheptou *et al.* 2008, for methodological details) using only dispersing seeds to minimize maternal seed-size effects. Measurements in the common environment revealed that patchy populations exhibit reduced dispersal as they produced a 50% increase in the number of non-dispersing achenes compared to rural continuous populations (Figure 24.3). The lower dispersal in urban habitat is in line with several recent empirical studies (Fresnillo and Ehlers 2008; Riba *et al.* 2009).

24.5 Evolutionary scenario in fragmented metapopulation

Detecting evolutionary processes in natural populations requires establishing causal relationships and eventually, quantitative predictions. The need to establish causality is a major limitation of correlative studies in plants. The urban system makes the task easier. The regularity of patch design facilitates modelling, and the extreme and recent fragmentation of the population suggests that adaptive change could be rapid. Quantitative genetics tools allow the estimation of the shift in dispersal strategy in response to high dispersal costs in fragmented environments. Initially in a rural population distribution R_d (μ=0.1; σ^2=0.01), the fitness W (R) of an R_d phenotype can be expressed as the sum of non-dispersing and dispersing achenes. During the colonization of urban patches from rural populations, only dispersing achenes contribute to colonization and W_c (R) is thus proportional to (1–R), which gives an advantage to low R-ratio phenotype. Conversely, in urban patches, the local fitness W_l (R) is driven by the high cost of dispersal in fragmented habitat; i.e. W_l (R) = R + (1–R) Pr (stay in the patch), where Pr (stay in the patch) is the probability of a dispersing achene staying in the patch. Given that patch surface is small compared to the total surface, we assume that all the dispersing achenes leaving a patch are lost and do not reach a suitable patch. The cost of dispersal will tend to increase the R_d. The selection differential, S, is the difference in the mean R_d before and after selection (Falconer 1981). When the traits under selection are heritable, the response to selection can be predicted by the breeder's equation: R = h^2S (Falconer 1981). The evolutionary scenario assumes one generation of colonization of urban patches from external (rural) populations followed by various number of generations in the city. Figure 24.4 showed the predicted shift in R_d as a consequence of the cost of dispersal. A substantial decrease in dispersal traits is expected as the cost of dispersal get stronger. Of interest, the R_d shift from 10% to about 15% observed

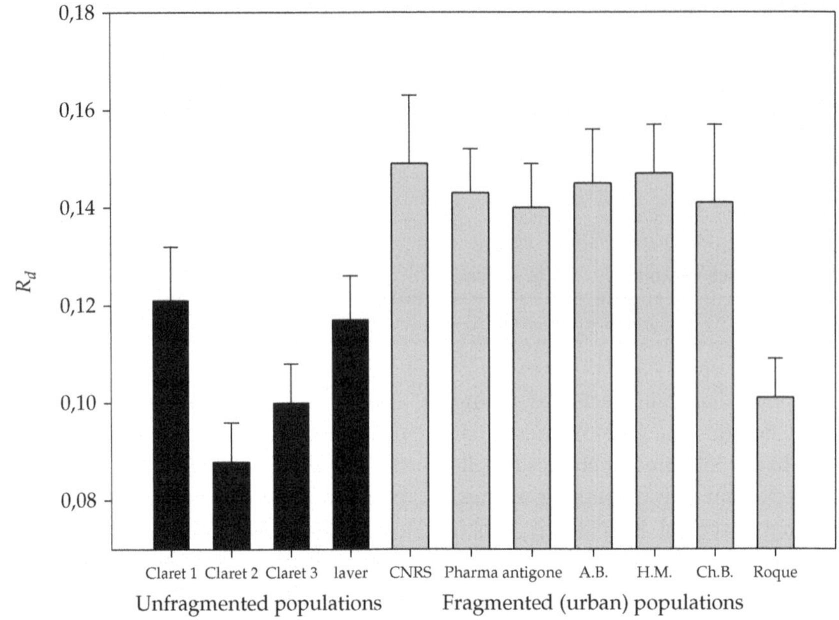

Figure 24.3 Mean (and standard error) for the proportion on of non-dispersing seeds (R_d) in unfragmented and fragmented 'patchy' populations measured in a common environment in a greenhouse.

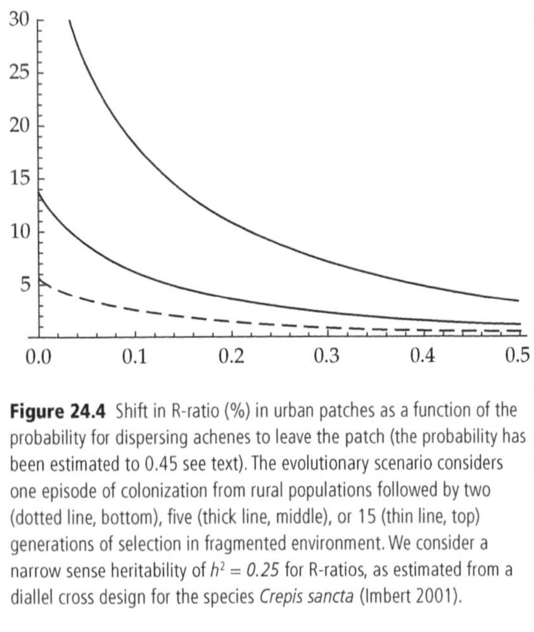

Figure 24.4 Shift in R-ratio (%) in urban patches as a function of the probability for dispersing achenes to leave the patch (the probability has been estimated to 0.45 see text). The evolutionary scenario considers one episode of colonization from rural populations followed by two (dotted line, bottom), five (thick line, middle), or 15 (thin line, top) generations of selection in fragmented environment. We consider a narrow sense heritability of $h^2 = 0.25$ for R-ratios, as estimated from a diallel cross design for the species *Crepis sancta* (Imbert 2001).

in our system (Figure 24.3) is consistent with less than 15 generations of selection. This shows that reduction of dispersal can be rapid once dispersal costs are high in fragmented environment.

24.6 Dispersal in plant metapopulation: lessons from simple natural systems

24.6.1 Dispersal and metapopulation dynamics

Dispersal in plants is sometimes been viewed as a mechanistic process from which dispersal kernels and then spatial structure could be deduced (Cousens *et al.* 2008). Application of the SPOM framework allows the estimation of metapopulation scenarios and inference of a phenomenological measure of dispersal from observed patterns of occupancy. Because the metapopulation approach captures the complexity of a species' habitat, the confrontation of phenomenological and morphological estimate of dispersal is meaningful in an evolutionary ecological perspective. The SPOM methodology is a flexible tool able to capture colonization patterns in plant populations. For instance, it has revealed its ability to detect signals of unidentified external sources and may thus be a powerful method to infer the dependence of populations to external sources (e.g. source/sink dynamics). Admittedly, the negative exponential kernel may provide a poor fit for dispersal processes (Chapter 15), which points out the need to add various dispersal kernels in such inferences. In spite of these limitations, dispersal inferences were close to direct estimates of seed dispersal in *Crepis sancta* (Imbert 1999). Some authors have questioned the applicability of metapopulation theory to plants (see Freckleton and Watkinson 2002). Admittedly, the urban *Crepis sancta* urban system is much simpler than typical natural systems (such as the absence of a seed bank or the definition of suitable habitat) that may be more difficult to overcome in plant systems in general. An important criticism made by Freckleton and Watkinson (2002) is the distinction between a 'true metapopulation' (where colonizations are rare enough to be disconnected from local demographic processes) and 'patchy populations' (where colonization relies on local processes and impacts local demography). Whether the urban system behaves as a true metapopulation or a patchy population is indeed questionable. This question may apply to natural populations in general (not only plant populations), and while Ovaskainen and Hanski (2004) concluded that metapopulation inferences are robust to this assumption, this question is an important issue in metapopulation studies. Compared to mobile organisms, sessile organisms such as plants may however make the task easier because colonization processes occur only once and over short episodes in the life cycle.

In a metapopulation context, dispersal limitation is often thought to create spatial genetic structuring of populations (due to inbreeding within patches). Importantly, while the low number of individuals of *C. sancta* within urban patches would have suggested high genetic drift and local inbreeding, microsatellites data have revealed very low F_{st} and an absence of significant inbreeding (Dornier and Cheptou 2012) probably due to important pollen flow among patches. Schematically, the set of patches is close to a panmictic unit. At the same time, we were able to detect limited seed dispersal in SPOMs. This shows that metapopulation dynamics *sensu* Levins do not necessarily imply significant genetic structure (Whitlock 1990), which points out the contrasted effect of dispersal on metapopulation genetics and metapopulation demography.

24.6.2 Dispersal evolution in fragmented environments

While dispersal reduction has been observed in several plants (e.g. Fresnillo and Ehlers 2008; Riba *et al.* 2009), and several empirical studies have revealed substantial genetic variance in dispersal traits (Venable and Burquez, 1989; Donohue *et al.* 2005), the main difficulty in natural system lies in the proper characterization of selection pressure. In the case of *Crepis sancta* in highly fragmented environment, (1) we have shown that the cost of dispersal was high in fragmented environments, (2) we predicted that shift to dispersal should be rapid due to substantial inheritance, and (3) the pattern observed in fragmented populations is consistent with five to 15 generations of selection driven by the cost of dispersal. Making of the cost of dispersal the driving force in fragmented environment is plausible but its strength may be highly variable depending on the sizes of habitat fragments. It is however possible that positive selection on dispersal arises from extinction in fragmented habitat (Gandon and Michalakis 2001) which may provide an advantage to dispersing phenotypes. However, the pattern observed suggests that extinction in not a major driver of adaptation in urban environment. Importantly, recent theoretical models indicate that positive selection on dispersal caused by extinction depends on the rapid saturation of local patches, which therefore gives a strong advantage to dispersers colonizing an empty patch. If patches do not saturate, as is the case in the urban system and probably in many natural systems, extinction can decrease local competition, thus favouring non-dispersing phenotypes so that the net effect of extinction is not trivial (see Ronce *et al.* 2000). This problem is similar to the discussion of whether natural populations fulfil true metapopulation assumptions or patchy population assumptions. The high cost of dispersal resulting from fragmentation is a consequence of passive dispersal as it is the rule in wind-dispersed plant. It is, however, important to note that oriented dispersal among fragments such as animal dispersal of animal-dispersed plant would probably not suffer from a high cost of dispersal in fragmented landscapes.

Overall, the metapopulation framework is both simple and useful in order to understand plant dispersal, at least in simple systems. Whether occupancy patterns can be captured by colonization/extinction dynamics and thus provide reliable information on plant dispersal, requires to analysis of the suitability of metapopulation models to each plant species' ecology (seed dormancy, habitat preference, etc.). We anticipate that plant dynamics issues do not fundamentally depart from species dynamics issues in general.

Acknowledgements

The TrameVerte ANR program (ANR-08-VILL-0003) provided financial support for this study. We thank one anonymous reviewer and Tim Benton for clarifying the text.

References

Baker, H.G. and Stebbins, G.L. (1965). *The genetics of colonizing species*. Academic Press, New York.

Carlquist, S. (1966). 'The biota of long-distance dispersal. II. Loss of dispersability in Pacific Compositae', *Evolution* **20**: 30–48.

Cheptou, P.O. and Avendano, L.G. (2006). 'Pollination processes and the Allee effect in highly fragmented populations: consequences for the mating system in urban environments', *New Phytologist* **172**: 774–83.

Cheptou, P.O., Carrue, O., Rouifed, S. *et al.* (2008). 'Rapid evolution of seed dispersal in an urban environment in the weed Crepis sancta', *Proceedings of the National Academy of Sciences of the USA* **105**: 3796–9.

Cheptou, P.O., Lepart J., Escarre J. (2002). 'Mating system variation along a successional gradient in the allogamous and colonizing plant Crepis sancta (Asteraceae)', *Journal of Evolutionary Biology* **15**: 753–62.

Cody, M.L. and Overton, J.M. (1996). 'Short-term evolution of reduced dispersal in island plant populations', *Journal of Ecology* **84**: 53–61.

Cordeiro, N.J. and Howe, H.F. (2003). 'Forest fragmentation severs mutualism between seed dispersers and an endemic African tree', *Proceedings of the National Academy of Sciences of the USA* **100**: 14052–6.

Cousens, R., Dytham, C. *et al.* (2008). *Dispersal in plants: a population perspective*. Oxford University Press, New York.

Donohue, K., Polisetty, C.R., Wender, N.J. (2005). 'Genetic basis and consequences of niche construction: Plasticity-induced genetic constraints on the evolution of seed dispersal in Arabidopsis thaliana', *American Naturalist* **165**: 537–50.

Dornier, A., Pons, V. Cheptou, P.-O. (2011). 'Colonization and extinction dynamics of an annual plant metapopulation in an urban environment', *Oikos* **120**: (8) 1240–46.

Dornier A. and Cheptou P.-O. 2012 Determinants of extinction in fragmented plant populations: Crepis sancta (asteraceae) in urban environments. *Oecologia* (online first)

Eriksson O. (1996). 'Regional dynamics of plants: A review of evidence for remnant, source-sink and metapopulations', *Oikos* **77**: 248–58.

Etienne R., ter Braak, C., and Vos, C. (2004), 'Applications of stochastic patch occupancy models to real metapopulations', in I. Hanski, O. Gaggiotti (eds), *Ecology, genetics and evolution of metapopulation*. Elsevier Academic Press, Amsterdam, 105—132.

Falconer D.S. (1981). *Introduction to quantitative genetics*, 3rd edition Longman Scientific and Technical, Essex, England.

Freckleton, R.P. Watkinson, A.R. (2002). 'Large-scale spatial dynamics of plants: metapopulation regional ensembles and patchy populations', *Journal of Ecology* **90**: 419–34.

Fresnillo, B. and Ehlers, B.K. (2008). 'Variation in dispersability among mainland and island populations of three wind dispersed plant species', *Plant Systematics and Evolution* **270**: 243–55.

Gandon, S. and Michalakis, Y. (2001). 'Multiple causes of the evolution of dispersal', in J. Colbert, E. Danchin, A.A. Dhondt *et al.* (eds), *Dispersal*. Oxford University Press, Oxford, 155–167.

Gotelli, N.J. (1991). 'Metapopulation Models—the Rescue Effect, the Propagule Rain, and the Core-Satellite Hypothesis', *American Naturalist* **138**: 768–76.

Hanski, I. (1994). 'A Practical Model of Metapopulation Dynamics', *Journal of Animal Ecology* **63**: 151–62.

Hanski, I. (1998). 'Metapopulation dynamics', *Nature* **396**: 41–9.

Husband, B.C. and Barrett, C.H. (1996). 'A metapopulation perspective in plant population biology', *Journal of Ecology* **84**: 461–9.

Imbert, E. (1999). 'The effects of achene dimorphism on the dispersal in time and space in Crepis sancta (Asteraceae)', *Canadian Journal of Botany-Revue Canadienne De Botanique* **77**: 508–13.

Imbert, E. (2001). 'Capitulum characters in a seed heteromorphic plant, *Crepis sancta* (Asteraceae): variance partitioning and inference for the evolution of dispersal rate', *Heredity* **86**: 78–86.

Imbert, E. (2002). 'Ecological consequences and ontogeny of seed heteromorphism', *Perspectives in plant ecology, evolution and systematics* **5**: 13–36.

Imbert, E., Escarre, J., Lepart, J. (1996). 'Achene dimorphism and among-population variation in Crepis sancta (Asteraceae)', *International Journal of Plant Science* **157**: 309–15.

Imbert, E. and Ronce, O. (2001). 'Phenotypic plasticity for dispersal ability in the seed heteromorphic Crepis sancta (Asteraceae)', *Oikos* **93**: 126–34.

Levins, R. (1969). 'Some demographic and genetic consequences of environment heterogeneity for biological control', *Bulletin of the Entomological Society of America* **15**: 237–40.

Moilanen, A. (1999). 'Patch occupancy models of metapopulation dynamics: Efficient parameter estimation using implicit statistical inference', *Ecology* **80**: 1031–43.

Moilanen, A. (2004). 'SPOMSIM: software for stochastic patch occupancy models of metapopulation dynamics', *Ecological Modelling* **179**: 533–50.

Moilanen, A. and Nieminen, M. (2002). 'Simple connectivity measures in spatial ecology', *Ecology* **83**: 1131–45.

Molofsky, J. and Ferdy, J.B. (2005). 'Extinction dynamics in experimental metapopulations', *Proceedings of the National Academy of Sciences of the USA* **102**: 3726–31.

Ouborg, N. and Eriksson, O. (2004). 'Toward a metapopulation concept for plants', in I. Hanski, *Ecology, genetics and evolution of metapopulation*. Elsevier Academic Press, Amsterdam, 447—469.

Ovaskainen, O. and Hanski, I. (2004). 'From individual behavior to metapopulation dynamics: Unifying the patchy population and classic metapopulation models', *American Naturalist* **164**: 364–77.

Riba, M., Mayol, M., Giles, B.E. *et al.* (2009). 'Darwin's wind hypothesis: does it work for plant dispersal in fragmented habitats?', *New Phytologist* **183**: 667–77.

Roff, D.A. (1990). 'The Evolution of Flightlessness in Insects', *Ecological Monographs* **60**: 389–421.

Ronce, O. (2007). 'How does it feel to be like a rolling stone? Ten questions about dispersal evolution', *Annual Review of Ecology and Systematics* **38**: 231–53.

Ronce, O., Perret F.,Olivieri, I. (2000). 'Evolutionary dispersal rates do not always increase with local extinction rates', *American Naturalist* **155**: 485–96.

Venable, D.L. and Brown, J.S. (1988). 'The Selective Interactions of Dispersal, Dormancy, and Seed Size as Adaptations for Reducing Risk in Variable Environments', *American Naturalist* **131**: 360–84.

Venable, D.L. and Burquez, A.M. (1989). 'Quantitative genetics of size, shape, life-history and fruit characteristics of the seed-heteromorphic composite Heterosperma pinnatum. I-Variation within and among populations', *Evolution* **43**: 113–24.

Whitlock, M.C. (1990). 'Some population genetic consequences of colony formation and extinction: genetic correlations within founding groups', *Evolution* **44**: 1717–24.

PART VI

Dispersal and Climate Change

CHAPTER 25

Dispersal and range dynamics in changing climates: a review

Jean François Le Galliard, Manuel Massot, and Jean Clobert

25.1 Introduction

Most animal and plant species inhabit complex environments characterized by substantial spatial and temporal heterogeneity. Dispersal (defined here to include both dispersal rate and dispersal distances) is often crucial for a species' ability to exploit this habitat heterogeneity at a variety of spatial scales, and to persist in a dynamic and heterogeneous environment over ecological timescales (Wiens *et al.* 1993). In turn, temporal and spatial variations in habitat quality have long been recognized as primary factors driving the evolution of dispersal rates and distances (Ronce *et al.* 2000). For example, species of planthoppers living in temporally unstable environments produce more dispersal forms than species of planthoppers living in stable environments, which allow them to escape from deteriorating local conditions and to colonize new habitats (Denno *et al.* 1991). Thus, some understanding of spatial and temporal variation in habitat quality is crucial to explain both the evolution of dispersal strategies and the ecological dynamics of spatially heterogeneous populations.

Generally, each species is characterized by a set of environmental conditions (water availability, nutrients, light, soil geochemistry, etc.) that enable its growth, survival, and reproduction (the fundamental ecological niche, *sensu* Hutchinson 1957). A large number of animals species are ectotherms that depend on external heat sources to warm their body, and the physiology of most plants and many endothermic animals is also strongly influenced by external temperatures (reviewed in Angilletta 2009). Most species can therefore exploit only a limited number of thermal or other climatic conditions (i.e. humidity, wind speed, solar radiation), which define their fundamental climate niche (Guisan and Thuiller 2005). In turn, the fundamental climate niche can have a substantial impact on the realized niche (i.e. observed spatial distribution of the species; Holt and Gomulkiewicz 1997) and therefore the costs and benefits of long-distance dispersal and colonization. Until recently, however, the effects of spatial and temporal climatic variation on dispersal strategies have received relatively little attention relative to other factors such as food availability, competition, or predation (Clobert *et al.* 2001; Bullock *et al.* 2002, but see Watkinson and Gill 2002).

The growing awareness that most ecosystems are being strongly impacted by global climate changes has fortunately led to a recent surge of interest in studies of dispersal aimed at understanding and predicting the ecological effects of climate change (Watkinson and Gill 2002; Parmesan, 2006; Berg *et al.* 2010). The two last decades have seen accumulating evidence that changing climate conditions, especially climate warming, can influence populations of single species, communities of interacting species, and whole ecosystems. One of the most common ecological responses to climate change involves range or habitat shifts due to spatial expansions at the cool range margins and spatial retractions at the warm range margins (Parmesan *et al.* 1999; Watkinson and Gill 2002; Parmesan and Yohe 2003; Parmesan 2006; Sinervo *et al.* 2010). Dispersal plays a key role in determining the rate at which these spatial range shifts occur and has a

Dispersal Ecology and Evolution. First Edition. Edited by Jean Clobert *et al.*

critical influence on the capacity of a species to survive climate change (Chapter 26). There is therefore an urgent need for a good understanding of the effects of climate change on dispersal. Natural history data hint at important effects of climate conditions on dispersal. For example, many animals disperse during episodic bouts determined by seasonal changes in climate conditions, the dispersal of plant and animal propagules requires some movements of air or water masses that can themselves be directly influenced by climate conditions, and some animal species are capable of long-distance movements when climate conditions are inappropriate for their survival. However, we must move beyond these observations to bridge the gap between climate change ecology and dispersal ecology, and develop predictive models of ecological responses to climate change. This chapter presents an attempt to summarize existing results and methods in this field of research with the aim of highlighting (i) existing patterns and trends in the current data, (ii) the most important knowledge gaps, and (iii) some fruitful avenues for future research.

25.2 Climate change and the spatial distribution of plants and animals

The spatial distribution of plants and animals is determined by multiple interacting processes including dispersal, biotic interactions (e.g. competition), abiotic factors (e.g. climate and geology), and adaptation (reviewed in Case and Taper 2000). The important role of climate factors for range dynamics is exemplified by the significant range shifts associated with past and present climate changes. In this section, we review patterns of range shifts from both the Quaternary period and from the last decades (see Table 25.1).

25.2.1 Inferences from range shifts during the Quaternary period

A major ecological footprint of climate changes is seen in long time series of palaeontological records gathered at multiple sites and encompassing the Quaternary period, which includes the Pleistocene that lasted from about 2.5 million years ago until 10 000 years ago, and the Holocene from 10 000 years ago to the present (Brewer *et al.* 2002). These epochs are characterized by cycles of glacial and interglacial climates lasting each approximately 100 000 years of which 10 000 years (including the current Holocene period) exhibit relatively warm climates. Climate changes have been dramatic during this time but on average slower relative to current warming trends. However, fast climatic changes of greater amplitude than projections for the near future have sometimes occurred—for example, periods of climate warming of 5 °C to 12 °C within less than a century (Brewer *et al.* 2002).

These past climate changes led to major distributional shifts in the Northern hemisphere for which very good fossil records are available (Delcourt and Delcourt 1988; Graham *et al.* 1996; Bradshaw 1999), although similar range dynamics are seen in the tropics and the Southern hemisphere (Hewitt 2000). Four major patterns have emerged where quantitative data on spatial range shifts are available. First, across species, there is a lack of evidence for highly synchronized dynamics of range contraction during cooling periods and of range expansion during warming periods. Many animal and plant species have moved northwards or southwards during the glacial and interglacial periods as intuition would predict, but other species have been able to survive for long periods without shifting their ranges and others have moved in unexpected eastward or westward directions (Graham *et al.* 1996; Bradshaw 1999; Davis and Shaw 2001; Lyons 2003). These individualistic responses to climate change have led to important changes in the structure of animal and plant communities during glacial and interglacial periods. It is assumed that species responded differently to climate change because of differences in climate tolerances (Davis and Shaw 2001), differences in dispersal abilities (Lyons *et al.* 2010; Roy *et al.* 2001), and the influence of geography on range shifting (Botkin *et al.* 2007). Second, historical spread rates of many plant species have been calculated with some accuracy from pollen and other palaeobotanical records (see Table 25.1 and Figure 25.1a). Some authors suggest that these

Table 25.1 Examples of range shifts during past and contemporary periods. Range shifts during the Quaternary were calculated from pollen or paleobotanical records for trees, and from records of mammalian fossils.

Reference	Study organisms	Study period	Spread rate
Observed historical range shifts during the late Quaternary period			
Cain *et al.* 1998	Woodland herbs during the Holocene	16 000 years BP	+ 0.03 km/year northward
Birks 1989	Tree species on the British Isles during Holocene	16 000 years BP	+ 0.05 to + 0.7 km/year northward
Lyons *et al.* 2010	Mammalian fauna during Pre-glacial to Glacial	30 000 years BP	0.01 to 0.32 km/year
	Mammalian fauna during Glacial to Holocene	15 000 years BP	0.002 to 0.32 km/year
	Mammalian fauna during Holocene	10 000 years BP	0.004 to 0.21 km/year
Observed contemporary range shifts during the last decades			
Parmesan *et al.* 1999	Meta-analysis of non-migratory temperate butterflies	1920–1990	+ 0.35 to + 2.4 km/year northward
Parmesan and Yohe 2003	Global meta-analysis of 99 temperate species of birds, butterflies and herbs	Decades to century	+ 0.61 km/year northward
Hickling *et al.* 2006	Meta-analysis of British flora and fauna (329 species)	1960–2000	+ 1.24 to + 2.4 km/year northward + 0.28 to + 2.48 m/year uphill
Beckage *et al.* 2008	Forest ecotone including several tree species	1962–2005	+ 2.11 to + 2.77 m/year uphill
Devictor *et al.* 2008	Assemblage of bird species in France	1989–2006	+ 5.35 km/year northward
Parolo and Rossi 2008	Vascular plant species	1956–2004	+ 3.42 m/year uphill
Hochkirch and Damerau 2009	Wing-dimorphic bush cricket during a warm spell	2003	+ 8 km northward in one year
Predicted spread rates needed to match the ongoing and predicted climate change			
Davis and Shaw 2001	Tree species in the Northern hemisphere	Next century	+ 3–5 km/year northward
Devictor *et al.* 2008	Assemblage of bird species in France	1989–2006	+ 16 km/year northward
Loarie *et al.* 2009	Global mean velocity of climate change	Next century	0.42 km/year
	Range of the mean velocity of climate change		0.08 to 1.26 km/year

historical spread rates have been much faster than those predicted on the basis of current dispersal abilities, measured by the dispersal distance of seeds in present-day species (Cain *et al.* 1998; Clark 1998). The underestimation of the current dispersal capacity of these species may be due to rare, long-distance dispersal events (Chapter 29). In addition, the calculation of historical spread rates is debated because some species might have been able to persist in cryptic refugia and so supposed spread rates would be lower (McLachlan *et al.* 2005; Provan and Bennett 2008). Yet this shows that it can be difficult to infer the rate and patterns of range shifts from a quantitative knowledge of the dispersal kernel. Third, there is some evidence from a few regional studies of trees that range dynamics have not always been uniform but have proceeded by a combination of rare dispersal events leading to long-distance colonization with more common short-distance dispersal allowing population spread within habitat patches (e.g. Delcourt and Delcourt 1988). Several species have persisted in topographically complex habitats by moving upward or downward in altitude (Lyons *et al.* 2010). These observations highlight interactions between climate change, landscape structure, and topography. Fourth, despite dramatic climate fluctuations and range shifts during the Quaternary, this period is also characterized by modest rates of extinction (Hewitt 2000; Davis and Shaw 2001; Botkin *et al.* 2007). This finding demonstrates that many species had the capacity to adapt to past climate fluctuations.

The analysis of the genetic structure of current populations yield further insights into the demographic events experienced during the Quaternary, which leave footprints in the spatial genetic data (Hewitt 2000). These genetic studies have now been conducted in several species and demonstrate, as do fossil records, the complexity of spatial responses to past climate changes. In Europe, the genetic structure of most populations is shaped by common patterns of range retractions into three refugia located in the south, range expansions towards the north during interglacial periods, declining genetic diversity in recently founded populations, and genetic adaptation to novel climate conditions after colonization (Taberlet *et al.* 1998; Hewitt 2000; Davis and Shaw 2001). However, species differ importantly in their colonization routes, sensitivity to dispersal barriers, and capacity to adapt to novel conditions. In lizards and snakes, for example, range expansions to the north during interglacial ages were more complex than a uniform expansion front (Figure 25.1B). Different snake and lizard species followed distinct expansion routes, some recolonization routes were constrained by the presence of mountainous barriers, and others were limited by adaptations to live in warmer climates, such as oviparity.

25.2.2 Evidence of contemporary ranges shifts

One of the major trends of recent global climate changes is that the mean global temperature increased by ca. 0.6 °C during the last century. This warming has been particularly marked at high latitudes, in some continental areas, and during the second half of the last century (IPCC 2007). Several long-term studies provide strong evidence for a pattern of range shifting coincident with these warming trends (Parmesan *et al.* 1999; Parmesan and Yohe 2003; Parmesan 2006). In a global meta-analysis using data for plants, birds, insects, fishes, marine invertebrates, and marine zooplankton collected during the last 30–80 years, Parmesan and Yohe (2006) concluded that:

Figure 25.1 Examples of historical and contemporary range dynamics. (a) Range dynamics of oak trees (*Quercus spp.*) during the Holocene illustrate a rapid range expansion poleward lasting approximately 10 000 years. Isolines give the approximate location of the cool margin of the distribution after pollen data collected across the British Isles (Birks 1989). (b) Expansion routes of the adder (*Vipera berus*) across Europe illustrate rapid colonization poleward (arrows) during the late Quaternary from three main refugia in southern Europe and recent contact zones (dashed lines). Data were inferred from a genetic study of contemporary populations across Europe (Ursenbacher *et al.* 2006). (c) Mobile species such as some butterflies have expanded quickly poleward during the last century. Data illustrate the twentieth-century changes in the range of the speckled wood (*Pararge aegeria*) in Great Britain (Parmesan *et al.*, 1999). A coloured grid cell indicates more than one population in 1915–1939 (black), 1940–1969 (dark grey) or 1970–1997 (light grey). (d) The treeline of the black spruce (*Picea mariana*) has shifted of about 12 kms along a warm–cold coastal gradient in the Hudson Bay since late 1800s. Data were obtained from transect surveys using tree ring analysis (Lescop-Sinclair and Payette 1995).

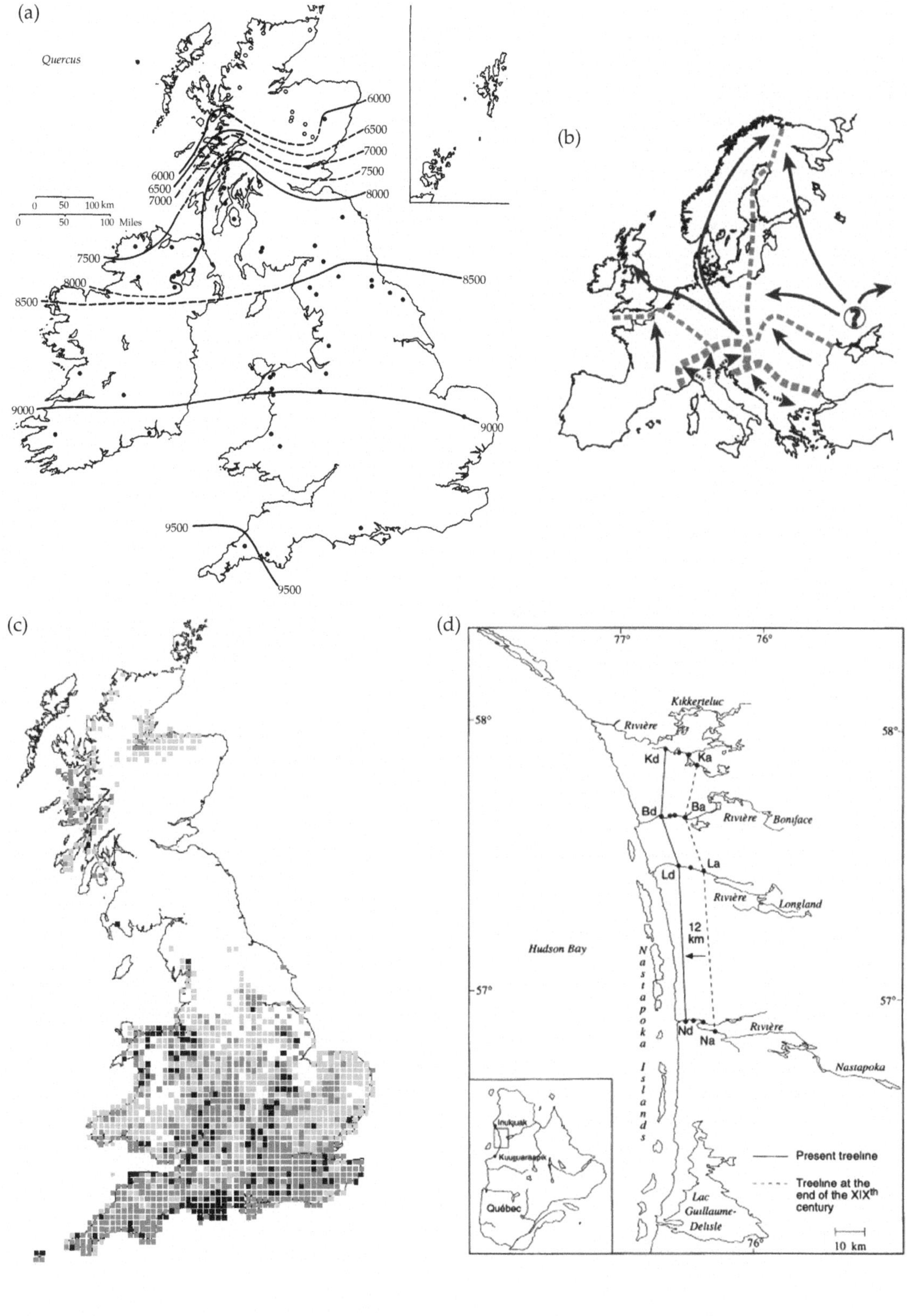
(a)
Quercus
6000
6500
7000
7500
8000
8500
9000
9500
0 50 100 km
0 50 100 Miles
(b)
(c)
(d)
77°
76°
58°
57°
Kikkerteluc
Rivière
Kd
Ka
Bd
Ba
Rivière Boniface
La
Ld
Rivière Longland
12 km
Hudson Bay
Nastapoka Islands
Nd
Na
Rivière Nastapoka
Inukjuak
Kuujjuarapik
Québec
Lac Guillaume-Delisle
Present treeline
Treeline at the end of the XIXth century
10 km

- range expansions towards formerly colder areas are very frequent with average spread rates of the order of several kilometres per year (see Table 25.1 and Figure 25.1C-D), though spread rates are highly variable across species (Parmesan and Yohe 2003);
- several species have responded weakly to recent climate warming;
- range contractions at warmer margins caused by population extinction are observed less frequently. For example, the southern boundaries of 35 northern temperate butterfly species was stable through time and only 22% displayed a range retraction in the south (Parmesan *et al.* 1999);
- very fast range shifts have been observed in the oceans where the passive drift of propagules can occur rapidly over very large spatial scales.

These range shifts could be confounded with changes in land use and are also consistent with an alternative hypothesis of non-equilibrium range dynamics that are continuing in response to past climate changes. Against these alternative explanations, long-term correlations between range shits and climate data indicate that many of the observed changes can be mechanistically linked to recent local or regional climate changes (Parmesan 2006). For example, there is evidence from a few highly mobile animals that rapid shifts in spatial distribution can occur immediately after a warm spell (e.g. Hochkirch and Damerau 2009), and marine zooplankton track very quickly their shifting thermal isoclines (e.g. Beaugrand *et al.* 2002).

Despite the rapid accumulation of substantial data on range shifts, few authors have attempted to quantify the relationship between the observed range shifts and traits of species. Generally, it is assumed that habitat generalists with high dispersal capacities, fast life histories, and low inter-specific competitive abilities would be able to expand their ranges faster to track changing climates (Hill *et al.* 1999a). To our knowledge, however, only two studies have examined these relationships. Pöyry *et al.* (2009) found that more mobile butterflies living in forest edges displayed faster range shifts towards the north in Finland. A similar positive correlation between dispersal capacity and range shifts was observed in vascular plants migrating upward in the Alps in response to climate warming (Parolo and Rossi 2008). We need more empirical studies that quantify differences in range shifts capacities among species.

25.3 Dispersal and resilience to climate change

The effects of climate change on the past and present spatial distributions of plants and animals are obvious from past and contemporary data (see Section 25.2 earlier). These effects should be strongly influenced by the dispersal capacity of a species. Under global climate change, the persistence of a species is indeed mediated by the interplay between dispersal and local adaptive responses (Chapter 26). Here, we review briefly evidence of ecological and evolutionary effects of dispersal in the context of spatial range shifts and climate change, and provide two contrasting case studies. We summarize predicted trends for the future and discuss major uncertainties in our predictions. Our analysis focuses on climate warming for which most data are available but results can be applied to the broader situation of climate change.

25.3.1 Dispersal enables ecological and evolutionary rescue to climate warming

Dispersal is important in allowing the colonization of novel habitat patches made available by climate warming, and is the key process by which species expand their ranges. This effect of dispersal can involve both dispersal rates and distances, and species with further dispersal distances and more frequent dispersal are therefore likely to be more capable at tracking their shifting climate niches. In addition, dispersal may be crucial for the persistence of a species at its contracting margin where climate warming will create larger gaps between formerly hospitable habitats. Good dispersal abilities are indeed important to compensate for decreased connectivity and to exploit opportunities of altitudinal shifts. Most models therefore predict that a high dispersal capacity diminishes the extinction risks from climate warming over ecological timescales (Chapter 26). Dispersal is also a major factor because it determines the spatial distribution of genetic diversity and therefore the capacity of a species to evolve in response to changing conditions at range margins. In addition, dispersal itself can be subject to evolutionary changes during range dynamics (Chapter 26). In turn, these evolutionary changes in dispersal can accelerate or slow down range expansion (Dytham 2009; Phillips *et al.* 2010). Thus, over evolutionary timescales, a high dispersal capacity will not necessarily rescue species against the extinction risks of climate warming.

25.3.2 Two contrasted case studies

A detailed study of the sachem skipper (*Atalopedes campestris*) highlights how good dispersal capacities can enable a species to respond quickly to a warming climate. This common generalist butterfly is found across the southern United States and Mexico and has expanded its range northward at high spread rates during the past 40 years (see Figure 25.2). Field and laboratory data indicate that this northward shift has not been constrained by dispersal limitations. The overwintering range of this species tracks a thermal cline of low winter temperature, which is associated with low larval survival in acute and chronic stress experiments in the laboratory (Crozier 2003). Recently colonized populations are characterized by a lower overwintering survival, which is predicted well from laboratory data linking temperature and survival, and there is no evidence that these populations are locally adapted to lower temperatures (Crozier 2003, 2004). The sachem skipper is characterized by high dispersal capacities with spread rates as fast as 19 km per year. The resultant range dynamics are therefore more strongly affected by landscape structure and physiological constraints than by dispersal distances (Crozier and Dwyer 2006).

On the contrary, many species are trapped inside a restricted range where they display limited dispersal capacities. This includes some tree species which do not seem to have filled their potential range since the last glaciations (Svenning and Skov 2004), and several endemic species occurring in limited areas such as mountain summits, isolated islands, or marginal distributions. Endemic species are already characterized by a syndrome of low dispersal, strong local adaptation, and potentially reduced genetic diversity, and are likely to be at high risks of extinction from climate change. A knapweed (*Centaurea corymbosa*), which occurs in a small area of 3 km^2 in southern France, is a good example of such a species (Colas *et al.* 2004). It is a cliff specialist that is out-competed by more dominant plants and scrubs on plateaus and depressions, and it has very limited dispersal abilities (Figure 25.2). Available cliff habitat, even that separated by less than 300 m, is therefore not well connected, and each cliff contains an independent panmictic population. In this species, assisted dispersal or *in situ* conservation will be needed to prevent extinction from climate warming in the near future.

25.3.3 Predicted range shifts under climate warming for the future

Most models forecast accelerations of climate warming over the next century even under some optimistic socio-economic scenarios (Table 25.1). On average, changes in climate conditions will include warmer conditions (likely range of + 0.2 °C to + 0.4 °C per decade for mean annual temperature during the next century), very likely precipitation increases in high latitudes but a decrease in water resources in semi-arid areas, as well as very likely increases in the frequency of hot extremes, heat waves, and heavy rainfall (IPCC 2007). Forecasts of range dynamics under these future climate projections have been obtained using two distinct methodological approaches. On the one hand, most predictive studies have used a 'niche-based approach' where one predicts the climatic niche of a species from spatial occurrence data and projects this niche into the future (Guisan and Thuiller 2005). This approach can be applied to a wide range of plant and animal species, and does not predict dynamics *per se* but the distribution of the future habitat. The species may or may not be able to occupy this habitat. On the other hand, mechanistic models have been developed, which may be used in conjunction with projections of the distribution of the future habitat. These models include details about the

Figure 25.2 Two species with contrasting dispersal abilities. (a) The sachem skipper *Atalopedes campestris* is a widespread and good dispersing butterfly species. Photograph © Will Cook. (b) The overwintering range of this species (shaded area) has expanded quickly northward (lighter shading) during the past 40 years and has tracked remarkably well the shifting thermal isocline of the January average minimum -4 °C isotherm, a temperature which is lethal in this species (Crozier 2003; used with permission from Springer Science + Business Media). Lines correspond to the isocline in the past (solid) and in the present (dotted). (c) The knapweed *Centaurea corymbosa* is rare plant endemic to a few cliffs in southern France. Photograph © Bruno Colas. (d) This species is a very poor disperser (data represent distribution of seed dispersal distances from the maternal plant), which we predict will be unable to track changing climates (Colas *et al*. 1997; Copyright (1997) National Academy of Sciences, USA).

relationships between physiological or life-history traits and climate conditions, as well as information on demographic processes such as dispersal and population dynamics. Mechanistic models cannot be applied to a wide range of species (Botkin *et al.* 2007), but can identify traits that contribute the most to distribution changes; for example, the relative importance of dispersal curves, density-dependence, and climate change. Examples of this approach include vegetation dynamics models (e.g. Iverson *et al.* 2004), physiologically structured models (e.g. Buckley 2008), and metapopulation models (e.g. Anderson *et al.* 2009).

Niche-based models predict that future climate trends would accelerate the rate at which the habitat of many species will shift poleward and upward, and mechanistic models suggest further that this could precipitate species extinctions worldwide (e.g. Thomas *et al.* 2004; Thuiller *et al.* 2005; Kearney and Porter 2009; Sinervo *et al.* 2010). However, there is a high error around these predictions because of temporal and spatial uncertainty about climate forecasts, uncertainty about ecological responses to future climate changes, and uncertainty about the ecological effects of evolutionary responses (Botkin *et al.* 2007). In particular, niche models and most mechanistic models tend to ignore biotic interactions, non-equilibrium demographic dynamics, and evolutionary responses. Also, most niche-based models neglect dispersal or do not account for differences in dispersal capacities and colonization routes among species (but see Fitzpatrick *et al.* 2008; Engler *et al.* 2009; Morin and Thuiller 2009). It has been demonstrated that models making slightly more realistic assumptions about dispersal mechanisms and ecological interactions can yield to strongly altered predictions (Chapter 26). Niche models are good at predicting current distributions of many species, but one of the few studies testing its accuracy for ongoing range shifts found that niche models developed at a broad scale cannot predict rapid range expansion dynamics at a fine scale (Mitikka *et al.* 2008).

25.3.4 Open questions

To summarize, we see that past and present changes in plant and animal distributions have proceeded through a process of colonization and expansion influenced by the environmental mosaic, the dispersal capabilities, and the evolutionary capacity of each species (see Sections 25.2 and 25.3.2). These idiosyncratic dynamics are generally not accounted for in niche models, nor in most mechanistic models of climate change (Section 25.3.3). It has been proposed that a combination of niche and mechanistic models would be extremely fruitful to address both large- and fine-scale spatial processes (del Barrio *et al.* 2006). However, we anticipate that these new models would not entirely solve uncertainties until a more detailed understanding of dispersal has been developed. Three major questions about dispersal are relevant to our capacity to predict future range shifts due to climate change:

- Should we account for the heterogeneity and plasticity of dispersal?
- Should we account for interactive effects between climate change, habitat fragmentation, and species interactions?
- Are dispersal effects similar at the expanding margin, contracting margin, and core of the range?

25.4 Dispersal heterogeneity

Most niche and mechanistic models tend to ignore dispersal or have unrealistic assumptions, such as complete philopatry or unlimited dispersal (Guisan and Thuiller 2005). It is becoming increasingly clear that dispersal is often restricted in space and time, and that dispersal is a plastic or condition-dependent trait that depends on individual, parental, and environmental attributes (Chapter 1). Including details about the heterogeneity and plasticity of dispersal into some predictive models would increase their realism and accuracy but also requires a substantial amount of work to understand and model the dispersal process (Engler *et al.* 2009). Here, we describe four aspects of the dispersal process that are relevant to this perspective and highlight recent results showing how extra information on the heterogeneity and plasticity of dispersal can make large differences to predicted migration rates and patterns of range shifts in changing climates.

25.4.1 Between-individual heterogeneity

An important feature of dispersal kernels for many plants and animals is their leptokurtic distribution; i.e. the presence of many short-distance dispersal events (relative to the mean) together with a tail of long-distance dispersal (LDD) events (relative to the mean; Chapter 15). The occurrence of a tail of LDD events strongly determines the speed at which species track their climate niche, as demonstrated in Chapter 29. Unfortunately, methodological and statistical problems render the detection of LDD problematic, though there are accumulating species for which dispersal kernels are known with certainty (Chapter 15).

Several hypotheses have been raised to explain the long-tailed distribution of dispersal distances, such as variable environmental conditions, seed polymorphism in plants, the existence of multi-vector dispersal, or behavioural polymorphism in animals (e.g. Higgins and Richardson 1999; Skalski and Gilliam 2000). In general, heterogeneity in dispersal abilities among propagules or individuals is likely to contribute significantly to the dispersal kernels of many species. In animals, dispersal kernels comprise a heterogeneous pool of individuals that move for distinct reasons and differ from each other in size, morphology and many other attributes (Chapter 17). For example, Kinlan and Gaines (2003) compared propagule dispersal across marine and terrestrial taxa and observed that many terrestrial species have flexible, adult-mediated dispersal. Individual variation has also been emphasized by field studies of movement distances in the lesser marsh grasshopper, *Chorthippus albomarginatus* (Walters *et al.* 2006), as well as in fishes (Fraser *et al.* 2001) and salamanders (Lowe 2009), and in a recent analysis of intraspecific variation in dispersal distances for butterflies (Stevens *et al.* 2010). According to Fraser *et al.* (2001), differences in behavioural or physiological incentives to move are generally large and stable enough in animal populations to generate a leptokurtic dispersal kernel. Such inter-individual variation in dispersal rates may facilitate colonization and range expansion (Duckworth 2008; Clobert *et al.* 2009), and should be better accounted for in predictive models. This could be done with a range of demographic models allowing inter-individual variation in dispersal such as metapopulation or agent-based approaches (see the model in Clobert *et al.* 2009).

25.4.2 Flexible changes in dispersal

Dispersal behaviour can also vary over time within individuals depending on climate conditions. Climate conditions can constrain movement abilities through direct effects on the drift of propagules or temporal opportunities for movement in active dispersers, and climate conditions have more complex physiological effects on energy metabolism and movement capacities. We refer to these as flexible changes in dispersal abilities because of their short-term duration and reversibility. Few studies have attempted to link directly climatic conditions and dispersal abilities, though recent methodological and conceptual advances in the field of movement ecology provide some promising results (Nathan *et al.* 2008). For example, detailed mechanistic models of LDD in plants have emphasized the importance of wind turbulence and micrometeorological patterns for seed dispersal (Thompson and Katul 2008; Nathan *et al.* 2002). Air flow must be uplifting with a sufficient coherence in space and time to promote long-distance seed dispersal. Strong, direct effects from climate change on seed dispersal distances are therefore expected. For example, Kuparinen *et al.* (2009) predicted that a temperature increase of + 3 °C would promote an average increase of the rate of LDD in various plants from boreal forest ecosystems. The generality of these predictions in other ecosystems and species remains untested.

Similar mechanistic approaches to describing dispersal are more difficult to implement in species that disperse actively (Walls *et al.* 2005), but might be relevant to predict the effects of climate change on dispersal in marine and terrestrial organisms that disperse more or less passively. For example, dispersal in some flying terrestrial arthropods can be explained by the interplay between passive movements in lifting winds and active movements during take-off and flight phases (Reynolds and Reynolds 2009). It is also not uncommon that insect species exhibit occasional LDD as well as opportunistic range expansions during warm events (e.g. Showers *et al.* 1989). In one species of grasshoppers, daily movements are enhanced in warmer climates, which should result in faster spread rates under climate warming (Walters *et al.* 2006). Several bird and mammal species are also capable of long-distance dispersal movements in response to changes in climate conditions (e.g. Figuerola 2007). State-space models of animal movements and mechanistic models of dispersal will provide fruitful avenues to quantify these direct effects of climatic conditions on dispersal distances and unravel consequences for range dynamics (Nathan *et al.* 2008).

25.4.3 Plastic changes in dispersal

Climate change will most likely cause plastic changes in many life-history traits because of changes in climatic factors that influence life-history decisions (Parmesan 2006). An often-studied source of plasticity in natural populations is temperature (Angilletta 2009). Thermal plasticity has been described for a variety of morphological and demographic traits, and contributes importantly to adaptation to climate change. In contrast, we know relatively little about the thermal plasticity of dispersal itself.

In animals, some levels of thermal plasticity of dispersal are expected because thermal conditions impacts locomotion and pay-offs of dispersal. Field studies of two endothermic bird species (Møller *et al.* 2006, Chapter 27) and one ectothermic lizard species (Massot *et al.* 2008; Lepetz *et al.* 2009) have documented long-term trends in dispersal tendencies that could be explained by thermal plasticity (see Table 25.2). For example, in the case of the common lizard (*Zootoca vivipara*), late spring temperatures in the field, when young of the year are still *in utero*, have increased concurrently with a decrease of natal dispersal probability. This pattern was explained by maternal effects or indirect effects of climate change on habitat quality and food availability (Massot *et al.* 2008). In fact, despite substantial evidence that developmental plasticity for dispersal occurs in many species (Chapter 1), there is only one truly convincing example to-date of thermal effects on the ontogeny of dispersal (Table 25.2). In a laboratory study, Bonte *et al.* (2008) demonstrated that thermal conditions experienced during juvenile development can influence the adult dispersal of the spider *Erigone atra*. Like many other spiders, this species can produce silk balloons to disperse over long distances up to several hundred metres (i.e. ballooning strategy) or take-off with a silk thread attached to the ground to disperse over shorter distances (i.e. rappelling strategy). Short-distance rappelling was less frequent at low temperatures and long-distance ballooning was less frequent at highest temperatures, in accordance with seasonal changes in dispersal observed in the field. Since developmental temperature often acts as a cue for habitat quality and time of the year, we speculate that adaptive thermal plasticity for dispersal will be discovered in a variety of other ectothermic species in the near future.

25.4.4 Evolutionary changes in dispersal

Several recent studies have emphasized the importance of evolutionary processes in populations expanding or contracting in space, and have demonstrated that a rapid evolution can be observed in populations at spatial disequilibrium (Phillips *et al.* 2010, Chapter 26). For example, evolution towards higher dispersal propensities will result in accelerated range expansion poleward and upward. However, these evolutionary and ecological responses also depend on other factors, such as proximate cues of dispersal, habitat turnover, costs of dispersal, life history, and the degree of local adaptation (Chapter 26). Substantial genetic variation exists for locomotor traits and these traits can evolve rapidly in the laboratory (reviewed in Feder *et al.* 2010). There are also now few examples from wing-monomorphic and wind-dimorphic insects that evolutionary changes in dispersal and habitat selection have facilitated range expansion in response to climate change (Table 25.2). However, it remains unclear which factors (climate change, population density, habitat quality, etc.) drive the evolution of dispersal in these expanding populations. Future studies should try to address the importance of these evolutionary changes in other species and also address the critical issue of adaptation in range contracting margins.

Table 25.2 Plastic and evolutionary changes in dispersal in response to climate conditions.

Study species	Study design	Pattern reported in the study	Reference
Plastic changes in dispersal			
Artic terns	Field study	Natal dispersal distance increases with temperature and humidity the year of hatching and increases with temperature and NAO# the year of breeding. Breeding dispersal distance decreases with temperature and increases with NAO and SOI* during the second breeding year	Møller *et al.*, 2006
Dipper	Field study	Immigration increases after a warm winter	Chapter 27
House sparrows	Field study	Natal dispersal increases with spring temperature in low-quality habitats Natal dispersal is independent of spring temperature in high-quality habitats	Chapter 27
Common lizards	Field study	Natal dispersal probability declines with rising temperature during embryogenesis but increases with temperature during dispersal Lower immigration into the study site with rising temperatures	Massot *et al.* 2008, Lepetz *et al.* 2009
Spiders	Laboratory study	Dispersal investment is stronger at intermediate developmental temperatures (20–25 °C), rappelling behaviour is less frequent at 15 °C and ballooning behaviour is less common at 30 °C	Bonte *et al.* 2008
Evolutionary changes in dispersal			
Speckled wood butterfly	Common garden	Evolution of a flight morphology after a recent range expansion	Hill *et al.* 1999a
Butterflies	Field study and common garden	Evolution of a generalist habitat selection after a recent range expansion	Thomas *et al.* 2001
Bush crickets	Field study	Increase in the frequency of macropterous individuals after a recent range expansion	Thomas *et al.* 2001
Bush crickets	Field study and common garden	Evolution of a flight morphology in species subject to a recent range expansion	Simmons and Thomas 2004
Speckled wood butterfly	Common garden	Evolution of a flight morphology with reduced fecundity in marginal populations	Hughes *et al.* 2003
Speckled wood butterfly	Common garden	Evolution of a flight morphology along a transect from the core to the expanding range margins	Hughes *et al.* 2007

North Atlantic Oscillation, * Southern Oscillation Index

25.5 Interactions with fragmentation and biotic interactions

The lessons drawn from studies of colonization in invading species is that there exists some degree of dependence of spatial range shifts upon species traits (e.g. dispersal ability, life history), landscape features (e.g. patchiness, non-climatic factors), and biotic interactions (e.g. predator and parasite release). Similar mechanisms may be important to consider in the context of spatial range shifts caused by climate change.

25.5.1 Dispersal and habitat fragmentation

Scaling down the approaches of global niche and mechanistic models to predict patterns at a regional or a local scale is often problematic because the dispersal of propagules is constrained at local scales by fine-grained habitat fragmentation (Mitikka *et al.* 2008). Attempts to include details about the landscape structure using metapopulation models have shown that habitat fragmentation can act as a barrier to colonization of newly available habitat patches and slow down range expansion from climate change (Anderson *et al.* 2009). The picture may be extremely pessimistic for many species that could suffer more from climate change in fragmented and degraded habitats than in pristine habitats (Chapter 26). Coupling models and data on land use, species dispersal, and climate change would allow the investigation of these 'deadly cocktails' in greater details. Also, climate change has direct effects on the habitat: it can generate transient landscape dynamics such as large gaps at range margins, increases in the breadth of habitat use by colonizing species, or temporary habitat corridors (Hill *et al.* 1999b). Finally, it is worth mentioning that changes in habitat fragmentation could have unexpected but rapid evolutionary consequences (Chapter 26).

25.5.2 Dispersal and species interactions

A common feature of niche and mechanistic models of range shifts is that they are fitted independently to single species and therefore assume that biotic interactions are expressed sufficiently in the calculation of the niche—i.e. that the fitted niche is the realized niche, which includes the effects of biotic interactions. The error of these models may be that biotic interactions change with climate and geography. For example, climate-dependent competitive interactions and dispersal determine jointly the spatial distribution across several scales for two oak species in France and the United Kingdom (Bullock *et al.* 2000). Perspectives in this field have recently emphasized the importance of focusing on community modules of strongly interacting species (Gilman *et al.* 2010), but we have still made surprisingly little progress in our understanding of shifting communities.

There are several reasons why dispersal responses to climate change could depend on species interactions. Firstly, the capacity of a species to track shifting climate niches depends not only on dispersal capacity but also on settlement success. Settlement success is strongly influenced by interactions with resident species; for example, a strong resident competitor could importantly slow down range shifting (Caplat *et al.* 2008). Second, species of different trophic levels and body size usually differ strongly in their dispersal capacities (see Figure 25.3). Thus, range shifts could create spatial mismatches between closely interacting species, which would cause pronounced ecological effects that depend on community structure and trophic interactions (Gilman *et al.* 2010). In terrestrial ecosystems, phytophagous insects have faster spread rates than plants (see Figure 25.3) and could be importantly constrained by the spatial spread of their host plants (Schweiger *et al.* 2008). Third, dispersal responses could modify co-evolutionary dynamics among species leading to novel community structures and the loss of many species and important ecosystem properties (Ackerly 2003). For example, plant–insect mutualisms are likely to be at great risks if one species is more likely to disperse than the other (but see Leotard *et al.* 2009).

25.6 Dispersal across the species' range

There is no reason to expect that species will react to climate change uniformly within their geographic range (Parmesan *et al.* 2005; Reich and Oleksyn 2008). Indeed, responses to climate change should vary across the species range depending on interactions between local climate conditions, dispersal propensities and local adaptation. For example, in temperate areas, climate warming in narrowly distributed species will lead to high risks of total extinction, while climate warming will be associated with high risks of local extinctions at warm margins in widely distributed species (Figure 25.4). For narrowly distributed species limited by thermal conditions in cold climates (Figure 25.4B), dispersal should allow range expansion, but predicting future species ranges in intermediate and

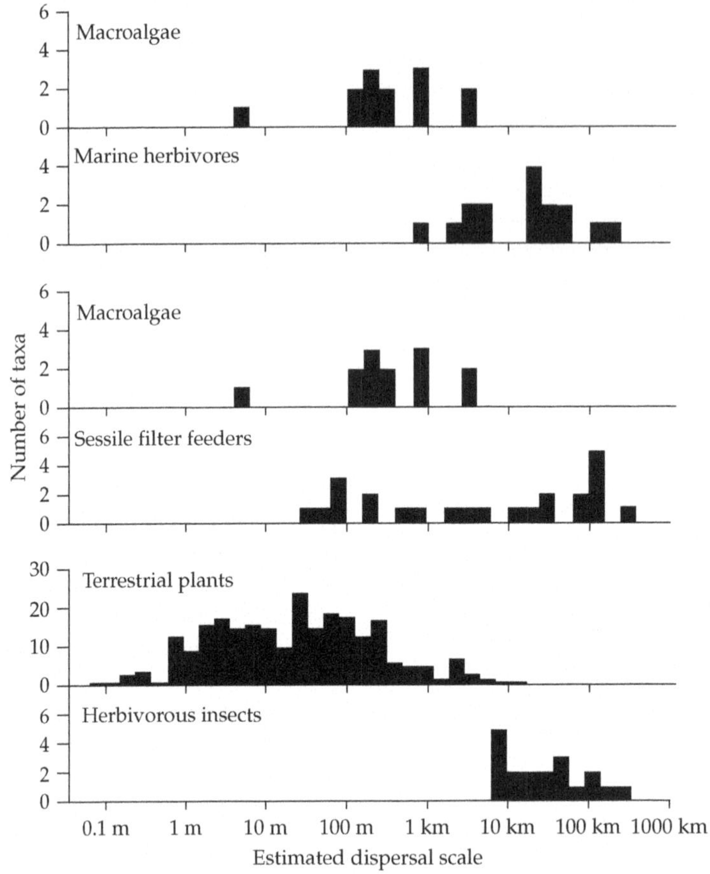

Figure 25.3 Species of interacting functional groups from marine and terrestrial ecosystems display very different dispersal scales. Dispersal scales were tabulated from genetic isolation-by-distance models except for terrestrial plants where direct estimates are shown. Upper: estimates for sessile macroalgae versus their herbivorous invertebrates and fishes; centre: estimates for macroalgae and their sessile filter feeders, which use macroalgae as substrate for fixation; lower: estimates for terrestrial plants and their phytophagous insects (Kinlan and Gaines 2003, used with permission from the Ecological Society of America).

warm regions is more complex. Indeed, the range limits of a species may be caused by another constraint than thermal conditions (e.g. inter-specific competition, predators, parasitism). When this constraint is not relaxed by climate warming, dispersal will not be efficient at shifting the species range. We therefore predict no expanding species range in intermediate regimes and species extinction in warm thermal regimes whatever the dispersal propensity (Figure 25.4B). In other words, the species' persistence in warm thermal regimes will be only possible with a concurrent change in the distributional limiting factors.

In widely distributed species, consequences of global change will probably more importantly depend on prevailing patterns of local adaptation to climate conditions (Davis and Shaw 2001). For example, climate warming shifts thermal regimes poleward and upward in temperate areas, resulting in unsuitable areas at warm margins and new suitable areas at cold margins. Climate warming also changes thermal regimes in other parts of species range (core area or mid-altitude in an altitudinal gradient). Thus, provided the species has adapted locally to different thermal regimes throughout its

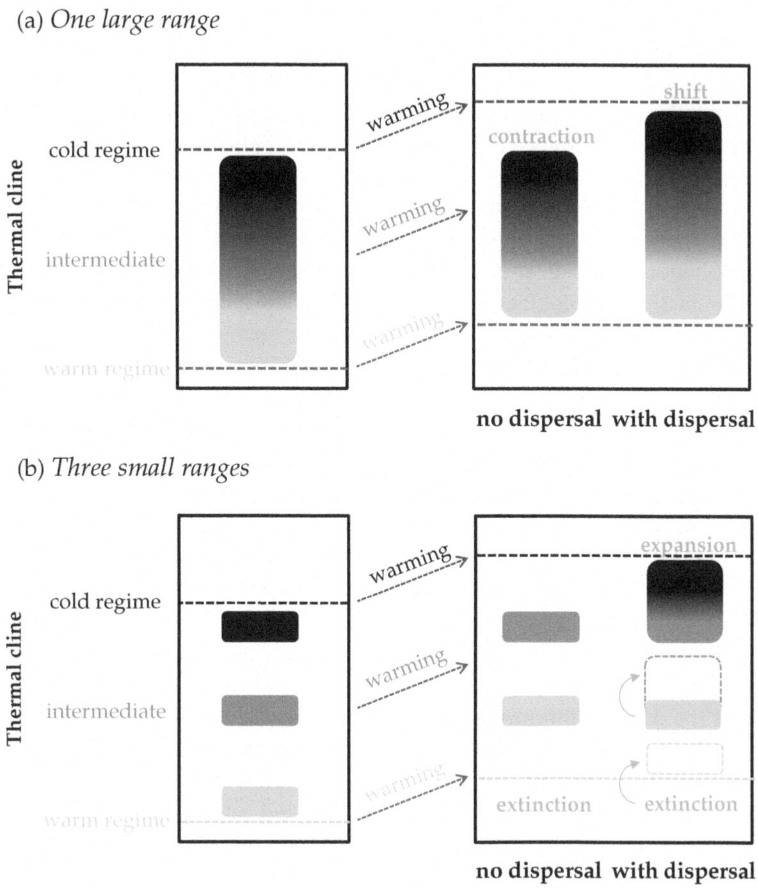

Figure 25.4 Schematic representation of the responses of species to climate warming across a thermal cline (altitudinal or latitudinal gradient). The warming effects depends on the interplay between the initial size of the species range, the initial local thermal regime, and the dispersal capacity. The thermal cline may depends on a combination of latitudinal and altitudinal factors. Dispersal can be a constant property of a given species or a changing trait with climate warming (see Section 25.4). The greyscale corresponds to thermal conditions within the species ranges (light grey: warm, dark grey: intermediate, black: cold). The right panels show the expected range changes induced by a spatial shift of the thermal cline due to climate warming. (a) Expected range changes for a species with a large initial range. (b) Expected range changes for three allopatric species with a small initial range. Solid arrows indicate expected range shift or expansion when other constraints on species range than climate are relatively unimportant.

range (e.g. Reich and Oleksyn 2008), climate warming should lead to deterioration of environmental conditions across all existing localities (Figure 25.4A). Genetic adaptation and dispersal could both facilitate range expansion at the warm margins and accelerate range contraction at the cold margins. This is because asymmetric dispersal during a range shift increases genetic variability at the leading edge, and also transfers genes from the trailing to the leading edge of the migrating front. On the contrary, fewer migrations can be expected around the cold margins where adaptation may even be slowed down by genetic dispersal from the core (Davis and Shaw 2001). This shows that ecological and evolutionary responses at expanding or contracting range margins are different and may depend on processes occurring within the species range. Yet it remains unknown whether adaptation will be fast enough relative to the current speed of climate change (Sexton *et al.* 2009).

In the same manner, we expect that ecological responses occurring at the cold, expanding, and warm, contracting, margins will involve two different types of dispersal processes. Range dynamics at the warm margins involve the invasion of newly available habitats. Recent studies have emphasized that the relative importance of behavioural flexibility and dispersal syndromes may be crucial to the invasion success (e.g. Duckworth and Badyaev 2007). There is also accumulating evidence that some phenotypic traits and life-history attributes may facilitate colonization and range expansion (Duckworth and Badyaev 2007; Hanski *et al.* 2004). In the common lizard, for example, individuals differ in social personality (i.e. behavioural attraction towards conspecifics), and social personality predicts whether some individuals disperse to join high-density rather than low-density patches (Cote and Clobert 2007). Thus, the colonization dynamics at cold margins will most likely be governed by the demographic structure of source populations. When new habitats become available, the capacity of a species to expand or invade will depend upon the populations structure at the margin of the distribution range (presence of colonizing types) and the dominant dispersal cause acting in those populations (Clobert *et al.* 2009). On the contrary, range dynamics at warm margins involve habitat degradation within and between local populations, which may select against dispersal. A decreased dispersal capacity is known to evolve in species occurring in isolated and highly fragmented landscapes (Colas *et al.* 2004) and dispersal inhibition in populations at warm margins could lead to an evolutionary trap (Massot *et al.* 2008).

25.7 Synthesis

Our review shows that the burgeoning fields of climate ecology and movement ecology have shed new light on our understanding of dispersal behaviour and spatial range dynamics. Our compilation of various data and models across a wide range of animal and plant species indicates that three important factors should be accounted for in the future to obtain more accurate prediction of ecological effects of climate change:

1. Climate conditions can induce flexible or plastic changes in dispersal behaviour, which may speed up or slow down range expansion. We need more data to understand the generality and relative importance of behavioural plasticity, but models are now available to describe links between climate conditions and dispersal processes.
2. Even though many species are sensitive to thermal conditions, their demographic responses to climate change will depend on other climate factors, such as humidity, and a complex interplay with the effects of landscape structure and biotic interactions with other species. So far, there has been a tendency to confound the realized niche of a species with its climate niche, which is in contradiction with theory and data from biogeography and community ecology. We believe this to be a major challenge for a future generation of climate-niche models.
3. The range of a species is usually highly heterogeneous and our understanding of ecological and evolutionary dynamics across a species range suggests that different processes may occur in expanding than in contracting margins. More attention should be paid to examining the contrasted causes and consequences of dispersal in expanding and contracting margins.

References

Ackerly, D.D. (2003). 'Community assembly, niche conservatism, and adaptive evolution in changing environments', *International Journal of Plant Sciences* **164**: S165–84.

Anderson, B.J., Akcakaya, H.R., Araujo, M.B. *et al.* (2009). 'Dynamics of range margins for metapopulations under climate change', *Proceedings of the Royal Society* **276**: 1415–20.

Angilletta, M.J. (2009). *Thermal adaptation: a theoretical and empirical synthesis.* Oxford University Press, Oxford.

Beaugrand, G., Reid, P.C., Ibanez, F. *et al.* (2002). 'Reorganization of North Atlantic marine copepod biodiversity and climate', *Science* **296**: 1692–4.

Beckage, B., Osborne, B., Gavin, D.G. *et al.* (2008). 'A rapid upward shift of a forest ecotone during 40 years of warming in the Green Mountains of Vermont', *Proceedings of the National Academy of Sciences of the USA* **105**: 4197–202.

Berg, M.P., Kiers, E.T., Driessen, G. *et al.* (2010). 'Adapt or disperse: understanding species persistence in a changing world', *Global Change Biology* **16**: 587–98.

Birks, H.J.B. (1989). 'Holocene isochrone maps and patterns of tree-spreading in the British Isles', *Journal of Biogeography* **16**: 503–40.

Bonte, D., Travis, J.M.J., De Clercq, N. *et al.* (2008). 'Thermal conditions during juvenile development affect adult dispersal in a spider', *Proceedings of the National Academy of Sciences USA* **105**: 17000–5.

Botkin, D.B., Saxe, H., Araujo, M.B. *et al.* (2007). 'Forecasting the effects of global warming on biodiversity', *Bioscience* **57**: 227–36.

Bradshaw, R.H.W. (1999). 'Spatial responses of animals to climate change during the Quaternary', *Ecological Bulletins* **47**: 16–21.

Brewer, S., Cheddadi, R., de Beaulieu, J.L. *et al.* (2002). 'The spread of deciduous Quercus throughout Europe since the last glacial period', *Forest Ecology and Management* **156**: 27–48.

Buckley, L.B. (2008). 'Linking traits to energetics and population dynamics to predict lizard ranges in changing environments', *American Naturalist* **171**: E1–19.

Bullock, J.M., Edwards, R.J., Carey, P.D. *et al.* (2000). 'Geographical separation of two *Ulex* species at three spatial scales: does competition limit species' ranges. *Ecography* **23**: 257–71.

Bullock, J.M., Kenward, R.E., Hails, R.S. (2002). *Dispersal ecology*, Blackwell Publishing, Oxford.

Cain, M.L., Damman, H., Muir, A. (1998). 'Seed dispersal and the Holocene migration of woodland herbs', *Ecological Monographs* **68**: 325–47.

Caplat, P., Anand, M., Bauch, C. (2008). 'Interactions between climate change, competition, dispersal, and disturbances in a tree migration model', *Theoretical Ecology* **1**: 209–20.

Case, T.J. Taper, M.L. (2000). 'Interspecific competition, environmental gradients, gene flow, and the coevolution of species' borders', *American Naturalist* **155**: 583–605.

Clark, J.S. (1998). 'Why trees migrate so fast: confronting theory with dispersal biology and the paleorecord', *American Naturalist* **152**: 204–24.

Clobert, J., Danchin, E., Dhondt, A.A. *et al.* (2001). *Dispersal*. Oxford University Press, Oxford.

Clobert, J., Le Galliard, J.-F., Cote, J. *et al.* (2009). 'Informed dispersal, heterogeneity in animal dispersal syndromes and the dynamics of spatially structured populations', *Ecology Letters* **12**: 197–209.

Colas, B., Olivieri, I., Riba, M. (1997). '*Centaurea corymbosa*, a cliff-dwelling species tottering on the brink of extinction: A demographic and genetic study', *Proceedings of the National Academy of Sciences of the USA* **94**: 3471–6.

Colas, B., Thomas, C.D. Hanski, I. (2004). 'Adaptive responses to landscape disturbances: empirical evidence', in R. Ferrière, U. Dieckmann, U.D. Couvet (eds),. *Evolutionary Conservation Biology*. Cambridge University Press, Cambridge.

Cote, J., Clobert, J. (2007). 'Social information and emigration: lessons from immigrants', *Ecology Letters* **10**: 411–17.

Crozier, L. (2003). 'Winter warming facilitates range expansion: cold tolerance of the butterfly Atalopedes campestris', *Oecologia* **135**: 648–56.

Crozier, L. (2004). 'Warmer winters drive butterfly range expansion by increasing survivorship', *Ecology* **85**: 231–41.

Crozier, L., Dwyer, G. (2006). 'Combining population-dynamic and ecophysiological models to predict climate-induced insect range shifts', *American Naturalist* **167**: 853–66.

Davis, M.B., Shaw, R.G. (2001). 'Range shifts and adaptive responses to Quaternary climate change', *Science* **292**: 673–9.

del Barrio, G., Harrison, P.A., Berry, P.M. *et al.* (2006). 'Integrating multiple modelling approaches to predict the potential impacts of climate change on species' distributions in contrasting regions: comparison and implications for policy', *Environmental Science and Policy* **9**: 129–47.

Delcourt, H.R., Delcourt, P.A. (1988). 'Quaternary landscape ecology: Relevant scales in space and time', *Landscape Ecology* **2**: 23–44.

Denno, R.F., Roderick, G.K., Olsmtead, K.L. *et al.* (1991). 'Density-related migration in planthoppers (Homoptera: Delphacidae): the role of habitat persistence', *American Naturalist* **138**: 1513–41.

Devictor, V., Julliard, R., Couvet, D. *et al.* (2008). 'Birds are tracking climate warming, but not fast enough', *Proceedings of the Royal Society B* **275**: 2743–8.

Duckworth, R.A. (2008). 'Adaptive dispersal strategies and the dynamics of a range expansion', *American Naturalist* **172**: S4–17.

Duckworth, R.A. Badyaev, A.V. (2007). 'Coupling of dispersal and aggression facilitates the rapid range expansion of a passerine bird', *Proceedings of the National Academy of Sciences USA* **104**: 15017–22.

Dytham, C. (2009). 'Evolved dispersal strategies at range margins', *Proceedings of the Royal Society B* **276**: 1407–13.

Engler, R., Randin, C.F., Vittoz, P. *et al.* (2009). 'Predicting future distributions of mountain plants under climate change: does dispersal capacity matter?', *Ecography* **32**: 34–45.

Feder, M.E., Garland, T., Marden, J.H. *et al.* (2010). 'Locomotion in response to shifting climate zones: not so fast', *Annual Review of Physiology* **72**: 167–190.

Figuerola, J. (2007). 'Climate and dispersal: black-winged stilts disperse further in dry springs', *PLoS One* **2**: 4.

Fitzpatrick, M.C., Gove, A.D., Sanders, N. J. *et al.* (2008). 'Climate change, plant migration, and range collapse in a global biodiversity hotspot: the Banksia (Proteaceae) of Western Australia', *Global Change Biology* **14**: 1337–52.

Fraser, D.F., Gilliam, J.F., Daley, M.J. *et al.* (2001). 'Explaining leptokurtic movement distributions: Intrapopulation variation in boldness and exploration', *American Naturalist* **158**: 124–35.

Gilman, S.E., Urban, M.C., Tewksbury, J.J. *et al.* (2010). 'A framework for community interactions under climate change', *Trends in Ecology and Evolution* **25**: 325–31.

Graham, R.W., Lundelius, E.L., Graham, M.A. *et al.* (1996). 'Spatial response of mammals to late quaternary environmental fluctuations', *Science* **272**: 1601–6.

Guisan, A. and Thuiller, W. (2005). 'Predicting species distribution: offering more than simple habitat models', *Ecology Letters* **8**: 993–1009.

Hanski, I., Eralahti, C., Kankare, M. *et al.* (2004). 'Variation in migration propensity among individuals maintained by landscape structure', *Ecology Letters* **7**: 958–66.

Hewitt, G. (2000). 'The genetic legacy of the Quaternary ice ages', *Nature* **405**: 907–13.

Hickling, R., Roy, D.B., Hill, J K. *et al.* (2006). 'The distributions of a wide range of taxonomic groups are expanding polewards', *Global Change Biology* **12**: 450–5.

Higgins, S.I. and Richardson, D.M. (1999). 'Predicting plant migration rates in a changing world: The role of long-distance dispersal', *American Naturalist* **153**: 464–75.

Hill, J.K., Thomas, C.D., Blakeley, D.S. (1999). 'Evolution of flight morphology in a butterfly that has recently expanded its geographic range', *Oecologia* **121**: 165–70.

Hill, J.K., Thomas, C.D., Huntley, B. (1999). 'Climate and habitat availability determine 20th century changes in a butterfly's range margin', *Proceedings of the Royal Society B* **266**: 1197–206.

Hochkirch, A. and Damerau, M. (2009). 'Rapid range expansion of a wing-dimorphic bush-cricket after the 2003 climatic anomaly', *Biological Journal of the Linnean Society* **97**: 118–27.

Holt, R.D., Gomulkiewicz, R. (1997). 'The evolution of species' niches: a population dynamic perspective, in H.Othmer, F.Adler, M. Lewis, J. Dallon, J, (eds) *Case Studies in Mathematical Modelling: Ecology, Physiology, and Cell Biology*. Prentice-Hall, Englewood Cliffs, New Jersey.

Hughes, C.L., Dytham, C., Hill, J.K. (2007). 'Modelling and analysing evolution of dispersal in populations at expanding range boundaries', *Ecological Entomology* **32**: 437–45.

Hughes, C.L., Hill, J.K., Dytham, C. (2003). 'Evolutionary trade-offs between reproduction and dispersal in populations at expanding range boundaries', *Biology Letters* **270**: S147–150.

Hutchinson, R.E. (1957). 'Concluding remarks', *Cold Spring Harbor. Symposium Quantitative Biology* **22**: 415–27.

IPCC (2007). 'Climate Change 2007: Synthesis Report. Summary for Policymakers', in R.K. Pachauri, A. Reisinger, (eds), *Contribution of Working Groups I, II and III to the Fourth Assessment Report of the Intergovernmental Panel on Climate Change.* Geneva, Switzerland. Cambridge University Press, Cambridge

Iverson, L.R., Schwartz, M.W., Prasad, A.M. (2004). 'How fast and far might tree species migrate in the eastern United States due to climate change?', *Global Ecology and Biogeography* **13**: 209–19.

Kearney, M. and Porter, W. (2009). 'Mechanistic niche modelling: combining physiological and spatial data to predict species' ranges', *Ecology Letters* **12**: 334–50.

Kinlan, B.P. and Gaines, S.D. (2003). 'Propagule dispersal in marine and terrestrial environments: A community perspective', *Ecology* **84**: 2007–20.

Kuparinen, A., Katul, G., Nathan, R. *et al.* (2009). 'Increases in air temperature can promote wind-driven dispersal and spread of plants', *Proceedings of the Royal Society B* **276**: 3081–7.

Leotard, G., Debout, G., Dalecky, A., *et al.* (2009). 'Range expansion drives dispersal evolution in an equatorial three-species symbiosis', *Plos One* 4(4), e5377.

Lepetz, V., Massot, M., Chaine, A. *et al.* (2009). 'Climate warming and the evolution of morphotypes in a reptile', *Global Change Biology* **15**: 454–66.

Lescop-Sinclair, K., Payette, S. (1995). 'Recent advance of the arctic treeline along the eastern coast of Hudson Bay', *Journal of Ecology* **83**: 929–36.

Loarie, S.R., Duffy, P.B., Hamilton, H. *et al.* (2009). 'The velocity of climate change. *Nature* **462**: 1052-U111.

Lowe, W.H. (2009). 'What drives long-distance dispersal? A test of theoretical predictions', *Ecology* **90**: 1456–62.

Lyons, S.K. (2003). 'A quantitative assessment of the range shifts of Pleistocene mammals', *Journal of Mammalogy* **84**: 385–402.

Lyons, S.K., Wagner, P.J., Dzikiewicz, K. (2010). 'Ecological correlates of range shifts of Late Pleistocene mammals', *Philosophical Transactions of the Royal Society B* **365**: 3681–93.

Massot, M., Clobert, J., Ferrière, R. (2008). 'Climate warming, dispersal inhibition and extinction risk', *Global Change Biology* **14**: 1–9.

McLachlan, J.S., Clark, J.S., Manos, P.S. (2005). 'Molecular indicators of tree migration capacity under rapid climate change', *Ecology* **86**: 2088–98.

Mitikka, V., Heikkinen, R.K., Luoto, M. *et al.* (2008). 'Predicting range expansion of the map butterfly in Northern Europe using bioclimatic models', *Biodiversity and Conservation* **17**: 623–41.

Møller, A.P., Flensted-Jensen, E., Mardal, W. (2006). 'Dispersal and climate change: a case study of the Arctic tern *Sterna paradisaea*', *Global Change Biology* **12**: 2005–13.

Morin, X. and Thuiller, W. (2009). 'Comparing niche- and process-based models to reduce prediction uncertainty in species range shifts under climate change', *Ecology* **90**: 1301–13.

Nathan, R., Getz, W.M., Revilla, E. *et al.* (2008). 'A movement ecology paradigm for unifying organismal movement research', *Proceedings of the National Academy of Sciences of the USA* **105**: 19052–9.

Nathan, R., Katul, G.G., Horn, H.S. *et al.* (2002). 'Mechanisms of long-distance dispersal of seeds by wind', *Nature* **418**: 409–13.

Parmesan, C. (2006). 'Ecological and evolutionary responses to recent climate change', *Annual Review of Ecology and Systematics* **37**: 637–69.

Parmesan, C., Gaines, S., Gonzalez, L. *et al.* (2005). 'Empirical perspectives on species borders: from traditional biogeography to global change', *Oikos* **108**: 58–75.

Parmesan, C., Ryrholm, N., Stefanescu, C. *et al.* (1999). 'Poleward shifts in geographical ranges of butterfly species associated with regional warming', *Nature* **399**: 579–83.

Parmesan, C. and Yohe, G. (2003). 'A globally coherent fingerprint of climate change impacts across natural systems', *Nature* **421**: 37–42.

Parolo, G. and Rossi, G. (2008). 'Upward migration of vascular plants following a climate warming trend in the Alps', *Basic and Applied Ecology* **9**: 100–7.

Phillips, B.L., Brown, G.P., Shine, R. (2010). 'Life-history evolution in range-shifting populations', *Ecology* **91**: 1617–27.

Pöyry, J., Luoto, M., Heikkinen, R.K. *et al.* (2009). 'Species traits explain recent range shifts of Finnish butterflies', *Global Change Biology* **15**: 732–43.

Provan, J. and Bennett, K.D. (2008). 'Phylogeographic insights into cryptic glacial refugia', *Trends in Ecology and Evolution* **23**: 564–71.

Reich, P.B. and Oleksyn, J. (2008). 'Climate warming will reduce growth and survival of Scots pine except in the far north', *Ecology Letters* **11**: 588–97.

Reynolds, A.M. and Reynolds, D.R. (2009). 'Aphid aerial density profiles are consistent with turbulent advection amplifying flight behaviours: abandoning the epithet 'passive'', *Proceedings of the Royal Society B* **276**: 137–43.

Ronce, O., Perret, F., Olivieri, I. (2000). 'Landscape dynamics and evolution of colonizer syndromes: interactions between reproductive effort and dispersal in a metapopulation. *Evolutionary Ecology* **14**: 233–60.

Roy, K., Jablonski, D., Valentine, J.W. (2001). 'Climate change, species range limits and body size in marine bivalves', *Ecology Letters* **4**: 366–70.

Schweiger, O., Settele, J., Kudrna, O. *et al.* (2008). 'Climate change can cause spatial mismatch of trophically interacting species', *Ecology* **89**: 3472–9.

Sexton, J. P., McIntyre, P. J., Angert, A. L. and Rice, K. J. (2009). 'Evolution and ecology of species range limits', *Annual Review of Ecology Evolution and Systematics* **40**: 415–36.

Showers, W.B., Whitford, F., Smelser, R.B. *et al.* (1989). 'Direct evidence for meteorologically driven long-range dispersal of an economically important moth', *Ecology* **70**: 987–92.

Simmons, A.D. and Thomas, C.D. (2004). 'Changes in dispersal during species' range expansions', *American Naturalist* **164**: 378–95.

Sinervo, B., Mendez-de-la-Cruz, F., Miles, D.B. *et al.* (2010). 'Erosion of lizard diversity by climate change and altered thermal niches', *Science* **328**: 894–9.

Skalski, G.T. and Gilliam, J.F. (2000). 'Modeling diffusive spread in a heterogeneous population: a movement study with stream fish', *Ecology* **81**: 1685–1700.

Stevens, V. M., Pavoine, S., Baguette, M. (2010). 'Variation within and between closely related species uncovers high intra-specific variability in dispersal', *PLoS One* **5 (6)**: e11123

Svenning, J.C. and Skov, F. (2004). 'Limited filling of the potential range in European tree species', *Ecology Letters* **7**: 565–73.

Taberlet, P., Fumagalli, L., Wust-Saucy, A.G. *et al.* (1998). 'Comparative phylogeography and postglacial colonization routes in Europe', *Molecular Ecology* **7**: 453–64.

Thomas, C.D., Bodsworth, E.J., Wilson, R.J. *et al.* (2001). 'Ecological and evolutionary processes at expanding range margins', *Nature* **411**: 577–81.

Thomas, C.D., Cameron, A., Green, R.E. *et al.* (2004). 'Extinction risks from climate change', *Nature* **427**: 145–8.

Thompson, S. and Katul, G. (2008). 'Plant propagation fronts and wind dispersal: An analytical model to upscale from seconds to decades using superstatistics', *American Naturalist* **171**: 468–79.

Thuiller, W., Lavorel, S., Araujo, M.B. *et al.* (2005). 'Climate change threats to plant diversity in Europe', *Proceedings of the National Academy of Sciences of the USA* **102**: 8245–50.

Ursenbacher, S., Carlsson, M., Helfer, V. *et al.* (2006). 'Phylogeography and Pleistocene refugia of the adder (*Vipera berus*) as inferred from mitochondrial DNA sequence data', *Molecular Ecology* **15**: 3425–37.

Walls, S.S., Kenward, R.E., Holloway, G.J. (2005). 'Weather to disperse? Evidence that climatic conditions influence vertebrate dispersal', *Journal of Animal Ecology* **74**: 190–7.

Walters, R.J., Hassall, M., Telfer, M.G. *et al.* (2006). 'Modelling dispersal of a temperate insect in a changing climate', *Proceedings of the Royal Society B* **273**: 2017–23.

Watkinson, A.R. and Gill, J.A. (2002). 'Climate change and dispersal', in J.M. Bullock, R.E. Kenward, R.S. Hails, (eds), *Dispersal ecology.* Blackwell Publishing, Oxford.

Wiens, J.A., Stenseth, N.C., Van Horne, B. *et al.* (1993). 'Ecological mechanisms and landscape ecology', *Oikos* **66**: 1–12.

CHAPTER 26

Dispersal and climate change: a review of theory

Justin M. J. Travis and Calvin Dytham

26.1 Introduction

The potential responses of species to climate change can be characterized as 'adapt, migrate or die'. Species will either adapt their climate tolerances within their current ranges to the new set of environmental conditions, or shift their biogeographic range to areas of newly-suitable climate or, if they are unable to do either, they will become extinct. A species' dispersal characteristics will play a major role in determining both its potential to migrate and its ability to adapt. The rate at which newly-available habitat beyond a species' existing range is colonized will depend upon dispersal as will the gene flow between local populations that will be crucial in facilitating adaptation. Additionally, as species shift their ranges, there will often be very different selection pressures exerted on dispersal, especially at an expanding range margin. While all the same selective forces that act on dispersal in stationary ranges including kin competition and inbreeding depression (Chapters 1, 2, 3) may still play a role, a new selective force that favours those dispersal strategies which maximize the likelihood that some of an individual's descendants remain at an expanding margin, will often take precedence (Travis and Dytham 2002; see Shine *et al.* 2011 for an interesting recent perspective). We can expect dispersal to evolve rapidly at expanding margins and this will have implications both for how rapidly a species is able to shift its range and for how it may subsequently adapt to changing environmental conditions.

In this chapter, we focus on how models of dispersal can improve our understanding, prediction and ultimately management of species' range shifts under environmental change. Most models of species distribution and spread represent dispersal quite crudely (Chapter 29) and we start this chapter with some thoughts on how we might integrate these with more sophisticated models of dispersal. Secondly, we consider the importance of inter-individual variability in dispersal and the role that dispersal evolution may play in range shifting. We then present an example of evolutionary entrapment that arises when species expand their ranges over fragmented landscapes. Finally, we consider potential management strategies that may be used to promote range shifting.

26.2 Incorporating dispersal into species distribution models

A recognition that under periods of climate change the realized geographical distribution of a species will often be limited by its dispersal capacity has motivated the development of a 'second generation' of species distribution models. The majority of the research effort focused on predicting how biodiversity will respond to climate change has made use of what are often termed 'climate envelope' or 'environmental niche' models (e.g. Thuiller *et al.* 2005; Araújo *et al.* 2005). These typically make use of statistical or machine-learning approaches to generate a relationship between a species' current distribution and current climate and then use this relationship to determine the spatial distribution of suitable climate for the species in the future. The limitations of this approach have been well-described

Dispersal Ecology and Evolution. First Edition. Edited by Jean Clobert *et al.*

(Heikkinen *et al.* 2006; Sinclair *et al.* 2010) and include the lack of population dynamics, interspecific interactions, and dispersal. Several recent modelling frameworks (Engler and Guisan 2009; Midgley *et al.* 2010) now link models of spatial population dynamics with statistical projections of future-suitable climate space.

Despite these advances, the treatment of dispersal is still extremely simplistic. In several cases the dispersal process is not modelled and instead a more aggregated parameter describing either the rate at which a front expands (e.g. DeVisser *et al* 2010), or patch colonisation propensity (e.g. Williams *et al.* 2008), is incorporated. BioMove (Midgley *et al.* 2010) and MigClim (Engler and Guisan 2009) are both designed to project future species ranges and both assume individuals are dispersed according to a statistical distribution. Dispersal kernels are available for many species and in these cases it is straightforward to parameterize the dispersal function of these models. However, these models treat dispersal phenomenologically and greater realism and greater understanding is likely to come from the inclusion of more complex dispersal functions that describe the mechanism of movement. One major advantage of taking a mechanistic approach is that it is less likely to suffer from problems related to extrapolation.

Sophisticated mechanistic models of seed dispersal have been developed for wind-dispersed plants (Chapter 15) and can now be coupled with dynamic population models to simulate range shifting dynamics under climate change scenarios (Jeltsch *et al.* 2008). Gaining parameter estimates becomes increasingly challenging as model complexity increases. One approach, taken by Cabral and Schurr (2010), is to fit a dynamic model including a sophisticated dispersal function to temporal data on a population's range expansion. Passively dispersed animals (see Chapter 15) can be modelled using mechanistic models similar to those used for seeds. These approaches are becoming increasingly sophisticated: including modelling of seed or larva release or when spiders jump (Bonte *et al.* 2007; Soons and Bullock 2008); and modelling of the total dispersal kernel resulting from the distribution of weather conditions under which all dispersal events occur (Skarpaas and Shea 2007).

For actively dispersing animals, and for animal-dispersed seeds, mechanistic models of dispersal are less well developed and have not been used in species distribution modelling. For many animals, an effective model of dispersal requires careful consideration of the three key phases (emigration, transition and settlement—Chapter 1). Below we outline each of these phases in turn, asking how the behavioural rules associated with each are likely to influence the rate of a species' range expansion.

Individuals are highly unlikely to have a fixed probability of emigrating, although there are exceptions (Ranius and Hedin 2001); rather, condition-dependent emigration is likely to be the norm (Chapter 1). In many species emigration rate is positively density-dependent (Bowler and Benton, 2005). This effect can be important at an expanding range front where densities will always be low, often well below the point where competition would typically drive emigration. If the probability of an individual emigrating is very low under these conditions, it could have the effect of slowing the rate of range expansion (Best *et al.* 2007). This effect may become even more pronounced in the situation where the distance that individuals disperse is dependent upon resource availability and, if the underlying population dynamics are cyclic, the outcome may even be a rate of spread that fluctuates through time (Dwyer and Morris 2006). How density-dependence in general, and especially in dispersal, interacts with spatial patterns of habitat suitability will be important in determining rates of population spread. However, with one recent exception (Pachepsky and Levine 2011), there remains a real lack of theory for this area. Further theory on the causes and consequences of context dependent dispersal during range shifting is required, and it is vital that animal dispersal modules in species distribution models begin to include this complexity.

The movement rules that an emigrant follows will, together with the structure of the landscape, determine the likelihoods of it moving different distances (Chapter 16). Due to this interaction between landscape structure and movement rules, there is unlikely to be a fixed dispersal kernel for a species. This implies that if, for example, range expansion takes a species into a region with a very different landscape structure from that where it originated then the distribution of realized dispersal distances may also be different. Using mechanistic movement rules avoids this particular extrapolation problem (although there can be other dangers if geographical variation in the movement rules exists). There is now a well developed set of mechanistic and statistical methods for 'movement ecology' (e.g. Nathan *et al.* 2008) which can be applied to dispersal (Morales *et al.* 2010). Here we highlight the potential for integrating ideas from least cost path modelling with those from stochastic movement modelling (see Palmer *et al.* 2010 for details).

Least cost path modelling has been widely used in landscape ecology and genetics to derive indices for connectivity between two points on a landscape (Adriaensen 2003; Coulon *et al.* 2004; Stevens *et al.* 2006); i.e. rather than taking Euclidean distance between the points, the effect of habitat type on the cost of movement is accounted for and the least costly route between the points is established. It is plausible that an animal moving within its home range will take the least costly route because it has good knowledge of the environment. However, this is unlikely to be the case for an organism dispersing into environments about which it has no prior knowledge. In this case an individual can only make movement decisions based on the landscape it has already passed through and that currently lies within its perceptual range. The stochastic movement simulator proposed by Palmer *et al.* (2010) incorporates a perceptual range and includes a parameter that accounts for the tendency for individuals to continue to move in roughly the same direction (movement correlation). As with the least cost path approach, the stochastic movement simulator operates on a gridded landscape and this makes it ideal for linking with existing species distribution models to incorporate functional connectivity such that the role of landscape structure in determining species range shifting behaviour can be explored.

It is important to realize that incorporating increasing realism in dispersal into the 2nd generation species distribution models is unlikely to reduce the uncertainty associated with the results. Instead, adding realism will reveal the true extent of the uncertainty that is present in our projections by introducing more sources of input uncertainty. Our own pessimistic view is that, at least for the foreseeable future, predicting the future distribution of most species will have a high level of uncertainty (Chapter 29). However, this does not mean that models coupling ecological dynamics with environmental niche models are not useful. They will show which processes have the greatest influence on the prediction and they have an important role to play in informing the development of conservation strategies that will be effective in preserving biodiversity under environmental change. We return to this issue and provide suggestions for potentially fruitful research below.

26.3 Inter-individual variability and dispersal evolution during climate change

As is demonstrated in Chapters 10–14, there can often be substantial inter-individual variability in dispersal characteristics. Some of this variability is heritable and there is thus potential for dispersal evolution. In this section we consider how dispersal is likely to come under selection pressures which are very different during climate-induced range expansions from those within a stationary range.

Using an individual-based, lattice model Travis and Dytham (2002) demonstrated that selection can favour increased emigration rate during range expansions. At the leading edge of a spreading population, those individuals (and their descendents) possessing higher emigration rates are more likely to benefit from the low densities at the front. Conversely, the descendents of less dispersive individuals are unlikely to keep pace with the front. Importantly, the evolution of elevated emigration rates can result in accelerating range shifting. The theoretical prediction of heightened dispersal during range expansion has support from a classic study on lodgepole pine (Cwynar and MacDonald, 1987) and there are now many further examples of increased dispersal ability evolving during range expansions from a range of taxa (e.g. Simmons and Thomas 2004; Darling *et al.* 2008; Léotard *et al.* 2009).

This model has been extended to explore the evolution of density dependent emigration (Travis *et al.* 2009). In a stationary range, theory suggests that the optimal emigration strategy is to remain in the natal patch until a threshold density that is close to the local carrying capacity (Hovestadt *et al.* 2010). Above this threshold density there is a gradual asymptotic increase in emigration rate. Range expansion conditions give a very different result (Travis *et al.* 2009); the strategy selected at the front is one whereby many individuals emigrate even at very low local density. Importantly, the evolution of a density dependent emigration strategy accelerates range shifting more than is possible when only a density independent strategy is allowed evolve. Burton *et al.* (2010) added two extra complexities that may be generally important in the evolution of emigration rate during range expansion. First, they assumed a trade-off between dispersal ability and investment in life-history traits that influence reproductive capacity or competitive ability and second, they explored the evolution of these traits for species invading empty space or space already occupied by a competing species. The first key result was that dispersal consistently evolves upwards on a range margin and this increase in dispersiveness comes mainly at a cost to competitive ability. Investment in reproduction sometimes increases and sometimes decreases at an expanding margin and this is likely to depend upon the shape of the trade-offs. The second important observation is that when the model species expands into already occupied space, the rate of expansion accelerates less, due to a constraint on the extent to which competitive ability can be traded-off to pay for increased dispersal ability. There is a need for much more theory both considering the range of different costs involved in dispersal (Bonte *et al.*, 2012) and exploring how inter-specific interactions may influence dispersal evolution.

In addition to influencing the evolution of emigration propensities, the novel selective environment of an expanding range margin may act to alter both movement and settlement rules. In terms of movement, our expectation would be for selection to favour straight line trajectories with a high degree of correlation. Such behaviour will tend to maximize displacement distances increasing the likelihood that an individual, and its descendents, remain at the expanding front. Phillips *et al.* (2008) demonstrated this outcome in a model developed to explain the incredible acceleration of invasion rate observed for cane toads spreading through northern Australia. The cane toad's rate of range expansion was around 5 km yr^{-1} in the initial decade following introduction but increased as it spread westwards, and the rate is now 50 km yr^{-1}. While a small part of this acceleration may be attributed to an increase in athleticism (longer legs facilitating faster movement), it is likely that it is mostly due to the evolution of straighter trajectories (Phillips *et al.* 2008). This emphasizes the need for a much greater understanding of movement rules, their heritability and their evolution.

To date, few models have explored the evolution of movement rules in the context of dispersal. However, two recent studies integrate the individual-based evolutionary modelling approach with movement methods to ask what the optimal movement strategies should be for inter-patch dispersal (Heinz and Strand 2006; Bartoń *et al.* 2009). In addition to exploring how the degree of movement correlation evolves, Bartoń *et al.* (2009) asked how much an individual should be expected to bias its movement towards a non-natal patch. Their simulation demonstrated that when movement between patches is highly risky, the evolved strategy is one where individuals rapidly bias their movement towards the closest non-natal patch. This behaviour minimizes mortality of emigrating individuals but can also result in a poorly connected habitat network. When inter-patch movement is made less risky, selection favours a strategy where individuals essentially ignore non-natal patches until they are much closer to them. This results in higher mortality than would occur if emigrants biased their movement towards a patch from greater distances, but it has the benefit of increasing the connectivity of the network. During range expansion we expect that selection will favour riskier strategies (that increase mortality or perhaps other dispersal costs), which make it more likely that an individual and its descendants are represented at the expanding front. Our expectation is that during range expansion, selection will favour strategies that increase net displacement even when it carries substantial risks to the dispersing individual (Bartoń *et al.* 2012); for example passing by perfectly suitable patches, or even for individuals decline to settle in suitable patches that they have entered in order to continue on to more distant habitat.

For wind-dispersed plants an excellent opportunity exists to begin linking evolutionary theory with the mechanistic models that describe seed movement. Ultimately including greater mechanism in dispersal evolution theory promises to facilitate the development of tactical models incorporating dispersal evolution which may, for example, provide estimates of how much range expansions may be expected to accelerate. The realized dispersal kernel for a wind-dispersed plant will depend on several factors which we expect to be under selection. These include both the distribution of heights from which seeds are released and abscission thresholds (i.e. seeds will be more likely to be released under certain weather conditions), see Chapter 16 for detailed discussion of how kernels emerge from mechanism. Travis *et al.* (2010) presented a first relatively simple example model reword to..., in which the height of seed release is a heritable trait...upon which selection can act. There is considerable potential for gaining improved insight into dispersal evolution by further developing this type of mechanistic approach and in the next section we present some results from the Travis *et al.* (2010) model exploring evolution of dispersal during range expansion occurring across a fragmented landscape.

26.4 An interaction between landscape structure and climate change

The potentially synergistic effects of habitat destruction and climate change are increasingly well understood from an ecological perspective. However, there have been few theoretical studies exploring evolutionary responses to both drivers acting simultaneously. Using the evolutionary, mechanistic seed-dispersal model described in the previous section, Travis *et al.* (2010) simulated range expansion across a patchy landscape, and found a substantial increase in release height and that the increased seed dispersal distances resulted in an accelerating rate of spread (Travis *et al.* 2010). However, in some cases, the population appeared to become trapped during the initial phase of range expansion and failed to spread. Inspection of these model realizations indicated that this effect occurred when an unusually large gap between habitat patches inhibited spread. Further simulations confirmed that larger gaps between habitat patches can delay range expansions (Figure 26.1). This result is not unexpected as we would expect seeds to be less likely to reach a more distant patch and hence the time for that patch to be colonized will be greater. However, the results highlight that the range entrapment also has an evolutionary component; it is much more likely to occur during the early stages of range expansion before selection has increased dispersal ability (in the

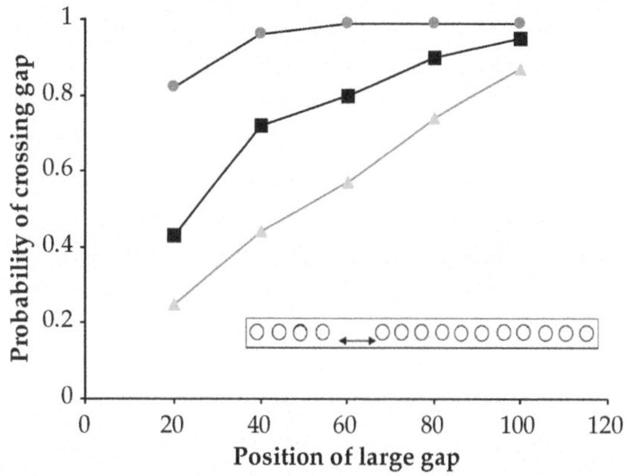

Figure 26.1 Unusually large gaps between patches of suitable habitat can block a range expansion. Here, we illustrate how the probability that a population successfully crosses a large habitat gap depends upon the position of the large gap relative to the initial position of the range and the size of the large gap (grey circles for gap size = 60; black square for gap size = 80; grey triangles for gap size = 100). The inset provides a schematic of the simulation design; the landscape is represented as a string of patches. The distance between adjacent patches is constant except for one larger gap. Both the size of that gap and its position relative to the start of range expansion is crucial. When the gap occurs further into the range expansion, dispersal evolution has had chance to occur, and a larger gap can then be more easily crossed. See Travis *et al.* (2010) for further details.

example here, greater dispersal is attained through evolving greater release height).

When a population with an expanding range encounters a large gap in suitable habitat, its best opportunity for crossing will be in the period immediately after reaching the gap. Edge effects are well known to select strongly against dispersal as many dispersing individuals will be lost to the unsuitable matrix. This effect is best known for island populations (Cody and Overton 1996), but is also seen in fragmented landscapes (Schitickzelle *et al.* 2006; Riba *et al.* 2009; Chapter 30). So, when an expanding range runs up against a large habitat gap, selection is likely to switch rapidly from increasing dispersal to acting against dispersive strategies.

26.5 Environmental gradients

Most of our discussion has concerned range expanding populations in general and the models discussed have, for simplicity, typically assumed homogeneous environmental conditions. While it may sometimes be the case that the invasion dynamics of recently introduced species occur across relatively homogenous conditions it will, almost by definition, not be the case for species expanding their ranges in response to environmental change. Species' natural biogeographic ranges are most often structured across gradients (Bridges *et al.* 2007), very often such that the populations towards the core of the range experience more suitable conditions than populations towards the margins—and it is these marginal populations that will be the ones leading climate-induced range expansions. The environmental gradients that structure a species' ranges do so through influencing one or more of the species' vital rates (e.g. Niewiarowski and Roosenburg 1993) or, in some cases, by influencing the amount of habitat that is available (for example, herbivores that are restricted by their food plant's range).

A growing number of modelling exercises incorporate environmental gradients (Holt and Keitt 2000; Brooker *et al.* 2007; Holt and Barfield 2009; Mustin *et al.* 2009) and they have provided some interesting findings for single species systems and for simple assemblages. For example, Brooker *et al.* (2007) explored the spatial dynamics of a competitor-mutualist system along a gradient. In a static climate, the mutualists are found in the harsher environmental conditions with competitors in the more productive area. Under climate change the distribution of the two interacting species depends upon an interaction between the proportion of long distance dispersal events and the rate of climate change. Mutualists are generally found to be the more vulnerable to climate change and they are most vulnerable when climate changes faster and there are more long distance dispersal events. Under these conditions, the competitors often leap-frog the mutualists' distributions and ultimately drive them to extinction. This result, from a very simple model, highlights that any predictions we make relating to how interacting species will respond to environmental change will depend critically on our assumptions of the type and strength of interspecific interactions and the dispersal of each of the species involved.

While there have been many studies focusing on ecological dynamics along gradients, far fewer have considered evolutionary dynamics. However, we highlight here two recent models that explore how selection on dispersal differs along a stationary and a shifting gradient. Dytham (2009) explored scenarios with ranges structured along gradients that influenced different parameters. In most cases, different dispersal strategies evolve at different positions across the range, although a gradient in death rate alone results in little spatial variation in dispersal. Increasing habitat turnover, reduced birth rate and reduced habitat quality all increase the evolved dispersal distance at the margin, while increased costs of dispersal and reduced habitat density resulted in reduced dispersal in the range margin populations. During climate change, the dispersal capabilities of the marginal populations will inevitably influence the rate of range expansion, so gaining an improved understanding of the selective pressures acting on dispersal in these populations is vital. These results also emphasize the danger of parameterising models developed to predict range expansions with data on dispersal obtained from core populations.

In an extension of this work, Kubisch *et al.* (2010) considered the evolution of dispersal during range shifting along environmental gradients. They employed a similar approach to Dytham (2009) but added simulated range expansion. The first important result is that dispersal consistently evolves upwards during range expansion regardless of the nature of the gradient. Most interestingly they sometimes observed an effect whereby the evolution of increased dispersal at an expanding range margin results in a range overshooting what will be its final position once environmental conditions stabilize. This effect, predicted by Holt (2003), comes about because once environment stabilizes, selection may substantially reduce dispersal in range margin populations, which can lead to some range contraction. Kubisch *et al.* (2010) demonstrate that this 'elastic range' effect is much more likely when the gradient is structured in a way that leads to substantially reduced dispersal in range margin populations under stable environmental conditions (for example, increasing dispersal costs towards the margins).

Finally for this section, we highlight the potential for dispersal to influence the ability of populations to adapt to novel or marginal conditions (see Kawecki *et al.* 2008; Bridle and Vines 2006 for reviews). Phillips (2012) modelled a population expanding its range across a homogenous landscape for some distance before it encounters an environmental gradient to which it can, in principle, adapt. The key result is that when dispersal has had longer to evolve upwards, adaptation to new conditions along the gradient occurs less readily. This is due to an increased flow of maladapted alleles from the homogenous conditions into the region of the environmental gradient. Whereas in the Travis *et al.* (2010) study, the evolution of increased dispersal ability at an expanding front facilitates the range expansion of a population across heterogenous space (by facilitating gap-crossing), in this second case, increased dispersal distance stymies the rate of local adaptation when an environmental gradient is encountered and hence slows spread.

26.6 Assisting migration by managing the landscape and dispersal

Gaining an improved understanding of the causes and consequences of dispersal behaviour during range shifting can help to inform management strategies aimed at facilitating range shifts. There are several potential conservation strategies that can be viewed as existing along a spectrum from least to most interventionist (Figure 26.2). We highlight three here, managing the landscape, assisted colonization and, what we will term, assisted gene flow.

By increasing the total amount of habitat, and by managing the preservation, or restoration, of habitat such that it has a particular spatial arrangement, we can change the expected rate at which a species will expand its range. Importantly, it is not necessarily the case that, for a given amount of habitat availability, increasing connectivity will lead to species being better able to spread across a landscape (McInerny *et al.* 2007). Depending upon a species' life-history traits, range expansion can be substantially faster when the available habitat is less aggregated; because, while clumps of habitat are typically smaller, the gaps between these clumps of habitat are, on average, also smaller. Whereas standard metapopulation theory, developed for species occupying stationary ranges, has resulted in conservation actions typically aimed at maximising connectivity, we now need to ask whether the management strategy needs to be shifted to find a balance between managing for connectivity and managing for permeability (the rate at which species spread across a landscape). A key task for modelling efforts will be to determine how well species with different dispersal characteristics will be able to shift their ranges on landscapes with different spatial structures. This modelling should not only consider how species with fixed dispersal kernels or fixed density-independent emigration rates will fare. Instead it needs to include proper consideration of the complexities of dispersal discussed above; emigration decisions, behavioural rules associated with transfer, settlement decisions and inter-individual variability are all key components that are likely to be important in determining the outcome.

An important result to emerge from theoretical work is that species' ranges can be 'pinned' by an Allee effect (Keitt *et al.* 2001) whereby low density populations at the range margins suffer reduced

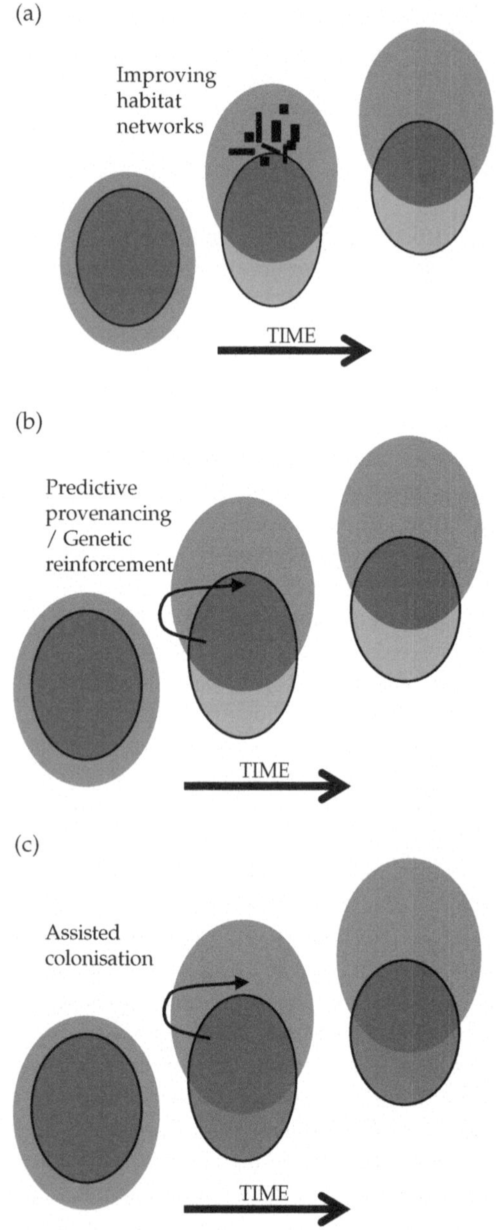

Figure 26.2 We illustrate three potential management actions that may assist migration of species' ranges. (a) The least interventionist method is to improve habitat connectivity; for example, with corridors and stepping stones. Genes can be artificially dispersed through a species range to facilitate local adaptation to changing conditions (b), or, probably most controversially, (c) individuals can be artificially dispersed ahead of the current range to facilitate colonization of newly suitable habitat.

growth rate. It has been suggested (e.g. Johnson *et al.* 2006) that for some invasive species one potential control strategy may be to suppress peaks in population abundance close to range borders, hence increasing the likelihood of invasion pinning by reducing the flow of individuals from higher density populations behind the front. In terms of managing species' responses to climate change we may want to consider the opposite; for species prone to Allee effects and hence liable to suffer invasion pinning, we might seek management actions that increase density behind the front (perhaps through habitat improvement) making it more likely that Allee effects at the front are overcome.

While managing the landscape to improve the likelihood that a species will successfully shift its range is unlikely to be particularly controversial (although it may be if it leads to a quite different spatial arrangement to that recommended by metapopulation theory), the idea that we should artificially disperse organisms to enhance their prospects of surviving climate change is fiercely debated (see McLachlan *et al.* 2007; Davidson and Simankin 2008; Hoegh-Guldberg *et al.* 2008; Minteer and Collins 2010). This debate has focused primarily on what may be viewed as the most interventionist of strategies, assisted colonization. Here, individuals are moved to suitable habitat that has become climatically suitable in a region ahead of their existing distribution. The objective is to establish populations ahead of the expanding range margin to accelerate the species' distributional shift. Quite apart from the debate on ethical issues and fears related to potential negative impacts akin to invasive species, we lack a sound scientific understanding of appropriate strategies should this policy be implemented. Generic questions that would benefit from the development of new theory include: From where and to where should we move individuals? How many should we move? When should we move them and how often? Assisted dispersal is likely to remain a hot topic and increasingly sophisticated, dynamic species distribution models offer a powerful tool to explore how we can optimize the probability of success, and also the potential consequences, of adopting this conservation strategy.

More interventionist than landscape management, but less than assisted colonization, is artificial dispersal with the objective of increasing gene flow (Figure 26.2). The idea of artificial dispersal is that by moving individuals from towards the core of a range to close to the expanding range limit, one can provide additional genetic variability that enables the marginal populations to adapt more rapidly to the changing climate. Individuals with 'warm genes' are introduced into populations that, having become adapted to colder conditions, possess 'cool genes', such that as the climate warms, the population will be able to adapt to that change and be more likely to thrive, and therefore be able to range shift more effectively. Recent theory suggests that genotypes conferring adaptation to range-core conditions may be vital to facilitating range shifting (Atkins and Travis 2010) and there is a need to develop a body of theory that will inform how, through artificial dispersal, we may promote local adaptation with the dual objectives of increasing the probability of local population survival and increasing a species' range-shifting ability.

Finally, we raise the possibility of assisting range shifting by selective, artificial dispersal of more dispersive genotypes. We have already highlighted that, under some conditions, individuals in range margin populations may have lower dispersal ability than individuals towards the core. For species exhibiting this pattern, a strategy of promoting range shifting could include translocation of more dispersive individuals from the core into the populations situated towards the potentially expanding margin. This would increase the rate of range expansion of these species and reduce the risk that they suffer evolutionary entrapment where poor dispersers are not able to push the range polewards (although Phillips *et al.*'s recent work (2011) indicates that, due to the interplay between dispersal and adaptation, increased dispersal at range fronts may not always be desirable).

26.7 Conclusion

Environmental change provides ecologists with a massive challenge: to apply our understanding of ecological systems effectively to inform the development of strategies to protect biodiversity under climate change. Over the last decade, substantial progress has been made in developing statistical approaches for charactering species' environmental niches and projecting future suitable climate space. These climate-matching models are beginning to incorporate ecological processes, including dispersal, into dynamic models. Over the same period, substantial progress has been made by theoretical modellers in understanding how dispersal interacts with population dynamics to determine range limits under stationary climatic conditions (see Keitt *et al.* 2001; Holt *et al.* 2005; Case *et al.* 2005), the consequences of dispersal for range shifting and how dispersal may evolve during range shifts. Additionally, major advances have been made in developing mechanistic models of dispersal (Katul *et al.* 2005; Kuparinen 2006; Bartoń *et al.* 2009). The need now is for greater integration of these approaches. The understanding gained from the theoretical models needs to be transferred to development of the species distribution models, which will benefit from incorporation of greater mechanistic detail in the dispersal functions. We have highlighted our scepticism of the predictive capability of species distribution models, but there is little doubt that they will provide a foundation for the development of conservation strategies that will prove effective under rapid environmental change. Species distribution models will certainly be used to assess the likely ability of species to track a changing climate over a fragmented landscape and to evaluate alternative landscape management options. They may also be employed to explore assisted migration strategies; while there may be adverse, and sometimes hard to predict, consequences of artificially manipulating biogeographic ranges in this way, for some species, these highly interventionist strategies may represent their only chance of survival (Minteer and Collins 2010) so it is important that we give them full and proper consideration.

References

Adriaensen, F. *et al.* (2003). 'The application of 'least-cost' modelling as a functional landscape model', *Landscape and Urban Planning* **64**: 233–47.

Araújo, M.B., Pearson, R.G., Thuiller, W. *et al.* (2005). 'Validation of species-climate impact models under climate change', *Global Change Biology* **11**: 1504–13.

Atkins, K.E. and Travis, J.M.J. (2010). 'Local adaptation and the evolution of species' ranges under climate change', *Journal of Theoretical Biology* **266**: 449–57.

Bartoń, K., Hovestadt, T., Phillips, B.L., Travis, J.M.J., (2012). 'Risky movement increases the rate of range expansion,' *Proc R Soc Lond B.* **279**: 1194–1202.

Bartoń, K.A., Phillips, B.L., Morales, J.M. *et al.* (2009). 'The evolution of an 'intelligent' dispersal strategy: Biased, correlated random walks in patchy landscapes', *Oikos* **118**: 309–19.

Berg, M.P., Kiers, E.T., Driessen, G. *et al.* and (2010). 'Adapt or disperse: understanding species persistence in a changing world', *Global Change Biology* **16**: 587–98.

Best, A.S., Johst, K., Münkemüller, T. *et al.* (2007). 'Which species will successfully track climate change? The influence of intraspecific competition and density dependent dispersal on range shifting dynamics', *Oikos* **116**: 1531–9.

Bonte, D., Bossuyt, B., Lens L. (2007). 'Aerial dispersal plasticity under different wind velocities in a salt marsh wolf spider', *Behavioral Ecology* **18**: 438–43.

Bonte, D., Van Dyck, H., Bullock, J.M. *et al.* (2012) 'The costs of dispersal', *Biological Reviews* **87**: 290–312.

Bowler, D.E. and Benton, T.G. (2005). 'Causes and consequences of animal dispersal strategies: relating individual behaviour to spatial dynamics', *Biological Reviews* **80**: 205–25.

Bridges, L.M. *et al.* 2007—Landscapes as gradients: the spatial structure of terrestrial ecosystem components in Southern Ontaria, Canada. *Ecological Complexity* **4**: 34–41.

Bridle, J.R. and Vines, T.H. (2007). 'Limits to evolution at range margins: when and why does adaptation fail?', *T.R.E.E.* **22**: 140–7.

Brooker, R.W., Travis, J.M.J., Clark, E.J. *et al.* (2007). 'Modelling species' range shifts in a changing climate: The impacts of biotic interactions, dispersal distance and the rate of climate change', *Journal of Theoretical Biology* **245**: 59–65.

Burton, O.J., Phillips, B.L., Travis, J.M.J. (2010). 'Trade-offs and the evolution of life-histories during range expansion', *Ecology Letters* **13**: 1210–20.

Dytham C. (2009). 'Evolved dispersal strategies at range margins', *Proceedings of the Royal Society of London Series B* **276**: 1407–13.

Cabral, J.S. and Schurr, F.M. (2010). 'Estimating demographic models for the range dynamics of plant species', *Global Ecology and Biogeography* **19**: 85–97.

Case, T.J., Holt, R.D., McPeek, M.A. *et al.* (2005). 'The community context of species' borders: ecological and evolutionary perspectives', *Oikos* **108**: 28–46.

Cody, M.L. and Overton, J.M. (1996). 'Short-term evolution of reduced dispersal in island plant populations', *Journal of Ecology* **84**: 53–61.

Coulon, A., Cosson, J.F., Angibault, J.M. *et al.* (2004). 'Landscape connectivity influences gene flow in a roe deer population inhabiting a fragmented landscape: an individual-based approach', *Molecular Ecology* **13**: 2841–50.

Cwynar L.C. and MacDonald, G.M. (1987). 'Geographical variation of lodgepole pine in relation to population history', *American Naturalist* **129**: 463–9.

Darling, E. *et al.* (2008). 'Increased seed dispersal potential towards geographic range limits in a Pacific coast dune plant', *New Phytologist* **178**: 424–35.

Davidson, I. and Simkanin, C. (2008). 'Sceptical of assisted colonization'. *Science* **322**: 1048–9.

DeVisser, M.H., Messina, J.P., Moore, N.J. *et al.* (2010). 'A dynamic species distribution model of *Glossina* subgenus *Morsitans*: The identification of tsetse reservoirs and refugia', *Ecosphere* **1**: article 6.

Dwyer, G. and Morris, W.F. (2006). 'Resource-dependent dispersal and the speed of biological invasions', *American Naturalist* **167**: 165–76.

Engler, R. and Guisan, A. (2009). 'MigClim: Predicting plant distribution and dispersal in a changing climate', *Diversity and Distributions* **15**: 590–601.

Heikkinen, R.K., Luoto, M., Araújo, M.B. *et al.* (2006). and 'Methods and uncertainties in bioclimatic envelope modelling under climate change', *Progress in Physical Geography* **6**: 751–77.

Heinz, S.K. and Strand, E. (2006). 'Adaptive patch searching strategies in fragmented landscapes', *Evolutionary Ecology* **20**: 113–30.

Hoegh-Guldberg, O., Hughes, L., McIntyre, S. *et al.* (2008). 'Assisted colonization and rapid climate change', *Science* **321**: 345–6.

Holt, R.D. (2003). 'On the evolutionary ecology of species' ranges', *Evolutionary Ecology Research* **5**: 159–78.

Holt, R.D. and Barfield, M. (2009). 'Trophic interactions and range limits: the diverse roles of predation', *Proceedings of the Royal Society B* **276**: 1435–42.

Holt, R.D. and Keitt, T.H. (2000). 'Alternative causes for range limits: a metapopulation perspective', *Ecology Letters* **3**: 41–7.

Holt, R.D., Keitt, T.H., Lewis, M.A. *et al.* (2005). 'Theoretical models of species' borders: single species approaches', *Oikos* **18**: 18–27.

Hovestadt, T.A., Kubisch, A., and Poethke, H.J. (2010). 'Information processing in models for density-dependent emigration: A comparison', *Ecological Modelling* **221**: 405–10.

Jeltsch, F., Moloney, K.A., Schurr, F.M. *et al.* (2008). 'The state of plant population modelling in light of environmental change', *Perspectives in Plant Ecology, Evolution and Systematics* **9**: 171–89.

Johnson, D.M., Liebhold, A.M., Tobin, P.C. *et al.* (2006). 'Allee effects and pulsed invasion by the gypsy moth', *Nature* **444**: 361–3.

Katul, G.G. *et al.* (2005). 'Mechanistic analytical models for long-distance seed dispersal by wind', *American Naturalist* **166**: 368–81.

Kawecki, T.J. (2008). 'Adaptation to marginal habitats', *Annual Review Ecological Evolutionary Systems* **39**: 321–42.

Keitt, T.H., Lewis, M.A., Holt, R.D. (2001). 'Allee effects, invasion pinning and species' borders', *American Naturalist* **157**: 203–16.

Kubisch, A., Hovestadt, T., Poethke, H-J. (2011). 'On the elasticity of range limits during periods of expansion', *Ecology* **91**: 3094–3099.

Kuparinen, A. (2006). 'Mechanistic models for wind dispersal', *Trends in Plant Science* **11**: 296–301.

Léotard, G., Debout, G., Dalecky, A. *et al.* (2009). 'Range expansion drives dispersal evolution in an Equatorial three-species symbiosis', *PLoS One* **4**:

McInerny, G., Travis, J.M.J. Dytham, C. (2007). 'Range shifting on a fragmented landscape', *Ecological Informatics* **2**: 1–8.

McLachlan, J.S., Hellmann, J.J., Schwartz, M.W. (2007). 'A framework for debate of assisted migration in an era of climate change', *Conservation Biology* **21**: 297–302.

Midgley, G.F., Davies I.D., Albert C.H. *et al.* (2010). 'BioMove—an integrated platform simulating the dynamic response of species to environmental change', *Ecography* **33**: 612–16.

Minteer, B.A. and Collins, J.P. (2010). 'Move it or lose it? The ecological ethics of relocating species under climate change', *Ecological Applications* **20**: 1801–4.

Morales, J.M., Moorcroft, P.R., Matthiopoulos, J. *et al.* (2010). 'Building the bridge between animal movement and population dynamics', *Philosphical Transactions of the Royal Society B*. **365**: 2289–301.

Mustin, K., Benton, T.G., Dytham, C. *et al.* (2009). 'The dynamics of climate-induced range shifting; perspectives from simulation modelling', *Oikos* **118**: 131–7.

Nathan, R., Getz, W.M., Revilla, E. *et al.* (2008). 'A movement ecology paradigm for unifying organismal movement research', *P.N.A.S.* **105**: 19052–9.

Niewiarowski, P.H. and Roosenburg, W.M. (1993). 'Reciprocal transplant reveals sources of variation in growth rates of the lizard, *Sceloporus undulatus*', *Ecology* **74**: 1992–2002.

Pachepsky, E. and Levine, J.M. (2011). 'Density dependence slows invader spread in fragmented landscapes', *American Naturalist* **177**: 18–28.

Palmer, S.C.F., Coulon, A., Travis, J.M.J. (2011). 'Introducing a 'Stochastic Movement Simulator' for estimating habitat connectivity', *Methods in Ecology and Evolution* **2**: 258–268.

Phillips, B.L. (2012). 'Range shift promotes the formation of stable range edges', *Journal of Biogeography* 39: 153–161.

Phillips, B.L., Brown, G.P., Travis, J.M.J. *et al.* (2008). 'Reid's paradox revisited: The evolution of dispersal kernels during range expansion', *American Naturalist* **172**: 34–8

Ranius, T. and Hedin, J. (2001). 'The dispersal rate of a beetle, *Osmoderma eremita*, living in tree hollows', *Oecologia*, **126**: 363–70.

Riba, M., Mayol, M., Giles, B. *et al.* and (2009). 'Darwin's wind hypothesis: does it work for plant dispersal in fragmented habitats?', *New Phytologist* **183**: 667–677.

Schtickzelle, N., Mennechez, G., Baguette, M. (2006). 'Dispersal depression with habitat fragmentation in the bog fritillary butterfly', *Ecology* **87**: 1057–65.

Shine, R., Brown, G.P., Phillips, B.L. (2011). 'An evolutionary process that assembles phenotypes through space rather than through time', *P.N.A.S* **108**: 5708–11.

Simmons, A.D. and Thomas, C.D. (2004). 'Changes in dispersal during species' range expansions', *American Naturalist* **164**: 378–95.

Sinclair, S.J., White, M.W., Newell, G.R. (2010). 'How useful are species distribution models for managing biodiversity under future climates?', *Ecology and Society* **15**: 8.

Skarpaas, O. and Shea, K., (2007). 'Dispersal patterns, dispersal mechanisms, and invasion wave speeds for invasive thistles', *American Naturalist* **170**: 421–30.

Soons, M.B. and Bullock, J.M., (2008). 'Non-random seed abscission, long-distance wind dispersal and plant migration rates', *Journal of Ecology* **96**: 581–90.

Stevens, V.M., Verkenne, C., Vandewoestijne, S. *et al.* (2006). 'Gene flow and functional connectivity in the natterjack toad', *Molecular Ecology* **15**: 2333–44.

Thuiller, W., Lavorel, S., Araújo, M.B. *et al.* (2005). 'Climate change threats to plant diversity in Europe', *P.N.A.S.* **23**: 8245–50.

Travis, J.M.J. and Dytham, C. (2002). 'Dispersal evolution during invasions', *Evolutionary Ecology Research*, **4**: 1119–29.

Travis, J.M.J., Mustin, K., Benton, T.G. *et al.* (2009). 'Accelerating invasion rates result from the evolution of density-dependent dispersal', *Journal of Theoretical Biology* **259**: 151–8.

Travis, J.M.J., Smith, H.S., Ranwala, S.M.W. (2010). 'Towards a mechanistic understanding of dispersal evolution in plants: conservation implications', *Diversity and Distributions* **16**: 690–702.

Williams, N.S.G., Hahs, A.K., Morgan, J.W. (2008). 'A dispersal-constrained habitat suitability model for predicting invasion of alpine vegetation', *Ecological Applications*, **18**, 347–59.

Wilson, J.R.U., Dormontt, E.E., Prentis, P.J. *et al.* (2009). 'Something in the way you move: dispersal pathways affect invasion success', *T.R.E.E.* **24**: 136–44.

CHAPTER 27

Influence of temperature on dispersal in two bird species

Henrik Pärn and Bernt-Erik Sæther

27.1 Introduction

The combined effects of climate change and habitat destruction represent serious threats to the long-term viability of many populations (Brook *et al.* 2008). Under such circumstances dispersal plays a key role for long-term persistence of populations by allowing individuals to track their suitable habitat in a fragmented landscape (Warren *et al.* 2001; Chapters 25, 26, 30, 31). Dispersal also affects gene flow and thereby contributes to the genetic variation and the capacity for adaptive evolution in response to changing environmental conditions (Chevin *et al.* 2010; Chapter 26). Thus, understanding the dispersal behaviour of a species is crucial when predicting the response of populations subject to climate change in a fragmented landscape (Best *et al.* 2007; Brooker *et al.* 2007).

There is a vast body of empirical work demonstrating that dispersal may vary considerably within species, often in response to environmental cues (see Chapter 1), such as patch size and patch configuration (Andreassen and Ims 2001), social structure (Le Galliard *et al.* 2003), food availability (Bowler and Benton 2009), and population size (Matthysen 2005; Enfjäll and Leimar 2009). Despite such apparent plasticity of dispersal, little is known about how the dispersal process is influenced by climate (but see Robinet and Roques 2010, and Chapter 28 for some insect examples), especially in vertebrates. Accordingly, several recent studies have stressed the need for more empirical studies on this topic (e.g. Stevens *et al.* 2010; Sæther and Engen 2010). In birds, the impact of local climate on demographic traits such as onset of breeding, survival, and fecundity is rather well-documented (Dunn and Winkler 2010; Sæther and Engen 2010). Whether also temporal variation in dispersal can be explained by climatic factors has received much less attention, and the underlying mechanisms are poorly understood (Sæther *et al.* 2000; Walls *et al.* 2005; Figuerola 2007; Balbontín *et al.* 2009; Grøtan *et al.* 2009).

In this chapter, we present two studies that have examined annual variation in dispersal in relation to climate. The first example, a long-term study on the dipper *Cinclus cinclus*, shows that fluctuations in winter temperature affect immigration, which in turn is an important contributor to fluctuations in population size. The second study addresses how the effect of climate on dispersal varies spatially and may depend on quality of the natal habitat in a metapopulation of house sparrows *Passer domesticus*.

27.2 Winter temperature, immigration, and impact on population dynamics in the dipper

The dipper is a 50–60 g passerine bird that depends on fast-flowing streams and rivers for food and nesting habitat (Tyler and Ormerod 1994). The *Cinclus* genus is widely distributed in the Palearctic region. The dipper has the ability to dive and swim, and it feeds on aquatic invertebrates and thus depends on open water. Dippers establish linear breeding territories along a river, and all activities take place within the territory. The size of the territories varies considerably and must include an area

Dispersal Ecology and Evolution. First Edition. Edited by Jean Clobert *et al.*

of water large enough to obtain sufficient amount of food. In the northern range of its distribution, the amount of ice strongly affects the availability of winter feeding habitats. The dipper is mainly sedentary across most of its range, but the Scandinavian (nominate) race *Cinclus cinclus cinclus* may migrate to areas with open water during winter (Lundberg *et al.* 1981). Seasonal movements, especially altitudinal migration in mountainous populations, also occur in other races (e.g. Gillis *et al.* 2008). It is strongly territorial in the breeding season but also during winter when population structure seems to depend on sex- and age-related dominance hierarchies (Lundberg *et al.* 1981). However, winter territoriality varies considerably between geographic locations and depends on winter severity (Tyler and Ormerod 1994). In a Welsh population of dippers, natal dispersal is common and takes place in late summer. About 30% of the birds ringed as nestlings or early summer juveniles dispersed more than 5 km, where females dispersed greater distances than males. In contrast, adults show strong fidelity to breeding sites and only 3.6% dispersed more than 2.5 km (Tyler *et al.* 1990). Dispersal typically occurs within a river basin, but may also include movements across watersheds between basins.

A long-term study of a dipper population in Norway has provided valuable data to investigate the link between climate, dispersal, and population dynamics (Sæther *et al.* 2000, Loison *et al.* 2002). The study area is located along the Lynga river in southern Norway (58° 15′ N, 7° 15′ E) and extends for 60 km from the mouth of the river to the inland. The same study area was surveyed from 1974 to 1997. Population size in terms of the number of breeding pairs has been estimated from 1978 by counting the number of active nests along the river bank. The fact that all the activities of the dipper take place within the spatially simple, linear habitat along the stream and its territoriality makes it possible to obtain precise population estimates, find nests, and capture individuals. Each year, nestlings were marked with metal rings, and adults were captured and marked with colour rings. A large proportion of the individuals was individually marked and the number of immigrants was estimated as the number of previously unringed birds. The number of immigrant recruits (i.e. an individual entering the breeding population) was larger following a warm winter (Figure 27.1). A possible proximate explanation is that the trend toward warmer winter temperature increases the number of ice-free days on the river, which allows access to foraging habitat. Juveniles that have performed natal dispersal to the area begin to establish territories during the autumn, and have to compete with adults who become territorial again after the moult. In a cold, ice-rich winter, the habitat and food availability is reduced, and aggression by dominant resident birds may prevent immigrant birds from establishing themselves and recruiting in vacant territories along the river. Thus, when ice reduces the availability of vacant territories, the establishment success of immigrants may be reduced. The number of immigrants was not related to the local population size (Sæther *et al.* 2000). An alternative explanation for the relationship between the number of immigrants and winter temperature is related to density-dependent emigration. A study of Welsh dippers suggested that dispersal distance in males increased when population size was larger (Tyler and Ormerod 1994). Possibly, such density-dependent emigration may contribute to a larger number of immigrant birds in warm years also in the Norwegian dipper population (see 27.3). However, the origin of the immigrants is unfortunately unknown and thereby the demographic properties of their natal populations. The increase in the number of immigrants after warm winters contributed significantly to the increase in population size. The winter temperature in the area is closely correlated with the North Atlantic Oscillation Index (NAO). Thus, the dipper example provides evidence of how changes in large-scale climatic factors affecting the local weather together with temperature-dependent dispersal affect the local dynamics of bird populations. Accordingly, in a recent study, Grøtan *et al.* (2009) showed how temperature-dependent immigration contributed to temporal variation in population growth rate of great tits *Parus major* in several different areas in the Netherlands.

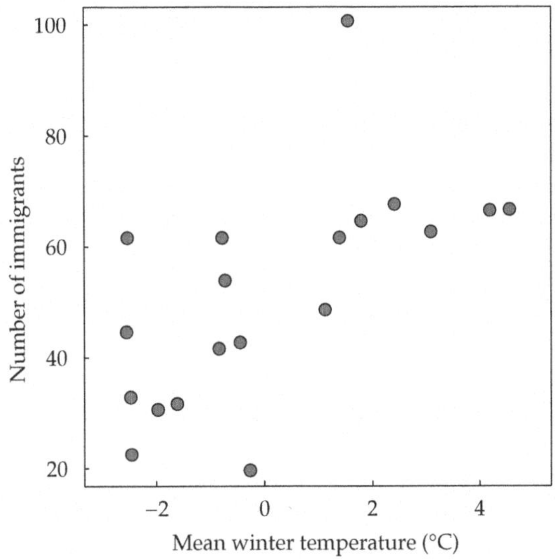

Figure 27.1 The relationship between the number of immigrants (estimated as the number of previously unringed birds) and winter temperature (°C) (January to March) in a dipper *Cinclus cinclus* population along the Lynga river in southern Norway (58° 15′ N, 7° 15′ E). Modified from Sæther *et al.* (2000).

27.3 Spring temperature, habitat quality, and natal dispersal in the house sparrow

Another important issue is that the effects of the same climate factor on demographic processes may differ between populations, even within regions experiencing the same variation in climate (Sæther and Engen 2010). Together with the fact that dispersal often is density-dependent, this implies that in order to understand the effect of climate variation on dispersal, one must consider fluctuations in population size and how the effects of density-dependence and climate differ between populations; for example, due to differences in habitat quality or habitat variability (Massot *et al.* 2002; Enfjäll and Leimar 2009).

To examine whether the relationship between dispersal rate and climate is influenced by spatial variation in habitat quality, we used data from a metapopulation of house sparrows (Pärn *et al.* 2012). The house sparrow is a small, non-migratory passerine bird and is one of the most widely distributed bird species on earth (Anderson 2006). It breeds in close association with man and preferred nest sites are usually crevices in buildings and other man-made structures, or in nest-boxes. The sparrows reach sexual maturity in their second calendar year, and in the study area sparrows may produce up to three clutches per season. The breeding season lasts from early May to mid-August. Most natal dispersal takes place during the autumn and breeding dispersal is negligible (Altwegg *et al.* 2000).

Our study area consists of several small island populations in an archipelago off the Norwegian coast (66° N, 13° E). The 18 islands in the study area can be categorized into two major types of habitat: islands where the house sparrows breed in close association with dairy farms, with access to cattle-food and shelter all year round, and islands without farms where sparrows are much more exposed to variation in food availability and weather conditions. Thus, we considered the farm islands to represent superior habitat for the house sparrows, whereas the non-farm islands represent inferior habitat for the house sparrows. The weather in the study area is highly spatially correlated (Ringsby *et al.* 2002). Together, this allowed comparison of responses of dispersal rates in several local populations exposed to more or less similar temporal fluctuations in temperature, but with spatial differences in habitat quality. A large proportion of the sparrows was individually marked as nestlings (Pärn *et al.* 2009) and the study area is large (1600 km^2) compared to the dispersal distances (Tufto *et al.* 2005). This implies two important advantages: it allows unambiguous identification of dispersal events between islands, and it reduces the bias in estimates of dispersal rates due to dispersal out of the study area (Doligez and Pärt 2008; Pärn *et al.* 2009). Because the origin of dispersing individuals is known, the ecological and demographical characteristics of the natal populations can be examined. Furthermore, in the study area, the spring temperature (mean temperature for April–May) in the study area has increased ca. 1 °C, from 4.5 °C to 5.5 °C during the last 50 years (Pärn *et al.* 2012).

Previous studies have shown that the laying date of first clutches is correlated with spring air temperature (Seel 1968). In years with warm springs, we expected early onset of breeding, which in turn may influence individual dispersal behaviour

(Altwegg *et al.* 2000). In addition, early onset breeding may increase the incidence of additional broods (Chastel *et al.* 2003), and thereby increase production of young. Similar to most other passerine birds in the temperate region, warm weather increases nestling and juvenile survival rate in the study area (Ringsby *et al.* 2002), and may influence total population density which shows large annual fluctuations. Accordingly, in years with warm springs, early onset of breeding and increased survival of young may lead to a high number of sparrows in the natal habitat, which in turn could affect individual dispersal decision.

Thus, we hypothesized that temperature-dependent production of juveniles and density-dependent dispersal may constitute the link between spring temperature and dispersal rate in our study system. Natal population size itself may induce emigration as a consequence of social interactions. However, if the effect of population size on emigration is mediated by competition over resources, the density-dependence of dispersal needs to be considered in relation to spatial variation in habitat quality (Enfjäll and Leimar 2009; Chapter 1). Accordingly, we expected that factors influencing production of juveniles and population size, e.g. temperature and onset of breeding, affected dispersal differently depending on quality of the natal habitat.

As an index of total population size on the natal island, we used the sum of number of adults and number of fledged juveniles. Nine of the islands, on which breeding was monitored more thoroughly than on the remaining islands, were included as natal islands and all the 18 islands were included as potential targets for dispersal. The influence of natal habitat quality on the effects of spring temperature and population size on dispersal was assessed by comparing dispersal from five high-quality and four low-quality natal islands, which differed in availability and predictability of resources (see earlier). Recaptures and observations of marked birds were used to determine whether fledglings survived until their second calendar year, i.e. recruited to the breeding population, and whether inter-island dispersal had occurred. Dispersers were defined as individuals which were marked as nestlings on one of the nine main study islands, and then performed inter-island dispersal and were recorded as recruits in their second calendar year on any of the other islands in the whole study system. Resident individuals were those which were marked as nestlings on any of the nine main study islands and then recruited on their natal island.

On the non-farm, low-quality islands the dispersal rate, i.e. the proportion of recruits that had performed inter-island dispersal, increased with spring temperature (Figure 27.2A), was higher in years of early breeding (Figure 27.2B), and increased with population size (Figure 27.2C). The increased dispersal rate also corresponded to an increase in the absolute number of dispersers from the low-quality islands. In contrast, on the high-quality, farm islands, no such relationships were found (Figure 27.2A–C). In a warm spring, early breeding and increased survival of young will increase the number of conspecifics in the natal habitat. If competition over resources is the proximate factor inducing emigration in warm years, the fact that the relationship between dispersal rate and population size depended on habitat quality suggests that the strength of competition depends not only on population size itself, but rather on population size in relation to habitat quality; e.g. availability and predictability of critical food resources and shelter. Emigration from the natal habitat is often triggered by competition over food, as shown experimentally in song sparrow *Melospiza melodia* (Arcese 1989) and in northern goshawk *Accipiter gentilis atricapillus* (Kennedy and Ward 2003; Chapter 1).

In our study area, it is likely that the strength of competition over food differs between islands with and without farms. On the farms, sparrows can feed on grass seeds in cattle-food and grain feed throughout the year, and cowsheds and barns provide a dry and warm shelter. In contrast, on the non-farm islands, the sparrows often have to rely on bird-feeders in the small settlements, which are much more variable food sources and also cause concentration of foraging sparrows to be higher than on the farms. Thus, in a warm spring, early breeding and increased survival of young will increase the number of conspecifics in the natal habitat. However,

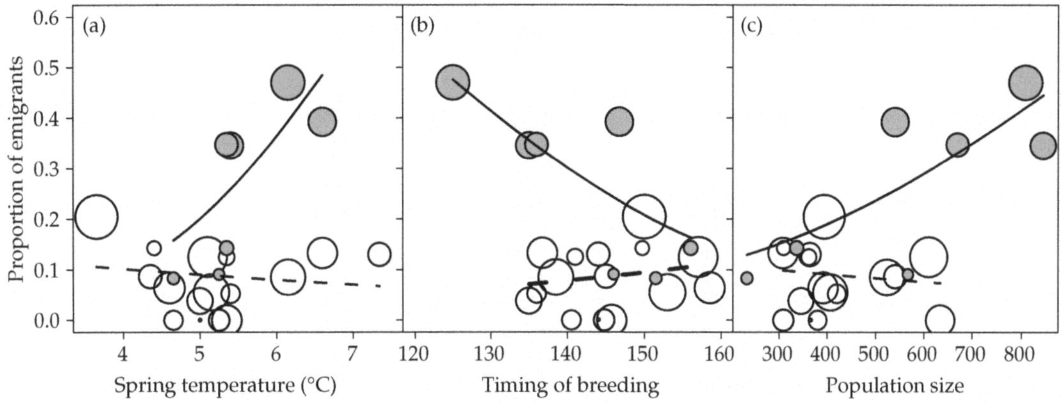

Figure 27.2 The relationship between the annual proportion of recruits that have performed inter-island natal dispersal in a house sparrow *Passer domesticus* metapopulation in northern Norway (66° N, 13° E) and (a) spring temperature (°C) (April to May), (b) timing of breeding estimated as mean day of first egg (1 = January 1), and (c) total population size (adults plus juveniles) on the natal islands. Islands were categorized as farm islands (open circles and dashed lines) or non-farm islands (filled circles and solid lines). The regression lines are predicted values from generalized linear models with binomial error and logit link function. Modified from Pärn *et al.* (2012).

the competition over food will increase to a larger extent on the non-farm islands (see Lowther 1983, and references therein). The role of habitat quality in density-dependent dispersal has been investigated in insects (e.g. Simmons and Thomas 2004; Chapter 1). However, in birds there are very few studies: the relationship between immigration and population density differed between Great Tit populations in Netherland (Grøtan *et al.* 2009), but did not vary between habitats in Wytham wood (Delestrade *et al.* 1996).

As described above, high spring temperature may cause density-dependent emigration in our study system. In addition, temperature may also influence density-independent emigration, via the effect of temperature on onset of breeding. We found that dispersal rate was higher in years of early breeding, but only on the non-farm islands. There is some evidence that individuals from early clutches are of low phenotypic quality (Ringsby *et al.* 1998), which potentially could affect their tendency to disperse. Although we have no strong *a priori* reason to expect that the effect of onset of breeding on dispersal should differ between high- and low-quality habitats, we cannot exclude a differential effect of early breeding on dispersal rate that is related to habitat quality.

27.4 Concluding remarks

In both our passerine examples, dispersal increased with local temperature; in the dipper the number of immigrants increased with winter temperature, and in the house sparrow the dispersal rate and number of dispersers increased with spring temperature. Because a dispersing house sparrow was defined as an individual which performed inter-island natal dispersal and was recorded as a recruit in its new breeding population, this implies that the increase in dispersal corresponded to an increase in the number of successful immigrants with temperature. Nevertheless, while the dipper study focused on conditions in the breeding habitat, the sparrow study focused on conditions in the natal habitat. The increase in dispersal with temperature has also been found in other bird species. For example, in great tits, warm springs led to a larger number of immigrants the following year (Grøtan *et al.* 2009), which in turn had a large effect on local population dynamics. In the non-migratory raptor, the common buzzard *Buteo buteo* (Walls *et al.* 2005), and in the migratory arctic tern *Sterna paradisaea* (Møller *et al.* 2006), dispersal distances increased with spring temperature. Although the number of studies so far is few, together these results suggest that, at least in bird species breeding in temperate areas, dispersal

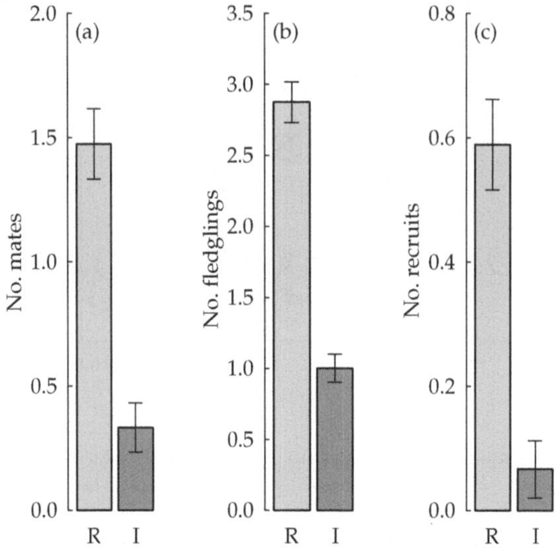

Figure 27.3 Lifetime number of (a) mates, (b) fledglings, and (c) recruits in resident (R) and immigrant (I) male house sparrows *Passer domesticus* in an insular metapopulation in northern Norway. Bars show means and error bars indicate ± 1 standard error. Modified from Pärn *et al.* (2009).

may be temperature-dependent, and that dispersal propensity often increases with temperature.

To what extent a climate-induced increase in dispersal contributes to future population growth rate, affects the genetic composition, maintains genetic diversity, or prevents inbreeding in the new population, depends not only on the absolute number of immigrants but also whether they reproduce successfully or not, and their relative contribution to the next generation (Chapter 20). Differences in lifetime reproduction between philopatric and immigrant individuals have been found in several bird species (e.g. Pärn *et al.* 2009, and references therein). In the house sparrow study system described earlier, the lifetime production of recruits by immigrant males is in fact less than a third than the LRS of resident individuals (Figure 27.3; Pärn *et al.* 2009). The overall effect of climate change on effective dispersal will thus be determined by both its influence on immigration rate and on the relative fitness of immigrants.

Furthermore, as our study on the house sparrow suggests (Figure 27.2), ecological differences between populations may modify how dispersal responds to climate change also within a limited geographic area. We focused on spatial heterogeneity of habitat characteristics and how such heterogeneity may cause differential density-dependence, and thereby affect the relationship between dispersal and temperature. However, our results also have relevance for temporal variation in temperature-sensitive dispersal, as climate change may modify the quality of natal habitats. So far little is known about how variation in habitat quality influences the relationship between dispersal rate and environmental changes. Nevertheless, the results emerging from our two studies of small passerines have some implications for the modelling of metapopulation dynamics. First, the dispersal process is affected by fluctuations in population size as well as stochastic environmental variation (Grøtan *et al.* 2009). Second, both the influence of density-dependence and environmental stochasticity differ among the local populations. We have shown that the response to the same climatic factor may differ between populations, also within a limited geographical area (Sæther *et al.* 2007). Thus, to understand how metapopulations respond to climate change, we need models that include local dynamics, i.e. that account for both dispersal and density-dependence (e.g. Lande *et al.* 1998), and that allow for spatial heterogeneity in environmental effects.

References

Altwegg, R., Ringsby, T.H., Sæther, B.-E. (2000). 'Phenotypic correlates and consequences of dispersal in a metapopulation of house sparrows *Passer domesticus*', *Journal of Animal Ecology* **69**: 762–70.

Anderson, T.R. (2006). *Biology of the ubiquitous house sparrow: from genes to populations*. Oxford University Press, New York.

Andreassen, H.P. and Ims, R.A. (2001). 'Dispersal in patchy vole populations: Role of patch configuration, density dependence, and demography', *Ecology* **82**: 2911–26.

Arcese, P. (1989). 'Intrasexual competition, mating system and natal dispersal in song sparrows', *Animal Behaviour* **38**: 958–79.

Balbontín, J., Møller, A.P., Hermosell, I.G. *et al.* (2009). 'Divergent patterns of impact of environmental conditions on life history traits in two populations of a long-distance migratory bird', *Oecologia* **159**: 859–72.

Best, A.S., Johst, K,. Münkemüller, T. *et al.* (2007). 'Which species will succesfully track climate change? The influence of intraspecific competition and density dependent dispersal on range shifting dynamics', *Oikos* **116**: 1531–9.

Bowler, D.E. and Benton, T.G. (2009). 'Variation in dispersal mortality and dispersal propensity among individuals: the effects of age, sex and resource availability', *Journal of Animal Ecology* **78**: 1234–41.

Brook, B.W., Sodhi, N.S., Bradshaw, C.J.A. (2008). 'Synergies among extinction drivers under global change', *Trends in Ecology and Evolution* **23**: 453–60.

Brooker, R.W., Travis, J.M.J., Clark, E.J. *et al.* (2007). 'Modelling species' range shifts in a changing climate: The impacts of biotic interactions, dispersal distance and the rate of climate change', *Journal of Theoretical Biology* **245**: 59–65.

Chastel, O., Lacroix, A., Kersten, M. (2003). 'Prebreeding energy requirements: thyroid hormone, metabolism and the timing of reproduction in house sparrows *Passer domesticus*', *Journal of Avian Biology* **34**: 298–306.

Chevin, L.-M., Lande, R., Mace, G.M. (2010). 'Adaptation, plasticity, and extinction in a changing environment: towards a predictive theory', *PLoS Biol* **8**: e1000357.

Delestrade, A., McCleery, R.H., Perrins, C.M. (1996). 'Natal dispersal in a heterogeneous environment: the case of the Great Tit in Wytham', *Acta Oecologica* **17**: 519–29.

Doligez, B. and Pärt, T. (2008). 'Estimating fitness consequences of dispersal: a road to 'know-where'? Non-random dispersal and the underestimation of dispersers' fitness', *Journal of Animal Ecology* **77**: 1199–211.

Dunn, P., Winkler, D.W. (2010). 'Effects of climate change on timing of breeding and reproductive success in birds', in A.P. Møller, W. Fiedler, P. Berthold (eds) *Effects of climate change on birds*. Oxford University Press, Oxford, 113—128.

Enfjäll, K.O. Leimar, O. (2009). 'The evolution of dispersal—the importance of information about population density and habitat characteristics', *Oikos* **118**: 291–9.

Figuerola, J. (2007). 'Climate and Dispersal: Black-winged stilts disperse further in dry springs', *PLoS One* **2**: e539.

Gillis, E.A., Green, D.J., Middleton, H.A. *et al.* (2008). 'Life history correlates of alternative migratory strategies in American Dippers', *Ecology* **89**: 1687–95.

Grøtan, V., Sæther, B.-E., Engen, S. *et al.* (2009). 'Spatial and temporal variation in the relative contribution of density dependence, climate variation and migration to fluctuations in the size of great tit populations', *Journal of Animal Ecology* **78**: 447–59.

Kennedy, P.L. and Ward, J.M.. (2003). 'Effects of experimental food supplementation on movements of juvenile northern goshawks (*Accipiter gentilis atricapillus*)', *Oecologia* **134**: 284–91.

Lande, R., Engen, S., Sæther, B.-E. (1998). 'Extinction times in finite metapopulation models with stochastic local dynamics', *Oikos* **83**: 383–9.

Le Galliard, J.F., Ferriere, R., Clobert, J. (2003). 'Mother-offspring interactions affect natal dispersal in a lizard', *Proceedings of the Royal Society B* **270**: 1163–9.

Loison, A., Sæther, B.-E., Jerstad, K. *et al.* (2002). 'Disentangling the sources of variation in the survival of the European dipper', *Journal of Applied Statistics* **29**: 289–304.

Lowther, P.E. (1983). 'Breeding biology of House Sparrows: Intercolony variation', *Occasional Papers of the Museum of Natural History, The University of Kansas, Lawrence, Kansas*. **107**: 1–17.

Lundberg, P., Bergman, A., Olsson, H. (1981). 'The ecology of wintering Dippers (*Cinclus cinclus*) in northern Sweden', *Journal für Ornithologie* **122**: 163–72.

Massot, M., Clobert, J., Lorenzon, P. *et al.* (2002). 'Condition-dependent dispersal and ontogeny of the dispersal behaviour: an experimental approach', *Journal of Animal Ecology* **71**: 253–61.

Matthysen, E. (2005). Density-dependent dispersal in birds and mammals. *Ecography* **28**: 403–16.

Møller, A.P., Flensted-Jensen, E., Mardal, W. (2006). Dispersal and climate change: a case study of the Arctic tern *Sterna paradisaea*', *Global Change Biology* **12**: 2005–13.

Pärn, H., Jensen, H., Ringsby, T.H. *et al.* (2009). 'Sex-specific fitness correlates of dispersal in a house sparrow metapopulation', *Journal of Animal Ecology* **78**: 1216–25.

Pärn, H., Ringsby, T.H., Jensen, H. *et al.* (2012). 'Spatial heterogeneity in the effects of climate and density-dependence on dispersal in a house sparrow metapopulation', *Proceedings of the Royal Society B* **279**: 144–52.

Ringsby, T.H., Sæther, B.-E., Solberg, E.J. (1998). 'Factors affecting juvenile survival in House Sparrow *Passer domesticus*', *Journal of Avian Biology* **29**: 241–7.

Ringsby, T.H., Sæther, B.E., Tufto, J. *et al.* (2002). 'Asynchronous spatiotemporal demography of a house sparrow metapopulation in a correlated environment', *Ecology* **83**: 561–9.

Robinet, C., and Roques, A. (2010). 'Direct impacts of recent climate warming on insect populations', *Integrative Zoology* **5**: 132–42.

Sæther, B.-E., and Engen, S. (2010). 'Population consequences of climate change', in A.P. Møller, W. Fiedler, P. Berthold (eds), *Effects of climate change on birds*. Oxford University Press, Oxford.

Sæther, B.-E., Engen, S. Grøtan, V. *et al.* (2007). 'The extended Moran effect and large-scale synchronous fluctuations in the size of great tit and blue tit populations', *Journal of Animal Ecology* **76**: 315–25.

Sæther, B.-E., Tufto, J., Engen, S. *et al.* (2000). 'Population dynamical consequences of climate change for a small temperate songbird', *Science* **287**: 854–6.

Seel, D.C. (1968). 'Breeding seasons of house sparrow and tree sparrow *Passer* spp. at Oxford', *Ibis* **110**: 129–44.

Simmons, A.D. and Thomas, C.D. (2004). 'Changes in dispersal during species' range expansions', *American Naturalist* **164**: 378–95.

Stevens, V.M., Turlure, C., Baguette, M. (2010). 'A meta-analysis of dispersal in butterflies', *Biological Reviews* **85**: 625–42.

Tufto, J., Ringsby, T.H., Dhondt, A.A. *et al.* (2005). 'A parametric model for estimation of dispersal patterns applied to five passerine spatially structured populations', *American Naturalist* **165**: E13–26.

Tyler, S.J., Ormerod, S.J. (1994). *The Dippers*. Poyser, London.

Tyler, S.J., Ormerod, S.J., Lewis, J.M.S. (1990). 'The postnatal and breeding dispersal of Welsh Dippers *Cinclus cinclus*', *Bird Study* **37**: 18–22.

Walls, S.S., Kenward, R.E., Holloway, G.J. (2005). 'Weather to disperse? Evidence that climatic conditions influence vertebrate dispersal', *Journal of Animal Ecology* **74**: 190–7.

Warren, M.S., Hill, J.K., Thomas, J.A. *et al.* (2001). 'Rapid responses of British butterflies to opposing forces of climate and habitat change', *Nature* **414**: 65–9.

CHAPTER 28

Dispersal under global change—the case of the Pine processionary moth and other insects

Hans Van Dyck

28.1 Temperature and mobility in a changing environment

Insects are known to be small, poikilothermous organisms that struggle to keep their body temperature within a suitable range for activity (Heinrich 1996). Although some insects (e.g. bees and moths) are able to elevate their body temperature by rapid muscle contractions, the basic metabolism of insects is a function of the temperature of the outside environment. Therefore, and in combination with their taxonomic dominance and multiple ecological associations and significance, insects have been popular study objects in the framework of climate and climate change (e.g. Dennis 1993; Musolin 2007). For several groups of insects, there is good evidence of ecological responses in space and/or time relative to recent climate change (Parmesan 2006; Altermatt 2010). Spatial responses include range expansion or shifts in latitude (e.g. Parmesan *et al.* 1999), and in altitude (e.g. Wilson *et al.* 2005). Responses in time include phenological shifts like earlier emergence in the spring (e.g. Stefanescu *et al.* 2003), and changes in the number of generations a year (voltinism, e.g. Tobin *et al.* 2008). There is also empirical evidence, and evidence from modelling, for extinctions of local isolated butterfly populations that were hastened by increasing variability in precipitation as a component of climate change (McLaughlin *et al.* 2002). Raised temperature has a direct effect on several ecological traits including development, survival, range and abundance of herbivorous insects (Bale *et al.* 2002). There is, however, also evidence of evolutionary responses (e.g. the pitcher-plant mosquito *Wyeomyia smithii*—Bradshaw and Holzapfel 2001). Ecology and evolution are no longer separate fields to our understanding of responses of short-living organisms (including insects) to climate change (e.g. Huey *et al.* 2000). Evolutionary ecology therefore is an essential, unifying field to improve our successes in applied biology to decrease population sizes of invasive species and to increase population sizes of endangered species.

Although several thermophilous insects have responded positively in terms of range expansion, there are also insects that are trapped in highly isolated boreo-alpine relict populations of cold vegetation types, like bog species (e.g. Turlure *et al.* 2010). In such cases, spontaneous dispersal is not going to be very helpful, and a much more specific habitat management focusing on thermal characteristics of the vegetation may be necessary. On the longer term, the conservation of species with such profile may become problematic at relatively low latitude, and will feed the debate on assisted dispersal (Hoegh-Guldberg *et al.* 2008). But even in species that appear to respond to regional warming by movements, there is evidence of range adjustments that considerably lag behind climate change. In British butterflies, only one-third of the predicted increase since the 1970s has taken place and successful introduction experiments of butterflies in climatically suitable areas beyond their current range confirm such time lag (Menéndez *et al.* 2006). Hence, dispersal and colonization ability are key factors to understand the impact of climate change

Dispersal Ecology and Evolution. First Edition. Edited by Jean Clobert *et al.*

on insects. Therefore, most current bioclimatic envelope models that do not account for dispersal to project changes in the distribution of species may give too optimistic impressions of species' capacities to track climate, and the next generation of these models need more sophisticated dispersal modules because dispersal and colonization rates will limit the realization of potential range shifts (Huntley *et al.* 2010).

The ability of insects to track suitable habitat and thermal conditions by flight is generally considered to be a key factor to their biological success. However, it is interesting to notice that the loss of the ability to fly has occurred in nearly all orders of winged insects (Speight *et al.* 2008). Flight is physiologically costly and it is thought that the stability of habitats in which insects live may select against costly phenotypic investments for displacement (Southwood 1962). In several insects, there is an evolutionary option to develop winged (macropterous) or wingless (brachypterous) forms that typically differ in their trade-off between flight capability and reproduction relative to environmental conditions (Zera and Denno 1997). In monomorphic, flying insects there can be more subtle variation in flight-related morphology between recently colonized populations at the expanding front of the range compared to populations at the core of the range (e.g. relative thorax size as a measure of investment in flight muscles in the butterfly *Pararge aegeria* in the UK; Hughes *et al.* 2003). However, the interpretation of flight morphology relative to dispersal ability under different climatic conditions may be confounded by the multifunctionality of flight (Van Dyck and Matthysen 1999; Berwaerts *et al.* 2008). Moreover, there is also evidence for the significance of physiological differences (e.g. relative to energy metabolism)—independent of morphological differences—among individuals and populations that differ in dispersal ability (e.g. Watt 1992; Mitikka and Hanski 2010).

In this chapter, I focus in a highly selective way on recent research on the Lepidoptera—moths and butterflies, with a particular emphasis on the case of the Pine processionary moth—to illustrate the multiple and complex relationships between climate change and dispersal in invertebrates. The impact of climate change is, however, not operating in isolation from other components of global change, including habitat fragmentation, urbanization, pollution, and invasive species. As elsewhere in this book, I will follow the conceptualization of dispersal as a three-step process (emigration—transfer—immigration and settlement). Climate change may have direct effects on both the propensity to move (emigration phase), the ability to cover particular distances in particular environments (transit phase), and the ability to settle and use, for example, novel host plants in colonized areas. But climate change may also have indirect effects on dispersal via altered habitat availability and community structures.

28.2 The case of the Pine processionary moth

The Pine processionary moth (*Thaumetopoea pityocampa* L.) is for several reasons an interesting study case in the context of insect dispersal and climate change (Netherer and Schopf 2010). This thermophilous moth has a Mediterranean distribution range, but over the last decades and years, it has expanded both to higher latitudes and altitudes. For example, its distribution has moved upwards in the Italian Alps by 110–230 m over the last 30 years (Battisti *et al.* 2005). In France, the northward expansion front moved *c.* 27 km per decade between 1972 and 2004, and has recently accelerated (last decade: > 55 km; Battisti *et al.* 2005).

This climate-related expansion is not only of biological interest, but also of economic and human-health interest (Hodar *et al.* 2003). *T. pityocampa* is a considerable forest pest as outbreaks cause heavy defoliations of pine trees (*Pinus* sp.) in forested areas and commercial plantations. Females lay their eggs in a single batch, and caterpillars continue feeding during autumn and winter making them particularly responsive to effects of climate change with higher frequencies of warmer winters. The gregarious caterpillars produce silk shelters in their host trees that contain hairy larval skins which provoke strong irritations and allergy responses on the skin and in the eyes of people who come in contact with them, like forest workers and recreationists (Hodar *et al.* 2003). The presence of the conspicuous silk tents in the canopy makes the species relatively easy to monitor by conventional visual count methods. Alternative methods to study their presence and abundance have been developed as well. Jactel *et al.* (2006) showed significant correlations between the mean number of males captured per day in pheromone-based traps, and the total number of nests in the stand. A pheromone-based trapping design is considered to be highly suitable for monitoring the populations of the Pine processionary moth across a wide range of European regions covering both the current core area as well as the expansion range (Jactel *et al.* 2006).

T. pityocampa is a striking insect example of the effect of climate warming on the release of thermal thresholds constraining species distribution (Robinet and Roques 2010). Winter temperature plays a key role in the life cycle of this moth, as the gregarious larvae feed on the coniferous needles during autumn and winter, which principally determines its range limits. As insects go through different life stages during the year, one needs to select carefully climatic data of the relevant periods—instead of using annual averages for temperatures—in order to improve understanding of the ongoing ecological processes that contribute to range expansion. In the same vein, Crozier (2004) showed, based on combined results from population and larval transplant analyses, that winter temperatures directly affected the persistence of the skipper butterfly *Atalopedes campestris* at its northern range edge, and that winter warming was a prerequisite for this butterfly's range expansion.

Besides this temporal aspect of temperature measures, working with insects like moths and butterflies also demonstrates the significance of the appropriate spatial extent and resolution to understand the effects of climate change on dispersal (and other life-history traits). Several studies use global or regional temperature data that are easily available in governmental databases. However, as shown by a recent study of Suggit *et al.* (2010) habitat type is major modifier of the temperature extremes experienced by an organism. Thermal heterogeneity has a direct impact on the thermal budget (e.g. Merckx *et al.* 2008), and hence metabolic performance, in flying insects.

More than 40 years ago, Huchon and Démonlin (1970) already developed a statistical distribution model assuming that the range of *T. pityocampa* was limited by the combination of the mean minimum temperature in January and the cumulative annual hours of sunshine. Robinet *et al.* (2007) re-tested the model with updated climatic data, but the historical model no longer successfully predicted the currently observed distribution. Instead, Robinet *et al.* (2007) developed a mechanistic model based on the ecophysiological data collected by Battisti *et al.* (2005) on thermal requirements for larval feeding. The mechanistic model showed that the range distribution of the Pine processionary moth in the Paris Basin area is since 2001, no longer limited by unfavourable larval feeding conditions; hence the pattern of range expansion will be determined by its dispersal capability and the distribution of host trees. Although winter temperatures have been recognized as a key factor, a recent analysis along a latitudinal gradient of *T. pityocampa* populations showed that winter temperatures were poorer predictors than latitude (Pimentel *et al.* 2010). The authors argue that this probably reflects a more complex relationship between life history and climate than has been appreciated before.

28.3 Flight, emigration, and ambient temperature

The number of insects in free flight depends on the population density and the amount of activity, which are both affected by climatic factors and especially by temperature (Taylor 1963). Temperature may directly affect flight propensity and ability in insects. Nocturnal flight activity of *T. pityocampa* females in outdoor cages increased with mean night temperature (Figure 28.1; Battisti *et al.* 2006). Ambient temperature of 14 °C was found to be a threshold for female flight. This behavioural parameter can be used as a predictor of the expansion in different years and regions as one could calculate flight activity windows based on the number of nights with a temperature above this threshold. Flight time may vary considerably among years. Moreover, although several studies focus on the effects of gradual changes in temperature, extreme

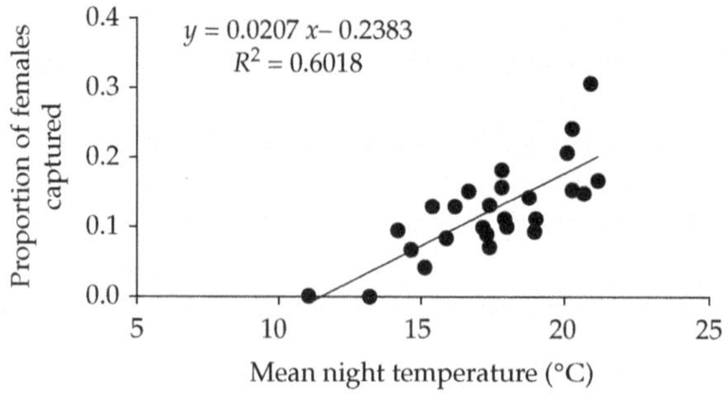

Figure 28.1 Relationship between mean night temperature and the proportion of females showing flight activity of *Thaumetopoea pityocampa* in an outdoor cage. Reproduced from Battisti et al. (2006) with permission from Wiley.

weather conditions can have a very significant effect on flight time, and hence potentially for range expansion. The European heat wave of the summer of 2003 was accompanied by a strong expansion of the Pine processionary moth to higher altitudes in the Italian Alps. Battisti *et al.* (2006) showed that during the female flight period of 2003, there were > 5 times more nights above the threshold of 14 °C compared with the two years before and the year after. Hence, rapid responses following climatic anomalies may generate significant acceleration of the expansion of insects. Predictions based on projected average changes may, hence, not fully capture the spread of such an organism.

However, it would be interesting to know whether temperature only increases the proportion of females in flight and their amount of time spent flying, or whether temperature also affects flight performance in terms of covered distance. Several dispersal studies interpret any flight activity to be relevant to dispersal and expansion. This is, however, questionable as routine, daily movements related to local resource search may differ from special displacement movements (Van Dyck and Baguette 2005). Flight activity experiments of the Peach fruit moth (*Carposina saskii*) in a flight mill did not show differences in flight duration relative to ambient temperature in a range of 14–26 °C (Ishiguri and Shirai 2004). Flight duration was only shorter at lower and higher temperatures. Flight speed was, however, significantly higher at high ambient temperature (20–29 °C) compared to low temperature (11–17 °C). Several studies on mobility of pest moths (e.g. Tu *et al.* 2010) but also of other flying insects (e.g. Orthoptera; Simmons and Thomas 2004), have estimated flight potential by flight mills with tethered flight in the laboratory. But it would be valuable to test the relevance of such laboratory estimates as proxies of (long-distance) dispersal in the field under natural conditions. Stimuli to fly and to continue flight may differ under spontaneous free flight compared to forced tethered flight. Females of the common cutworm (*Spodoptera litura*) were, for example, observed to fly on average for 24 h during a 72 h session under tethered conditions which would correspond to a distance of > 105 km (Tu *et al.* 2010). Although flight performance measures have often been taken at the species level, there is significant intraspecific variation in flight behaviour. The temperature-related sensitivity to show active flight may vary among population of the same species as shown in the Glanville fritillary *Melitaea cinxia* (e.g. Wang *et al.* 2011).

Besides the influence on the number of individuals in flight and the distance covered, ambient temperature may also have an effect on the energetic cost of flight and dispersal at different thermal conditions. From a life-history viewpoint, the cost of dispersal may also operate via altered trade-offs with other life-history traits. In the Light brown apple moth (*Epiphyas postvittana*) flight capacity was negatively correlated with fecundity (Gu and Danthanarayana 1992) which is often considered to be a classic trade-off, but recent work has questioned the universal nature, even within the Lepidoptera (e.g. Hanksi *et al.* 2006).

The direct impact of warm weather on regional mobility is an ecological effect in the first place, but in relatively short-living organisms we cannot exclude an additional effect of rapid evolutionary change on flight ability. I am not aware of such studies in the Pine processionary moth, but there is relevant work in other moths. Free flight activity responded rapidly to artificial selection in both sexes of the Oriental fruit moth (*Grapholita molesta*) (Torriani *et al.* 2010). Flight activity in a laboratory setting was found to be a good proxy of the mobility in the field (Keil *et al.* 2001). In line with the cost of flight and trade-offs, selection for flight activity typically shows correlated responses of other life-history traits that are traded off against mobility. In the study of Torriani *et al.* (2010) upward-selected moths showed reduced survival in both males and females, although there was no effect on female reproductive output. Wing loading did not differ in response to the induced selection regime. Wing loading—i.e. body mass relative to wing surface—is one of the classical functional morphological traits studied in flying animals, including the Lepidoptera (e.g. Chai and Srygley 1990). However, the relationship with (long-distance) dispersal may not be unequivocal.

28.4 The transit stage: barriers to expansion?

Ecological and genetic studies suggest sex-biased dispersal in the Pine processionary moth; females are poorer fliers than males (Kerdelhué *et al.* 2009). Although *T. pityocampa* clearly is an expanding species, a genetic study using AFLP and mitochondrial DNA markers found low estimates of the absolute number of migrants between populations (Nm) suggesting low gene flow (Salvato *et al.* 2002). But they also showed evidence of gender-related dispersal with adult males dispersing more than females. Females would typically fly for only a few hours during the first night after emergence covering on average *c.* 300 m only, although at much lower frequency individuals will fly up to a few kilometers. Consequently, long-distance colonization by adult females is considered to be a rare event (Salvato *et al.* 2002). Kerdelhué *et al.* (2009) confirmed that the actual dispersal of the females is highly limited over most of the distribution range. Males fly more and also cover longer distances than females. Hence, dispersing males make more or less constant contributions to the genetic diversity of populations on a relatively wide range, reducing the risk of genetic problems like inbreeding and genetic erosion. Rather than functioning as barriers to dispersal, mountains appear to have served as refugia for *T. pityocampa* during glaciations, and current distributions still largely reflect expansion from these refugial populations (Rousselet *et al.* 2010). Their phylogeographical study did not detect any role of physical barriers to dispersal.

Climate change may regionally improve landscape connectivity in thermophilous insects when more vegetation patches become thermally accessible or even suitable for reproduction. There is also evidence of changing habitat associations in thermally constrained species and changes in habitat specificity at the climatic range boundaries (e.g. Davies *et al.* 2006, Oliver *et al.* 2009). Anthropogenic environments are usually interpreted as barriers to the movements of several wild organisms (e.g. Mader *et al.* 1990). However, the case of the Pine processionary moth is interesting to illustrate that the opposite situation may occur as well. Urban areas are known to be warmer than more rural areas; this is particularly true during the night and during winter periods (Shochat *et al.* 2006). So this makes much sense to *T. pityocampa* as warm urban areas can be particularly favourable for winter feeding, as has been shown for Paris (Robinet *et al.* 2007). Warm urban climates may also affect adult flight behaviour. However, in urban areas other anthropogenic factors may also interfere with moth movements. Light pollution is likely to be a significant factor to guide moths outside forest fragments into the landscape matrix (Longcore and Rich 2004). Dispersing moths may be attracted to artificial lights which can be sites with high bat predation (Rydell 1992). There is still much to learn on the behavioural mechanisms that explain dispersal and expansion patterns, including research on perception in changing environments.

Insects—particularly the immobile life cycle stages like eggs and pupae—can be easy targets of passive dispersal by human-mediated transport. This may be particularly significant for range expansion in forest-pest species as plant material and some soil may be moved from the current range of the species to new areas. Moreover, the pupae of *T. pityocampa* are soil-borne. So, in principle also soil transport has some potential of spreading individuals regionally, but the impact is probably limited. The role of human trade and travel routes for accelerating the involuntary spread has been well documented in another moth species, the non-native Gypsy moth (*Lymantria dispar*) (Tobin *et al.* 2010).

28.5 Settlement: the advantage of trophic plasticity

With climate-related range expansion, insect herbivores may enter regions where their preferred host plant is relatively rare or even absent. Successful use of second or novel hosts may facilitate colonization of areas without the primary host or with a low density of the latter. Experiments by Stastny *et al.* (2006) in the laboratory and in the field showed that relative growth rate and mortality of *T. pityocampa* caterpillars did not differ significantly among the primary host *Pinus nigra*, the secondary host *P. sylvestris*, and a novel host *P. mugo*. So the moth is not limited in its spread by the lack of a host plant. However, the use of host plants is not only a function of larval host plant suitability (Thompson and Pellmyr 1991). Female behaviour and host acceptance are also key factors to the successful use of host plants, and hence to the colonization of stands. The study by Statsny *et al.* (2006) showed differences in the frequency of host acceptance among the trees, even in the absence of alternative choices. The primary host was preferred consistently. Since the female lays a single batch of eggs only, host choice is essential in this time-limited species and bet hedging by spreading the eggs across different *Pinus* trees is no option. So, range expansion may imply selection on continued host search performance or efficiency in such a time-limited insect. Alternatively, there could be selection to alter the innate host recognition behaviour as the insect moves into areas with different host plants frequencies. Limited gene flow across populations could facilitate adaptation to local host plants as well as to local climatic regimes. Translocation experiments are an interesting approach to test levels of local adaptation in the field (e.g. Pelini *et al.* 2009). Divergence of local *T. pityocampa* populations driven by the host plant is a factor that has to be taken into account (Salvato *et al.* 2002), and the species has a great trophic plasticity. This plasticity may be of significance in this context as the moth sometimes attacks exotic conifer hosts (Robinet *et al.* 2007). A successful switch to a novel host plant that facilitates colonizations in anthropogenic environments has been documented in several insects; for instance, the North American butterfly *Euphydryas editha* that incorporated a European weed into its diet (e.g. Thomas *et al.* 1987).

28.6 Research perspectives and challenges

The issues reviewed in this chapter have illustrated the multiple direct and indirect impacts of climate change on the different stages of the dispersal process. Moreover, climate change is only one complex component among the other components of global change (habitat fragmentation, urbanization, invasive species, human mobility) that have additive or interactive effects on dispersal and other correlated life-history traits. Climate change is often, but incorrectly, narrowed to gradual climate warming only. Several other correlated or non-correlated climatic variables can be the subject of significant climate change (e.g. cloudiness, relative humidity). There is still much to learn about the influence of such climatic factors on insect activity and mobility. Clouds reduce the efficiency of heliothermous butterflies (diurnal Lepidoptera) to regulate their body temperature by solar basking, whereas cloudy nights may create warmer conditions and hence more efficient flight conditions in endothermous moths (nocturnal Lepidoptera). Biologists tend to think differently when focusing on vertebrates or invertebrates, but major different thermal physiological strategies (ectotherm versus endotherm, poikilotherm versus homeotherm) may even be of larger significance to understand the mechanisms of climate change on the ecology and evolution of insects and other organisms. Lepidoptera are from that perspective an interesting group.

There is interesting and challenging scope for improving the integration of thermal ecology, behavioural ecology, eco-physiology, and life-history work including phenotypic plasticity in dispersal studies within a global change context (Helmuth *et al.* 2004; Walthers *et al.* 2006; Chown *et al.* 2007; Ward and Masters 2007; Berg *et al.* 2010; Bardsen *et al.* 2011; Reed *et al.* 2011). This multidisciplinary work is essential in order to improve our understanding of the mechanisms beyond the observed patterns of climate-related responses of insects in time and space.

Acknowledgements

HVD is supported by an ARC research grant of the Académie Louvain (Grant no 10/15-031). This is publication BRC 266 of the Biodiversity Research Centre (UCL, Louvain-la-Neuve).

References

Altermatt, F. (2010). 'Climatic warming increases voltinism in European butterflies and moths', *Proceedings of the Royal Society B* **277:** 1281–7.

Bale, J.S., Masters, G.J., Hodkinson, I.D. *et al.* (2002). 'Herbivory in global climate change research: direct effects of rising temperature on insect herbivores', *Global Change Biology* **8**: 1–16.

Bardsen, B.-J., Henden, J.-A., Fauchald, P., *et al.* (2011). 'Plastic reproductive allocation as a buffer against environmental stochasticity—linking life history and population dynamics to climate', *Oikos* **120:** 245–57.

Battisti, A., Stastny, M., Buffo, E. *et al.* (2006). 'A rapid altitudinal range expansion in the pine processionary moth produced by the 2003 climatic anomaly', *Global Change Biology* **12:** 662–71.

Battisti, A., Stastny, M., Netherer, S. *et al.* (2005). 'Expansion of geographic range in the pine processionary moth caused by increased winter temperatures', *Ecological Applications* **15:** 2084–96.

Berg, M.P., Kiers, E.T., Driessen, G. *et al.* (2010). 'Adapt or disperse: understanding species persistence in a changing world', *Global Change Biology* **16:** 587–98.

Berwaerts, K., Matthysen, E. and Van Dyck, H. (2008). 'Take-off flight performance in the butterfly *Pararge aegeria* relative to sex and morphology: a quantitative genetic assessment', *Evolution* **62:** 2525–33.

Bradshaw, W.E. and Holzapfel, C.M. (2001). 'Genetic shift in photoperiodic response correlated with global warming', *Proceedings of the National Academy of Sciences USA* **98:** 14509–11.

Chai, P. and Srygley, R.B. (1990). 'Predation and the flight, morphology, and temperature of Neotropical rain-forest butterflies', *American Naturalist* **135:** 748–65.

Chown, S.L., Slabber, S., McGeoch, M.A. *et al.* (2007). 'Phenotypic plasticity mediates climate change responses among invasive and indigenous arthropods', *Proceedings of the Royal Society B* **274:** 2531–7.

Crozier, L. (2004). 'Warmer winters drive butterfly range expansion by increasing survivorship', *Ecology* **85:** 231–41.

Davies, Z.G., Wilson, R.J., Coles, S. *et al.* (2006). 'Changing habitat associations of a thermally constrained species, the silver-spotted skipper butterfly, in response to climate warming', *Journal of Animal Ecology* **75:** 247–56.

Dennis, R.L.H. (1993). *Butterflies and Climate Change*. Manchester University Press, Manchester.

Gu, H. and Danthanarayana, W. (1992). 'Quantitative genetic analysis of dispersal in *Epiphyas postvittana*. II. Genetic covariations between flight capacity and life-history traits', *Heredity* **68:** 61–9.

Hanksi, I., Saastamoinen, M. and Ovaskainen, O. (2006). 'Dispersal-related life-history trade-offs in a butterfly metapopulation', *Journal of Animal Ecology* **75:** 91–100.

Heinrich, B. (1996). *The Thermal Warriors. Strategies of Insect Survival*. Harvard University Press, Cambridge, Mass.

Helmuth, B, Kingsolver, J.G. and Carrington, E. (2004). 'Biophysics, physiological ecology, and climate change: does mechanism matter', *Annual Review of Physiology* **67:** 177–201.

Hodar, J.A., Castro, J. and Zamora, R. (2003). 'Pine processionary caterpillar *Thaumetopoea pityocampa* as a new threat for relict Mediterranean Scots pine forests under climate warming', *Biological Conservation* **110:** 123–9.

Hoegh-Guldberg, O., Hughes, L., McIntyre, S. *et al.* (2008). 'Assisted colonization and rapid climate change', *Science* **321:** 345–6.

Huchon, H. and Démonlin, G. (1970) 'La bioécologie de la processionaire du pin. Dispersion potentielle. Dispersion actuelle', *Phytoma* **23:** 11–20.

Huey, R.B., Gilchrist, G.W., Carlsson, M.L. *et al.* (2000). 'Rapid evolution of a geographic cline in size in an introduced fly', *Science* **287:** 308–9.

Hughes, C.L., Hill, J.K. and Dytham, C. (2003). 'Evolutionary trade-offs between reproduction and dispersal in populations at expanding range boundaries', *Proceedings of the Royal Society B* **270:** S147–50.

Huntley, B., Barnard, P., Altwegg, R. *et al.* (2010). 'Beyond bioclimatic envelopes: dynamic species' range and abundance modelling in the context of climatic change', *Ecography* **33:** 621–6.

Ishiguri, Y. and Shirai, Y. (2004). 'Flight activity of the peach fruit moth, *Carposina sasakii* (Lepidoptera: Carposinidae), measured by a flight mill', *Applied Entomology and Zoology* **39:** 127–31.

Jactel, H., Menassieu, P., Vétillard, F. *et al.* (2006) 'Population monitoring of the pine processionary moth(Lepidoptera: Thaumetopoeidae) with pheromone-baited traps', *Forest Ecology and Management* **235:** 96–106.

Keil, S., Gu, H. and Dorn, S. (2001). 'Response of *Cydia pomonella* to selection on mobility: laboratory evaluation and field verification', *Ecological Entomology* **26:** 495–501.

Kerdelhué, C., Zane, L., Simonato, M. *et al.* (2009). 'Quaternary history and contemporary patterns in a currently expanding species', *BMC Evolutionary Biology* **9**: 220 (e-paper).

Longcore, T. and Rich, C. (2004). 'Ecological light pollution', *Frontiers in Ecology and Environment* **2**: 191–8.

Mader, H.J., Shell, C. and Kornacker, P. (1990). 'Linear barriers to arthropod movements in the landscape', *Biological Conservation* **54:** 209–22.

McLaughlin, J.F., Hellmann, J.J., Boggs, C.L. *et al.* (2002). 'Climate change hastens population extinctions', *Proceedings of the National Academy of Sciences USA* **99:** 6070–4.

Menéndez, R., Gonzalez Megias A., Hill, J.K. *et al.* (2006). 'Species richness changes lag behind climate change', *Proceedings of the Royal Society B* **273:** 1465–70.

Merckx, T., Van Dongen, S., Matthysen, E. *et al.* (2008). 'Thermal flight budget of a woodland butterfly in woodland versus agricultural landscapes: an experimental assessment', *Basic and Applied Ecology* **9:** 433–42.

Mitikka, V. and Hanski, I. (2010). 'Pgi genotype influences flight metabolism at the expanding range margin of the European map butterfly', *Annales Zoologici Fennici* **47:** 1–14.

Musolin, D.L. (2007). 'Insects in a warmer world: ecological, physiological and life-history responses of true bugs (Heteroptera) to climate change', *Global Change Biology* **13:** 1565–85.

Netherer, S. and Schopf, A. (2010). 'Potential effects of climate change on insect herbivores in European forests—General aspects and the pine processionary moth as specific example', *Forest Ecology and Management* **259:** 831–8.

Oliver, T., Hill, J.K., Thomas, C.D. *et al.* (2009). 'Changes in habitat specificity of species at their climatic range boundaries', *Ecology Letters* **12:** 1091–102.

Parmesan, C. (2006). 'Ecological and evolutionary responses to recent climate change', *Annual Review of Ecology, Evolution, and Systematics* **37:** 637–69.

Parmesan, C., Ryrholm, N., Stefanescu, C. *et al.* (1999). 'Poleward shifts in geographical ranges of butterfly species associated with regional warming', *Nature* **399:** 579–83.

Pelini, S.L., Dzurisin, J.D.K., Prior, K.M., *et al.* (2009). 'Translocation experiments in butterfly species reveal limitations to range shifts under climate change', *Proceedings of the National Academy of Sciences USA* **106:** 11160–5.

Pimentel, C., Ferreira, C. and Nilsson, J.-A. (2010). 'Latitudinal gradients and the shaping of life-history traits in a gregarious caterpillar', *Biological Journal of the Linnean Society* **100:** 224–36.

Reed, T.E., Schindler, D.E. and Waples, R.S. (2011). 'Interacting effects of phenotypic plasticity and evolution on population persistence in a changing climate', *Conservation Biology* **25:** 56–63.

Robinet, C., Baier, P., Pennerstorfer, J. *et al.* (2007). 'Modelling the effects of climate change on the potential feeding activity of *Thaumetopoea pityocampa* (Den. and Schiff.) (Lep., Notodontidae) in France', *Global Ecology and Biogeography* **16:** 460–71.

Robinet, C. and Roques, A. (2010). 'Direct impacts of recent climate warming on insect populations', *Integrative Zoology* **5**: 132–42.

Rousselet, J., Zhao, R., Argal, D. *et al.* (2010). 'The role of topography in structuring the demographic history of the pine processionary moth, *Thaumetopoea pityocampa* (Lepidoptera: Notodontidae)', *Journal of Biogeography* **37:** 1478–90.

Rydell, J. (1992). 'Exploitation of insects around streetlamps by bats in Sweden', *Functional Ecology* **6**: 747–59.

Salvato, P., Battisti, A., Concato, S. *et al.* (2002). 'Genetic differentiation in the winter pine processionary moth (*Thaumetopoea pityocampa—wilkinsoni* complex), inferred by AFLP and mitochondrial DNA markers', *Molecular Ecology* **11:** 2435–44.

Shochat, E., Warren, P.S., Faeth, S.H. *et al.* (2006). 'From patterns to emerging processes in mechanistic urban ecology', *Trends in Ecology and Evolution* **21:** 186–91.

Simmons, A.D. and Thomas, C.D. (2004). 'Changes in dispersal during species' range expansions', *American Naturalist* **164:** 378–95.

Southwood, T.R.E. (1962). 'Migration of terrestrial arthropods in relation to habitat', *Biological Reviews* **37:** 171–211.

Speight, M.R., Hunter, M.D. and Watt, A.D. (2008). *Ecology of Insects: Concepts and Applications* (2nd Edition). Wiley-Blackwell, Oxford.

Stastny, M., Battisti, A., Petrucco-Toffolo, E. *et al.* (2006). 'Host-plant use in the range expansion of the pine processionary moth, *Thaumetopoea pityocampa*', *Ecological Entomology* **31:** 481–90.

Stefanescu, C., Peñuelas, J. and Filella, I. (2003). 'Effects of climatic change on the phenology of butterflies in the northwest Mediterranean Basin', *Global Change Biology* **9**: 1494–506.

Suggitt, A.J., Gillingham, P.K., Hill, J.K. *et al.* (2011). 'Habitat microclimates drive fine-scale variation in extreme temperatures', *Oikos* **120:** 1–8.

Taylor, L.R. (1963). 'Analysis of the effect of temperature on insects in flight', *Journal of Animal Ecology* **32:** 99–117.

Thomas, C.D., Ng, D., Singer, M.C. *et al.* (1987). 'Incorporation of a European weed into the diet of a North American herbivore', *Evolution* **41:** 892–901.

Thompson, J.N. and Pellmyr, O. (1991). 'Evolution of oviposition behaviour and host preference in Lepidoptera', *Annual Review of Entomology* **36:** 65–89.

Tobin, P.C., Nagarkatti, S., Loeb, G. and Saunders, M.C. (2008). 'Historical and projected interactions between climate change and insect voltinism in a multivoltine species', *Global Change Biology* **14:** 951–7.

Tobin, P.C., Van Stappen, J. and Blackburn, L.M. (2010). 'Human visitation rates to the Apostle Islands National Lakeshore and the introduction of the non-native species *Lymantria dispar* (L.)', *Journal of Environmental Management* **91:** 1991–6.

Torriani, M.V.G., Mazzi, D., Hein, S. *et al.* (2010). 'Directed and correlated responses to artificial selection on flight activity in the oriental fruit moth (Lepidoptera: Tortricidae)', *Biological Journal of the Linnean Society* **100:** 879–89.

Tu, Y-G., Wu, K-M., Xue, F-S. *et al.* (2010). 'Laboratory evaluation of flight activity of the common cutworm, *Spodoptera litura* (Lepidoptera: Noctuidae)', *Insect Science* **17:** 53–9.

Turlure, C., Choutt, J., Baguette, M. *et al.* (2010). 'Microclimate buffering and resource-based habitat of a relict butterfly species: significance for conservation under climate change', *Global Change Biology* **16:** 1883–93.

Van Dyck, H. and Baguette, M. (2005). 'Dispersal behaviour in fragmented landscapes: routine or special movements?', *Basic and Applied Ecology* **6:** 535–45.

Van Dyck, H. and Matthysen, E. (1999). 'Habitat fragmentation and insect flight: a changing 'design' in a changing landscape?', *Trends in Ecology and Evolution* **14:** 172–4.

Walthers, R.J., Hassall, M., Telfer, M.G. *et al.* (2006). 'Modelling dispersal of a temperate insect in a changing climate', *Proceedings of the Royal Society B* **273:** 2017–23.

Wang, R., Ovaskainen, O., Cao, Y. *et al.* (2011). 'Dispersal in the Glanville fritillary butterfly in fragmented versus continuous landscapes: comparison between three methods', *Ecological Entomology* **36:** 251–60.

Ward, N.L. and Masters, G.J. (2007). 'Linking climate change and species invasion: an illustration using insect herbivores', *Global Change Biology* **13:** 1605–15.

Watt, W.B. (1992). 'Eggs, enzymes, and evolution: natural genetic variants change insect fecundity', *Proceedings of the National Academy of Sciences USA* **89:** 10608–12.

Wilson, R.J., Gutiérrez, D., Gutiérrez, J. *et al.* (2005). 'Changes to the elevational limits and extent of species ranges associated with climate change', *Ecology Letters* **8**: 1138–46.

Zera, A.J. and Denno, R.F. (1997). 'Physiology and ecology of dispersal polymorphism in insects', *Annual Review of Entomology* **42:** 207–30.

CHAPTER 29

Plant dispersal and the velocity of climate change

James M. Bullock

29.1 Projecting climate change impacts on plants

Attempts to predict plant responses to climate change comprise a major research effort at this time. Curiously, the majority of this research has not considered dispersal in any great detail. Many published modelling studies seek to understand the extent to which different plant species will be able to survive and redistribute in future climates. Clearly, it is vital that dispersal is represented in such models (Midgley *et al.* 2006; Engler *et al.* 2006). Indeed, there is empirical evidence that dispersal limitation has led to incomplete range filling of some plants (Svenning and Skov 2004). However, the major focus of early research has been the characterization of the climatic tolerances of species, with little realism in representations of dispersal; for example, by treating dispersal as alternatively absent or unlimited (e.g. Thomas *et al.* 2004). Some recent modelling of species migration under climate change has attempted to represent dispersal more realistically by using some functions to describe the kernels (Iverson *et al.* 2004; Midgley *et al.* 2006; Engler *et al.* 2009; Hillyer and Silman 2010). While these are important advances, the kernels used are not based on actual dispersal data and the functions used are not based on a good understanding of dispersal processes. This is despite the fact that, for a long time, more general modelling of species spread has shown that the form of the dispersal kernel is critical and should be based on actual data (Clark *et al.* 2003; Higgins *et al.* 2003a). In an excellent review, Thuiller *et al.* (2008) considered the range of processes important in understanding species migration in responses to climate change, which include: detailed demography and dispersal data, and spatial and temporal variation in these; biotic interactions; spatial habitat structure, evolution, and temporal stochasticity.

In this chapter, I want to take a step back from such complexity, and consider only one process in detail; that is dispersal and its realistic characterization in models of plant spread under climate change. Simple models which treat space as homogeneous and spread rate as unvarying can allow an understanding of the potential spread rates of plants. Inclusion of spatial and temporal variation in analytical models or using simulation models tends to result in lower projected spread rates (Bullock *et al.* 2008; Travis *et al.* 2011). This greater realism is also gained at the expense of generality and increased data needs (Travis *et al.* 2011). Therefore, if we treat all space as suitable habitat and allow high local population growth rates, then we can consider such models as best-case scenarios, with which the maximum reasonable spread rate of a plant species can be projected given what we know of its dispersal ability (e.g. Nathan *et al.* 2011a).

An important paper by Loarie *et al.* (2009) supplies estimates of the velocity of climate change, against which such projected plant spread rates can be compared. Using ensemble climate model projections for the period 2000–2100 and spatial climatic gradients, Loarie *et al.* (2009) derived a measure of the local velocity of movement across the Earth's surface needed to maintain a constant temperature. These suggest minimum rates of spread needed for species to track climate change.

Dispersal Ecology and Evolution. First Edition. Edited by Jean Clobert *et al.*

The global average (geometric mean) velocity is 0.42 km yr−1, but the value varies across biomes, from 0.11–1.46 km yr−1 (Figure 29.1), according to local climate change projections and topography (i.e. spatial temperature gradients). Nathan *et al.* (2011a) recently used a simple model—assuming spatial and temporal homogeneity—to explore the potential spread rates of 12 wind-dispersed North American trees. They used realistic dispersal kernels, projected using the mechanistic WALD model. Under reasonable parameter values, they concluded that the spread rates would not be fast enough to allow these trees to track climate change.

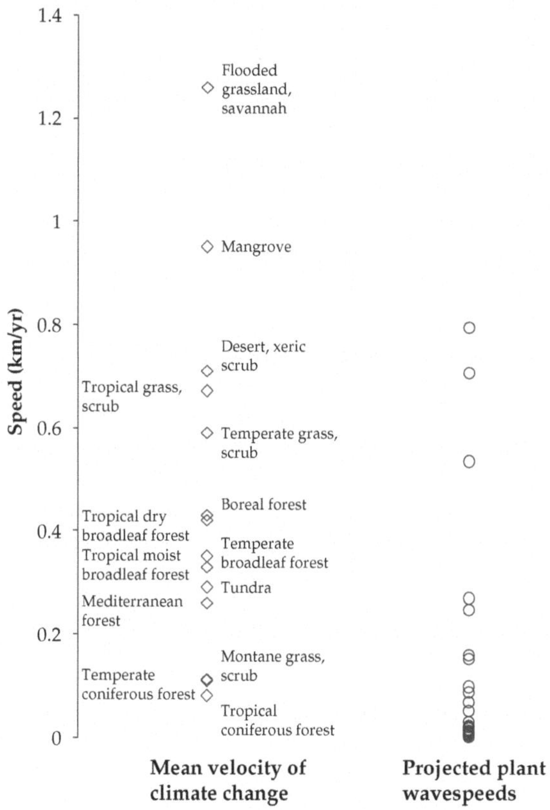

Figure 29.1 A comparison of projected rates of spread of selected plant species with estimated velocities of climate change in the world's biomes. The climate change velocities were calculated by Loarie *et al.* (2009) and the values given are the geometric means for each biome for the period 2000–2100. The projected rates of spread are for each of the 32 plant species in Table 29.1, and were calculated with Equation 29.2 using published dispersal data and an assumed λ value of 2.

29.2 Modelling plant spread

The modelling of plant spread has followed two divergent paths. Complex simulation approaches can simulate the spread process in explicit space and allow detailed and stochastic representations of demography and dispersal (Grimm and Railsback 2005). While such approaches may be appealing due to their apparent realism, they are data-hungry, are often poorly validated and model structure varies from study to study. Analytical models of population spread are based on well-developed mathematical theory. While these analytical models represent a highly simplified version of reality, their assumptions are well characterized, and the models have been shown to make accurate predictions in some circumstances (Bullock *et al.* 2008). These two types of modelling can inform each other, and recent research is seeking to develop hybrid approaches (Travis *et al.* 2011).

For the purposes of this study, the use of an analytical modelling method allows a focus on the dispersal process and projection of population spread under best-case scenarios. The approach used here is within the class of wavespeed models, which represent the spread of a population as a travelling wave in one dimension. The theory and application of wavespeed models is explained in Kot *et al.* (1996), Neubert and Caswell (2001), and Lewis *et al.* (2006). Dispersal is assumed to be isotropic (having no directional bias) in a homogeneous environment so that the distribution of dispersal distances r, i.e. the dispersal kernel $k(r)$, depends only on the relative locations of the dispersal start y and finish points x;

$$k(r) = k(x - y). \qquad \text{(Equation 29.1)}$$

Assuming population dynamics are simple and show no Allee effects nor overcompensation, λ represents the per capita population growth rate at low density. A local population then grows and spreads in one dimension in the form of a travelling wave, which has an asymptotic speed c^*,

$$c^* = \min_{s>0}\left(\frac{1}{s}\ln[\lambda M(s)]\right), \qquad \text{(Equation 29.2)}$$

where $M(s)$ is the moment generating function (mgf) of the dispersal kernel $k(r)$ in one dimension and Equation 29.2 is solved according to the auxiliary variable s. This wavespeed c^*, is therefore the modelled rate of spread of the population, which can be derived analytically using data on demography and dispersal.

Many functions which are used to describe dispersal kernels have an mgf, such as the Laplace or inverse Gaussian (Wald) functions (Forbes *et al.* 2011). However, other dispersal functions (e.g. lognormal, Weibull) do not have a simple mgf. More critically, many dispersal kernels are not well described by mathematical functions (Bullock *et al.* 2006; Chapter 15). Clark *et al.* (2001a; see Lewis *et al.* 2006 for a user's guide) show a method for deriving an empirical, non-parametric estimate of the mgf from raw dispersal data, which can be either those measured or those generated from a dispersal model (e.g. Skarpaas and Shea 2007; Soons and Bullock 2008). The exact method used will depend on the form of the dispersal dataset (Lewis *et al.* 2006).

29.3 Real dispersal data: sourcing and analysis

To determine what the best available dispersal data can tell us about potential spread rates, I searched the peer-reviewed literature for studies providing dispersal kernels for plant species representing a variety of dispersal mechanisms and life forms, and from a range of biomes. This search was not systematic, nor was it designed to make use of all published dispersal data. Rather, I aimed to find studies with good quality data, and which measured dispersal over the longest distances currently available for each dispersal mechanism. Thus, I used my own eclectic bibliography of dispersal studies, and supplemented this with a search over all dates on Web of Science using the term: topic = 'dispersal' AND 'seed*' AND 'distance OR kernel OR curve OR shadow' AND 'measure* OR method* OR sampl*'. I sifted the resulting citations to discover extractable data conforming to the criteria described. I included only studies which measured dispersal by the following means. 1) Studies in which dispersal distances had been measured by tracking seeds or locating their final positions, including following animals carrying seeds (e.g. Whitney 2002), the use of seed traps (e.g. Bullock and Clarke 2000), or utilizing markers such as gene sequences (e.g. Jordano *et al.* 2007) or stable isotopes (e.g. Carlo *et al.* 2009). 2) Studies in which the movement data for the animal vector was combined with measures of seed retention on or in the animal to derive seed dispersal distances (e.g. Lenz *et al.* 2010; Koike *et al.* 2010).

Mechanistic models provide a means to accurately determine dispersal kernels, especially long distance events (Nathan *et al.* 2011b). As such, they tend to predict longer dispersal tails than measured kernels (Skarpaas and Shea 2007). Mechanistic models are available only for wind dispersal, and so to achieve comparability of dispersal kernels among the different vectors I did not include mechanistic kernels.

I extracted data for 32 species from 25 studies, representing grasses, herbs, shrubs, and trees, and a range of biomes from temperate grassland to tropical forest (Table 29.1). In most cases the data were graphical, and I extracted them using a drawing board. These provided kernels for dispersal by dehiscing pods (ballistic), wind (including an example of tumble dispersal along the ground), ants, birds, a range of mammals, and a rare example for seeds carried by a flowing river (Table 29.1; Figure 29.2). Some studies graphed dispersal distances of individual seeds (Table 29.1), but the majority of studies graphed the data as proportions, numbers, or densities of seeds in distance bins. Where necessary, I converted the data into bin-specific densities, which I used to derive probability densities, from which I then simulated a large number (10^6) of individual dispersal distances. In most cases the dispersal measures were taken in two dimensions and so the set of N distances, either taken directly from the paper or derived as just explained, were used to calculate an empirical estimation (denoted by the superscript E) of the mgf using

$$M^E(s)=\frac{1}{N}\sum_{i=1}^{N} I_0(sr_i), \qquad \text{(Equation 29.3)}$$

where I_0 is the modified Bessel function of the first kind. In some cases (Table 29.1), dispersal was measured along linear features (e.g. along a river or a track). I treated these dispersal kernels as one-dimensional and so calculated the empirical mgf using

$$M^E(s)=\frac{1}{N}\sum_{i=1}^{N} \exp(sr_i). \qquad \text{(Equation 29.4)}$$

Table 29.1 Characteristics of plant species for which dispersal data were extracted from published studies, and used to model population spread. Percentiles of the 1-dimensional dispersal kernels are derived from the extracted data which were analysed as described in the text. Wavespeeds are given for three values of λ, calculated as described in the text.

Species	Life form	Habitat	Dispersal mechanism	Dispersal percentiles (m)		Wavespeed (myr^{-1})			Study
				50th	99th	λ = 1.1	λ = 2	λ = 4	
Geranium maculatum	Perennial herb	Wet woodland, USA	Ballistic	1.62	4.20	0.9	2.2	3.0	Stamp and Lucas (1983)
G. carolinum	Perennial herb	Open herbaceous vegetation, USA	Ballistic	1.98	3.88	1.0	2.4	3.1	Stamp and Lucas (1983)
Asclepias syriaca	Perennial herb	Tall-grass prairie, USA	Wind	7.16	24.9	4.0	11.1	16.4	Platt and Weiss (1977)
Apocynum sibiricum	Perennial herb	Tall-grass prairie, USA	Wind	12.9	49.2	7.3	20.2	29.4	Platt and Weiss (1977)
Cirsium undulatum	Short-lived herb	Tall-grass prairie, USA	Wind	8.73	25.0	4.7	12.3	16.8	Platt and Weiss (1977)
Mirabilis hirsuta	Perennial herb	Tall-grass prairie, USA	Wind	0.16	0.72	0.1	0.3	0.5	Platt and Weiss (1977)
Verbena stricta	Perennial herb	Tall-grass prairie, USA	Wind	0.54	1.87	0.3	0.8	1.2	Platt and Weiss (1977)
Oenethera biennis	Short-lived herb	Tall-grass prairie, USA	Wind	1.06	3.29	0.5	1.5	2.1	Platt and Weiss (1977)
Calluna vulgaris	Low shrub	Temperate grassland, England	Wind	0.20	0.82	1.3	7.0	12.9	Bullock and Clarke (2000)
Erica cinerea	Low shrub	Temperate grassland, England	Wind	0.19	0.93	1.2	6.9	12.7	Bullock and Clarke (2000)
Protea repens	Low shrub	Fynbos, South Africa	Wind + tumble	4.26	41.5	4.5	15.7	25.4	Higgins *et al.* (2003b)
Vincetoxicum rossicum†	Vine (invasive)	Mown grassland, Canada	Wind	2.55	16.3	1.5	4.4	6.8	Cappuccino *et al.* (2002)
Lonchocarpus pentaphyllus	Canopy tree	Tropical forest, Barro Colorado Island	Wind	1.58	27.7	3.2	10.0	15.5	Augspurger and Hogan (1983)
Tilia americana	Canopy tree	Temperate forest, Canada	Wind	4.60	56.4	6.5	20.4	31.5	Greene and Calogeropoulos (2002)
Fraxinus excelsior	Canopy tree	Temperate forest, Germany	Wind	4.17	54.7	6.5	20.0	30.6	Wagner (1997)
Adriana quadripartita	Tall shrub	Australian temperate coastal scrub	Ant	0.12	2.66	0.3	1.0	1.6	Beaumont *et al.* (2009)
Melampyrum lineare	Annual herb	Temperate herbaceous vegetation, USA	Ant	0.10	1.71	0.2	0.8	1.2	Gibson (1993)
Acacia suaveolens	Medium shrub	Temperate sclerophyllous heath, Australia	Ant	0.26	3.65	0.4	1.5	2.5	Andersen (1988)
Acacia holosericea	Tall shrub	Tropical savanna woodland, Australia	Ant	0.14	5.14	0.6	2.2	3.8	Parr *et al.* (2007)
Acacia ligulata†	Tall shrub	Arid red-earth dunes, Australia	Ant	87.0	180	32.7	86.3	119	Whitney (2002)
Capsicum anuum†	Annual herb	Desert grassland, USA	Bird	12.5	105	9.5	28.6	43.5	Carlo *et al.* (2009)
Unspecified	-	Fragmented temperate forest, S. Africa	Bird	1.64	21.2	25.1	151	289	Lenz *et al.* (2010)
Berberis darwinii	Tall shrub (invasive)	Regenerating forest, New Zealand	Bird	1.57	34.3	10.0	49.8	89.4	McAlpine and Jesson (2008)
Prunus mahaleb	Tall shrub	Open dry woodland, Spain	Bird, mammal	8.60	249	108	706	1,376	Jordano *et al.* (2007)
Simarouba amara	Canopy tree	Tropical forest, Barro Colorado Island	Bird, mammal	47.0	611	82.9	247	374	Hardesty *et al.* (2006)
Quercus ilex	Canopy tree	Mediterranean fields and forests, Spain	Bird	6.64	116	15.9	66.8	116	Pons and Pausas (2007)
Aglaia flavida	Canopy tree	Tropical hill forest, New Guinea	Bird	15.9	441	44.2	158	259	Mack (1995)
Aesculus turbinata	Canopy tree	Temperate floodplain forest, Japan	Rodent	2.19	43.4	5.5	19.6	32.2	Hoshizaki *et al.* (1999)
Trillium grandiflorum	Perennial herb	Temperate forest, USA	Deer	77.7	1365	161	535	882	Vellend *et al.* (2003)
Virola calophylla	Canopy tree	Tropical floodplain forest, Peru	Monkey	43.5	836	84.1	269	421	Russo *et al.* (2006)
Prunus jamasakura	Tall shrub	Temperate forest, Japan	Bear	54.5	1469	196	794	1,487	Koike *et al.* (2011)
*Ailanthus altissima**	Pioneer tree (invasive)	Urban Germany	River	2.81	99.7	22.2	97.0	168	Säumel and Kowarik (2010)

The studies supplied 2-dimensional dispersal data in distance bins, except for the studies marked: *, where data were in 1-dimension, and, † where dispersal distances were given

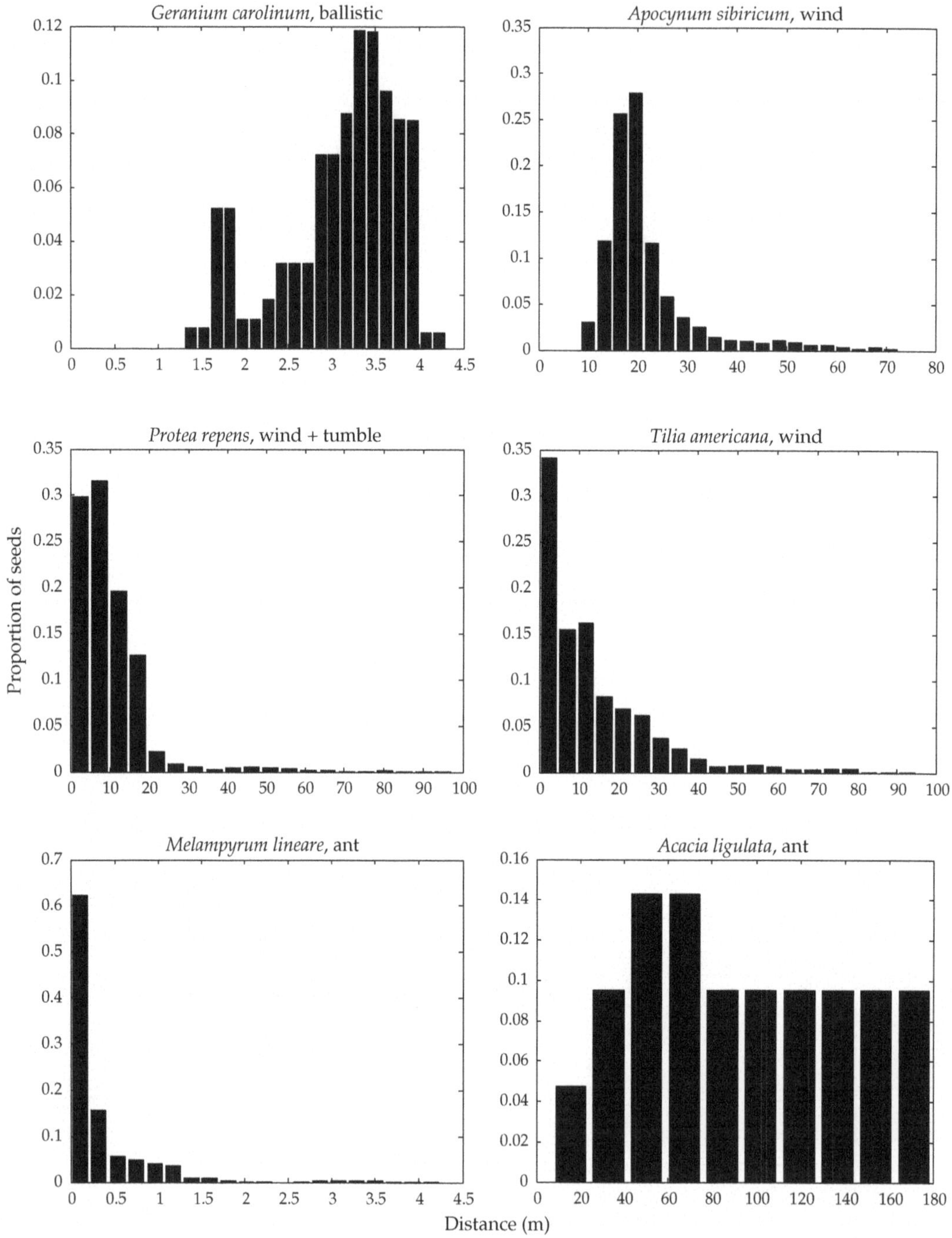

Figure 29.2 Dispersal kernels for a selection of the species listed in Table 29.1, illustrating a range of dispersal mechanisms. Dispersal data were extracted from published studies in a variety of forms. To allow comparison, all kernels show dispersal distances for a single direction (i.e. one dimension) and the proportion of seeds in each distance bin.

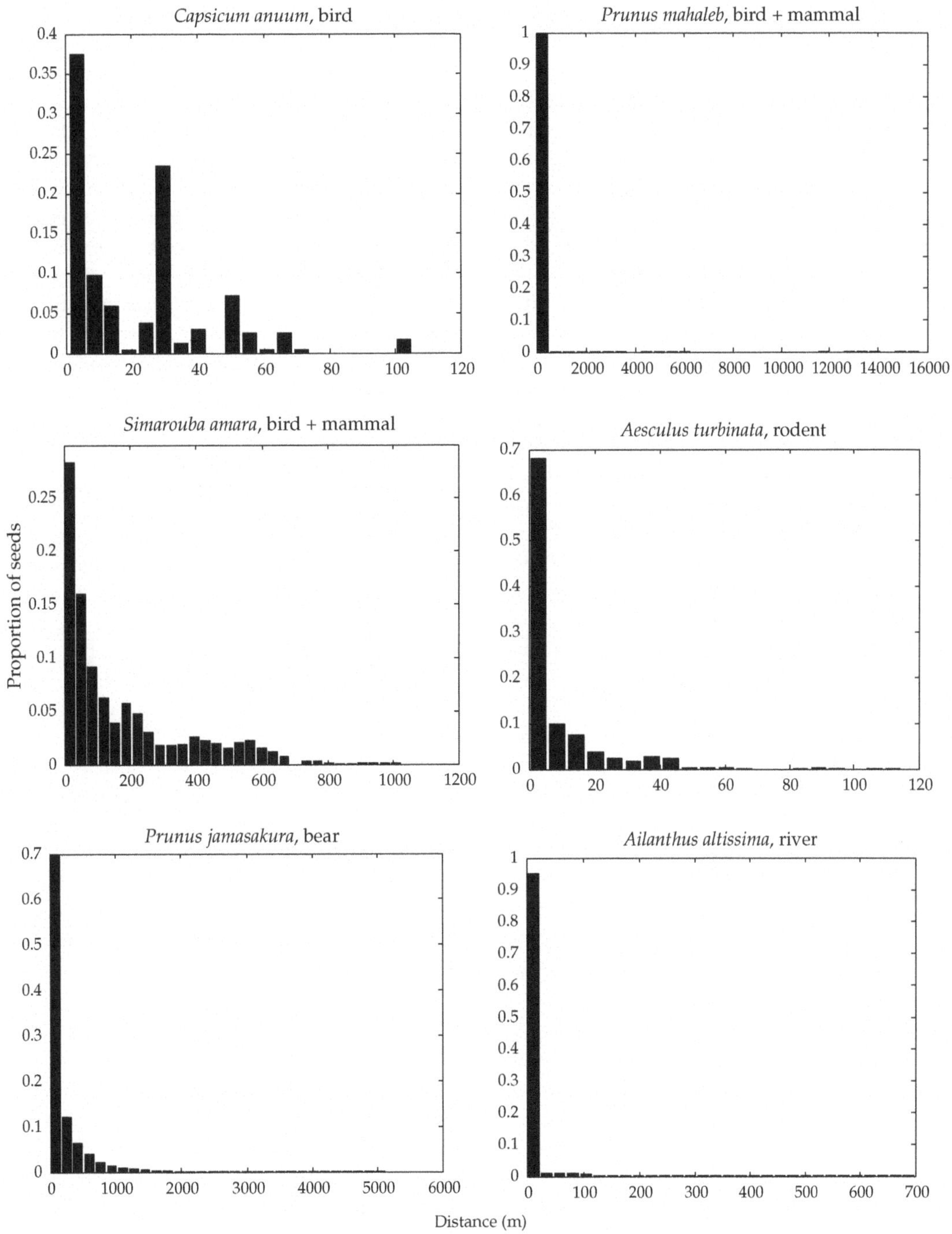
Capsicum anuum, bird
0.4
0.35
0.3
0.25
0.2
0.15
0.1
0.05
0
0 20 40 60 80 100 120
Prunus mahaleb, bird + mammal
1
0.9
0.8
0.7
0.6
0.5
0.4
0.3
0.2
0.1
0
0 2000 4000 6000 8000 10000 12000 14000 16000
Simarouba amara, bird + mammal
0.25
0.2
0.15
0.1
0.05
0
0 200 400 600 800 1000 1200
Aesculus turbinata, rodent
0.7
0.6
0.5
0.4
0.3
0.2
0.1
0
0 20 40 60 80 100 120
Proportion of seeds
Prunus jamasakura, bear
0.7
0.6
0.5
0.4
0.3
0.2
0.1
0
0 1000 2000 3000 4000 5000 6000
Ailanthus altissima, river
1
0.9
0.8
0.7
0.6
0.5
0.4
0.3
0.2
0.1
0
0 100 200 300 400 500 600 700
Distance (m)

More explanation of this method is given by Lewis *et al.* (2006).

The extracted dispersal kernels, once transformed into comparable one-dimensional distance distributions (Figure 29.2), show a variety of shapes and extents (Table 29.1). In particular, many kernels show a rapid descent to a long and quite flat tail (Figure 29.2). Thus, the 99th percentile of the kernel is usually much greater than the median (Table 29.1). Indeed, in several cases, the tail is so long and flat that the 99th percentile describes the extent of dispersal poorly; for example, compare this percentile in Table 29.1 with the maximum dispersal distances in Figure 29.2 for *Prunus mahaleb, P. jamasakura*, and *Ailanthus altissima*.

These kernels represent the longest measured dispersal distances for each life form/dispersal mechanism combination, and they show great variation even within dispersal mechanisms (Table 29.1). Ballistic dispersal is known to achieve only short distances (Willson 1993), but wind and ant dispersal can achieve quite long distances, depending on the system. Dispersal by birds and mammals has been extremely well studied over the last decade, and the examples here show some very long distances. Other vectors, such as dispersal by water, have been much less studied, but the river example shows the potential for appreciable distances.

29.4 Projected spread rates of selected species

Detailed demographic information can be used in wavespeed models, for example by using structured population models (Neubert and Caswell 2000; Soons and Bullock 2008). To allow a focus on the dispersal data, I used Equation 29.2, which has a simple, scalar representation of demography, λ. This rate of population increase is specified for low population density, as it represents demography at the front of the travelling wave. To allow straightforward comparison among the species, I used the same λ values for all, but used a range. Among real plant populations, the calculated λ can be much greater than one; in a collation of demographic studies, Silvertown *et al.* (1993) report a maximum of 11.8. Allowing that the wavespeed modelling requires an increasing population ($\lambda > 1$), I chose arbitrary values for λ at low density of 1.1, 2, and 4.

Solving Equation 29.2 using the estimated mgf for each species and the chosen λ values gave, unsurprisingly, a wide range of wavespeeds, from a very few metres/year, up to over a kilometre/year (Figure 29.1). Taking $\lambda = 2$ as a reasonable value for population growth at low density, seven of the 32 species had projected wavespeeds above the lowest climate change velocity of 0.11 km yr^{-1} (Figure 29.1; Table 29.1), and these were all dispersed by vertebrates. However, about half of the species had very low speeds, at most an order of magnitude less than 0.11 km yr^{-1}.

29.5 Enhancing dispersal and spread under climate change

My spread modelling using actual dispersal data gives good and bad news. It suggests the potential for some plant species to track climate change, while also showing that many species have dispersal abilities that fall extremely short of those necessitated by the predicted velocities of climate change. Even the good news must be tempered. These models suggest best-case scenarios. The introduction of processes into wavespeed models such as temporal stochasticity (Caswell *et al.* 2010; Travis *et al.* 2011), spatial variation in habitat quality (Dewhirst and Lutscher 2009), or demographic stochasticity (Clark *et al.* 2001b) will tend to slow the modelled rates of spread. However, in the following I will concentrate on the majority of cases where dispersal seems too limited to allow climate tracking.

29.5.1 Better dispersal data

Dispersal is difficult to measure, and it is especially problematic to detect rare and long distance events. This issue is critical, because spread rates are highly sensitive to the extent and frequency of the furthest dispersal distances (Caswell *et al.* 2003; Bullock *et al.* 2008). The last 15 years has seen a great deal of research in characterizing empirical plant dispersal kernels, and an increase in reported maximum distances for most vectors (compare with the review in Willson 1993). Novel methods to improve dispersal measurements are appearing all the time, including: refining experimental design (Skarpaas *et al.* 2005, Bullock *et al.* 2006), molecular and statistical methods for seed parentage analysis (Jones *et al.* 2008), new approaches for marking and detecting seeds (Carlo *et al.* 2009; Lemke *et al.* 2009), and combining tracking with measures of seed retention for animal vectors (Lenz *et al.* 2010; Bullock *et al.* 2011). Thus, as dispersal data gathering improves, estimates of spread rates may increase. One can only speculate about the extent of these increases, but I suggest that there is unlikely to be much increase in the measured distances for ballistic dispersal, dispersal by ants or wind dispersal of low-growing plants. Dispersal kernels in such circumstances are relatively easy to measure because distances reached are somewhat limited. The measurement of dispersal by vertebrates has advanced greatly in recent years through the use of molecular methods and the improved ability to track animals. It is probable that the longest of the distances reported in Table 29.1 and Figure 29.2 for vertebrate dispersal of seeds is unlikely to be surpassed greatly.

Wind dispersal is more problematic, because it is difficult to follow air-borne seeds. Thus empirical studies are likely to result in underestimates of dispersal distances. Happily, wind dispersal is increasingly studied using a set of accurate mechanistic models (e.g. Soons *et al.* 2004; Katul *et al.* 2005; Nathan *et al.* 2011b), which represent the physical process of dispersal independently of any dispersal data and are parameterised using information on plant and vector characteristics (Chapter 15). In some cases, by allowing a better estimation of the tail than that measured, mechanistic models can increase the projected wavespeed (Skarpaas and Shea 2007). However, while mechanistic models sometimes compare well against measured dispersal data (Soons *et al.* 2004; Skarpaas and Shea 2007), they have also been found to give a variable match (Stephenson *et al.* 2007), or even to underestimate the measured tail (Katul *et al.* 2005; Soons and Bullock 2008).

Other vectors are extremely poorly studied. Dispersal in water—along rivers or in ocean currents—is known to be a major process for many plants (e.g. Wadsworth *et al.* 2000; Bond *et al.* 2005), but kernels are not well characterized (Säumel and Kowarik 2010). It is likely that a focus of research on dispersal by water will provide dispersal kernels covering extremely long distances.

29.5.2 Different dispersal mechanisms—human-mediated dispersal

It is increasingly clear that the dispersal mechanism deduced from morphology of the seed or fruit does not represent the sole mechanism by which many seeds are dispersed (Bullock *et al.* 2006). In some cases, such 'non-standard' mechanisms may take seeds further than the assumed standard mechanism (Higgins *et al.* 2003b). Perhaps the most important non-standard seed dispersal mechanism is that by humans. There are plenty of indications of the potential for dispersal of seed and other organisms over massive distances by human activities; for example, by aeroplanes (Tatem *et al.* 2009), or shipping (Kaluza *et al.* 2010). Quantification of human-mediated dispersal kernel is still rare. Kernels have been measured for dispersal directly on humans (Bullock and Primack 1977; Wichmann *et al.* 2009; Pickering *et al.* 2011), on vehicles (Bullock *et al.* 2003, and on livestock driven by humans (Manzano and Malo 2006). Wichmann *et al.* (2009) found a maximum wind dispersal distance of the herb *Brassica oleracea* of 70 m, but dispersal on the boots of hikers extended the maximum to at least 5 km. Using the spread modelling approach described above, with $\lambda = 2$, gives a projected wavespeed of 5.5 m yr^{-1} for wind dispersal, but of 38.8 m yr^{-1} for human dispersal (assuming humans pick up only 1% of seeds). Manzano and Malo (2006) found that sheep being shepherded between distant seasonal pastures dispersed several herbs up to 400 km. Taking these as the full dispersal kernels (i.e. assuming all seeds are dispersed in this way), and $\lambda = 2$, results in extremely fast spread rates of over 200 km yr^{-1}.

These results suggest that human-mediated dispersal may be important in increasing spread rates of plants under climate change, and so research into dispersal should concentrate more on humans as vectors. Accidental dispersal of seeds by humans is most likely to affect those species growing in human-impacted or created environments such as urban areas and roadsides (Hodkinson *et al.* 1997; von der

Lippe and Kowarik 2008). However, because humans have penetrated most of the world biomes, they act as significant dispersal agents even in remote areas (Hughes *et al.* 2010; Pickering *et al.* 2011).

29.5.3 Changes in the behaviour of dispersal vectors

As well as affecting the growing conditions for plants, climate change may alter the dispersal process directly. Some global circulation and regional climate models provide projections of changes in surface windspeeds under climate change, although these are rather uncertain. Both increases and decreases in mean windspeed are projected, depending on the model used, the scenario, and the region considered. Windspeeds were projected by Räisänen *et al.* (2004) to increase in northern Europe by up to 12% by the late twenty-first century, but to decrease by up to 12% in southern Europe. Nathan *et al.* (2011a) showed that rates of spread of wind-dispersed trees could be quite sensitive to altered horizontal windspeeds and their effect on dispersal distances. However, the uncertainty in projections currently makes it difficult to determine whether windspeeds and thus wavespeeds will increase or decrease in future climates.

One can only speculate about changes in the behaviour of other vectors under climate change. Increased flooding (especially in the northern hemisphere; Milly *et al.* 2008) could increase dispersal by water. Where seeds are dispersed by large herbivores, it is possible that the increased movement of these vectors, as they change their biogeographic ranges in response to climate change (e.g. Parmesan and Yohe 2003), will aid the spread of the associated plants. Conversely, if major animal dispersal agents decline or go extinct due to climate change, then spread of the affected plant species will be slowed. The impacts of climate change on human society are multifarious and hard to predict (Moss *et al.* 2010), but these are likely to have ramifications for seed dispersal by humans.

29.5.4 Evolution of dispersal ability

Modelling studies have shown that climate change, by gradually shifting the location of suitable habitat, should select for increased dispersal ability and lead to further dispersal distances (Dytham 2009; Travis *et al.* 2010). Empirical studies have shown that alterations in landscape spatial structure can lead to evolutionary changes over very few generations; for example, by modifying the ratio of plumed to non-plumed seeds (Cheptou *et al.* 2008), or by altering seed and pappus volumes (Cody and Overton 1996). Such examples tend to demonstrate secondary loss of dispersal ability in island conditions. In reality, we know very little about the potential for plants to evolve increased dispersal ability, in terms of both how much evolution in the short-term can change realized dispersal distances to the extent needed, and how rapidly such evolution can take place.

29.5.5 Conclusions and a comment on assisted migration

The analyses carried out here provide support for the general idea that the dispersal kernels of many plant species are too limited to allow them to spread at a rate sufficient to match the velocity of climate change. Dispersal may be sufficient in some cases, especially those involving dispersal by vertebrates, and certain factors, especially human-mediated dispersal, may further enhance dispersal and spread rates. However, despite the many uncertainties detailed in this section, we should be prepared for the fact that the response of many plants to climate change will be primarily limited by dispersal. Conservationists have come up with few practicable responses to this problem. Increasing landscape connectivity (e.g. Vos *et al.* 2008) does not address the issue that dispersal, even in homogeneous environments, may not be sufficient to track climate change. Assisted migration or colonization is a highly controversial proposition (e.g. Ricciardi and Simberloff 2009; Vitt *et al.* 2009) designed to overcome the dispersal limitation highlighted in this chapter, whereby humans would move selected species into newly available habitat. A consideration of plants highlights the impracticality of this strategy, which necessarily targets charismatic and high-profile species. It would be virtually impossible to assist the migration of the many thousands of plant species (most of which are poorly studied) which will be unable to track climate change unaided.

References

Andersen, A.N. (1988). 'Dispersal distance as a benefit of myrmecochory' *Oecologia* **75**: 7507–11.

Augspurger, C. and Hogan, K.P. (1983). 'Wind dispersal of fruits with variable seed number in a tropical tree', *American Journal of Botany* **70**: 1031–7.

Beaumont, K.P., Mackay, D.A., Whalen, M.A. (2009). 'Combining distances of ballistic and myrmecochorous seed dispersal in *Adriana quadripartita* (Euphorbiaceae)', *Acta Oecologica-International Journal of Ecology* **35**: 3429–36.

Bond, J.M., Daniels, R., Bioret, F. (2005). 'Genetic diversity in *Crambe maritima* along the English Channel: the role of ocean currents in determining population structure', *Ecography* **28**: 374–84.

Bullock, J.M. and Clarke, R.T. (2000). 'Long distance seed dispersal by wind: measuring and modelling the tail of the curve', *Oecologia* **124**: 506–21.

Bullock, J.M., Galsworthy, S.J., Manzano, P. *et al.* and (2011). 'Process-based functions for seed retention on animals: a test of improved descriptions of dispersal using multiple data sets', *Oikos* **120**: 1201–1208.

Bullock, J M., Moy, I.L., Coulson, S.J. *et al.* (2003). 'Habitat-specific dispersal: environmental effects on the mechanisms and patterns of seed movement in a grassland herb *Rhinanthus minor*', *Ecography* **26**: 692–704.

Bullock, J.M., Pywell, R.F., Coulson-Phillips, S.J. (2008). 'Managing plant population spread: Prediction and analysis using a simple model', *Ecological Applications* **18**: 945–53.

Bullock, J.M., Shea, K., Skarpaas, O. (2006). 'Measuring plant dispersal: an introduction to field methods and experimental design', *Plant Ecology* **186**: 217–34.

Bullock, S.H. and Primack, R.B. (1977). 'Comparative experimental study of seed dispersal on mammals', *Ecology* **58**: 681–6.

Cappuccino, N., Mackay, R., Eisner, C. (2002). 'Spread of the invasive alien vine *Vincetoxicum rossicum*: Tradeoffs between seed dispersability and seed quality'. *American Midland Naturalist* **148**: 263–70.

Carlo, T.A., Tewksbury, J.J. del Rio, C.M. (2009). 'A new method to track seed dispersal and recruitment using N-15 isotope enrichment', *Ecology* **90**: 3516–25.

Caswell, H., Lensink, R., Neubert, M.G. (2003). 'Demography and dispersal: Life table response experiments for invasion speed', *Ecology* **84**: 1968–78.

Caswell, H., Neubert, M., Hunter, C. (2010). 'Demography and dispersal: invasion speeds and sensitivity analysis in periodic and stochastic environments', *Theoretical Ecology* **4**: 1–15.

Cheptou, P.O., Carrue, O., Rouifed, S. *et al.* (2008). 'Rapid evolution of seed dispersal in an urban environment in the weed *Crepis sancta*', *Proceedings of the National Academy of Sciences of the USA* **105**: 3796–9.

Clark, J., Horvath, L., Lewis, M. (2001). 'On the estimation of spread rate for a biological population', *Statistics and Probability Letters* 51: 225–34.

Clark, J.S., Lewis, M., Horvath, L. (2001). 'Invasion by extremes: Population spread with variation in dispersal and reproduction', *American Naturalist* **157**: 537–54.

Clark, J.S., Lewis, M., McLachlan, J.S. *et al.* (2003). 'Estimating population spread: What can we forecast and how well?', *Ecology* **84**: 1979–88.

Cody, M.L. and Overton, J.M. (1996). 'Short-term evolution of reduced dispersal in island plant populations', *Journal of Ecology* **84**: 53–61.

Dewhirst, S. and Lutscher, F. (2009). 'Dispersal in heterogeneous habitats: thresholds, spatial scales, and approximate rates of spread', *Ecology* **90**: 1338–45.

Dytham, C. (2009). 'Evolved dispersal strategies at range margins', *Proceedings of the Royal Society B* **276**: 1407–13.

Engler, R., Randin, C.F., Vittoz, P. *et al.* (2009). 'Predicting future distributions of mountain plants under climate change: does dispersal capacity matter?', *Ecography* **32**: 34–45.

Forbes, C., Evans, M., Hastings, N. *et al.* and (2011). *Statistical distributions*, 4th edition, Hoboken, Wiley-Blackwell.

Gibson, W. (1993). 'Selective advantages to hemi-parasitic annuals, genus *Melampyrum*, of a seed-dispersal mutualism involving ants. 1. Favorable nest sites;, *Oikos* **67**: 334–44.

Greene, D. F. and Calogeropoulos, C. (2002). 'Measuring and modelling seed dispersal of terrestrial plants', in J.M. Bullock, R.E. Kenward, R. Hails (eds)and *Dispersal ecology*. Backwells, Oxford.

Grimm, V. and Railsback, S.F. (2005). '*Individual-based modeling and ecology*. Princeton University Press, Oxford

Hardesty, B.D., Hubbell, S.P., Bermingham, E. (2006). 'Genetic evidence of frequent long-distance recruitment in a vertebrate-dispersed tree', *Ecology Letters* **9**: 516–25.

Higgins, S.I., Clark, J.S., Nathan,R. *et al.* and (2003). 'Forecasting plant migration rates: managing uncertainty for risk assessment', *Journal of Ecology* **91**: 341–7.

Higgins, S.I., Nathan, R., Cain, M.L. (2003). 'Are long-distance dispersal events in plants usually caused by non-standard means of dispersal?', *Ecology* **84**: 1945–56.

Hillyer, R. and Silman, M.R. (2010). 'Changes in species interactions across a 2.5 km elevation gradient: effects on plant migration in response to climate change', *Global Change Biology* **16**: 3205–14.

Hodkinson, D.J. and Thompson, K. (1997). 'Plant dispersal: the role of man. *Journal of Applied Ecology* **34**: 1484–96.

Hoshizaki, K., Suzuki, W., Nakashizuka, T. (1999). 'Evaluation of secondary dispersal in a large-seeded tree *Aesculus turbinata*: a test of directed dispersal', *Plant Ecology* **144**: 167–76.

Hughes, K.A., Lee, J.E., Ware, C. *et al.* (2010). 'Impact of anthropogenic transportation to Antarctica on alien seed viability', *Polar Biology* **33**: 1125–30.

Iverson, L.R., Schwartz, M.W., Prasad, A.M. (2004). 'How fast and far might tree species migrate in the eastern United States due to climate change?', *Global Ecology and Biogeography* **13**: 209–19.

Jones, F.A. and Muller-Landau, H.C. (2008). 'Measuring long-distance seed dispersal in complex natural environments: an evaluation and integration of classical and genetic methods', *Journal of Ecology* **96**: 642–52.

Jordano, P., Garcia, C., Godoy, J.A. *et al.* (2007). 'Differential contribution of frugivores to complex seed dispersal patterns', *Proceedings of the National Academy of Sciences of the USA* **104**: 3278–82.

Kaluza, P., Kolzsch, A., Gastner, M.T. *et al.* (2010). 'The complex network of global cargo ship movements', *Journal of the Royal Society Interface* **7**: 1093–103.

Katul, G.G., Porporato, A., Nathan, R *et al.* (2005). 'Mechanistic analytical models for long-distance seed dispersal by wind', *American Naturalist* **166**: 368–81.

Koike, S., Masaki, T., Nemoto, Y. *et al.* (2011). 'Estimate of the seed shadow created by the Asiatic black bear Ursus thibetanus and its characteristics as a seed disperser in Japanese cool-temperate forest', *Oikos* **120**: 280–90.

Kot, M., Lewis, M.A., van den Driessche, P. (1996). 'Dispersal data and the spread of invading organisms', *Ecology* **77**: 2027–42.

Lemke, A., von der Lippe, M., Kowarik, I. (2009). 'New opportunities for an old method: using fluorescent colours to measure seed dispersal', *Journal of Applied Ecology* **46**: 1122–8.

Lenz, J., Fiedler, W., Caprano, T. *et al.* (2010). 'Seed-dispersal distributions by trumpeter hornbills in fragmented landscapes', *Proceedings of the Royal Society B* **278**: 2257–64.

Lewis, M.A., Neubert, M.G., Caswell, H. *et al.* (2006). 'A guide to calculating discrete-time invasion rates from data', in M. Cadotte, S. Mcmahon, T. Fukami, (eds), *Conceptual ecology and invasions biology: reciprocal approaches to nature*. Kluwer Academic Press, Dordrecht, 169–92.

Loarie, S.R., Duffy, P.B., Hamilton, H. *et al.* (2009). 'The velocity of climate change', *Nature* **462**: 1052–5.

Mack, A. L. (1995). 'Distance and non-randomness of seed dispersal by the dwarf cassowary Casuarius bennetti', *Ecography* **18**: 286–95.

Manzano, P. and Malo, J.E. (2006). 'Extreme long-distance seed dispersal via sheep', *Frontiers in Ecology and the Environment* **4**: 244–8.

McAlpine, K.G. and Jesson, L.K. (2008). 'Linking seed dispersal, germination and seedling recruitment in the invasive species *Berberis darwinii* (Darwin's barberry)', *Plant Ecology* **197**: 119–29.

Midgley, G.F., Hughes, G.O., Thuiller, W. *et al.* (2006). 'Migration rate limitations on climate change-induced range shifts in Cape Proteaceae', *Diversity and Distributions* **12**: 555–62.

Milly, P.C.D., Betancourt, J., Falkenmark, M. *et al.* (2008). 'Climate change—Stationarity is dead: Whither water management?', *Science* **319**: 573–4.

Moss, R.H., Edmonds, J.A., Hibbard, K.A. *et al.* (2010). 'The next generation of scenarios for climate change research and assessment', *Nature* **463**: 747–56.

Nathan, R., Horvitz, N., He, Y.P. *et al.* and (2011). Spread of North American wind-dispersed trees in future environments', *Ecology Letters* **14**: 211–19.

Nathan, R., Katul, G., Bohrer, G. *et al.* and (2011). 'Mechanistic models of seed dispersal by wind', *Theoretical Ecology*, **4**: 113–32.

Neubert, M.G. and Caswell, H. (2000). 'Demography and dispersal: calculation and sensitivity analysis of invasion speed for structured populations', *Ecology* **81**: 1613–28.

Parmesan, C. and Yohe, G. (2003). 'A globally coherent fingerprint of climate change impacts across natural systems', *Nature* **421**: 37–42.

Parr, C.L., Andersen, A.N., Chastagnol, C. *et al.* (2007). 'Savanna fires increase rates and distances of seed dispersal by ants', *Oceologia* **151**: 33–41.

Pickering, C.M.A.M., Wichmann, M.C., Bullock, J.M. (2011). 'Quantifying human-mediated dispersal of seeds within an Australian protected area', *Biological Invasions* **13**: 1869–1880.

Platt, W.J. and Weiss, I.M. (1977). 'Resource partitioning and competition within a guild of fugitive prairie plants', *American Naturalist* **111**: 479–513.

Pons, J. and Pausas, J.G. (2007). 'Acorn dispersal estimated by radio-tracking', *Oceologia* **153**: 903–11.

Räisänen, J., Hansson, U., Ullerstig, A. *et al.* (2004). 'European climate in the late twenty-first century: regional simulations with two driving global models and two forcing scenarios', *Climate Dynamics* **22**: 13–31.

Ricciardi, A. and Simberloff, D. (2009). 'Assisted colonization is not a viable conservation strategy', *Trends in Ecology and Evolution* **24**: 248–53.

Russo, S.E., Portnoy, S., Augspurger, C.K. (2006). 'Incorporating animal behavior into seed dispersal models: Implications for seed shadows', *Ecology* **87**: 3160–74.

Säumel, I. and Kowarik, I. (2010). 'Urban rivers as dispersal corridors for primarily wind-dispersed invasive tree species', *Landscape and Urban Planning* **94**: 244–9.

Silvertown, J., Franco, M., Pisanty, I. *et al.* (1993). 'Comparative plant demography—relative importance of life-cycle components to the finite rate of increase in woody and herbaceous perennials', *Journal of Ecology* **81**: 465–76.

Skarpaas, O. and Shea, K. (2007). 'Dispersal patterns, dispersal mechanisms, and invasion wave speeds for invasive thistles', *American Naturalist* **170**: 421–30.

Skarpaas, O., Shea, K., Bullock, J.M. (2005). 'Optimizing dispersal study design by Monte Carlo simulation', *Journal of Applied Ecology* **42**: 731–9.

Soons, M.B. and Bullock, J.M. (2008). 'Non-random seed abscission, long-distance wind dispersal and plant migration rates', *Journal of Ecology* **96**: 581–90.

Soons, M.B., Heil, G.W., Nathan, R. *et al.* (2004). 'Determinants of long-distance seed dispersal by wind in grasslands', *Ecology* **85**: 3056–68.

Stamp, N.E. and Lucas, J.R. (1983). 'Ecological correlates of explosive seed dispersal', *Oecologia* **59**: 272–8.

Stephenson, C.M., Kohn, D.D., Park, K.J. *et al.* (2007). 'Testing mechanistic models of seed dispersal for the invasive Rhododendron ponticum (L.)', *Perspectives in Plant Ecology Evolution and Systematics* **9**: 15–28.

Svenning, J.C. and Skov, F. (2004). 'Limited filling of the potential range in European tree species', *Ecology Letters* **7**: 565–73.

Tatem, A.J. (2009). 'The worldwide airline network and the dispersal of exotic species: 2007–2010', *Ecography* **32**: 94–102.

Thomas, C.D., Cameron, A., Green, R.E. *et al.* (2004). 'Extinction risk from climate change', *Nature* **427**: 145–8.

Thuiller, W., Albert, C., Araujo, M.B. *et al.* (2008). 'Predicting global change impacts on plant species' distributions: Future challenges', *Perspectives in Plant Ecology Evolution and Systematics* **9**: 137–52.

Travis, J.M.J., Harris, C.M., Park, K.J. *et al.* (2011). 'Improving prediction and management of range expansions by combining analytical and individual-based modelling approaches', *Methods in Ecology and Evolution* **2**: 477–488.

Travis, J.M.J., Smith, H.S,. Ranwala, S.M.W. (2010). 'Towards a mechanistic understanding of dispersal evolution in plants: conservation implications', *Diversity and Distributions* **16**: 690–702.

Vellend, M., Myers, J.A., Gardescu, S. *et al.* (2003). 'Dispersal of *Trillium* seeds by deer: Implications for long-distance migration of forest herbs', *Ecology* **84**: 1067–72.

Vitt, P., Havens, K., Hoegh-Guldberg, O. (2009). 'Assisted migration: part of an integrated conservation strategy', *Trends in Ecology and Evolution* **24**: 473–4.

von der Lippe, M. and Kowarik, I. (2008). 'Do cities export biodiversity? Traffic as dispersal vector across urban-rural gradients', *Diversity and Distributions* **14**: 18–25.

Vos, C.C., Berry, P., Opdam, P. *et al.* (2008). 'Adapting landscapes to climate change: examples of climate-proof ecosystem networks and priority adaptation zones', *Journal of Applied Ecology* **45**: 1722–31.

Wadsworth, R.A., Collingham, Y.C., Willis, S.G. *et al.* and (2000). 'Simulating the spread and management of alien riparian weeds: are they out of control?', *Journal of Applied Ecology* **37**: 28–38.

Wagner, S. (1997). 'Ein Modell zur Fruchtausbreitung der Esche (*Fraxinus excelsior* L.) unter Berücksichtigung von Richtungseffekten', *Allgemeine Forst und Jagdzeitung* **168**: 149–55.

Whitney, K.D. (2002). 'Dispersal for distance? *Acacia ligulata seeds and meat ants Iridomyrmex viridiaeneus', Austral Ecology* **27**: 589–95.

Wichmann, M.C., Alexander, M.J., Soons, M.B. *et al.* (2009). 'Human mediated dispersal of seeds over long distances', *Proceedings of the Royal Society B* **276**: 523–32.

Willson, M.F. (1993). 'Dispersal mode, seed shadows, and colonization patterns', *Vegetatio* **108**: 261–80.

PART VII

Dispersal and Habitat Fragmentation

CHAPTER 30

Evolutionary ecology of dispersal in fragmented landscape

Michel Baguette, Delphine Legrand, Hélène Fréville, Hans Van Dyck, and Simon Ducatez

30.1 Habitat fragmentation is a non-linear process

Anthropogenic environmental changes result in habitat loss for many living organisms, which translates into the ongoing biodiversity extinction crisis (Foley *et al.* 2005; Rockstrom *et al.* 2009): direct habitat loss is recognized as the main cause of species extinction (Tilman *et al.* 1994; Dobson *et al.* 1997; Fahrig 2003). Habitat fragmentation is an inherent consequence of habitat loss: the progressive disappearance of a given habitat entails changes in both the area of remnant fragments and their spatial configuration, with consequences on the structural connectivity among remnants. This causal relationship has complicated the analysis of the relative impact of habitat loss and fragmentation on biodiversity (Lindenmayer and Fischer 2007), their effects being hard to distinguish. Fahrig's observation that 'to correctly interpret the influence of habitat fragmentation on biodiversity, the effects of these two components of fragmentation must be measured independently' (Fahrig 2003), is merely a pious wish in natural populations, given the causal relationship between habitat loss and the rupture of continuity of habitat remnants. Model-based approaches nonetheless highlight the necessity of considering the effect of both habitat loss and fragmentation on genetic variation and dynamics of metapopulations (Wiegand *et al.* 2005; Bruggeman *et al.* 2010), and further experimental designs under controlled conditions should help validate such predictions. It is indeed possible to disentangle those effects by using (quasi-)experimental designs in controlled landscapes, a method that was not yet widely applied given the inherent technical difficulties in controlling the environment at the landscape scale (but see e.g. Bierregaard *et al.* 1992; Gonzalez *et al.* 1998; Mortelliti *et al.* 2011).

The voluminous review material that addressed the consequences of habitat fragmentation on biodiversity repeatedly identified the decrease of connectivity among habitat remnants associated with habitat fragmentation as a major driver of extinction (e.g. Saunders *et al.* 1991; Ewers and Didham 2006—but see Fahrig 2003). At the same time, several modelling studies using percolation theory showed how the fragmentation effects on the degree of connectivity among habitat and the number of fragments are non-linear. These studies highlighted key thresholds during the fragmentation process: (1) when the proportion of suitable habitat in the study area falls below 60%, the larger habitat fragment generally breaks apart and no longer exists; (2) when the proportion of suitable habitat in the study area falls below ca. 40%, the distances among the habitat fragments suddenly increase; and (3) the number of fragments is maximal when the proportion of suitable habitat is around 30% (e.g. Gustafson and Parker 1992; Andren 1994; Bascompte and Sole 1996; Swift and Hannon 2010). Besides, the rate of extinction of local populations—a corollary of small population size and isolation—also responds in a non-linear way to habitat fragmentation, and plays a major role in the evolution of dispersal (Ronce *et al.* 2000; Poethke *et al.* 2003). Hence we can predict that if dispersal is sensitive to these structural

Dispersal Ecology and Evolution. First Edition. Edited by Jean Clobert *et al.*

aspects of fragmentation, the evolution of dispersal will also be non-linear in response to habitat fragmentation (Chapter 31).

30.2 Models of dispersal evolution along gradients of fragmentation

Several modelling approaches used either simulations or analytical models to address how dispersal rates evolve in spatially structured populations according to the level of habitat fragmentation. Overall, these models confirmed a non-monotonic relationship between habitat availability and dispersal rates along the gradient of fragmentation (Travis and Dytham 1999; Heino and Hanski 2001; Mathias *et al.* 2001).

All these models basically showed that the cost-benefit balance of dispersal is the main driver of its evolution in fragmented landscapes. This balance is influenced by biotic and abiotic parameters of the environment (habitat geometry and quality, habitat persistence, habitat predictability, landscape matrix suitability), and by population features (kin competition and inbreeding, variance in reproductive success) (Fahrig 2007; Poethke *et al.* 2007; Ronce 2007; Bonte *et al.* 2012; Chapter 31). Besides this general pattern, these models provided interesting insights into the emergence of dispersal polymorphism according to the configuration and the quality of habitat remnants. At intermediate levels of fragmentation and medium time of habitat persistence, the grain of the landscape (i.e. the spatial organization of the resource and habitat patches, Baguette and Van Dyck 2007) becomes important: low-dispersal phenotypes are advantaged in isolated local populations, whereas high-dispersal phenotypes are advantaged in clustered local populations (Travis and Dytham 1999). Moreover, habitat quality also modifies the selection on dispersal, with low-dispersal phenotypes occurring in high-quality habitats only, and high-dispersal phenotypes occurring in habitats of both good and poor quality (Mathias *et al.* 2001). A high-dispersal phenotype therefore may be regarded as part of a generalist strategy, with individuals opting for more risk spreading in space, whereas the low-dispersal phenotype is a specialist that utilizes the locally high quality of the habitat (Mathias *et al.* 2001).

However, these models address only emigration (or dispersal propensity); that is, the first step of dispersal. If dispersal is considered as an integrated process, including emigration, transfer, and settlement, dispersal distances will reflect the two latter steps. Using individual-based, simulation models, Bonte *et al.* (2010) simulated the evolution of two strategies of dispersal distance: short- and long-distance dispersal, respectively, along habitat fragmentation gradients. Empirical data indeed document the co-existences of these strategies in metapopulations (Hovestadt *et al.* 2011). The model predicted that both strategies were counter-selected when habitat fragmentation increased. However, for a given fragmentation level, when habitat fragments were more clumped, the short-dispersal strategy was favoured, whereas the long-dispersal distance strategy was counter-selected (Bonte *et al.* 2010).

Besides their heuristic value in revealing the non-linearity of the evolution of dispersal rates along gradients of fragmentation, or the diverging evolution of dispersal distance strategies according to habitat configuration, these models unambiguously predict that dispersal is under strong selective pressure in fragmented landscapes, which can entail the emergence of dispersal polymorphism under certain conditions (Travis and Dytham 1999; Mathias *et al.* 2001).

30.3 Spatial strategies, dispersal timing, and habitat fragmentation

In mobile animal species, the timing and the frequency of dispersal are contrasted mainly according to individual strategies of space use. The interplay between resource dynamics and life-history evolution indeed has shaped two broad, contrasted spatial strategies that co-exist among species (Mueller and Fagan 2008), but also within species (Cote *et al.* 2010). Sedentary individuals are those animals that spent most of their lifetime in the same area (home range or territory), in which they find all the resources required to complete their life cycle, including mates. In contrast, nomadic individuals are always on the move, constantly sampling the environment to acquire the resources they need. In sedentary individuals, dispersal is a rare event that occurs usually during well-defined periods of the life cycles. In nomadic individuals, dispersal is susceptible to occur any time in the individual lifetime. We suggest that the sensitivity of these two groups to habitat fragmentation could markedly differ, as sedentary individuals being more vulnerable than nomads, with potential consequences on their respective patterns of dispersal evolution under habitat fragmentation. Indeed, we expect that those nomadic individuals that need to be always on the move to sustain their daily requirements will be more able to rapidly adapt one or several components of movements (motion or navigation capacity, movement decision, information acquisition and storage; Nathan *et al.* 2008) to cope with changes in dispersal costs entailed by habitat fragmentation.

30.4 Dispersal costs in fragmented landscapes

Habitat fragmentation is thus likely to alter the costs, and hence also the cost-benefit ratio, of dispersal. Dispersal costs were classified into energetic, time, risk, and opportunity costs that may be levied either directly or deferred, during (or before) emigration, transfer, or immigration (or after settlement) (Bonte *et al.* 2012). The collection of accurate empirical data on changes in dispersal costs relative to fragmentation is challenging because mortality associated with dispersal is logistically difficult to detect, and the demonstration of the lower fitness of successful dispersers needs a thorough comparison with the fitness of the same individuals without dispersal (Belichon *et al.* 1996; Baguette and Van Dyck 2007; Doligez and Part 2008). Two different approaches (detailed following) can be used to bypass these difficulties. Firstly, experimental systems or analytical procedures may be used to directly or indirectly assess dispersal mortality. Secondly, dispersal costs can be revealed by changes in (suite of) phenotypic traits associated with, or traded off against, dispersal when comparing dispersing and resident individuals in fragmented and continuous landscapes.

30.4.1 Direct and indirect assessment of dispersal mortality

Bowler and Benton (2009a; 2009b) experimentally investigated how inter-habitat patch distances affect dispersal mortality. They showed in a two-patch laboratory set-up with mites that (1) emigration decreased and dispersal mortality increased with the distance between patches, (2) the proportion of juveniles (versus adults) in colonists was positively related with the distance between patches, and (3) the habitat quality at the patch of departure impacted both the probability of emigration and the dispersal mortality. Another interesting example of how dispersal mortality may be assessed experimentally is the study of Cheptou *et al.* (2008) with the plant *Crepis sancta* (Case Study IV; Chapter 14). In this species, individuals may either produce dispersing fruits with a pappus, or non-dispersing fruits without the pappus. Cheptou *et al.* (2008) experimentally assessed the cost of dispersal in a fragmented urban landscape by quantifying the probability of a dispersing fruit to fall within its patch, relatively to a non-dispersing fruit. They showed that dispersing fruits had a 55% lower chance of settling in tiny patches of the urban landscape, compared to non-dispersing fruits. Given the low availability of suitable habitats in the urban landscape, dispersing fruits thus were more likely to fall within unsuitable areas (in the urban matrix) compared to an unfragmented landscape. Investments in dispersal-related structures may also limit the allocation of resources to other functions which may indirectly affect progeny fitness. For instance, a negative correlation between dispersal capability and germination success has been evidenced in some wind-dispersed plant species (Morse and Schmitt 1985; Gravuer *et al.* 2003).

Besides such a direct, experimental approach to assess dispersal mortality, one can also use capture-mark-recapture (CMR) data and appropriate simulation models based on reliable assumptions. This makes it possible to disentangle dispersal mortality rates from mortality rates within natal or breeding patch, and hence to assess the mean dispersal cost within a spatially structured population. The virtual migration model (VM) (Hanski *et al.* 2000) uses individual capture histories obtained from CMR data, and allows assessing mortality during dispersal using a maximum likelihood estimation. Several assumptions of the model correspond to organisms with an insect-like life history. For example, Matter (2006) applied VM to CMR data of the herbivorous beetle *Tetraopes tetraophthalmus* collected in an experiment with suitable patches that were changed in size during a five-year study period. He showed that the increase in the size of suitable patches over the five years reduced the mean dispersal distance by half, and was associated with a decrease of dispersal mortality of several orders of magnitude. In another study, Schtickzelle *et al.* (2006) applied VM to CMR data collected in four spatially structured populations (each with several distinct habitat patches) of the butterfly *Boloria eunomia* organized along a gradient of fragmentation, from a continuous landscape with ca. 50% of suitable habitat to a highly fragmented landscape with only 0.4% of suitable habitat. Dispersal events leading to the death of the dispersing individuals increased from less than 5% in the continuous landscape to ca. 30% in the highly fragmented landscape (Schtickzelle *et al.* 2006).

30.4.2 Changes in emigration, transfer, and immigration

Although we may expect, from theoretical studies, fragmentation to alter all three steps of dispersal (i.e. emigration, transfer and immigration: Stenseth and Lidicker 1992; Ims and Yoccoz 1997; Clobert *et al.* 2009), empirical works mostly look at (and report) changes in dispersal traits associated with emigration. The dispersal propensity, i.e. the decision taken by an individual to cross the border of its habitat to initiate the transfer phase, was repeatedly shown to respond to the level of habitat fragmentation, both in actively dispersing species and in passive dispersers. Some evidence is now increasingly accumulating to show that habitat fragmentation may also change traits associated to other phases of the dispersal process, particularly with transfer, but also, to a lesser extent, to settlement.

In the butterfly *Boloria eunomia*, dispersal propensity was one order of magnitude lower in a highly fragmented landscape than in a continuous one (Schtickzelle *et al.* 2006; Chapter 33). Precise individual trajectories of butterflies were compared in those landscapes, in order to explain this decrease in dispersal propensity, and thereby show how dispersal costs entail changes in dispersal traits. This comparison revealed that individuals in the fragmented landscape were more reluctant to cross habitat patch borders than conspecifics in the more continuous landscape (Schtickzelle *et al.* 2007). The former were more likely to make U-turns at the habitat-matrix edge, whereas the latter readily crossed edges to continue their flight in the continuous landscape.

Gap-crossing avoidance that is documented in many forest animals is another striking example of reduced dispersal propensity related to fragmentation. Forest specialists are most often totally reluctant to cross even small gaps in the forest matrix, as demonstrated by the experimental translocation of breeding birds from their breeding territory to forests interspersed by open habitats, and the subsequent radio-tracking of their homing trajectory (e.g. Bakker and Van Vuren 2004; Awade and Metzger 2008; Ibarra-Macias *et al.* 2011). For such species, the functional grain of the landscape, i.e. the smallest spatial scale at which an individual recognizes spatial heterogeneity according to its perceptual range (Baguette and Van Dyck 2007), is thus extremely tight.

Even in species with passive dispersal, there are changes in dispersal propensity with habitat fragmentation. In ballooning spiders, long-distance dispersal events are dependent upon wind as individuals produce silk threads that serve as parachutes transported by air currents. Nonetheless, the initiation of flight is controlled by the dispersing individual and preceded by a special behaviour (tiptoe behaviour), which differs according to contrasted habitat fragmentation level (Bonte *et al.* 2006; Case Study II). In a common-garden experiment, the comparison of offspring from females of the ballooning spider *Pardosa monticola* originating from three landscapes with different levels of fragmentation (continuous, fragmented, and highly fragmented) showed that the proportion of individuals displaying the tiptoe behaviour correlated with the degree of habitat fragmentation: highest in individuals from the continuous landscape and lowest in individuals from the highly fragmented landscape. The comparisons of offspring from females taken in four patches of the fragmented landscape showed moreover that dispersal propensity was related to the size and the structural connectivity (i.e. the distance between populations) of the population of origin (Bonte *et al.* 2006).

Rather than documenting dispersal costs, these examples and others (review in Ries *et al.* 2004) basically show how dispersal costs are mediated by decreasing dispersal propensity. But dispersal propensity is not the only step of the dispersal process that is affected by habitat fragmentation. Comparative studies of movements within and between habitats or resource patches indicate significant changes in the topology of individual trajectories. In actively dispersing species (i.e. those species that disperse by their own locomotory apparatus), dispersal may occur either as a by-product of daily or routine movements (e.g. foraging), or as the result of movements shaped for net displacement, depending on the grain of the landscape (i.e. the configuration and the density of resource and habitat patches in the landscape) (Van Dyck and Baguette 2005). By increasing the distance among such patches, landscape fragmentation is expected to increase dispersal costs, including loss of time and energy and the risk of getting lost. A higher risk of predation out of suitable habitat is also possible (Vanvuren and Armitage 1994; Sakai and Noon 1997; Aars *et al.* 1999; Smith and Batzli 2006), especially when the contrast between the habitat and the outside world increases the probability of detection of the dispersing individuals.

The adoption of direct and rapid movements is interpreted as an adaptive response to minimize the time spent in the matrix. In fragmented landscapes where the grain of suitable habitats is loose, the mean flight speed of *B. eunomia* butterflies changes dramatically if individuals fly within habitats or in the matrix. Mean flight speed in the matrix was 1–1.47 km/h in the habitat, versus 1.93–2.07 km/h and > 3 km/h in open surrounding areas and scrubs, respectively (Turlure *et al.* 2011). Such acceleration after crossing habitat boundaries corresponds to the shift from those routine movements with many turns in the habitat that are characteristic of foraging behaviour (Bell 1991), to special movements with more rectilinear flight trajectories used for net displacements out of the habitat (Van Dyck and Baguette 2005). Flight behaviour (speed and linearity of flights), a trait characteristic of the transfer phase of dispersal, may thus also respond to fragmentation. In the butterfly *Maniola jurtina*, flight between habitat patches occurs either using foray search (i.e. loops corresponding to foraging, routine movements) or rectilinear movements optimizing net displacements (Conradt *et al.* 2000; Delattre *et al.* 2010). The relative frequency of each type of movements is related to the mean inter-patch distance in the landscape, the frequency of rectilinear movements increasing with habitat fragmentation (Delattre *et al.* 2010). Similar patterns have been found in other organisms including vertebrates and invertebrates. In a study of woodland caribous (*Rangifer tarandus*) collared with global positioning system (GPS) collars, Johnson *et al.* (2002) differentiated intra-patch from inter-patch movements. Intra-patch movements had a lower cost of movement and were associated with cover types where foraging behaviours occurred. Besides, experimental translocation of hedgehogs (*Erinaceus europaeus*) showed that translocated individuals moved substantially further and faster on average in the matrix than in suitable habitats (Doncaster *et al.* 2001). In another experimental study, Goodwin and Fahrig (2002) showed that individuals of goldenrod beetle (*Trirhabda borealis*) moved slowly and sinuously in habitat patches, whereas they turned to directed movements in an open matrix. In this experiment, an artificial increase of the complexity of the matrix resulted in the adoption of meandering but faster movements than the directed movements observed in the open matrix.

Such changes in dispersal behaviour during the transfer phase associated with habitat fragmentation are coherent with the assumption that individuals will try to minimize their dispersal costs. However, the adaptation of movements to the various challenges associated with fragmentation implies that individuals have the basic ability to move between fragments, which is not always guaranteed. A striking example of limitation of dispersal ability was reported on tropical birds. The ability to fly even short distances (< 100 m) between habitat fragments varies dramatically and consistently among species of forest specialist birds (Moore *et al.* 2008). In a release experiment, some individuals of five species out of a sample of ten showed dispersal limitations when released in the matrix at 100 m from forest fragments, while any individual from these five species was able to cross a gap of 300 m.

The final part of the dispersal chain process, i.e. immigration, has been the least studied. Immigration in fragmented landscape systems implies the detection of habitats, arriving, and finally settle in (un) occupied habitat. We might expect that in highly fragmented systems with small proportion of habitats and hence high dispersal costs, any increase in perceptual range, i.e. the distance at which an individual is able to detect suitable habitat using its sensory organs, should represent a benefit for the dispersing individual. Using a simulation model Vuilleumier and Perrin (2006) suggest that when energy is limited, the ability to detect suitable habitat from a distance is the best strategy in comparison to random search or even to the use of information collected nearby.

Changes in dispersal behaviour associated with habitat fragmentation are expected to induce modifications of immigration and emigration rates within spatially structured populations, and hence variation in metapopulation dynamics. Theory suggests that the shift of (correlated) random walks associated with foraging to directed dispersal displacements will have significant consequences for predictions about the dynamics of populations and communities (Armsworth and Roughgarden 2005). Empirical and experimental studies comparing fragmented and continuous study systems indeed confirm these expectations. Using an experimental design mimicking either dispersal failure by manually re-integrating dispersing common lizards (*Zootoca vivipara*) in their population of origin in some cases, or successful dispersal in other cases, Boudjemadi *et al.* (1999) nicely showed how the loss of connectivity between pairs of populations entailed a suite of consequences that ultimately modified the relative demographic population trajectories. Matthysen and Currie (1996) showed that the establishment in summer territories by first-year nuthatches *Sitta europaea* varied considerably according to forest fragmentation. In small forest fragments, settlers arrived at a slower rate, more of them remained unpaired, and more territories remained vacant at the end of the dispersal period than in continuous larger forests. Furthermore, territories were taken up in a highly predictable order related to territory quality in the large forest, but not in the fragments. Costly dispersal in a highly fragmented habitat reduces the number of settlers at the population level, and reduces opportunities for pair formation and habitat selection at the individual level, with obvious potential consequences for population dynamics.

Finally, the selection associated with dispersal in fragmented landscapes is expected to filter individuals with particular suite of traits that are associated with successful dispersal and immigration when the total amount of habitats within the landscape reaches a certain threshold. How this filtering process and the resulting selection of *dispersotypes* (i.e. dispersing and resident individuals) will influence metapopulation dynamics is an open and intriguing research question (Clobert *et al.* 2009; Cote *et al.* 2010).

30.5 Habitat fragmentation and the selection of dispersotypes

Proximal explanations of changes in dispersal observed in the wild relative to landscape fragmentation remain scarce. Heterotypic species develop particular morphologies that promote dispersal, which may result in a sharp phenotypic distinction between dispersotypes. Such differences can be more gradual, or more cryptic, like, for instance, differences in physiological or behavioural types. Heterotypic populations are thus excellent models to understand the relative importance of phenotypic plasticity and genetic assimilation in shaping such dispersotypes.

As mentioned earlier, Cheptou *et al.* (2008) showed that seed dispersal of the heteromorphic plant *Crepis sancta* was selected against in a highly fragmented urban landscape, due to the high cost of dispersal. To elucidate the mechanism underlying this selective process, a common-garden experiment showed that plants from fragmented populations indeed produced significantly more seeds of low mobility (without a pappus) than did plants from unfragmented populations, and a quantitative genetic model confirmed that this pattern is consistent with short-term evolution occurring during 5–12 generations of natural selection against dispersal in the fragmented landscape.

Behavioural changes in the wingless water strider *Aquarius najas* was also related to landscape functional grain (Ahlroth *et al.* 2010). In this species that requires running water, laboratory-born individuals that were released in the field had a lower dispersal propensity if their parents were caught in small, isolated streams than those individuals whose parents came from large and continuous habitats. This common-garden experiment shows that dispersal propensity is inheritable, and thus is controlled either by genetic effects, by maternal effects, or by their interaction, rather than by differences in learning or experiences between the populations dealing with the two levels of fragmentation (Ahlroth *et al.* 2010).

Besides the decrease of dispersal propensity relative to habitat fragmentation, behaviours observed during the two other steps of the dispersal process (i.e. transience and settlement) are also affected by connectivity loss (Chapter 33). But here again, the processes at work behind these patterns (phenotypic plasticity or genetic assimilation) are most often unclear.

We showed that dispersal costs in fragmented landscapes caused the selection for phenotype-dependent dispersal. Can we find cues of such 'dispersotypes' in homotypic species where variation in morphological traits associated with dispersal variation is not observed? Recent studies indeed revealed a high level of individual variation in dispersal performances within natural populations of species without dispersal-related polymorphism in morphology, both in mobile and in sessile organisms. Hanski *et al.* (2004) and Saastamoinen (2007) documented for a metapopulation of the butterfly *Melitaea. cinxia* in the Åland islands in Finland, a system with frequent deterministic extinctions of local populations, that colonist females were more mobile and had a higher fecundity and a longer lifespan than resident females. These super-colonists bear specific alleles of the Phosphoglucoisomerase enzyme. This enzyme is involved in the glycolysis, and hence, in the energy metabolism that is significant to flight performance (Case Study III; Chapter 8). Variation in flight performances is thus here correlated with physiological polymorphism rather than morphological polymorphism.

Before going further in the investigation of the evolution of such dispersotypes (either morphological, physiological, or other), the confirmation of a clear relationship between mobility estimates and dispersal is missing. Ducatez *et al.* (accepted) designed three original experimental tests to assess in the laboratory different measures of butterfly mobility, reflecting different components of performance exploration, and decision to move. They showed that individual responses to these tests were consistent across time and that there were strong correlations of individual responses across tests, which reflect the existence of a mobility syndrome as a particular case of the widespread behavioural syndromes (Sih 2004). The test of performance showed that mobility was costly to the individual, both in terms of survival and fecundity, and revealed the co-existence of two broad strategies: highly mobile individuals having higher survival and fecundity, and less mobile individuals with lower survival and fecundity. Movement performance was also in this case partly heritable ($h^2 = 0.37$), positively correlated with lifetime and fecundity, and related with morphology (wing shape and length), a character also highly heritable in the species ($h^2 = 0.50$) (Ducatez 2011). The co-existence of two cryptic strategies corresponds to the finding that less mobile individuals can take advantage of stable resources to develop local adaptations, whereas more mobile individuals are able to track dynamic resources in the landscape (Ducatez *et al.* accepted). To confirm that more mobile individuals in the laboratory were also the most dispersive under fragmented conditions, individuals of known mobility (as assessed by the laboratory tests) were introduced into the Metatron, a large experimental device consisting of paired cages joined by a narrow corridor (Legrand *et al.* accepted). In each pair, one cage was initially left empty of butterflies and the other was occupied by a small butterfly population. After their introduction into the cage, the butterflies were free to disperse through the corridor and their position was regularly recorded. Legrand *et al.* (accepted) showed that highly mobile individuals crossed the corridors significantly more often than less mobile butterflies. Besides, eggs were laid and offspring hatched into those cages that were empty at the beginning of the experiment, which confirm that corridor crossing resulted in effective dispersal between paired cages. This experimental test of metapopulation dynamics showed that indeed, more mobile individuals were also those dispersers that realized effective dispersal events in laboratory.

Among sessile homomorphic species, plants have been by far the most studied species regarding the evolution of dispersal in response to habitat fragmentation. Effects of habitat fragmentation on pollen dispersal have been widely investigated by studying pollinator abundance, pollinator visits and pollinator behaviour in biotic pollinated species (e.g. Aguirre *et al.* 2011), individual seed set (Vandewoestijne *et al.* 2009), or pollen dispersal kernels inferred from parentage analysis (Dick *et al.* 2003; Bacles and Ennos 2008), in relation to the spatial configuration of plant populations and their size or density (see Kearns *et al.* 1998 and Ghazoul, 2005 for reviews). However, none of these empirical studies elucidates the ecological and evolutionary processes that drive pollen dispersal change in response to fragmentation.

Evidence for seed dispersal evolution in homomorphic species in response to habitat fragmentation comes from the comparison of the presence/absence of morphological traits associated with dispersal among populations from islands versus mainland (Cody and Overton 1996; Gravuer *et al.* 2003; Fresnillo and Ehlers 2008), from landscapes with varying degree of fragmentation (Soons and Heil 2002; Riba *et al.* 2009), or from populations with different sizes (Mix *et al.* 2006). For instance, in the wind-dispersed species *Mycelis muralis*, where all achenes bear a pappus, Riba *et al.* (2009) showed that achene dispersal ability decreased significantly with decreasing landscape connectivity. Although most of those studies assessed population differentiation on dispersal ability from plants grown in a common-garden experiment, it remains, however, unclear whether such phenotypic variation in dispersal-related morphological traits has a genetic basis or may only reflect non-genetic inheritance (e.g. parental effects). Looking for differences in dispersal gene polymorphism and expression will surely provide new insights into this pending question (Chapter 6).

30.6 Conclusions and perspectives

Landscape fragmentation is a key factor moulding individual strategies in dispersal and associated traits; to what extent these processes will differently challenge resident and dispersal strategies and will affect metapopulation dynamics is a crucial question that deserves further research efforts (e.g. Clobert *et al.* 2009; Cote *et al.* 2010). Many empirical studies confirm that habitat fragmentation drives changes in phenotypic traits related to dispersal, and may promote dispersal polymorphism, resulting from either continuous or discrete variation in morphological, behavioural, or physiological attributes. The relative role of phenotypic plasticity and genetic assimilation in these changes, however, still remains unclear. Moreover, empirical studies are still lacking that investigate whether the evolution of dispersal in response to habitat fragmentation is indeed non-linear, and how suites of traits might be affected by the selection acting on dispersal through habitat fragmentation. A recent quantitative analysis shows that in butterflies, dispersal is strongly integrated into life histories (Stevens *et al.* 2012). However, the direction of the association between dispersal and life-history traits varies among species, depending among other things on the ability of butterflies to cope with changes in temperature. Species completing their life cycle under narrow temperature ranges are significantly less dispersive than generalists. Thermal specialists face thus double jeopardy from landscape fragmentation and climate change. The evolution of increased dispersal could allow such specialists to track their shifting climate envelope (Berg *et al.* 2010). The evolution of decreased dispersal, resulting in increased genetic isolation among local populations, could allow local adaptations (Berg *et al.* 2010) to cope with climate changes. Understanding how dispersal will evolve under such contrasted selective pressures is probably the most urgent research question in the current era of species extinction.

Acknowledgements

We acknowledge Virginie M. Stevens for constructive review and helpful discussions. MB and DL are funded by the Agence Nationale de la Recherche, via the programs (1) open call DIAME (DIspersal and MEtapopulation), (2) 6th extinction MOBIGEN (with HF), and (3) Biodiversa ERANET TenLamas. HVD is funded by an ARC-research grant of the Académie Louvain (ARC grant no. 10-15/031). MB and DL are part of the 'Laboratoire d'Excellence' (LABEX) entitled TULIP (ANR-10-LABX-41).

References

Aars, J., Johannesen, E., Ims, R.A. (1999). 'Demographic consequences of movements in subdivided root vole populations', *Oikos* **85**: 204–16.

Aguirre, A., Guevara, R., Dirzo, R. (2011). 'Effects of forest fragmentation on assemblages of pollinators and floral visitors to male- and female-phase inflorescences of Astrocaryum mexicanum (Arecaceae) in a Mexican rain forest', *Journal of Tropical Ecology* **27**: 25–33.

Ahlroth, P., Alatalo, R.V., Suhonen, J. (2010). 'Reduced dispersal propensity in the wingless waterstrider Aquarius najas in a highly fragmented landscape', *Oecologia* **162**: 323–30.

Andren, H. (1994). 'Effects of habitat fragmentation on birds and mammals in landscapes with different proportions of suitable habitat—a review', *Oikos* **71**: 355–66.

Armsworth, P.R. and Roughgarden, J.E. (2005). 'The impact of directed versus random movement on population dynamics and biodiversity patterns', *American Naturalist* **165**: 449–65.

Awade, M. and Metzger, J P. (2008). 'Using gap-crossing capacity to evaluate functional connectivity of two Atlantic rainforest birds and their response to fragmentation', *Austral Ecology* **33**: 863–71.

Bacles, C. E. F. and Ennos, R. A. (2008). 'Paternity analysis of pollen-mediated gene flow for Fraxinus excelsior L. in a chronically fragmented landscape', *Heredity* **101**: 368–80.

Baguette, M. and Van Dyck, H. (2007). 'Landscape connectivity and animal behavior: functional grain as a key determinant for dispersal', *Landscape Ecology* **2:** 1117–29.

Bakker, V. J. and Van Vuren, D.H. (2004). 'Gap-crossing decisions by the red squirrel, a forest-dependent small mammal', *Conservation Biology* **18**: 689–97.

Bascompte, J. and Sole, R.V. (1996). 'Habitat fragmentation and extinction thresholds in spatially explicit models', *Journal of Animal Ecology* **65**: 465–73.

Belichon, S., Clobert, J., Massot, M. (1996). 'Are there differences in fitness components between philopatric and dispersing individuals?', *Acta Oecologica-International Journal of Ecology*: **17**: 503–17.

Bell, W.J. (1991). *Searching Behaviour. The Behavioural Ecology of Finding Resources*. Chapman and Hall, London.

Berg, M.P., Kiers, E.T., Driessen, G. *et al.* (2010). 'Adapt or disperse: understanding species persistence in a changing world', *Global Change Biology* **16**: 587–98.

Bierregaard, R.O., Lovejoy, T.E., Kapos, V. *et al.* (1992). 'The biological dynamics of tropical rain-forest fragments.' *Bioscience* **42**: 859–66.

Bonte, D., Hovestadt, T., Poethke, H.J. (2010). 'Evolution of dispersal polymorphism and local adaptation of dispersal distance in spatially structured landscapes', *Oikos* **119**: 560–6.

Bonte, D., Van Dyck, H., Bullock, J.M. *et al.* (2012). 'The costs of dispersal', *Biological Reviews* **87**: 290–312.

Bonte, D., Vanden Borre, J., Lens, L. *et al.* (2006). 'Geographical variation in wolf spider dispersal behaviour is related to landscape structure', *Animal Behaviour* **72**: 655–62.

Boudjemadi, K., Lecomte, J, Clobert, J. (1999). 'Influence of connectivity and dispersal in two contrasting habitats: an experimental approach', *Journal of Animal Ecology* **68**: 1207–24.

Bowler, D. E. and Benton, T.G. (2009a). 'Impact of dispersal on population growth: the role of inter-patch distance', *Oikos* **118**: 403–12.

Bowler, D.E. and Benton, T.G. (2009b). 'Variation in dispersal mortality and dispersal propensity among individuals: the effects of age, sex and resource availability', *Journal of Animal Ecology* **78**: 1234–41.

Bruggeman, D.J., Wiegand, T., Fernandez, N. (2010). 'The relative effects of habitat loss and fragmentation on population genetic variation in the red-cockaded woodpecker (Picoides borealis)' *Molecular Ecology* **19**: 3679–91.

Cheptou, P.O., Carrue, O., Rouifed, S. *et al.* (2008). 'Rapid evolution of seed dispersal in an urban environment in the weed Crepis sancta', *Proceedings of the National Academy of Sciences of the USA* **105**: 3796–9.

Clobert, J., Le Galliard, J.F., Cote, J. *et al.* (2009). 'Informed dispersal, heterogeneity in animal dispersal syndromes and the dynamics of spatially structured populations', *Ecology Letters* **12**: 197–209.

Cody, M.L. and Overton, J.M. (1996). 'Short-term evolution of reduced dispersal in island plant populations', *Journal of Ecology* **84**: 53–61.

Conradt, L., Bodsworth, E.J., Roper, T.J. *et al.* (2000). 'Non-random dispersal in the butterfly Maniola jurtina: implications for metapopulation models', *Proceedings of the Royal Society B* **267**: 1505–10.

Cote, J., Clobert, J., Brodin, T. *et al.* (2010). 'Personality-dependent dispersal: characterization, ontogeny and consequences for spatially structured populations', *Philosophical Transactions of the Royal Society B-* **365**: 4065–76.

Delattre, T., Burel, F., Humeau, A. *et al.* (2010). 'Dispersal mood revealed by shifts from routine to direct flights in the meadow brown butterfly Maniola jurtina', *Oikos* **119**: 1900–8.

Dick, C. W., Etchelecu, G., Austerlitz, F. (2003). 'Pollen dispersal of tropical trees (Dinizia excelsa: Fabaceae) by native insects and African honeybees in pristine and fragmented Amazonian rainforest', *Molecular Ecology* **12**: 753–64.

Dobson, A.P., Bradshaw, A.D., Baker, A.J.M. (1997). 'Hopes for the future: Restoration ecology and conservation biology', *Science* **277**: 515–22.

Doligez, B. and Part, T. (2008). 'Estimating fitness consequences of dispersal: a road to 'know-where'? Non-random dispersal and the underestimation of dispersers' fitness', *Journal of Animal Ecology* **77**: 1199–211.

Doncaster, C.P., Rondinini, C., Johnson, P.C.D. (2001). 'Field test for environmental correlates of dispersal in hedgehogs Erinaceus europaeus', *Journal of Animal Ecology* **70**: 33–46.

Ducatez, S., Legrand, D., Chaput-Bardy, A., Stevens, V.M., Fréville, H., and Baguette, M. 'Inter-individual variation in movement: is there a mobility syndrome in the large white butterfly Pieris brassicae?' *Ecological Entomology*, accepted.

Ducatez, S. (2011). "Evolution of mobility in butterflies: constraints and adaptations" PhD thesis. Paris, Museum National d'Histoire Naturelle, 271 p.

Ewers, R.M. and Didham, R.K. (2006). 'Confounding factors in the detection of species responses to habitat fragmentation', *Biological Reviews* **81**: 117–42.

Fahrig, L. (2003). 'Effects of habitat fragmentation on biodiversity', *Annual Review of Ecology, Evolution and Systematics* **34**: 487–515.

Fahrig, L. (2007). 'Non-optimal animal movement in human-altered landscapes', *Functional Ecology* **21**: 1003–15.

Foley, J.A., DeFries, R., Asner, G.P. *et al.* (2005). 'Global consequences of land use', *Science* **309**: 570–4.

Fresnillo, B. and Ehlers, B.K. (2008). 'Variation in dispersability among mainland and island populations of three wind dispersed plant species', *Plant Systematics and Evolution* **270**: 243–55.

Ghazoul, J. (2005). 'Pollen and seed dispersal among dispersed plants', *Biological Reviews* **80**: 413–43.

Gonzalez, A., Lawton, J.H., Gilbert, F.S. *et al.* (1998). 'Metapopulation dynamics, abundance, and distribution in a microecosystem', *Science* **281**: 2045–7.

Goodwin, B. J. and Fahrig, L. (2002). 'Effect of landscape structure on the movement behaviour of a specialized goldenrod beetle, Trirhabda borealis', *Canadian Journal of Zoology-Revue Canadienne de Zoologie* **80**: 24–35.

Gravuer, K., von Wettberg, E.J., Schmitt, J. (2003). 'Dispersal biology of Liatris scariosa var. novae-angliae (Asteraceae), a rare New England grassland perennial', *American Journal of Botany* **90**: 1159–67.

Gustafson, E.J. and Parker, G.R. (1992). 'Relationships between landcover proportion and indexes of landscape spatial pattern', *Landscape Ecology* **7**: 101–10.

Hanski, I., Alho, J. and Moilanen, A. (2000). 'Estimating the parameters of survival and migration of individuals in metapopulations', *Ecology* **81**: 239–51.

Hanski, I., Eralahti, C., Kankare, M. *et al.* (2004). 'Variation in migration propensity among individuals maintained by landscape structure', *Ecology letters* **7**: 958–66.

Heino, M. and Hanski, I. (2001). 'Evolution of migration rate in a spatially realistic metapopulation model', *American Naturalist* **157**: 495–511.

Hovestadt, T., Binzenhofer, B., Nowicki, P. *et al.* (2011). 'Do all inter-patch movements represent dispersal? A mixed kernel study of butterfly mobility in fragmented landscapes', *Journal of Animal Ecology* **80**: 1070–7.

Ibarra-Macias, A., Robinson, W.D., Gaines, M.S. (2011). 'Experimental evaluation of bird movements in a fragmented Neotropical landscape', *Biological Conservation* **144**: 703–12.

Ims, R.A. and Yoccoz, N. (1997). 'Studying transfer processes in metapopulations: emigration, migration and colonization', in I.Hanski., M.Gilpin (eds), *Metapopulation biology. Ecology, Genetics, and Evolution.* and Academic Press, San Diego, 247–65.

Johnson, C.J., Parker, K.L., Heard, D.C. *et al.* (2002). 'Movement parameters of ungulates and scale-specific responses to the environment', *Journal of Animal Ecology* **71**: 225–35.

Kearns, C. A., Inouye, D. W., Waser, N. M. (1998). 'Endangered mutualisms: the conservation of plant-pollinator interactions', *Annual Review of Ecology and Systematics* **29**: 83-112.

Legrand, D., Guillaume, O., Baguette, M., Cote, J., Trochet, A., Calvez, O., Zajitschek, S., Zajitschek, F., Lecomte, J., Bénard, Q., Le Galliard, J.-F., and Clobert, J. 'The Metatron: an experimental system to study dispersal and metaecosystem for terrestrial organisms'. *Nature Methods*, accepted.

Lindenmayer, D.B. and Fischer, J. (2007). 'Tackling the habitat fragmentation panchreston', *Trends in Ecology and Evolution* **22**: 127–32.

Mathias, A., Kisdi, E., Olivieri, I. (2001). 'Divergent evolution of dispersal in a heterogeneous landscape', *Evolution* **55**: 246–59.

Matter, S.F. (2006). 'Changes in landscape structure decrease mortality during migration', *Oecologia* **150**: 8–16.

Matthysen, E. and Currie, D. (1996). 'Habitat fragmentation reduces disperser success in juvenile nuthatches Sitta europaea: Evidence from patterns of territory establishment', *Ecography* **19**: 67–72.

Mix, C., Picó, F.X., van Groenendael, J.M. *et al.* (2006). 'Inbreeding and soil conditions affect dispersal and components of performance of two plant species in fragmented landscapes', *Basic and Applied Ecology* **7**: 59–69.

Moore, R.P., Robinson, W.D., Lovette, I.J.and Robinson, T.R. (2008). 'Experimental evidence for extreme dispersal limitation in tropical forest birds', *Ecology Letters* **11**: 960–968.

Morse, D.H. and Schmitt, J. (1985). 'Propagule size, dispersal ability, and seedling performance in Asclepias syriaca', *Oecologia* **67**: 372–9.

Mortelliti, A., Amori, G., Capizzi, D. *et al.* (2011). 'Independent effects of habitat loss, habitat fragmentation and structural connectivity on the distribution of two arboreal rodents', *Journal of Applied Ecology* **48**: 153–62.

Mueller, T., and Fagan, W.F. (2008). 'Search and navigation in dynamic environments—from individual behaviours to population distributions', *Oikos* **117**: 654–64.

Nathan, R., Getz, W.M., Revilla, E. *et al.* (2008). 'A movement ecology paradigm for unifying organismal movement research', *Proceedings of the National Academy of Sciences of the USA* **105**: 19052–9.

Poethke, H.J., Hovestadt, T., Mitesser, O. (2003). 'Local extinction and the evolution of dispersal rates: Causes and correlations', *American Naturalist* **161**: 631–40.

Poethke, H.J., Pfenning, B., Hovestadt, T. (2007). 'The relative contribution of individual and kin selection to the evolution of density-dependent dispersal rates', *Evolutionary Ecology Research* **9**: 41–50.

Ries, L., Fletcher, R.J., Battin, J. *et al.* (2004). 'Ecological responses to habitat edges: mechanisms, models, and variability explained', *Annual Review of Ecology, Evolution, and Systematics* **35**: 491–522.

Riba, M., Mayol, M., Giles, B.E. *et al.* (2009). 'Darwin's wind hypothesis: does it work for plant dispersal in fragmented habitats?' *New Phytologist* **83**: 667–77.

Rockstrom, J., Steffen, W., Noone, K. *et al.* (2009). 'A safe operating space for humanity', *Nature* **461**: 472–5.

Ronce, O. (2007). 'How does it feel to be like a rolling stone? Ten questions about dispersal evolution', *Annual Review of Ecology, Evolution, and Systematics* **38**: 231–53.

Ronce, O., Perret, F., Olivieri, I. (2000). 'Landscape dynamics and evolution of colonizer syndromes: interactions between reproductive effort and dispersal in a metapopulation', *Evolutionary Ecology* **14**: 233–60.

Saastamoinen, M. (2007). 'Mobility and lifetime fecundity in new versus old populations of the Glanville fritillary butterfly', *Oecologia* **153**: 569–78.

Sakai, H.F. and Noon, B.R. (1997). 'Between-habitat movement of dusky-footed woodrats and vulnerability to predation', *Journal of Wildlife Management* **61**: 343–50.

Saunders, D.A., Hobbs, J.A., Margules, C.R. (1991). 'Biological consequences of ecosystem fragmentation: a review', *Conservation Biology* **5**: 18–32.

Schtickzelle, N., Joiris, A., Van Dyck, H. *et al.* (2007). 'Quantitative analysis of changes in movement behaviour within and outside habitat in a specialist butterfly', *BMC Evolutionary Biology* **7**: 4:

Schtickzelle, N., Mennechez, G. Baguette, M. (2006). 'Dispersal depression with habitat fragmentation in the bog fritillary butterfly', *Ecology* **87**: 1057–65.

Sih, A., Bell, A., Chadwick Johnson, J. (2004). 'Behavioral syndromes: an ecological and evolutionary overview', *Trends in Ecology and Evolution* **19**: 372-8.

Smith, J.E. and Batzli, G.O. (2006). 'Dispersal and mortality of prairie voles (Microtus ochrogaster) in fragmented landscapes: a field experiment', *Oikos* **112**: 209–17.

Soons, M.B. and Heil, G.W. (2002). 'Reduced colonization capacity in fragmented populations of wind-dispersed grassland forbs', *Journal of Ecology* **90**: 1033–43.

Stenseth, N.C. and Lidicker, W.Z. (1992). *Animal Dispersal*. Chapman and Hall, London.

Stevens, V.M., Trochet, A., Van Dyck, H. *et al.* (2012). How is dispersal integrated in life histories: a quantitative analysis using butterflies', *Ecology Letters* **15**: 65–73.

Swift, T.L. and Hannon, S.J. (2010). 'Critical thresholds associated with habitat loss: a review of the concepts, evidence, and applications', *Biological Reviews* **85**: 35–53.

Tilman, D., May, R.M., Lehman, C.L. *et al.* (1994). 'Habitat destruction and the extinction debt', *Nature* **371**: 65–6.

Travis, J.M.J. and Dytham, C. (1999). 'Habitat persistence, habitat availability and the evolution of dispersal', *Proceedings of the Royal Society B-Biological Sciences* **266**: 723–8.

Turlure, C., Baguette, M., Stevens, V.M. *et al.* (2011). 'Species- and sex-specific adjustments of movement behaviour to landscape heterogeneity in butterflies', *Behavioral Ecology* **63**: 1581–91.

Vandewoestijne, S., Róis, A.S., Caperta, A. *et al.* (2009). 'Effects of individual and population parameters on reproductive success in three sexually deceptive orchid species', *Plant Biology* **11**: 454–63.

Van Dyck, H. and Baguette, M. (2005). 'Dispersal behaviour in fragmented landscapes: Routine or special movements?', *Basic and Applied Ecology* **6**: 535–45.

Vanvuren, D. and Armitage, K.B. (1994). 'Survival of dispersing and philopatric yellow-bellied marmots—what is the cost of dispersal', *Oikos* **69**: 179–81.

Vuilleumier, S. and Perrin, N. (2006). 'Effects of cognitive abilities on metapopulation connectivity', *Oikos* **113**: 139–47.

Wiegand, T., Revilla, E., Moloney, K.A. (2005). 'Effects of habitat loss and fragmentation on population dynamics', *Conservation Biology* **19**: 108–21.

CHAPTER 31

Modelling the effects of habitat fragmentation

Calvin Dytham and Justin M. J. Travis

31.1 Introduction

Habitat can be naturally fragmented, but human activity has already massively increased the degree of fragmentation. The persistence of populations within fragmented landscapes will often be critically dependent upon the species' dispersal characteristics. Also, the dispersal characteristics of populations existing in recently fragmented landscapes may come under strong selection, often for reduced dispersal ability. Thus, understanding the role of dispersal in determining spatial population dynamics across fragmented landscapes, understanding the consequences of habitat fragmentation on the evolution of dispersal, and exploring the interplay between the evolutionary and ecological dynamics remain key issues in dispersal biology. Our primary objective in this chapter is to describe approaches to modelling spatial population dynamics in fragmented landscapes. We consider three of the main aspects of modelling the effects of habitat fragmentation on biological populations: the creation of landscapes allowing for the flexibility and accessibility for habitat destruction scenarios, methods for adding agents representing either individuals or groups of interacting individuals to the landscape, and the encapsulation of output from complex models by simple, interpretable metrics. We give a schematic representation of the decisions that a modeller needs to make (Figure 31.1) and many of the decisions are considered in more detail below. After considering the model building process, we present some examples where habitat fragmentation influences model construction and model output. We conclude by outlining some important areas for future research.

31.2 Landscapes

The world is characterized by spatial heterogeneity at all scales. To make analysis tractable, models often simplify the world making assumptions about panmixis and spatial homogeneity. However, these assumptions don't allow questions about the effect of the extent and pattern of habitat loss or creation to be addressed. Here, we detail a range of alternative modelling approaches for incorporating habitat heterogeneity with differing levels of complexity and realism.

Two-patch models are the simplest spatially extended approach and allow for the simplest level of habitat heterogeneity to be examined (e.g. McPeek and Holt 1992). The effects of habitat fragmentation can only be examined very crudely by applying a mortality cost to patch transfers, or reducing the quality of one of the patches, or removing it completely and then looking at the population response.

Spatially implicit models: The original metapopulation model of Levins (1969) assumes a very large number of identical and equally accessible patches. Colonization is affected by the proportion of occupied and unoccupied patches in the landscape. The seminal Hamilton-May evolution of dispersal model uses a very similar modelling framework, accounting for the patchy nature of space but not considering the explicit spatial arrangement of patches (Hamilton and May 1977). In spatial models set up in this way, habitat fragmentation can only be modelled as a fraction of patches that are unavailable for colonization (e.g. Nee and May 1992); there is no

Dispersal Ecology and Evolution. First Edition. Edited by Jean Clobert *et al*.

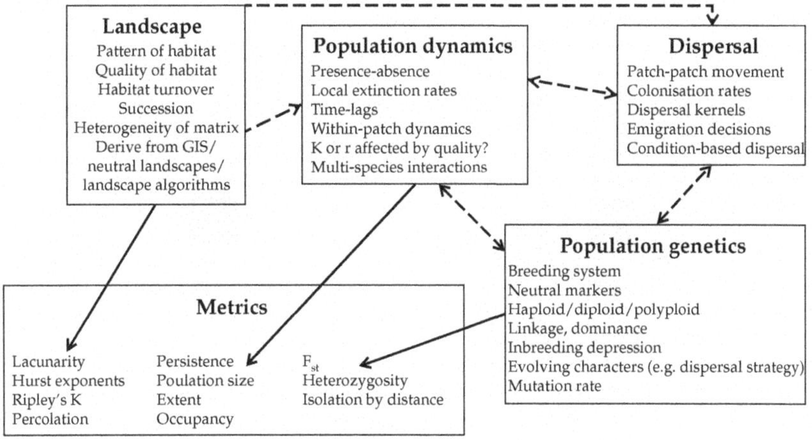

Figure 31.1 A summary of the decisions that need to be made when constructing a model of the effects of habitat fragmentation. The main areas are i) the landscape, ii) the population: its population dynamics, population genetics and dispersal, iii) the metrics that are used to describe the landscape and population (indicated by solid lines). The dashed lines with arrows indicate when areas influence each other. For example the landscape affects dispersal, while population dynamics and population genetics affect each other. No metrics are given for dispersal although there are decisions to be made there too: emigration rate, immigration rate and dispersal kernel are all commonly used.

way to explore the effect of the physical arrangement of patches. Analytical methods for accounting for the spatial arrangement of patches are possible by using moment closure techniques, and these can be used to explore the effect of loss of patches.

Spatially explicit models: The simplest way to describe space explicitly is to use an array of tessellated squares (Hogeweg 1988). Spatially explicit models will include many additional processes that are excluded from the spatially implicit approach and therefore can give rather different insights. For example, Dytham (1994) showed that a spatially explicit version of the spatially implicit model of Nee and May (1992) gave different results, especially at low levels of habitat availability. Grids of squares suit the way that data is held as a two-dimensional array in a computer model, and easily incorporate the sort of GIS raster data that is now readily available describing landscapes and species' distributions. Further information on landscape suitability or permeability can be extracted from GIS layers, and data can be rescaled so that all are based on squares or pixels with the same dimensions. Landscapes can be matched to existing distributions statistically, although that makes the assumptions that a population is currently matched up with the habitat that suits it (i.e. where it is absent the habitat is assumed to be unsuitable in some way) and, as described below, there are difficulties in choosing the appropriate statistic to describe the match of simulation output with species distribution data.

Square grids have the problem that there are two types of immediate neighbour—those connected along an edge and those only joined at the corner. One simple solution to this is to use tessellated hexagons that only have one class of neighbour (see Gonzalez-Andujar and Perry 1995, for an implementation). The main problem with this approach is that GIS data are rarely collected in a hexagonal way. More complex methods of dividing the landscape into groups of square cells that come from the GIS can use other rules for tessellation. One commonly-used method is the Dirichilet tessellation where centroids of patches are located and then divisions are drawn exactly half way between the centroids. This generates irregular polygons with an average of six borders that are much more like natural situations than the regular grid from which they are derived which could describe available territories of different shapes and sizes

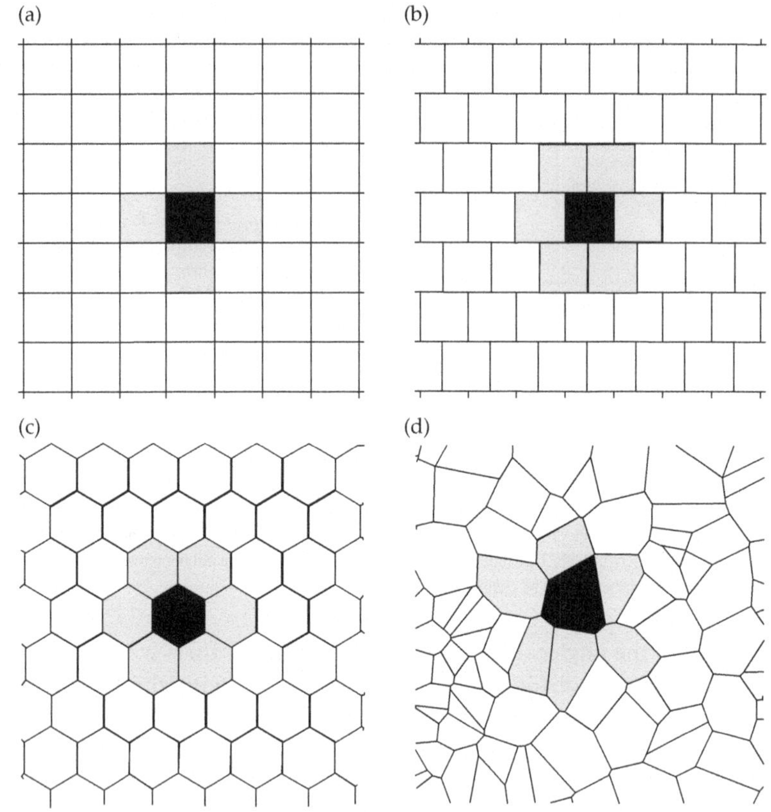

Figure 31.2 Dividing the landscape into cells: (a) shows a simple square grid where each cell has four first-order neighbours (the von Neuman neighbourhood), (b) shows that if the squares are arranged such that alternate rows are offset by half a side length, it gives a landscape where each square has six neighbours, (c) shows regular hexagons where there are six first-order neighbours, (d) is an example of a Dirichilet tesselation where borders are drawn as far as possible from randomly-place centroids (cells have an average of six neighbours). Figure modified from Holland *et al*. (2007).

(Holland *et al.* 2007). Some of these different arrangements of patches are shown in Figure 31.2.

Stochastic Patch Occupancy Models (SPOMs) describe networks where the patches (nodes) are arranged in a configuration such that inter-patch distances and patch sizes are used to generate the rates of colonizations of patches (Hanski 1999; Moilanen 2004). This approach can be used to show key patches in a network (Ovaskainen and Hanski 2003). It can also draw on a huge theoretical underpinning for network analysis which can help identify the effects of management. SPOMs have a big advantage in that large networks of patches can be described, but their disadvantage is that the matrix between patches is effectively ignored, as is any variation within patches (see Kindlmann *et al.* 2005 for a discussion of the effects of variation in the quality of the matrix).

Methods for generating artificial landscapes: It is possible to use simple algorithms to generate particular configurations of habitat within a landscape. For example, Dytham and Simpson (2007) used a sequential algorithm to generate a landscape of islands and reefs where islands were located at random and then expanded from seed foci, before reef foci were located next to some of the islands and expanded with a restriction of maximum and minimum distance from land. While such algorithms are often useful for specific scenarios, they have unpredictable spatial properties that vary with scale.

Fractals are widely used to generate realistic looking 'neutral' landscapes, although it is important to remember that this resemblance may be superficial and that the world isn't really fractal (Halley *et al.* 2004). Fractals are patterns which retain spatial characteristics at all scales. However, this is not what makes them attractive to modellers, rather it is the availability of simple algorithms that can be used to generate landscapes with different characteristics of habitat density and critically the spatial autocorrelation of habitat; this is the Hurst exponent which ranges from 0 for a random landscape to 1 for a very highly spatially autocorrelated landscape. A simple method for generating fractal landscapes is the 'mid-point displacement' method of Pietgen and Saupe (1988), which is very easy to implement, although it only generates a true fractal landscape at a Hurst exponent of 0.5. More complex methods, such as the algorithm of Chipperfield *et al.* (2011), use spectral synthesis to generate a model landscape using a Fourier series (a combination of a series of waves of decreasing wavelengths with different amplitudes). Fractal landscape algorithms result in a value being assigned to each pixel. Often a particular landscape is divided into a habitat/non-habitat dichotomy using a threshold, but there is scope for using continuous values to describe, for example, 'pixel' habitat quality (Figure 31.3). Complications can arise with edges if periodic boundaries are required and periodic fractals can be generated (as implemented in Dey *et al.* 2006).

Fractal landscapes can be combined with, for example, gradients of habitat suitability mimicking a climate window (Travis and Dytham 2004). Using this approach with a moving gradient allows complex landscapes to change through time in a realistic manner, enabling simulations to be developed in order to explore the ecological and evolutionary dynamics of populations responding to climate change as their ranges migrate across a fragmented landscape (see Figure 31.3 and Chapter 26 for further discussion on this general topic). It should be remembered that making more naturalistic landscapes may obscure some of the mechanisms of interest and simplistic landscapes will often suffice.

31.3 Populating the landscapes

The simplest models take the Levins (1969) metapopulation approach and have a patch focus with only empty or occupied patches. There are no within patch population dynamics (i.e. the local process of population growth after colonization is assumed to occur very rapidly). Habitat loss can be modelled by adding a further category of 'unavailable' patches. More complicated models, such as SPOMs, also use a simple empty/occupied dichotomy with the assumption that the population dynamics are working on a much shorter timescale than the extinction-colonization dynamics, so it is assumed that once colonized a patch will immediately reach its equilibrium density. However, some SPOMs assume that the size of the patch affects the rate of extinction and colonization of a patch (Moilanen 2004).

Incorporation of population dynamics in spatially extended models can be achieved in a wide variety of ways. If there are few patches, and simplifying assumptions about equal properties of patches and populations are made, then matrix models can be used, as it is possible to combine population processes and dispersal processes into matrices. Reaction-diffusion methods have a long history in ecology (Fisher 1937; Skellam 1951). They use partial differential equations to account for the rate of change in density in a population in continuous space and time. This approach is only really suitable for situations where the assumption of a homogeneous habitat can be met. Furthermore, they are often cast as one-dimensional systems (e.g. a shore-line) to avoid the problems associated with extending the logic to two dimensions. Reaction-diffusion models predict that populations will expand equally in all directions, and that the rate of expansion is determined by a combination of the intrinsic rate of population increase and the dispersal rate. There have been attempts to add spatial heterogeneity to reaction-diffusion models in a one-dimensional (Shigesada *et al.* 1986) and two-dimensional (Kinezaki *et al.* 2003) system. It is generally agreed that diffusion models do not capture the rate of spread of populations as they underestimate rare, long-distance dispersal events (e.g. Clark 1998).

Nearly all biological populations comprise unique, discrete individuals each of which has a different history and different location. Models ignore this complexity to make modelling tractable. With

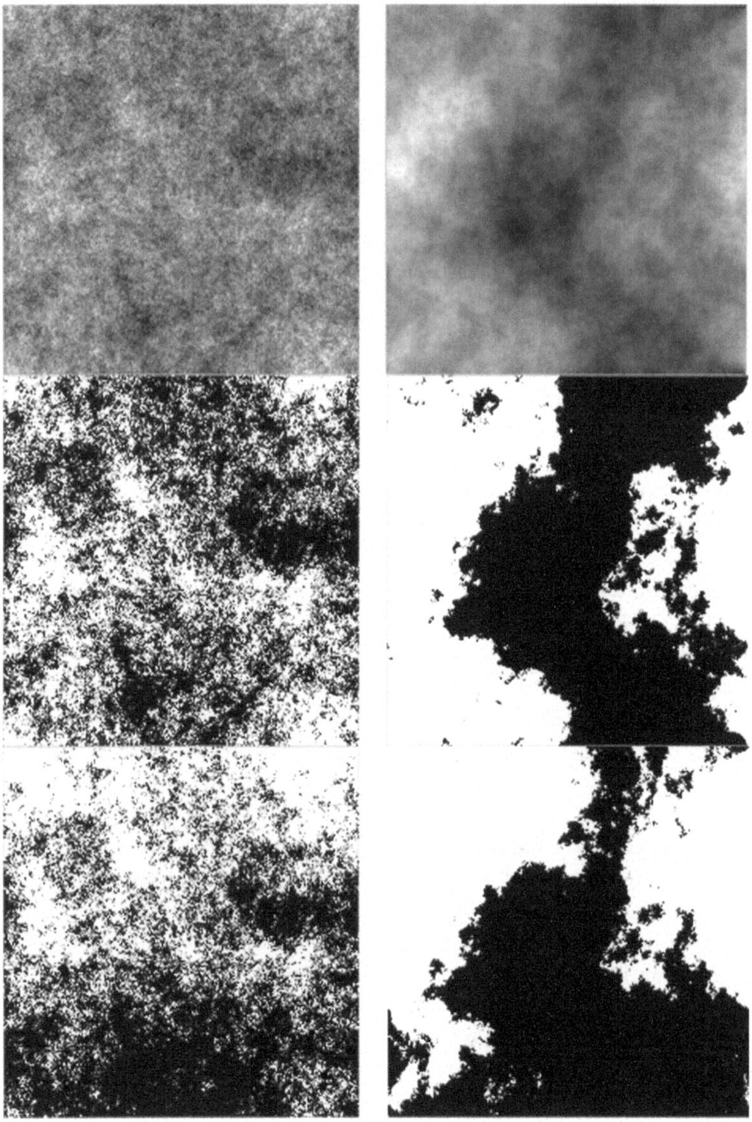

Figure 31.3 Representations of three fractal landscapes. The two columns have different Hurst exponents (a measure of spatial autocorrelation) of 0.2 and 0.7. The three rows represent: top—the landscape with continuous variation shown as a grey-scale; middle—shown as a binary landscape so that half the cells are shaded; bottom—a binary landscape with half the cells shaded but with a gradient of suitability added such that cells are likely to be shaded at the bottom of the grid than the top.

very large populations and the incorporation of stochasticity, population-level models can be an adequate descriptor of the behaviour of populations. However, in small populations, at range margins, at low density, and especially where selection and drift lead to variations in phenotype in space, individual-based models (IBMs) are more appropriate (e.g. Judson 1994). Individuals can be modelled occupying sub-populations in a grid (the coupled-map-lattice approach), or can be modelled as agents located in continuous space. The advantages of IBMs are clear in that the unique nature of individuals, perhaps having different phenotypes, is captured, and selection can be very easily incorporated.

They also include demographic stochasticity and kin competition by default (Poethke *et al.* 2007). Patterns often emerge from IBMs, such as the clear benefits to increased dispersal distances at range margins (Travis and Dytham 2002).

However, there are drawbacks including the computational time required, and the increase in state-space and stochasticity. Each time a model is developed the modeller should consider carefully the purpose of the model being constructed and weigh up the costs and benefits of the different approaches available.

31.4 Implementing dispersal

Simple metapopulation models assume infinite dispersal: there is an equal chance of moving to any patch. Which means dispersal can only be controlled by the rate of leaving a patch. While this has been very effective at revealing the forces driving the evolution of dispersal (e.g. Hamilton and May 1977) and can account for habitat loss (e.g. Nee and May 1992) it does not allow for any concept of habitat fragmentation pattern or any change in distance dispersed.

Spatially explicit patch models often have simple rules about single step dispersal. There are severe assumptions based on the type of neighbourhood considered. For example, a von Neumann neighbourhood (four cardinal neighbours) only has dispersal steps of 1, while a Moore neighbourhood (eight nearest cells) has two dispersal distances which, unless weighting rules are applied, are equally likely to occur: 1 and 1.41. Other commonly used rules for dispersal to 12 or 24 or 48 neighbours have similar problems. Holland *et al.* (2007) have a review of the effects of different assumptions of dispersal through regular grids.

Individual-based models in grids have problems when a continuous dispersal kernel is imposed on a grid. Short-distance moves will have no effect, while those that just leave the cell will be augmented. Obviously, as the size of the kernel expands relative to the size of the imposed grid the effect declines (Collingham *et al.* 1996). Chesson and Lee (2005) and Chipperfield *et al.* (2011) attempt to overcome this problem and provide methods for implementing continuous dispersal in a lattice.

As described above, reaction-diffusion models work in continuous space and use mean dispersal distance and population growth rate to give expansion rate. The problem is that the mean dispersal distance isn't the most important factor at expanding margins. This is Reid's paradox: expansion is much faster than predicted from mean dispersal (Reid 1899), and is solved when rare long-distance events considered.

Many other complications arise when considering dispersal in continuous space as consideration of interaction kernels (i.e. how effects decline with distance) and dispersal kernels often become intractable when habitat fragmentation is added. However, this may be important in the future as it is clear that continuous space models don't always give the same results as discrete space ones (e.g. Murrell 2005).

The interactions of intelligent dispersal strategies and a fragmented landscape have been highlighted elsewhere in this volume. It is clear that any consideration of edges (i.e. whether and when individuals choose to cross habitat boundaries or not) will have a big effect on the population dynamics (e.g. Chapman *et al.* 2007). It also follows that if edges have an impact on dispersal, then the shape of a patch has an effect on the population dynamics as, for example, irregularly shaped patches have much more edge to encounter than circular ones (e.g. Ewers and Didham 2007 discuss the effect of patch shape on population size). The interaction of habitat fragmentation pattern and condition-dependent dispersal strategies (Chapters 10–14) will also need to be considered if the full effect of habitat fragmentation and loss is to be established.

31.5 Metrics

Landscapes: One of the problems facing modellers is that once the complexities of landscapes and populations are described in sufficient detail the resulting system has a vast number of parameters and these need to be reduced to simple metrics. Commonly used metrics are total number of suitable patches or mean quality of patches (statistically, they only describe the first spatial moment). However, these ignore the spatial arrangement of patches; so simple metrics to encapsulate the spatial variance are required (statistically, something related to the second spatial moment). The Hurst exponent is a useful measure of the spatial clumpiness of fractal landscapes, but is not a metric that can be extracted from any landscape. Quadrat-variance methods (where, for example, the variance-mean ratio is calculated for a range of different quadrat sizes) have a long history in ecology (Bliss 1941; Greig-Smith 1952) and give a useful indication of the spatial scale of clumps (Wu *et al.* 2006 for a more recent variant). More useful, now computation power is not an issue, are measures of the spatial clumpiness based on Ripley's K (Ripley 1976) where the number of habitat patches encountered up to a certain distance from each patch are accumulated (see Gustafson 1998). This approach has the advantage of providing an individual's eye view of the density of habitat which will provide a meaningful measure of crowding and will also reveal the scale of clumping. Related measures, such as the 'O-ring' statistic use the density of habitat at a particular distance from a habitat patch (Wiegand and Moloney 2004). Other measures of landscapes include lacunarity (Allain and Cloitre 1991, and its variants Dale 2000; Holland *et al.* 2009), which have the advantage of allowing measures of landscape structure which allow for continuous variation in habitat rather than just a dichotomy. See Hargis *et al.* (1998) for a discussion of the performance of various landscape metrics.

Species: There are several simple metrics that can be used to score the success of a species in a landscape. These can be simply the number of patches occupied, the total population size, the mean population size over a period, or a rate or extent of range expansion (McInerny *et al.* 2007). If the species is declining, the rate of contraction, probability of extinction (Fahrig 1998), or time to extinction may be more appropriate. Metrics using the spatial arrangement of individuals can be captured with quadrat-variance, Ripley's K, or O-rings (as described for landscape metrics above).

Combining the effects of landscape structure with a consideration of the dispersal abilities or distances can be achieved with permeability measures (With and King 1999; King and With 2002). These metrics measure how likely a population will be to cross a particular landscape. Permeability is rarely linearly related to habitat availability and there will usually be distinct thresholds below which a landscape is impermeable.

Diversity: Once more than one species is considered, or individuals within a species are genetically different, a further layer of metrics is required; for multi-species systems, the species richness at a location or some measure of the alpha-diversity (diversity at a point). However, as spatial arrangement is important, the beta-diversity (a measure of the complementarity of two point diversity measures; see Magurran (2004) for discussion of metrics of beta-diversity) is a more useful metric of the effect of landscape on diversity. There is a range of population genetic equivalents (F statistics, isolation-by-distance measures). Note that there is ongoing controversy for both ecological and genetic measures of spatial variation and no truly settled measures have been agreed.

Statistical fitting: One of the issues facing modellers trying to match results to species distribution data is the choice of statistic for comparing the model output to the real data. A simple statistic, such as the 'area under curve' (AUC) uses the proportion of correct matches and mismatches in cells to give a measure of the accuracy (Fawcett 2007). Unfortunately, this type of approach loses the spatial information giving equal weight to a mismatched cell close to a match and one very far from any match (see Wisz and Guisan 2009 for an assessment of the effectiveness of AUC statistics and Lobo *et al.* 2008 for a criticism of the method). Furthermore it assumes that the species distribution is 'real'—i.e. accurate and complete. More sophisticated approaches use 'observation models' to account to the inaccuracies in the 'real' data, and likelihood approaches or Bayesian statistics which attempt to overcome some of the problems with the simple approach.

31.6 Implementing models

Once a landscape has been created, a method for simulating populations chosen and metrics selected the process of modelling is still not complete.

Whatever the modelling approach used, a sensitivity analysis will be revealing. This is where parameters are varied in a systematic fashion and the effect on the metrics noted. Sensitivity analysis, if conducted carefully, can reveal which parameters have most influence on the results, and which combinations of parameters interact. Machine-learning algorithms such as boosted regression trees can be used to aid interpretation of sometimes complex simulation outputs (Munkemuller *et al.* 2011). There are further steps required if a stochastic modelling approach has been used. Unlike analytical approaches, where scenarios can be solved, stochastic models need to have many realizations and need the importance of starting conditions to be explored. If a disturbance, such as a habitat reduction event, is being modelled, it is important that the realization has reached an equilibrium before the disturbance is applied. However, because the model is stochastic there will not be a true equilibrium rather a dynamic, quasi-equilibrium, and this can be difficult to establish formally. Often, the pragmatic approach is to produce, for a small number of parameter combinations, time-series of ecological or evolutionary indices and visually determine when the dynamics stabilize. Then, in the sensitivity analyses, the time period allowed for equilibrium to be obtained (prior to a perturbation) is set to be somewhat longer than the longest time observed in the set of time-series.

31.7 What happens to populations in fragmented landscapes?

Some species will be more prone to extinction than others and, to a large extent, this will be determined by a species' dispersal characteristics. Importantly, dispersal is not a fixed property of a species. Instead, there is within-species variability. Some of this is heritable and thus selection can act on dispersal as the spatial environment is changed. As habitat is fragmented, we may expect evolution of reduced dispersal rates in dumb dispersers (i.e. dispersers that take no account of landscape, habitat, population density, or internal condition) because of the increased cost of dispersal (e.g. Schtickzelle *et al.* 2006). This reduction in dispersal can have knock-on effects for the probability of regional persistence as colonization of empty habitat is reduced. Some models can even predict evolutionary suicide as a response to fragmentation (e.g. Gyllenberg *et al.* 2002). Fragmentation can also have genetic impacts: limited dispersal leads to stronger kin interactions and an increase in the problems of inbreeding (which usually has a negative effect on fitness, although Reed (2010) gives some counter-examples), but might result in increased local adaptation as there is less dilution of location-specific genes. Hanski *et al.* (2011) extend the SPOM framework to include local adaptation, revealing interactions between the strength of local adaptation and the rate of inter-patch movements.

We highlight in the following some of the important issues related to habitat fragmentation and dispersal, and the consequences for population dynamics. We start with an empirical example that neatly illustrates several of the key processes and effects.

Rockall is a very isolated islet in the North Atlantic between Britain and Iceland. There are seven species of winkles (intertidal gastropod molluscs in the genus *Littorina*) that live in the British Isles with a range of dispersal modes from planktonic larvae to direct development. Of the seven, only one has been recorded from Rockall. Counter-intuitively, this species is the one that is least dispersive. This has been dubbed the 'Paradox of Rockall' (Johannesson 1988) where a really isolated patch can only be reached by a very dispersive species, but only exploited by a non-dispersive species. This effect is likely to occur in models of habitat fragmentation where, for dispersive species, there will be sporadic occupancy of isolated areas of habitat resulting from frequent colonization and extinction while less dispersive species will only colonize very rarely, but may have much lower extinction rates once established. Cody and Overton (1996) suggest that there is strong selection for reduced dispersal on islands (i.e. isolated patches of habitat). The location of small islands of habitat is critical. Small patches close to larger patches might not be big enough to have viable populations but may be frequently 'rescued' by immigration from bigger patches and so be continuously occupied.

There will be circumstances where moderate dispersal is the worst strategy—low dispersal keeps a population going in pockets where habitat is dense while high dispersal is enough to jump from one area to another. Dorset butterflies seem to show this effect where the habitat specialists and generalists are both faring better than intermediate species (Thomas 2000). Within a species, this disruptive selection might be enough for the evolution of polymorphisms in dispersal strategy at some levels of habitat fragmentation (Travis and Dytham 1999). In situations where there is temporal variation in fragmentation perhaps due to seasons or inter-annual variation in climate, there will be a shift in the optimum dispersal strategy. A flexible, or shifting, dispersal strategy is common in nature (e.g. pea aphids in a season move from almost no dispersal and clonal reproduction when food is plentiful, to lots of dispersal and sexual reproduction when they shift host plants to overwinter on a tree). Models of dispersal will rarely have the flexibility to describe a complex scenario like this, and will result in predictions of intermediate or bet-hedging dispersal strategies.

It is not loss of habitat *per se* that is the most important element of fragmentation. Dispersal strategies will interact with patterns of loss in space, and it is this combination of dispersal strategy with the scale and pattern of habitat fragmentation that will determine the outcome for the species. Consider three extreme scenarios of habitat loss in a region: a general degradation of all habitat to 50% quality, the loss of 50% of habitat at random, or the loss of all the habitat from half the region. Although these all have the same effect on the regional 'carrying capacity', they will have very different effects on the evolution of dispersal. The first will tend to increase dispersal rates as smaller populations have more kin interactions and more stochasticity. The second will tend to lower dispersal rates as inter-patch distances will have inevitably increased, and the third will have almost no effect as the landscape in half of the region is unaltered.

In some cases, a block of continuous habitat with loss elsewhere is best. This will be especially for species with constrained dispersal and no ability to find habitat. Some species, in contrast, may benefit from moderate fragmentation as their predators are reduced and they gain release (Jeffries and Lawton 1984). Moloney and Levin (1996) modelled the effect of auto-correlated disturbance in neutral landscapes, and showed that the effect of disturbance on their community could only be understood by understanding the spatial pattern of disturbance. Humans generate a lot of disturbance to habitats through logging, road building, agriculture, and mining. Consequently, ruderal species (those with rapid maturation, high reproductive rates, and good dispersal ability; Grime 1979) tend to do very well when humans are around. Humans have spread a small number of weeds and pests around the world. Rapid fragmentation, whether human-mediated or not, leads to a lot of species 'playing the wrong strategy'. In rainforests where the dispersal strategies are evolved to exploit a slowly shifting mosaic of gaps in a relatively continuous habitat, the abrupt loss of forest for agriculture has left many species unable to recruit as suitable habitats are beyond their dispersal range. These have been called the 'living dead' by Janzen (1986) and are part of a slow erosion of diversity that follows habitat loss termed 'extinction debt'.

As habitat is slowly lost, it is loss of 'connectivity' that can be crucial for the survival of a species. This effect will be non-linear as there will be a threshold of habitat loss that will break the connectivity of a region (Figure 31.4). Conservation activities and habitat restoration are often aimed at creating corridors to allow interactions between patches. Unfortunately, it has been very difficult to implement targeted corridor generation in practice, except tactically for single species, because the permeability of the matrix and the willingness of individuals to exploit corridors varies so much both between and within species.

Dispersal kernels which match the pattern of habitat in the landscape will evolve as this maximizes the success of dispersers (Chapters 15, 16). However, this will take a long time and a population in a dynamic environment will not be able to react quickly to changes in habitat pattern.

All the predictions in this section are assuming that habitat patches are equivalent and dispersal is dumb. Once these assumptions are relaxed the

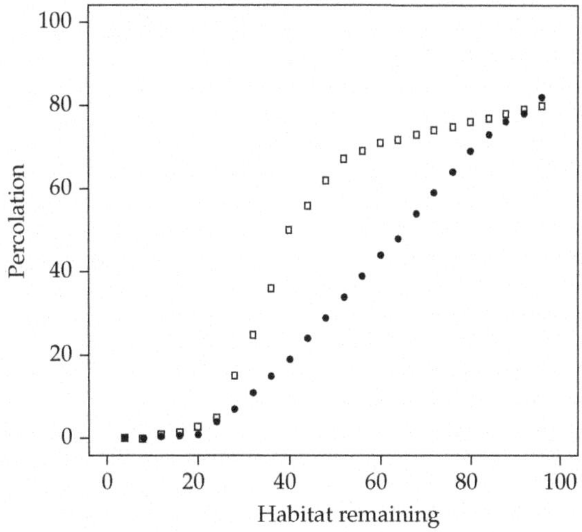

Figure 31.4 Non-linear effects of fragmentation on percolation (chance of moving from one side of the landscape to the other within a time limit) using a landscape of size 512 × 512 pixels. When habitat is removed at random there is a linear decline in percolation probability (solid symbols). However, as habitat is removed in a more aggregated way (open symbols) there is a clear threshold level of habitat loss below which the probability of percolation is greatly reduced.

conclusions are likely to change. If patches vary in quality evolved dispersal strategies will respond. As high-quality patches are more important than low-quality ones it is possible to reduce regional population size by improving the quality of some low quality patches (Poethke *et al.* 2011). If individuals are not dumb and disperse towards habitat there may be selection for increased, rather than reduced, dispersal as habitat becomes fragmented (e.g. Heino and Hanski 2001). This has been demonstrated in an extended SPOM of the Glanville fritillary butterfly (Zheng *et al.* 2009).

31.8 Future directions

Information: Many organisms are able to acquire information about the environment and use this information to modify dispersal decisions (Chapter 10). One example is the foray searching behaviour of butterflies (Conradt *et al.* 2003), when individuals finding themselves away from suitable habitat use a looping search strategy. This foraging behaviour will affect the oviposition opportunities available and therefore affect dispersal (Chapter 32). Intelligent dispersal will mitigate the effects of fragmentation—individuals in the 'matrix' can continue dispersal or detect suitable habitat at range. It is not always the case that information has to be acquired by an individual as decision rules can be passed between individuals either genetically or culturally (Clobert *et al.* 2009). For example, organisms can traverse long distances of bad habitat if they know that there is good habitat to be found (e.g. elephants moving around deserts to specific locations of water that are known about culturally). This could be the result of gradual reduction in habitat quality over large areas (e.g. Sahel expansion; Chapter 25).

Age and stage structure: The majority of spatially extended models applied to predicting the effects of habitat fragmentation make gross simplifying assumptions about population dynamics. These assumptions will typically assume that individuals become mature immediately, or always mature after one time-step. This means that all the age structure in a population is ignored, and while that might be a fair assumption for a strictly annual organism, for most organisms, age will affect size, fecundity, survival, and many other aspects of ecology. Stage structure is equally overlooked. Endopterygous insects, for example, all have four distinct life stages: egg, larva, pupa, adult, and within those stages there may be further complexity with several larval instars or delayed maturity within the adult stage. There are detailed models of complex life histories but they have not been implemented in the spatially extended context required for modelling the effects of habitat fragmentation. Buckley *et al.* (2010) demonstrate the effectiveness of a statistical approach combining climate parameters with a detailed life-history model to predict where species would be expected to persist. However, as their approach had no dispersal, it would not be a useful approach to modelling habitat fragmentation.

Inter-individual variation: Not all individuals have the same genetic background and experience even when they have the same parents and occupy the same patch. It is a dangerous assumption to take the mean dispersal distance or mean life expectancy, especially during a range expansion where the extreme individuals will have a disproportionate influence on the movements of a front. Within fragmented landscapes there is clearly variation in habitat qualities, and such landscapes may be more fully exploited by a population with a range of strategies than a population of identical individuals.

Multi-species: Interspecific interactions will affect dispersal patterns and it is known that fragmentation patterns will mediate those effects. Currently, most models incorporating both interspecific effects and habitat fragmentation are tactical and focused on a single species (e.g. water voles, Rushton *et al.* 2000; spread of rabbit haemorrhagic fever, White *et al.* 2003). Holzkamper *et al.* (2006) address the problem of habitat modification when there are species with different habitat requirements sharing the same landscape. The complexity inherent in multi-species modelling will mean that progress will be slow, but there is a growing need for a landscape-level consideration of the effect of management on biodiversity so the pressure to expand models from the tactical will be intense.

Climate change: The combined effects of environmental change (Chapter 25) and habitat fragmentation on ecological communities provide rich opportunities for the modeller. As with the effects of habitat fragmentation, climate change will have idiosyncratic effects on species, or even individual phenotypes, and as each species is embedded in a complex network of interactions, modelling both habitat fragmentation and climate change will be a challenge indeed.

References

Allain, C. and Cloitre, M. (1991). 'Characterizing the lacunarity of random and deterministic fractal sets', *Physical Review A* 44: 3552–8.

Bliss, C.I. (1941). 'Statistical problems in estimating populations of Japanese beetle larvae', *Journal of Economic Entomology* 34: 221–32.

Buckley, L.B., Urban, M.C., Angilletta, M.J. *et al.* (2010). 'Can mechanism inform species' distribution models?', *Ecology Letters* 13: 1041–54.

Chapman, D.S., Dytham, C., Oxford, G.S. (2007). 'Landscape and fine-scale movements of a leaf beetle: The importance of boundary behaviour', *Oecologia* 154: 55–64.

Chesson, P. and Lee, C.T. (2005). 'Families of discrete kernels for modeling dispersal', *Theoretical Population Biology* 67: 241–56.

Chipperfield, J.D., Dytham, C., Hovestadt, T. (2011). 'An updated algorithm for the generation of neutral landscapes by spectral synthesis', *PLoS ONE* 6: e17040.

Chipperfield, J.D., Holland, E.P., Dytham, C. *et al.* (in press) 'On the approximation of continuous dispersal kernels in discete-space models', *Methods in Ecology and Evolution* DOI: 10.1111/j.2041-210X.2011.00117.x

Clark, J.S. (1998). 'Why trees migrate so fast: confronting theory with dispersal biology and the paleorecord', *American Naturalist* 152: 204–24.

Clobert, J., Le Galliard, J-F., Cote, J. *et al.* (2009). 'Informed dispersal, heterogeneity in animal dispersal syndromes and the dynamics of spatially structured populations', *Ecology Letters* 12: 197–209.

Cody, M.L. and Overton, J.M. (1996). 'Short-term evolution of reduced dispersal in island plant populations', *Journal of Ecology* 84: 53–61.

Collingham, Y.C., Hill, M.O., Huntley, B. (1996). 'The migration of sessile organisms: a simulation model with measureable parameters', *Journal of Vegetation Science* 7: 831–46.

Conradt, L., Zollner, P.A., Roper, T.J. *et al.* (2003). 'Foray search: an effective systematic dispersal strategy in fragmented landscapes', *American Naturalist* 161: 905–15.

Dale, M.R.T. (2000). 'Lacunarity analysis of spatial pattern: A comparison', *Landscape Ecology* 15: 467–78.

Dey, S., Dabholkar, S., Joshi, A. (2006). 'The effect of migration on metapopulation stability is qualitatively unaffected by demographic and spatial heterogeneity', *Journal of Theoretical Biology* 238: 78–84.

Dytham, C. (1994). 'Habitat destruction and competitive coexistence: a cellular model', *Journal of Animal Ecology* **63**: 490–91.

Dytham, C. and Simpson, S.D. (2007). 'Evolutionary models of active movement by larval coral reef fishes in response to mortality rates on the reef', *Marine Ecology Progress Series* 346: 255–64.

Ewers, R.M. and Didham, R.K. (2007). 'The effect of fragment shape and species' sensitivity to habitat edges on animal population size', *Conservation Biology* 21: 926–36.

Fahrig, L. (1998). 'When does fragmentation of breeding habitat affect population survival?', *Ecological Modelling* 105: 273–92.

Fawcett, T. (2006). 'An introduction to ROC analysis', *Pattern Recognition Letters* 27: 861–74.

Fisher, R.A. (1937). 'The wave of advance of advantageous genes', *Annals of Eugenics* 7: 355–69.

González-Andujar, J.L., Perry, J.N. (1995). 'Models for the herbicidal control of the seed bank of *Avena sterilis*: the effects of spatial and temporal heterogeneity and of dispersal', *Journal of Applied Ecology* 32: 578–87.

Greig-Smith, P. (1952). 'The use of random and contiguous quadrats in the study of plant communities', *Annals of Botany* 16: 293–316.

Grime, J.P. (1979). *Plant Strategies and Vegetation Processes.* John Wiley and Sons, Chichester.

Gustafson, E.J. (1998). 'Quantifying landscape spatial pattern: What is the state of the art?', *Ecosystems* 1: 143–56.

Gyllenberg, M., Parvinen, K. and Dieckmann, U. (2002). 'Evolutionary suicide and evolution of dispersal in structured metapopulations', *Journal of Mathematical Biology* 45: 79–105.

Halley, J.M., Hartley, S., Kallimanis, A.S. *et al.* (2004). 'Uses and abuses of fractal methodology in ecology', *Ecology Letters* 7: 254–71.

Hamilton, W.D. and May, R. (1977). 'Dispersal in stable habitats', *Nature* 269: 578–81.

Hanski, I. (1999). *Metapopulation Ecology*. Oxford University Press.

Hanski, I., Mononen, T., Ovaskainen, O. (2011). 'Eco-evolutionary metapopulation dynamics and the spatial scale of adaptation', *American Naturalist* 177: 29–43.

Hargis, C.D., Bissonette, J.A., David, J.L. (1998). 'The behavior of landscape metrics commonly used in the study of habitat fragmentation', *Landscape Ecology* 13: 167–86.

Heino, M. and Hanski, I. (2001). 'Evolution of migration rate in a spatially realistic metapopulation model. *American Naturalist* 157: 495–511.

Hogeweg, P. (1988). 'Cellular automata as a paradigm for ecological modeling', *Applied Mathematics and Computation* 27: 81–100.

Holland, E.P., Aegerter, J.N., Dytham, C. (2009) 'Comparing resource representations and choosing scale in heterogeneous landscapes', *Landscape Ecology* 24: 213–27.

Holland, E.P., Aegerter, J.N., Dytham, C. *et al.* (2007). 'Landscape as a model: the importance of geometry', *PLoS Computational Biology* 3: e200.

Holzkämper, A., Lausch, A., Seppelt, R. (2006). 'Optimizing landscape configuration to enhance habitat suitability for species with contrasting habitat requirements', *Ecological Modelling* 198: 277–92.

Janzen, D.H. (1986). 'The Future of Tropical Ecology', *Annual Review of Ecology and Systematics* 17: 305–24.

Jeffries, M.J. and Lawton, J.H. (1984). 'Enemy free space and the structure of ecological communities', *Biological Journal of the Linnean Society* 23: 269–86.

Johannesson, K. (1988). 'The paradox of Rockall: why is a brooding gastropod (*Littorina saxatilis*) more widespread than one having a planktonic larval dispersal stage (*L. littorea*)?', *Marine Biology* 99: 507–13.

Judson, O.P. (1994). 'The rise of the individual-based model in ecology', *Trends in Ecology and Evolution* 9: 9–14.

Kindlmann, P., Aviron, S., Burel, F. (2005). 'Simulating the spatial and temporal dynamics of landscapes using generic and complex models', *Ecological Complexity* 2: 150–8.

Kinezaki, N., Kawasaki, K., Takasu, F. *et al.* (2003). 'Modeling biological invasions into periodically fragmented environments', *Theoretical Population Biology* 64: 291–302.

King, A.W. and With, K.A. (2002). 'Dispersal success on spatially structured landscapes: when do spatial pattern and dispersal behavior really matter? *Ecological Modelling* 145: 211–24.

Levins, R. (1969). 'Some demographic and genetic consequences of environmental heterogeneity for biological control. *Bulletin of the Entomological Society of America* 15: 237–40.

Lobo, J.M., Jiménez-Valverdel, A., Real, R. (2008). 'AUC: a misleading measure of the performance of predictive distribution models', *Global Ecology and Biogeography* 17: 145–51.

Magurran, A.E. (2004). *Measuring Biological Diversity.* Blackwell Publishing, Oxford.

McInerny, G., Travis, J.M.J., Dytham, C. (2007). 'Range shifting on a fragmented landscape', *Ecological Informatics* 2: 1–8.

McPeek, M.A. and Holt, R.D. (1992). 'The evolution of dispersal in spatially and temporally varying environments', *American Naturalist* 140: 1010–27.

Moilanen, A. (2004). 'SPOMSIM: software for stochastic patch occupancy models of metapopulation dynamics', *Ecological Modelling* 179: 533–50.

Moloney K.A. and Levin S.A. (1996). 'The effects of disturbance architecture on landscape-level population dynamics', *Ecology* 77: 375–94.

Münkemüller, T., Travis, J.M.J., Burton, O.J. *et al.* (2011). 'Density-regulated population dynamics and conditional dispersal alter the fate of mutations occurring at the front of an expanding population', *Heredity* 106: 678–89.

Murrell, D.J. (2005). 'Local spatial structure and predator-prey dynamics: counterintuitive effects of prey enrichment', *American Naturalist* 166: 354–67.

Nee, S. and May, R.M. (1992). 'Dynamics of metapopulations—habitat destruction and competitive coexistence', *Journal of Animal Ecology* 61: 37–40.

Ovaskainen, O. and Hanski, I. (2003). 'How much does an individual habitat fragment contribute to metapopulation dynamics and persistence?', *Theoretical Population Biology* 64: 481–95.

Pietgen, H.O. and Saupe, D. (1988). *The Science of Fractal Images*. Springer-Verlag.

Poethke, H.J., Dytham, C. Hovestadt, T. (2011). 'A metapopulation paradox: partial improvement of habitat may reduce metapopulation persistence', *American Naturalist* 177: 792–9.

Poethke, H.J., Pfenning, B. Hovestadt, T. (2007). 'The relative contributions of individual- and kin-selection in the evolution of density-dependent dispersal rates', *Evolutionary Ecology Research* 9: 41–50.

Reed, D.H. (2010). 'Albatrosses, eagles and newts, oh my!: Exceptions to the prevailing paradigm concerning genetic diversity and population viability?', *Animal Conservation* 13: 448–57.

Reid, C. (1899). *The Origin of the British Flora*. Dulau, London.

Ripley, B.D. (1976). 'The second-order analysis of stationary point processes', *Journal of Applied Probability* 13: 225–66.

Rushton, S.P., Barreto, G.W., Cormack, R.M. *et al.* (2000). 'Modelling the effects of mink and habitat fragmentation on the water vole', *Journal of Applied Ecology* 37: 475–90.

Schtickzelle, N., Mennechez, G., Baguette, M. (2006). 'Dispersal depression with habitat fragmentation in the bog fritillary butterfly', *Ecology* 87: 1057–65.

Shigesada, N., Kawasaki, K., Teramoto, E. (1986). 'Traveling periodic waves in heterogeneous environments', *Theoretical Population Biology* 30: 143–60.

Skellam, J.G. (1951). 'Random dispersal in theoretical populations', *Biometrika* 38: 196–218.

Thomas, C.D. (2000). 'Dispersal and extinction in fragmented landscapes', *Proceedings of the Royal Society B* 267: 139–45.

Travis, J.M.J. and Dytham, C. (1999). 'Habitat persistence, habitat availability and the evolution of dispersal', *Proceedings of the Royal Society B* 266: 723–8.

Travis, J.M.J. and Dytham, C. (2002). 'Dispersal evolution during invasions', *Evolutionary Ecology Research* 4: 1119–29.

Travis, J.M.J. and Dytham, C. (2004). 'A method for simulating patterns of habitat availability at static and dynamics range margins', *Oikos* 104: 410–6.

White, P.C.L., Newton-Cross, G., Gray, M. *et al.* (2003). 'Spatial interactions and habitat use of rabbits on pasture and implications for the spread of rabbit haemorrhagic disease in New South Wales', *Wildlife Research* 30: 49–58.

Wiegand, T. and Moloney, K. (2004). 'Rings, circles and null-models for point pattern analysis in ecology', *Oikos* 104: 209–29.

Wisz, M.S. and Guisan, A. (2009). 'Do pseudo-absence selection strategies influence species distribution models and their predictions? An information-theoretic approach based on simulated data', *BMC Ecology* 9: 8.

With, K.A. and King, A.W. (1999). 'Dispersal success on fractal landscapes: a consequence of lacunarity thresholds', *Landscape Ecology* 14: 73–82.

Wu, B.M., Subbarao, K.V., Ferrandino, F.J. *et al.* (2006). 'Spatial analysis based on variance of moving averages', *Journal of Phytopathology* 154: 349–60.

Zheng, C., Ovaskainen, O., Hanski, I. (2009). 'Modelling single nucleotide effects in phosphoglucose isomerase on dispersal in the Glanville fritillary butterfly: coupling of ecological and evolutionary dynamics', *Proceedings of the Royal Society B* 364: 1519–32.

CHAPTER 32

High connectivity despite high fragmentation: iterated dispersal in a vertebrate metapopulation

Xavier Lambin, Diane Le Bouille, Matthew K. Oliver, Chris Sutherland, Edoardo Tedesco, and Alex Douglas

32.1 Introduction

A dispersal strategy is defined by the frequency and triggers of emigration, the characteristics of movement trajectories during the transience phase (step length, degree of correlation of dispersal paths, and perceptual ranges), and the influence of conspecifics on immigration. Identifying the selective pressures on each of those traits is the topic of a sustained research effort (e.g. Barton *et al.* 2009). Dispersal strategies and the context dependency of dispersal decisions remain poorly characterized for most systems. In order to understand the costs and benefits of dispersal, the empirical challenges include 1) the requirement for sufficiently large study areas to encompass the whole distribution of realized movements, 2) the difficulty in observing the transience phase, and 3) the characterization of the landscape experienced by dispersers, especially with respect to the distribution of relatives present in the natal area, as well as more distant patches. Because of these obstacles, empirical evidence lags behind theoretical developments, and too many theories remain unchallenged by data.

Whilst experimental approaches are clearly best suited for testing the relevance of mechanisms putatively causing dispersal (Chapters 3, 4, 13, 14, 23, 33), these are by necessity performed over small, tractable scales. Descriptive studies of patterns and correlates of dispersal at more ecologically relevant scales are thus paramount, because small-scale processes do not necessarily predict large-scale patterns. Indeed, studies of dispersal that have checked the consistency of pattern emerging at behavioural, individual, and population scale are rare or altogether lacking.

In this chapter, we review published and unpublished studies on dispersal by water voles (see Figure 32.1) inhabiting fragmented habitats and organized as metapopulations. We describe how we have drawn inferences from exceptionally large-scale, but largely descriptive, studies of dispersal through the use of molecular markers, combined with small-scale individual-level experiments. These studies reveal a high degree of connectivity through dispersal between geographically isolated water vole colonies, and experiments with 'enforced dispersers' show how water vole behaviour during the transience phase of dispersal might bring this about, if dispersal takes place over a long time through multiple stepping-stone movements.

32.2 Study system

Water voles are large rodents (200–300 g) which, in the UK, occupy margins of slow-flowing stretches of waterways with high grass coverage (Woodroffe *et al.* 2008). Individuals construct burrows in the plant rooting zone and residents do not normally move more than a few meters away from water. Both sexes mark range boundaries with latrines, and consequently the absence of latrines makes vacant sites easily identifiable. Voles born in the preceding year (overwintered adults) produce most

Dispersal Ecology and Evolution. First Edition. Edited by Jean Clobert *et al.*

Figure 32.1 Water voles are large rodents up to 300 g that are semi-aquatic and restricted to riparian habitats in the UK. They form small colonies in grass-rich sections of waterways with slow flowing water. They show a colour polymorphism, with black, brown, or black-brown coloration. Photo Alexandre Millon.

of the young (Telfer *et al.* 2003). Some juveniles from the first litter of the year tend to mature in the year of birth in lowland areas, and typically disappear from their natal range when they reach 80 to 130 g, equivalent to 1–2 months old. Juveniles from later litters in the lowland, and virtually all juveniles born in upland areas where the growing season is shorter, do not reproduce in the year of birth, and their dispersal takes place over a more protracted period. Only a small fraction of adults survive to breed in two successive years, such that, except for the rare early maturing individuals, generations are nearly non-overlapping at the breeding stage.

Water vole populations described in Scotland closely conform to a metapopulation structure (reviewed in Lambin *et al.* 2004). In the Scottish uplands, suitable sections of waterways (habitat patches hereafter) are imbedded in a matrix of unsuitable habitat dominated by prostrate dwarf heath, sedge and rush, moss heaths, and bogs in which they never settle. These suitable stretches are typically 505 m apart (range: 201 m–1507 m) and amount to 11% of the 860 km of waterways in the study area. They are separated by fast-flowing water, rocky or heather margined sections. Discrete breeding colonies use sections of riverbank typically 200 m long (range 10 m–1845 m). In lowland Scotland, the matrix consists of arable land and pastures. Because of overall higher productivity, a greater fraction of the waterway network is suitable for occupancy. Upland colonies typically include a single breeding female and her progeny. Thus, in the absence of dispersal, there would be a high potential for kin competition and close inbreeding between siblings, but virtually none between parents and offspring.

Our approach has been to study the demography, dispersal, and spatial dynamics of water voles over an exceptionally large spatial scale (140 km^2) relative to the size of the organism and that which is normally attempted with microtine rodents. We aimed to census all individuals once per year. Thus all patches ever identified as potentially suitable for occupancy are visited in summer, and where voles are present, live trapping is conducted over 3–5 days. While such snapshot sampling yields information on the status and composition of colonies at a given time, rich additional information on individual breeding success and dispersal can be derived through genotyping of all trapped water voles (details in Aars *et al.* 2006; Stewart *et al.* 1998; Telfer *et al.* 2003).

32.3 High connectivity between fragmented populations

The limitations of classical population genetics approaches to make inferences about dispersal in populations are well known (Lowe and Allendorf 2010) but, critically, such approaches do not require exhaustive sampling over a large area to reveal large-scale dispersal (Ehrich *et al.* 2001). In our empirical studies addressing the theory that genetic diversity should become eroded in fragmented populations (Aars *et al.* 2006), evidence of the mixing power of dispersal at the largest scale in our upland water vole metapopulation is somewhat indirect, but nevertheless compelling. Aars *et al.* (2006) reported that temporal genetic drift (whether measured by the slope of the relationship between F_{st} or G'_{st} estimates derived from 12 microsatellite loci and time between sampling occasion) amongst groups of water vole colonies occurred at similar rates at different spatial scales up to 20 km (Oliver *et al.* 2009). By inference, dispersal must take place at a sufficient rate and frequency to achieve effective panmixia over a few generations, and preclude any divergence between sub-groups of colonies due to independent drift up to at least such a distance. The lack of any evidence of isolation by distance at the scale of metapopulation networks covering > 100s km^2 was also indicative of frequent dispersal up to a few kilometres. As with any inference drawn from population genetics data, a degree of care must be exercised when interpreting these strands of evidence (Lowe and Allendorf 2010). Firstly, high degrees of genetic cohesiveness may arise from cumulative smaller scale dispersal movements by multiple individuals or less frequent long-distance dispersal (Ehrich *et al.* 2001). Second, the degree of fragmentation, frequency of extinction, and fluctuations in abundance of populations may interact in complex ways with dispersal strategies in shaping the emerging genetic structure. Thus, while it might seem reasonable to assume that the dispersal ability of a given species is invariant, it is striking that similar analyses to those we performed, but conducted on the fossorial form and cyclically fluctuating populations of the same vole species, revealed clear genetic discontinuities between areas with contrasting density (Berthier *et al.* 2005). Similarly, two studies with *Microtus arvalis*, a related vole species, both in agricultural habitats, revealed a persistent overall genetic structure among populations separated by 0.4–2.5 km in one instance (Schweizer *et al.* 2007), and a single genetic unit covering the entire 500 km^2 area in the other instance where the focal population experienced multi-annual cycles (Gauffre *et al.* 2008). Thus, differences observed in the genetic patterns of populations of a given species must be attributed to variation in demographic processes and flexible dispersal strategies rather than to differences in inherent dispersal abilities.

Aars *et al.* (2006) used population genetics approaches to draw inferences on the age-specific prevalence of dispersal by water voles over ecological timescales using the distinctive age structure visible in water vole colonies as they are sampled after the emergence, but typically before the dispersal of juveniles from the first annual litter. F_{IS} values among juveniles sampled in the year of birth were negative, reflecting heterozygosity in excess of Hardy Weinberg equilibrium, whereas they clustered around zero among adults born in previous years. F_{ST} values among colonies were high for juveniles, but moderate and often insignificant for parents. The excess heterozygosity seen within colonies was taken to reflect the few individuals dispersing from beyond the range of the closest neighbour patch to form discrete breeding colonies. Thus rapid reproduction and the formation of local family groups apparently cause an increase in differentiation, but this is cancelled by subsequent dispersal.

Comparisons of F_{IS} between years at the metapopulation level show that excess heterozygosity (across both adults and juveniles) increases with the average level of isolation between colonies (Oliver *et al.* 2009). This illustrates a situation where individuals dispersing from distant colonies, characterized by different genetic frequencies, locate mates in a sparsely occupied landscape. The resulting mixing of genotypes sourced from distant colonies (with intervening colonies having gone extinct) results in levels of heterozygosity in excess of Hardy Weinberg equilibrium. Interestingly this demonstrates how excess heterozygosity, which may mistakenly be attributed to the action of selection or inbreeding avoidance, both of which would be expected to lead to a general excess of heterozygosity, rather than one that correlates specifically with increasing isolation, can be generated purely by metapopulation effects combined with long-distance dispersal events.

32.4 Natal dispersal of individuals linking populations

In addition to making inferences from population genetic parameters, genetic information can also be used to infer dispersal through the identification of genealogically related individuals. Here, we present an example where we infer the connectivity between patches by reconstructing pedigree relationships between pairs of water voles using the software KINSHIP 1.3.1 (Goodnight and Queller 1999). Other softwares (e.g. COLONY Jones and Wang 2010) can similarly be used, although computational requirements can vary substantially. KINSHIP calculates the genetic relatedness, *r*, between pairs of individuals which, when combined with likelihood tests on competing relatedness hypotheses (e.g. parent offspring versus non-related), can be used to assign full sibling or parent offspring pairs without requiring any other prior information. Including age data can help improve discrimination between parent offspring and full sibling pairs. Using this method, dispersers are those individuals who are located in a different patch from that where their putative close relatives are located. Identifying such individuals was facilitated by the very high proportion of related individuals present in the study area being sampled. Note, however, that in those instances where two close relatives both left their natal patch, dispersal distance may be under- or overestimated.

Water voles disperse over exceptionally long distances. Of 1498 individuals assigned and who survived to reproduce, 757 (51%) dispersed and the mean dispersal distance was 3.5 km (STD 4.7, max 23.9 km). The average dispersal distance was at least four times higher than predicted from allometric relationships based on other studies of mammalian dispersal (Sutherland *et al.* 2000), and substantially further than estimated using similar techniques for raccoon, which are at least tenfold larger (Cullingham *et al.* 2008). Dispersal in water voles shows no evidence of sex bias in either distance or frequency of dispersal (Telfer *et al.* 2003; Aars *et al.* 2006).

Dispersal movement binds water vole colonies in a highly connected network (Figure 32.2), the structure of which approaches randomness with very little dependency of connectivity on distance. Median dispersal distance doubled from within (1.02 km) to between year (1.97 km) timescales. When considering natal dispersal and resulting networks, connectedness markedly increased with time such that by the spring following birth, even patches on either side of major topographical obstacles were connected. Contrasting Figure 32.2A and 32.1B implies that water voles are sufficiently mobile to bypass obstacles, such that if considering the ecologically relevant annual timescale, the impact of intervening topographical obstacles on connectedness between patches would not be apparent. In such circumstances, least-cost surface distances would not outperform Euclidean distances as predictors of dispersal, reflecting the mismatch between observations at short (behavioural) and ecological (annual) timescales. Network statistics indicate that some patches, including those bounded by topographical obstacles, have high centrality, especially those used as transit patches in the year of birth.

32.5 Short-term experiments and the behaviour of transient water voles

While our study is otherwise wholly descriptive, we used simulated dispersal experiments of sub-adults fitted with small VHF radio-transmitters attached to collars to assess disperser's ability to immigrate to occupied or presently vacant but previously occupied sites (Fisher *et al.* 2009). Sub-adults aged 1.5–2 months that would often naturally disperse were used to minimize any bias arising from forced dispersal or behavioural differences in hormonally primed or motivated dispersal. Individuals were located hourly during daytime for the first three days and at least daily thereafter. Performing such experiments is challenging owing to the number of sites required, the care needed in selecting appropriate candidates for enforced dispersal, and the large scale of movements taking place relative to the limited range of radio-transmitters when dealing with a partly fossorial organism. Despite achieving only modest levels of replication (ten and seven voles moved to different occupied and vacant sites respectively, and three newly mated females radio-tracked but not translocated), we learned a great deal about how the behaviour of voles during the transient phase might translate into the extremely long movements observed in the uplands. The salient points are summarized here, though more details are available in Fisher *et al.* (2009).

The fate of the sub-adult voles that survived a few days post-movement (three were immediately predated) varied according to the context of their release. Of the six surviving voles translocated to a suitable but vacant patch, all rejected the site and dispersed; a vacant site where a young male was released was colonized by a natural (non-experimental) opposite-sex disperser and both settled there, and, half of those voles translocated to occupied sites settled (n = 8 non-predated). Dispersing voles also rejected all vacant sites they encountered

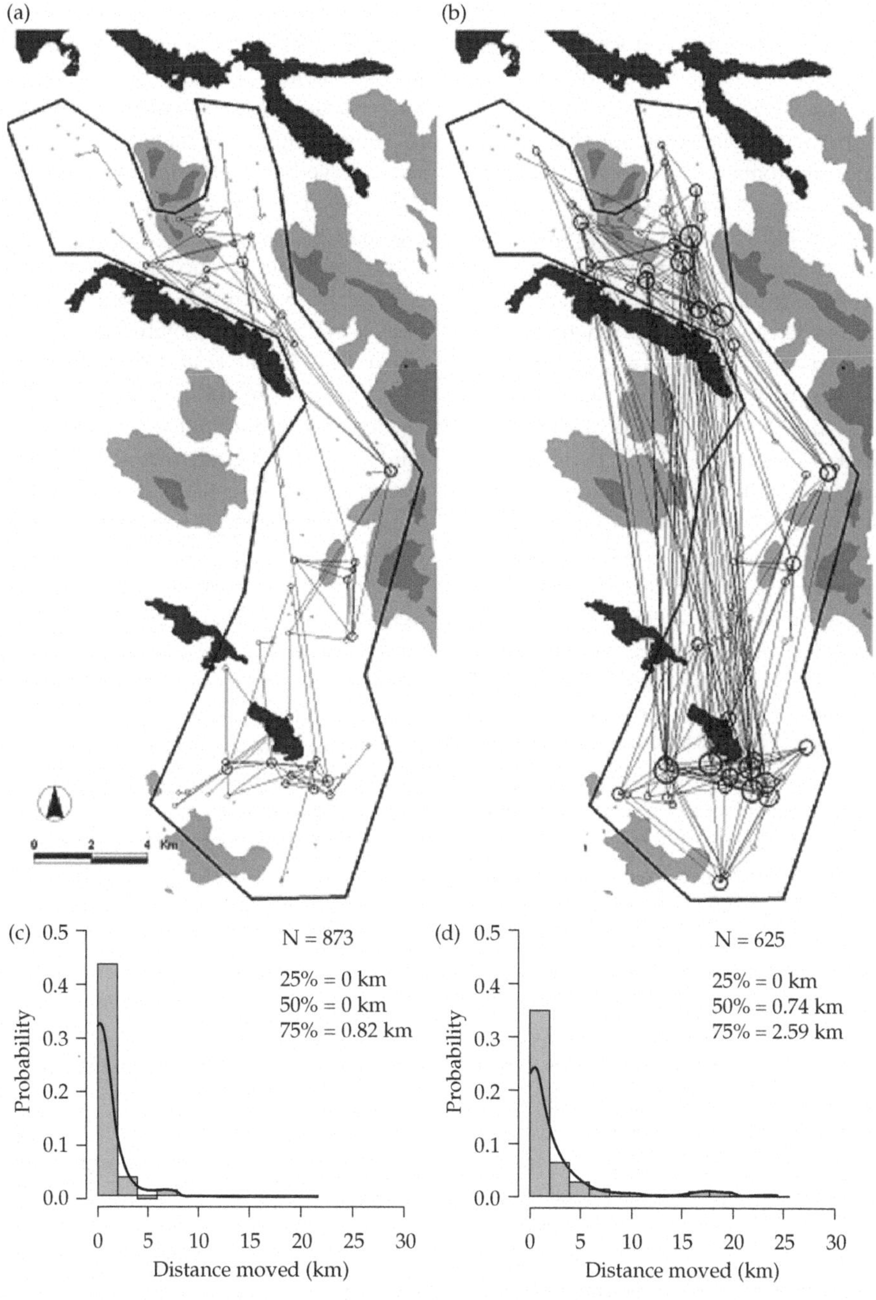

Figure 32.2 Network of dispersal movements by first-degree relative water voles detected through parentage assignment. Lines connect patches linked by dispersal movements that took place (a) in the first three months of life of the youngest vole in the pair; (b) up to 11 months after the birth of the youngest vole. The size of each circle reflects the centrality of a patch (number of connections to other patches). Solid grey areas represent areas above 400 and 600 m, including cliff faces and sparsely vegetated stony ground. Dark stippled areas represent water bodies (sea to the north) and lochs in the study area. The area where all patches are monitored yearly is delineated by the black line. Annual patch occupancy and colony extinction probability both average 35% but no account is taken of occupancy or population size in these network representations. C and D. Frequency distribution of dispersal distances for 873 and 625 individuals assigned pairs within year and between years respectively.

after leaving their release patch and often followed a 'stepping-stone' dispersal trajectory. Dispersing voles remained in the transient phase of dispersal for up to 16 days. Dispersers stayed for up to ten days at sites that were considered suitable (previously occupied) and contained no other water voles but, with the exception of two instances when they were joined by an individual of the opposite sex, they eventually abandoned vacant sites and moved on. There was some evidence of density dependent immigration; vole density at the sites where translocated voles settled (0.88 adult/100 m) was substantially and significantly lower than the density at the four occupied sites rejected by the translocated voles (2.35 adults/100 m).

Voles that rejected their release sites for settlement dispersed out of their own volition either along waterways (n = 4) or overland (n = 5). Thus, they readily entered the assumed matrix of unsuitable habitats where no vole ever settles. On four occasions voles were recorded as having dug a burrow in a transit area. Of these, one transient individual dug a substantial burrow in a field well away from water, and remained in this shelter for one night before resuming its exploration. Thus not only are these herbivorous rodents able to sustain themselves in a relatively permeable matrix, but they might also be able to reduce the cost of transience by creating temporary refuges. They do, however, suffer an elevated mortality risk relative to non-dispersers, estimated previously by comparing daily mortality risk to predation between translocated individual and the disappearance rate of individuals from the same juvenile spring cohort followed by capture recapture in their natal patches, to be fourfold higher for sub-adults of the same age and general area (Lambin *et al.* 2004). This predation cost rather than any depletion of energetic resources is also believed to account for the death of failed dispersal in Vancouver marmot (Brashares *et al.* 2010). Whereas one contributing factor to elevated marmot mortality during dispersal is their use of inadequate shelters such as fallen trees instead of burrows when away from colonies, the ability of water voles to create effective temporary shelters during dispersal is a likely contributory factor to their success as a species inhabiting highly fragmented habitats.

The collection of anecdotes reported above demonstrates that water voles occasionally leave their normal habitat, and embark on one-way dispersal movements. Each movement bout displaces them by some hundreds of meters per day; for example, we observed an individual moving 1800 m over 13 days. Such stepping-stone movements may be repeated multiple times and are seemingly interrupted by either an encounter with an opposite sex conspecific in a suitable site or the death of the disperser. Whereas early-maturing spring-born voles tracked by Fisher *et al.* (2009) may lose valuable breeding time in the year of birth if they fail to encounter a mate rapidly, later cohorts in the lowlands and probably all juveniles in upland areas do not face this time constraint. Instead, they have the option of dispersing over many months between adolescence and their first breeding opportunity in the spring following their birth. If they followed the iterated dispersal strategy described above, they would have the opportunity to sample a large number of habitat patches sequentially, hence to optimize patch or mate choice.

Using an iterated dispersal strategy between habitat patches, including waiting for the arrival of an opposite-sex conspecific seems effective for encountering a potential mate despite very low conspecific density, although little existing theory tackles this aspect of dispersal. Indeed, population genetic evidence (Oliver *et al.* 2009) suggests that pairs are formed by individuals of distant provenance at low density. However, whereas dispersal of the spring cohort that was the focus of Fisher *et al.*'s (2009) work is mostly synchronized, coinciding with sexual maturation, dispersal over the summer and autumn is probably more asynchronous, reducing the likelihood of encountering a mate during dispersal.

While there is no evidence of mating limitation amongst water voles (dispersers or not) that survived until the spring (Aars *et al.* 2006), failure to encounter an opposite-sex conspecific may amount to a depensatory process at low conspecific density, such as on the edge of a species range where the transience phase may be protracted (Chapter 26). A greater perceptual range of dispersers and higher ability to detect both habitat patches and the presence of conspecifics would obviously increase their success in dispersal.

32.6 Conclusions

We presented evidence, collected at daily, monthly, and annual scales that reveal a water vole patch network highly connected by dispersal despite long inter-patch distances. Individuals performed multiple stepping-stone dispersal movements over many days in search of opposite sex conspecifics. Multiple individuals dispersed between patches, mostly irrespective of the intervening distance, and colonies were thus genetically homogeneous over extended distances being formed by individuals originating from larger areas at low density. We conclude that in water voles, but also probably in many other species for which dispersal may take place over time through repeated stepping-stone movements, between-patch connectivity in a network brought about by dispersal may be higher than expected from properties of single dispersal events.

The pattern of movement employed by voles appears to be in contrast to the foray search pattern described for some butterflies (Conradt *et al.* 2003), but its effective straightness is not dissimilar to that predicted for species inhabiting sparsely populated environments (Barton *et al.* 2009; Heinz and Strand 2006; Zollner and Lima 2005). Here the effective displacement achieved by an individual employing such a 'move and wait' strategy should reflect its dispersal lifespan, which for water voles able to access food resources while transient, should in turn be influenced by the mortality risk during dispersal (Chapter 16). While pikas (Ochotona princeps) have a similar size and probably similar intrinsic mobility as water voles, their effective vagility is constrained, especially at low altitudes, by the need to have completed the process of dispersal in time for collecting the hay stores on which they rely in winter, and their inability to tolerate high diurnal temperature. These constraints limit the dispersal lifespan and contribute to a relatively low vagility of this species (Smith 1974). Dispersal lifespan is much shorter for butterflies living in appropriate climatic conditions for only a few days and relying on onboard energy stores (Chapter 18), though is much longer for species with delayed maturation able to acquire information from direct sampling of patches. While the patterns observed at the individual and population scales imply that water vole dispersal behaviour is protracted and probably spans many weeks or months, our observation that one individual achieved the median natal dispersal distance in only 13 days and the substantial mean dispersal of individuals sampled in their year of birth indicate that mobility *per se* might not be a constraint. Crucially, the information gleaned by dispersers concerning their environment and conspecifics is acquired up to breeding. As such, in contrast to common lizards making dispersal decisions based on maternal quality in the first days of life but years before they reproduce (Case Study I), the reliability of the information gleaned by dispersers should not be severely eroded by changing circumstances such as conspecific mortality.

The lack of clear theoretical predictions for how the behaviour of transients translates into overall effective dispersal stands in contrast with the plethora of predictions of the evolution of emigration (Chapter 10). Informing theory with relevant empirical data such as those presented here is likely to be a fruitful endeavor.

Acknowledgements

We thank Stuart Piertney for supervising genotyping, land owners for allowing access to their land, Diana Fisher for comments on an earlier version, and multiple field workers who contributed to this study over many years.

References

Aars J., Dallas, J.F, Piertney, S.B. *et al.* (2006). 'Widespread gene flow and high genetic variability in populations of water voles *Arvicola terrestris* in patchy habitats', *Molecular Ecology* **15**: 1455–66.

Barton, K.A, Phillips, B.L, Morales, J.M. *et al.* (2009). 'The evolution of an 'intelligent' dispersal strategy: biased, correlated random walks in patchy landscapes', *Oikos* **118**: 309–19.

Berthier, K., Galan M., Foltête, J.C. *et al.* (2005). 'Genetic structure of the cyclic fossorial water vole (*Arvicola terrestris*): landscape and demographic influences', *Molecular Ecology* **14**: 2861–71.

Brashares, J.S., Werner, J.R., Sinclair, A.R.E. (2010). 'Social 'meltdown' in the demise of an island endemic: Allee effects and the Vancouver Island marmot', *Journal of Animal Ecology* **79**: 965–73.

Conradt, L., Zollner, P.A., Roper, T.J. *et al.* (2003). 'Foray search: An effective systematic dispersal strategy in fragmented landscapes', *American Naturalist* **161**: 905–15.

Cullingham, C.I., Pond, B.A., Kyle, C.J. *et al.* (2008). 'Combining direct and indirect genetic methods to estimate dispersal for informing wildlife disease management decisions', *Molecular Ecology* **17**: 4874–86.

Ehrich, D., Jorde, P.E., Krebs, C.J. *et al.* (2001). 'Spatial structure of lemming populations (*Dicrostonyx groenlandicus*) fluctuating in density', *Molecular Ecology* **10**: 481–95.

Fisher, D.O., Lambin, X., Yletyinen, S.M. (2009). 'Experimental translocation of juvenile water voles in a Scottish lowland metapopulation', *Population Ecology* **51: 289–95.**

Gauffre, B., Estoup, A., Bretagnolle, V. *et al.* (2008). 'Spatial genetic structure of a small rodent in a heterogeneous landscape', *Molecular Ecology* **17**: 4619–29.

Goodnight, K. and Queller, D. (1999). 'Computer software for performing likelihood tests of pedigree relationship using genetic markers', *Molecular Ecology* **8**: 1231–4.

Heinz, S.K. and Strand, E. (2006). 'Adaptive patch searching strategies in fragmented landscapes', *Evolutionary Ecology* **20**: 113–30.

Jones, O.R., Wang, J.L. (2010). 'COLONY: a program for parentage and sibship inference from multilocus genotype data', *Molecular Ecology Resources* **10**: 551–5.

Lambin, X., Aars, J., Piertney, S.B. *et al.* (2004). 'Inferring pattern and process in small mammal metapopulations: insights from ecological and genetic data, in *Ecology, Genetics and Evolution of Metapopulations*. Academic Press 515–40.

Lowe, W.H. and Allendorf, F.W. (2010). 'What can genetics tell us about population connectivity?', *Molecular Ecology* **19**: 3038–51.

Oliver, M.K., Lambin, X., Cornulier, T. *et al.* (2009). 'Spatio-temporal variation in the strength and mode of selection acting on major histocompatibility complex diversity in water vole (*Arvicola terrestris*) metapopulations', *Molecular Ecology* **18**: 80–92.

Schweizer, M., Excoffier, L., Heckel, G. (2007). 'Fine-scale genetic structure and dispersal in the common vole (Microtus arvalis)', *Molecular Ecology* **16**: 2463–73.

Stewart, W.A., Piertney, S.B., Dallas, J.F. (1998). 'Isolation and characterization of highly polymorphic microsatellites in the water vole, *Arvicola terrestris*', *Molecular Ecology* **7**: 1258–9.

Sutherland, G.D., Harestad, A.S., Price, K. *et al.* (2000). Scaling of natal dispersal distances in terrestrial birds and mammals', *Conservation Ecology* **4**

Telfer, S., Piertney, S.B., Dallas, J.F. *et al.* (2003). 'Parentage assignment reveals widespread and large-scale dispersal in water voles', *Molecular Ecology* **12**: 1939–51.

Woodroffe, G., Lambin, X., Strachan, R. (2008). 'Water voles *Arvicola terrestris*', in S. Harris, D. Yalden (ed), *Mammals of the the British Isles*, 4th edition. The Mammal Society, Southhampton, 110–16.

Zollner, P.A., Lima, S.L., (2005). 'Behavioral tradeoffs when dispersing across a patchy landscape', *Oikos* **108**: 219–30.

CHAPTER 33

Dispersal and habitat fragmentation in invertebrates—examples from widespread and localized butterflies

Hans Van Dyck and Michel Baguette

33.1 Introduction

Habitat fragmentation through the action of intensive human land use is widely recognized as a key problem for biodiversity. The problem of habitat fragmentation has catalyzed the study of spatial ecology of species, including metapopulation dynamics and dispersal. The significance of mobility to understanding population dynamics and hence population viability in fragmented systems has long been recognized in butterflies with, amongst other methods, several decades of mark-release-recapture studies (e.g. Ehrlich 1961). This interest is surely not independent of their worldwide use as conservation targets (e.g. Schultz and Hammond 2003; Thomas *et al.* 2004). In the era of high extinction risks for many species, several researchers hunt for biological attributes or correlates that can be used to predict extinction. In this context, several studies considered differences in mobility relative to extinction risk. It is interesting to note that different studies on butterflies examined mobility in different ways. Koh *et al.* (2004) used a binary factor reflecting whether species had some records of 'migratory flights' or not. This mobility factor did not contribute to their regression model to estimate relative extinction proneness. Kotiaho *et al.* (2005) used the average dispersal ability of each species based on scores from questionnaires to lepidopterologists. Extinction risk increased with lower mobility. Finally, Shahabuddin and Riveros (2005) used the average flight speed of butterfly species as a surrogate of their ability to disperse. They found fast-flying species to be more vulnerable to extinction. So, interestingly, this sample of studies on a single 'well-studied' taxonomic group already offers all possible outcomes: dispersal did not contribute, was positively, or negatively, related to extinction risk. However, several studies have ignored questioning the assumptions related to the use of simple surrogates for complex multidimensional traits as dispersal (Baguette and Van Dyck 2007).

However, more recently, butterflies have been studied in much more behavioural detail to improve our understanding, from a mechanistic viewpoint, of the 'black box' of dispersal in fragmented systems. A recent meta-analysis of butterfly dispersal studies clearly demonstrated that the intraspecific variation in dispersal traits can be as large as the interspecific variation (Stevens *et al.* 2010). The significant intraspecific variation component has not been fully appreciated yet, as several basic and applied studies assume dispersal to be a species-specific fixed trait (e.g. Kotiaho *et al.* 2005). Dispersal should be considered as not just a fixed trait in a species (intraspecific variation), but also as a multivariate trait, since dispersal has several components. Moreover, several traits related to dispersal are likely to be either correlated or not. Hence it makes sense to understand dispersal relative to other (correlated) life-history traits. That is particularly true for studies on movements in fragmented habitats and landscapes.

In the same landscape with a network of meadow habitats, a habitat-specialist species and a habitat-

Dispersal Ecology and Evolution. First Edition. Edited by Jean Clobert *et al.*

generalist species (*Lycaena virgaureae* and *Maniola jurtina*, respectively) were found to cover similar distances, but the generalist did so at higher frequency (Schneider *et al.* 2003). So dispersal differences are then not captured by covered distances and the study suggests a difference between dispersal ability and dispersal propensity, at least if we assume no differential predation effects. In principle, we could find a similar pattern if *L. vigaureae* would be more prone to predation during the transience stage, but there is to our knowledge little or no evidence for differential predation in adult butterflies among these two species within the same area. Conceptually, it has been a fruitful approach to recognize dispersal as a three-stage process separating the emigration stage, the transition stage, and the settlement or immigration stage. Here we show the interactions between the organism's behaviour and the environment with different degrees of habitat fragmentation at each dispersal stage referring to recent work on a highly successful, expanding butterfly species, and a highly threatened species of conservation concern. The expanding species is the speckled wood butterfly (*Pararge aegeria*). Its European range has moved significantly to the north over recent years (e.g. Hill *et al.* 1999). In north-west Europe, *P. aegeria* has not only increased in distribution and abundance, but it has also expanded its habitat use. In this part of the range, *P. aegeria* is no longer limited to woodland, but it also occurs on agricultural land with some fragmented woodland aspect like hedgerows (e.g. Vandewoestijne and Van Dyck 2010). Woodlands represent cooler, moister, and thermally better buffered environments compared to hedgerows and small wood-lots in agricultural landscape (Karlsson and Van Dyck 2005). Here we discuss some of the dispersal-related differences between individuals of both landscape types that differ in degree of habitat fragmentation. The other focal species we selected for this chapter is the Bog fritillary butterfly (*Boloria eunomia*). It is a localized specialist species as it is a glacial relict species of bog habitat. Therefore, it has a highly fragmented distribution across Europe with strongly isolated populations and larger continuous populations in the north of Europe. It is a species of conservation concern that requires an adjusted management (Schtickzelle *et al.* 2007b). The general aim of this chapter is not to compare dispersal between this widespread and localized butterfly species, but rather illustrates principles based on the extensive recent work on these (and other) butterfly species.

33.2 Dispersal propensity—habitat edge-crossing

Several butterfly studies have addressed the first dispersal stage by analysing habitat edge-crossing as emigration behaviour (Ries and Debinski 2001; Schultz and Crone 2001). Crossing a habitat edge, and hence leaving a local habitat unit, may mean different things in different landscapes. Such an intraspecific effect was found in the peatbog specialist butterfly *B. eunomia* as individuals from more fragmented landscapes were more reluctant to cross habitat edges compared to conspecifics from less fragmented or continuous habitat (Schtickzelle and Baguette 2003). Butterflies were tracked individually and the movement behaviour was parameterized (move lengths and distribution of turning angles) within and outside habitat patches.

Merckx *et al.* (2003) tested for higher habitat edge-crossing tendencies in speckled woods *P. aegeria* that originated from less fragmented woodland landscape compared to conspecifics from fragmented agricultural landscapes. In woodland landscape, crossing a woodland edge is probably only under weak selection as the butterfly is very likely to end up in another woodland part of the landscape. In other words, there is less selection pressure for being reluctant to cross habitat edges compared to the highly fragmented agricultural landscape. The test was done under experimental conditions in large outdoor cages as a common arena providing closed 'woodland' conditions alternated by a zone of open grassland. Positions and hence movements of the marked individuals were followed in two cages which each had a mixture of F2-butterflies of woodland landscape and agricultural landscape origin with a fixed density. F2 refers to the second generation under standardized conditions in the laboratory. Moving from one woodland compartment to the other by crossing the open grassland compartment was used as the measure to

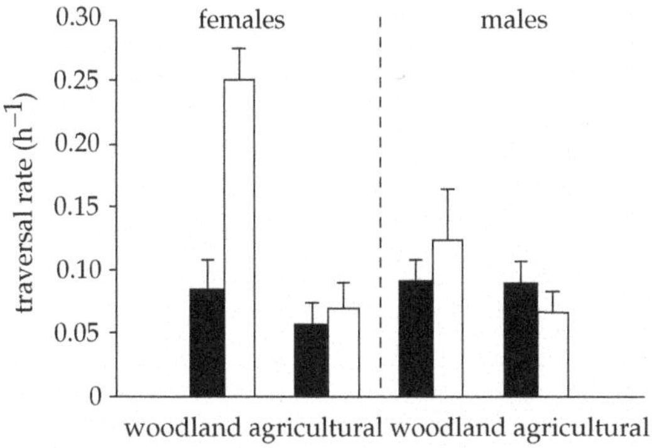

Figure 33.1 Landscape structure and habitat edge crossing. Difference in habitat edge crossing propensity between female and male speckled wood butterflies (*Pararge aegeria*) from woodland landscape origin and agricultural landscape origin under warm (open bars) and cooler weather conditions (filled bars). F2-generation butterflies were tested simultaneously in large outdoor cages. Habitat boundary crossing is expressed as a traversal rate between two shady 'woodland' compartments of the cage that are interrupted by open grassland conditions. Values are weighted means (with corresponding time present as weight) ± s.e. (by bootstrapping). Reproduced from Merckx *et al*. (2003) with permission of the Royal Society.

test their propensity or willingness to cross a habitat edge. Figure 33.1 shows that the results supported the predicted difference in habitat edge crossing between groups of different landscape-type origin. In speckled woods, females are the most dispersive sex. Accordingly, there was a much stronger effect of edge crossing in females than in males. Several studies on butterfly dispersal ignore sexual differences in movement behaviour, but differences between males and females can be very significant in fragmented systems as they vary in thermal flight physiology (e.g. Merck *et al*. 2006).

33.3 Flight types and movement during the transition stage

As butterflies use flight for almost all adult behaviours (including foraging, egg laying, mate searching, or escaping from predators), the assumption that any movement is relevant to dispersal may not hold. We earlier discussed the significance of distinguishing between routine movements (like those during foraging) within a habitat patch and displacement movements (Van Dyck and Baguette 2005). Both types of movement may contribute to dispersal, but the contribution of the latter type is likely to be predominant in fragmented landscapes. Movement paths have been carefully analysed in *B. eunomia* (Schtickzelle *et al*. 2007); outside the habitat in the landscape matrix, movement is significantly straighter (i.e. smaller turning angles and longer step lengths) than within the habitat where the movement path typically is highly tortuous (Figure 33.2). This result agrees with predictions of simulation studies as the straight type of movement is able to reduce the costs (time spent in the transition phase, predation risk) of travelling through a hostile matrix (Zollner and Lima 1999). In *P. aegeria*, movements in the landscape matrix have not been studied in detail yet, but work on a related satyrine species (*Maniola jurtina*) is compatible with the conclusions on *B. eunomia*, as *M. jurtina* also used a straight flight type as soon as they left the habitat in a fragmented agricultural landscape (Delattre *et al*. 2010). In both study cases, there was a clearcut structural difference between habitat zones with a concentration of all essential ecological resources (Dennis *et al*. 2003) and the landscape matrix (arable fields in *M. jurtina*, and peat bog lacking the host plant in *B. eunomia*). In species for whom essential ecological resources are not distributed in compact patches but rather scattered across the landscape (e.g. in White *Pieris* butterflies), individuals may more easily switch between explorative and displacement types of movement (Van Dyck and Baguette 2005).

33.4 Habitat detection—perceptual range

The detection of habitat during transition is essential to entering the final stage of immigration and settlement. In most organisms, the perceptual dimensions and abilities are poorly understood, and, moreover, there might be some bias in the interpretation through the eyes of the human observer. As any other organism, a butterfly will have its very own suite of environmental stimuli to which it is sensitive and to which it will respond. In ethology, the species-specific particular sub-sample of environmental information is often referred to as 'Umwelt' (Dyer and Brockmann 1996). Perceptual

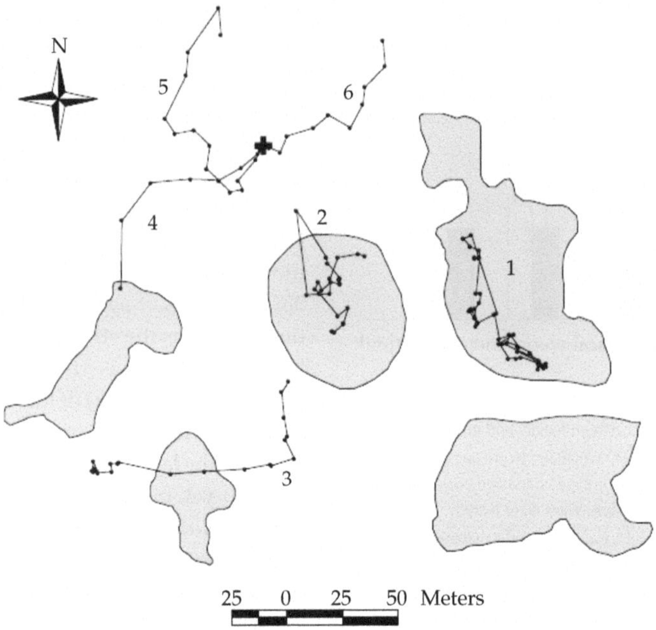

Figure 33.2 Flight paths during transition may differ from explorative movement within the habitat. Schematic drawing of several representative flight paths of the peatbog specialist butterfly *B. eunomia* as observed by individual tracking after release or after spontaneous flights (Schtickzelle *et al.* 2007a). Butterflies made tortuous paths (small moves interrupted by frequent and large U-turns) when flying within the habitat patch (Path 1), but flew straighter (long moves, small turns) in the matrix (Paths 4, 5, and 6). Path 2 shows the tendency to return into the patch after having crossed the habitat edge. Path 3 resulted from a spontaneous flight. Path 4 illustrates the return to a suitable patch, with the last turn at a distance of 25 m from the patch, which approaches the estimated perceptual range of this butterfly. Reproduced from Schtickzelle *et al.* (2007a) with permission from BioMed Publishers.

ranges relative to landscape structure have been particularly studied in small mammals (e.g. Zolner 2000), but some first steps have been made for the two focal butterfly species of this chapter.

Merckx and Van Dyck (2007) released *P. aegeria* butterflies at different distances from a target habitat in open agricultural land. They studied different movement parameters, including the distance at which the butterflies switched from a searching flight to a fast, directed flight. Distance from switch points to the target habitat is interpreted as perceptual range. The field experiment allowed testing the hypothesis of improved perceptual range for *P. aegeria* butterflies of fragmented systems. In line with this prediction, the experiment showed that individuals of agricultural landscape populations were on average directed to the target at a distance twice as far as individuals of woodland landscape populations. Most of the thinking about corridors, barriers, and functional connectivity of landscapes has been implicitly biased towards visual information, but other sensory mechanisms (including olfaction) may play an additional or even major role. This offers intriguing perspectives for future work on dispersal behaviour.

Field observations and release experiments outside habitat patches suggest a perceptual range < 30 m for being attracted to the habitat patch in the habitat specialist *B. eunomia* (Schtickzelle *et al.* 2007a). In *P. aegeria* the average perceptual ranges appear to be larger (50–100 m with higher average values for agricultural populations) (Merckx and Van Dyck 2007).

33.5 Cost of dispersal in differently fragmented landscapes

Dispersal in fragmented landscapes with movements through a hostile landscape matrix is assumed to be costly, but the empirical demonstration of such costs remains scarce. By comparing the survival of dispersing individuals of *B. eunomia* in four landscapes with increasing fragmentation levels (from *c.* 50% of suitable habitat to < 1%), Schtickzelle *et al.* (2006) estimated that dispersal mortality increased from 5–30%.

However, in order to understand to what extent behaviours and dispersal strategies may be altered in fragmented landscapes, one needs to consider costs relative to benefits and one needs to develop a more integrated approach to life-history strategies. But several ecological studies have addressed dispersal independent of other life-history traits. Trade-offs between dispersal and fecundity have, nevertheless, been a classical theme in insect evolutionary biology and life-history theory for several decades (e.g. Johnson 1963). The application of life-history thinking in habitat fragmentation studies is a fruitful example of the integration of ecology, evolutionary biology, and developmental biology (the so-called 'eco-devo-evo' field; Sultan 2007).

In laboratory experiments with recently established breeding stocks of woodland landscape populations and agricultural landscape *P. aegeria* populations, Gibbs and Van Dyck (2010) tested for different trade-offs between flight performance and other life-history traits with landscape of origin. Fragmented landscapes require higher flight performance levels in females to spread their eggs on wider scattered and suitable host plants. Therefore it was expected that females from agricultural landscape populations would invest more in flight at the expense of other life-history traits. Figure 33.3 demonstrates just such a landscape-specific effect. Under a control treatment, woodland-landscape-origin females lived for longer than did agricultural-landscape females. However, if we compare these results with a flight treatment in which females of both landscape groups were forced to fly in a standardized way in the laboratory, we see a strong decrease in longevity in woodland-origin females, but no further difference in agricultural-landscape females.

When forced to fly, both woodland and agricultural females reduced their resource allocation to egg size, but no changes in allocation to daily fecundity were observed (Gibbs and Van Dyck 2010). Females may compensate for excessive flight by re-allocating resources away from egg provisioning which in turn may affect offspring fitness (Gibbs *et al.* 2010a, 2010b). From such flight experiments in the laboratory, we learn that an increase in flight during the oviposition period alters resource allocation to reproduction, which generates cross-generational maternal effects that have the potential to influence population dynamics in fragmented landscapes that require higher flight performances. However, one needs to stay cautious about transferring experimental laboratory results to complex reality in the field.

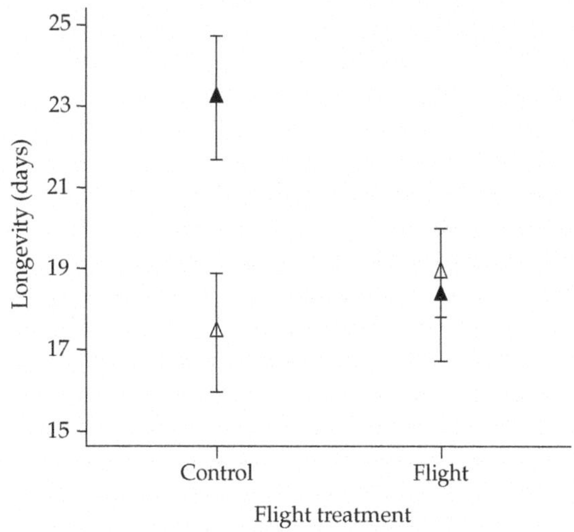

Figure 33.3 Flight performance and other life-history traits (i.e. longevity) in differently fragmented landscapes—Differences in mean (± s.e.) longevity (days) in females that were forced to fly versus control females for two landscape types of origin: woodland landscape (filled symbols) and agricultural landscape (open symbols). Reproduced from Gibbs and Van Dyck (2010) with permission from Springer.

33.6 Mechanisms for dealing with fragmented anthropogenic landscapes

Eco-evolutionary responses of species to habitat fragmentation will be better understood if we approach the dispersal process relative to other key life-history traits in an evolutionary-ecological way along the different components of the three life stages. As in the case of *P. aegeria*, some differences between populations that have to deal with different levels of habitat fragmentation result from either genetic differentiation or phenotypic plasticity. However, the real issue is to focus on G × E interactions and hence adopting reaction norm approaches for different dispersal components in different landscape settings. More consideration needs to be given to the role that flight-induced maternal effects may play in these ecological processes (Gibbs *et al.* 2010a). In line with the review by Gibbs and Van Dyck (2009), we argue that plasticity in ovipostion-site selection and maternal effects on offspring phenotypes may be an important, yet underexplored mechanism by which environmental conditions have cross-generational consequences in butterflies, and surely several other organisms. The impact of habitat fragmentation on oviposition site selection, and the related life-history trade-offs (e.g. number versus size or quality of eggs) may have important consequences for the patterns of both routine movements and special displacement movements (Van Dyck and Baguette 2005).

Over the next few years, the study of dispersal in these butterfly systems will be further integrated into the 'eco-devo-evo' study of alternative lifestyles with specific combinations of movement traits and several other life-history traits under particular environmental conditions. Combining such studies on declining localized and expanding widespread species among butterfly and other species will help us to get a better understanding about changing lifestyles in changing environments. Such an approach will also make a contribution to turning conservation biology into evolution in action (Carroll and Fox 2008).

Acknowledgements

HVD is funded by an ARC-research grant of the Académie Louvain (ARC grant no. 10-15/031). MB is part of the Laboratoire d'Excellence TULIP (ANR-10-LABX-41), and is funded by the Agence Nationale de la Recherche, via the programs (1) open call DIAME (DIspersal and MEtapopulation), (2) 6th extinction MOBIGEN and (3) EU FP6 Biodiversa ERANET TenLamas, and by the EU FP7 SCALES project (project no. 226852).

References

Baguette, M. and Van Dyck, H. (2007). 'Landscape connectivity and animal behavior: functional grain as a key determinant for dispersal', *Landscape Ecology* **22:** 1117–29.

Carroll, S.P. and Fox, C.W. (2008). *Conservation Biology: Evolution in Action*. Oxford University Press, Oxford, New York.

Ehrlich, P.R. (1961). 'Intrinsic barriers to dispersal in checkerspot butterfly', *Science* **134:** 108–9.

Delattre, T., Burel, F., Humeau, A *et al.* (2010). 'Dispersal mood revealed by shifts from routine to direct flights in the meadow brown butterfly *Maniola jurtina*', *Oikos* **119**: 1900–1908.

Dennis, R.L.H., Shreeve, T.G., Van Dyck, H. (2003). 'Towards a functional resource-based concept for habitat: a butterfly biology viewpoint', *Oikos* **102:** 417–26.

Dyer, F.C. and Brockmann, H.J. (1996). 'Biology of the Umwelt' in L.C.Drickamer, ed, *Foundations of Animal Behavior: Classic Papers with commentaries*. University of Chicago Press, Chicago, 529–38.

Gibbs, M., Breuker, C.J., Hesketh, H. *et al.* (2010a). 'Maternal effects, flight versus fecundity trade-offs and offspring immune defence in the Speckled Wood butterfly, *Pararge aegeria*', *BMC Evolutionary Biology* **10:** 345.

Gibbs, M., Breuker, C.J., Van Dyck, H. (2010b). Flight during oviposition reduces maternal egg provisioning and influences offspring development in *Pararge aegeria* (L.). *Physiological Entomology*, **35**, 29–39.

Gibbs, M. and Van Dyck, H. (2009). 'Reproductive plasticity, oviposition site selection and maternal effects in fragmented landscapes', *Behavioral Ecology and Sociobiology* **64:** 1–11.

Gibbs, M. and Van Dyck, H. (2010). 'Butterfly flight activity affects reproductive performance and longevity relative to landscape structure', *Oecologia* **163:** 341–50.

Hill, J.K., Thomas, C.D., Huntley, B. (1999). 'Climate and habitat availability determine 20th century changes in a butterfly's range margin', *Proceedings of the Royal Society B* **266:** 1197–206.

Johnson, C.G. (1963). 'Physiological factors in insect migration by flight', *Nature* **198:** 423–7.

Karlsson, B. and Van Dyck, H. (2005). 'Does habitat fragmentation affect temperature-related life-history traits? A laboratory test with a woodland butterfly', *Proceedings of the Royal Society B* **272:** 1257–63.

Koh, L.P., Sodhi, N.S.,Brook, B.W. (2004). 'Ecological correlates of extinction proneness in tropical butterflies', *Conservation Biology* **18:** 1571–8.

Kotiaho, J.S., Kaitala, V., Kolmonen, A., *et al.* (2005). 'Predicting the risk of extinction from shared ecological characteristics', *Proceedings of the National Academy of Sciences USA* **102:** 1963–7.

Merckx, T., Karlsson, B., Van Dyck, H. (2006). 'Sex- and landscape-related differences in flight ability under suboptimal temperatures in a woodland butterfly', *Functional Ecology* **20:** 436–41.

Merckx, T. and Van Dyck, H. (2007). 'Habitat fragmentation affects habitat-finding ability of the speckled wood butterfly (*Pararge aegeria* L.)', *Animal Behaviour* **74:** 1029–37.

Merckx, T., Van Dyck, H., Karlsson, B. *et al.* (2003). 'The evolution of movements and behaviour at boundaries in different landscapes: a common arena experiment with butterflies', *Proceedings of the Royal Society of London B* **270:** 1815–21.

Ries, L. and Debinski, D.M. (2001). 'Butterfly responses to habitat edges in the highly fragmented prairies of Central Iowa', *Journal of Animal Ecology* **70:** 840–52.

Schneider, C., Dover, J., Fry, G.L.A. (2003). 'Movement of two grassland butterflies in the same habitat network: the role of adult resources and size of the study area', *Ecological Entomology* **28:** 219–27.

Schtickzelle, N. and Baguette, M. (2003). 'Behavioural responses to habitat patch boundaries restrict dispersal and generate emigration-patch area relationships in fragmented landscapes', *Journal of Animal Ecology* **72:** 533–45.

Schtickzelle, N., Joiris, A., Van Dyck, H. *et al.* (2007a). 'Quantitative analysis of changes in movement behaviour within and outside habitat in a specialist butterfly', *BMC Evolutionary Biology* **7:** 4 (e-paper).

Schtickzelle, N., Mennechez, G., Baguette, M. (2006). Dispersal depression with habitat fragmentation in the bog fritillary butterfly. *Ecology* **87**: 1057–1065.

Schtickzelle, N., Turlure, C., Baguette, M. (2007b). 'Grazing management impacts on the viability of the threatened bog fritillary butterfly *Proclossiana eunomia*', *Biological Conservation* **136:** 651–60.

Schultz, C.B. and Crone, E.E. (2001). 'Edge-mediated dispersal behavior in a prairie butterfly', *Ecology* **82:** 1879–92.

Schultz, C.B. and Hammond, P.C. (2003). 'Using population viability analysis to develop recovery criteria for endangered insects: case study of the Fender's blue butterfly', *Conservation Biology* **17:** 1372–85.

Shahabuddin, G. and Riveros, C.A. (2005). 'Frugivorous butterfly species in tropical forest fragments: correlates of vulnerability to extinction', *Biodiversity and Conservation* **14:**, 1137–52.

Stevens, V.M., Pavoine, S., Baguette, M. (2010). 'Variation within and between closely related species uncovers high intra-specific variability in dispersal' *PLoS One* **5:** e11123.

Sultan, S.E. (2007). Development in context: the timely emergence of eco-devo. *Trends in Ecology & Evolution* **22**: 575–582.

Thomas, J.A., Telfer, M.G., Roy, D.B. *et al.* (2004). 'Comparative losses of British butterflies, birds, and plants and the global extinction crisis', *Science* **303:** 1879–81.

Van Dyck, H. and Baguette, M. (2005). 'Dispersal behaviour in fragmented landscapes: routine or special movements?', *Basic and Applied Ecology* **6:** 535–45.

Vandewoestijne, S. and Van Dyck, H. (2010). 'Population genetic differences along a latitudinal cline between original and recently colonized habitat in a butterfly', *PLoS One* **5 (11):** e13810.

Zollner, P.A. (2000). 'Comparing the landscape level perceptual abilities of forest sciurids in fragmented agricultural landscapes', *Landscape Ecology* **15:** 523–33.

Zollner, P.A. and Lima, S.L. (1999). 'Search strategies for landscape-level interpatch movements', *Ecology* **80:** 1019–30.

CHAPTER 34

Gene flow allows persistence of a perennial forest herb in a dynamic landscape

Olivier Honnay and Hans Jacquemyn

34.1 Consequences of habitat fragmentation for plant populations in remnant patches

The loss and fragmentation of habitats are major drivers of biodiversity loss world-wide (Butchart *et al.* 2010). How habitat fragmentation affects plant population viability and genetic diversity in remnant habitat patches has been the focus of a vast body of research in conservation biology and ecology (Honnay *et al.* 2005; Lindenmayer and Fischer 2006). Much of the habitat fragmentation research has focused on the plant population genetic consequences as it is known that decreasing genetic diversity reduces short- and long-term fitness of fragmented populations (Keller and Waller 2002; Frankham *et al.* 2010). Meta-analyses of this work have clearly shown that habitat fragmentation generally lowers within-population genetic diversity of the remnant plant populations through genetic drift, and that spatial isolation of habitat fragments does not allow replenishment of alleles which are lost through drift by seed or pollen flow (Leimu *et al.* 2006; Honnay and Jacquemyn 2007). However, the genetic response to habitat fragmentation is strongly mediated by plant traits such as the breeding system (Aguilar *et al.* 2008).

Habitat fragmentation has also been shown to negatively affect insect pollinator abundance and pollinator diversity in small habitat fragments (Ashman *et al.* 2004). Fragments can become too small to sustain pollinator communities or too isolated to attract a large diversity of pollinators (Tscharntke and Brandl 2004), both of which reduce pollinator abundance and efficiency, and therefore reproductive success of plant species (Aguilar *et al.* 2008). In addition to the direct effects of fragmentation through changes in pollinator guilds, indirect effects associated with altered pollinator flight patterns within patches have been shown to affect reproductive success and population viability of plant species in fragmented habitats (Lennartsson 2002).

Finally, habitat fragmentation also implies a relative increase in edge habitat; i.e. habitat prone to edge effects. Edge effects result from the matter, species, and energy flowing from patches with high-intensity land use into low-intensity land use patches. Edge effects have been shown to affect plant regeneration, interspecies competition and plant-animal interactions such as predation, seed dispersal, and pollination (Lindenmayer and Fischer 2006).

34.2 Challenges for plant survival in landscapes with patch turnover

Almost all empirical research so far has concentrated on the consequences of habitat fragmentation on persistence of plant populations in remnant habitat patches. The situation where many qualitatively similar habitat fragments stably co-exist in a landscape matrix of agricultural or urban land is a considerable simplification of reality, however. Even in natural landscapes that are prone to succession, new patches constantly arise and old patches may disappear (Picket and White 1985).

Plant species occurring in landscapes with habitat patch turnover face a range of specific challenges

Dispersal Ecology and Evolution. First Edition. Edited by Jean Clobert *et al.*

to survive in the long term. First, continuous seed dispersal from established populations into patches that recently originated in the landscape is required because of the recurrent destruction of seed source patches. Successful seed addition experiments in recently established forest patches have confirmed the importance of seed dispersal limitation during forest plant colonization (Eriksson and Ehrlén 1992; Verheyen and Hermy 2004). Because most plant species appear to be to some extent dispersal-limited, the distance between existing source populations in old patches and recently established habitat patches is a key factor mediating plant colonization rates. In forest herbs, species with seeds lacking specific dispersal mechanisms were less likely to colonize recent patches than species having specific adaptations that facilitate long-distance seed dispersal, suggesting that seed dispersal syndromes and colonization capacity are significantly related to each other (e.g. Honnay *et al.* 2002).

Second, after a successful seed dispersal event, the establishment of a viable population within recent patches must occur fast, relative to the rate of habitat clearance. This is not a trivial requirement for different reasons. Very recently established patches often show environmental conditions that strongly differ from those of older patches. For forest plant species, for example, the colonization of recently established forests on former agricultural land may be hampered by high levels of soil nutrients, in particular phosphorus, and by inter-specific competition between colonizing species and highly competitive plant species that very effectively exploit these nutrients (Honnay *et al.* 1999; Flinn 2007; Flinn and Marks 2007). Next to lower habitat quality, recently established populations may be susceptible to various kinds of genetic and demographic stochasticities because they are likely to be founded by a limited number of individuals. In the absence of continuous gene flow into these founder populations, both genetic drift and inbreeding can be expected to become important, causing a rapid loss of alleles and decreasing heterozygosity, respectively (Young *et al.* 1996; Keller and Waller 2002; Dlugosh *et al.* 2008). Demographic stochasticity may lead to reduced mate availability in founding populations of, in particular, self-incompatible and/or heteromorphic (e.g. heterostylous) plant species. Low mate availability, in turn, may result in low fruit and seed set (Baker 1955; Brys *et al.* 2007; Elam *et al.* 2007), which may translate in smaller population growth rates.

34.3 Insights from metapopulation models

Almost all current insights in the effects of patch turnover on the persistence of plant species in dynamic landscapes are based on models which are rooted in metapopulation theory (Johst *et al.* 2002; Ellner and Fussman 2003; Vellend 2003; Verheyen *et al.* 2004; Bossuyt and Honnay 2006). Basically, these approaches model patch occupancy of a species through time as a function of the proportion of ancient and recent habitat present in the landscape, and the habitat turnover rate. Model results suggest that the timing of habitat patch clearance can have dramatic effects on species persistence in the landscape (Vellend 2004; Verheyen *et al.* 2004). The smaller the period between patch clearance and recovery, the more likely the species will colonize new patches, and the higher the equilibrium patch occupancy rate, and the lower the extinction probability of the metapopulation.

From a population genetic perspective, the consequences of recurrent colonization and extinction have also received attention. The population genetic diversity of species in habitat patches under recurrent extinction and colonization dynamics is generally predicted to decrease due to founder events and genetic drift (Barton and Whitlock 1997; Pannell and Charlesworth 2000; but see Honnay *et al.* 2009). Other work has focused on predicting the effects of population extinction and recolonisation on the genetic differentiation between populations (Pannell and Charlesworth 2000).

Clearly, models based on metapopulation theory have increased our understanding of the effects of patch turnover and recurrent extinction and colonization on the long-term persistence of species at the landscape scale, and on the structuring of genetic diversity within and between populations. More empirical studies are required, however, to understand species' response to landscape dynamics. In what follows, we synthesize the results of a long-lasting study on the population dynamics and genetics of the perennial forest herb *Primula elatior* (Oxlip) in a fragmented and dynamic forest landscape in Central Belgium. More specifically, we aim at integrating data on i) forest age, ii) colonization rates of recent forests, iii) genetic diversity in founder populations, and iv) demography of founder and established populations to evaluate the potential for long-term persistence of the species in the landscape.

34.4 The forest herb *Primula elatior* in a dynamic landscape

Primula elatior is a distylous perennial forest herb showing a clear affinity to moist habitats (Whale 1984; Taylor and Woodell 2008). It appears as a late successional species that typically occurs in ancient forest habitats, although colonization of recent forests has frequently been observed (Jacquemyn *et al.* 2002, 2004). Like most other *Primula* species, it has a distylous breeding system, with two genetically determined style morphs (long-styled (pins) and short-styled morphs (thrums)) (Richards 1997). Only cross-pollination between pin and thrum morphs results in seed set. The species is mainly pollinated by bumblebees, bees, and beeflies (Taylor and Woodell 2008). Once they have reached the flowering stage, individual plants can survive for a long time (> 30 years). Seeds of *P. elatior* lack specific dispersal mechanisms (barochory). However, as the inflorescences are frequently eaten by roe deer, it is not unlikely that infrequent long-distance seed dispersal occurs through seed transfer by roe deer.

The study was conducted in a highly fragmented and dynamic landscape in Belgium with high forest turnover (Jacquemyn *et al.* 2002). Currently 241 different forest patches can be distinguished. Reconstruction of the land use history using nine available historical maps showed that the total forest area declined dramatically during the eighteenth and nineteenth centuries with yearly deforestation percentages varying between 0.17% and 0.56%. As a result, total forest area declined from 1947 ha in 1775 to 464 ha in 1970. In 1991, total forest area had increased again to 510 ha. At the same time, the total number of forests increased from 119 in 1775 to 195 in 1982. Mean forest patch area, on the other hand, decreased from 16.4 ha in 1775 to 2.5 ha in 1982. At the time of our surveys, forest occupied only 6% of the total area. Most forest patches were also small: 68% had an area smaller than 1 ha, and only six (2.5%) had an area larger than 5 ha. A map of the study area and a visual presentation of the deforestation rates are available in Jacquemyn *et al.* (2003).

34.5 Seed dispersal and colonization of recently established forests

Within the study area, *P. elatior* occurred in 69 of the 241 (29% of the total occurrence) surveyed forest patches, and 16 times (23%) in young forest patches. Young patches were defined as patches with an age < 30 years, whereas old patches were all patches with and age > 30 years. This suggests that the species is able to colonize recent forest habitats relatively easily. However, the probability of colonizing a recent forest patch strongly depended on its location relative to an old source patch. Using data on the geographic distribution of *P. elatior*, the probability of colonization of empty patches in relationship to the distance to the nearest old occupied forest patch was determined using logistic regression analysis. The results clearly showed that the probability of recent patches being colonized significantly decreased with increasing distance from occupied older patches ($P < 0.001$) (Figure 34.1). Recent patches adjacent to older occupied forests had a 40% chance of becoming colonized within a period of 30 years, whereas patches further than 400 m apart from occupied older patches had less than 5% chance of becoming colonized (Figure 34.1). In general, colonizing populations were also small (between 13 and 112 individuals), making them susceptible to demographic stochasticity that may jeopardize sexual reproduction.

34.6 Demographic stochasticity and reproductive success

The distylous breeding system of *P. elatior* can be expected to affect reproductive success and thus

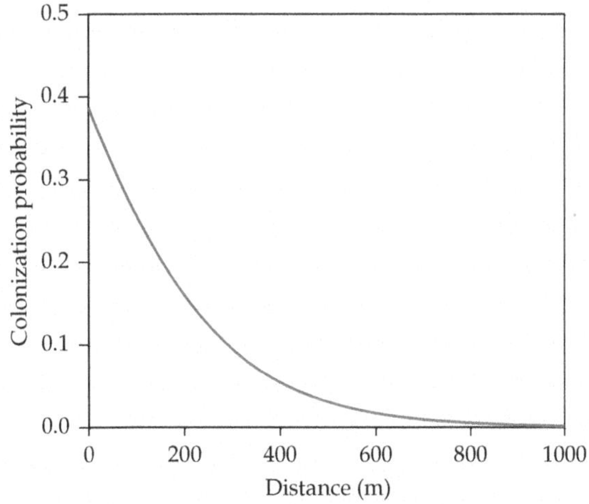

Figure 34.1 Colonization probability of *Primula elatior* in function of the distance to occupied, older forest patches. Redrawn from Jacquemyn *et al.* (2002) with kind permission from Springer Science and Business Media.

population viability, because of stochastic variation in pin:thrum ratios. The probability of finding equal ratios of pin and thrum morphs is expected to depend on the size of populations. In 19 populations that varied in size between four and > 2000 flowering plants, the morph ratio was determined. When population size was smaller than 200 individuals (the typical size of young colonizing populations), the morph bias quickly increased (Figure 34.2a). This increase in morph bias was associated with strong decreases in seed set (Figure 34.2b). Besides lower availability of pollinators in small populations, the lack of suitable mates has likely contributed to the lower reproductive success in small populations (Jacquemyn *et al.* 2002). Because colonizing populations are by nature small and are likely to have unbalanced morph ratios, these results thus clearly show that small colonizing populations may be suffering from lowered seed production, particularly if morph ratios are strongly skewed.

Because of the lower seed production in small colonizing populations, it can be expected that populations occurring in recently established forest patches have smaller long-term population growth rates than populations occurring in older patches. Additionally, it has been repeatedly reported that many forest plant species also suffer from recruitment limitation (Honnay *et al.* 1999; Baeten *et al.* 2009). High soil nutrient levels are expected to stimulate growth of strongly competitive species and as a result may hamper establishment and recruitment of new individuals. Lack of recruitment and competition with vigorously growing plant species are likely to affect the population dynamics of *P. elatior* in recently colonized patches.

34.7 Population dynamics in young versus old forests

To get better insights into the processes that drive the population dynamics of *P. elatior* in forest patches of different ages, and into the specific life cycle transitions that contributed most to variation in population growth rates, population projection matrices were constructed from demographic data that were collected between 2004 and 2007 (three transitions) (Jacquemyn and Brys 2008). Demographic data were collected in seven populations that occurred in patches along an age gradient varying between 22 and > 225 years. The studied populations varied in size between 39 and > 2000 flowering individuals (mean: 760), whereas the proximity to occupied patches (mean distance to the nearest five populations) ranged from 18–1231 m (mean: 413 m).

Five different life stages were considered: seedlings, juveniles, non-flowering adults, small flowering adults, and large flowering adults. No seed stage was modelled, as the species has no

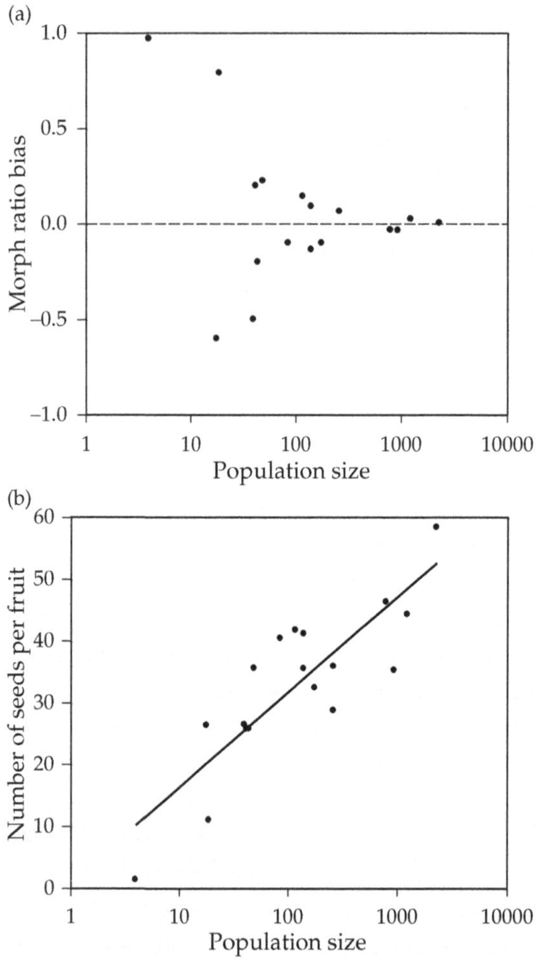

Figure 34.2 Morph ratio bias in relationship to population size in nineteen populations of the distylous forest herb *Primula elatior*. Redrawn from Jacquemyn *et al.* (2002) with kind permission from Springer Science and Business Media.

permanent seed bank. Transition probabilities between life stages were obtained by following the fate of each individual from one year to the next. A total of 2487 individuals were monitored. The 'anonymous reproduction method' (Caswell 2001) was used to estimate stage-specific fertility. The asymptotic growth rate of each population was calculated as the dominant eigenvalue of the transition matrices. Confidence intervals of 95% around λ were calculated based on a bootstrap procedure. Note that the asymptotic growth rates and demographic projections were obtained from a deterministic model, assuming that stochastic factors did not influence population dynamics, and environmental conditions were the same in the past, as from 2004 to 2007.

Average population growth rates (λ) increased significantly with increasing forest age, from 0.909 in the most recent forest, to 1.102 in the oldest forest (Figure 34.3). In the youngest forests (22 years), λ was significantly lower than 1, indicating that the populations in these forests are not viable and that populations are decreasing in size. Starting from a population size of 100 individuals, it can be projected that, without seed inflow, populations in these recent forests will disappear within the next 50 years. A significantly positive population growth rate was only obtained in forests that were older

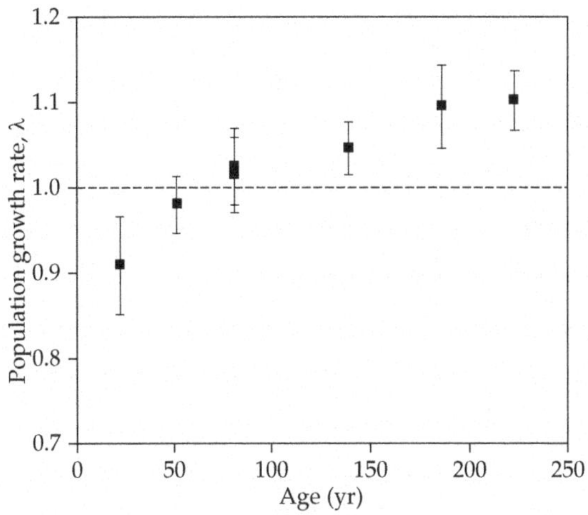

Figure 34.3 Relationship between forest age and asymptotic population growth rate, λ ($y = 0.821 \times \log_{10}x + 0.6575$, $r^2 = 0.938$, $P < 0.001$). Error bars show 95% bias-corrected bootstrap confidence intervals.

than 150 years (Figure 34.3). These results imply that continuous seed inflow from other established populations is necessary to allow long-term viability of populations that colonized recent forests. Seed inflow is required until these recently established forest patches reach an age of 150 years. These findings have important consequences for the persistence of the species at the landscape scale. If ancient forests are cleared before a sufficiently large amount of recently established forests has reached an age high enough to support viable populations of *P. elatior*, the species will go extinct at the landscape scale (Vellend 2003, 2004).

The question remains why population growth rates were significantly lower in recent forests than in ancient forests. Using a fixed one-way design LTRE analysis with the population in the oldest forest fragment as the reference matrix, Jacquemyn and Brys (2008) showed that the increase in λ with increasing forest age was mainly due to increased seedling and juvenile growth, and increased juvenile and adult stasis. Surprisingly, lower population growth rates in recent forest patches were not due to lower seedling recruitment. This suggests that reduced seed set, for example resulting from demographic stochasticity through a morph bias, was not limiting population growth rates in recent forests. On the contrary, the major bottleneck hampering establishment of large viable populations in recent forests appeared to be related to survival and growth of later life stages, in particular survival of seedlings and juveniles (Figure 34.4). These results thus indicate that population growth in recent forests is more restricted by establishment limitation than by seed limitation. The high recruitment rates in recent forests patches can be explained by the high proportion of open soil combined with relatively high proportions of light penetrating to the soil early in the growing season. In older patches, the forest ground is often densely carpeted with other forest herbs that flower at the same time (e.g. *Anemone nemorosa*), and that limit germination possibilities of other forest herbs. On the other hand, the low persistence of seedlings and juveniles in recent forests is caused by vigorous growth later in the season of strongly competitive plant species (most often *Urtica dioica*) that take advantage of the high levels of soil nutrients in the soil. To conclude, our results indicate that seed dispersal from older forest patches is needed to counterbalance high mortality rates of seedlings and juveniles.

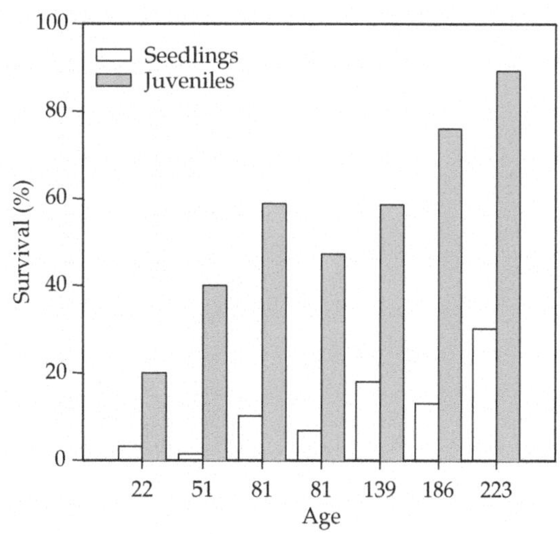

Figure 34.4 Survival of seedlings and juveniles of *Primula elatior* in forest patches of different ages.

34.8 Genetic diversity of colonizing populations

Next to demographic stochasticity, small founder populations can experience a reduction in genetic diversity through genetic drift and founder effects (Lande 1988; Dlugosh and Parker 2008). Founder effects are specific examples of random genetic drift, and are defined as genetic changes in populations that occur when few founders establish a new population. The loss of genetic diversity may jeopardize the ability of plant populations to adapt to changing environmental conditions. If mating between related individuals increases due to small population sizes, lower fitness can be expected in the relatively short term through inbreeding depression (Keller and Waller 2002).

Here, we aimed at gaining insight into the genetic diversity of a colonizing population. More specific, we wanted to know whether a colonizing population is losing genetic diversity through time. We identified a population in a forest patch that was not older than 20 years and that was only recently colonized by the study species (Jacquemyn *et al.* 2009). All flowering individuals (n = 112) in the population were meticulously mapped, and the size of each plant was determined by counting the number of rosettes. Large plants (i.e. plants with > 5 rosettes) were considered as the first plants to have colonized the forest (oldest plants), whereas plants with only one rosette were thought to have established only very recently (youngest plants). Genetic analyses using dominant Amplified Fragment Length Polymorphisms (AFLPs) were performed on all 112 individuals. Band richness (a substitute for allelic richness) and expected heterozygosity were calculated for each of the three *Primula* size classes.

First, this multigenerational analysis showed a rapid decline in genetic diversity due to drift, from the cohort of large (old) individuals to the cohort of recently established individuals (Figure 34.5). The genetic diversity of the largest, and likely oldest, individuals was also comparable to the genetic diversity of the populations in the surrounding ancient forests (Jacquemyn *et al.* 2004). Second, the analysis showed that pollen dispersal distances are relatively restricted (< 50 m), and that most fathers could be located within the forest patch. It is interesting to see that these results were recently con-

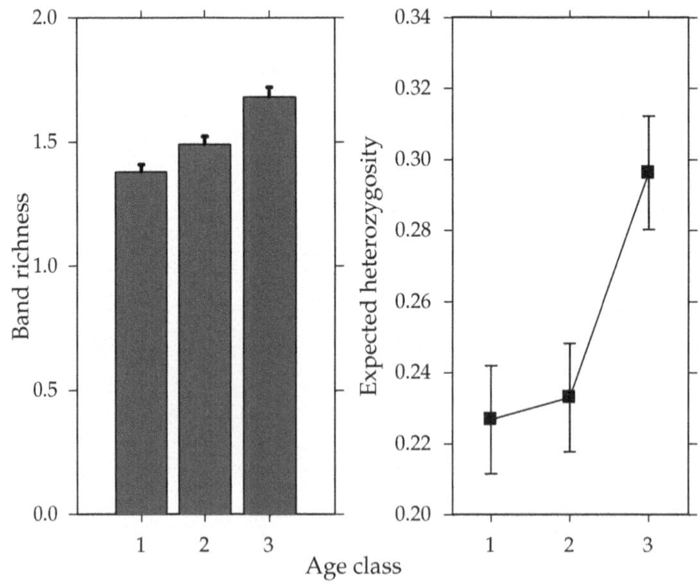

Figure 34.5 Band richness and genetic diversity (± SE) within different age classes in a founding population of *Primula elatior*. 1: young individuals (1 rosette) (n = 57), 2: medium-aged individuals (2–5 rosettes) (n = 34), 3: old individuals (> 5 rosettes) (n = 10).

firmed in other *P. elatior* populations near Brussels (Van Rossum *et al.* 2010). Pollen dispersal between forest fragments seems thus unlikely.

All this indicates that the inflow of seeds from other populations is not sufficient to replenish the alleles that are lost through genetic drift in the founding population. Given that the distance to the five nearest ancient patches was over 250 m, these results corroborate our earlier conclusion that seed dispersal across larger distances does occur. It is too limited, however, to wipe out the genetic signature of colonization and extinction dynamics. Low but consistent gene flow between patches is confirmed by a genetic differentiation (F_{ST}) of 0.04 ($p < 0.001$) between all *P. elatior* populations in the study area (Jacquemyn *et al.* 2004). The parentage analysis also showed that not all possible parents within the population contributed to gene flow. To conclude, without continuous seed flow into founding populations, they will lose genetic diversity relatively quickly.

34.9 Conclusion

Most habitat fragmentation research so far has focused on plant populations in remnant habitat patches of comparable habitat quality. Many landscapes, however, are dynamic and characterized by a mix of habitat patches of different age and quality. The continuous clearance of old patches and the addition of new patches, and especially the rate at which this happens, can have major consequences for the persistence of species at the landscape level. The effects of these landscape changes are complicated by the fact that species colonizing newly established habitats face different challenges to establish viable populations. We have quantified these challenges, and shown that small colonizing populations of *P. elatior* in recent forests have lower seed set, partly due to unequal proportions of pin and thrum morphs within populations. We have also shown that population growth rates of populations establishing in recent forests were significantly lower than 1 and that it takes more than a century before environmental conditions reach a state that allow persistence of *Primula* populations. The main conclusion we can draw from all these results is that recently established populations of *P. elatior* in recent forests are not viable in the long term, and that they act as sinks in the landscape. Without permanent seed influx from nearby source populations, they are bound to go extinct within a few decades. Seed dispersal is therefore of paramount importance to maintain these populations. This has important conservation implications. If patch destruction occurs too fast, the necessary seed sources to maintain recent populations will disappear. Ultimately, this will result in the extinction of the species from the landscape. The mere presence of *P. elatior* in recent forest patches is misleading in this sense, as these populations can only exist due to permanent seed dispersal from source populations.

Although we have no hard data on asymmetrical gene flow, our results provide some evidence for source-sink dynamics in dynamic landscapes. There has been theoretical and conceptual work on this topic (e.g. Pulliam 1988; Eriksson 1996), but evidence of the phenomenon remains rather scarce in plant species (but see Anderson and Geber 2009). In general, our results may indicate that the role of dispersal in maintaining populations in fragmented habitats may be underestimated. Dispersal is not only crucial for the initial recolonization of new habitats, it remains of major importance for the viability of founder populations, possibly decennia after the colonization event. This implies that many 'ghost' populations may be present in our fragmented landscapes.

References

Aguilar, R., Quesada, M., Ashworth L. *et al.* (2008). 'Genetic consequences of habitat fragmentation in plant populations: susceptible signals in plant traits and methodological approaches', *Molecular Ecology* **17**: 5177–88.

Anderson, J.T. and Geber M.A. (2009). 'Demographic source-sink dynamics restrict local adaptation in Elliott's blueberry (*Vaccinium elliottii*)', *Evolution* **64**: 370–84.

Ashman, T., Knight T.M., Steets J.A. *et al.* (2004). 'Pollen limitation of plant reproduction: ecological and evolutionary causes and consequences', *Ecology* **85**: 2408–21.

Baeten, L., Hermy, M.,Verheyen, K. (2009). 'Environmental limitation contributes to the differential colonization capacity of two forest herbs', *Journal of Vegetation Science* **20**: 209–23.

Baker, H.G. (1955). 'Self-compatibility and establishment after 'long-distance' dispersal', *Evolution* **9**: 347–9.

Barton, N.H. and Whitlock, M.C. (1997). 'The evolution of metapopulations', in: I. Hanski, M. Gilpin, eds. *Metapopulation Dynamics: Ecology, genetics and evolution*, Academic Press, San Diego. 183–210.

Bossuyt, B. and Honnay, O. (2006). 'Trade-offs between plant life span, seed dispersal capacity and fecundity determine metapopulation viability in a dynamic landscape', *Landscape Ecology* **21**: 1195–205.

Bossuyt, B., Honnay, O., Hermy, M. (2003). 'A landscape ecological view on the successional pathway in wet dune slacks', *Journal of Vegetation Science* **14**: 781–8.

Brys, R., Jacquemyn, H., De Bruyn, L. *et al.* (2007). 'Pollination success and reproductive output in experimental populations of the self-incompatible *Primula vulgaris*', *International Journal of Plant Science* **168**: 571–8.

Butchart, S.H.M., Walpole, L., Collen, B. *et al.* (2010). 'Global Biodiversity: Indicators of recent declines', *Science* **328**: 1164–8.

Caswell, H. (2001). *Matrix population models. Construction, analysis, and interpretation.* Sinauer Associates, Sunderland, Mass.

Dlugosh, K.M. and Parker, I.M. (2008). 'Founding events in species invasions: genetic variation, adaptive evolution, and the role of multiple introductions', *Molecular Ecology* **17**: 431–49.

Elam D.R., Ridley C.E., Goodell, K. *et al.* (2007). 'Population size and relatedness affect fitness of a self-incompatible invasive plant', *Proceedings of the National Academy of Science USA* **104**: 549–52.

Ellner, S.P. and Fussmann, G. (2003). 'Effects of successional dynamics on metapopulation persistence', *Ecology* **84**: 882–9.

Eriksson, O. (1996). 'Regional dynamics of plants: a review of evidence for remnant, source-sink and metapopulations', *Oikos* **77**: 248–58.

Eriksson, O. and Ehrlèn, J. (1992). 'Seed mass and microsite limitation of recruitment in plant populations', *Oecologia* **91**: 360–4.

Flinn, K.M. (2007). 'Microsite-limited recruitment controls fern colonization of post-agricultural forests', *Ecology* **88**: 3103–14.

Flinn, K.M. and Marks, P.L. (2007). 'Agricultural legacies in forest environments: Tree communities, soil properties and light availability', *Ecological Applications* **17**: 452–63.

Flinn, K.M. and Vellend, M. (2005). 'Recovery of forest plant communities in post-agricultural landscapes', *Frontiers in Ecology and the Environment* **3**: 243–50.

Frankham, R., Balloux, J.D., Briscoe, D.A. (2010). *Introduction to conservation genetics.* Cambridge University Press, Cambridge.

Harrison, S. and Bruna, E. (1999). 'Habitat fragmentation and large-scale conservation: what do we know for sure?', *Ecography* **22**: 225–32.

Honnay, O., Hermy, M., Coppin, P. (1999). 'Habitat quality factors affecting the colonisation of forest plant species', *Forest Ecology and Management* **115**: 157–70.

Honnay, O. and Jacquemyn, H. (2007). 'Susceptibility of rare and common plant species to the genetic consequences of habitat fragmentation', *Conservation Biology* **24**: 824–31.

Honnay, O., Jacquemyn, H., Bossuyt, B. *et al.* (2005). 'Forest fragmentation effects on patch occupancy and population viability of herbaceous plant species', *New Phytologist* **166:** 723–36.

Honnay, O., Jacquemyn, H., Van Looy, K. *et al.* (2009). 'Temporal and spatial genetic variation in a metapopulation of the annual *Erysimum cheiranthoides* on stony river banks', *Journal of Ecology* **97:** 131–41.

Honnay, O., Verheyen, K., Butaye, J. *et al.* (2002). 'Possible effects of climate change and habitat fragmentation on the range of forest plant species', *Ecology Letters* **5**: 525–30.

Jacquemyn, H. and Brys, R. (2008). 'Population growth rates of the forest herb *Primula elatior* increase with stand age in post-agricultural forests', *Ecology* **89**: 3480–9.

Jacquemyn, H., Brys, R.,Hermy, M. (2002). 'Patch occupancy, population size and reproductive success of a forest herb (*Primula elatior*) in a fragmented landscape', *Oecologia* **130**: 617–25.

Jacquemyn, H., Butaye, J., Hermy, M. (2003). 'Influence of environmental and spatial variables on regional distribution of forest plant species in a fragmented and changing landscape', *Ecography* **26**: 768–76.

Jacquemyn, H., Honnay, O., Galbusera, P. *et al.* (2004). 'Genetic structure of the forest herb *Primula elatior* in a changing landscape', *Molecular Ecology* **13**: 211–19.

Jacquemyn, H., Vandepitte K., Roldán-Ruiz, I *et al.* (2009). 'Rapid loss of genetic variation in a founding population of *Primula elatior* (Primulaceae) after colonization', *Annals of Botany* **103**: 777–83.

Johst, K., Brandl, R., Eber, S. (2002). 'Metapopulation persistence in dynamic landscapes: the role of dispersal distance', *Oikos* **98:** 263–70.

Keller, L.F. and Waller, D.M. (2002). 'Inbreeding effects in wild populations', *Trends in Ecology and Evolution* **17**: 230–41.

Kolb, A. and Diekmann, M. (2004). 'Effects of environment, habitat configuration and forest continuity on the distribution of forest plant species', *Journal of Vegetation Science* **15**: 199–208.

Lande, R. (1988). 'Genetics and demography in biological conservation', *Science* **241**: 1455–60.

Leimu, R., Mutikainen, P., Koricheva, J. *et al.* (2006). 'How general are positive associations between population size, fitness, and genetic variation?', *Journal of Ecology* **94**: 942–52.

Lennartsson, T. (2002). 'Extinction thresholds and disrupted plant–pollinator interactions in fragmented plant populations', *Ecology* **83**: 3060–72.

Lindenmayer, D.B. and Fischer, J. (2006). *Habitat fragmentation and landscape change. An ecological and conservation synthesis.* Island Press, Washington.

Pannell, J.R. and Charlesworth, B. (2000). 'Effects of metapopulation processes on measures of genetic diversity', *Philosophical Transactions of the Royal Society B* **355**: 1851–64.

Picket, S.T.A. and White P.S. (1985). *The ecology of natural disturbance and patch dynamics.* Academic Press, San Diego, CA.

Pulliam, H.R. (1988). 'Sources, sinks, and population regulation', *American naturalist* **132**: 652–61.

Richards, A.J. (1997). *Plant Breeding Systems.* Allen and Unwyn, London.

Taylor, K. and Woodell, S.R.J. (2008). 'Biological Flora of the British Isles: *Primula elatior* (L.) Hill', *Journal of Ecology* **96**: 1098–116.

Tscharntke, T. and Brandl, R. (2004). 'Plant-insect interactions in fragmented landscapes', *Annual Review of Entomology* **49**: 405–30.

Van Rossum, F., Stiers, I., Van Geert, A. *et al.* (2010). 'Fluorescent dye particles as pollen analogues for measuring pollen dispersal in an insect-pollinated forest herb', *Oecologia* **165**: 663–74.

Vellend, M. (2003). 'Habitat loss inhibits recovery of plant diversity as forests regrow', *Ecology* **84**: 1158–64.

Vellend, M. (2004). 'Metapopulation dynamics following habitat loss and recovery', in O Honnay, K. Verheyen, B. Bossuyt, M Hermy (eds), *Forest Biodiversity –Lessons from history for conservation*, CABI, Wallingford, 163–177.

Verheyen, K. and Hermy, M. (2004). 'Recruitment and growth of herb-layer species with different colonizing capacities in ancient and recent forest', *Journal of Vegetation Science* **15**: 125–34.

Verheyen, K., Vellend, M., Van Calster, H. *et al.* (2004). 'Metapopulation dynamics in changing landscapes: a new spatially realistic model for forest plants', *Ecology* **85**: 3302–12.

Whale, D.M. (1984). 'The response of *Primula* species to soil waterlogging and soil drought', *Oecologia* **58**: 272–7.

Young, A., Boyle, T., Brown, T. (1996). 'The population genetic consequences of habitat fragmentation for plants', *Trends in Ecology and Evolution* **11**: 413–8.

Conclusion

CHAPTER 35

Human expansion: research tools, evidence, mechanisms

Francesco d'Errico, William E. Banks, and Jean Clobert

35.1 Introduction

Paradoxical as it may appear to some, the processes at work during human expansion parallel, to a large degree, those observed in the expansion of plant and animal species. As with all other species on the planet, the expansion of past human populations (and present-day patterns of dispersal) have been influenced by numerous parameters such as climate and geography (Chapters 25 and 30), vegetation regimes and food availability (Chapters 20 and 30), as well as intra- and inter-species competition (Chapters 1, 20, and 21), to name just a few.

When viewed from a diachronic perspective, expansion by members of our lineage shows a great degree of variability, a pattern that is likely to be correlated with the changes in social behaviour, emergence of innovations, and the inferred complexity of means of communication. For the most part, these changes occurred in our lineage at a higher rate than has been observed in the remainder of the living world (although numerous instances of rapid, human-assisted invasions have been recorded: Chapters in Parts VI and VII). Understanding the numerous human expansions that are known to have occurred in prehistory is therefore a challenge: driving factors may have been different or varied in their influence based on the time period and the hominin species in questions. Thus, when studying human dispersals, each case must be examined in such a way that all possible mechanisms are taken into account, since one cannot assume that a single mechanism consistently played an exclusive role, and it is likely that different mechanisms were interrelated, a situation which is parallel to many other species (see Part I: the multiple causes of the dispersal process). Our aim here is threefold: 1) we briefly review and critically evaluate the tools that anthropologists and archaeologists have used to tackle these issues; 2) we highlight what we know, and don't know, about the main steps of hominin expansion; and 3) we discuss the possible mechanisms that have stimulated expansion and determined their success or failure, in the light of what we know about plant and animal dispersal.

35.2 Proxies of expansion

Geographic distributions and dated occurrences of hominin remains may be considered the most obvious indicators of the dispersal of members of our lineage. Unfortunately for more ancient periods, human remains are scarce, their recovery often accidental, and too fragmentary for unambiguous taxonomic determinations. There exists, however, sufficient paleontological evidence to establish, on a general timescale, when members of our lineage expanded out of Africa to other regions of the Old World. Also, the last 20 years of paleoanthropological research has significantly enlarged and improved, through the application of new analytical methods, the criteria used to address skeletal variability and make taxonomic attributions (Wood 2010). However, there still exists considerable debate surrounding taxonomic classification and phylogenetic significance of key hominin fossils (Wood and Lonergan 2008; White *et al.* 2010). This is complicated by differential preservation, meaning that sites with well-preserved human remains may

Dispersal Ecology and Evolution. First Edition. Edited by Jean Clobert *et al.*

not necessarily be situated in a region that was at the heart of an ancient dispersal corridor. Skeletal morphology has been used to infer the structure and variation in ancient populations (Gunz *et al.* 2009), as well as adaptation to specific environmental conditions or ecological niches and a hominin's ability to migrate to and settle in new regions (Potts and Teague 2010). However, cases are recorded in which there is not a straightforward match between morphology and function (Ungar *et al.* 2008). For more recent periods for which there are large samples of preserved human remains, morphological variability of specific features, such as cranial shape, allow scenarios of demic population diffusions to be proposed (González-José *et al.* 2008; Pinhasi and von Cramon-Taubadel 2009).

As in studies of plant and animal phylogeography, the analysis of genetic data is becoming an increasingly important means to address human expansion. These studies fall into three broad classes. First, present human genetic variability is used to infer the timing, geographic origin, route of dispersal, rate of genetic exchange, and in some cases, the likely size of the migrant population (Cavalli-Sforza *et al.* 1994; Li *et al.* 2008). For example, present-day genetic data demonstrate that anatomically modern humans originated in Africa approximately 200 000 years before present (BP) and has been used to infer the timing and paths of human dispersal through Asia, Europe, Australia, and the Americas (see Henn *et al.* 2011 for a review; Figure 35.1a). Based on the same data, it has been recently proposed that modern humans intermixed with archaic human populations in Africa circa 35 000 BP (Hammer *et al.* 2011). The Neolithic expansion into Europe from the Near East has been repeatedly addressed using the same research philosophy (Chikhi *et al.* 2009). Second, the dramatic improvements in extraction and sequencing technology of ancient DNA seen in recent years, has profoundly impacted our ability to study past human variability and dispersal. The reconstruction of a large portion of the Neanderthal genome (Green *et al.* 2010) and the recent sequencing of the Denosovian genome (Krause *et al.* 2010) has demonstrated that modern humans leaving Africa interbred to varying degrees with two separate archaic human populations in the Near East and Asia. This has had profound impacts on our understanding of past human dispersals, and the speed at which such methods are improving suggest that this may be one of the most promising approaches with which to study past human expansions. Lastly, human expansions can be traced by examining genetic data from species that accompanied or encountered expanding human populations. Present-day human dispersals are used to trace long-distance animal and plant dispersals (see Chapter 15). Similarly, paleogenetic data of domesticated species have been used to trace the expansion of the first herding societies. A case in point is the use of ancient *Sus* DNA to trace human expansion in South-east Asia and Oceania during the Neolithic (Larson *et al.* 2007). For Europe, genetic data from goats, cattle, and dogs are used to trace Neolithic human expansions (see Deguilloux *et al.* 2009; Larson 2011; Tresset *et al.* 2009). Another interesting example is the use of chicken paleogenetic data to understand prehistoric sources of migration into South America (Storey *et al.* 2007). The genetics of human parasites have proven to be extremely informative in tracing human expansion and contacts between ancient human populations (see Rinaldi 2007 for a review). Lice have been used to indicate direct contacts between modern human colonizers and resident archaic populations in Asia (Reed *et al.* 2004). Other examples (Figure 35.1b) include viruses and bacteria such as leprosy (Dominguez-Bello and Blaser 2011; Monot *et al.* 2005).

Strontium, oxygen, carbon, and nitrogen isotopes, as well as concentrations of major, minor, and trace elements, have also been used to examine paleodiet and infer paleomobility (see Knudson *et al.* 2011 for a review of methods). However, these techniques are more useful at regional scales than they are for tracing geographically expansive human migrations. With respect to other disciplines, such methods have been used to examine bird migrations and dispersals (Sellick *et al.* 2009).

Expansions can also be inferred by examining material remains in the archaeological record. Unlike other animal species, prehistoric humans produced chipped stone-tool industries, one of the most traditional proxies used in archaeology to

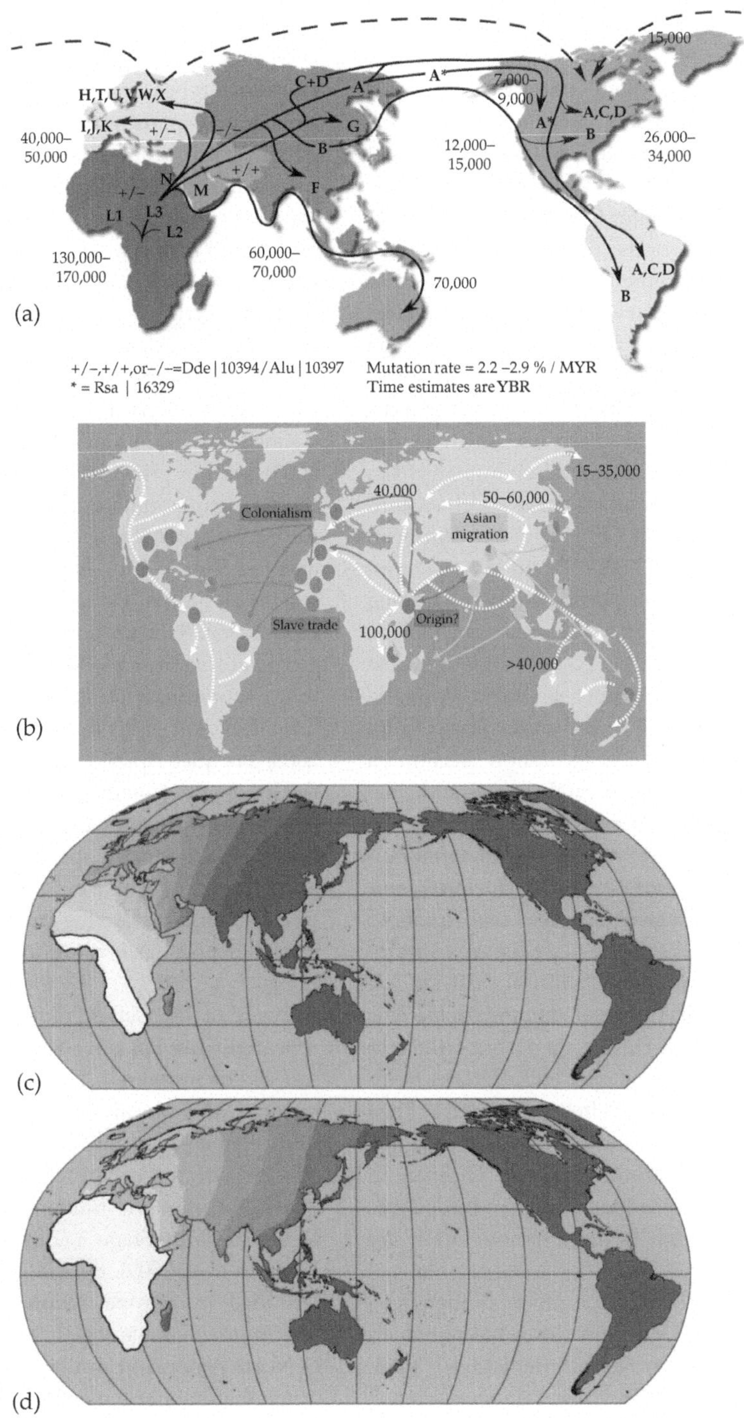

Figure 35.1 (a) *Homo sapiens* migrations based on the modern variability of mitochondrial DNA; (b) dissemination of leprosy in the world, based on the analysis of single-nucleotide polymorphisms. White dotted arrows indicate dispersal of modern humans based on genetic and archaeological evidence. Plain arrows indicate direction of human migration predicted by the polymorphism of the leprosy bacteria; (c)–(d) likely area of language origin under a founder effect model of phonemic diversity inferred from individual languages (a) and mean diversity across language families. (a) modified after <http://www.mitomap.org/mitomap/WorldMigrations.pdf>; (b) from Monot *et al*. 2005, used with permission from Science; (c)–(d) from Atkinson 2011, used with permission from Science.

infer past expansions and contractions of geographic ranges. Stone tool morphology and the processes of their manufacture are thought to reflect the maintenance and transmission of cultural traditions (Bar-Yosef and Belfer-Cohen 2012). For most recent prehistoric periods, during which cultural behaviour becomes more complex, lithic technological data are combined with other cultural markers to reconstruct vectors and extents of expansion events. For example, Mellars (2006) uses the presence of ostrich egg shell beads, engraved objects, and crescent-shaped chipped stone tools to identify an expansion of anatomically modern humans from southern Africa into southern and south-eastern Asia (between 60–40 000 BP). For the European Upper Paleolithic (40–12 000 BP), bone tool styles, mobiliary art, and personal ornament types, among others, have been used trace expansion such as the recolonization of Northern Europe after the Last Glacial Maximum (Gamble *et al.* 2005). With the Holocene archaeological record (after 10 000 BP), ceramic morphology and decoration become a major proxy for defining cultural entities and tracing their possible expansions.

One problem with the reliance on material culture and specific artefact styles is that during expansion events, technologies and styles can undergo change or be abandoned as a consequence of adaptation to new environments, cultural drift, or cultural exchanges with encountered populations. This makes it difficult, particularly for early periods, to associate specific technologies with certain hominins (see Bar-Yosef and Belfer-Cohen 2012; Shea 2010). Additionally, with the appearance of complex societies and social stratification we observe an increased frequency of artefacts being moved across the landscape through trade or made locally by travelling specialists.

Tracing and modelling dispersals of human populations via archaeological proxies is strengthened by the incorporation of chronological data. Biochronology (faunal and vegetal associations), paleomagnetic stratigraphy, and radiometric dating are used to trace the expansion of the first member of our lineage to leave Africa (*Homo erectus*). Such methods have improved considerably during recent decades, but in the absence of clearly diagnostic stone-tool types, archaeologists are unable to determine if an expansion occurred as a single dispersal event or if it represents multiple waves. For early dispersals of anatomically modern humans (AMH), the above methods are supplemented by more precise dating methods such as OSL, TL, and radiocarbon (^{14}C). Examples include dispersals across northern Africa (Barton *et al.* 2009), possible corridors out of Africa (Armitage *et al.* 2011), through southern Asia and into Australia (Balme 2011). For expansions occurring after 40 000 BP, radiocarbon dating and its updated calibration (Reimer *et al.* 2009) are the principal means used to reconstruct the timing of population movements across Eurasia, the arrival of AMH into the European cul-de-sac and the New World, as well as island colonization (e.g. Polynesian chains, Mediterranean islands, Japan, etc.). Radiocarbon chronologies have played a crucial role in modelling the Neolithic expansion into Europe from the Near East (Ammerman and Cavalli-Sforza 1984; Bocquet-Appel and Bar-Yosef 2008; Bocquet-Appel *et al.* 2009). Chronologies at the limit of the method (50 000–30 000 BP) can be tested and verified by the presence of tephra, which serves to fingerprint specific volcanic events of a known age. For example, the Campanian Ignimbrite has been used to trace the arrival of modern humans into Italy and Eastern Europe (Blockley *et al.* 2008; Longo *et al.* 2012).

In and of themselves, radiometric ages have been repeatedly used as proxies by which human dispersals can be measured. The increasing precision and number of more ancient ages, along with recent improvements in radiocarbon calibration, have allowed for the construction of scenarios for the arrival of modern humans into Europe during Marine Isotope Stage 3 (Zilhão and d'Errico 1999; Higham *et al.* 2009), the recolonization of northern latitudes in Europe following the Last Glacial Maximum (Gamble *et al.* 2005), and the colonization of North America (Hamilton and Buchanan 2007).

As in animal and plant species, paleogeographic reconstructions based on tectonics (e.g. Schattner and Lazar 2009) and glacioeustatic and sea level changes (Bailey *et al.* 2011) are also used to constrain the dispersal and development of hominin populations in some regions such as *Homo erectus* in Java or

Homo florensis on Flores Island (Zaim 2010). Information on sea level changes is combined with archaeological information on sea-faring navigation to reconstruct potential dispersal corridors. In examining sea level reconstructions, it becomes clear that parts of south-eastern Asia could only have been populated by *Homo erectus* if they crossed bodies of water (see Balme 2012), but these water barriers would have been less than 20 km due to lowered sea levels. Purposeful and long-distance dispersals by sea crossing are attested to in some regions of the world, such as Sahul, at around 50 000 BP. Much later, the Neolithic expansion along the Mediterranean coast is thought to have been facilitated by an intense use of boats that were of sufficient size to transport both people and domesticated animals (Broodbank 2006). It is important to point out that while human populations used sea-faring technology to settle distant regions and islands, there are several instances in which these technologies were subsequently lost or abandoned by later generations.

As with bird song (Seddon *et al.* 2008), historical linguistic means have been developed to establish language phylogenies which in cases of migration may be used as tracers of the related population expansions. A well-known, but also controversial case is the history of Indo-European languages associated with a number of population movements into Europe from the East (Renfrew 1987). Linguistic variability has been a key element in the study of the Bantu expansion in Africa, which refers to the large population movement throughout sub-Saharan Africa ca. 5000 BP (see Berniell-Lee *et al.* 2009 for a review). In the Pacific, language phylogenies have served to identify periods of expansion and stasis in human settlement (Gray *et al.* 2009). Recently, attempts have been made in using language diversity to predict the African region where modern languages originated (Figure 35.1c–d; Atkinson 2011). For the recent historic period, surnames serve to study mobility and temporal demographic variability (Darlu *et al.* 2011). Finally, we must not forget that humans are the only species to tell and share stories about where they come from, although a social group's history of its origin tends to be influenced by a number of biases and may not accurately reflect a demographic reality.

The introduction of modelling techniques to dispersal studies, with the aim of creating scenarios and evaluating them against empirical data (e.g. Ackland *et al.* 2007; Banks *et al.* 2008a; Bocquet-Appel *et al.* 2009; Currat and Excoffier 2011; Field *et al.* 2007; Hughes *et al.* 2007), can allow one to determine if deviations from expectations can be accounted for by including additional parameters or different model structures (Steele 2009). For example, with modelling methods, one can include key parameters similar to those used to study plant and animal dispersals such as carrying capacity, population density, and population growth.

35.3 The record of hominin expansions

What have all these different approaches highlighted or revealed about dispersals in general, and more specifically, the successive expansions of members of our lineage since our separation from the common ancestor that we share with chimpanzees?

Between 2 and 3 million years ago (mya) early hominins appear to have exploited ecological situations that were coincident with transitions between wooded and open environments present in both the Rift valley of East Africa and southern Africa. It had been proposed that a shift in climate that occurred in the late Miocene may have favoured evolutionary changes in early hominins, which would have led to bipedalism and their expansion into open savannah environments (Bonnefille *et al.* 2004). This hypothesis has now been largely abandoned after the discovery of earlier hominin remains to the west of the Rift Valley as well as in Chad (Brunet 2009). This does not rule out the idea, however, that climate change led to changes in adaptation and geographic expansion of early hominins (de Menocal 2011), as is currently the case with plants and animals due to global warming (see chapters in Part VI). Selection on the habitat margin (like social adaptations, morphological adaptations, and selection on dispersal capacity which can be viewed as innovation) may have played important roles (Chapters 25 and 26).

The rarity of associated hominin remains makes it difficult to establish the authors of the first unmistakable knapped stone tool industries in East Africa around 2.6 mya (de la Torre 2011). By ca. 1.77 mya the archaeological record indicates a first dispersal of *Homo erectus* bearing a core and flake technology into various regions of Africa (see Bar-Yosef and Belfer-Cohen 2012) as well as into southern Eurasia (site of Dmanisi; Agusti and Lordkipanidze 2011). Almost immediately afterwards, around 1.76 mya, we see in East Africa the earliest instances of a biface-dominated stone tool industry termed the Acheulian (Lepre *et al.* 2011). By 1.5–1.4 mya, the bearers of Acheulian technology had dispersed out of Africa, across western Asia, and into South Asia. It is still unclear whether the Acheulian occupied the whole of Africa before 0.7 mya (Raynal *et al.* 2001). Europe does not seem to have been populated during these expansions and it is not until much later, ca. 1.2 mya, that we find hominin remains and flake/core industries in southern Europe (Carbonell *et al.* 2008). The Acheulian appears in the Iberian Peninsula around 0.9 mya (Scott and Gibert 2009), but does not extend into northern Europe until after 0.5 mya (Monnier 2006). The British Isles do not have a record of occupation before 0.78 mya, and that occupation is not associated with Acheulian technology (Parfitt *et al.* 2010). Recent genetic evidence suggests that the archaic hominins who colonized Europe and Asia became largely isolated by 1.0 mya and gave rise to at least two separate lineages that ultimately led to Neanderthals and Denisovans, respectively (Krause *et al.* 2010; Reich *et al.* 2010). The latter is named after the eponymous site located in the Altai Mountains of southern Siberia from which human remains that preserved ancient DNA were recovered. Paleoanthropological evidence suggests that archaic hominins living in Europe slowly acquired typical Neanderthal features during the late Middle and early Upper Pleistocene, i.e. 400 000–40 000 BP. During this period, pre-Neanderthal and Neanderthal ranges fluctuated such that they intermittently occupied regions such as the British Isles, the Near East, western Asia, and possibly more northern latitudes near the arctic circle (Slimak *et al.* 2011; Pavlov *et al.* 2001), although the latter still needs further evaluation to be confirmed.

Genetic and paleoanthropological evidence suggests that modern human features appear in West Africa around 200 000 BP (White *et al.* 2003). Due to a lack or fragmentary nature of human remains over large regions of Africa, it is, however, uncertain when expansions of AMH occurred within Africa and whether there was a single expansion or multiple expansions, although the latter is more likely (Gunz *et al.* 2009). Evidence has recently been presented that indicates a genetic contribution of African archaic hominins to the modern African gene pool (Hammer *et al.* 2011). By ca. 130 000 BP modern human populations, with some archaic features and bearing a technology similar to that of coeval Neanderthal populations occupying the Near East, are recognized in the Maghreb and in the Near East. The presence of these populations outside of Africa is limited to the Near East and is not recognized in neighboring regions of Western Asia (Garcea 2012). This suggests that this expansion was not successful and may only represent populations following the natural expansion of their ecological range during this period of climatic amelioration. It is interesting to note that similar differences in colonizing capacity among populations in different refuges are documented in the plant and animal kingdom after the last glacial episode (Hewitt 2008).

It is not until after the cold and rigorous period known as Marine Isotope Stage 4, which terminated around 60 000 BP, that we see a significant expansion of anatomically modern humans out of Africa. The path and timing of this expansion are highly debated, but it was quite rapid towards the East as we recognize a modern human presence in Australia by ca. 50 000 BP (Balme 2012). Genetic evidence indicate encounters between modern humans and resident Neanderthal populations probably occurred just after the former's exit from Africa since Neanderthal genes are present (ca. 2–4%) in all non-African modern human populations (Green *et al.* 2010). During their dispersal through Asia, modern humans also interbred with Denisovan populations as indicated by a relatively large contribution to the modern human genome of populations in Melanesia (Reich *et al.* 2010). Often considered to be a single event, this major dispersal may have taken a variety of forms with multiple exits from Africa during a period that covered several millennia.

Evidence for the first arrival of modern humans in Europe dates to ca. 42–42 000 BP. The timing of this expansion, Neanderthal disappearance and the mechanisms behind this founder population event are highly debated (Pinhasi *et al.* 2011; Banks *et al.* 2008a). There is evidence suggesting a late survival of Neanderthals in the Iberian Peninsula (ca. 34 000 BP) which supports the idea that modern human dispersal into Europe was not a single, rapid event and may have taken different forms in different regions.

It is debated whether the cultural changes within the Upper Paleolithic are the result of cultural drift (local cultural evolution) or population dispersals. The latter has been proposed for the Gravettian based principally on genetic data (Semino *et al.* 2000) and more convincingly, combining genetic and archaeological data, for the Magdalenian (Gamble *et al.* 2006). It is clear, however, that millennial-scale climatic variability of MIS 3 and 2, as well as the impact in Europe of the Last Glacial Maximum, must have had profound influences on human ranges (Banks *et al.* 2008b). During the Bølling-Allerød climatic amelioration (14 700–12 700 BP) and more intensively at the beginning of the Holocene, ca. 10 000 BP, human populations recolonized northern Europe. Mesolithic hunter-gatherers are recognized in the British Isles and Ireland around 9 000 BP and northern Scandinavia after 8 000 BP. Recent genetic evidence suggests that similar recolonizations may have occurred in Asia well before the spread of the Neolithic (Zheng *et al.* 2011).

Early dates from a number of archaeological sites indicates that the first human colonization of the Americas, favoured by the use of watercraft along the southern coast of the Bering Land Bridge, occurred between 15 000 and 13 500 BP. Pre-Clovis hunter-gatherers were present in the state of Washington by at least 13 700 BP (Waters *et al.* 2011). Radiocarbon ages from South American sites suggest the process was extremely rapid (Figure 35.2a) and may have taken less than a millennium (Rothhammers and Dillehay 2009; Meltzer 2009).

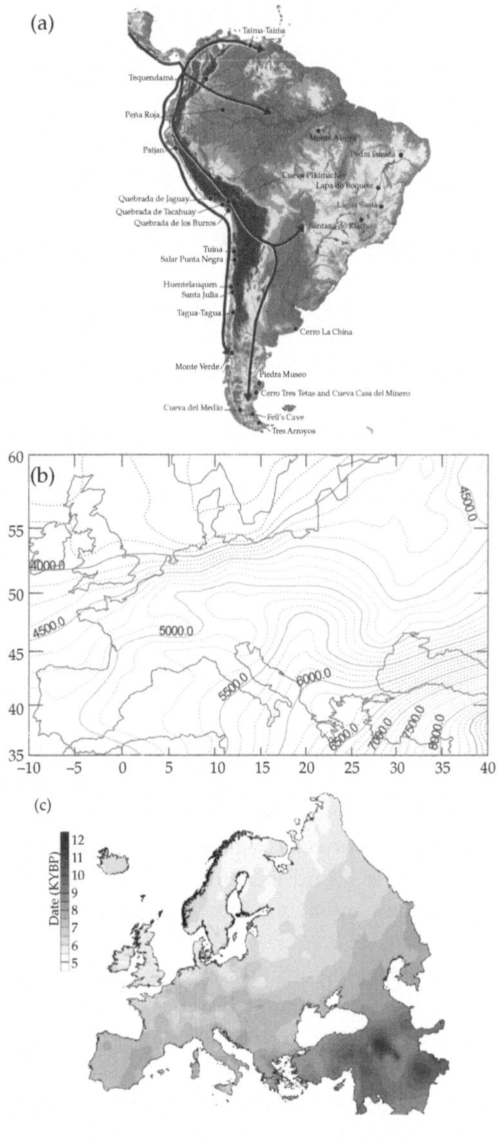

Figure 35.2 (a) Hypothesized migration routes into South America and major archaeological sites dating to the late Pleistocene; (b)–(c) representations of the isochrones of the Neolithic farming expansion into Europe; (a) after Rothhammer and Dillehay 2009, used with permission from John Wiley and Sons; (b) after Bocquet-Appel *et al.* 2012, used with permission from Elsevier; (c) after Balaresque *et al.* 2010.

The transition from hunter-gatherer to farming/herding economies has also led, in many areas of the world, to population dispersals. For many years, it was questioned whether the emergence of agriculture and animal domestication in Europe was, at least in some regions, the result of independent local invention, the spread and adoption of this adaptation by resident hunter-gather populations, or the actual physical dispersals of populations that used such production-oriented economies. It is now widely accepted that the latter scenario is likely the case for most of Europe and that early Neolithic adaptations spread westward from the Near East (Figure 35.2b, c) at around 10 000 BP across the Mediterranean and reached peripheral zones in northwestern Europe by ca. 4000 BP (Bocquet-Appel *et al.* 2009). The nature of cultural and genetic exchanges between invading Neolithic populations and local hunter-gatherers are still the subject of intense debate fueled by the input of a variety of disciplines (Bocquet-Appel and Bar-Yosef 2008). The process that we observe in Europe was not the rule in all other areas of the world. In North America, the transition to full-scale agriculture was relatively slow and was for the most part characterized by the diffusion of horticultural and agricultural technologies rather than large scale movements or expansion of people. It is possible that differences in dispersal syndrome (Chapter 10), linking colonization ability and agricultural skills, might have differed between regions. Similar instances are recurrently found in animal species, although obviously not involving the same traits (Clobert *et al.* 2009).

More remote regions of the globe were not systematically settled until well after the spread of agricultural economies in mid- and tropical latitudes of continental regions of Eurasia and the Americas. This is the case for the pre-Dorset colonization of Arctic North America by 4500 years ago. This population originated in Siberia and carried with it an adaptation that allowed for the permanent occupation of an arctic landscape. An adaptation to a full arctic landscape is not seen until the Thule expansion from western Alaska around 800 years ago and that reached the southern coast of Greenland 300 years later (Friesen and Arnold 2008).

Human dispersal and colonization has taken various forms, was dependent on the population and species in question, and was influenced by multiple abiotic and biotic factors. Alone, human dispersal illustrates well the multicausal and multifactorial nature of this complex behaviour (see chapters in Part I).

35.4 Factors behind human expansion

35.4.1 The climate hypothesis

For most species, when expansion is the focus of investigation, climatic and resultant environmental changes are often evoked as the key factors (Chapters 25 and 26). Indeed, models predict that the evolution of dispersal should be more sensitive to environmental stochasticity when all potential causes for its evolution are combined (Gandon *et al.* 2001). For humans, the basic assumption here is that shifts in climate alter the environmental structure, and therefore resource availability, of a given region. Climatic changes can alter the geographic footprint of a given ecological niche, and a population may expand to track this changing footprint (Chapter 25). A good example of this latter scenario is the repopulation of northern Europe by hunter-gatherers during a period of climatic amelioration following the Last Glacial Maximum (Gamble *et al.* 2005). Potential outcomes of environmental change on a population include extinction, migration, isolation, and eventually speciation. The new species may benefit from mutations that allow it to expand its ecological niche (Chapter 26), which can be manifested by a physical expansion. Some have advanced the idea that the key to understanding the successful adaptations leading to the multiple exits from Africa is the process of speciation and the resulting cognitive advantages (McBrearty and Brooks 2000). This would explain the expansion out of Africa by *Homo erectus* as well as by modern humans. Attributing human expansion solely to climate change and speciation, however, can hardly explain why, compared to other species, members of our lineage, spread so quickly across a wide variety of ecosystems.

Clearly other factors were at play during the evolution of our lineage and influenced human expansions to varying degrees depending on the period in question. Many authors have explored the role of culture in its various forms. Such a role can be compared, to some extent, to that of maternal effects and phenotypic plasticity in plants and animals (see chapters in Part III). Richerson and Boyd (2005; Richerson *et al.* 2005) argue that cumulative culture is the means by which our species has escaped the rules of natural selection in order to cope with rapid-scale and high-amplitude climatic changes that increasingly characterized our planet's climate during the last million years. We have created a costly tool that allows us to accumulate and transmit knowledge across generations (Boyd *et al.* 2011). From this perspective, human dispersal is at once the outcome of environmental change and the accumulation of successful cultural adaptations.

Chipped-stone tools, hunting technologies, containers, systematic use of fire, clothing, and sea-faring technology are among the cultural innovations that have certainly played a role in facilitating human expansions by reducing dispersal cost (Bonte *et al.* 2012). In order for expansions to be viable in challenging environments and for complex cultural adaptations to be developed and maintained, the establishment and maintenance of communication systems and social networks is essential. The production of symbolic material culture, including body decoration, art, imposed style in manufactured objects, etc. is the archaeological reflection of the communication systems developed by human groups. Such mechanisms may also serve to distinguish differences between groups and result in the establishment of local or more regional cultural traditions. When seen from an ecological perspective, such processes echo the interplay between dispersal and local adaptation following a colonization event, which we see with the expansion of other animal and plant species.

In turn, the creation of these advantageous cultural adaptations likely favoured organic evolution that allowed and facilitated human expansion. The development of more complex communication systems may have stimulated changes in the brain that led to increased language capacity.

35.4.2 The cultural hypothesis

Recently, the involvement of climatic variability in the process of expansion has been challenged by researchers who stress the importance of cultural and genetic features associated with different populations to account for success and failure of dispersal events. Each hominin group expanding inside and outside Africa would be characterized, according to them, by an ability to cope with environmental stress that would have conditioned their ability to occupy new territories and have that expansion be viable over the long term (Bar-Yosef and Belfer-Cohen 2012; Chudek and Henrich 2011). A similar situation may have played out in plant and animal expansions, where novelties or selection would have reduced dispersal costs or enhanced settlement success of the dispersers (Bonte *et al.* 2012; Chapters in Part III).

The integration of ecological niche and species distribution modelling methods into archaeological investigations has allowed these issues to be addressed directly via an approach termed eco-cultural niche modelling (Banks *et al.* 2006). The utility of eco-cultural niche modelling is that it allows one to evaluate quantitatively whether links exist between a given adaptive system and ecological constraints, or if the characteristics and geographic distribution of a cohesive cultural system may have been influenced more by non-ecological (i.e. cultural) processes (Banks *et al.* 2009, 2011). Additionally, one can evaluate culture-environment relationships across time and determine if dispersal events were associated with ecological niche shifts or if they represent simple range changes that tracked fluctuating environmental footprints. The interest of this and similar approaches is that: (1) they allow one to disentangle and evaluate the various mechanisms behind specific human dispersal events; and (2) they can be used to create the foundations of a 'unified theory' of human expansion that combines the inputs from modelling techniques, genetic and ecological thinking, and archaeological and paleoanthropological evidence.

35.4.3 The social hypothesis

While climatic factors certainly played a role in past human expansion, in conjunction with the use of culture to reduce dispersal costs, there is a dimension, which can be termed the social dimension, whose factors might also have played an important role. Evidence of social factors motivating human dispersal is found in most modern human expansions. Social-based factors encompass all factors which arise from interactions either among kin or among congeners (Chapter 1), and as an extension among tribes or groups of humans or tribes (Towner 1999; Kok 2010). Potential examples are numerous. There is a tradition in oceanic island populations of sending selected sons/daughters from different families to colonize new islands, even in the absence of any information about the existence and accessibility of these new islands (Edwards 2003, Finney 1996). This resembles a potential link between kin competition and colonization, a pattern that is found in some animals populations (Sinervo and Clobert 2003; Cote *et al.* 2007). In aristocratic families, often only the oldest male child will inherit the title and domain, the youngest having to develop their own way with a minimum of parental investment (Block 1864; Kok 2010). A similar uneven inheritance principle was applied in farmer families (Furby 1896; Kok 2010). Not surprisingly, the youngest children of aristocratic families have often played an important role in colonizing new territories during the colonization by Europeans of other continents in the last 500 years or so. Competition among congeners for food, territory, or other resources has also play an important role in the same colonization process. The early colonization of North America is a good example of this with numerous people taking the decision of leaving their natal country in the hope of having a better life in those newly available areas. Competition for goods is also playing a major role in the present-day colonization of European countries by people from countries with depressed economies. Sadly enough, the above example also serves to illustrate that colonization comes with some cost: many individuals die during the dispersal process (boat people, ICMC Europe report 2011), or during initial settlement with an increased mortality rate (up to the failing of the colonization) due to small population size, illness, or lack of local adaptation (Greenland colonization by Eric the Red in McGovern 1980; Christopher Columbus' first implantation in Columbus and Fuson 1987).

With respect to animals and plants, the colonization process often involves individuals with a different phenotype than that of the resident individual (chapters in Part III). Although no systematic studies have been conducted on humans (Arango 2000), the actual behavioural profile of human pioneers (Laland and Brown 2011; Massey 1990, Whybrow 2005) suggests such a possibility. To what extent this might explain the current structure and reference values of nations arising from such colonization (such as in North America) is open to debate, especially if the later waves of colonization were achieved by people with different behavioural profiles than those of the first (the colonizer-joiner process, Clobert *et al.* 2009). A comparison between Australia and North America might be of help here since these colonization events were achieved in very different ways.

How social factors may have played an important role in early human expansions is difficult to envision, although it is possible that these forces increased in importance when the adoption of agriculture led to human populations becoming more sedentary (Larsen 1995).

35.5 Conclusions

Our review shows that the mechanisms behind dispersals throughout the history of the human lineage are diverse and serves to illustrate the multiple causes and origins of the dispersal processes and patterns that are observed in other species (see chapters in Parts I and IV). The examples of different dispersal events described above serve to show that there is no common denominator when considering human populations, and that this fact is likely to be true for most other species as well. For example, early dispersals within the African continent and the earliest dispersals of *Homo* out of Africa appear to be related to speciation events and their success tied to biological adaptations (Chapters 25 and 26). Changing environmental conditions seem to have strongly conditioned certain expansions by facilitating movement through geographic corridors during specific climatic events. The movement into regions devoid of human populations, and therefore relatively free from competition, likely favoured early dispersals out of Africa across the Old World, and also later when populations first entered the Americas. The pioneer syndrome should then have been heavily selected for. Likewise, competition was also a key factor when one sees dispersals into regions that already had resident human populations, as is the case with the earliest, but unsuccessful, expansions of anatomically modern humans out of Africa into the Near East where archaic human populations were present. It is likely that social factors increased in importance when human populations became more sedentary and that the pioneer syndrome was selected against.

In human populations, it is evident that in many instances cultural adaptations played a key role in successful dispersal events, a role which is, to some extent, comparable to phenotypic plasticity and maternal effects in other species. The development of technologies and the appearance of certain social behaviours appear to have conferred slight competitive advantages to expanding populations, as is the case with the second migration of modern humans out of Africa around 50 000 years ago into regions of the Old World that had resident archaic human populations, as well as during the Neolithic some 40 000 years later. However, dispersal comes with a cost, which has probably increased as human populations increased their social structures. Therefore, when one considers human dispersal, each case must be examined individually and all possible factors behind such events must be considered—one cannot assume *a priori* that a single mechanism consistently plays a major role. In other words, causes and consequences of dispersal in humans and other living beings not surprisingly share many common characteristics. Only the relative importance of some mechanisms, such as cultural ones, might have permitted humans to colonize a larger array of habitat types with great success.

Acknowledgements

We thank the editors for inviting us to contribute a chapter to this monograph. Financial support was provided to the authors by the European Research Council Advanced Grant TRACSYMBOLS (FP7 No. 249587).

References

Ackland, G.J., Signitzer, M., Stratford, K. *et al.* (2007). 'Cultural hitchhiking on the wave of advance of beneficial technologies', *Proceedings of the National Academy of Sciences* **104**: 8714–9.

Agusti, J., Lordkipanidze, D. (2011). 'How 'African' was the early human dispersal out of Africa?', *Quaternary Science Reviews* **30**: 1338–42.

Ammerman, A.J., Cavalli-Sforza, L.L. (1984). 'The Neolithic Transition and the Genetics of Populations in Europe', Princeton University Press: Princeton.

Arango, J. (2000). 'Explaining the migrations: a critical review', *Revue internationale des Sciences Sociales* **165**: 329–42.

Armitage, S.J., Jasim, S.A., Marks, A.E. *et al.* (2011). 'The Southern Route "Out of Africa": Evidence for an Early Expansion of Modern Humans into Arabia', *Science* **331**: 453–6.

Atkinson, Q.D. (2011). 'Phonemic Diversity Supports a Serial Founder Effect Model of Language Expansion from Africa', *Science* **332**: (6027), 346–9.

Bailey, G.N., Reynolds, S.C., King, G.C.P. (2011). 'Landscapes of human evolution: models and methods of tectonic geomorphology and the reconstruction of hominin landscapes', *Journal of Human Evolution* **60**: 257–80.

Balaresque, P., Bowden, G.R., Adams, S.M. *et al.* (2010). 'Predominantly Neolithic Origin for European Paternal Lineages', *PLoS Biology* **8**: (1) e1000285. DOI:10.1371/journal.pbio.1000285

Balme, J. (2012). 'Of boats and string: The maritime colonisation of Australia', *Quaternary International* (in press).

Banks, W.E., Aubry, T., d'Errico, F. *et al.* (2011). 'Eco-cultural niches of the Badegoulian: Unraveling links between cultural adaptation and ecology during the Last Glacial Maximum in France', *Journal of Anthropological Archaeology* **30**: 359–74.

Banks, W.E., d'Errico, F., Dibble, H.L. *et al.* (2006). 'Eco-cultural niche modelling: New tools for reconstructing the geography and ecology of past human populations', *PaleoAnthropology* **4**: 68–83.

Banks, W.E., d'Errico, F., Peterson, A.T. *et al.* (2008a). 'Neanderthal Extinction by Competitive Exclusion', *PLoS One* **3**: (12) e3972.

Banks, W.E., d'Errico, F., Peterson, A.T. *et al.* (2008b). 'Human ecological niches and ranges during the LGM in Europe derived from an application of eco-cultural niche modelling', *Journal of Archaeological Science* **35**: 481–91.

Banks, W.E., Zilhão, J., d'Errico, F. *et al.* (2009). 'Investigating links between ecology and bifacial tool types in Western

Europe during the Last Glacial Maximum', *Journal of Archaeological Science* **36**: 2853–67.

Bar-Yosef, O. and Belfer-Cohen, A. (2012). 'Following Pleistocene road signs of human dispersals across Eurasia', *Quaternary International* (in press) doi: 10.1016/j.quaint.2011.07.043.

Barton, R.N.E., Bouzouggar, A., Collcutt, S.N. *et al.* (2009). 'OLS dating of the Aterian levels at Dar es-Soltan I (Rabat, Morocco) and implications for the dispersal of modern *Homo sapiens*', *Quaternary Science Reviews* **28**: 1914–31.

Berniell-Lee, G., Calafell, F., Bosch, E. *et al.* (2009). 'Genetic and Demographic Implications of the Bantu Expansion: Insights from Human Paternal Lineages', *Molecular Biology and Evolution* **26** (7): 1581–9.

Block, M.M. (1864). *Dictionnaire general de la politique.* Lorenz O. Libraire-Editeur, Paris, France.

Blockley, S.P.E., Ramsey, C.B., Higham, T.F.G. (2008). 'The Middle to Upper Paleolithic transition: dating, stratigraphy, and isochronous markers', *Journal of Human Evolution* **55**: 764–1.

Bocquet-Appel, J.-P. and Bar-Yosef, O. (eds), (2008). *The Neolithic Demographic Transition and Its Consequences.* Springer: New York.

Bocquet-Appel, J.-P., Naji, S., Vander Linden, M. *et al.* (2012). 'Understanding the rates of expansion of the farming system in Europe', *Journal of Archaeological Science* **39**: 531–46.

Bocquet-Appel, J.-P., Naji, S., Vander Linden M. *et al.* (2009). 'Detection of diffusion and contact zones of early farming in Europe from the space-time distribution of 14C dates', *Journal of Archaeological Science* **36**: 807–20.

Bonnefille R., Potts, R., Chalié, F. *et al.* (2004). 'High-resolution vegetation and climate change associated with Pliocene *Australopithecus afarensis*', *Proceedings of the National Academy of Sciences* **101**(33): 12125–9.

Bonte, D., Van Dyck, H., Bullock, J. *et al.* (2012). 'Costs of dispersal', *Biological Reviews* DOI:10.1111/j.1496- 185X.2011.00201.x.

Boyd, R., Richerson, P.J., Henrich, J. (2011). 'The cultural niche: Why social learning is essential for human adaptation', *Proceedings of the National Academy of Sciences* **108**: 10918–25.

Broodbank, C. (2006). 'The Origins and Early Development of Mediterranean Maritime Activity', *Journal of Mediterranean Archaeology* **19(2)**: 199–230.

Brunet, M. (2009). 'Origine et évolution des hominidés: Toumaï, une confirmation éclatante de la prédiction de Darwin', *Comptes Rendus Palevol* **8**: 311–9.

Carbonell, E., Bermudez de Castro, J.M., Parees J.M. *et al.* 2008. 'The first hominin of Europe', *Nature* **452**: 465–70.

Cavalli-Sforza, L.L., Menozzi, P., Piazza, A. (1994). *The History and Geography of Human Genes.* Princeton University Press: Princeton.

Chikhi, L., Destro-Bisol, G., Pascali, V. *et al.* (2009). 'Clinal Variation in the Nuclear DNA of Europeans', *Human Biology* **81** (5–6): 625–38.

Chudek, M. and Henrich, J. (2011). 'Culture–gene coevolution, norm-psychology and the emergence of human prosociality', *Trends in Cognitive Sciences* **15**: 218–26.

Clobert, J., Le Galliard, J.-F., Cote, J. *et al.* (2009). 'Informed dispersal, heterogeneity in animal dispersal syndromes and the dynamics of spatially structured populations', *Ecology Letters* **12**:197–209.

Columbus, C. and Fuson, F.R. (1987). *The log of Chistopher Columbus.* International Marine Publishing Company, Camden, Me.

Cote, J., Clobert, J., Fitze, P.S. (2007). 'Kin competition promotes colonization success', *Proceedings of National Academy of Sciences* USA **104**: 9703–8.

Currat, M. and Excoffier, L. (2011). 'Strong reproductive isolation between humans and Neanderthals inferred from observed patterns of introgression', *PNAS* **108** (37): 15129–34.

Darlu, P., Brunet, G., Barbero, D. (2011). 'Spatial and Temporal Analyses of Surname Distributions to Estimate Mobility and Changes in Historical Demography: The Example of Savoy (France) from the Eighteenth to the Twentieth Century', in M.P. Gutmann *et al.* (eds) *Navigating Time and Space in Population Studies.* Springer: Berlin, 99–113,

De la Torre, I. (2011). 'The origins of stone tool technology in Africa: a historical perspective', *Philosophical Transactions of the Royal Society B* **366** (1567): 1028–37.

de Menocal, P.B. (2011). 'Climate and Human Evolution', *Science* **331** (6017): 540–2.

Deguilloux, M.F., Moquel, J., Pemonge, M.H. *et al.* (2009). 'Ancient DNA supports lineage replacement in European dog gene pool: insight into Neolithic South-East France', *Journal of Archaeological Science* **36**: 513–19.

Dominguez-Bello, M.G., Blaser, M.J. (2011). 'The Human Microbiota as a Marker for Migrations of Individuals and Populations', *Annual Review of Anthropology* **40**: 451–74.

Edwards, P. ed. (2003). *James Cook: The Journals.* Penguin Books: London.

Field, J.S., Petraglia, M.D., Lahr, M.M. (2007). 'The southern dispersal hypothesis and the South Asian archaeological record: Examination of dispersal routes through GIS analysis', *Journal of Anthropological Archaeology* **26**: 88–108.

Finney, B. (1996). 'Colonizing an Island World', *Transactions of the American Philosophical Society* **86**: 71–116.

Friesen, T.M. and Arnold, C.D. (2008). 'The timing of the Thule migration: new dates from the western Canadian Arctic', *American Antiquity* **73**: 527–38.

Furby, C. (1896). *Essai sur le droit d'ainese en Angleterre.* Durand & Pedone-Lauriel Edition, Paris France. http://www.archive.org/stream/essaisurledroit00furbgoog/essaisurledroit00furbgoog_djvu.txt

Gamble, C., Davies, W., Pettitt, P. *et al.* (2005). 'The Archaeological and Genetic Foundations of the European Population during the Late Glacial: Implications for "Agricultural Thinking"', *Cambridge Archaeological Journal* **15**: 193–223.

Gamble, C., Davies, W., Pettitt, P. *et al.* (2006). 'The Late Glacial ancestry of Europeans: Combining genetic and archaeological evidence', *Documenta Praehistorica* **33**: 1–10.

Gandon, S. and Michalakis, Y. (2001). 'Multiple causes of the evolution of dispersal', in J. Clobert. E. Danchin, A.A. Dhindt *et al.* (eds), *Dispersal.* Oxford University Press, Oxford, 155–67.

Garcea, E.A. (2012). 'Successes and failures of human dispersals from North Africa', *Quaternary International* (in press).

González-José, R., Bortolini, M.C., Santos, F.R. *et al.* (2008). 'The Peopling of America: Craniofacial Shape Variation on a Continental Scale and Its Interpretation From an Interdisciplinary View', *American Journal of Physical Anthropology* **137**: 175–87.

Gray, R.D., Drummond, A.J., Greenhill, S.J. (2009). 'Language Phylogenies Reveal Expansion Pulses and Pauses in Pacific Settlement', *Science* **323** (5913): 479–83.

Green, R.E., Krause, J., Briggs, A.W. *et al.* (2010). 'A Draft Sequence of the Neandertal Genome', *Science* **328**: 710–22.

Gunz, P., Bookstein, F.L., Mitteroecker, P. *et al.* (2009). 'Early modern human diversity suggests subdivided population structure and a complex out-of-Africa scenario', *Proceedings of the National Academy of Sciences USA* **106**: 6094–8.

Hamilton, M.J. and Buchanan, B. (2007). 'Spatial gradients in Clovis-age radiocarbon dates across North America suggest rapid colonization from the north', *PNAS* **104**(40): 15625–30

Hammer, M.F., Woerner, A.E., Mendez, F.L. *et al.* (2011). 'Genetic evidence for archaic admixture in Africa', *Proceedings of the National Academy of Sciences USA* **108**: 15123–8.

Henn, B.M., Gignoux, C.R., Jobin, M. *et al.* (2011). 'Hunter-gatherer genomic diversity suggests a southern African origin for modern humans', *Proceedings of the National Academy of Sciences USA* **108**: 5154–62.

Hewitt, G.M. (2008). 'Post-glacial re-colonization of European biota', *Biological Journal of the Linnean Society* **68**: 87–112.

Higham, T., Brock, F., Peresani, M. *et al.* (2009). 'Problems with radiocarbon dating the Middle to Upper Palaeolithic transition in Italy', *Quaternary Science Reviews* **28** (13–14): 1257–67.

Hughes J.K., Haywood, A., Mithen, S.J. *et al.* (2007). 'Investigating early hominin dispersal patterns: developing a framework for climate data integration', *Journal of Human Evolution* **53**(5): 465–74.

ICMC Europe (2011) '"MAYDAY!" Strengthening responses of assistance and protection to boat people and other migrants arriving in Southern Europe', *International Catholic Migration Commission publication report*.

Knudson, K.J., O'Donnabhain, B., Carver, C. *et al.* (2011). 'Migration and Viking Dublin: paleomobility and paleodiet through isotopic analyses', *Journal of Archaeological Science* **32**: 903–13.

Kok, J. (2010). 'The family factor in migration decisions', in J. Lugassen, L. Lucassen, P. Manning (eds), *Migration History in World History: Multidisciplinary Approaches.* Koninklijke Brill NV, Leiden, The Netherlands, 215–50.

Krause, J., Fu, Q., Good, J.M. *et al.* (2010). 'The complete mitochondrial DNA genome of an unknown hominin from southern Siberia', *Nature* **464**: 894–7.

Laland, K.L. and Brown, G. (2011). *Sense and non sense: evolutionary perspective on human behaviour.* Oxford University Press: Oxford.

Larsen, C.S. (1995). 'Biological changes in human populations with agriculture', *Annual Review of Anthropology* **24**: 185–213.

Larson, G. (2011). Genetics and Domestication: Important Questions for New Answers', *Current Anthropology* **52**(S4): S485–95.

Larson, G., Dobney K., Albarella, U. *et al.* (2007). 'Phylogeny and ancient DNA of *Sus* provides insights into Neolithic expansion in Island Southeast Asia and Oceania', *Proceedings of the National Academy of Sciences USA* **104**: 4834–9.

Lepre, C.J., Roche, H., Kent, D.V. *et al.* (2011). 'An earlier origin for the Acheulian', *Nature* **477**: 82–5.

Li, J. Z., Absher, D. M., Tang, H. *et al.* (2008). Worldwide 'Human Relationships Inferred from Genome-Wide Patterns of Variation', *Science* **319** (5866): 1100–4.

Longo, L., Boaretto, E., Giunti, P. *et al.* (2012). 'Did Neandertals and anatomically modern humans coexist in northern Italy during the late MIS 3?', *Quaternary International* (in press).

Massey, D.S. (1990). 'Social structure, household strategies, and the cumulative causation of migration', *Population Index* **56**: 3–26.

Mcbrearty, S. and Brooks A.S. (2000). 'The revolution that wasn't: a new interpretation of the origin of modern human behaviour', *Journal of Human Evolution* **39**(5): 453–563.

McGovern, T.H. (1980). 'Cows, Harp Seals, and Churchbells: Adaptation and Extinction in Norse Greenland', *Human Ecology* **10**: 245–75.

Mellars, P. (2006). 'Going East: New Genetic and Archaeological Perspectives on the Modern Human Colonization of Eurasia', *Science* **313**: 796–800.

Meltzer, D.J. (2009). 'First people in a new world', University of California Press, Berkeley.

Monnier, L. (2006). 'Les premiers peuplements de l'ouest de la France: Cadre chronostratigraphique et paléoenvironnemental', *Bulletin du Musée d'anthropologie préhistorique de Monaco* **46**: 3–20.

Monot H., Honore N., Garnier T. *et al.* (2005). 'On the origin of leprosy', *Science* **308**: 1040–2.

Parfitt, S.A., Ashton, N.M., Lewis, S. *et al.* (2010). 'Early Pleistocene human occupation at the edge of the boreal zone in northwest Europe', *Nature* **466**: 229–33.

Pavlov, P., Svendsen, J.I., Indrelid, S. (2001). 'Human presence in the European Arctic nearly 40,000 years ago', *Nature* **413**: 64–7.

Pinhasi, R., Gasparian, B., Nahapetyan, S. *et al.* (2011). 'Middle Palaeolithic human occupation of the high altitude region of Hovk-1, Armenia', *Quaternary Science Reviews* **30**: 3846–57.

Pinhasi, R. and von Cramon-Taubadel, N. (2009). 'Craniometric Data Supports Demic Diffusion Model for the Spread of Agriculture into Europe', *PLoS One* **4**(8): e6747.

Potts, R. and Teague, R. (2010). 'Behavioural and Environmental Background to 'Out-of-Africa I' and the Arrival of *Homo erectus* in East Asia', in J.G. Fleagle, J.J. Shea, F.E. Grine *et al.* (eds), *Out of Africa I: The First Hominin Colonization of Eurasia.* Springer: London, 67–85.

Raynal, J.-P., Sbihi Alaoui, F.Z., Geraads, D. *et al.* (2001). 'The earliest occupation of North-Africa: the Moroccan perspective', *Quaternary International* **75**: 65–75.

Reed, D.L., Smith, V.S., Hammond, S.L. *et al.* (2004). 'Genetic Analysis of Lice Supports Direct Contact between Modern and Archaic Humans', *PLoS Biology* **2**(11): e340.

Reich, D., Green, R.E., Kircher, M. *et al.* (2010). 'Genetic history of an archaic hominin group from Denisova Cave in Siberia', *Nature* **468**: 1053–60.

Reimer, P.J., Baillie, M.G.L., Bard, E. *et al.* (2009). 'IntCal09 and Marine09 radiocarbon age calibration curves, 0–50,000 years cal BP', *Radiocarbon* **51**: 1111–50.

Renfrew, C. (1987). *Archaeology and language: the puzzle of Indo-European origins*. Cambridge University Press: Cambridge.

Richerson, P.J., Boyd, R. (2005). *Not by genes along: How culture transformed human evolution*. University of Chicago Press: Chicago.

Richerson P. J., Boyd R., Bettinger R. L. (2005). 'Evolution on a restless planet: were environmental variability and environmental change major drivers of human evolution?', in F.M. Wuketits and F.J. Ayala (eds), *Handbook of evolution, Vol. 2*.John Wiley & Sons: Weinheim, 223–42.

Rinaldi, A. (2007). 'Tiny travel companions', *European Molecular Biology Organization Reports* **8**: 121–5.

Rothhammer, F. and Dillehay, T.D. (2009). 'The Late Pleistocene Colonization of South America: An Interdisciplinary Perspective', *Annals of Human Genetics* **73**: 540–9.

Schattner, U. and Lazar, M. (2009). 'Subduction, collision and initiation of hominin dispersal', *Quaternary Science Reviews* **28** (19–20): 1820–4.

Scott, G.R. and Gibert, L. (2009). 'The oldest hand-axes in Europe. *Nature* **461**: 82–5.

Seddon, N., Richard M., Merrill, R.M. *et al.* (2008). 'Sexually Selected Traits Predict Patterns of Species Richness in a Diverse Clade of Suboscine Birds', *American Naturalist* **171**: 620–31.

Sellick, M. J., Kyser, T. K., Wunder, M. B. *et al.* (2009). 'Geographic Variation of strontium and hydrogen isotopes in Avian tissue: implications for tracking migration and dispersal', Plos One: **4** (3): e4735.

Semino, O., Passarino, G., Oefner, P.J., Lin, A. A., Arbuzova, S. *et al.* (2000). 'The Genetic Legacy of Paleolithic Homo sapiens sapiens in Extant Europeans: A Y Chromosome Perspective', *Science* **290**: 1155–9

Shea, J.J. (2010). 'Stone Age Visiting Cards Revisited: A Strategic Perspective on the Lithic Technology of Early Hominin Dispersal', in J.G. Fleagle, J.J. Shea, F.E. Grine et al. (eds), *Out of Africa I: The First Hominin Colonization of Eurasia* Springer: London, 47–64.

Sinervo, B. and Clobert, J. (2003). 'A lizard color polymorphism drives settlement behaviour owing to selective benefits of cooperation', *Science* **300**: 1949–51.

Slimak, L., Svendsen, J.I., Mangerud, J. *et al.* (2011). 'Late Mousterian Persistence near the Arctic Circle', *Science* **332** (6031): 841–5.

Steele, J. (2009). 'Human Dispersals: Mathematical Models and the Archaeological Record', *Human Biology* **81**: 121–40.

Storey, A.A., Ramírez, J.M., Quiroz, D. *et al.* (2007). 'Radiocarbon and DNA evidence for a pre-Columbian introduction of Polynesian chickens to Chile', *Proceedings of the National Academy of Sciences USA* **104**: 10335–9.

Towner, M.C. (1999). 'A dynamic model of human dispersal in a land-based economy', *Behavioural Ecology and Sociobiology* **46**: 82–94.

Tresset, A., Bollongino, R., Edwards, C.J. *et al.* (2009). 'Early diffusion of domestic bovids in Europe: An indicator for human contacts, exchanges and migrations?', in F. d'Errico, J.-M. Hombert (eds), *Becoming Eloquent: Advances in the emergence of language, human cognition, and modern cultures.* John Benjamins Publishing Company, Amsterdam, 69–90.

Ungar, P.S., Grine, F.E., Teaford, M.F. (2008). 'Dental Microwear and Diet of the Plio-Pleistocene Hominin *Paranthropus boisei*', *PLoS One* **3(4)**: e2044.

Waters, M.R., Stafford Jr. T.W., McDonald, H.G. *et al.* (2011). 'Pre-Clovis Mastodon Hunting 13,800 Years Ago at the Manis Site, Washington', *Science* **334** (6054): 351–3.

White, T., Suwa, G., Lovejoy, C.O. (2010). 'Response to Comment on the Paleobiology and Classification of *Ardipithecus ramidus*', *Science* **328** (5982): 1105.

White, T.D., Asfaw, B., DeGusta, D. *et al.* (2003). 'Pleistocene Homo sapiens from Middle Awash, Ethiopia', *Nature* **423**: 742–7.

Whybrow, P.C. (2005). *American Mania: when more is enough.* Norton W.W. & Company, New York.

Wood, B.A. (2010). 'Systematics, Taxonomy, and Phylogenetics: Ordering Life, Past and Present', in C.S. Larsen (ed.), *A Companion to Biological Anthropology.*AQ: 2 Wiley-Blackwell: Oxford.

Wood, B. and Lonergan, N. (2008). 'The hominin fossil record: taxa, grades and clades', *Journal of Anatomy* **212**: 354–76.

Zaim, Y. (2010). 'Geological Evidence for the Earliest Appearance of Hominins in Indonesia', in J.G. Fleagle, J.J. Shea, F.E. Grine *et al.* (eds.). Out of Africa I. Vertebrate Paleobiology and Paleoanthropology, Part 2, Dordrecht: Springer, pp. 97–110,

Zheng, H-X., Yan, S., Qin, Z-D. *et al.* (2011). 'Major Population Expansion of East Asians Began before Neolithic Time: Evidence of mtDNA Genomes', *PLoS ONE* **6**(10): e25835. doi:10.1371/journal.pone.0025835

Zilhão J.and d'Errico F. (1999). 'The Chronology and Taphonomy of the Earliest Aurignacian and Its Implications for the Understanding of Neandertal Extinction', Journal of World Prehistory **13** (1), 1–68.

Index

Note: page numbers in **bold** refer to definitions of terms, and those in *italics* refer to figures and tables.

The manufacturer's authorised representative in the EU for product safety is Oxford University Press España S.A. of El Parque Empresarial San Fernando de Henares, Avenida de Castilla, 2 - 28830 Madrid (www.oup.es/en or product.safety@oup.com). OUP España S.A. also acts as importer into Spain of products made by the manufacturer.

Printed and bound by CPI Group (UK) Ltd, Croydon, CR0 4YY

12/09/2026

02215930-0008